EXPERIMENTAL SOLID MECHANICS

Arun Shukla
University of Rhode Island

James W. Dally
University of Maryland
College Park

College House Enterprises, LLC
Knoxville, Tennessee

The manuscript was prepared using Microsoft Word with 10 point Cambria font. Publishing and Printing Inc., Knoxville, TN printed this book from pdf files of each chapter.

College House Enterprises, LLC.
5713 Glen Cove Drive
Knoxville, TN 37919, U. S. A.
Phone or FAX (865) 558 6111
e mail jwd@collegehousebooks.com
http://www.collegehousebooks.com

10 Digit ISBN 0-9792581-8-9
13 Digit ISBN 978-0-9792581-8-3

CONTENTS

CHAPTER 8 STRAIN ANALYSIS METHODS

PART III OPTICAL METHODS OF STRESS, STRAIN AND DISPLACEMENT ANALYSIS

CHAPTER 9 BASIC OPTICS

CHAPTER 10 PHOTOELASTICITY

CHAPTER 11 APPLIED PHOTOELASTICITY

CHAPTER 12 INTERFEROMETRY AND HOLOGRAPHY

CHAPTER 13 MOIRÉ AND MOIRÉ INTERFEROMETRY

CHAPTER 14 SPECKLE METHODS

CHAPTER 15 DIGITAL IMAGE CORRELATION

CHAPTER 16 OPTICAL METHODS FOR DETERMINING FRACTURE PARAMETERS

PART IV MEASUREMENTS UNDER HIGH STRAIN RATES AND SMALL LENGTH SCALES

CHAPTER 17 DYNAMIC MEASUREMENTS

CHAPTER 18 NANOSCALE MEASUREMENTS

PART V MEASUREMENT AND ANALYSIS METHODS

CHAPTER 19 DIGITAL RECORDING SYSTEMS

CHAPTER 20 STATISTICAL ANALYSIS OF EXPERIMENTAL DATA

PREFACE

Experimental Solid Mechanics has evolved from the 4th edition of Experimental Stress Analysis. The title change reflects the fact that the field today is much broader than it was in 1965 when the first edition of Experimental Stress Analysis was published. Experimental Solid Mechanics describes methods used to measure forces, pressure, displacements, stresses, strains and fracture mechanics parameters. Measurements described include electrical and optical methods.

The text is intended for upper-division undergraduate students or graduate students beginning to study experimental methods. Extensive revisions have been incorporated, which reflects many of the changes in experimental mechanics that have occurred during the past decade. A significant amount of new content has been added by expanding existing chapters and by introducing new chapters. Some material covering outdated methods have been removed. The organization of the textbook includes five parts, which are briefly described below:

Part I: Elementary Elasticity and Fracture Mechanics contains three chapters on elasticity and an introductory chapter on elementary fracture mechanics. A section has been added to Chapter 2 to provide the stress-strain relations for composite materials.

Part II: Displacement and Strain Measurement Methods contains four chapters describing up to date techniques for measuring these quantities. Chapter 5 defines the properties of strain-gage systems. It provides descriptions of several strain-gage systems as well as an introduction to five different types of sensors that are used in measuring mechanical quantities. Chapter 6 describes electrical resistance strain gages in considerable detail. Strain-gage circuits and parameters affecting their performance are covered in Chapter 7. Chapter 8 describes analysis methods and illustrates techniques for determining principal stresses from rosettes. The chapter also describes torque and stress gages, and covers techniques for measuring stress intensity factors, crack initiation toughness and residual stresses.

Part III: Optical-Methods of Stress, Strain and Displacement Analysis, the core of this textbook, includes eight chapters. Chapter 9 covers basic optics that serves as a foundation for subsequent chapters. The coverage of photoelasticity is divided into two chapters. Chapter 10 provides the theoretical basis for the method. Chapter 11 provides practical information on the application of photoelasticity including two- and three-dimensional methods and birefringent coatings. Chapter 12 introduces interferometric methods which serve as a basis for describing techniques used in holographic interferometery. Chapter 13 covers both classical moiré methods, the more modern procedures of moiré interferometry and e beam moiré. Chapter 14 describes speckle methods including speckle interferometry and electronic speckle pattern interferometry. Chapter 15 describes digital image correlation methods, which have surged in their applications during the past decade. Chapter 16 covers methods for determining fracture parameters and extends the textbook's coverage to include fracture analysis.

Part IV: Measurements under High Strain Rates and Small Length Scales consists of two chapters that are new and they reflect the growing interest in high-rate loading in the measurement of material properties and in the behavior of nanoscale materials. Chapter 17 describes seven different loading methods used today to provide a wide range of strain-rates to relatively small specimens. Techniques are also introduced to measure stress and strain during these dynamic experiments. Chapter 18 describes nanoscale measurement, which includes description of three different microscopes that enable observation of the substructures so important in the understanding of material behavior at these scales.

Part V: Measurement and Analysis Methods contains two chapters. Chapter 19 describes digital recording systems covering the important topic of analog to digital conversion and aliasing. Data logging systems and well as PC based data acquisition systems are covered. Chapter 20 deals with the application of statistics in enhancing experimental accuracy and in improving the method of reporting experimental results that show variation.

Each part of the book is essentially independent so that instructors can be quite flexible in selecting course content. For instance, a two- or three-credit course on strain gages can be offered by using three chapters of Part I and all of Part II. Parts I and III can be combined to provide a thorough three- or four-credit course on optical methods of experimental analysis. Selected chapters from the first four parts can be organized to introduce the broader field of experimental stress analysis. Chapters selected from Part II, Part III and PartIV can be combined to teach an experimental measurements course. Part IV provides the basis for an advanced course on modern methods in Experimental Mechanics. A complete detailed treatment of the subject matter covered in the text and supplemented with laboratory exercises on strain gages, photoelasticity, moiré interferometry, speckle methods and digital image correlation will require six- to eight-credit hours.

The essential feature of the text is its completeness in introducing the entire range of experimental methods to the student. A reasonably deep coverage is presented of the theory required to understand experimental stress analysis and of the five primary methods employed: strain gages, photoelasticity, moiré, interferometry (including holography, coherent gradient sensing and speckle) and digital image correlation. While primary emphasis is placed on the theory of experimental stress analysis, the important experimental techniques associated with each of the four major methods are covered in sufficient detail to permit the student to begin laboratory work with a firm understanding of experimental procedures. Exercises designed to support and extend the treatment and to show the application of the theory have been placed at the end of all of the chapters.

Laboratory exercises have not been included, because laboratory work will depend strongly on local conditions such as the equipment and supplies available, the instructor's interests, the number of students in the class and research activities of current interest. It is believed that the instructor is best qualified to specify the associated laboratory exercises on the basis of interest, equipment, supplies, and time available for this important supplement to the course.

A significant amount of new material has been added to edition; however, space limitations did not permit coverage of the many modern research topics such as, bio-mechanics, smart structures, MEMs, etc. It is anticipated that the instructor will, in certain instances, treat these topics by using his or her lecture notes or by using recent papers published in the technical journals. The authors hope that most instructors will find the fundamental material required to present a complete and practical course on the theory of experimental solid mechanics in this text.

The material presented here has been assembled by both authors. Courses have been developed on Experimental Stress Analysis, Photoelasticity, and Photomechanics at Illinois Institute of Technology, Cornell University, The U. S. Air Force Academy, University of Rhode Island and the University of Maryland. The material has been shown to be interesting by the students participating in these courses. The mathematics employed in this treatment can easily be understood by senior undergraduates. Cartesian notation and/or vector notation has been used to enhance the student's understanding of the field equations. A great deal of effort was devoted to the selection and preparation of the illustrations employed. These illustrations complement the text and should aid appreciably in presenting the material to the student.

ACKNOWLEDGEMENTS

The contributions of many investigators working in experimental mechanics should be acknowledged. This edition represents a summary of many of the more mature methods which have been developed by the combined efforts of many diligent researchers and instructors. In particular, we wish to thank Drs. A. J. Durelli and G. R. Irwin our mentors in experimental stress analysis and fracture mechanics. Next, we acknowledge the excellent illustrations provided by Drs. Fu-pen Chiang, R. Chona, Gary Cloud, A. J. Durelli, Y. Guo, W. P. T. North, D. Post, R. J. Sanford, W.N. Sharpe, Jr., C. W. Smith, Karl Stetson, C. E. Taylor and the suppliers of commercial products. Finally, the review of Chapter 14 by Gary Cloud and the review of Chapter 15 by Bill Chao were extremely helpful. We also acknowledge the assistance of graduate students Michael Heeder, Erheng Wang and Puneet Kumar in carefully proof reading the text, and in aiding with the preparation of Chapters 16, 17 and 18.

Arun Shukla
University of Rhode Island

James W. Dally
College Park, MD

LIST OF SYMBOLS

a acceleration, amplitude, crack dimension
A area
A_n coefficients
A_0 amplitude of light
b aperture width
B magnetic field, specimen thickness, parameter
B_m coefficient
B_L lateral boundary
c stress optical coefficient, velocity of light
c_1, c_2 stress optical coefficients, wave velocities
cc center to center dimension
C capacitance, constant, count, correlation coefficient
C_c coating coefficient of sensitivity
C_e equivalent capacitance
C_J junction capacitance
C_s coefficient of sensitivity, stray capacitance
C_v coefficient of variation
C_ν Poisson's ratio mismatch correction factor
CMRR common mode rejection ratio
d distance, dimension, diameter, pitch
d_x mean deviation
D damping coefficient, diameter
D volume dilatation ($\varepsilon_1 + \varepsilon_2 + \varepsilon_3$)
D_p difference due to pressure effects
D_0 fog density of a photographic film
DA area dose
DL line dose
e electron charge
E energy, error, error function
E magnitude of a light vector, photon energy
E modulus of elasticity, potential gradient
E_b bias voltage
E^c modulus of elasticity of a coating
E_k kinetic energy
E_0 exposure inertia of a photographic film
E_R reference voltage
E* modulus of elasticity of a calibrating beam
f focal length, frequency, function
f_a aliasing frequency
f_{bw} bandwidth
f_c cut-off frequency
f_f fringe frequency
f_m frequency of moiré fringes
f_r reference grating frequency
f_s sampling rate
f_t theoretical frequency
f_ε material strain-fringe value
f_σ material stress-fringe value
F force
F_{CB} bending correction factor
F_{CR} reinforcing correction factor
$\mathbf{F_n}$ resultant force
F_r, F_θ, F_z polar components—body-force intensity
F_x, F_y, F_z Cartesian body-force intensity components
FSV full-scale voltage
g gap dimension, gravitational constant, function
g gray scale values of intensity
G amplifier gain, gray level, shear modulus
G energy release rate or crack extension force
G_{Ic} critical energy release rate
h bi-cubic spline function, Planck's constant
h height, gap, thickness
H Hamaker constant
H(ω) frequency response function
i imaginary number = $\sqrt{-1}$
i, j, k unit vectors
I current, intensity of light, moment of inertia
I' current density
I_e intensity of emerging light
I_g gage current
I_i intensity of incident light
I_0 object beam intensity
I_r intensity of reflected light
I_s supply current
I_B electron beam current
I_R reference beam intensity
I_1, I_2, I_3 first, second, and third invariants of stress
j imaginary number = $\sqrt{-1}$
J integral about crack tip, polar moment of inertia
J_{Ic} critical value of J integral
J_1, J_2, J_3 first, second and third strain invariants
J_0 zero order Bessel function
k Boltzmann's constant, multiplication factor
k proportionality constant, spring rate
K bulk modulus, dielectric constant,
K strength of a light source, optical strain coefficient
K_{ij} coefficient of elasticity
K_t transverse-sensitivity factor
K_I stress intensity factor (opening mode)
K_{Ic} crack initiation toughness
K_{II} stress intensity factor (shear mode)
K_{III} stress intensity factor (tearing mode)
K_T compressibility constant
l, m, n direction cosines
l length
l_g gage length
l_0 gage length
L length, dimension
L_0 gage length
LSB least significant bit
$\mathscr{L}$ loss factor
m mass, fitting parameter
M bending moment, magnification factor
n index of refraction, integer
n normal vector

n_0 index of refraction in an unstressed medium
n_1, n_2, n_3 indices of refraction, principal directions
N cycles of relative retardation, fringe order
N number of charge carriers, of cycles
NF noise factor
OPL optical path length
p pitch, pressure, probability
p_m pitch of moiré fringes
p_r pitch of reference grating
p_T transducer power
P applied load, power, force
P_g power dissipated by a gage
P_s load shedding due to yielding
P_D power density
P_R load range of transducer
q electric charge, resistance ratio
Q figure of merit
r position vector
r radial coordinate, dimension, resistance ratio
r^*_p apparent distance
R radius, crack growth resistance, range
R reflection coefficient, resistance, resolution
R responsivity, rotation
R_c contact resistance, capacitive reactance
R_d dynamic range
R_e equivalent resistance
R_f feedback resistance
R_g gage resistance
R_p parallel resistance
R_s series resistance
R_{sh} shunt resistance
R_x external resistance
R_M resistance of measuring circuit
R_T transducer resistance
RL record length
$\mathcal{R}$ impact resistance
s displacement, distance, spacing
s_1, s_2 curvilinear coordinates
S sensitivity index, flow stress, stiffness
S_a axial strain sensitivity of a gage
S_c circuit sensitivity
S_{cc} sensitivity constant current circuit
S_{cv} sensitivity of a constant voltage circuit
S_g strain sensitivity of a gage, gage factor
S_q charge sensitivity
S_s subjective speckle size
S_{sc} strain sensitivity of a semiconductor material
S_t gage sensitivity to time and transverse strain sensitivity of a gage
S_v voltage sensitivity
S_x standard deviation
$S_{\bar{x}}$ standard error
S_x^2 variance
S_ε strain sensitivity
S_A strain sensitivity of a material
$S_{A/B}$ sensitivity of materials A and B
S_0 objective speckle size
S_T gage sensitivity to temperature
S_σ stress sensitivity
$S_\sigma{}^c$ coating sensitivity to stress
S/N signal to noise ratio
SR sampling rate
t time and thickness
t_d duration of waveform
t_e exposure time
t_r rise time
t^* integration time
T time, period, temperature, threshold intensity
T torque, transmission coefficient
$\mathbf{T_n}$ resultant stress
T_{nx}, T_{ny}, T_{nz} Cartesian components—resultant stress
T_x, T_y, T_z Cartesian components—surface tractions
T_0 reference temperature
T_s sampling time
u unit vector
u distance, displacement, dimension, velocity
u, v, w Cartesian components of displacement
u_r, u_θ, u_z polar components of displacement
U strain energy
v velocity, voltage and distance
v′ potential gradient
v_b bias voltage
v_i input voltage
v_0 output voltage
v_s supply voltage
v_u unknown voltage
v_R reference voltage
v_T transducer voltage
v_Z Zener diode breakdown voltage
V volume
V_B bridge excitation voltage
w width
w_0 gage width
W beam depth, strain energy density
x, y, z Cartesian coordinates
$\bar{x}$ sample mean
z coordinate along optical axis, on complex plane
Z impedance, Westergaard complex stress function
α angle of incidence, mapping parameter, multiplier
α coefficient of thermal expansion, mismatch factor
α_p polarizing angle
$\alpha, \beta, \gamma, \phi, \theta$ angles
β angle of reflection, thermal coefficient expansion
γ angle of refraction, shear strain component
γ surface energy, temp. coefficient of resistivity
$\gamma_{r\theta}$ shear strain component in polar coordinates
$\left.\begin{array}{l}\gamma_{xy} = \gamma_{yx}\\ \gamma_{yz} = \gamma_{zy}\\ \gamma_{zx} = \gamma_{xz}\end{array}\right\}$ Cartesian shear strain components
δ displacement, linear phase difference
Δ relative phase difference, relative retardation
ΔR_T resistance change due to temperature
ΔR_ε resistance change due to strain
ε normal strain
ε_a axial strain

ε_c calibration strain
ε_n normal strain component
$\varepsilon_{rr}, \varepsilon_{\theta\theta}, \varepsilon_{zz}$ normal strain components in polar coordinates
ε_{sh} shunt calibration strain
ε_t transverse strain
$\varepsilon_{xx}, \varepsilon_{yy}, \varepsilon_{zz}$ Cartesian components of normal strain
$\varepsilon_1, \varepsilon_2, \varepsilon_3$ principal normal strains
$\varepsilon^c{}_1, \varepsilon^c{}_2$ principal normal strains in a coating
$\varepsilon^s{}_1, \varepsilon^s{}_2$ principal normal strains in a specimen
η nonlinear term
θ angular deflection
λ aspect ratio, Lame's constant, wavelength
λ_g wavelength of light propagating in glass
μ mobility of charge carriers, shear modulus
μ true arithmetic mean of a population
ν Poisson's ratio
ν^c Poisson's ratio of a coating
ν^s Poisson's ratio of a specimen
ξ wave number
ξ, η elliptic coordinates
π_{ijkl} piezoresistive tensor
ρ radius of curvature, mass density, specific resistance
ρ_{ij} resistivity coefficient
σ normal stress true standard deviation
σ_e effective stress
σ_n normal component of the resultant stress
σ_{pl} proportional limit
$\sigma_{rr}, \sigma_{\theta\theta}, \sigma_{zz}$ polar components of normal stresses
$\sigma_{xx}, \sigma_{yy}, \sigma_{zz}$ Cartesian components of normal stresses
σ_{ys} yield strength
σ_0 applied stress
σ^R residual stress
$\sigma_1, \sigma_2, \sigma_3$ principal normal stresses
$\sigma^c{}_1, \sigma^c{}_2$ principal normal stresses in a coating
$\sigma^s{}_1, \sigma^s{}_2$ principal normal stresses in a specimen
σ'_1, σ'_2 secondary principal stresses
σ_η normal stress component in elliptic coordinates
σ^* normal stress in a calibrating beam
τ shear stress component, time constant
τ_n shear stress component of the resultant stress
τ_{kl} stress tensor
$\tau_{r\theta}$ shear stress component in polar coordinates

$$\left.\begin{array}{l}\tau_{xy} = \tau_{yx} \\ \tau_{yz} = \tau_{zy} \\ \tau_{zx} = \tau_{xz}\end{array}\right\}$$ Cartesian shear stresses components

ϕ Airy's stress function, phase angle
ω angular frequency, rigid body rotation
Ω body-force function
χ dimensionless constant

PART I

ELEMENTARY ELASTICITY
AND
FRACTURE MECHANICS

CHAPTER 1

STRESS

1.1 INTRODUCTION

An experimental stress analyst must have a thorough understanding of stress, strain, and the laws relating stress to strain. For this reason, the first three chapters of this text have been devoted to the elementary concepts of the theory of elasticity. The first chapter deals with stresses produced in a body due to external and body-force loadings. The second chapter deals with deformations and strains produced by the loadings and with relations between the stresses and strains. The third chapter covers plane problems in the theory of elasticity, important because a large part of a first course in experimental stress analysis deals with two-dimensional problems. Also treated is the stress-function approach to the solution of plane problems. Upon completing the subject matter of the first three chapters of the text, the student should have a firm understanding of stress and strain and should be able to solve some of the more elementary two-dimensional problems in the theory of elasticity by using the Airy's-stress-function approach.

Failures under loading conditions well below the yield stress often occur in structures with small crack or crack like flaws. Such failures show that conventional strength studies, no matter how accurately conducted, are not always sufficient to guarantee structural integrity under operational conditions. The field of study, which considers crack-extension behavior as a function of applied loads is known as **fracture mechanics**. As a result of research during the past 20 years, fracture mechanics is now used to solve many practical engineering problems in failure analysis, material selection, and structural-life prediction. For this reason, Chapter 4 was introduced in the third edition to provide students with a brief treatment of some of the fundamental concepts of fracture mechanics.

1.2 DEFINITIONS

Two basic types of force act on a body to produce stresses. Forces of the first type are called **surface forces** for the simple reason that they act on the surfaces of the body. Surface forces are generally exerted when one body comes in contact with another. Forces of the second type are called **body forces** because they act on each element of the body. Body forces are commonly produced by centrifugal, gravitational, or other force fields. The most common body forces are gravitational, being present to some degree in almost all cases. For many practical applications, however, they are so small compared with the surface forces present that they can be neglected without introducing serious error. Body forces are included in the following analysis for the sake of completeness.

Consider an arbitrary internal or external surface, which may be plane or curvilinear, as shown in Fig. 1.1. Over a small area ΔA of this surface in the neighborhood of an arbitrary point P, a system of forces acts which has a resultant represented by the vector $\mathbf{\Delta F_n}$, in the figure. It should be noted that the line of action of the resultant force vector $\mathbf{\Delta F_n}$ does not necessarily coincide with the outer normal $\mathbf{n}$ associated with the element of ΔA. If the resultant force $\mathbf{\Delta F_n}$ is divided by the increment of area ΔA, the average stress which acts over the area is obtained. In the limit as ΔA approaches zero, a quantity defined as the resultant stress $\mathbf{T_n}$ acting at the point P is obtained. This limiting process is illustrated in equation form below:

$$\mathbf{T_n} = \lim_{\Delta A \to 0} \frac{\Delta F_n}{\Delta A} \tag{1.1}$$

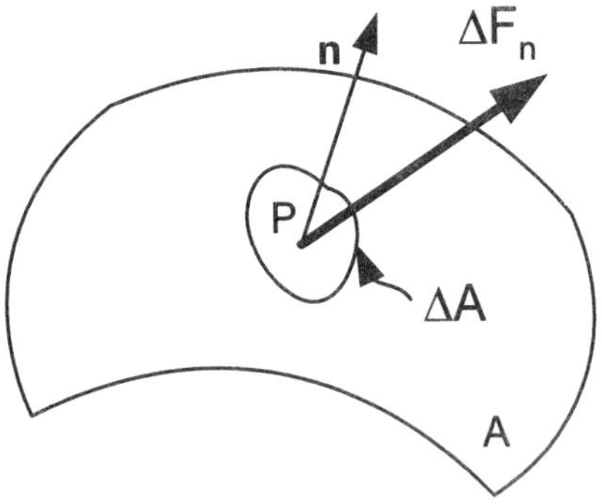

Figure 1.1 Arbitrary surface (either internal or external) showing the resultant of all forces acting over the element ΔA.

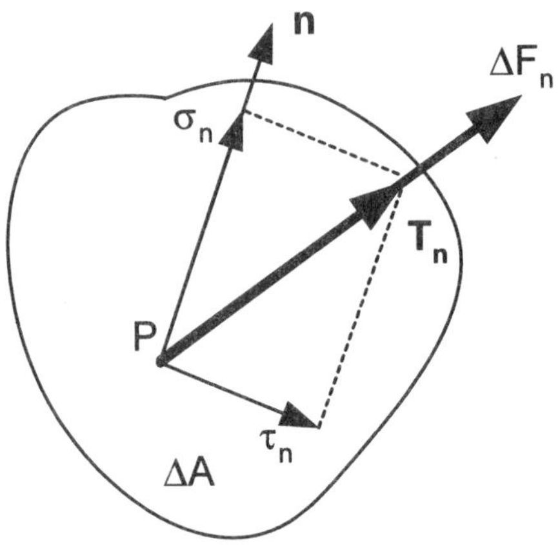

Figure 1.2 Resolution of the resultant stress $\mathbf{T_n}$ into normal and tangential components σ_n and τ_n.

The line of action of this resultant stress $\mathbf{T_n}$ coincides with the line of action of the resultant force $\mathbf{\Delta F_n}$, as illustrated in Fig. 1.2. It is important to note at this point that the resultant stress $\mathbf{T_n}$ is a function of both the position of the point P in the body and the orientation of the plane which is passed through the point and identified by its outer normal **n**. In a body subjected to an arbitrary system of loads, both the magnitude and the direction of the resultant stress $\mathbf{T_n}$ at any point P change as the orientation of the plane under consideration is changed.

As illustrated in Fig. 1.2, it is possible to resolve $\mathbf{T_n}$ into two components: one σ_n normal to the surface is known as the resultant normal stress, while the component τ_n is known as the resultant shearing stress.

Cartesian components of stress for any coordinate system can also be obtained from the resultant stress. Consider first a surface whose outer normal is in the positive z direction, as shown in Fig. 1.3. If the resultant stress $\mathbf{T_n}$ associated with this particular surface is resolved into components along the x, y, and z-axes, the Cartesian stress components τ_{zx}, τ_{zy}, σ_{zz} are obtained. The components τ_{zx} and τ_{zy} are shearing stresses because they act tangent to the surface under consideration. The component σ_{zz} is a normal stress because it acts normal to the surface.

If the same procedure is followed using surfaces with outer normals in the positive x and y directions, two more sets of Cartesian components, τ_{xy}, τ_{xz}, σ_{xx} and τ_{yx}, τ_{yz}, σ_{yy}, respectively, can be obtained. The three different sets of three Cartesian components for the three selections of the outer normal are summarized in the array below:

σ_{xx}	τ_{xy}	τ_{xz}	outer normal parallel to x-axis
τ_{yx}	σ_{yy}	τ_{yz}	outer normal parallel to y-axis
τ_{zx}	τ_{zy}	σ_{zz}	outer normal parallel to z-axis

From this array, it is clear that nine Cartesian components of stress exist. These components can be arranged on the faces of a small cubic element, as shown in Fig. 1.4. The sign convention employed in placing the Cartesian stress components on the faces of this cube is as follows: If the outer normal defining the cube face is in the direction of increasing x, y, or z, then the associated normal and shear stress components are also in the direction of positive x, y, or z. If the outer normal is in the direction of negative x, y, or z, then the normal and shear stress components are also in the direction of negative x, y, or z. As for subscript convention, the first subscript refers to the outer normal and defines the plane upon which the stress component acts, whereas the second subscript gives the direction in which the stress acts. Finally, for normal stresses, positive signs indicate tension and negative signs indicate compression.

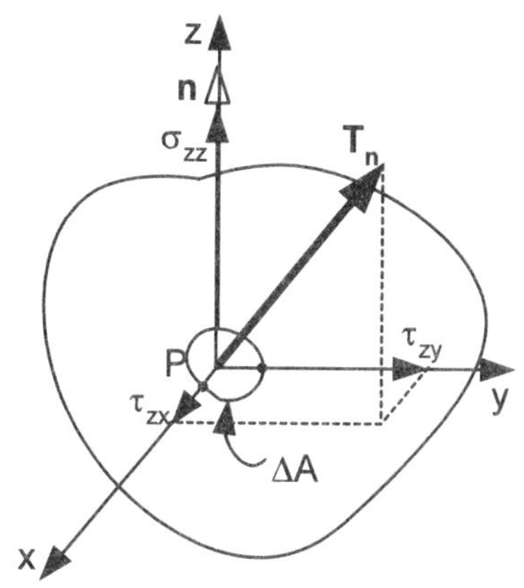

Figure 1.3 Resolution of the resultant stress $\mathbf{T_n}$ into its three Cartesian components τ_{zx}, τ_{zy} and σ_{zz}.

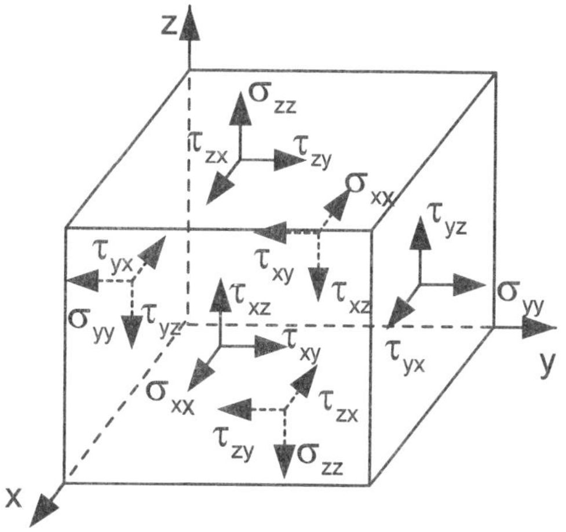

Figure 1.4 Cartesian components of stress acting on the faces of a small cubic element.

1.3 STRESS AT A POINT

At a given point of interest within a body, the magnitude and direction of the resultant stress $\mathbf{T_n}$ depend upon the orientation of the plane passed through the point. Thus an infinite number of resultant-stress vectors can be used to represent the resultant stress at each point because an infinite number of planes can be passed through each point. It is easy to show, however, that the magnitude and direction of each of these resultant $\mathbf{T_n}$ stress vectors can be specified in terms of the nine Cartesian components of stress acting at the point. This fact can be established by considering equilibrium of the elemental tetrahedron shown in Fig. 1.5. In this figure, the stresses acting over the four faces of the tetrahedron are represented by their average values. The average value is denoted by placing a ~ sign over the stress symbol. In order for the tetrahedron to be in equilibrium, the following condition must be satisfied. First consider equilibrium in the x direction.

$$\tilde{T}_{nx}A - \tilde{\sigma}_{xx}A\cos(\mathbf{n},\, x) - \tilde{\tau}_{yx}A\cos(\mathbf{n},\, y) - \tilde{\tau}_{zx}A\cos(\mathbf{n},\, z) + \tilde{F}_x hA/3 = 0$$

where h = altitude of tetrahedron; A = area of base of tetrahedron; $\tilde{F}_x$ = average body-force intensity in x direction; and $\tilde{T}_{nx}$ = component of resultant stress in the x direction, and A cos(**n**, x), A cos(**n**, y), and A cos(**n**, z) are the projections of the area A on the yz, xz, and xy planes, respectively.

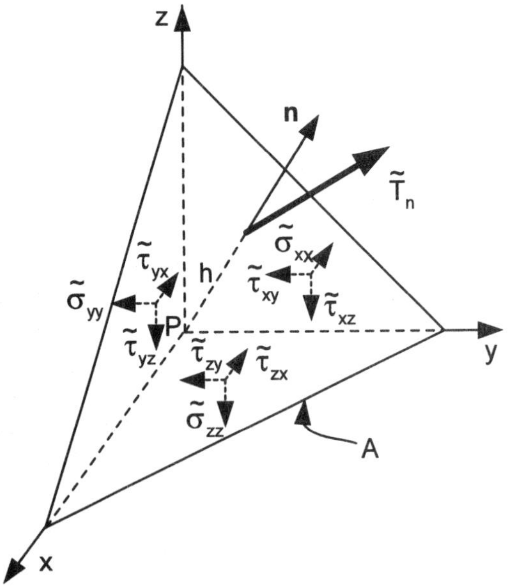

Figure 1.5 Elemental tetrahedron at a point P showing the average stresses that act over its Cartesian faces

By letting the altitude h ⇒ 0, after eliminating the common factor A from each term of the expression, it can be seen that the body-force term vanishes, the average stresses become exact stresses at the point P, and the previous expression becomes:

$$T_{nx} = \sigma_{xx} \cos(\mathbf{n}, x) + \tau_{yx} \cos(\mathbf{n}, y) + \tau_{zx} \cos(\mathbf{n}, z) \qquad (1.2a)$$

Two similar expressions are obtained by considering equilibrium in the y and z directions:

$$T_{ny} = \tau_{xy} \cos(\mathbf{n}, x) + \sigma_{yy} \cos(\mathbf{n}, y) + \tau_{zy} \cos(\mathbf{n}, z) \qquad (1.2b)$$

$$T_{nz} = \tau_{xz} \cos(\mathbf{n}, x) + \tau_{yz} \cos(\mathbf{n}, y) + \sigma_{zz} \cos(\mathbf{n}, z) \qquad (1.2c)$$

After the three Cartesian components of the resultant stress have been determined from Eqs. (1.2), the resultant stress $\mathbf{T_n}$ can be determined by using the expression:

$$|\mathbf{T_n}| = \sqrt{T_{nx}^2 + T_{ny}^2 + T_{nz}^2}$$

The three direction cosines, which define the line of action of the resultant stress $\mathbf{T_n}$ are:

$$\cos(T_n, x) = \frac{T_{nx}}{|\mathbf{T_n}|} \qquad \cos(T_n, y) = \frac{T_{ny}}{|\mathbf{T_n}|} \qquad \cos(T_n, z) = \frac{T_{nz}}{|\mathbf{T_n}|}$$

The normal stress σ_n and the shearing stress τ_n which act on the plane under consideration can be obtained from the expressions:

$$\sigma_n = |\mathbf{T_n}|\cos(T_n, \mathbf{n}) \quad \text{and} \quad \tau_n = |\mathbf{T_n}|\sin(T_n, \mathbf{n})$$

The angle between the resultant-stress vector $\mathbf{T_n}$ and the normal to the plane $\mathbf{n}$ can be determined by using the well-known relationship:

$$\cos(T_n, \mathbf{n}) = \cos(T_n, x)\cos(\mathbf{n}, x) + \cos(T_n, y)\cos(\mathbf{n}, y) + \cos(T_n, z)\cos(\mathbf{n}, z)$$

It should also be noted that the normal stress σ_n could be determined by considering the projections of T_{nx}, T_{ny}, and T_{nz} onto the normal to the plane under consideration. Thus,

$$\sigma_n = T_{nx} \cos(\mathbf{n}, x) + T_{ny} \cos(\mathbf{n}, y) + T_{nz} \cos(\mathbf{n}, z)$$

When σ_n has been determined, τ_n can easily be found because:

$$\tau_n = \sqrt{T_n^2 - \sigma_n^2}$$

1.4 STRESS EQUATIONS OF EQUILIBRIUM

In a body subjected to a general system of body and surface forces, stresses of variable magnitude and direction are produced throughout the body. The distribution of these stresses must be such that the overall equilibrium of the body is maintained; furthermore, equilibrium of each element in the body must be maintained. This section deals with the equilibrium of the individual elements of the body. On the element shown in Fig. 1.6, only the stress and body-force components, which act in the x direction, are shown. Similar components exist and act in the y and z directions. The stress values shown are average stresses over the faces of an element, which is assumed to be very small. A summation of forces in the x direction gives:

$$\left(\sigma_{xx} + \frac{\partial \sigma_{xx}}{\partial x} dx - \sigma_{xx}\right) dydz + \left(\tau_{yx} + \frac{\partial \tau_{yx}}{\partial y} dy - \tau_{yx}\right) dxdz$$

$$+\left(\tau_{zx} + \frac{\partial \tau_{zx}}{\partial z} dz - \tau_{zx}\right) dxdy + F_x\, dx\, dy\, dz = 0$$

Dividing through by dx dy dz gives,

$$\frac{\partial \sigma_{xx}}{\partial x} + \frac{\partial \tau_{yx}}{\partial y} + \frac{\partial \tau_{xz}}{\partial z} + F_x = 0 \tag{1.3a}$$

By considering the force and stress components in the y and z directions, it can be established in a similar fashion that:

$$\frac{\partial \tau_{xy}}{\partial x} + \frac{\partial \sigma_{yy}}{\partial y} + \frac{\partial \tau_{zy}}{\partial z} + F_y = 0 \tag{1.3b}$$

$$\frac{\partial \tau_{xz}}{\partial x} + \frac{\partial \tau_{yz}}{\partial y} + \frac{\partial \sigma_{zz}}{\partial z} + F_z = 0 \tag{1.3c}$$

where F_x, F_y, and F_z are body-force intensities (in lb/in^3 or N/m^3) in the x, y, and z directions, respectively.

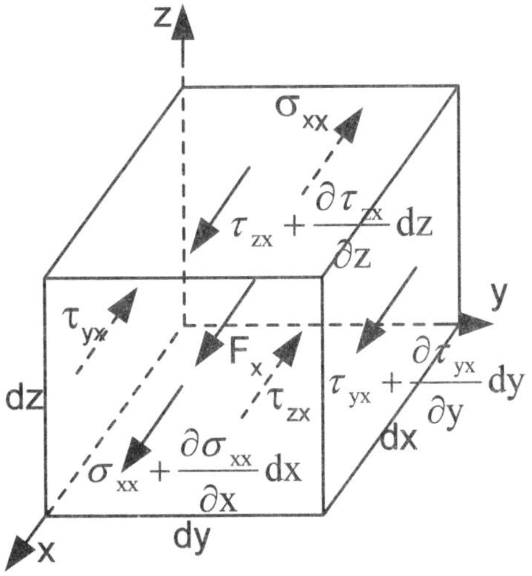

Figure 1.6 A small element removed from a body, showing stresses acting in the x direction only.

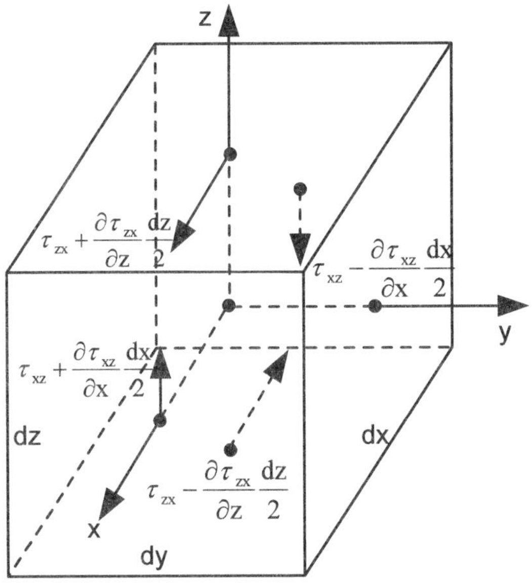

Figure 1.7 A small element removed from a body, showing the stresses that produce a moment about the y-axis.

Equations (1.3) are the well-known stress equations of equilibrium, which any theoretically or experimentally obtained stress distribution must satisfy. In obtaining these equations, three of the six equilibrium conditions have been employed. The three remaining conditions can be utilized to establish additional relationships between the stresses.

Consider the element shown in Fig. 1.7. Only those stress components that will produce a moment about the y-axis are shown. Because the coordinate system has been selected with its origin at the centroid of the element, the normal stress components and the body forces do not produce moments.

A summation of moments about the y-axis gives the following expression:

$$\left(\tau_{zx} + \frac{\partial \tau_{zx}}{\partial z}\frac{dz}{2}\right)dxdy\frac{dz}{2} + \left(\tau_{zx} - \frac{\partial \tau_{zx}}{\partial z}\frac{dz}{2}\right)dxdy\frac{dz}{2}$$

$$-\left(\tau_{xz} + \frac{\partial \tau_{xz}}{\partial x}\frac{dx}{2}\right)dydz\frac{dx}{2} - \left(\tau_{xz} - \frac{\partial \tau_{xz}}{\partial x}\frac{dx}{2}\right)dydz\frac{dx}{2} = 0$$

this reduces to:

$$\tau_{zx}\, dx\, dy\, dz - \tau_{xz}\, dx\, dy\, dz = 0$$

Therefore,

$$\tau_{zx} = \tau_{xz} \tag{1.4a}$$

The remaining two equilibrium conditions can be used in a similar manner to establish that:

$$\tau_{xy} = \tau_{yx} \tag{1.4b}$$

$$\tau_{yz} = \tau_{zy} \tag{1.4c}$$

The equalities given in Eqs. (1.4) reduce the nine Cartesian components of stress to six independent components, which may be expressed in the following array:

$$\begin{matrix} \sigma_{xx} & \tau_{xy} & \tau_{zx} \\ \tau_{xy} & \sigma_{yy} & \tau_{yz} \\ \tau_{zx} & \tau_{yz} & \sigma_{zz} \end{matrix}$$

1.5 LAWS OF STRESS TRANSFORMATION

It has previously been shown that the resultant-stress vector $\mathbf{T_n}$ acting on an arbitrary plane defined by the outer normal **n** can be determined by substituting the six independent Cartesian components of stress into Eqs. (1.2). However, it is often desirable to make another transformation—that from the stress components σ_{xx}, σ_{yy}, σ_{zz}, τ_{xy}, τ_{yz}, and τ_{zx}, which refer to an Oxyz coordinate system, to the stress components $\sigma_{x'x'}$, $\sigma_{y'y'}$, $\sigma_{z'z'}$, $\tau_{x'y'}$, $\tau_{y'z'}$, and $\tau_{z'x'}$ that refer to an Ox'y'z' coordinate system. The transformation equations commonly used to perform this operation will be developed in this section.

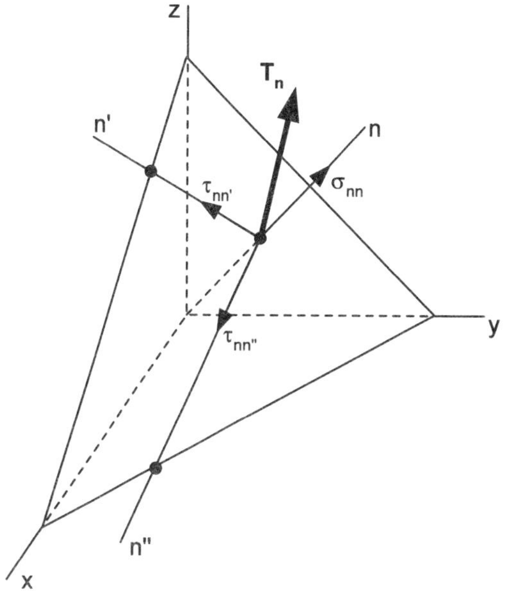

Figure 1.8 Resolution of $\mathbf{T_n}$ into three Cartesian components σ_{nn}, $\tau_{nn'}$ and $\tau_{nn''}$.

Consider an element similar to Fig. 1.5 with an inclined face having an outer normal **n**. Two mutually perpendicular directions **n'** and **n''** can then be denoted in the plane of the inclined face, as shown in Fig. 1.8. The resultant stress $\mathbf{T_n}$ acting on the inclined face can be resolved into components along the directions **n**, **n'** and **n''** to yield the stresses σ_{nn}, $\tau_{nn'}$, $\tau_{nn''}$. This resolution of the resultant stress into components can be accomplished most easily by utilizing the Cartesian components T_{nx}, T_{ny} and T_{nz}. Thus:

$$\sigma_{nn} = T_{nx} \cos(\mathbf{n}, x) + T_{ny} \cos(\mathbf{n}, y) + T_{nz} \cos(\mathbf{n}, z)$$

$$\tau_{nn'} = T_{nx} \cos(\mathbf{n}', x) + T_{ny} \cos(\mathbf{n}', y) + T_{nz} \cos(\mathbf{n}', z)$$

$$\tau_{nn''} = T_{nx} \cos(\mathbf{n}'', x) + T_{ny} \cos(\mathbf{n}'', y) + T_{nz} \cos(\mathbf{n}'', z)$$

If the results from Eqs. (1.2) and (1.4) are substituted into these expressions, the following important equations are obtained:

$$\sigma_{nn} = \sigma_{xx} \cos^2(\mathbf{n}, x) + \sigma_{yy} \cos^2(\mathbf{n}, y) + \sigma_{zz} \cos^2(\mathbf{n}, z) + 2\tau_{xy} \cos(\mathbf{n}, x) \cos(\mathbf{n}, y) + 2\tau_{yz} \cos(\mathbf{n}, y)\cos(\mathbf{n}, z) + 2\tau_{zx} \cos(\mathbf{n}, z) \cos(\mathbf{n}, x) \quad (1.5a)$$

$$\tau_{nn'} = \sigma_{xx}\cos(\mathbf{n},x)\cos(\mathbf{n}',x) + \sigma_{yy}\cos(\mathbf{n},y)\cos(\mathbf{n}',y) + \sigma_{zz}\cos(\mathbf{n},z)\cos(\mathbf{n}',z)$$
$$+\tau_{xy}[\cos(\mathbf{n},x)\cos(\mathbf{n}',y) + \cos(\mathbf{n},y)\cos(\mathbf{n}',x)]$$
$$+\tau_{yz}[\cos(\mathbf{n},y)\cos(\mathbf{n}',z) + \cos(\mathbf{n},z)\cos(\mathbf{n}',y)]$$
$$+\tau_{zx}[\cos(\mathbf{n},z)\cos(\mathbf{n}',x) + \cos(\mathbf{n},x)\cos(\mathbf{n}',z)] \quad (1.5b)$$

$$\tau_{nn''} = \sigma_{xx}\cos(\mathbf{n},x)\cos(\mathbf{n}'',x) + \sigma_{yy}\cos(\mathbf{n},y)\cos(\mathbf{n}'',y) + \sigma_{zz}\cos(\mathbf{n},z)\cos(\mathbf{n}'',z)$$
$$+\tau_{xy}[\cos(\mathbf{n},x)\cos(\mathbf{n}'',y) + \cos(\mathbf{n},y)\cos(\mathbf{n}'',x)]$$
$$+\tau_{yz}[\cos(\mathbf{n},y)\cos(\mathbf{n}'',z) + \cos(\mathbf{n},z)\cos(\mathbf{n}'',y)]$$
$$+\tau_{zx}[\cos(\mathbf{n},z)\cos(\mathbf{n}'',x) + \cos(\mathbf{n},x)\cos(\mathbf{n}'',z)] \quad (1.5c)$$

Equations (1.5) provide the means for determining normal- and shear-stress components at a point associated with any set of Cartesian reference axes provided the stresses associated with one set of axes are known.

Expressions for the stress components $\sigma_{x'x'}$, $\sigma_{y'y'}$, $\sigma_{z'z'}$, $\tau_{x'y'}$, $\tau_{y'z'}$, and $\tau_{z'x'}$ can be obtained directly from Eq. (1.5a) or Eq. (1.5b) by employing the following procedure. To determine $\sigma_{x'x'}$ select a plane having an outer normal **n** coincident with x'. A resultant stress $\mathbf{T_n} = \mathbf{T}_{x'}$ is associated with this plane. The normal stress $\sigma_{x'x'}$ associated with this plane is obtained directly from Eq. (1.5a) by substituting x' for **n**. Thus:

$$\sigma_{x'x'} = \sigma_{xx}\cos^2(x',x) + \sigma_{yy}\cos^2(x',y) + \sigma_{zz}\cos^2(x',z) + 2\tau_{xy}\cos(x',x)\cos(x',y)$$
$$+ 2\tau_{yz}\cos(x',y)\cos(x',z) + 2\tau_{zx}\cos(x',z)\cos(x',x) \quad (1.6a)$$

By selecting **n** coincident with the y' and z' axes and following the same procedure, expressions for $\sigma_{y'y'}$ and $\sigma_{z'z'}$ can be obtained as follows:

$$\sigma_{y'y'} = \sigma_{yy}\cos^2(y',y) + \sigma_{zz}\cos^2(y',z) + \sigma_{xx}\cos^2(y',x) + 2\tau_{yz}\cos(y',y)\cos(y',z)$$
$$+ 2\tau_{zx}\cos(y',z)\cos(y',x) + 2\tau_{xy}\cos(y',x)\cos(y',y) \quad (1.6b)$$

$$\sigma_{z'z'} = \sigma_{zz}\cos^2(z',z) + \sigma_{xx}\cos^2(z',x) + \sigma_{yy}\cos^2(z',y) + 2\tau_{zx}\cos(z',z)\cos(z',x)$$
$$+ 2\tau_{xy}\cos(z',x)\cos(z',y) + 2\tau_{yz}\cos(z',y)\cos(z',z) \quad (1.6c)$$

The shear-stress component $\tau_{x'y'}$ is obtained by selecting a plane having an outer normal **n** coincident with x' and the in-plane direction **n'** with y', as shown in Fig. 1.9. The shear stress $\tau_{x'y'}$ is then obtained from Eq. (1.5b) by substituting x' for **n** and y' for **n'**. Thus:

$$\tau_{x'y'} = \sigma_{xx}\cos(x',x)\cos(y',x) + \sigma_{yy}\cos(x',y)\cos(y',y) + \sigma_{zz}\cos(x',z)\cos(y',z)$$
$$+\tau_{xy}[\cos(x',x)\cos(y',y) + \cos(x',y)\cos(y',x)]$$
$$+\tau_{yz}[\cos(x',y)\cos(y',z) + \cos(x',z)\cos(y',y)]$$
$$+\tau_{zx}[\cos(x',z)\cos(y',x) + \cos(x',x)\cos(y',z)] \quad (1.6d)$$

By selecting **n** and **n'** coincident with the y' and z', and z' and x' axes, additional expressions can be developed for $\tau_{y'z'}$ and $\tau_{z'x'}$, respectively, as follows:

$$\tau_{y'z'} = \sigma_{yy}\cos(y',y)\cos(z',y) + \sigma_{zz}\cos(y',z)\cos(z',z) + \sigma_{xx}\cos(y',x)\cos(z',x)$$
$$+\tau_{yz}[\cos(y',y)\cos(z',z) + \cos(y',z)\cos(z',y)]$$
$$+\tau_{xz}[\cos(y',z)\cos(z',x) + \cos(y',x)\cos(z',z)]$$
$$+\tau_{xy}[\cos(y',x)\cos(z',y) + \cos(y',y)\cos(z',x)] \quad (1.6e)$$

$$\tau_{z'x'} = \sigma_{zz}\cos(z',z)\cos(x',z) + \sigma_{xx}\cos(z',x)\cos(x'\,x) + \sigma_{yy}\cos(z',y)\cos(x',y)$$
$$+\tau_{zx}[\cos(z',z)\cos(x',x) + \cos(z',x)\cos(x',z)]$$
$$+\tau_{xy}[\cos(z',x)\cos(x',y) + \cos(z',y)\cos(x',x)]$$
$$+\tau_{yz}[\cos(z',y)\cos(x',z) + \cos(z',z)\cos(x',y)] \quad (1.6f)$$

These six equations permit the six Cartesian components of stress relative to the Oxyz coordinate system to be transformed into a different set of six Cartesian components of stress relative to an Ox'y'z' coordinate system.

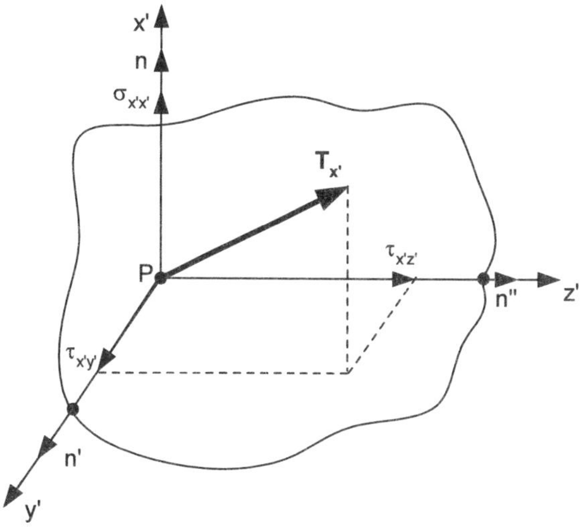

Figure 1.9 Resolution of $\mathbf{T_n}$ into three Cartesian components $\sigma_{x'x'}$, $\tau_{x'y'}$ and $\tau_{x'z'}$.

1.6 PRINCIPAL STRESSES

In Section 1.2 it was noted that the resultant-stress vector $\mathbf{T_n}$ at a given point P depended upon the choice of the plane upon which the stress acted. If a plane is selected such that $\mathbf{T_n}$ coincides with the outer normal **n**, as shown in Fig. 1.10, it is clear that the shear stress τ_n vanishes and that $\mathbf{T_n}$, σ_n and **n** are coincident.

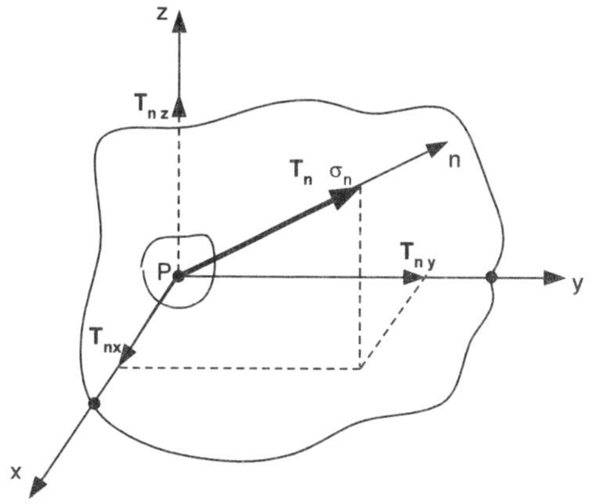

Figure 1.10 Coincidence of T_n with the outer normal n indicates that the shear stresses vanish and that σ_n is equal in magnitude to T_n.

If **n** is selected so that it coincides with $\mathbf{T_n}$ then the plane defined by **n** is known as a principal plane. The direction given by **n** is a principal direction, and the normal stress acting on this particular plane is a principal stress. In every state of stress there exist at least three principal planes, which are mutually perpendicular, and associated with these principal planes there are at most three distinct principal stresses. These statements can be established by referring to Fig. 1.10 and noting that:

$$T_{nx} = \sigma_n \cos(\mathbf{n}, x) \qquad T_{ny} = \sigma_n \cos(\mathbf{n}, y) \qquad T_{nz} = \sigma_n \cos(\mathbf{n}, z) \tag{a}$$

If Eqs. (1.2) are substituted into Eqs. (a), the following expressions are obtained:

$$\begin{aligned}
\sigma_{xx} \cos(\mathbf{n}, x) + \tau_{yx} \cos(\mathbf{n}, y) + \tau_{zx} \cos(\mathbf{n}, z) &= \sigma_n \cos(\mathbf{n}, x) \\
\tau_{xy} \cos(\mathbf{n}, x) + \sigma_{yy} \cos(\mathbf{n}, y) + \tau_{zy} \cos(\mathbf{n}, z) &= \sigma_n \cos(\mathbf{n}, y) \\
\tau_{xz} \cos(\mathbf{n}, x) + \tau_{yz} \cos(\mathbf{n}, y) + \sigma_{zz} \cos(\mathbf{n}, z) &= \sigma_n \cos(\mathbf{n}, z)
\end{aligned} \tag{b}$$

Rearranging Eqs. (b) gives:

$$(\sigma_{xx} - \sigma_n) \cos(\mathbf{n}, x) + \tau_{yx} \cos(\mathbf{n}, y) + \tau_{zx} \cos(\mathbf{n}, z) = 0$$

$$\tau_{xy} \cos(\mathbf{n}, x) + (\sigma_{yy} - \sigma_n) \cos(\mathbf{n}, y) + \tau_{zy} \cos(\mathbf{n}, z) = 0 \qquad \text{(c)}$$

$$\tau_{xz} \cos(\mathbf{n}, x) + \tau_{yz} \cos(\mathbf{n}, y) + (\sigma_{zz} - \sigma_n) \cos(\mathbf{n}, z) = 0$$

Solving for any of the direction cosines, say cos (**n**, x), by determinants gives:

$$\cos(\mathbf{n}, x) = \frac{\begin{vmatrix} 0 & \tau_{yx} & \tau_{zx} \\ 0 & \sigma_{yy} - \sigma_n & \tau_{zy} \\ 0 & \tau_{yz} & \sigma_{zz} - \sigma_n \end{vmatrix}}{\begin{vmatrix} \sigma_{xx} - \sigma_n & \tau_{yx} & \tau_{zx} \\ \tau_{xy} & \sigma_{yy} - \sigma_n & \tau_{zy} \\ \tau_{xz} & \tau_{yz} & \sigma_{zz} - \sigma_n \end{vmatrix}} \qquad \text{(d)}$$

It is clear that nontrivial solutions for the direction cosines of the principal plane will exist only if the determinant in the denominator is zero. Thus:

$$\begin{vmatrix} \sigma_{xx} - \sigma_n & \tau_{yx} & \tau_{zx} \\ \tau_{xy} & \sigma_{yy} - \sigma_n & \tau_{zy} \\ \tau_{xz} & \tau_{yz} & \sigma_{zz} - \sigma_n \end{vmatrix} = 0 \qquad \text{(e)}$$

Expanding the determinant after substituting Eqs. (1.4) gives the following important cubic equation:

$$\sigma_n^3 - (\sigma_{xx} + \sigma_{yy} + \sigma_{zz})\sigma_n^2 + (\sigma_{xx}\sigma_{yy} + \sigma_{yy}\sigma_{zz} + \sigma_{zz}\sigma_{xx} - \tau_{xy}^2 - \tau_{yz}^2 - \tau_{zx}^2)\sigma_n$$

$$- (\sigma_{xx}\sigma_{yy}\sigma_{zz} - \sigma_{xx}\tau_{yz}^2 - \sigma_{yy}\tau_{zx}^2 - \sigma_{zz}\tau_{xy}^2 + 2\tau_{xy}\tau_{yz}\tau_{zx}) = 0 \qquad (1.7)$$

The roots of this cubic equation are the three principal stresses. By substituting the six Cartesian components of stress into this equation, one can solve for σ_n and obtain three real roots. Three possible solutions exist.

1. If σ_1, σ_2 and σ_3 are distinct, then n_1, n_2 and n_3 are unique and mutually perpendicular.
2. If $\sigma_1 = \sigma_2 \neq \sigma_3$, then n_3 is unique and every direction perpendicular to n_3 is a principal direction associated with $\sigma_1 = \sigma_2$.
3. If $\sigma_1 = \sigma_2 = \sigma_3$, then a hydrostatic state of stress exists and every direction is a principal direction.

When the three principal stresses have been established, they can be substituted individually into Eqs. (c) to give three sets of simultaneous equations which together with the relation:

$$\cos^2(\mathbf{n}, x) + \cos^2(\mathbf{n}, y) + \cos^2(\mathbf{n}, z) = 1$$

can be solved to give the three sets of direction cosines defining the principal planes. A numerical example of the procedure used in computing principal stresses and directions is given in the exercises at the end of the chapter.

In treating principal stresses it is often useful to order them so that $\sigma_1 > \sigma_2 > \sigma_3$. When the stresses are ordered in this fashion, σ_1 is the normal stress having the largest algebraic value at a given point and σ_3 is the normal stress having the smallest algebraic value. It is important to recall in this ordering process that tensile stresses are considered positive and compressive stresses are considered negative.

Another important concept is that of stress invariants. It was noted in Section 1.5 that a state of stress could be described by its six Cartesian stress components with respect to either the Oxyz coordinate system or the Ox'y'z' coordinate system. Furthermore, Eqs. (1.6) were established to give the relationship between these two systems. In addition to Eqs. (1.6), three other relations exist which are called the three invariants of stress. To establish these invariants, refer to Eq. (1.7) that is the cubic equation in terms of the principal stresses σ_1, σ_2 and σ_3. By recalling that σ_1, σ_2 and σ_3 are independent of the Cartesian coordinate system employed, it is clear that the coefficients of Eq. (1.7), which contain Cartesian components of the stresses, must also be independent or invariant of the coordinate system. Thus, from Eq. (1.7) it is clear that:

$$I_1 = \sigma_{xx} + \sigma_{yy} + \sigma_{zz} = \sigma_{x'x'} + \sigma_{y'y'} + \sigma_{z'z'}$$

$$\begin{aligned} I_2 &= \sigma_{xx}\sigma_{yy} + \sigma_{yy}\sigma_{zz} + \sigma_{zz}\sigma_{xx} - \tau_{xy}^2 - \tau_{yz}^2 - \tau_{zx}^2 \\ &= \sigma_{x'x'}\sigma_{y'y'} + \sigma_{y'y'}\sigma_{z'z'} + \sigma_{z'z'}\sigma_{x'x'} - \tau_{x'y'}^2 - \tau_{y'z'}^2 - \tau_{z'x'}^2 \end{aligned} \tag{1.8}$$

$$\begin{aligned} I_3 &= \sigma_{xx}\sigma_{yy}\sigma_{zz} - \sigma_{xx}\tau_{yz}^2 - \sigma_{yy}\tau_{zx}^2 - \sigma_{zz}\tau_{xy}^2 + 2\tau_{xy}\tau_{yz}\tau_{zx} \\ &= \sigma_{x'x'}\sigma_{y'y'}\sigma_{z'z'} - \sigma_{x'x'}\tau_{y'z'}^2 - \sigma_{y'y'}\tau_{z'x'}^2 - \sigma_{z'z'}\tau_{x'y'}^2 + 2\tau_{x'y'}\tau_{y'z'}\tau_{z'x'} \end{aligned}$$

where I_1, I_2 and I_3 are the first, second, and third invariants of stress, respectively. If the Oxyz coordinate system is selected coincident with the principal directions, Eqs. (1.8) reduce to:

$$I_1 = \sigma_1 + \sigma_2 + \sigma_3 \qquad I_2 = \sigma_1\sigma_2 + \sigma_2\sigma_3 + \sigma_3\sigma_1 \quad I_3 = \sigma_1\sigma_2\sigma_3 \tag{1.9}$$

1.7 MAXIMUM SHEAR STRESS

In developing equations for maximum shear stresses, the special case will be considered in which $\tau_{xy} = \tau_{yz} = \tau_{zx} = 0$. No loss in generality is introduced by considering this special case because it involves only a reorientation of the reference axes to coincide with the principal directions. In the following development n_1, n_2 and n_3 will be used to denote the principal directions. In Section. 1.3, the resultant stress on an oblique plane was given by:

$$T_n^2 = T_{nx}^2 + T_{ny}^2 + T_{nz}^2 \tag{a}$$

Substituting values for T_{nx}, T_{ny} and T_{nz} from Eqs. (1.2), with principal normal stresses and zero shearing stresses, yields:

$$T_n^2 = \sigma_1^2 \cos^2(\mathbf{n}, n_1) + \sigma_2^2 \cos^2(\mathbf{n}, n_2) + \sigma_3^2 \cos^2(\mathbf{n}, n_3) \tag{b}$$

Also from Eq. (1.5a):

$$\sigma_n = \sigma_1 \cos^2(\mathbf{n}, n_1) + \sigma_2 \cos^2(\mathbf{n}, n_2) + \sigma_3 \cos^2(\mathbf{n}, n_3) \tag{c}$$

Because $\tau_n^2 = T_n^2 - \sigma_n^2$, an expression for the shear stress τ_n on the oblique plane is obtained from Eqs. (b) and (c) after substituting $l = \cos(\mathbf{n}, n_1)$, $m = \cos(\mathbf{n}, n_2)$ and $n = \cos(\mathbf{n}, n_3)$ as:

$$\tau_n^2 = \sigma_1^2\, l^2 + \sigma_2^2\, m^2 + \sigma_3^2\, n^2 - (\sigma_1\, l^2 + \sigma_2\, m^2 + \sigma_3\, n^2)^2 \tag{d}$$

The planes on which maximum and minimum shearing stresses occur can be obtained from Eq. (d) by differentiating with respect to the direction cosines l, m and n . One of the direction cosines, n for example, in Eq. (d) can be eliminated by solving the expression:

$$l^2 + m^2 + n^2 = 1 \tag{e}$$

for n and substituting into Eq. (d). Thus:

$$\tau_n^2 = (\sigma_1^2 - \sigma_3^2)l^2 + (\sigma_2^2 - \sigma_3^2)m^2 + \sigma_3^2 - [(\sigma_1 - \sigma_3)l^2 + (\sigma_2 - \sigma_3)m^2 + \sigma_3]^2 \tag{f}$$

By taking the partial derivatives of Eq. (f), first with respect to l and then with respect to m, and equating to zero, the following equations are obtained for determining the direction cosines associated with planes having maximum and minimum shearing stresses:

$$l[\tfrac{1}{2}(\sigma_1 - \sigma_3) - (\sigma_1 - \sigma_3)l^2 - (\sigma_2 - \sigma_3)m^2] = 0 \tag{g}$$

$$m[\tfrac{1}{2}(\sigma_2 - \sigma_3) - (\sigma_1 - \sigma_3)l^2 - (\sigma_2 - \sigma_3)m^2] = 0 \tag{h}$$

One solution of these equations is obviously $l = m = 0$. Then from Eq. (e), $n = \pm 1$ (a principal plane with zero shear). Solutions, different from zero, are also possible for this set of equations. Consider first that $m = 0$; then from Eq. (g), $l = \pm(\tfrac{1}{2})^{1/2}$ and from Eq. (e), $n = \pm(\tfrac{1}{2})^{1/2}$. Also if $l = 0$, then from Eq. (h), $m = \pm(\tfrac{1}{2})^{1/2}$ and from Eq. (e), $n = \pm(\tfrac{1}{2})^{1/2}$. Repeating the above procedure by eliminating l and m in turn from Eq. (f) yields other values for the direction cosines that make the shearing stresses maximum or minimum. Substituting the values $l = \pm(\tfrac{1}{2})^{1/2}$ and $n = \pm(\tfrac{1}{2})^{1/2}$ into Eq. (d) yields:

$$\tau_n^2 = \tfrac{1}{2}\sigma_1^2 + 0 + \tfrac{1}{2}\sigma_3^2 - (\tfrac{1}{2}\sigma_1 + 0 + \tfrac{1}{2}\sigma_3)^2$$

from which:

$$\tau_n = \tfrac{1}{2}(\sigma_1 - \sigma_3)$$

Similarly, using the other values for the direction cosines, which maximize the shearing stresses gives:

$$\tau_n = \tfrac{1}{2}(\sigma_1 - \sigma_2) \quad \text{and} \quad \tau_n = \tfrac{1}{2}(\sigma_2 - \sigma_3)$$

Of these three possible results, the largest magnitude will be obtained from $\sigma_1 - \sigma_3$ if the principal stresses are ordered such that $\sigma_1 \geq \sigma_2 \geq \sigma_3$. Thus:

$$\tau_{max} = \tfrac{1}{2}(\sigma_{max} - \sigma_{min}) = \tfrac{1}{2}(\sigma_1 - \sigma_3) \tag{1.10}$$

A useful aid for visualizing the complete state of stress at a point is the three-dimensional Mohr's circle shown in Fig. 1.11. This representation, which is similar to the familiar two-dimensional Mohr's circle, shows the three principal stresses, the maximum shearing stresses, and the range of values within which the normal and shear stress components must lie for a given state of stress.

Figure 1.11 Mohr's circle for the three-dimensional state of stress.

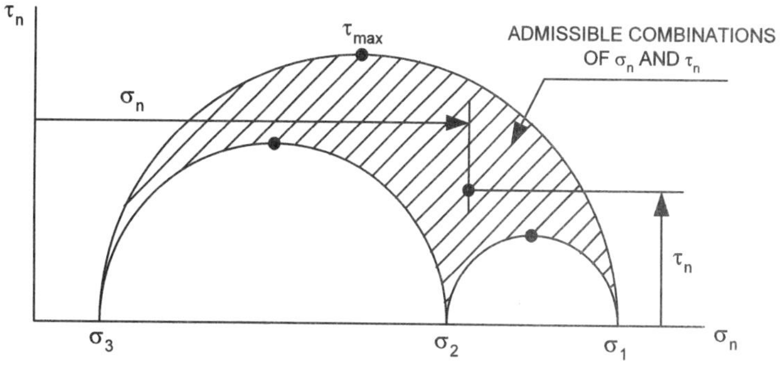

1.8 A TWO-DIMENSIONAL STATE OF STRESS

For two-dimensional stress fields where $\sigma_{zz} = \tau_{zx} = \tau_{yz} = 0$, z' is coincident with z, and θ is the angle between x and x', then Eqs. (1.6a) to (1.6f) reduce to:

$$\begin{aligned}\sigma_{x'x'} &= \sigma_{xx}\cos^2\theta + \sigma_{yy}\sin^2\theta + 2\tau_{xy}\sin\theta\cos\theta \\ &= ½(\sigma_{xx} + \sigma_{yy}) + ½(\sigma_{xx} - \sigma_{yy})\cos 2\theta + \tau_{xy}\sin 2\theta \qquad (1.11a)\end{aligned}$$

$$\begin{aligned}\sigma_{y'y'} &= \sigma_{yy}\cos^2\theta + \sigma_{xx}\sin^2\theta - 2\tau_{yz}\sin\theta\cos\theta \\ &= ½(\sigma_{yy} + \sigma_{xx}) + ½(\sigma_{yy} - \sigma_{xx})\cos 2\theta - \tau_{xy}\sin 2\theta \qquad (1.11b)\end{aligned}$$

$$\begin{aligned}\tau_{x'y'} &= \sigma_{yy}\cos\theta\sin\theta - \sigma_{xx}\cos\theta\sin\theta + \tau_{xy}(\cos^2\theta - \sin^2\theta) \\ &= ½(\sigma_{yy} - \sigma_{xx})\sin 2\theta + \tau_{xy}\cos 2\theta \qquad (1.11c)\end{aligned}$$

$$\sigma_{z'z'} = \tau_{y'z'} = \tau_{z'x'} = 0 \qquad (1.11d)$$

The relationships between the stress components, given in Eqs. (1.11), can be represented by using Mohr's circle of stress, as indicated in Fig. 1.12.

In this diagram, normal-stress components σ are plotted horizontally, while shear-stress components τ are plotted vertically. Tensile stresses are plotted to the right of the τ axis. Compressive stresses are plotted to the left. Shear-stress components, which tend to produce a clockwise rotation of a small element surrounding the point, are plotted above the σ axis. Those tending to produce a counterclockwise rotation are plotted below. When plotted in this manner, the stress components associated with each plane through the point are represented by a point on the circle. The diagram thus gives an excellent visual picture of the state of stress at a point. Mohr's circle and Eqs. (1.11) are often used in experimental stress-analysis work when stress components are transformed from one coordinate system to another. These relationships will be used frequently in later sections of this text, where strain gage and photoelasticity methods of analysis are discussed. Because two-dimensional stress systems are often considered in subsequent chapters, it will be useful to consider the principal stresses which occur in a two-dimensional stress system. If a coordinate system is chosen so that $\sigma_{zz} = \tau_{zx} = \tau_{yz} = 0$, then a state of plane stress exists and Eq. (1.7) reduces to:

$$\sigma_n[\sigma_n^2 - (\sigma_{xx} + \sigma_{yy})\sigma_n + (\sigma_{xx}\sigma_{yy} - \tau_{xy}^2)] = 0 \qquad (a)$$

Solving this equation for the three principal stresses yields:

$$\sigma_1,\ \sigma_2 = \frac{\sigma_{xx} + \sigma_{yy}}{2} \pm \sqrt{\left(\frac{\sigma_{xx} - \sigma_{yy}}{2}\right)^2 + \tau_{xy}^2} \qquad \sigma_3 = 0 \qquad (1.12)$$

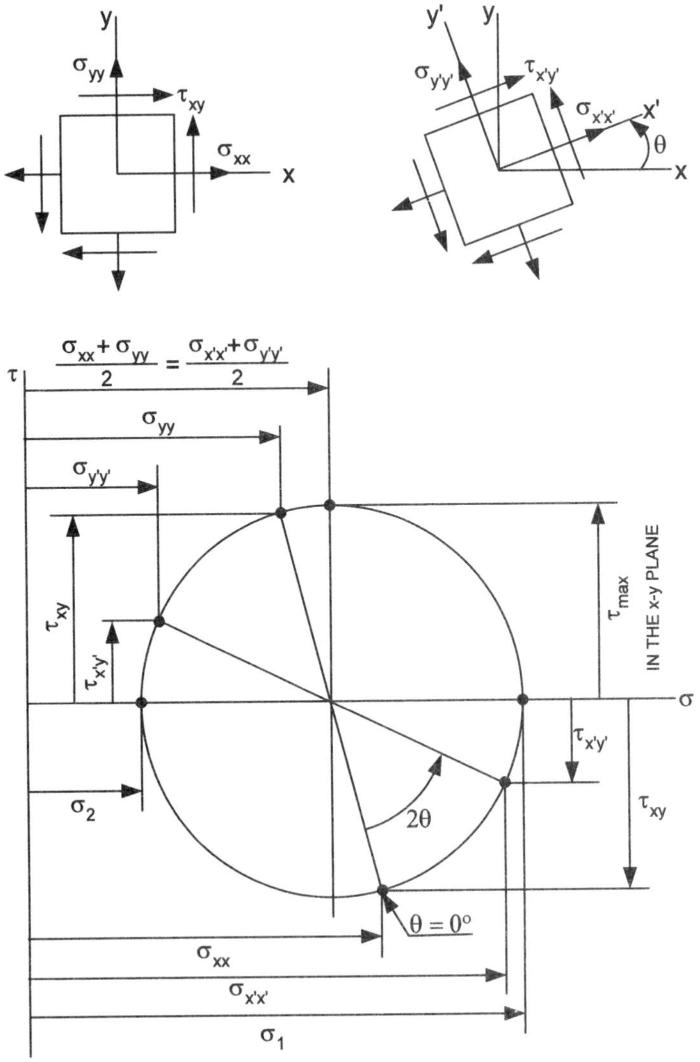

Figure 1.12 Mohr's circle of stress.

The two direction cosines, which define the two principal planes, can be determined from Eq. (1.11c), which gives $\tau_{x'y'}$ in terms of σ_{xx}, σ_{yy}, τ_{xy} and the angle θ between x and x'. If x' and y' are selected so that x'= $\mathbf{n_1}$ and y' = $\mathbf{n_2}$, then $\tau_{x'y'}$ must vanish because no shearing stresses can exist on principal planes. Thus the following equation can be written:

$$\frac{\sigma_{yy}-\sigma_{xx}}{2}\sin 2(n_1, x) + \tau_{xy}\cos 2(n_1, x) = 0 \tag{1.13}$$

Dividing through by cos $2(n_1, x)$ and simplifying gives:

$$\tan 2(n_1, x) = \frac{2\tau_{xy}}{\sigma_{xx}-\sigma_{yy}} \tag{1.14a}$$

and hence:

$$\cos 2(n_1, x) = \frac{\sigma_{xx} - \sigma_{yy}}{\sqrt{(\sigma_{xx} - \sigma_{yy})^2 + 4\tau_{xy}^2}} \tag{1.14b}$$

$$\sin 2(n_1, x) = \frac{2\tau_{xy}}{\sqrt{(\sigma_{xx} - \sigma_{yy})^2 + 4\tau_{xy}^2}} \tag{1.14c}$$

Equations (1.14) are used in solving for the direction of $\mathbf{n_1}$ if the Cartesian stress components σ_{xx}, σ_{yy} and τ_{xy} are known.

1.9 STRESSES RELATIVE TO PRINCIPAL COORDINATES

If the coordinate system Oxyz is selected to coincide with the three principal directions $\mathbf{n_1}$, $\mathbf{n_2}$ and $\mathbf{n_3}$, then $\sigma_1 = \sigma_{xx}$, $\sigma_2 = \sigma_{yy}$, $\sigma_3 = \sigma_{zz}$ and $\tau_{xy} = \tau_{yz} = \tau_{zx} = 0$. This reduces the six components of stress to three, which permits a considerable simplification in some of the previous results. Equations (1.2) become:

$$T_{nx} = \sigma_1 \cos(\mathbf{n}, x) \qquad T_{ny} = \sigma_2 \cos(\mathbf{n}, y) \qquad T_{nz} = \sigma_3 \cos(\mathbf{n}, z) \tag{1.15}$$

and Equations (1.6) reduce to:

$$\sigma_{x'x'} = \sigma_1 \cos^2 (x', x) + \sigma_2 \cos^2 (x', y) + \sigma_3 \cos^2 (x', z)$$

$$\sigma_{y'y'} = \sigma_1 \cos^2 (y', x) + \sigma_2 \cos^2 (y', y) + \sigma_3 \cos^2 (y', z)$$

$$\sigma_{x'x'} = \sigma_1 \cos^2 (z', x) + \sigma_2 \cos^2 (z', y) + \sigma_3 \cos^2 (z', z) \tag{1.16}$$

$$\tau_{x'y'} = \sigma_1 \cos (x', x) \cos (y', x) + \sigma_2 \cos (x', y) \cos (y', y) + \sigma_3 \cos (x', z) \cos (y', z)$$

$$\tau_{y'z'} = \sigma_1 \cos (y', x) \cos (z', x) + \sigma_2 \cos (y', y) \cos (z', y) + \sigma_3 \cos (y', z) \cos (z', z)$$

$$\tau_{z'x'} = \sigma_1 \cos (z', x) \cos (x', x) + \sigma_2 \cos (z', y) \cos (x', y) + \sigma_3 \cos (z', z) \cos (x', z)$$

Often experimental methods yield principal stress directly, and in these cases Eqs. (1.16) are frequently used to obtain the stresses acting on other planes.

1.10 SPECIAL STATES OF STRESS

Two states of stress occur so frequently in practice that they have been classified. They are the state of pure shearing stress and the hydrostatic state of stress. Both are defined below.

1. A state of pure shear stress exists if one particular set of axes Oxyz can be found such that $\sigma_{xx} = \sigma_{yy} = \sigma_{zz} = 0$. It can be shown that this particular set of axes Oxyz exists if and only if the first invariant of stress $I_1 = 0$. The proof of this condition is beyond the scope of this text. Two of the infinite number of arrays which represent a state of pure shearing stress are given below:

$$\begin{Vmatrix} 0 & \tau_{xy} & \tau_{xz} \\ \tau_{xy} & 0 & \tau_{yz} \\ \tau_{xz} & \tau_{yz} & 0 \end{Vmatrix} \quad \text{or} \quad \begin{Vmatrix} \sigma_{xx} & \tau_{xy} & \tau_{xz} \\ \tau_{xy} & \sigma_{yy} = -\sigma_{xx} & \tau_{yz} \\ \tau_{xz} & \tau_{yz} & 0 \end{Vmatrix}$$

Pure shear

This array can be converted to the form shown on the left by a suitable rotation of the co-ordinate system.

2. A state of stress is said to be hydrostatic if $\sigma_{xx} = \sigma_{yy} = \sigma_{zz} = p$ and all of the shearing stresses vanish. In photoelastic studies a hydrostatic state of stress is often called an isotropic state of stress. The stress array for this case is given by:

$$\begin{Vmatrix} -p & 0 & 0 \\ 0 & -p & 0 \\ 0 & 0 & -p \end{Vmatrix}$$

One particularly important property of these two states of stress is that they can be combined to form a general state of stress. Of more importance, however, is the fact that any state of stress can be separated onto a state of pure shear plus a hydrostatic state of stress. This is easily seen from the three arrays shown below:

$$\begin{Vmatrix} \sigma_{xx} & \tau_{xy} & \tau_{xz} \\ \tau_{xy} & \sigma_{yy} & \tau_{yz} \\ \tau_{xz} & \tau_{yz} & \sigma_{zz} \end{Vmatrix} = \begin{Vmatrix} -p & 0 & 0 \\ 0 & -p & 0 \\ 0 & 0 & -p \end{Vmatrix} + \begin{Vmatrix} \sigma_{xx}+p & \tau_{xy} & \tau_{xz} \\ \tau_{xy} & \sigma_{yy}+p & \tau_{yz} \\ \tau_{xz} & \tau_{yz} & \sigma_{zz}+p \end{Vmatrix} \tag{1.17}$$

General State = Hydrostatic State + State of Pure Shear

These arrays indicate that a general state of stress can be represented by a hydrostatic state of stress plus a state of pure shearing stress. It is immediately clear that the array on the left represents a general state of stress and that the center array represents a hydrostatic state of stress; however, the right-hand array represents a state of pure shear if and only if its first stress invariant is zero. This fact implies that:

$$(\sigma_{xx} + p) + (\sigma_{yy} + p) + (\sigma_{zz} + p) = 0$$

Hence:

$$p = -(1/3)(\sigma_{xx} + \sigma_{yy} + \sigma_{zz}) \tag{1.18}$$

If the p represented in the hydrostatic state of stress satisfies Eq. (1.18) then the separation of the state of stress given in Eq. (1.17) is valid. In the study of plasticity, the effect of the hydrostatic stresses is usually neglected; consequently, the principle illustrated above is quite important.

EXERCISES

1.1 Prepare a scale drawing representing the stresses associated with a resultant stress $\mathbf{T_n}$ where $|T_n|$ = 10 units and its direction is described by cos (**n**, x) = 0.6 and cos (**n**, y) = 0.7. The normal to the surface upon which $\mathbf{T_n}$ acts is given by **n** = (1/3) **i** + (2/3) **j** + (2/3) **k**. In this drawing show σ_n and τ_n to scale.

1.2 Derive the relation:

$$l^2 + m^2 + n^2 = 1$$

where l = cos (**n**, x), m = cos (**n**, y), and n = cos (**n**, z).

1.3 At a point in a stressed body, the Cartesian components of stress are σ_{xx} = 70 MPa, σ_{yy} = −35 MPa, σ_{zz} = 35 MPa, τ_{xy} = 40 MPa, τ_{yz} = 0 and τ_{zx} = 0. Determine the normal and shear stresses on a plane whose outer normal has the direction cosines:

cos (**n**, x) = (15/35) cos (**n**, y) = (18/35) cos (**n**, z) = (26/35)

1.4 At a point in a stressed body, the Cartesian components of stress are σ_{xx} = 75 MPa, σ_{yy} = 60 MPa, σ_{zz} = 50 MPa, τ_{xy} = 25 MPa, τ_{yz} = −25 MPa and τ_{zx} = 30 MPa. Determine the normal and shear stresses on a plane whose outer normal has the direction cosines:

cos (**n**, x) = (12/25) cos (**n**, y) = (15/25) cos (**n**, z) = (16/25)

1.5 At a point in a stressed body, the Cartesian components of stress are σ_{xx} = 45 MPa, σ_{yy} = 90 MPa, σ_{zz} = 45 MPa, τ_{xy} = 108 MPa, τ_{yz} = 54 MPa and τ_{zx} = 36 MPa. Determine (a) the normal and shear stresses on a plane whose outer normal has the direction cosines:

cos (**n**, x) = (4/9) cos (**n**, y) = (4/9) cos (**n**, z) = (7/9)

and (b) the angle between $\mathbf{T_n}$ and the outer normal **n**.

1.6 At a point in a stressed body, the Cartesian components of stress are σ_{xx} = 60 MPa, σ_{yy} = − 40 MPa, σ_{zz} = 20 MPa, τ_{xy} = − 40 MPa, τ_{yz} = 20 MPa and τ_{zx} = 30 MPa. Determine (a) the normal and shear stresses on a plane whose outer normal has the direction cosines:

cos (**n**, x) = 0.429 cos (**n**, y) = 0.514 cos (**n**, z) = 0.743

and (b) the angle between $\mathbf{T_n}$ and the outer normal **n**.

1.7 Determine the normal and shear stresses on a plane whose outer normal makes equal angles with the x, y, and z axes if the Cartesian components of stress at the point are $\sigma_{xx} = \sigma_{yy} = \sigma_{zz}$ = 0, τ_{xy} = 75 MPa, τ_{yz} = 0 and τ_{zx} = 100 MPa.

1.8 At a point in a stressed body, the Cartesian components of stress are $\sigma_{xx} = \sigma_{yy}$ = 75 MPa, σ_{zz} = − 30 MPa, τ_{xy} = 0, τ_{yz} = 45 MPa and τ_{zx} = 75 MPa. Determine the normal and shear stress on a plane whose outer normal has the direction cosines:

cos (**n**, x) = (2/3) cos (**n**, y) = (2/3) cos (**n**, z) = (1/3)

1.9 The following stress distribution has been determined for a machine component:

$\sigma_{xx} = 3x^2 - 3y^2 - z$ $\sigma_{yy} = 3y^2$ $\sigma_{zz} = 3x + y - z + 1.25$

$\tau_{xy} = z - 6xy - 0.75$ $\tau_{yz} = 0$ $\tau_{zx} = x + y - 1.5$

Is equilibrium satisfied in the absence of body forces?

1.10 If the state of stress at any point in a body is given by the equations:

$\sigma_{xx} = ax + by + cz$ $\quad$ $\sigma_{yy} = dx^2 + ey^2 + fz^2$ $\quad$ $\sigma_{zz} = gx^3 + hy^3 + iz^3$

$\tau_{xy} = k$ $\quad$ $\tau_{yz} = Ly + mz$ $\quad$ $\tau_{zx} = nx^2 + pz^2$

What equations must the body-force intensities F_x, F_y and F_z satisfy?

1.11 For a two dimensional state of stress with $\sigma_{zz} = \tau_{yz} = \tau_{zx} = 0 = 0$ and zero body forces, show that:

$$\int \frac{\partial \sigma_{xx}}{\partial x} dy = \int \frac{\partial \sigma_{yy}}{\partial y} dx$$

1.12 At a point in a stressed body, the Cartesian components of stress are $\sigma_{xx} = 100$ MPa, $\sigma_{yy} = 50$ MPa, $\sigma_{zz} = 30$ MPa, $\tau_{xy} = 30$ MPa, $\tau_{yz} = -30$ MPa and $\tau_{zx} = 60$ MPa. Transform this set of Cartesian stress components into a new set of Cartesian stress components relative to an Ox'y'z' set of coordinates where the Ox'y'z' axes are defined as:

θ	**Case 1**	**Case 2**	**Case 3**	**Case 4**
x – x'	π/4	π/2	0	π/2
y – y'	π/4	π/2	π/2	0
z – z'	0	0	π/2	π/2

1.13 At a point in a stressed body, the Cartesian components of stress are $\sigma_{xx} = 100$ MPa, $\sigma_{yy} = 60$ MPa, $\sigma_{zz} = 50$ MPa, $\tau_{xy} = 40$ MPa, $\tau_{yz} = -30$ MPa and $\tau_{zx} = 60$ MPa. Transform this set of Cartesian stress components into a new set of Cartesian stress components relative to an Ox'y'z' set of coordinates where the direction angles associated with the Ox'y'z' axes are:

	x	y	z
x'	π/3	π/3	π/4
y'	3π/4	π/4	π/2
z'	π/3	π/3	3π/4

1.14 At a point in a stressed body, the Cartesian components of stress are $\sigma_{xx} = 90$ MPa, $\sigma_{yy} = 54$ MPa, $\sigma_{zz} = -36$ MPa, $\tau_{xy} = 27$ MPa, $\tau_{yz} = -36$ MPa and $\tau_{zx} = 45$ MPa. Transform this set of Cartesian stress components into a new set of Cartesian stress components relative to an Ox'y'z' set of coordinates where the Ox'y'z' axes are defined by the following direction cosines:

(a)

	x	y	z
x'	2/3	2/3	−1/3
y'	−2/3	1/3	−2/3
z'	−1/3	2/3	2/3

(b)

	x	y	z
x'	1/9	−8/9	4/9
y'	4/9	4/9	7/9
z'	−8/9	1/9	4/9

1.15 At a point in a stressed body, the Cartesian components of stress are $\sigma_{xx} = 90$ MPa, $\sigma_{yy} = 60$ MPa, $\sigma_{zz} = -30$ MPa, $\tau_{xy} = 45$ MPa, $\tau_{yz} = -15$ MPa and $\tau_{zx} = 45$ MPa. Transform this set of Cartesian stress components into a new set of Cartesian stress components relative to an Ox'y'z' set of coordinates where the Ox'y'z' axes are defined by the following direction cosines:

(a)

	x	y	z
x'	2/15	10/15	11/15
y'	14/15	−5/15	2/15
z'	−5/15	−10/15	10/15

(b)

	x	y	z
x'	2/11	6/11	9/11
y'	6/11	7/11	−6/11
z'	9/11	−6/11	2/11

1.16 At a point in a stressed body, the Cartesian components of stress are σ_{xx} = 100 MPa, σ_{yy} = 50 MPa, σ_{zz} = − 25 MPa, τ_{xy} = 50 MPa, τ_{yz} = − 25 MPa and τ_{zx} = 75 MPa. Transform this set of Cartesian stress components into a new set of Cartesian stress components relative to an Ox'y'z' set of coordinates where the Ox'y'z' axes are defined by the following direction cosines:

(a)

	x	y	z
x'	1/17	12/17	12/17
y'	12/17	−9/17	8/17
z'	12/17	8/17	−9/17

(b)

	x	y	z
x'	9/25	12/25	20/25
y'	12/25	16/25	−15/25
z'	20/25	−15/25	0

1.17 For the state of stress at the point of Exercise 1.3, determine the principal stresses and the maximum shear stress at the point.

1.18 For the state of stress at the point of Exercise 1.4, determine the principal stresses and the maximum shear stress at the point.

1.19 For the state of stress at the point of Exercise 1.5, determine the principal stresses and the maximum shear stress at the point.

1.20 For the state of stress at the point of Exercise 1.6, determine the principal stresses and the maximum shear stress at the point.

1.21 For the state of stress at the point of Exercise 1.12, determine the principal stresses and the maximum shear stress at the point.

1.22 For the state of stress at the point of Exercise 1.14, determine the principal stresses and the maximum shear stress at the point.

1.23 Determine the principal stresses and the maximum shear stress at the point x = ½, y = 1, z = ¾ for the stress distribution given in Exercise 1.9.

1.24 At a point in a stressed body, the Cartesian components of stress are σ_{xx} = 50 MPa, σ_{yy} = 60 MPa, σ_{zz} = 70 MPa, τ_{xy} = 100 MPa, τ_{yz} = 75 MPa and τ_{zx} = 50 MPa. Determine (a) the principal stresses and the maximum shear stress at the point and (b) the orientation of the plane on which the maximum tensile stress acts.

1.25 At a point in a stressed body, the Cartesian components of stress are $\sigma_{xx} = \sigma_{yy} = \sigma_{zz} = 0$, τ_{xy} = 75 MPa, τ_{yz} = 0 and τ_{zx} = − 75 MPa. Determine (a) the principal stresses and the maximum shear stress at the point and (b) the orientation of the plane on which the maximum tensile stress acts.

1.26 At a point in a stressed body, the Cartesian components of stress are $\sigma_{xx} = \sigma_{yy} = \sigma_{zz}$ = 50 MPa, τ_{xy} = 200 MPa, τ_{yz} = 0 and τ_{zx} = 150 MPa. Determine the principal stresses and the associated principal directions. Check on the invariance of I_1, I_2 and I_3.

1.27 At a point in a stressed body, the Cartesian components of stress $\sigma_{xx} = \sigma_{yy} = \sigma_{zz} = 0$, $\tau_{xy} = \tau_{yz} = \tau_{zx}$ = 60 MPa. Determine the principal stresses and the associated principal directions. Check on the invariance of I_1, I_2 and I_3.

1.28 A machine component is subjected to loads which produce the following stress field in a region where an oil hole must be drilled: σ_{xx} = 100 MPa, σ_{yy} = − 60 MPa, σ_{zz} = − 30 MPa, τ_{xy} = 50 MPa, τ_{yz} = 0 and τ_{zx} = 0. To minimize the effects of stress concentrations, the hole must be drilled along a line parallel to the direction of the maximum tensile stress in the region. Determine the direction cosines associated with the centerline of the hole with respect to the reference Oxyz coordinate system.

1.29 A two-dimensional state of stress ($\sigma_{zz} = \tau_{zx} = \tau_{zy} = 0$) exists at a point on the free surface of a machine component. The remaining Cartesian components of stress are σ_{xx} = 90 MPa, σ_{yy} = – 80 MPa, τ_{xy} = – 30 MPa. Determine (a) the principal stresses and their associated directions at the point and (b) the maximum shear stress at the point.

1.30 A two-dimensional state of stress ($\sigma_{zz} = \tau_{zx} = \tau_{zy} = 0$) exists at a point on the surface of a loaded member. Determine the principal stresses and the maximum shear stress at the point if the remaining Cartesian components of stress are σ_{xx} = 45 MPa, σ_{yy} = 30 MPa, τ_{xy} = 20 MPa.

1.31 A two-dimensional state of stress ($\sigma_{zz} = \tau_{zx} = \tau_{zy} = 0$) exists at a point on the surface of a loaded member. The remaining Cartesian components of stress are σ_{xx} = 100 MPa, σ_{yy} = 50 MPa, τ_{xy} = – 20 MPa. Determine the principal stresses and the maximum shear stress at the point.

1.32 A two-dimensional state of stress ($\sigma_{zz} = \tau_{zx} = \tau_{zy} = 0$) exists at a point on the surface of a loaded member. The remaining Cartesian components of stress are σ_{xx} = 90 MPa, σ_{yy} = 40 MPa, τ_{xy} = 60 MPa. Determine the principal stresses and the maximum shear stress at the point.

1.33 Solve Exercise 1.31 by means of Mohr's circle.

1.34 Solve Exercise 1.32 by means of Mohr's circle.

1.35 Write a computer program to determine (a) The principal stresses, (b) The principal directions, (c) The maximum shear stresses, and (d) The von Mises stress if the Cartesian components of stresses are given.

1.36 At the point of Exercise 1.31, determine the normal and shear stresses on a plane whose outer normal has the direction cosines:

$$\cos(\mathbf{n}, x) = 3/5 \qquad \cos(\mathbf{n}, y) = 4/5 \qquad \cos(\mathbf{n}, x) = 0$$

1.37 At the point of Exercise 1.32, determine the normal and shear stresses on a plane whose outer normal has the direction cosines:

$$\cos(\mathbf{n}, x) = 1/3 \qquad \cos(\mathbf{n}, y) = 2/3 \qquad \cos(\mathbf{n}, x) = 2/3$$

1.38 At a point in a machine part the principal stresses are σ_1 = 100 MPa, σ_2 = 60 MPa, and σ_3 = 30 MPa. Determine the normal and shear stresses on a plane whose outer normal has the direction cosines:

$$\cos(\mathbf{n}, n_1) = \sqrt{3}/2 \qquad \cos(\mathbf{n}, n_2) = 0 \qquad \cos(\mathbf{n}, n_3) = ½$$

1.39 If the three principal stresses relative to the Oxyz reference system are $\sigma_1 = \sigma_{xx}$ = 50 MPa, $\sigma_2 = \sigma_{yy}$ = 40 MPa and $\sigma_3 = \sigma_{zz}$ = – 20 MPa, determine the six Cartesian components of stress relative to the Ox'y'z' reference system where Ox'y'z' is defined as:

θ	Case 1	Case 2	Case 3	Case 4
x – x'	$\pi/4$	$\pi/2$	0	$\pi/4$
y – y'	$\pi/4$	$\pi/2$	$\pi/4$	0
z – z'	0	0	$\pi/4$	$\pi/4$

1.40 If the three principal stresses relative to the Oxyz reference system are $\sigma_1 = \sigma_{xx}$ = 75 MPa, $\sigma_2 = \sigma_{yy}$ = 60 MPa and $\sigma_3 = \sigma_{zz}$ = 50 MPa, determine the six Cartesian components of stress relative to an Ox'y'z' reference system where the direction angles associated with the Ox,y,z, axes are:

	x	y	z
x'	$\pi/3$	$\pi/3$	$\pi/4$
y'	$3\pi/4$	$\pi/4$	$\pi/2$
z'	$\pi/3$	$\pi/3$	$3\pi/4$

1.41 If the three principal stresses relative to the Oxyz reference system are $\sigma_1 = \sigma_{xx} = 135$ MPa, $\sigma_2 = \sigma_{yy} = 90$ MPa and $\sigma_3 = \sigma_{zz} = 45$ MPa, determine the six Cartesian components of stress relative to an Ox'y'z' reference system where the Ox'y'z' axes are defined by the following direction cosines:

(a)					(b)			
	x	y	z			x	y	z
x'	2/3	2/3	− 1/3		x'	1/9	− 8/9	4/9
y'	− 2/3	1/3	− 2/3		y'	4/9	4/9	7/9
z'	− 1/3	2/3	2/3		z'	− 8/9	1/9	4/9

1.42 Resolve the general state of stress given in Exercise 1.3 into a hydrostatic state of stress and a state of pure shearing stress.

1.43 Resolve the general state of stress given in Exercise 1.4 into a hydrostatic state of stress and a state of pure shearing stress.

1.44 Resolve the general state of stress given in Exercise 1.5 into a hydrostatic state of stress and a state of pure shearing stress.

1.45 Resolve the general state of stress given in Exercise 1.6 into a hydrostatic state of stress and a state of pure shearing stress.

1.46 Determine the octahedral normal and shearing stresses associated with the principal stresses σ_1, σ_2 and σ_3. Octahedral normal and shearing stresses occur on planes whose outer normal makes equal angles with the principal directions $\mathbf{n}_1$, $\mathbf{n}_2$ and $\mathbf{n}_3$.

REFERENCES

1. Boresi, A. P., and P. P. Lynn: Elasticity in Engineering Mechanics, Prentice-Hall, Inc., Englewood Cliffs, N.J., 1974.
2. Chou, P. C., and N. J. Pagano: Elasticity, 1, D. Van Nostrand Company, Inc., Princeton, N.J., 1967.
3. Durelli, A. J., E. A. Phillips, and C. H. Tsao: Introduction to the Theoretical and Experimental Analysis of Stress and Strain, McGraw-Hill Book Company, New York, 1958.
4. Love, A. E. H.: A Treatise on the Mathematical Theory of Elasticity, Dover Publications, Inc., New York, 1944.
5. Atanackovic and A. Guran: Theory of Elasticity for Scientists and Engineers, Birkhauser, Boston, MA. 2000.
6. Slaughter, W. S.: Linearized Theory of Elasticity, Birkhauser, Boston, MA. 2002.
7. Fung, Y. C. and P. Tang, Classical and Computational Solid Mechanics, World Scientific, Singapore, 2001.
8. Atkin, R. J. and N. Fox: An Introduction to the Theory of Elasticity, Longman, London, 1980.
9. Sechler, E. E.: Elasticity in Engineering, John Wiley & Sons, Inc., New York, 1952.
10. Sokolnikoff, I. S.: Mathematical Theory of Elasticity, 2d ed. McGraw-Hill Book Company, New York, 1956.
11. Southwell, R. V.: An Introduction to the Theory of Elasticity, Oxford University Press, Fair Lawn, N.J., 1953.
12. Timoshenko, S. P., and J. N. Goodier: Theory of Elasticity, 2d ed., McGraw-Hill Book Company, New York, 1951.

CHAPTER 2

STRAIN AND THE STRESS STRAIN RELATIONS

2.1 INTRODUCTION

In the preceding chapter, the state of stress, which develops at an arbitrary point within a body as a result of surface- or body-force loadings, was discussed. The relationships obtained were based on the conditions of equilibrium, and because no assumptions were made regarding body deformations or physical properties of the material of which the body was composed, the results are valid for any material and for any amount of body deformation. In this chapter, the subject of body deformation and associated strain will be discussed. Because strain is a pure geometric quantity, no restrictions on body material will be required. However, in order to obtain linear equations relating displacement to strain, restrictions must be placed on the magnitude of the allowable deformations. In a later section, when the stress-strain relations are developed, the elastic constants of the body material must be considered.

2.2 DEFINITIONS OF DISPLACEMENT AND STRAIN

If a given body is subjected to a system of forces, individual points of the body will, in general, move. This movement of an arbitrary point is a vector quantity known as a **displacement**. If the various points in the body undergo different movements, each can be represented by its own unique displacement vector. Each vector can be resolved into components parallel to a set of Cartesian coordinate axes such that u, v, and w are the displacement components in the x, y, and z directions, respectively. Motion of the body may be considered as the sum of two parts:

1. The translation and/or rotation of the body as a whole.
2. The movement of points within the body relative to each other.

The translation or rotation of the body as a whole is known as **rigid-body motion**. This type of motion is applicable to either the idealized rigid body or the real deformable body. The movement of the points of the body relative to each other is known as a **deformation** and is obviously a property of real bodies only. Rigid-body motions can be large or small. Deformations, in general, are small except when rubber-like materials or specialized structures such as long, slender beams are involved.

Strain is a geometric quantity, which depends on the relative movements of two or three points in the body and therefore is related only to the deformation displacements. Because rigid-body displacements do not produce strains, they will be neglected in all further developments in this chapter. In the preceding chapter, two types of stresses were discussed—normal stress and shear stress. This same classification will be used for strains. A normal strain is defined as the change in length of a line segment between two points divided by the original length of the line segment. A shearing strain is defined as the angular change between two line segments that were originally perpendicular. The relationships between strains and displacements can be determined by considering the deformation of an arbitrary cube in a body as a system of loads is applied. This deformation is illustrated in Fig. 2.1, in which a general point P is moved through a distance u in the x direction, v in the y direction and w in the z direction. The other corners of the cube are also displaced and, in general, they will be displaced by amounts that differ from those at point P. For example the displacements u*, v* and w* associated with point Q can be expressed in terms of the displacements u, v and w at point P by means of a Taylor-series expansion. Thus:

$$u^* = u + \frac{\partial u}{\partial x}\Delta x + \frac{\partial u}{\partial y}\Delta y + \frac{\partial u}{\partial z}\Delta z + \cdots$$

$$v^* = v + \frac{\partial v}{\partial x}\Delta x + \frac{\partial v}{\partial y}\Delta y + \frac{\partial v}{\partial z}\Delta z + \cdots \qquad (2.1)$$

$$w^* = w + \frac{\partial w}{\partial x}\Delta x + \frac{\partial w}{\partial y}\Delta y + \frac{\partial w}{\partial z}\Delta z + \cdots$$

The terms shown in the above expressions are the only significant terms if it is assumed that the cube is sufficiently small for higher-order terms such as $(\Delta x)^2$, $(\Delta y)^2$, $(\Delta z)^2$, ... to be neglected. Under these conditions, planes will remain plane and straight lines will remain straight lines in the deformed cube, as shown in Fig. 2.1.

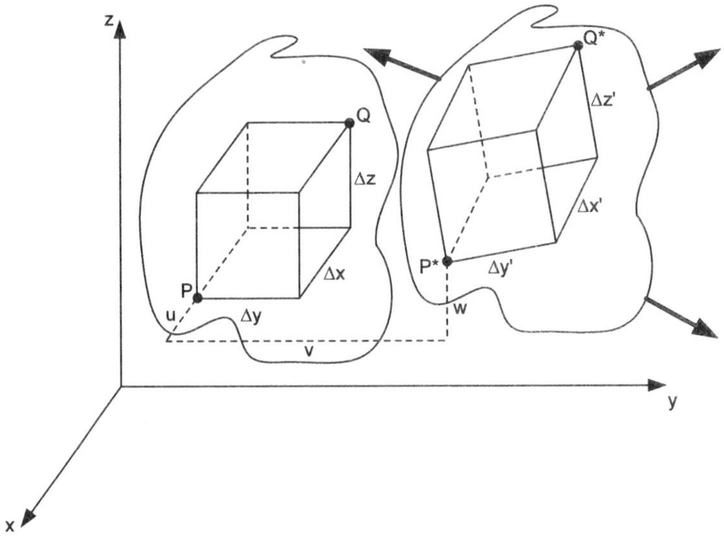

Figure 2.1 The distortion of an arbitrary cube in a body after applying a system of forces.

The average normal strain along an arbitrary line segment was previously defined as the change in length of the line segment divided by its original length. This normal strain can be expressed in terms of the displacements experienced by points at the ends of the segment. For example, consider the line PQ originally oriented parallel to the x-axis, as shown in Fig.2.2. Because y and z are constant along PQ, Eqs. (2.1) yield the following displacements for point Q if the displacements for point P are given by u, v, and w:

$$u^* = u + \frac{\partial u}{\partial x}\Delta x \qquad v^* = v + \frac{\partial v}{\partial x}\Delta x \qquad w^* = w + \frac{\partial w}{\partial x}\Delta x$$

From the definition of normal strain:

$$\varepsilon_{xx} = \frac{\Delta x' - \Delta x}{\Delta x} \qquad (a)$$

which is equivalent to:

$$\Delta x' = (1 + \varepsilon_{xx})\Delta x \qquad (b)$$

As shown in Fig. 2.2, the deformed length $\Delta x'$ can be expressed in terms of the displacement gradients as:

$$(\Delta x')^2 = \left[\left(1+\frac{\partial u}{\partial x}\right)\Delta x\right]^2 + \left(\frac{\partial v}{\partial x}\Delta x\right)^2 + \left(\frac{\partial w}{\partial x}\Delta x\right)^2 \qquad \text{(c)}$$

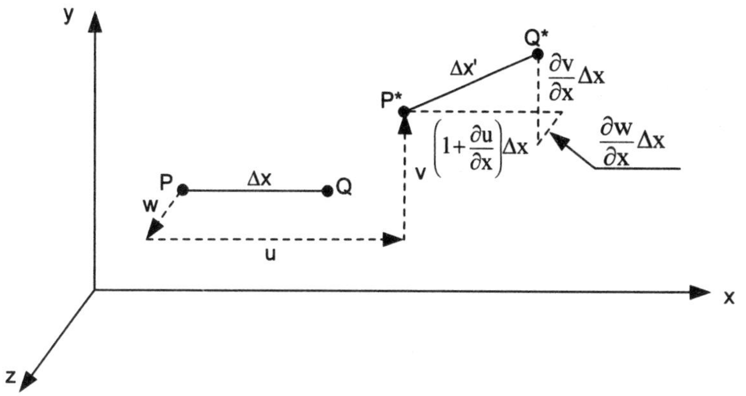

Figure 2.2 Displacement gradients associated with the normal strain ε_{xx}.

Squaring Eq. (b) and substituting Eq. (c) yields:

$$(1+\varepsilon_{xx})^2(\Delta x)^2 = \left[1+2\frac{\partial u}{\partial x}+\left(\frac{\partial u}{\partial x}\right)^2+\left(\frac{\partial v}{\partial x}\right)^2+\left(\frac{\partial w}{\partial x}\right)^2\right](\Delta x)^2$$

or

$$\varepsilon_{xx} = \sqrt{1+2\frac{\partial u}{\partial x}+\left(\frac{\partial u}{\partial x}\right)^2+\left(\frac{\partial v}{\partial x}\right)^2+\left(\frac{\partial w}{\partial x}\right)^2} - 1 \qquad \text{(2.2a)}$$

In a similar manner, considering line segments originally oriented parallel to the y and z-axes leads to:

$$\varepsilon_{yy} = \sqrt{1+2\frac{\partial v}{\partial y}+\left(\frac{\partial v}{\partial y}\right)^2+\left(\frac{\partial w}{\partial y}\right)^2+\left(\frac{\partial u}{\partial y}\right)^2} - 1 \qquad \text{(2.2b)}$$

$$\varepsilon_{zz} = \sqrt{1+2\frac{\partial w}{\partial z}+\left(\frac{\partial w}{\partial z}\right)^2+\left(\frac{\partial u}{\partial z}\right)^2+\left(\frac{\partial v}{\partial z}\right)^2} - 1 \qquad \text{(2.2c)}$$

The shear strain components can also be related to the displacements by considering the changes in right angle experienced by the edges of the cube during deformation. For example, consider lines PQ and PR, as shown in Fig. 2.3. The angle θ^* between P*Q* and P*R* in the deformed state can be expressed in terms of the displacement gradients because the cosine of the angle between any two intersecting lines in space is the sum of the pair wise products of the direction cosines of the lines with respect to the same set of reference axes. Thus:

$$\cos\theta^* = \left[\left(1+\frac{\partial u}{\partial x}\right)\frac{\Delta x}{\Delta x'}\right]\left(\frac{\partial u}{\partial y}\frac{\Delta y}{\Delta y'}\right)+\left(\frac{\partial v}{\partial x}\frac{\Delta x}{\Delta x'}\right)\left[\left(1+\frac{\partial v}{\partial y}\right)\frac{\Delta y}{\Delta y'}\right]+\left(\frac{\partial w}{\partial x}\frac{\Delta x}{\Delta x'}\right)\left(\frac{\partial w}{\partial y}\frac{\Delta y}{\Delta y'}\right) \qquad \text{(d)}$$

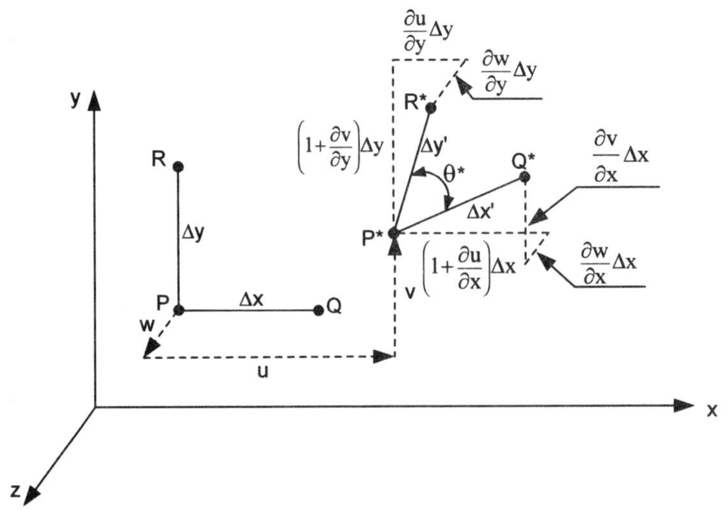

Figure 2.3 Displacement gradients associated with the shear strain γ_{xy}.

From the definition of shear strain,

$$\gamma_{xy} = [(\pi/2) - \theta^*] \tag{e}$$

therefore:

$$\sin \gamma_{xy} = \sin [(\pi/2) - \theta^*] = \cos \theta^* \tag{f}$$

Substituting Eq. (d) into Eq. (f) and simplifying yields:

$$\sin\gamma_{xy} = \left[\left(1+\frac{\partial u}{\partial x}\right)\frac{\partial u}{\partial y} + \left(1+\frac{\partial v}{\partial x}\right)\frac{\partial v}{\partial y} + \frac{\partial w}{\partial x}\frac{\partial w}{\partial y}\right]\left(\frac{\Delta x}{\Delta x'}\frac{\Delta y}{\Delta y'}\right)$$

From Eq. (b)

$$\Delta x' = (1 + \varepsilon_{xx})\Delta x \qquad \text{and} \qquad \Delta y' = (1 + \varepsilon_{yy})\Delta y$$

therefore:

$$\gamma_{xy} = \arcsin\frac{\dfrac{\partial u}{\partial y} + \dfrac{\partial v}{\partial x} + \dfrac{\partial u}{\partial x}\dfrac{\partial u}{\partial y} + \dfrac{\partial v}{\partial x}\dfrac{\partial v}{\partial y} + \dfrac{\partial w}{\partial x}\dfrac{\partial w}{\partial y}}{(1+\varepsilon_{xx})(1+\varepsilon_{yy})} \tag{2.3a}$$

In a similar manner, by considering two line segments originally oriented parallel to the y and z-axes and the z and x axes, one obtains:

$$\gamma_{yz} = \arcsin\frac{\dfrac{\partial v}{\partial z} + \dfrac{\partial w}{\partial y} + \dfrac{\partial v}{\partial y}\dfrac{\partial v}{\partial z} + \dfrac{\partial w}{\partial y}\dfrac{\partial w}{\partial z} + \dfrac{\partial u}{\partial y}\dfrac{\partial u}{\partial z}}{(1+\varepsilon_{yy})(1+\varepsilon_{zz})} \tag{2.3b}$$

$$\gamma_{zx} = \arcsin\frac{\dfrac{\partial w}{\partial x} + \dfrac{\partial u}{\partial z} + \dfrac{\partial w}{\partial z}\dfrac{\partial w}{\partial x} + \dfrac{\partial u}{\partial z}\dfrac{\partial u}{\partial x} + \dfrac{\partial v}{\partial z}\dfrac{\partial v}{\partial x}}{(1+\varepsilon_{zz})(1+\varepsilon_{xx})} \tag{2.3c}$$

Equations (2.2) and (2.3) represent a common engineering description of strain in terms of positions of points in a body before and after deformation. In the development of these equations, no limitations were imposed on the magnitudes of the strains. One restriction was introduced, however, when the higher-order terms in the Taylor-series expansion for displacement were neglected. This restriction has the effect of limiting the length of the line segment (gage length) used for strain determinations unless displacement gradients ($\partial u/\partial x$, $\partial v/\partial y$, ...) in the region of interest are essentially constant. If displacement gradients change rapidly with position in the region of interest, very short gage lengths will be required for accurate strain measurements.

In a wide variety of engineering problems, the displacements and strains produced by the applied loads are very small. Under these conditions, it can be assumed that products and squares of displacement gradients will be small with respect to the displacement gradients and therefore can be neglected. With this assumption, Eqs. (2.2) and (2.3) reduce to the strain-displacement equations frequently encountered in the theory of elasticity. The reduced form of each equation is:

$$\begin{aligned}
\varepsilon_{xx} &= \frac{\partial u}{\partial x} \\
\varepsilon_{yy} &= \frac{\partial v}{\partial y} \\
\varepsilon_{zz} &= \frac{\partial w}{\partial z} \\
\gamma_{xy} &= \frac{\partial v}{\partial x} + \frac{\partial u}{\partial y} \\
\gamma_{yz} &= \frac{\partial w}{\partial y} + \frac{\partial v}{\partial z} \\
\gamma_{zx} &= \frac{\partial u}{\partial z} + \frac{\partial w}{\partial x}
\end{aligned} \tag{2.4}$$

Equations (2.4) indicate that it is a simple matter to convert a displacement field into a strain field. However, as will be emphasized later, an entire displacement field is rarely determined experimentally. Usually, strains are determined at a number of small areas on the surface of the body by using strain gages. In certain problems, however, the displacement field can be determined analytically, and in these instances, Eqs. (2.4) become very important.

2.3 STRAIN EQUATIONS OF TRANSFORMATION

Now that the normal and shearing strains in the x, y, z, directions have been determined, consider the normal strain in an arbitrary direction. Refer to Fig. 2.4 and consider the elongation of the diagonal PQ. By definition the normal strain along PQ is:

$$\varepsilon_{PQ} = \frac{P^*Q^* - PQ}{PQ} \tag{a}$$

From geometric considerations, as illustrated in Fig. 2.4:

$$(PQ)^2 = (\Delta x)^2 + (\Delta y)^2 + (\Delta z)^2 \tag{b}$$

$$(P^*Q^*)^2 = (\Delta x^*)^2 + (\Delta y^*)^2 + (\Delta z^*)^2 \tag{c}$$

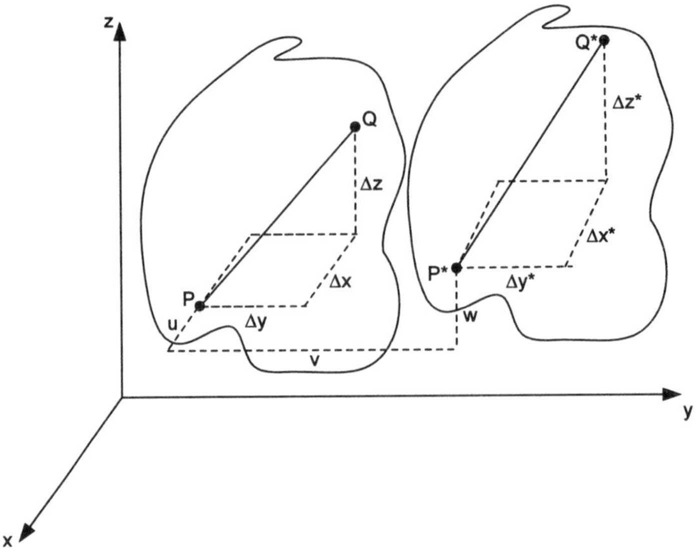

Figure 2.4 Displacements of points P and Q in a body that result from a system of loads.

In general, the component Δx^* will have a different length than the component Δx because of the deformation of the body in the x direction. From Fig. 2.4 it can easily be seen that:

$$\Delta x^* = \left(1+\frac{\partial u}{\partial x}\right)\Delta x + \frac{\partial u}{\partial y}\Delta y + \frac{\partial u}{\partial z}\Delta z$$

$$\Delta y^* = \frac{\partial v}{\partial x}\Delta x + \left(1+\frac{\partial v}{\partial y}\right)\Delta y + \frac{\partial v}{\partial z}\Delta z \qquad \text{(d)}$$

$$\Delta z^* = \frac{\partial w}{\partial x}\Delta x + \frac{\partial w}{\partial y}\Delta y + \left(1+\frac{\partial w}{\partial z}\right)\Delta z$$

If Eqs. (d) are substituted into Eq. (c), the length of the deformed line segment P*Q* can be computed. In the substitution, because the deformations are extremely small, the products and squares of derivatives can be neglected. Thus:

$$\begin{aligned}(P^*Q^*)^2 &= \left(1+2\frac{\partial u}{\partial x}\right)(\Delta x)^2 + \left(1+2\frac{\partial v}{\partial y}\right)(\Delta y)^2 + \left(1+2\frac{\partial w}{\partial z}\right)(\Delta z)^2 \\ &+ 2\left(\frac{\partial u}{\partial y}+\frac{\partial v}{\partial x}\right)\Delta x\Delta y + 2\left(\frac{\partial v}{\partial z}+\frac{\partial w}{\partial y}\right)\Delta y\Delta z + 2\left(\frac{\partial w}{\partial x}+\frac{\partial u}{\partial z}\right)\Delta z\Delta x\end{aligned} \qquad \text{(e)}$$

Equation (a) can be rearranged in the following form:

$$\varepsilon_{PQ} = \frac{P^*Q^*}{PQ} - 1$$

or

$$(\varepsilon_{PQ}+1)^2 = \left(\frac{P^*Q^*}{PQ}\right)^2$$

If Eqs. (b) and (e) are substituted into this rearranged form of Eq. (a), the following relation can be obtained after some rearrangement of terms:

$$\begin{aligned}(\varepsilon_{PQ}+1)^2 &= \cos^2(x,PQ)+\cos^2(y,PQ)+\cos^2(z,PQ)\\ &+2\frac{\partial u}{\partial x}\cos^2(x,PQ)+2\frac{\partial v}{\partial y}\cos^2(y,PQ)+2\frac{\partial w}{\partial z}\cos^2(z,PQ)\\ &+2\left(\frac{\partial u}{\partial y}+\frac{\partial v}{\partial x}\right)\cos(x,PQ)\cos(y,PQ)+2\left(\frac{\partial v}{\partial z}+\frac{\partial w}{\partial y}\right)\cos(y,PQ)\cos(z,PQ)\\ &+2\left(\frac{\partial w}{\partial x}+\frac{\partial u}{\partial z}\right)\cos(z,PQ)\cos(x,PQ)\end{aligned} \tag{f}$$

If the left side of Eq. (f) is expanded, the ε_{PQ}^2 term can be neglected because it is of the same order of magnitude as the products and squares of the displacement derivatives, which were neglected, in a previous step of this development. Recall also that:

$$\cos^2(x, PQ) + \cos^2(y, PQ) + \cos^2(z, PQ) = 1$$

Thus, the basic equation for the strain along an arbitrary line segment is:

$$\begin{aligned}\varepsilon_{PQ} &= \frac{\partial u}{\partial x}\cos^2(x,PQ)+\frac{\partial v}{\partial y}\cos^2(y,PQ)+\frac{\partial w}{\partial z}\cos^2(z,PQ)\\ &+\left(\frac{\partial u}{\partial y}+\frac{\partial v}{\partial x}\right)\cos(x,PQ)\cos(y,PQ)+\left(\frac{\partial v}{\partial z}+\frac{\partial w}{\partial y}\right)\cos(y,PQ)\cos(z,PQ)\\ &+\left(\frac{\partial w}{\partial x}+\frac{\partial u}{\partial z}\right)\cos(z,PQ)\cos(x,PQ)\end{aligned} \tag{2.5a}$$

Equations (2.4) and (2.5a) can be used to determine $\varepsilon_{x'x'}$ by choosing the direction of PQ parallel to the x' axis. Then:

$$\begin{aligned}\varepsilon_{x'x'} &= \varepsilon_{xx}\cos^2(x,x')+\varepsilon_{yy}\cos^2(y,x')+\varepsilon_{zz}\cos^2(z,x')\\ &+\gamma_{xy}\cos(x,x')\cos(y,x')+\gamma_{yz}\cos(y,x')\cos(z,x')\\ &+\gamma_{zx}\cos(z,x')\cos(x,x')\end{aligned} \tag{2.6a}$$

In a similar manner $\varepsilon_{y'y'}$ and $\varepsilon_{z'z'}$ can be determined by choosing the direction of PQ parallel to the y' and z' axes, respectively:

$$\begin{aligned}\varepsilon_{y'y'} &= \varepsilon_{yy}\cos^2(y,y')+\varepsilon_{zz}\cos^2(z,y')+\varepsilon_{xx}\cos^2(x,z')\\ &+\gamma_{yz}\cos(y,y')\cos(z,y')+\gamma_{zx}\cos(z,y')\cos(x,y')\\ &+\gamma_{xy}\cos(x,y')\cos(y,y')\end{aligned} \tag{2.6b}$$

$$\begin{aligned}\varepsilon_{z'z'} &= \varepsilon_{zz}\cos^2(z,z') + \varepsilon_{xx}\cos^2(x,z') + \varepsilon_{yy}\cos^2(y,z') \\ &+ \gamma_{zx}\cos(z,z')\cos(x,z') + \gamma_{xy}\cos(x,z')\cos(y,z') \\ &+ \gamma_{yz}\cos(y,z')\cos(z,z')\end{aligned} \tag{2.6c}$$

A similar but somewhat more involved derivation can be used to establish the shearing strains. Consider the angular change in an arbitrary right angle formed by two line segments PQ_1 and PQ_2. The shearing strain $\gamma_{PQ1,\,PQ2}$ can be shown to be given by:

$$\begin{aligned}\gamma_{PQ1,\,PQ2} &= 2\varepsilon_{xx}\cos(x, PQ_1)\cos(x, PQ_2) + 2\varepsilon_{yy}\cos(y, PQ_1)\cos(y, PQ_2) + 2\varepsilon_{zz}\cos(z, PQ_1)\cos(z, PQ_2) \\ &+ \gamma_{xy}[\cos(x, PQ_1)\cos(y, PQ_2) + \cos(x, PQ_2)\cos(y, PQ_1)] \\ &+ \gamma_{yz}[\cos(y, PQ_1)\cos(z, PQ_2) + \cos(y, PQ_2)\cos(z, PQ_1)] \\ &+ \gamma_{zx}[\cos(z, PQ_1)\cos(x, PQ_2) + \cos(z, PQ_2)\cos(x, PQ_1)]\end{aligned} \tag{2.5b}$$

By choosing PQ_1 parallel to x' and PQ_2 parallel to y', an expression for $\gamma_{x'y'}$ is obtained as follows:

$$\begin{aligned}\gamma_{x'y'} &= 2\varepsilon_{xx}\cos(x, x')\cos(x, y') + 2\varepsilon_{yy}\cos(y, x')\cos(y, y') + 2\varepsilon_{zz}\cos(z, x')\cos(z, y') \\ &+ \gamma_{xy}[\cos(x, x')\cos(y, y') + \cos(x, y')\cos(y, x')] \\ &+ \gamma_{yz}[\cos(y, x')\cos(z, y') + \cos(y, y')\cos(z, x')] \\ &+ \gamma_{zx}[\cos(z, x')\cos(x, y') + \cos(z, y')\cos(x, x')]\end{aligned} \tag{2.6d}$$

Similarly:

$$\begin{aligned}\gamma_{y'z'} &= 2\varepsilon_{yy}\cos(y, y')\cos(y, z') + 2\varepsilon_{zz}\cos(z, y')\cos(z, z') + 2\varepsilon_{xx}\cos(x, y')\cos(x, z') \\ &+ \gamma_{yz}[\cos(y, y')\cos(z, z') + \cos(y, z')\cos(z, y')] \\ &+ \gamma_{zx}[\cos(z, y')\cos(x, z') + \cos(z, z')\cos(x, y')] \\ &+ \gamma_{xy}[\cos(x, y')\cos(y, z') + \cos(x, z')\cos(y, y')]\end{aligned} \tag{2.6e}$$

$$\begin{aligned}\gamma_{z'x'} &= 2\varepsilon_{zz}\cos(z, z')\cos(z, x') + 2\varepsilon_{xx}\cos(x, z')\cos(x, x') + 2\varepsilon_{yy}\cos(y, z')\cos(y, x') \\ &+ \gamma_{zx}[\cos(z, z')\cos(x, x') + \cos(z, x')\cos(x, z')] \\ &+ \gamma_{xy}[\cos(x, z')\cos(y, x') + \cos(x, x')\cos(y, z')] \\ &+ \gamma_{yz}[\cos(y, z')\cos(z, x') + \cos(y, x')\cos(z, z')]\end{aligned} \tag{2.6f}$$

Equations (2.6a) to (2.6f) are the strain equations of transformation and can be used to transform the six Cartesian components of strain ε_{xx}, ε_{yy}, ε_{zz}, γ_{xy}, γ_{yz} and γ_{zx} relative to the Oxyz reference system to six other Cartesian components of strain relative to the Ox'y'z' reference system.

A comparison of Eqs. (2.6) with the stress equations of transformation, Eq. (1.6), shows remarkable similarities:

$$\sigma_{xx} \Leftrightarrow \varepsilon_{xx} \qquad 2\tau_{xy} \Leftrightarrow \gamma_{xy}$$

$$\sigma_{yy} \Leftrightarrow \varepsilon_{yy} \qquad 2\tau_{yz} \Leftrightarrow \gamma_{yz} \qquad (2.7)$$

$$\sigma_{zz} \Leftrightarrow \varepsilon_{yy} \qquad 2\tau_{zx} \Leftrightarrow \gamma_{zx}$$

Here the symbol $\Leftrightarrow$ indicates an interchange. This interchange is important because many of the derivations given in the preceding chapter for stresses can be converted directly into strains. Some of these conversions are indicated in the next section.

2.4 PRINCIPAL STRAINS

From the similarity between the laws of stress and strain transformation it can be concluded that there exist at most three distinct principal strains with their three associated principal directions. By substituting the conversions indicated by Eqs. (2.7) into Eq. (1.7), the cubic equation whose roots give the principal strains is obtained:

$$\varepsilon_n^3 - (\varepsilon_{xx} + \varepsilon_{yy} + \varepsilon_{zz})\varepsilon_n^2 + (\varepsilon_{xx}\varepsilon_{yy} + \varepsilon_{yy}\varepsilon_{zz} + \varepsilon_{zz}\varepsilon_{xx} - \gamma_{xy}^2/4 - \gamma_{yz}^2/4 - \gamma_{zx}^2/4)\varepsilon_n$$

$$- (\varepsilon_{xx}\varepsilon_{yy}\varepsilon_{zz} - \varepsilon_{xx}\gamma_{yz}^2/4 - \varepsilon_{yy}\gamma_{zx}^2/4 - \varepsilon_{zz}\gamma_{xy}^2/4 + \gamma_{xy}\gamma_{yz}\gamma_{zx}/4) = 0 \qquad (2.8)$$

As with principal stresses, three situations exist:

$$\varepsilon_1 \neq \varepsilon_2 \neq \varepsilon_3 \qquad \varepsilon_1 = \varepsilon_2 \neq \varepsilon_3 \qquad \varepsilon_1 = \varepsilon_2 = \varepsilon_3$$

The significance of these three cases is described in the discussion presented in Section 1.6.

Similarly, there are three strain invariants that are analogous to the three stress invariants. By substituting Eqs. (2.7) into Eqs. (1.8), the following expressions are obtained for the strain variants:

$$J_1 = \varepsilon_{xx} + \varepsilon_{yy} + \varepsilon_{zz}$$

$$J_2 = \varepsilon_{xx}\varepsilon_{yy} + \varepsilon_{yy}\varepsilon_{zz} + \varepsilon_{zz}\varepsilon_{xx} - \gamma_{xy}^2/4 - \gamma_{yz}^2/4 - \gamma_{zx}^2/4 \qquad (2.9)$$

$$J_3 = \varepsilon_{xx}\varepsilon_{yy}\varepsilon_{zz} - \varepsilon_{xx}\gamma_{yz}^2/4 - \varepsilon_{yy}\gamma_{zx}^2/4 - \varepsilon_{zz}\gamma_{xy}^2/4 + \gamma_{xy}\gamma_{yz}\gamma_{zx}/4$$

It is clear that other equations derived in Chap. 1 for stresses could easily be converted into equations in terms of strains. A few more will be covered in the exercises at the end of the chapter, and others will be converted as the need arises.

2.5 COMPATIBILITY

From a given displacement field—three equations expressing u, v, and w as functions of x, y, and z—a unique strain field can be determined by using Eqs. (2.4). However, an arbitrary strain field may yield an impossible displacement field, i.e., one in which the body might contain voids after deformation. A valid displacement field can be ensured only if the body under consideration is simply connected and if the strain field satisfies a set of equations known as the compatibility relations. The six equations of compatibility, which must be satisfied, are:

$$\frac{\partial^2 \gamma_{xy}}{\partial x \partial y} = \frac{\partial^2 \varepsilon_{xx}}{\partial y^2} + \frac{\partial^2 \varepsilon_{yy}}{\partial x^2} \quad (2.10a)$$

$$\frac{\partial^2 \gamma_{yz}}{\partial y \partial z} = \frac{\partial^2 \varepsilon_{yy}}{\partial z^2} + \frac{\partial^2 \varepsilon_{zz}}{\partial y^2} \quad (2.10b)$$

$$\frac{\partial^2 \gamma_{zx}}{\partial z \partial x} = \frac{\partial^2 \varepsilon_{zz}}{\partial x^2} + \frac{\partial^2 \varepsilon_{xx}}{\partial z^2} \quad (2.10c)$$

$$2\frac{\partial^2 \varepsilon_{xx}}{\partial y \partial z} = \frac{\partial}{\partial x}\left(-\frac{\partial \gamma_{yz}}{\partial x} + \frac{\partial \gamma_{zx}}{\partial y} + \frac{\partial \gamma_{xy}}{\partial z}\right) \quad (2.10d)$$

$$2\frac{\partial^2 \varepsilon_{yy}}{\partial z \partial x} = \frac{\partial}{\partial y}\left(\frac{\partial \gamma_{yz}}{\partial x} - \frac{\partial \gamma_{zx}}{\partial y} + \frac{\partial \gamma_{xy}}{\partial z}\right) \quad (2.10e)$$

$$2\frac{\partial^2 \varepsilon_{zz}}{\partial x \partial y} = \frac{\partial}{\partial z}\left(\frac{\partial \gamma_{yz}}{\partial x} + \frac{\partial \gamma_{zx}}{\partial y} - \frac{\partial \gamma_{xy}}{\partial z}\right) \quad (2.10f)$$

In order to derive Eq. (2.10a), begin by recalling:

$$\gamma_{xy} = \frac{\partial u}{\partial y} + \frac{\partial v}{\partial x} \quad (a)$$

Differentiating γ_{xy} once with respect to x and then again with respect to y gives:

$$\frac{\partial^2 \gamma_{xy}}{\partial x \partial y} = \frac{\partial^3 u}{\partial x \partial y^2} + \frac{\partial^3 v}{\partial x^2 \partial y} \quad (b)$$

Note that:

$$\frac{\partial^2 \varepsilon_{xx}}{\partial y^2} = \frac{\partial^3 u}{\partial x \partial y^2} \quad \text{and} \quad \frac{\partial^2 \varepsilon_{yy}}{\partial x^2} = \frac{\partial^3 v}{\partial x^2 \partial y} \quad (c)$$

Substituting Eqs. (c) into Eq. (b) gives:

$$\frac{\partial^2 \gamma_{xy}}{\partial x \partial y} = \frac{\partial^2 \varepsilon_{xx}}{\partial y^2} + \frac{\partial^2 \varepsilon_{yy}}{\partial x^2} \quad (d)$$

and establishes Eq. (2.10a). By using the same approach, Eqs. (2.10b) and (2.10c) can be verified. The proof of Eq. (2.10d) is obtained by considering four identities:

$$\frac{\partial^2 \varepsilon_{xx}}{\partial y \partial z} = \frac{\partial^3 u}{\partial x \partial y \partial z} \quad (e)$$

$$\frac{\partial^2 \gamma_{xy}}{\partial x \partial z} = \frac{\partial^3 u}{\partial x \partial y \partial z} + \frac{\partial^3 v}{\partial x^2 \partial z} \qquad \text{(f)}$$

$$\frac{\partial^2 \gamma_{zx}}{\partial x \partial y} = \frac{\partial^3 w}{\partial x^2 \partial y} + \frac{\partial^3 u}{\partial x \partial y \partial z} \qquad \text{(g)}$$

$$\frac{\partial^2 \gamma_{yz}}{\partial x^2} = \frac{\partial^3 w}{\partial x^2 \partial y} + \frac{\partial^3 v}{\partial x^2 \partial z} \qquad \text{(h)}$$

Now by forming:

$$2(e) = (f) + (g) - (h) \qquad \text{(i)}$$

thus obtaining:

$$2\frac{\partial^2 \varepsilon_{xx}}{\partial y \partial z} = \frac{\partial}{\partial x}\left(\frac{\partial \gamma_{xy}}{\partial z} + \frac{\partial \gamma_{zx}}{\partial y} - \frac{\partial \gamma_{yz}}{\partial x}\right) \qquad \text{(j)}$$

Eq. (2.10d) is verified. The remaining two compatibility relations can be established in an identical manner.

In order to gain a better physical understanding of the compatibility relations, consider a two-dimensional body made up of a large number of small, square elements. When the body is loaded, the elements deform. By measuring angle changes and length changes, the strains that develop in each element can be determined. This procedure is accomplished theoretically by differentiating the displacement field. Consider now the inverse problem. Suppose a large number of small, deformed elements are given which must be fitted together to form a body free of voids and discontinuities. **If and only if** each element is properly strained can the body be reassembled without voids. The deformed elements correspond to the case of a prescribed strain field. The check if the elements are all properly strained, and hence compatible with each other, is determined with compatibility relations. If these relations are satisfied, the elements will fit together properly, thus guaranteeing a satisfactory displacement field.

2.6 STRESS-STRAIN RELATIONS

Thus far stress and strain have been discussed individually, and no assumptions have been required regarding the behavior of the material except that it was a continuous medium.[1] In this section, stress will be related to strain; therefore, certain restrictive assumptions regarding the body material must be introduced. The first of these assumptions pertains to the linearity of the stress with strain in the body. With a linear stress-strain relationship, it is possible to write the general stress-strain expressions as follows:

$$\sigma_{xx} = K_{11}\,\varepsilon_{xx} + K_{12}\,\varepsilon_{yy} + K_{13}\,\varepsilon_{zz} + K_{14}\,\gamma_{xy} + K_{15}\,\gamma_{yz} + K_{16}\,\gamma_{zx}$$

$$\sigma_{yy} = K_{21}\,\varepsilon_{xx} + K_{22}\,\varepsilon_{yy} + K_{23}\,\varepsilon_{zz} + K_{24}\,\gamma_{xy} + K_{25}\,\gamma_{yz} + K_{26}\,\gamma_{zx}$$

$$\sigma_{zz} = K_{31}\,\varepsilon_{xx} + K_{32}\,\varepsilon_{yy} + K_{33}\,\varepsilon_{zz} + K_{34}\,\gamma_{xy} + K_{35}\,\gamma_{yz} + K_{36}\,\gamma_{zx}$$

(2.11)

---

[1] Actually most metals are not strictly continuous because they are composed of a large number of rather small grains. However, the grains are in almost all cases small enough in comparison with the size of the body for the body to behave as if it were a continuous medium.

$$\tau_{xy} = K_{41}\,\varepsilon_{xx} + K_{42}\,\varepsilon_{yy} + K_{43}\,\varepsilon_{zz} + K_{44}\,\gamma_{xy} + K_{45}\,\gamma_{yz} + K_{46}\,\gamma_{zx}$$

$$\tau_{yz} = K_{51}\,\varepsilon_{xx} + K_{52}\,\varepsilon_{yy} + K_{53}\,\varepsilon_{zz} + K_{54}\,\gamma_{xy} + K_{55}\,\gamma_{yz} + K_{56}\,\gamma_{zx}$$

$$\tau_{zx} = K_{61}\,\varepsilon_{xx} + K_{62}\,\varepsilon_{yy} + K_{63}\,\varepsilon_{zz} + K_{64}\,\gamma_{xy} + K_{65}\,\gamma_{yz} + K_{66}\,\gamma_{zx}$$

where K_{11} to K_{66} are the coefficients of elasticity of the material and are independent of the magnitudes of both the stress and the strain, provided the elastic limit of the material is not exceeded. If the elastic limit is exceeded, the relationship between stress and strain is no longer linear, and Eqs. (2.11) are not valid.

There are 36 coefficients of elasticity in Eqs. (2.11); however, they are not all independent. By strain energy considerations, which are beyond the scope of this book, the number of independent coefficients of elasticity can be reduced to 21. This reduction is quite significant; however, even with 21 constants, Eqs. (2.11) may be considered rather long and involved. By assuming that the material is isotropic, i.e., that the elastic constants are the same in all directions and hence independent of the choice of a coordinate system, the 21 coefficients of elasticity reduce to two constants. The stress-strain relationships then reduce to:

$$\sigma_{xx} = \lambda J_1 + 2\mu\varepsilon_{xx} \quad \sigma_{yy} = \lambda J_1 + 2\mu\varepsilon_{yy} \quad \sigma_{zz} = \lambda J_1 + 2\mu\varepsilon_{zz}$$
$$\tau_{xy} = \mu\gamma_{xy} \qquad \tau_{yz} = \mu\gamma_{yz} \qquad \tau_{zx} = \mu\gamma_{zx} \tag{2.12}$$

where J_1 = first invariant of strain ($\varepsilon_{xx} + \varepsilon_{yy} + \varepsilon_{zz}$); λ = Lame's constant; μ = shear modulus

Equations (2.12) can be solved to give the strains as a function of stress:

$$\varepsilon_{xx} = \frac{\lambda+\mu}{\mu(3\lambda+2\mu)}\sigma_{xx} - \frac{\lambda}{2\mu(3\lambda+2\mu)}(\sigma_{yy}+\sigma_{zz})$$

$$\varepsilon_{yy} = \frac{\lambda+\mu}{\mu(3\lambda+2\mu)}\sigma_{yy} - \frac{\lambda}{2\mu(3\lambda+2\mu)}(\sigma_{xx}+\sigma_{zz})$$

$$\varepsilon_{zz} = \frac{\lambda+\mu}{\mu(3\lambda+2\mu)}\sigma_{zz} - \frac{\lambda}{2\mu(3\lambda+2\mu)}(\sigma_{yy}+\sigma_{xx}) \tag{2.13}$$

$$\gamma_{xy} = \tau_{xy}/\mu \qquad \gamma_{yz} = \tau_{yz}/\mu \qquad \gamma_{zx} = \tau_{zx}/\mu$$

The elastic coefficients μ and λ shown in Eqs. (2.12) and (2.13) arise from a mathematical treatment of the general linear stress-strain relations. In experimental work, Lame's constant λ is rarely used because it has no physical significance; however, as will be shown later, the shear modulus has physical significance and can easily be measured.

Consider a two-dimensional case of pure shear where:

$$\sigma_{xx} = \sigma_{yy} = \sigma_{zz} = \tau_{yz} = \tau_{zx} = 0 \qquad \tau_{xy} = \text{applied shearing stress}$$

From Eqs. (2.13),

$$\mu = \tau_{xy}/\gamma_{xy} \tag{2.14a}$$

Hence, the shear modulus μ is the ratio of the shearing stress to the shearing strain in a two-dimensional state of pure shear.

In a conventional tension test, which is often used to determine the mechanical properties of materials, a long, slender bar is subjected to a state of uniaxial stress in, say, the x direction. In this instance:

$\sigma_{yy} = \sigma_{zz} = \tau_{xy} = \tau_{yz} = \tau_{zx} = 0$ $\qquad\qquad$ σ_{xx} = applied normal stress

From Eqs. (2.13),

$$\varepsilon_{xx} = \frac{\lambda + \mu}{\mu(3\lambda + 2\mu)}\sigma_{xx} \qquad \text{(a)}$$

$$\varepsilon_{yy} = \varepsilon_{zz} = -\frac{\lambda}{2\mu(3\lambda + 2\mu)}\sigma_{xx} \qquad \text{(b)}$$

In elementary strength-of-materials texts, the stress-strain relations for the case of uniaxial stress are often written as:

$$\varepsilon_{xx} = \sigma_{xx}/E \qquad \text{(c)}$$

$$\varepsilon_{yy} = \varepsilon_{zz} = -\nu\sigma_{xx}/E \qquad \text{(d)}$$

By equating the coefficients in Eqs. (a) and (b) to those in Eqs. (c) and (d),

$$E = \frac{\mu(3\lambda + 2\mu)}{\lambda + \mu} \qquad \text{(2.14b)}$$

$$\nu = \frac{\lambda}{2(\lambda + \mu)} \qquad \text{(2.14c)}$$

where E is the modulus of elasticity and ν is Poisson's ratio, defined as:

$$\nu = -\frac{\varepsilon_{yy}}{\varepsilon_{xx}} \qquad \text{(2.14d)}$$

Equations (2.14b) and (2.14c) indicate the conversion from Lame's constant λ and the shear modulus μ to the more commonly used modulus of elasticity E and Poisson's ratio ν.

To establish the definition and physical significance of a fifth elastic constant, consider a state of hydrostatic stress given by:

$\sigma_{xx} = \sigma_{yy} = \sigma_{zz} = -p$ $\qquad\qquad$ $\tau_{xy} = \tau_{yz} = \tau_{zx} = 0$

where p is the uniform pressure acting on the body.

Adding together the first three of Eqs. (2.12) gives:

$$-3p = (3\lambda + 2\mu)J_1$$

or

$$p = -[(3\lambda + 2\mu)J_1]/3 = -KJ_1 = -KD$$

$$K = \frac{(3\lambda + 2\mu)}{3} = -\frac{p}{D} \tag{2.14e}$$

The constant K is known as the **bulk modulus** and is the ratio of the applied hydrostatic pressure to the volume dilatation.

Five elastic constants λ, μ, E, ν and K have been discussed. The constant λ has no physical significance and is employed because it simplifies, mathematically speaking, the stress-strain relations. The constant μ has both mathematical and physical significance. It is used extensively in torsion problems. The constants E and ν are the most widely recognized of the five constants considered and are used in almost all areas of stress analysis. The rather specialized bulk modulus K is used primarily for computing volume changes in a given body subjected to hydrostatic pressure. As indicated previously, there are two and only two independent elastic constants. The five constants discussed are related to each other as shown in Table 2.1.

Because the constants E and ν will be used almost exclusively throughout the remainder of this text, Eqs. (2.15b) and (2.15c) have been substituted into Eqs. (2.12) and (2.13) to obtain expressions for strain in terms of stress and the constants.

$$\begin{aligned}
\varepsilon_{xx} &= (1/E)[\sigma_{xx} - \nu(\sigma_{yy} + \sigma_{zz})] \\
\varepsilon_{yy} &= (1/E)[\sigma_{yy} - \nu(\sigma_{xx} + \sigma_{zz})] \\
\varepsilon_{zz} &= (1/E)[\sigma_{zz} - \nu(\sigma_{yy} + \sigma_{xx})]
\end{aligned} \tag{2.15}$$

$$\gamma_{xy} = \frac{2(1+\nu)}{E}\tau_{xy} \qquad \gamma_{yz} = \frac{2(1+\nu)}{E}\tau_{yz} \qquad \gamma_{zx} = \frac{2(1+\nu)}{E}\tau_{zx}$$

and for stress in terms of strain and the constants.

$$\begin{aligned}
\sigma_{xx} &= \frac{E}{(1+\nu)(1-2\nu)}\left[(1-\nu)\varepsilon_{xx} + \nu(\varepsilon_{yy} + \varepsilon_{zz})\right] \\
\sigma_{yy} &= \frac{E}{(1+\nu)(1-2\nu)}\left[(1-\nu)\varepsilon_{yy} + \nu(\varepsilon_{xx} + \varepsilon_{zz})\right] \\
\sigma_{zz} &= \frac{E}{(1+\nu)(1-2\nu)}\left[(1-\nu)\varepsilon_{zz} + \nu(\varepsilon_{xx} + \varepsilon_{yy})\right]
\end{aligned} \tag{2.16}$$

$$\tau_{xy} = \frac{E}{2(1+\nu)}\gamma_{xy} \qquad \tau_{yz} = \frac{E}{2(1+\nu)}\gamma_{yz} \qquad \tau_{zx} = \frac{E}{2(1+\nu)}\gamma_{zx}$$

Table 2.1
Relationships among the elastic constants

	λ equals	μ equals	E equals	ν equals	K equals
λ, μ			$\frac{\mu(3\lambda+2\mu)}{\lambda+\mu}$	$\frac{\lambda}{2(\lambda+\mu)}$	$\frac{3\lambda+2\mu}{3}$
λ, E		$\frac{A^*+(E-3\lambda)}{4}$		$\frac{A^*-(E+\lambda)}{4\lambda}$	$\frac{A^*+(3\lambda+E)}{6}$
λ, ν		$\frac{\lambda(1-2\nu)}{2\nu}$	$\frac{\lambda(1+\nu)(1-2\nu)}{\nu}$		$\frac{\lambda(1+\nu)}{3\nu}$
λ, K		$\frac{3(K-\lambda)}{2}$	$\frac{9K(K-\lambda)}{3K-\lambda}$	$\frac{\lambda}{3K-\lambda}$	
μ, E	$\frac{\mu(2\mu-E)}{E-3\mu}$			$\frac{(E-2\mu)}{2\mu}$	$\frac{\mu E}{3(3\mu-E)}$
μ, ν	$\frac{2\mu\nu}{1-2\nu}$		$2\mu(1+\nu)$		$\frac{2\mu(1+\nu)}{3(1-2\nu)}$
μ, K	$\frac{3K-2\mu}{3}$		$\frac{9K\mu}{3K+\mu}$	$\frac{3K-2\mu}{2(3K+\mu)}$	
E, ν	$\frac{\nu E}{(1+\nu)(1-2\nu)}$	$\frac{E}{2(1+\nu)}$			$\frac{E}{3(1-2\nu)}$
K, E	$\frac{3K(3K-E)}{9K-E}$	$\frac{3KE}{9K-E}$		$\frac{3K-E}{6K}$	
ν. K	$\frac{3K\nu}{1+\nu}$	$\frac{3K(1-2\nu)}{2(1+\nu)}$	$3K(1-2\nu)$		

$A^* = [E^2 + 2\lambda E + 9\lambda^2]^{1/2}$

2.7 STRAIN-TRANSFORMATION EQUATIONS AND STRESS-STRAIN RELATIONS FOR A TWO-DIMENSIONAL STATE OF STRESS

Simplified forms of the strain-transformation equations and the stress-strain relations, which will be extremely useful in later chapters when electrical-resistance strain-gage and other experimental methods are discussed, are the equations applicable to the strain field associated with a two-dimensional state of stress ($\sigma_{zz} = \tau_{zx} = \tau_{zy} = 0$).

The strain-transformation equations can be obtained from Eqs. (2.6) selecting z' coincident with z and noting from Eqs. (2.15) that $\gamma_{zx} = \gamma_{zy} = 0$. The notation can also be simplified by denoting the angle between x' and x as θ. The equations obtained are:

$$
\begin{aligned}
\varepsilon_{x'x'} &= \varepsilon_{xx}\cos^2\theta + \varepsilon_{yy}\sin^2\theta + \gamma_{xy}\sin\theta\cos\theta \\
\varepsilon_{y'y'} &= \varepsilon_{yy}\cos^2\theta + \varepsilon_{xx}\sin^2\theta - \gamma_{xy}\sin\theta\cos\theta \\
\gamma_{x'y'} &= 2(\varepsilon_{yy} - \varepsilon_{xx})\sin\theta\cos\theta + \gamma_{xy}(\cos^2\theta - \sin^2\theta) \\
\varepsilon_{z'z'} &= \varepsilon_{zz} \qquad \gamma_{y'z'} = \gamma_{z'x'} = 0
\end{aligned}
\tag{2.17}
$$

The stress-strain relations for a two-dimensional state of stress are obtained by substituting $\sigma_{zz} = \tau_{zx} = \tau_{zy} = 0$ into Eqs. (2.15). Thus:

$$
\begin{aligned}
&\varepsilon_{xx} = \frac{1}{E}\left(\sigma_{xx} - \nu\sigma_{yy}\right) \qquad \varepsilon_{yy} = \frac{1}{E}\left(\sigma_{yy} - \nu\sigma_{xx}\right) \qquad \varepsilon_{xx} = -\frac{\nu}{E}\left(\sigma_{xx} + \sigma_{yy}\right) \\
&\gamma_{xy} = \frac{2(1+\nu)}{E}\tau_{xy} \qquad \gamma_{yz} = \gamma_{zx} = 0
\end{aligned}
\tag{2.18}
$$

In a similar manner the equations for stress in terms of strain for the two-dimensional state of stress are obtained from Eqs. (2.16). Thus:

$$
\begin{aligned}
&\sigma_{xx} = \frac{E}{(1-\nu^2)}\left(\varepsilon_{xx} + \nu\varepsilon_{yy}\right) \qquad \sigma_{yy} = \frac{E}{(1-\nu^2)}\left(\varepsilon_{yy} + \nu\varepsilon_{xx}\right) \\
&\sigma_{zz} = \tau_{yz} = \tau_{zx} = 0 \qquad \tau_{xy} = \frac{E}{2(1+\nu)}\gamma_{xy}
\end{aligned}
\tag{2.19}
$$

One additional relationship, which relates the strain ε_{zz} to the measured strains ε_{xx} and ε_{yy} in experimental analyses, is obtained from Eq. (2.16) by substituting $\sigma_{zz} = 0$. Thus:

$$
\varepsilon_{zz} = -\left[\frac{\nu}{1-\nu}\right]\left(\varepsilon_{xx} + \varepsilon_{yy}\right) \tag{2.20}
$$

This equation can be used to establish the magnitude of the third principal strain associated with a two-dimensional state of stress. This information is useful for maximum shear-strain determinations.

2.8 STRESS-STRAIN RELATIONS FOR COMPOSITE MATERIALS

In recent years, composite materials fabricated from strong continuous fibers bonded together with a plastic matrix have been employed in many engineering applications because of their high strength to weight ratios. Composite materials are often employed in high performance structures to reduce weight and in sports equipment for their toughness. The high strength of a composite material is achieved by orienting high-strength fibers (carbon or glass fibers are commonly employed) in the direction of the applied stress. Because of the selective orientation of their fibers, composite materials are anisotropic. In many applications, composite materials are relatively thin when they are used as skin on aircraft and as panels for flooring, etc. In these applications, the composite materials are fabricated with a cross ply structure as illustrated by the cross-section shown in Fig. 2.5. With the fibers arranged in alternating plies in orthogonal directions, the composite is classified as an orthotropic material.

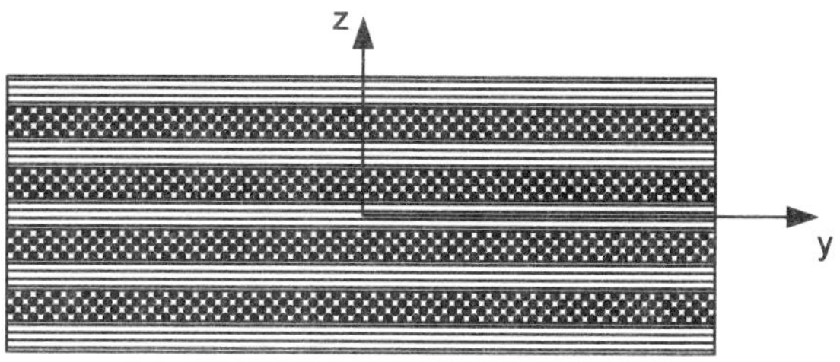

Figure 2.5 Cross section of a cross-ply material showing alternating orthogonal plies of fibers. The x axis is perpendicular to the y-z axis.

In other applications, where composite materials are employed in fabricating long thin members, all of the fibers are oriented along their axis. The cross section showing the fiber orientation in these uniaxial members is presented in Fig. 2.6. With the fibers all aligned with the axis of the uniaxial member, this type of composite can be classified as a transversely isotropic material. The ends of the fibers shown in Fig. 2.6 lie in the transversely isotropic plane.

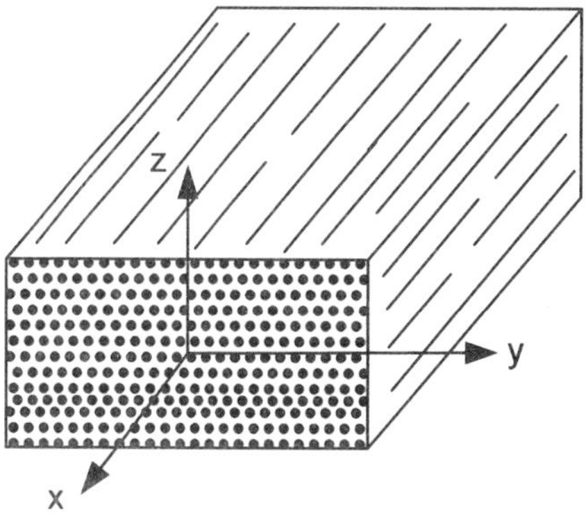

Figure 2.6 Cross section of a uniaxial-ply material showing all of the fibers oriented along the axis of the structural member.

In defining the constitutive equations for the two classes of composite materials described above, it is necessary to model the composite as **effectively homogeneous**. Of course, composites are heterogeneous because they consist of fibers embedded in a matrix material. It is possible to treat these materials as effectively homogeneous only if the local stresses in the matrix and fibers are ignored. Instead average stresses over a representative volume element are determined when the composite is treated as effectively homogeneous. A representative volume element for an effectively homogeneous material, presented in Fig. 2.7, includes many fibers and the matrix material that surrounds them. Individual fibers and small volumes of matrix material are not representative volume elements and cannot be studied with the analysis methods that are described in the subsequent paragraphs. In the following sections, stress strain relations for uniaxial and cross-ply composites will be presented.

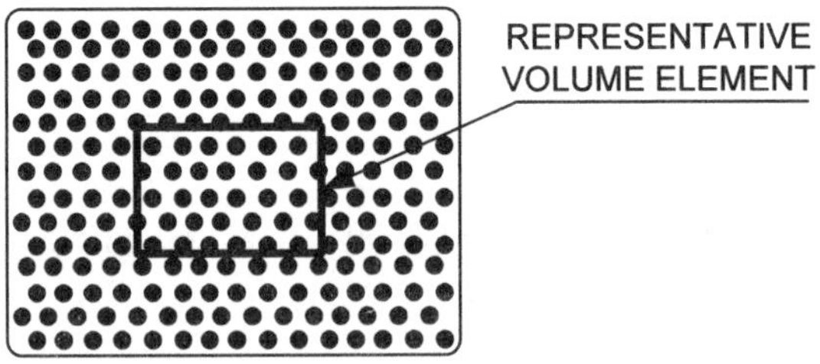

Figure 2.7 A representative volume element contains many fibers and surrounding matrix material.

2.8.1 Stress-strain Relations for Orthotropic Materials

Consider a composite material with a cross ply orientation as shown in Fig. 2.5. It is assumed that the directions of the fibers in the alternating lamina correspond to the x and y coordinates and that the z coordinate is perpendicular to the planes containing these fibers. In this case, the symmetrical stiffness matrix contains nine independent constants as indicated in Eq. (2.21).

$$[K]=\begin{vmatrix} K_{11} & K_{12} & K_{13} & 0 & 0 & 0 \\ K_{12} & K_{22} & K_{23} & 0 & 0 & 0 \\ K_{13} & K_{23} & K_{33} & 0 & 0 & 0 \\ 0 & 0 & 0 & K_{44} & 0 & 0 \\ 0 & 0 & 0 & 0 & K_{55} & 0 \\ 0 & 0 & 0 & 0 & 0 & K_{66} \end{vmatrix} \tag{2.21}$$

The stress strain relations corresponding to this stiffness matrix can be written as:

$$\sigma_{xx} = K_{11}\,\varepsilon_{xx} + K_{12}\,\varepsilon_{yy} + K_{13}\,\varepsilon_{zz}$$

$$\sigma_{yy} = K_{12}\,\varepsilon_{xx} + K_{22}\,\varepsilon_{yy} + K_{23}\,\varepsilon_{zz} \tag{2.22}$$

$$\sigma_{zz} = K_{13}\,\varepsilon_{xx} + K_{23}\,\varepsilon_{yy} + K_{33}\,\varepsilon_{zz}$$

$$\tau_{yz} = K_{44}\,\gamma_{yz} \qquad \tau_{xz} = K_{55}\,\gamma_{xz} \qquad \tau_{xy} = K_{66}\,\gamma_{xy}$$

The constants K_{ij} in the stiffness matrix are not measured directly in the usual tensile tests in a materials laboratory. Engineering elastic constants are measured in these tensile tests. To establish the relations among the constants K_{ij} in the stiffness matrix and these engineering elastic constants, it is assumed that the orientation of the fibers in the composite material coincide with the coordinate system. When the direction of the fibers coincide with the x and y axes for a cross ply composite, it is possible to write the orthotropic strain-stress relations in terms of the usual elastic constants as:

$$\varepsilon_{xx} = \frac{\sigma_{xx}}{E_x} - \frac{\nu_{yx}}{E_y}\sigma_{yy} - \frac{\nu_{zx}}{E_z}\sigma_{zz}$$

$$\varepsilon_{yy} = -\frac{\nu_{xy}}{E_x}\sigma_{xx} + \frac{\sigma_{yy}}{E_y} - \frac{\nu_{zy}}{E_z}\sigma_{zz}$$

$$\varepsilon_{zz} = -\frac{\nu_{xz}}{E_x}\sigma_{xx} - \frac{\nu_{yz}}{E_y}\sigma_{yy} + \frac{\sigma_{zz}}{E_z}$$

$$\gamma_{xy} = \frac{\tau_{xy}}{G_{xy}} \qquad \gamma_{yz} = \frac{\tau_{yz}}{G_{yz}} \qquad \gamma_{xz} = \frac{\tau_{xz}}{G_{xz}} \tag{2.23}$$

where E_x, E_y and E_z are the moduli of elasticity in the three coordinate directions. Because the fiber directions coincide with the coordinate axes (principal material coordinates), E_x and E_y are the elastic moduli in the fiber directions for a cross ply material and E_z is the transverse modulus.

Poisson's ratio is defined in terms of the strains measured in a tensile test where, for example, the stress is applied in the x direction and the lateral strain is measured in the y and z directions. This tensile test provides the data to measure ν_{xy} and ν_{xz} as:

$$\nu_{xy} = -\varepsilon_{yy}/\varepsilon_{xx} \qquad \nu_{xz} = -\varepsilon_{zz}/\varepsilon_{xx} \qquad \text{x-axis loading} \tag{2.24a}$$

By conducting additional tests with tensile specimens loaded in the y and z directions and measuring strains in the lateral directions, the remaining values of Poisson's ratio can be determined from:

$$\nu_{yz} = -\varepsilon_{zz}/\varepsilon_{yy} \qquad \nu_{yx} = -\varepsilon_{xx}/\varepsilon_{yy} \qquad \text{y-axis loading} \tag{2.24b}$$

$$\nu_{zx} = -\varepsilon_{xx}/\varepsilon_{zz} \qquad \nu_{zy} = -\varepsilon_{yy}/\varepsilon_{zz} \qquad \text{z-axis loading} \tag{2.24c}$$

In general $\nu_{xy} \neq \nu_{yx}$, $\nu_{yz} \neq \nu_{zy}$ and $\nu_{zx} \neq \nu_{xz}$,

The shear moduli G_{xy}, G_{yz} and G_{zx} are measured by applying shear stresses τ in the three different planes. One simplification is possible because of the symmetry in the stiffness matrix that enables one to write:

$$E_x \nu_{yx} = E_y \nu_{xy} \qquad E_x \nu_{zx} = E_z \nu_{xz} \qquad E_z \nu_{yz} = E_y \nu_{zy} \tag{2.25}$$

From Eqs. (2.23), (2.24) and (2.25), it is clear that nine independent elastic constants (E_x, E_y, E_z, G_{xy}, G_{xz}, G_{yz}, ν_{xy}, ν_{xz} and ν_{zy}) are necessary to express the stress strain relations for an orthotropic material.

2.8.2 Stress-strain Relations for Orthotropic Materials Under Plane Stress Conditions

The relations presented in the previous section pertained to a three dimensional state of stress imposed on composite materials with orthotropic properties. These relations simplify if the body is plane and the stresses $\sigma_{zz} = \tau_{yz} = \tau_{zx} = 0$. Again it is assumed that the x and y directions define the fiber directions and that the z axis is perpendicular to the plane of the composite. In this case, the stiffness matrix Eq. (2.21) reduces to:

$$[K] = \begin{vmatrix} K_{11} & K_{12} & 0 \\ K_{12} & K_{22} & 0 \\ 0 & 0 & K_{66} \end{vmatrix} \tag{2.26}$$

The stress-strain relations corresponding to this reduced stiffness matrix can be written as:

$$\sigma_{xx} = K_{11}\,\varepsilon_{xx} + K_{12}\,\varepsilon_{yy}$$

$$\sigma_{yy} = K_{12}\,\varepsilon_{xx} + K_{22}\,\varepsilon_{yy}$$
$$\tau_{xy} = K_{66}\,\gamma_{xy} \tag{2.27}$$

Examination of Eq. (2.27) indicates that there are five constants in the stiffness matrix for orthotropic materials subjected to plane states of stress. However due to symmetry of the matrix it is possible to show that only four of the five constants are independent.

The constants from the stiffness matrix can be expressed in terms of the elastic constants that were defined previously. For the plane stress conditions, these relations are written as:

$$K_{11} = \frac{E_x}{1-\nu_{xy}\nu_{yx}} \qquad K_{12} = \frac{\nu_{xy}E_y}{1-\nu_{xy}\nu_{yx}} = \frac{\nu_{yx}E_x}{1-\nu_{xy}\nu_{yx}} \qquad K_{22} = \frac{E_y}{1-\nu_{xy}\nu_{yx}}$$
$$K_{66} = G_{xy} \tag{2.28}$$

Again symmetry of the stiffness matrix enables one to write:

$$E_x\nu_{yx} = E_y\nu_{xy} \tag{2.29}$$

Substituting Eqs. (2.28) into Eqs. (2.27) yields:

$$\sigma_{xx} = \frac{E_x}{1-\nu_{xy}\nu_{yx}}\varepsilon_{xx} + \frac{\nu_{xy}E_y}{1-\nu_{xy}\nu_{yx}}\varepsilon_{yy} = \frac{E_x}{1-\nu_{xy}\nu_{yx}}\left[\varepsilon_{xx} + \nu_{yx}\varepsilon_{yy}\right]$$
$$\sigma_{yy} = \frac{\nu_{xy}E_y}{1-\nu_{xy}\nu_{yx}}\varepsilon_{xx} + \frac{E_y}{1-\nu_{xy}\nu_{yx}}\varepsilon_{yy} = \frac{E_y}{1-\nu_{xy}\nu_{yx}}\left[\nu_{xy}\varepsilon_{xx} + \varepsilon_{yy}\right] \tag{2.30}$$
$$\tau_{xy} = G_{xy}\gamma_{xy}$$

It is clear from an inspection of Eqs. (2.30) that coupling does not occur between the normal stresses and the shearing strain when the coordinate axes x and y are coincident with the fiber directions.

2.8.3 Stress-strain Relations for Transversely Isotropic Materials

A composite with all of its fibers oriented in the x direction is a transversely isotropic material because its effective elastic constants are isotropic on the plane with its outer normal in the x direction. In Fig. 2.6, which represents a body with uniaxial fibers, the y-z plane is an isotropic plane. This isotropy results in a reduction of the number of independent constants in the stiffness matrix [Eq. (2.21)] because of the dependence of some of the constants as listed below:

$$K_{22} = K_{33} \qquad K_{12} = K_{13} \qquad K_{55} = K_{66} \qquad K_{44} = (K_{22} - K_{23})/2 \tag{2.31}$$

Noting the relations listed in Eq. (2.31) enables one to write the stiffness matrix as:

$$[K] = \begin{vmatrix} K_{11} & K_{12} & K_{12} & 0 & 0 & 0 \\ K_{12} & K_{22} & K_{23} & 0 & 0 & 0 \\ K_{12} & K_{23} & K_{22} & 0 & 0 & 0 \\ 0 & 0 & 0 & \frac{K_{22} - K_{23}}{2} & 0 & 0 \\ 0 & 0 & 0 & 0 & K_{66} & 0 \\ 0 & 0 & 0 & 0 & 0 & K_{66} \end{vmatrix} \tag{2.32}$$

The stress strain relations corresponding to this stiffness matrix can be written as:

$$\sigma_{xx} = K_{11}\,\varepsilon_{xx} + K_{12}\,\varepsilon_{yy} + K_{12}\,\varepsilon_{zz}$$

$$\sigma_{yy} = K_{12}\,\varepsilon_{xx} + K_{22}\,\varepsilon_{yy} + K_{23}\,\varepsilon_{zz}$$

$$\sigma_{zz} = K_{12}\,\varepsilon_{xx} + K_{23}\,\varepsilon_{yy} + K_{22}\,\varepsilon_{zz} \tag{2.33}$$

$$\tau_{yz} = [(K_{22} - K_{23})/2]\gamma_{yz} \qquad \tau_{xz} = K_{66}\,\gamma_{xz} \qquad \tau_{xy} = K_{66}\,\gamma_{xy}$$

From these relations it is evident that only four elastic constants are required to characterize a transversely isotropic composite. It is also evident that with all of the fibers aligned with the x-axis, one can write:

$$E_x >> E_y \qquad E_y = E_z \qquad \nu_{xy} = \nu_{xz} \qquad \nu_{yz} = \nu_{zy} \qquad \nu_{yx} << \nu_{xy} \qquad G_{xy} = G_{xz} \tag{2.34}$$

From Eqs. (2.23) and (2.34), one obtains:

$$\varepsilon_{xx} = \frac{\sigma_{xx}}{E_x} - \frac{\nu_{yx}}{E_y}\sigma_{yy} - \frac{\nu_{zx}}{E_y}\sigma_{zz}$$

$$\varepsilon_{yy} = -\frac{\nu_{xy}}{E_x}\sigma_{xx} + \frac{\sigma_{yy}}{E_y} - \frac{\nu_{yz}}{E_y}\sigma_{zz}$$

$$\varepsilon_{zz} = -\frac{\nu_{xy}}{E_x}\sigma_{xx} - \frac{\nu_{yz}}{E_y}\sigma_{yy} + \frac{\sigma_{zz}}{E_y} \tag{2.35}$$

$$\gamma_{xy} = \frac{\tau_{xy}}{G_{xy}} \qquad \gamma_{yz} = \frac{\tau_{yz}}{G_{yz}} \qquad \gamma_{xz} = \frac{\tau_{xz}}{G_{xy}}$$

For plane stress conditions where $\sigma_{zz} = \tau_{yz} = \tau_{zx} = 0$, and for transversely isotropic composites where $E_y = E_z$, the strain-stress relations reduce to:

$$\sigma_{xx} = \frac{E_x}{1-\nu_{xy}\nu_{yx}}\varepsilon_{xx} + \frac{\nu_{xy}E_y}{1-\nu_{xy}\nu_{yx}}\varepsilon_{yy} = \frac{E_x}{1-\nu_{xy}\nu_{yx}}\left[\varepsilon_{xx} + \nu_{yx}\varepsilon_{yy}\right]$$

$$\sigma_{yy} = \frac{\nu_{xy}E_y}{1-\nu_{xy}\nu_{yx}}\varepsilon_{xx} + \frac{E_y}{1-\nu_{xy}\nu_{yx}}\varepsilon_{yy} = \frac{E_y}{1-\nu_{xy}\nu_{yx}}\left[\nu_{xy}\varepsilon_{xx} + \varepsilon_{yy}\right] \tag{2.36}$$

$$\tau_{xy} = G_{xy}\gamma_{xy}$$

From Eq. (2.36) it is clear that only four elastic constants[2] (E_x, E_y, G_{xy} and ν_{xy}) are required to write the stress strain relations for an orthotropic material subjected to plane stress. If $\nu_{yx} \Rightarrow 0$ for a uniaxial composite material with fibers oriented in the x direction, then Eqs. (2.36) reduce to:

$$\sigma_{xx} = E_x\varepsilon_{xx}$$

$$\sigma_{yy} = E_y\left[\nu_{xy}\varepsilon_{xx} + \varepsilon_{yy}\right] \qquad (2.37)$$

$$\tau_{xy} = G_{xy}\gamma_{xy}$$

EXERCISES

2.1 Given the displacement field:

$u = (3x^4 + 2x^2y^2 + x + y + z^3 + 3)(10^{-3})$
$v = (3xy + y^3 + y^2z + z^2 + 1)(10^{-3})$
$w = (x^2 + xy + yz + zx + y^2 + z^2 + 2)(10^{-3})$

Compute the associated strains at point (1,1,1). Compare the results obtained by using Eqs. (2.2) and (2.3) with those obtained by using Eqs. (2.4).

2.2 Repeat Exercise 2.1 but change the multiplier from 10^{-3} to 10^{-1}.

2.3 Given the displacement field:

$u = (x^2 + y^4 + 2y^2z + yz)(10^{-3})$
$v = (xy + xz + 3x^2z)(10^{-3})$
$w = (y^4 + 4y^3 + 2z^2)(10^{-3})$

Compute the associated strains at point (2,2,2). Compare the results obtained by using Eqs. (2.2) and (2.3) with those obtained by using Eqs. (2.4).

2.4 Repeat Exercise 2.3 but change the multiplier from 10^{-3} to 10^{-2}.

2.5 Transform the set of Cartesian stain components:

$\varepsilon_{xx} = 300\ \mu\varepsilon$ $\quad$ $\varepsilon_{yy} = 200\ \mu\varepsilon$ $\quad$ $\varepsilon_{zz} = 100\ \mu\varepsilon$
$\gamma_{xy} = 200\ \mu\varepsilon$ $\quad$ $\gamma_{yz} = 100\ \mu\varepsilon$ $\quad$ $\gamma_{zx} = 150\ \mu\varepsilon$

into a new set of Cartesian strain components relative to a set of coordinates, where the direction angles associated with the Ox'y'z' axes are:

θ	**Case 1**	**Case 2**	**Case 3**	**Case 4**
x-x'	π/4	π/2	0	π/2
y-y'	π/4	π/2	π/2	0
z-z'	0	0	π/2	π/2

[2] Note that ν_{yx} is not independent because $\nu_{yx} = (E_y/E_x)\ \nu_{xy}$.

2.6 Transform the set of Cartesian stain components:

$\varepsilon_{xx} = 400\ \mu\varepsilon$ $\varepsilon_{yy} = 250\ \mu\varepsilon$ $\varepsilon_{zz} = 125\ \mu\varepsilon$
$\gamma_{xy} = 275\ \mu\varepsilon$ $\gamma_{yz} = 175\ \mu\varepsilon$ $\gamma_{zx} = 225\ \mu\varepsilon$

into a new set of Cartesian strain components relative to an Ox'y'z' set of coordinates, where the direction angles associated with the Ox'y'z' axes are:

	x	y	z
x'	$\pi/3$	$\pi/3$	$\pi/4$
y'	$3\pi/4$	$\pi/4$	$\pi/2$
z'	$\pi/3$	$\pi/3$	$3\pi/4$

2.7 At a point in a stressed body, the Cartesian components of strain are:

$\varepsilon_{xx} = 300\ \mu\varepsilon$ $\varepsilon_{yy} = 450\ \mu\varepsilon$ $\varepsilon_{zz} = 300\ \mu\varepsilon$
$\gamma_{xy} = 600\ \mu\varepsilon$ $\gamma_{yz} = 375\ \mu\varepsilon$ $\gamma_{zx} = 450\ \mu\varepsilon$

Transform this set of Cartesian components into a new set of Cartesian strain components relative to an Ox'y'z' set of coordinates where the Ox'y'z' axes are defined by the following direction cosines:

(a)					(b)			
	x	y	z			x	y	z
x'	2/3	2/3	−1/3		x'	1/9	−8/9	4/9
y'	− 2/3	1/3	−2/3		y'	4/9	4/9	7/9
z'	−1/3	2/3	2/3		z'	−8/9	1/9	4/9

2.8 At a point in a stressed body, the Cartesian components of strain are:

$\varepsilon_{xx} = 450\ \mu\varepsilon$ $\varepsilon_{yy} = 300\ \mu\varepsilon$ $\varepsilon_{zz} = 150\ \mu\varepsilon$
$\gamma_{xy} = 150\ \mu\varepsilon$ $\gamma_{yz} = 150\ \mu\varepsilon$ $\gamma_{zx} = 300\ \mu\varepsilon$

Transform this set of Cartesian components into a new set of Cartesian strain components relative to an Ox'y'z' set of coordinates where the Ox'y'z' axes are defined by the following direction cosines:

(a)					(b)			
	x	y	z			x	y	z
x'	1/17	12/17	12/17		x'	9/25	12/25	20/25
y'	12/17	−9/17	8/17		y'	12/25	16/25	−15/25
z'	12/17	8/17	−9/17		z'	20/25	−15/25	0

2.9 At a point in a stressed body, the Cartesian components of strain are:

$\varepsilon_{xx} = 990\ \mu\varepsilon$ $\varepsilon_{yy} = 825\ \mu\varepsilon$ $\varepsilon_{zz} = 550\ \mu\varepsilon$
$\gamma_{xy} = 330\ \mu\varepsilon$ $\gamma_{yz} = 660\ \mu\varepsilon$ $\gamma_{zx} = 500\ \mu\varepsilon$

Transform this set of Cartesian components into a new set of Cartesian strain components relative to an Ox'y'z' set of coordinates where the Ox'y'z' axes are defined by the following direction cosines:

(a)	x	y	z		(b)	x	y	z
x'	2/15	10/15	11/15		x'	2/11	6/11	9/11
y'	14/15	−5/15	2/15		y'	6/11	7/11	-6/11
z'	−5/15	−10/15	10/15		z'	9/11	−6/11	2/11

2.10 Write a computer program to determine the Cartesian strain components relative to the Ox'y'z' in terms of the Cartesian strain components referred to Oxyz and the appropriate direction cosines.

2.11 Determine the three principal strains and the maximum shearing strain at the point having the Cartesian strain components given in Exercise 2.5. Check the three strain invariants.

2.12 Determine the three principal strains and the maximum shearing strain at the point having the Cartesian strain components given in Exercise 2.6. Check the three strain invariants.

2.13 Determine the three principal strains and the maximum shearing strain at the point having the Cartesian strain components given in Exercise 2.7.

2.14 Determine the three principal strains and the maximum shearing strain at the point having the Cartesian strain components given in Exercise 2.8.

2.15 Determine whether the following strain fields are compatible:

(a)
$\varepsilon_{xx} = 2x^2 + 3y^2 + z + 1$
$\varepsilon_{yy} = 2y^2 + x^2 + 3z + 2$
$\varepsilon_{zz} = 3x + 2y + z^2 + 1$
$\gamma_{xy} = 10\,xy$
$\gamma_{yz} = 0$
$\gamma_{zx} = 0$

(b)
$\varepsilon_{xx} = 3y^2 + xy$
$\varepsilon_{yy} = 2y + 4z + 3$
$\varepsilon_{zz} = 3zx + 2xy + 3yz + 2$
$\gamma_{xy} = 5\,xy$
$\gamma_{yz} = 2x + z$
$\gamma_{zx} = 2y + x$

2.16 Determine whether the following strain fields are compatible:

(a)
$\varepsilon_{xx} = 3x^2 + 4xy - y^2$
$\varepsilon_{yy} = x^2 + xy + 3y^2$
$\varepsilon_{zz} = 0$
$\gamma_{xy} = -x^2 - 6xy - 4y^2$
$\gamma_{yz} = 2x + y$
$\gamma_{zx} = z + 3$

(b)
$\varepsilon_{xx} = 12x^2 - 6y^2 - 4z$
$\varepsilon_{yy} = 12\,y^2 - 6x^2 + 4z$
$\varepsilon_{zz} = 12x + 4y - z + 5$
$\gamma_{xy} = 4z - 24xy - 3$
$\gamma_{yz} = y + z - 4$
$\gamma_{zx} = 4x + 4y - 6$

2.17 Given the strain field:

$\varepsilon_{xx} = ay$ $\quad \varepsilon_{yy} = by$ $\quad \varepsilon_{zz} = by$
$\gamma_{xy} = 0$ $\quad \gamma_{yz} = 0$ $\quad \gamma_{zx} = 0$

Compute the displacement fields. What physical problem does this strain field represent?

Determine λ, μ and K for steel, brass, aluminum, plastic and magnesium, using the following values for E and ν:

Material	E (GPa)	ν
Steel	207	0.30
Brass	106	0.33
Aluminum	71	0.33
Plastic	3	0.40
Magnesium	45	0.35

2.18 A cube of steel (E = 207 GPa and ν = 0.30) is loaded with a uniformly distributed pressure of 300 MPa on the four faces having outward normals in the x and y directions. Rigid constraints limit the total deformation of the cube in the z direction to 0.05 mm. Determine the normal stress, if any, which develops in the z direction. The length of a side of the cube is 190 mm.

2.19 Determine the change in volume of a 10-mm cube of aluminum (E = 71 GPa and ν = 0.33) when dropped a distance of 8 km to the ocean floor.

2.20 Determine the stresses at a point in a steel (E = 207 GPa and ν = 0.30) machine component if the Cartesian components of strain at the point are as listed in Exercise 2.5.

2.21 Determine the stresses at a point in an aluminum (E = 71 GPa and ν = 0.33) machine component if the Cartesian components of strain at the point are as listed in Exercise 2.6.

2.22 The Cartesian components of stress at a point in a steel (E = 207 GPa and ν = 0.30) machine part are:

σ_{xx} = 220 MPa $\quad$ σ_{yy} = 77 MPa $\quad$ σ_{zz} = 154 MPa
τ_{xy} = 110 MPa $\quad$ τ_{yz} = 55 MPa $\quad$ τ_{zx} = 66 MPa

Determine the principal strains at that point.

2.23 At a point on the free surface of an alloy-steel (E = 207 GPa and ν = 0.30) machine part normal strains of 1000 με, 2000 με and 1200 με were measured at angles of 0°, 60° and 120°, respectively, relative to the x axis. Design considerations limit the maximum normal stress to 510 MPa, the maximum shearing stress to 275 MPa, the maximum normal strain 2200 με, and the maximum shear strain to 2500 με. What is your evaluation of the design?

2.24 A thick-walled cylindrical pressure vessel will be used to store gas under a pressure of 100 MPa. During initial pressurization of the vessel, axial and hoop components of strain were measured on the inside and outside surfaces. On the inside surface, the axial strain was 500 με and the hoop strain was 750 με. On the outside surface, the axial strain was 500 με, and the hoop strain was 100 με. Determine the axial and hoop components of stress associated with these strains if E = 207 GPa and ν = 0.30.

2.25 The Cartesian components of stress at a point in a steel (E = 207 GPa and ν = 0.30) machine part are as follows:

σ_{xx} = 280 MPa $\quad$ σ_{yy} = −120 MPa $\quad$ σ_{zz} = 140 MPa
τ_{xy} = 280 MPa $\quad$ τ_{yz} = 0 $\quad$ τ_{zx} = 0

Determine the three principal strains, the principal-strain directions, and the maximum shearing strain.

2.26 Mohr's circle for stress and Mohr's circle for strain are convenient graphical methods for visualizing three-dimensional states of stress and strain at points in a stressed body. At a particular point in a body fabricated from steel (E = 207 GPa and ν = 0.30), the three principal stresses are:

$\sigma_1 = 120$ MPa $\qquad \sigma_2 = 60$ MPa $\qquad \sigma_3 = -40$ MPa

(a) Sketch the three-dimensional Mohr's circle for stress at the point.
(b) Sketch the three-dimensional Mohr's circle for strain at the point.
(c) On a plane through the point, the shearing stress $\tau = 70$ MPa. What normal stress must exist on this plane?
(d) On another plane through the point, the shearing stress $\tau = 50$ MPa. Within what range of values must the normal stress associated with this plane fall?
(e) Along a line through the point (say the x-axis), the normal strain is zero. What range of values may the shearing strain γ assume for different orientations of the x-axis?

2.27 A thin rubber membrane is stretched in such a manner that the following uniform strain field is produced:

$\varepsilon_{xx} = 5000\ \mu\varepsilon \qquad \varepsilon_{yy} = -6000\ \mu\varepsilon \qquad \gamma_{xy} = 2000\ \mu\varepsilon$

A rectangle is drawn on the membrane before stretching. How should the rectangle be oriented if the angles are to remain 90° during stretching?

2.28 A thin rectangular aluminum (E = 71 GPa and ν = 0.33) plate 75 by 100 mm is acted upon by a two-dimensional stress distribution which produces the following uniform distribution of strains in the plate:

$\varepsilon_{xx} = 2000\ \mu\varepsilon \qquad \varepsilon_{yy} = -500\ \mu\varepsilon \qquad \gamma_{xy} = 2000\ \mu\varepsilon$

(a) Determine the changes in length of the diagonals of the plate.
(b) Determine the maximum shearing strain in the plate. Prepare a sketch of two initially perpendicular lines in the plate associated with this maximum shearing strain.

2.35 For an aluminum (E = 71 GPa and ν = 0.33) body under plane-stress conditions with $\sigma_{zz} = \tau_{yz} = \tau_{zx} = 0$, strains on the surface of the body at a given point are:

$\varepsilon_{xx} = 1200\ \mu\varepsilon \qquad \varepsilon_{yy} = 900\ \mu\varepsilon$

Determine the strain ε_{zz}.

2.36 Determine the stresses σ_{xx}, σ_{yy} and σ_{zz} in a material with ν = ½ if

$\varepsilon_{xx} = \varepsilon_{yy} = \varepsilon_{zz} = -1000\ \mu\varepsilon$

Explain your results.

2.37 Write a computer program using the relations given in Table 2.1 so that the five elastic constants λ, μ, E, ν and K can be determined from any two of the constants that are given.

2.38 Write the strain equations of transformation for a composite material for a state of plane stress.

2.39 Write a test procedure for measuring E_x and ν_{xy} if a tension specimen fabricated from a cross ply material is available. Note the longitudinal axis of the tension specimen coincides with the x-axis fiber orientation in the cross ply material. List the data required and the equations that will be used in the determination.

2.40 Write a test procedure for measuring E_y and ν_{yx} if a tension specimen fabricated from a cross ply material is available. Note the longitudinal axis of the tension specimen coincides with the y-axis fiber orientation in the cross ply material. List the data required and the equations that will be used in the determination.

2.41 Write a test procedure for determining G_{xy} from a tension specimen fabricated with its axis at ±45° relative to the fiber orientations. List the data required and the equations that will be used in the determination.

2.42 A thin-walled tube fabricated from a cross ply material is subjected to torsion. Write a test procedure for determining G_{xy} by measuring strains on the outside surface of the tube. Describe the gage placement, list the data required and the equations that will be used in the determination.

REFERENCES

1. Boresi, A. P., and P. P. Lynn: Elasticity in Engineering Mechanics, Prentice-Hall, Inc., Englewood Cliffs, N.J., 1974.
2. Chou, P. C., and N. J. Pagano: Elasticity, D. Van Nostrand Company, Inc., Princeton, N.J., 1967.
3. Durelli, A. J., E. A. Phillips, and C. H. Tsao: Introduction to the Theoretical and Experimental Analysis of Stress and Strain, McGraw-Hill Book Company, New York, 1958.
4. Love, A. E. H.: A Treatise on the Mathematical Theory of Elasticity, Dover Publications, Inc., New York, 1944.
5. Sechler, E. E.: Elasticity in Engineering, John Wiley & Sons, Inc., New York, 1952.
6. Sokolnikoff, I. S.: Mathematical Theory of Elasticity, 2nd ed., McGraw-Hill Book Company, New York, 1956.
7. Southwell, R. V.: An Introduction to the Theory of Elasticity, Oxford University Press, Fair Lawn, N.J., 1953.
8. Timoshenko, S. P., and J. N. Goodier: Theory of Elasticity, 2nd ed., McGraw-Hill Book Company, New York, 1951.
9. Herakovich, C. T.: Mechanics of Fibrous Composites, John Wiley & Sons, New York, NY, 1998.
10. Agarwal, B. D. and Broutman, L. J.: Analysis and Performance of Fiber Composites, 2nd Edition, John Wiley & Sons, New York, NY, 1990.
11. Lekhnitskii, S. G.: Theory of Elasticity of an Anisotropic Body, Holden-Day, San Francisco, CA, 1963.
12. Atanackovic and A. Guran: Theory of Elasticity for Scientists and Engineers, Birkhauser, Boston, MA. 2000.
13. Slaughter, W. S.: Linearized Theory of Elasticity, Birkhauser, Boston, MA. 2002.
14. Fung, Y. C. and P. Tang: Classical and Computational Solid Mechanics, World Scientific, Singapore, 2001.
15. Atkin, R. J. and N. Fox: An Introduction to the Theory of Elasticity, Longman, London, 1980.

CHAPTER 3

BASIC EQUATIONS AND PLANE-ELASTICITY THEORY

3.1 FORMULATION OF THE PROBLEM

In the general three-dimensional elasticity problem there are 15 unknown quantities, which must be determined at every point in the body—the 6 Cartesian components of stress, the 6 Cartesian components of strain, and the 3 components of displacement. Attempts can be made to obtain a solution to a given problem after the following quantities have been adequately defined:

1. The geometry of the body
2. The boundary conditions
3. The body-force field as a function of position
4. The elastic constants

In order to solve for the above-mentioned 15 unknown quantities, 15 independent equations are required. Three are provided by the stress equations of equilibrium Eqs. (1.3), six are provided by the strain-displacement relations Eqs. (2.4), and the remaining six can be obtained from the stress-strain expressions Eqs. (2.16).

A solution to an elasticity problem, in addition to satisfying these 15 equations, must also satisfy the boundary conditions. In other words, the stresses acting over the surface of the body must produce tractions that are equivalent to the loads being applied to the body. Boundary conditions are often classified to define the four different types of boundary-value problems listed below:

Type 1. If the displacements are prescribed over the entire boundary, the problem is classified as a type 1 boundary-value problem. As an example, consider a long, slender rod, which is given an axial displacement, say, u and transverse displacements v and w. In this instance, displacements are prescribed over the entire boundary of the rod.

Type 2. The most frequently encountered boundary-value problem is the type where normal and shearing forces are given over the entire surface. For instance, a sphere subjected to a uniform hydrostatic pressure has zero shearing stress and a normal stress equal to $-p$ on the surface and hence is a type 2 boundary-value problem.

Type 3. This type is a mixed boundary-value problem where the normal and shearing forces are given over a portion of the boundary and the displacements are given over the remainder of the body. To illustrate this type of problem consider the shrinking of a sleeve over a shaft. In the shrinking process, a radial displacement is given to the sleeve at the interface between the shaft and the sleeve. On all other surfaces of the sleeve, both the normal and the shearing components of stress are zero.

Type 4. This type of boundary-value problem is the most general of the four considered. Over one portion of the surface, displacements are prescribed. Over a second portion of the surface, normal and shearing stresses are prescribed. Over a third portion of the surface, the normal component of displacement and the shearing component of stress are prescribed. Over a fourth portion of the surface, the shearing component of displacement and the normal component of stress are prescribed. Obviously, the first three types of problems can be regarded as special cases of this general fourth type.

One of the most difficult problems encountered in any experimental study is the design and construction of the loading fixture for applying the required displacements or tractions to the model being studied. The classifications given previously should be kept in mind when the loading fixture is designed. In general, it has been found that tractions cannot be adequately simulated by applying a displacement field to the model and vice versa. Type 1 and type 2 boundary-value problems are usually the easiest to approach experimentally. In general, type 3 and type 4 problems offer more difficulties in correctly loading the model.

3.2 FIELD EQUATIONS

Thus far in the development, four sets of field equations have been discussed—the stress equations of equilibrium, the strain-displacement relations, the stress-strain expressions, and the equations of compatibility. Quite often two or more of these sets of equations can be combined to give a new set that may be more applicable to a specific problem. As an example, consider the six stress-displacement equations that can be derived from the six stress-strain relations and the six strain-displacement equations by substituting Eqs. (2.4) into Eqs. (2.16).

$$\frac{\partial u}{\partial x} = \frac{1}{E}\left[\sigma_{xx} - \nu(\sigma_{yy} + \sigma_{zz})\right]$$

$$\frac{\partial v}{\partial y} = \frac{1}{E}\left[\sigma_{yy} - \nu(\sigma_{zz} + \sigma_{xx})\right] \tag{3.1}$$

$$\frac{\partial w}{\partial z} = \frac{1}{E}\left[\sigma_{zz} - \nu(\sigma_{xx} + \sigma_{yy})\right]$$

$$\frac{\partial u}{\partial y} + \frac{\partial v}{\partial x} = \frac{\tau_{xy}}{\mu} \qquad \frac{\partial v}{\partial z} + \frac{\partial w}{\partial y} = \frac{\tau_{yz}}{\mu} \qquad \frac{\partial w}{\partial x} + \frac{\partial u}{\partial z} = \frac{\tau_{zx}}{\mu}$$

It is interesting to note that the set of equations consisting of the stress equations of equilibrium Eqs.(1.3) and the stress-displacement relations Eqs. (3.1) are expressed as nine equations in terms of nine unknowns. The reduction in the number of unknowns from 15 to 9 was made possible by eliminating the strains.

The problem can be reduced further (from nine to three unknowns) if the stress equations of equilibrium Eqs. (1.3) are combined with the stress-displacement equations Eqs. (3.1). The displacement equations of equilibrium obtained can be written as follows:

$$\nabla^2 u + \frac{1}{1-2\nu}\frac{\partial}{\partial x}\left(\frac{\partial u}{\partial x} + \frac{\partial v}{\partial y} + \frac{\partial w}{\partial z}\right) + \frac{1}{\mu}F_x = 0$$

$$\nabla^2 v + \frac{1}{1-2\nu}\frac{\partial}{\partial y}\left(\frac{\partial u}{\partial x} + \frac{\partial v}{\partial y} + \frac{\partial w}{\partial z}\right) + \frac{1}{\mu}F_y = 0 \tag{3.2}$$

$$\nabla^2 w + \frac{1}{1-2\nu}\frac{\partial}{\partial z}\left(\frac{\partial u}{\partial x} + \frac{\partial v}{\partial y} + \frac{\partial w}{\partial z}\right) + \frac{1}{\mu}F_z = 0$$

where ∇^2 is the operator $\partial^2/\partial x^2 + \partial^2/\partial y^2 + \partial^2/\partial z^2$.

It is clear that a solution of the displacement equations of equilibrium will yield the three displacements u, v, and w. When the displacements are known, the six strains and the six stresses can easily be obtained by using Eqs. (2.4) to obtain the strains and Eqs. (2.16) to obtain the stresses.

Analytical solutions for three-dimensional elasticity problems are quite difficult to obtain, and the number of problems, which have been solved in an exact fashion to date, is surprisingly small. The most successful approach has been through the use of the Boussinesq-Popkovich stress functions, which are defined so as to satisfy Eq. (3.2). The development of this approach is somewhat involved and is therefore beyond the scope and objectives of this elementary treatment of the theory of elasticity. The interested reader should consult the references at the end of the chapter for a detailed development of the Boussinesq-Popkovich stress-function approach.

Before this section is completed, the stress equations of compatibility will be developed because they are the basis for an important theorem regarding the dependence of stresses on the elastic constants. If the stress-strain relations Eqs (2.16), the stress equations of equilibrium Eqs.(1.3), and the strain compatibility equations Eqs. (2.10) are combined, the six stress equations of compatibility are obtained as follows:

$$\nabla^2\sigma_{xx} + \frac{1}{1+\nu}\frac{\partial^2}{\partial x^2}I_1 = -\frac{\nu}{1-\nu}\left(\frac{\partial F_x}{\partial x} + \frac{\partial F_y}{\partial y} + \frac{\partial F_z}{\partial z}\right) - 2\frac{\partial F_x}{\partial x}$$

$$\nabla^2\sigma_{yy} + \frac{1}{1+\nu}\frac{\partial^2}{\partial y^2}I_1 = -\frac{\nu}{1-\nu}\left(\frac{\partial F_x}{\partial x} + \frac{\partial F_y}{\partial y} + \frac{\partial F_z}{\partial z}\right) - 2\frac{\partial F_y}{\partial y}$$

$$\nabla^2\sigma_{zz} + \frac{1}{1+\nu}\frac{\partial^2}{\partial z^2}I_1 = -\frac{\nu}{1-\nu}\left(\frac{\partial F_x}{\partial x} + \frac{\partial F_y}{\partial y} + \frac{\partial F_z}{\partial z}\right) - 2\frac{\partial F_z}{\partial z}$$

$$\nabla^2\tau_{xy} + \frac{1}{1+\nu}\frac{\partial^2}{\partial x\partial y}I_1 = -\left(\frac{\partial F_x}{\partial y} + \frac{\partial F_y}{\partial x}\right) \tag{3.3}$$

$$\nabla^2\tau_{yz} + \frac{1}{1+\nu}\frac{\partial^2}{\partial y\partial z}I_1 = -\left(\frac{\partial F_y}{\partial z} + \frac{\partial F_z}{\partial y}\right)$$

$$\nabla^2\tau_{zx} + \frac{1}{1+\nu}\frac{\partial^2}{\partial z\partial x}I_1 = -\left(\frac{\partial F_x}{\partial z} + \frac{\partial F_z}{\partial x}\right)$$

where the first invariant of stress $I_1 = \sigma_{xx} + \sigma_{yy} + \sigma_{zz}$, and F_x, F_y, F_z are the body-force intensities in the x, y, z directions, respectively.

If this system of six equations is solved for the six Cartesian stress components, and if the boundary conditions are satisfied, the problem can be considered solved. Of great importance to the experimentalist is the appearance of the elastic constants in Eqs. (3.3). Recall that the stress equations of equilibrium did not contain elastic constants. Because only Poisson's ratio ν appears in Eqs. (3.3), it follows that the stresses are independent of the modulus of elasticity E of the model material and can at most depend upon Poisson's ratio alone. Of course, this is true only for a simply connected body because the strain compatibility equations are valid only for this condition.

This independence of the stresses on the elastic modulus E is very important in three-dimensional photoelasticity, where a low-modulus plastic model is used to simulate a metal prototype. Only the difference in Poisson's ratio between the model and the prototype is a source of error. The very large difference between the moduli of elasticity of the model and the prototype does not produce any significant errors in the determination of stresses using a three-dimensional photoelastic approach, provided the strains induced in the photoelastic model remain sufficiently small.

3.3 THE PLANE ELASTIC PROBLEM

In the theory of elasticity there exists a special class of problems, known as plane problems, which can be solved more readily than the general three-dimensional problem because certain simplifying assumptions can be made in their treatment. The geometry of the body and the nature of the loading on the boundaries, which permit a problem to be classified as a plane problem, are as follows:

By definition a plane body consists of a region of uniform thickness bounded by two parallel planes and by any closed lateral surface B_L, as indicated by Fig. 3.1. Although the thickness of the body must be uniform, it need not be limited. It may be very thick or very thin; in fact, these two extremes represent the most desirable cases for this approach, as will be described later.

In addition to the restrictions on the geometry of the body, the following restrictions are imposed on the loads applied to the plane body.

1. Body forces, if they exist, cannot vary through the thickness of the region, that is, $F_x = F_x(x,y)$ and $F_y = F_y(x,y)$. Furthermore, the body force F_z must equal zero.
2. The surface tractions or loads on the lateral boundary B_L must be in the plane of the model and must be uniformly distributed across the thickness, i.e., constant in the z direction. Hence, $T_x = T_x(x,y)$, $T_y = T_y(x,y)$ and $T_z = 0$.
3. No loads can be applied on the parallel planes bounding the top and bottom surfaces, that is, $T_n = 0$ on $z = \pm t$.

When the geometry and loading have been defined, stresses can be determined by using either the plane-strain or the plane-stress approach. Usually, the plane-strain approach is used when the body is very thick relative to its lateral dimensions. The plane-stress approach is employed when the body is relatively thin in relation to its lateral dimensions.

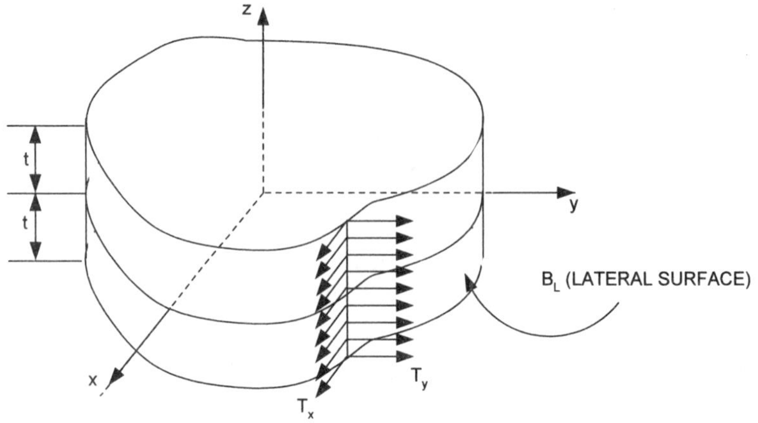

Figure 3.1 A body can be considered for the plane-elasticity approach if it is bounded on the top and bottom by two parallel planes and laterally by any surface normal to these two planes.

3.4 THE PLANE-STRAIN APPROACH

If it is assumed that the strains in the body are plane, i.e., the strains in the x and y directions are functions of x and y alone and the strains in the z directions are equal to zero, the strain-displacement relations Eqs. (2.4) can be simplified as follows:

$$\begin{aligned} &\varepsilon_{xx} = \frac{\partial u}{\partial x} \qquad \varepsilon_{yy} = \frac{\partial v}{\partial y} \qquad \varepsilon_{zz} = \frac{\partial w}{\partial z} = 0 \\ &\gamma_{xy} = \frac{\partial u}{\partial y} + \frac{\partial v}{\partial x} \qquad \gamma_{yz} = \frac{\partial v}{\partial z} + \frac{\partial w}{\partial y} \qquad \gamma_{zx} = \frac{\partial w}{\partial x} + \frac{\partial u}{\partial z} = 0 \end{aligned} \tag{3.4}$$

Similarly, if Eqs. (3.4) are substituted into Eqs. (2.13), a reduced form of the stress-strain relations for the case of plane strain is obtained:

$$\begin{aligned} &\sigma_{xx} = \lambda J_1 + 2\mu\varepsilon_{xx} \quad \sigma_{yy} = \lambda J_1 + 2\mu\varepsilon_{yy} \quad \sigma_{zz} = \lambda J_1 \\ &\tau_{xy} = \mu\gamma_{xy} \qquad \tau_{yz} = \tau_{zx} = 0 \end{aligned} \tag{3.5}$$

where $J_1 = \varepsilon_{xx} + \varepsilon_{yy} + \varepsilon_{zz}$. In addition, the stress equations of equilibrium, Eqs. (1.3), reduce to

$$\frac{\partial \sigma_{xx}}{\partial x} + \frac{\partial \tau_{xy}}{\partial y} + F_x = 0 \qquad \frac{\partial \tau_{xy}}{\partial x} + \frac{\partial \sigma_{yy}}{\partial y} + F_y = 0 \tag{3.6}$$

Any solution for a plane-strain problem must satisfy Eqs. (3.4) to (3.6) in addition to the boundary conditions on the lateral boundary B_L and the bounding planes. The boundary conditions on B_L can be expressed in terms of the stresses by referring to Eqs. (1.2), which give the x, y, and z components of the resultant-stress vector in terms of the Cartesian components of stress. Thus, on B_L the following relations must be satisfied:

$$\begin{aligned} &T_{nx} = \sigma_{xx}\cos(n, x) + \tau_{xy}\cos(n, y) \\ &T_{ny} = \tau_{xy}\cos(n, x) + \sigma_{yy}\cos(n, y) \\ &T_{nz} = 0 \end{aligned} \tag{3.7}$$

where T_{nx}, T_{ny} and T_{nz} are the x, y, z components of the stresses applied to the body on surface B_L. Finally, on the two parallel bounding planes,

$$\mathbf{T}_n = 0 \tag{3.8}$$

i.e., no tractions are applied to these surfaces; hence, τ_{yz}, τ_{zx} and σ_{zz} must be zero on these surfaces.

It is clear from Eqs. (3.5) that σ_{zz} will be equal to zero, as demanded by Eq. (3.8), only when the dilatation J_1 is equal to zero. In most problems, J_1 will not be equal to zero; therefore, the solution will not be exact because the boundary conditions on the parallel planes are violated. In many problems, this violation of the boundary conditions can be cleared by superimposing an equal and opposite distribution of σ_{zz} (residual solution) onto the original solution.

It is possible to obtain an exact solution to the residual problem only when σ_{zz} is a linear function of x and y. When σ_{zz} is nonlinear, an approximate solution based on Saint-Venant's principle[1] is often utilized. When the nonlinear distribution of σ_{zz} on the parallel boundaries is replaced by a linear distribution that is statically equivalent, the solution will be valid only in regions well removed from the parallel bounding planes. Thus, it is clear that the plane-strain approach is necessarily limited to the central regions of bodies such as shafts or dams that are very long, i.e., thick, relative to their lateral dimensions. In the central region of such a long body, the stresses σ_{xx}, σ_{yy} and τ_{xy} can be found from the solution of the original problem because the superposition of the residual solution onto the original problem does not influence these stresses but only serves to make σ_{zz} vanish.

In this section, the plane-strain approach has been discussed without indicating a method for solving for σ_{xx}, σ_{yy} and τ_{xy}. This problem will be treated later in this chapter when the Airy's-stress-function is discussed. In this plane-strain section, it is important for the reader to understand the plane-strain assumption, why it usually leads to a violation of the boundary conditions on the two parallel planes, and finally how these undesired stresses can be removed from the planes by superimposing a statically equivalent linear stress system. Also quite important is Saint-Venant's principle, because an experimentalist in simulating loads often relies on this principle to permit simplification in the design of the loading fixtures.

3.5 PLANE STRESS

In the preceding section, it was noted that the plane-strain method is limited to very long or thick bodies. In those cases where the body thickness is small relative to its lateral dimensions, it is advantageous to assume that

$$\sigma_{zz} = \tau_{yz} = \tau_{zx} = 0 \tag{3.9}$$

throughout the thickness of the plate. With this assumption, the stress equations of equilibrium again reduce to

$$\frac{\partial \sigma_{xx}}{\partial x} + \frac{\partial \tau_{xy}}{\partial y} + F_x = 0 \qquad \frac{\partial \tau_{xy}}{\partial x} + \frac{\partial \sigma_{yy}}{\partial y} + F_y = 0 \tag{3.10}$$

and the stress-strain relations, Eqs. (2.13), become

$$\begin{aligned} &\sigma_{xx} = \lambda J_1 + 2\mu\varepsilon_{xx} \quad \sigma_{yy} = \lambda J_1 + 2\mu\varepsilon_{yy} \quad \sigma_{zz} = \lambda J_1 + 2\mu\varepsilon_{zz} = 0 \\ &\tau_{xy} = \mu\gamma_{xy} \qquad \tau_{yz} = \mu\gamma_{yz} = 0 \qquad \tau_{zx} = \mu\gamma_{zx} = 0 \end{aligned} \tag{3.11}$$

From the third equation of Eqs. (3.11), the following relationship can be obtained:

$$\varepsilon_{zz} = -\frac{\lambda}{\lambda + 2\mu}\left(\varepsilon_{xx} + \varepsilon_{yy}\right) \tag{a}$$

With this value of ε_{zz} the first strain invariant J_1 becomes:

[1] Saint-Venant's principle states that a system of forces acting over a small region of the boundary can be replaced by a statically equivalent system of forces without introducing appreciable changes in the distribution of stresses in regions well removed from the area of load application.

$$J_1 = \frac{2\mu}{\lambda + 2\mu}\left(\varepsilon_{xx} + \varepsilon_{yy}\right) \tag{b}$$

Substituting the value for J_1 given in Eq. (b) into Eqs. (3.11) yields:

$$\begin{aligned}
\sigma_{xx} &= \frac{2\lambda\mu}{\lambda + 2\mu}\left(\varepsilon_{xx} + \varepsilon_{yy}\right) + 2\mu\varepsilon_{xx} \\
\sigma_{yy} &= \frac{2\lambda\mu}{\lambda + 2\mu}\left(\varepsilon_{xx} + \varepsilon_{yy}\right) + 2\mu\varepsilon_{yy} \\
\tau_{xy} &= \mu\gamma_{xy} \qquad \sigma_{zz} = \tau_{yz} = \tau_{zx} = 0
\end{aligned} \tag{3.12}$$

Unfortunately, in the general case, σ_{xx}, σ_{yy} and τ_{xy} are not independent of z, and thus the boundary conditions imposed on the boundary B_L cannot be rigorously satisfied. To overcome this difficulty, average stresses and displacements over the thickness are commonly used. If the body is relatively thin, these averages closely approximate the true boundary conditions on B_L. Average values for the stresses and displacements over the thickness of the body are obtained as follows:

$$\begin{gathered}
\tilde{\sigma}_{xx} = \frac{1}{2t}\int_{-t}^{t} \sigma_{xx}\,dz \qquad \tilde{\sigma}_{yy} = \frac{1}{2t}\int_{-t}^{t} \sigma_{yy}\,dz \qquad \tilde{\tau}_{xy} = \frac{1}{2t}\int_{-t}^{t} \tau_{xy}\,dz \\
\tilde{u} = \frac{1}{2t}\int_{-t}^{t} u\,dz \qquad \tilde{v} = \frac{1}{2t}\int_{-t}^{t} v\,dz
\end{gathered} \tag{3.13}$$

The symbol ~ over the stresses and displacements indicates average values. Substituting the average values of the stresses into Eqs. (1.2) gives the boundary conditions that must be satisfied on B_L:

$$\begin{aligned}
T_{nx} &= \tilde{\sigma}_{xx}\cos(n,\ x) + \tilde{\tau}_{xy}\cos(n,\ y) \\
T_{ny} &= \tilde{\tau}_{xy}\cos(n,\ x) + \tilde{\sigma}_{yy}\cos(n,\ y)
\end{aligned} \tag{3.14}$$

If the equations that the plane-strain and the plane-stress solutions must satisfy are compared, it can be observed that they are identical except for the comparison between Eqs. (3.5) and (3.11). An examination of a typical equation from each of these sets,

$$\sigma_{xx} = \begin{cases} \lambda(\varepsilon_{xx} + \varepsilon_{yy}) + 2\mu\varepsilon_{xx} & \text{plane stress} \\ \dfrac{2\lambda\mu}{\lambda + 2\mu}(\varepsilon_{xx} + \varepsilon_{yy}) + 2\mu\varepsilon_{xx} & \text{plane strain} \end{cases}$$

indicates that they are identical except for the coefficients of the $\varepsilon_{xx} + \varepsilon_{yy}$ term. Because all other equations for the plane-stress and plane-strain solutions are identical, results from plane strain can be transformed into plane stress by letting:

$$\lambda \rightarrow \frac{2\lambda\mu}{\lambda + 2\mu}$$

that is equivalent to letting:

$$\frac{\nu}{1-\nu} \rightarrow \nu \tag{3.15}$$

In a similar manner, a plane-stress solution can be transformed into a plane-strain solution by letting

$$\frac{2\lambda\mu}{\lambda + 2\mu} \rightarrow \lambda$$

or

$$\nu \rightarrow \frac{\nu}{1-\nu} \tag{3.16}$$

In the plane-stress approach, it is generally assumed that

$$\sigma_{zz} = \tau_{yz} = \tau_{zx} = 0$$

and the unknown stresses σ_{xx}, σ_{yy} and τ_{xy} will have a z dependence. As a result of this z dependence, the boundary conditions on B_L are violated. This difficulty can be eliminated and an approximate solution to the problem can be obtained by using average values for the stresses and displacements. Finally, it was shown that plane-stress and plane-strain solutions can be transformed from one case into the other by a simple replacement involving Poisson's ratio in one solution or the other, as indicated in Eqs. (3.15) and (3.16).

3.6 AIRY'S STRESS FUNCTION

In the plane problem, three unknowns σ_{xx}, σ_{yy} and τ_{xy} must be determined that will satisfy the required field equations and boundary conditions. The most convenient sets of field equations to use in this determination are the two equations of equilibrium and one stress equation of compatibility.

The equilibrium equations in two dimensions are:

$$\frac{\partial \sigma_{xx}}{\partial x} + \frac{\partial \tau_{xy}}{\partial y} + F_x = 0 \tag{3.17a}$$

$$\frac{\partial \tau_{xy}}{\partial x} + \frac{\partial \sigma_{yy}}{\partial y} + F_y = 0 \tag{3.17b}$$

The stress compatibility equation for the case of plane strain is

$$\nabla^2(\sigma_{xx} + \sigma_{yy}) = -\frac{2(\lambda+\mu)}{\lambda+2\mu}\left(\frac{\partial F_x}{\partial x} + \frac{\partial F_y}{\partial y}\right) \tag{3.17c}$$

Suppose the body-force field is defined by $\Omega(x, y)$ so that the body-force intensities are given by

$$F_x = -\frac{\partial \Omega}{\partial x} \qquad F_y = -\frac{\partial \Omega}{\partial y} \tag{3.18}$$

Then, by substituting Eqs. (3.18) into Eqs. (3.17) and noting that

$$\frac{2(\lambda+\mu)}{(\lambda+2\mu)}=\frac{1}{(1-\nu)}$$

it is apparent that

$$\frac{\partial\sigma_{xx}}{\partial x}+\frac{\partial\tau_{xy}}{\partial y}=\frac{\partial\Omega}{\partial x}\qquad\frac{\partial\tau_{xy}}{\partial x}+\frac{\partial\sigma_{yy}}{\partial y}=\frac{\partial\Omega}{\partial y}$$
$$\nabla^2\left(\sigma_{xx}+\sigma_{yy}-\frac{\Omega}{1-\nu}\right)=0 \tag{3.19}$$

Equations (3.19) represent the three field equations that σ_{xx}, σ_{yy} and τ_{xy} must satisfy. Assume that the stresses can be represented by a stress function ϕ such that

$$\sigma_{xx}=\frac{\partial^2\phi}{\partial y^2}+\Omega\qquad\sigma_{yy}=\frac{\partial^2\phi}{\partial x^2}+\Omega\qquad\tau_{xy}=-\frac{\partial^2\phi}{\partial x\partial y} \tag{3.20}$$

If Eqs. (3.20) are substituted into Eqs. (3.19), it can be seen that the two equations of equilibrium are exactly satisfied, and the last of Eqs. (3.19) gives:

$$\nabla^4\phi=-\frac{1-2\nu}{1-\nu}\nabla^2\Omega \tag{3.21}$$

Thus, equilibrium and compatibility are immediately satisfied if ϕ satisfies Eq. (3.21). The expression ϕ is known as **Airy's stress function**. If Eq. (3.21) is solved for ϕ, an expression containing x, y, and a number of constants will be obtained. The constants are evaluated from the boundary conditions given in Eqs. (3.17), and the stresses are computed from ϕ according to Eqs. (3.20). Of course, evaluation of ϕ from Eq. (3.21) produces a solution for stresses for the plane-strain case. A solution for the plane-stress case can be obtained by letting $\nu/(1-\nu)\Rightarrow\nu$, as indicated in Eq. (3.15). This substitution leads to

$$\nabla^4\phi=-(1-\nu)\nabla^2\Omega \tag{3.22}$$

that is valid for plane-stress problems.

It is important to note that if the body-force intensities are zero or constant, such as those encountered in a gravitational field, then

$$\nabla^2\Omega=0$$

and Eqs. (3.21) and (3.22) both become

$$\nabla^4\phi=0 \tag{3.23a}$$

This is a biharmonic equation, which can also be written in the form

$$\frac{\partial^4\phi}{\partial x^4}+2\frac{\partial^4\phi}{\partial x^2\partial y^2}+\frac{\partial^4\phi}{\partial y^4}=0 \tag{3.23b}$$

Examination of this equation shows that ϕ and thus σ_{xx}, σ_{yy} and τ_{xy} are independent of the elastic constants. This consideration is very important in two-dimensional photoelasticity because it indicates that the stresses obtained from a plastic model are identical to those in a metal prototype if the model is simply connected and subjected to a zero or a uniform body-force field. Differences in the values of the modulus of elasticity and Poisson's ratio between model and prototype do not influence the results for the stresses. There are exceptions to the simply connected restriction, however, which will be covered in a later chapter on photoelasticity.

3.7 AIRY'S STRESS FUNCTION IN CARTESIAN COORDINATES

Any Airy's stress function used in the solution of a plane problem must satisfy Eqs. (3.23a) and (3.23b) and provide stresses via Eqs. (3.20) that satisfy the defined boundary conditions. Some Airy's stress functions commonly used are polynomials in x and y. In this section, polynomials from the first to the fifth degree will be discussed.

3.7.1 Airy's Stress Function in Terms of a First-Degree Polynomial

$$\phi_1 = a_1x + b_1y$$

It is clear from Eqs. (3.20) that

$$\sigma_{xx} = \sigma_{yy} = \tau_{xy} = 0 \tag{3.24}$$

and that Eqs. (3.23a) and (3.23b) are satisfied. This function is suitable only for indicating a zero stress field and therefore is of little use in the solution of any problem.

3.7.2 Airy's Stress Function in Terms of a Second-Degree Polynomial

$$\phi_2 = a_2x^2 + b_2xy + c_2y^2$$

From Eqs. (3.20) the stresses are

$$\sigma_{xx} = 2c_2 \qquad \sigma_{yy} = 2a_2 \qquad \tau_{xy} = -b_2 \tag{3.25}$$

Note that Eqs. (3.23a) and (3.23b) are satisfied and that the stress function ϕ_2 gives a uniform stress field over the entire body that is independent of x and y.

3.7.3 Airy's Stress Function in Terms of a Third-Degree Polynomial

$$\phi_3 = a_3x^3 + b_3x^2y + c_3xy^2 + d_3y^3$$

Again, by use of Eqs. (3.20) the stresses are given by

$$\sigma_{xx} = 2c_3x + 6d_3y \qquad \sigma_{yy} = 6a_3x + 2b_3y \qquad \tau_{xy} = -2b_3x - 2c_3y \tag{3.26}$$

Equations (3.23a) and (3.23b) are satisfied unconditionally, and the stress function ϕ_3 provides a linearly varying stress field over the body.

3.7.4 Airy's Stress Function in Terms of a Fourth-Degree Polynomial

$$\phi_4 = a_4x^4 + b_4x^3y + c_4x^2y^2 + d_4xy^3 + e_4y^4$$

From Eqs. (3.20) it is apparent that:

$$\sigma_{xx} = 2c_4x^2 + 6d_4xy + 12e_4 y^2$$

$$\sigma_{yy} = 12a_4x^2 + 6b_4xy + 2c_4y^2 \qquad (3.27)$$

$$\tau_{xy} = -\ 3b_4x^2 - 4c_4xy - 3d_4y^2$$

When ϕ_4 is substituted into Eq. (3.23b), it should be noted that ϕ_4 is not unconditionally satisfied. In order for $\nabla^4 \phi = 0$ it is necessary that

$$e_4 = -\ [a_4 + (c_4/3)]$$

Substituting this equation into the relations for the stresses gives

$$\sigma_{xx} = 2c_4x^2 + 6d_4xy - 12a_4y^2 - 4c_4y^2$$

and σ_{yy} and τ_{xy} are unchanged. Thus, ϕ_4 yields a stress field that is a second-degree polynomial in x and y.

3.7.5 Airy's Stress Function in Terms of a Fifth-Degree Polynomial

$$\phi_5 = a_5x^5 + b_5x^4y + c_5x^3y^2 + d_5x^2y^3 + e_5xy^4 + f_5y^5$$

Employing Eqs. (3.20) to solve for the stresses gives

$$\sigma_{xx} = 2c_5x^3 + 6d_5x^2y + 12e_5xy^2 + 20f_5y^3$$

$$\sigma_{yy} = 20a_5x^3 + 12b_5\, x^2y + 6c_5xy^2 + 2d_5y^3$$

$$\tau_{xy} = -\ 4b_5x^3 - 6c_5x^2y - 6d_5xy^2 - 4e_5y^3$$

Again, note that ϕ_5 must be subjected to certain conditions involving the constants e_5 and f_5. For Eqs. (3.23a) and (3.23b) to be satisfied, these conditions are

$$e_5 = -\ (5a_5 + c_5) \qquad f_5 = -\ (1/5)(b_5 + d_5)$$

Subject to the restrictive conditions listed above, the Cartesian stress components become

$$\sigma_{xx} = 2c_5x^3 + 6d_5x^2y - 12(5a_5 + c_5)xy^2 - 4(b_5 + d_5)y^3$$

$$\sigma_{yy} = 20a_5x^3 + 12b_5\, x^2y + 6c_5xy^2 + 2d_5y^3 \qquad (3.28)$$

$$\tau_{xy} = -\ 4b_5x^3 - 6c_5x^2y - 6d_5xy^2 + 4(5a_5 + c_5)y^3$$

Thus, it is clear that ϕ_5 yields a stress field that is a third-degree polynomial in x and y. It is possible to continue this procedure to include ϕ_6, ϕ_7, etc., provided Eqs. (3.23a) and (3.23b) are satisfied. It is also possible to add together two or more stress functions to form another function, for example, $\phi^* = \phi_2 + \phi_3$. Thus, by simply adding terms or by eliminating terms from the stress function, it is theoretically possible to develop any stress field that can be expressed as a function of x and y.

3.8 EXAMPLE PROBLEM

Airy's stress function, expressed in Cartesian coordinates, can be employed to solve a particular class of two-dimensional problems where the boundaries of the body can be adequately represented by the Cartesian reference frame. As an example, consider the simply supported beam with a uniform loading shown in Fig. 3.2. An examination of the loading conditions indicates that

$$\sigma_{yy} = \tau_{xy} = 0 \qquad \text{at} \qquad y = -h/2 \tag{a}$$

$$\sigma_{yy} = -q \text{ and } \qquad \tau_{xy} = 0 \qquad \text{at} \qquad y = h/2 \tag{b}$$

Also at $x = \pm L/2$

$$\int_{-h/2}^{h/2} \tau_{xy} dy = R = \frac{qL}{2} \tag{c}$$

$$\int_{-h/2}^{h/2} \sigma_{xx} dy = 0 \tag{d}$$

$$\int_{-h/2}^{h/2} \sigma_{xx} y dy = 0 \tag{e}$$

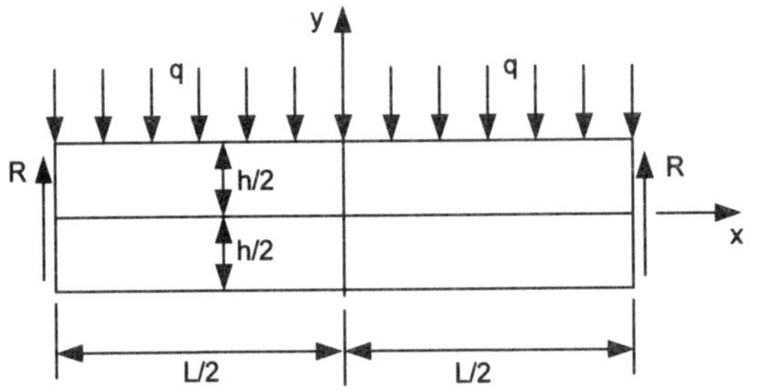

Figure 3.2 A simply supported beam of length L, height h and unit depth subjected to a uniformly distributed load.

Note that the bending moment (and consequently σ_{xx}) is a maximum at position x = 0 and decreases with a change in x in either the positive or the negative direction. This is possible only if the stress function contains even functions of x. Note also that σ_{yy} varies from zero at $y = -h/2$ to a maximum value of $-q$ at $y = +h/2$; thus, the stress function must contain odd functions of y. From the stress functions listed in Section 3.7, the following even and odd functions can be selected to form a new stress function ϕ that satisfies the previously listed conditions.

$$\phi = a_2x^2 + b_3x^2y + d_3y^3 + a_4x^4 + b_5x^4y + d_5x^2y^3 + f_5y^5 \tag{f}$$

The stress function ϕ must satisfy the equation $\nabla^4 \phi = 0$; hence,

$$a_4 = 0 \qquad f_5 = -(1/5)(b_5 + d_5) \tag{g}$$

From Eqs. (3.20) the Cartesian stress components are

$$\begin{aligned} \sigma_{xx} &= 6d_3y + 6d_5x^2y - 4(b_5 + d_5)y^3 \\ \sigma_{yy} &= 2a_2 + 2b_3y + 12b_5\, x^2y + 2d_5y^3 \\ \tau_{xy} &= -2b_3x - 4b_5x^3 - 6d_5xy^2 \end{aligned} \tag{h}$$

Examination of the boundary conditions listed in Eq. (a) indicates that σ_{yy} must be independent of x; hence the coefficient $b_5 = 0$. Consequently, Eqs. (h) reduce to

$$\sigma_{xx} = 6d_3y + 6d_5x^2y - 4d_5y^3$$

$$\sigma_{yy} = 2a_2 + 2b_3y + 2d_5y^3 \qquad \text{(i)}$$

$$\tau_{xy} = -2b_3x - 6d_5xy^2$$

The problem can be solved if the coefficients a_2, b_3, d_3 and d_5 can be selected so that the boundary conditions given in Eqs. (a) to (e) are satisfied. From Eqs. (a)

$$\sigma_{yy} = 0 = 2a_2 + 2b_3(-h/2) + 2d_5(-h/2)^3 \qquad a_2 - b_3h/2 - d_5h^3/8 = 0 \qquad \text{(j)}$$

and from Eqs. (b)

$$\sigma_{yy} = -q = 2a_2 + 2b_3h/2 + 2d_5(h/2)^3 \qquad a_2 + b_3h/2 + d_5h^3/8 = -q/2 \qquad \text{(k)}$$

Adding Eqs. (j) and (k) gives

$$a_2 = -q/4 \qquad \text{(l)}$$

From Eqs. (a) and (b)

$$\tau_{xy} = 0 = -2x[b_3 + 3d_5(\pm h/2)^2] \qquad b_3 = -(3/4)(h^2d_5) \qquad \text{(m)}$$

Substituting Eqs. (m) into Eqs. (j),

$$d_5h^3/8 - 3h^3d_5/8 = -q/4 \qquad d_5 = q/h^3 \qquad \text{(n)}$$

and

$$b_3 = -(3/4)(q/h) \qquad \text{(o)}$$

With the values of a_2, b_3, and d_5 given by Eqs. (l), (o), and (n), respectively, Eqs. (c) and (d) are identically satisfied. Equation (e) can be used to solve for the remaining unknown d_3.

$$\int_{-h/2}^{h/2}\left(6d_3y^2 + \frac{3}{2}\frac{q}{h^3}L^2y^2 - 4\frac{qy^4}{L^3}\right)dy = 0$$

$$\left[2d_3y^3 + \frac{q}{2h^3}L^2y^3 - \frac{4}{5}\frac{qy^5}{h^3}\right]_{-h/2}^{h/2} = 0 \qquad \text{(p)}$$

Solving Eq. (p) for d_3 gives

$$d_3 = [q/(240I)](2h^2 + 5L^2) \qquad \text{(q)}$$

where $I = h^3/12$ is the second moment of the cross sectional area (moment of inertia) of the unit-width beam. Substituting Eqs. (q), (o), (n), and (l) into Eqs. (i) gives the final equations for the Cartesian components of stress:

$$\sigma_{xx} = [q/(8I)](4x^2 - L^2)y + (q/60I)(3h^2y - 20\,y^3)$$

$$\sigma_{yy} = [q/(24I)](4y^3 - 3h^2y - h^3) \quad \text{(r)}$$

$$\tau_{xy} = [qx/(8I)](h^2 - 4y^2)$$

The conventional strength-of-materials solution for this simple beam problem, namely, that $\sigma_{xx} = My/I$, gives

$$\sigma_{xx} = [qx/(8I)](4x^2 - L^2)y \quad \text{(s)}$$

that is identical with the first term of the relation given for σ_{xx} in Eqs. (r). The second term,

$$[q/(60I)](3h^2 y - 20y^3) \quad \text{(t)}$$

is a correction term for the strength-of-materials approach. Recall in the conventional strength-of-materials solution that it is assumed that plane sections remain plane after bending. This assumption is not exactly true except for the case of pure bending, and as a consequence, the strength-of-materials solution obtained lacks the correction term shown above. It is clear that the correction term is small when L >> h; therefore, the strength-of-materials solution is sufficiently accurate for most engineering work.

This simple example illustrates how elementary elasticity theory can be employed to extend the reader's understanding of the distribution of stresses in simple two-dimensional problems. Other examples are included in the exercises at the end of this chapter.

3.9 TWO-DIMENSIONAL PROBLEMS IN POLAR COORDINATES

In Section 3.6 the Airy's-stress-function approach to the solution of two-dimensional elasticity problems in Cartesian coordinates was developed. This method was then applied to solve an elementary problem that was well suited to the Cartesian reference frame. In many problems, however, the geometry of the body does not lend itself to the use of a Cartesian coordinate system, and it is more expeditious to work with a different system. A large class of problems (such as circular rings, curved beams, and half-planes) can be solved by employing a polar coordinate system. In any elasticity problem, the proper choice of the coordinate system is extremely important because this choice establishes the complexity of the mathematical expressions employed to satisfy the field equations and the boundary conditions.

In order to solve two-dimensional elasticity problems by employing a polar-coordinate reference frame, the equations of equilibrium, the definition of Airy's stress function, and one of the stress equations of compatibility must be reestablished in terms of polar coordinates. On the following pages, the equations of equilibrium will be derived by considering a polar element instead of a Cartesian element. The equations for the polar components of stress in terms of Airy's stress function as well as the stress equation of compatibility will be transformed from Cartesian to polar coordinates. Finally, a set of stress functions will be developed that satisfies the stress equation of compatibility.

The stress equations of equilibrium in polar coordinates can be derived from the free-body diagram of the polar element shown in Fig. 3.3. The element is assumed to be very small. The average values of the normal and shearing stress, which act on surface 1, are denoted by σ_{rr} and $\tau_{r\theta}$, respectively. Because the stresses may vary as a function of r, values of the normal and shearing stresses on surface 3 are given by:

$$\sigma_{rr} + (\partial\sigma_{rr}/\partial r)dr \qquad \text{and} \qquad \tau_{r\theta} + (\partial\tau_{r\theta}/\partial r)dr.$$

Similarly, the average values of the normal and shearing stresses that act on surface 2 are given by $\sigma_{\theta\theta}$ and $\tau_{r\theta}$. Because the stresses may also vary as a function of θ, values of the normal and shearing stresses on surface 4 are:

$$\sigma_{\theta\theta} + (\partial\sigma_{\theta\theta}/\partial\theta)d\theta \qquad \text{and} \qquad \tau_{r\theta} + (\partial\tau_{r\theta}/\partial\theta)d\theta.$$

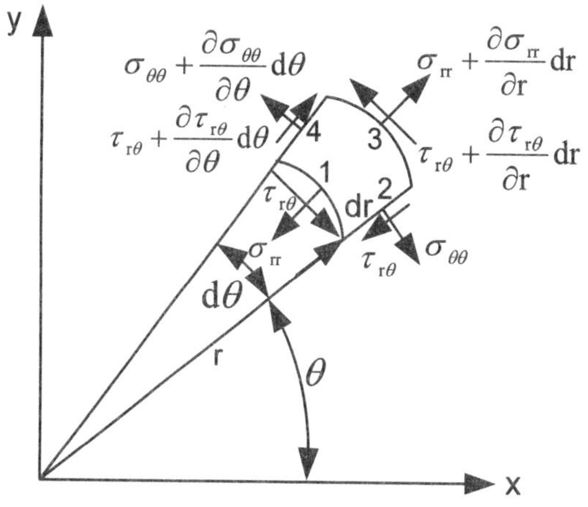

Figure 3.3 Polar element of unit depth showing the stresses acting on the four faces.

For a polar element of unit thickness to be in a state of equilibrium, the sum of all forces in the radial r and tangential θ directions must equal zero. Summing forces first in the radial direction and considering the body-force intensity F_r gives the equation of equilibrium as:

$$\left(\sigma_{rr}+\frac{\partial\sigma_{rr}}{\partial r}dr\right)(r+dr)d\theta-\sigma_{rr}rd\theta-\left[\sigma_{\theta\theta}dr+\left(\sigma_{\theta\theta}+\frac{\partial\sigma_{\theta\theta}}{\partial\theta}d\theta\right)dr\right]\frac{d\theta}{2}$$
$$+\left(\tau_{r\theta}+\frac{\partial\tau_{r\theta}}{\partial\theta}d\theta-\tau_{r\theta}\right)dr+F_r rd\theta\,dr=0 \qquad \text{(a)}$$

Dividing Eq. (a) by dr dθ and simplifying gives:

$$\frac{\partial\sigma_{rr}}{\partial r}dr-\frac{\partial\sigma_{\theta\theta}}{\partial\theta}\frac{d\theta}{2}+\sigma_{rr}+\frac{\partial\sigma_{rr}}{\partial r}r-\sigma_{\theta\theta}+\frac{\partial\tau_{r\theta}}{\partial\theta}d\theta+F_r r=0 \qquad \text{(b)}$$

If the element is made infinitely small by permitting dr and dθ each to approach zero, the first two terms in Eq. (b) also approach zero and the expression can be rewritten as:

$$\frac{\partial\sigma_{rr}}{\partial r}+\frac{1}{r}\frac{\partial\tau_{r\theta}}{\partial\theta}+\frac{1}{r}(\sigma_{rr}-\sigma_{\theta\theta})+F_r=0 \qquad \text{(3.29a)}$$

The equation of equilibrium in the tangential direction can be derived in the same manner if the forces acting in the θ direction on the polar element are summed and set equal to zero. Hence:

$$\frac{1}{r}\frac{\partial\sigma_{\theta\theta}}{\partial\theta}+\frac{\partial\tau_{r\theta}}{\partial r}+\frac{2\tau_{r\theta}}{r}+F_\theta=0 \qquad \text{(3.29b)}$$

Equations (3.29a) and (3.29b) represent the equations of equilibrium in polar coordinates. They are analogous to the equations of equilibrium in Cartesian coordinates presented in Eqs. (3.17a) and (3.17b). Any solution to an elasticity problem must satisfy these field equations.

3.10 TRANSFORMATION OF $\nabla^4\phi = 0$ INTO POLAR COORDINATES

In the coverage of Airy's stress function given in Section 3.6, it was shown that the stress function ϕ had to satisfy the biharmonic equation $\nabla^4\phi = 0$, provided the body forces are zero or constants. In polar coordinates, the stress function must satisfy this same equation; however, the definition of the ∇^4 operator must be modified to suit the polar-coordinate system. This modification may be accomplished by transforming the ∇^4 operator from the Cartesian system to the polar system.

In transforming from Cartesian coordinates to polar coordinates, recall that:

$$r^2 = x^2 + y^2 \qquad \theta = \arctan(y/x) \qquad (3.30)$$

where r and θ are defined in Fig. 3.3.

Differentiating Eqs. (3.30) gives:

$$(\partial r/\partial x) = x/r = \cos\theta \qquad (\partial r/\partial y) = y/r = \sin\theta$$
$$(\partial\theta/\partial x) = -y/r^2 = -(\sin\theta)/r \qquad (\partial\theta/\partial y) = x/r^2 = (\cos\theta)/r \qquad (3.31)$$

The form of the ∇^4 operator in Cartesian coordinates is

$$\nabla^4\phi = \left(\frac{\partial^2}{\partial x^2} + \frac{\partial^2}{\partial y^2}\right)\left(\frac{\partial^2\phi}{\partial x^2} + \frac{\partial^2\phi}{\partial y^2}\right)$$

Individual elements of this expression can be transformed by employing Eqs. (3.30) and (3.31) as follows. If it is assumed that ϕ is a function of r and θ,

$$\frac{\partial\phi}{\partial x} = \frac{\partial\phi}{\partial r}\frac{\partial r}{\partial x} + \frac{\partial\phi}{\partial\theta}\frac{\partial\theta}{\partial x} \qquad (a)$$

$$\frac{\partial^2\phi}{\partial x^2} = \frac{\partial\phi}{\partial r}\frac{\partial^2 r}{\partial x^2} + \left(\frac{\partial r}{\partial x}\right)^2\frac{\partial^2\phi}{\partial r^2} + 2\frac{\partial^2\phi}{\partial r\partial\theta}\frac{\partial r}{\partial x}\frac{\partial\theta}{\partial x} + \frac{\partial\phi}{\partial\theta}\frac{\partial^2\theta}{\partial x^2} + \left(\frac{\partial\theta}{\partial x}\right)^2\frac{\partial^2\phi}{\partial\theta^2} \qquad (b)$$

Substituting the equalities given in Eqs. (3.31) into Eq. (b) yields:

$$\frac{\partial^2\phi}{\partial x^2} = \frac{\sin^2\theta}{r}\frac{\partial\phi}{\partial r} + \cos^2\theta\frac{\partial^2\phi}{\partial r^2} - \frac{\sin 2\theta}{r}\frac{\partial^2\phi}{\partial r\partial\theta} + \frac{\sin 2\theta}{r^2}\frac{\partial\phi}{\partial\theta} + \frac{\sin^2\theta}{r^2}\frac{\partial^2\phi}{\partial\theta^2} \qquad (3.32a)$$

By following the same procedure, it is clear that

$$\frac{\partial^2\phi}{\partial y^2} = \frac{\cos^2\theta}{r}\frac{\partial\phi}{\partial r} + \sin^2\theta\frac{\partial^2\phi}{\partial r^2} + \frac{\sin 2\theta}{r}\frac{\partial^2\phi}{\partial r\partial\theta} - \frac{\sin 2\theta}{r^2}\frac{\partial\phi}{\partial\theta} + \frac{\cos^2\theta}{r^2}\frac{\partial^2\phi}{\partial\theta^2} \qquad (3.32b)$$

$$\frac{\partial^2\phi}{\partial x\partial y} = -\frac{\sin\theta\cos\theta}{r}\frac{\partial\phi}{\partial r} + \sin\theta\cos\theta\frac{\partial^2\phi}{\partial r^2} + \frac{\cos 2\theta}{r}\frac{\partial^2\phi}{\partial r\partial\theta} - \frac{\cos 2\theta}{r^2}\frac{\partial\phi}{\partial\theta} - \frac{\sin\theta\cos\theta}{r^2}\frac{\partial^2\phi}{\partial\theta^2} \qquad (3.32c)$$

Adding Eqs. (3.32a) and (3.32b) gives

$$\frac{\partial^2 \phi}{\partial x^2}+\frac{\partial^2 \phi}{\partial y^2}=\frac{\partial^2 \phi}{\partial r^2}+\frac{1}{r}\frac{\partial \phi}{\partial r}+\frac{1}{r^2}\frac{\partial^2 \phi}{\partial \theta^2} \tag{3.33}$$

Furthermore, it is evident that

$$\begin{aligned}\nabla^4\phi &= \left(\frac{\partial^2}{\partial x^2}+\frac{\partial^2}{\partial y^2}\right)\left(\frac{\partial^2 \phi}{\partial x^2}+\frac{\partial^2 \phi}{\partial y^2}\right)\\ &= \left(\frac{\partial^2}{\partial r^2}+\frac{1}{r}\frac{\partial}{\partial r}+\frac{1}{r^2}\frac{\partial^2}{\partial \theta^2}\right)\left(\frac{\partial^2 \phi}{\partial r^2}+\frac{1}{r}\frac{\partial \phi}{\partial r}+\frac{1}{r^2}\frac{\partial^2 \phi}{\partial \theta^2}\right)=0\end{aligned} \tag{3.34}$$

Equation (3.34) is the stress equation of compatibility in terms of Airy's stress function referred to a polar-coordinate system.

3.11 POLAR COMPONENTS OF STRESS IN TERMS OF AIRY'S STRESS FUNCTION

By referring to the two-dimensional equations of stress transformation Eqs. (1.11), expressions can be obtained that relate the polar stress components σ_{rr}, $\sigma_{\theta\theta}$ and $\tau_{r\theta}$ to the Cartesian stress components σ_{xx}, σ_{yy} and τ_{xy} as follows:

$$\begin{aligned}\sigma_{rr} &= \sigma_{xx}\cos^2\theta + \sigma_{yy}\sin^2\theta + \tau_{xy}\sin 2\theta\\ \sigma_{\theta\theta} &= \sigma_{yy}\cos^2\theta + \sigma_{xx}\sin^2\theta - \tau_{xy}\sin 2\theta\\ \tau_{r\theta} &= (\sigma_{yy}-\sigma_{xx})\sin\theta\cos\theta + \tau_{xy}\cos 2\theta\end{aligned} \tag{3.35}$$

If Eqs. (3.20) are substituted into Eqs. (3.35) and Ω set equal to zero (that is equivalent to setting both F_x and F_y equal to zero), then

$$\begin{aligned}\sigma_{rr} &= \frac{\partial^2 \phi}{\partial y^2}\cos^2\theta+\frac{\partial^2 \phi}{\partial x^2}\sin^2\theta-\frac{\partial^2 \phi}{\partial x\partial y}\sin 2\theta\\ \sigma_{\theta\theta} &= \frac{\partial^2 \phi}{\partial x^2}\cos^2\theta+\frac{\partial^2 \phi}{\partial y^2}\sin^2\theta+\frac{\partial^2 \phi}{\partial x\partial y}\sin 2\theta\\ \tau_{r\theta} &= \left(\frac{\partial^2 \phi}{\partial x^2}-\frac{\partial^2 \phi}{\partial y^2}\right)\sin\theta\cos\theta-\frac{\partial^2 \phi}{\partial x\partial y}\cos 2\theta\end{aligned} \tag{3.36}$$

If the results from Eqs. (3.32a) and (3.32c) are substituted into Eqs. (3.36), the polar components of stress in terms of Airy's stress function are obtained:

$$\sigma_{rr} = \frac{1}{r}\frac{\partial\phi}{\partial r} + \frac{1}{r^2}\frac{\partial^2\phi}{\partial\theta^2} \qquad \sigma_{\theta\theta} = \frac{\partial^2\phi}{\partial r^2} \qquad \tau_{r\theta} = \frac{1}{r^2}\frac{\partial\phi}{\partial\theta} - \frac{1}{r}\frac{\partial^2\phi}{\partial r\partial\theta} \tag{3.37}$$

When Airy's stress function ϕ in polar coordinates has been established, these relations can be employed to determine the stress field as a function of r and θ.

3.12 FORMS OF AIRY'S STRESS FUNCTION IN POLAR COORDINATES

The equation $\nabla^4\phi = 0$ is a fourth-order biharmonic partial differential equation that can be reduced to an ordinary fourth-order differential equation by using a separation-of variables technique, where

$$\phi^{(n)} = R_n(r)\begin{Bmatrix}\cos n\theta \\ \sin n\theta\end{Bmatrix}$$

The resulting differential equation is an Euler type that yields four different stress functions upon solution. These stress functions are tabulated, together with the stress and displacement distributions, which they provide, on the following pages.

One of the stress functions obtained can be expressed in the following form:

$$\phi^{(0)}(r) = a_0 + b_0 \ln r + c_0 r^2 + d_0 r^2 \ln r \tag{3.38a}$$

By using Eq. (3.37), the stresses associated with this particular stress function can be expressed as

$$\sigma_{rr} = (b_0/r^2) + 2c_0 + d_0(1 + 2\ln r)$$

$$\sigma_{\theta\theta} = -(b_0/r^2) + 2c_0 + d_0(3 + 2\ln r) \qquad \tau_{r\theta} = 0 \tag{3.38b}$$

The displacements associated with this function can be determined by integrating the stress displacement relations, giving:

$$u_r = (1/E)[-(1+\nu)(b_0/r) + 2(1-\nu)c_0 r + 2(1-\nu)d_0 r \ln r - (1+\nu)d_0 r] + \alpha_2\cos\theta + \alpha_3\sin\theta$$

$$u_\theta = (1/E)(4d_0 r\theta) - \alpha_1 r - \alpha_2\sin\theta + \alpha_3\cos\theta \tag{3.38c}$$

where u_r and u_θ are the displacements in the radial and circumferential directions, respectively. The terms containing α_1, α_2 and α_3 are associated with rigid-body displacements. It should be noted that the stresses in this solution are independent of θ; hence, the stress function $\phi^{(0)}$ should be employed to solve problems that have rotational symmetry.

Another stress function and the stresses and displacements associated with it can be expressed as follows:

$$\phi^{(1)} = \left(a_1 r + \frac{b_1}{r} + c_1 r^3 + d_1 r\ln r\right)\begin{Bmatrix}\sin\theta \\ \cos\theta\end{Bmatrix}$$

$$\sigma_{rr} = \left(-\frac{2b_1}{r^3} + 2c_1 r + \frac{d_1}{r}\right)\begin{Bmatrix}\sin\theta \\ \cos\theta\end{Bmatrix}$$

(3.39)

$$\sigma_{\theta\theta} = \left(\frac{2b_1}{r^3} + 6c_1 r + \frac{d_1}{r}\right)\begin{Bmatrix}\sin\theta \\ \cos\theta\end{Bmatrix}$$

$$\tau_{r\theta} = \left(-\frac{2b_1}{r^3} + 2c_1 r + \frac{d_1}{r}\right)\begin{Bmatrix}-\cos\theta \\ \sin\theta\end{Bmatrix}$$

$$u_r = \frac{1}{E}\left\{\left[(1+\nu)\frac{b_1}{r^2} + (1-3\nu)c_1 r^2 - (1+\nu)d_1 + (1-\nu)d_1 \ln r\right]\begin{Bmatrix}\sin\theta \\ \cos\theta\end{Bmatrix} - (2d_1\theta)\begin{Bmatrix}\cos\theta \\ -\sin\theta\end{Bmatrix}\right\}$$

$$+\alpha_2\cos\theta + \alpha_3\sin\theta$$

$$u_\theta = \frac{1}{E}\left\{\left[-(1+\nu)\frac{b_1}{r^2} - (5+\nu)c_1 r^2 + (1-\nu)d_1 \ln r\right]\begin{Bmatrix}\cos\theta \\ -\sin\theta\end{Bmatrix} + (2d_1\theta)\begin{Bmatrix}\sin\theta \\ \cos\theta\end{Bmatrix}\right\}$$

$$-\alpha_1 r - \alpha_2\sin\theta + \alpha_3\cos\theta$$

The third stress function of interest and its associated stresses and displacements are as follows:

$$\phi^{(n)} = \left(a_n r^n + b_n r^{-n} + c_n r^{2+n} + d_n r^{2-n}\right)\begin{Bmatrix}\sin n\theta \\ \cos n\theta\end{Bmatrix}$$

$$\sigma_{rr} = \left[a_n(n-n^2)r^{n-2} - b_n(n+n^2)r^{-n-2} + c_n(2+n-n^2)r^n + d_n(2-n-n^2)r^{-n}\right]\begin{Bmatrix}\sin n\theta \\ \cos n\theta\end{Bmatrix}$$

$$\sigma_{\theta\theta} = \left[a_n(n^2-n)r^{n-2} + b_n(n^2+n)r^{-n-2} + c_n(2+3n+n^2)r^n + d_n(2-3n+n^2)r^{-n}\right]\begin{Bmatrix}\sin n\theta \\ \cos n\theta\end{Bmatrix}$$

$$\tau_{r\theta} = \left[a_n(n^2-n)r^{n-2} - b_n(n+n^2)r^{-n-2} + c_n(n+n^2)r^n + d_n(n-n^2)r^{-n}\right]\begin{Bmatrix}-\cos n\theta \\ \sin n\theta\end{Bmatrix}$$

$$u_r = \frac{1}{E}\left\{-a_n(1+\nu)nr^{n-1} + b_n(1+\nu)nr^{-n-1} + c_n[4-(1+\nu)(2+n)]r^{n+1} + d_n[4-(1+\nu)(2-n)]r^{-n+1}\right\}\begin{Bmatrix}\sin n\theta \\ \cos n\theta\end{Bmatrix}$$

$$+\alpha_2\cos\theta + \alpha_3\sin\theta$$

$$u_\theta = \frac{1}{E}\left\{-a_n(1+\nu)nr^{n-1} - b_n(1+\nu)nr^{-n-1} - c_n[4+(1+\nu)n]r^{n+1} + d_n[4-(1+\nu)n]r^{-n+1}\right\}\begin{Bmatrix}\cos n\theta \\ -\sin n\theta\end{Bmatrix}$$

$$-\alpha_1 r - \alpha_2\sin\theta + \alpha_3\cos\theta$$

(3.40)

For the stress function $\phi^{(n)}$, the value of n is given by $n \geq 2$.

The fourth stress function of interest and the associated stresses and displacements are expressed as follows:

$$\phi^{(*)} = a^*\theta + b^* r^2\theta + c^* r\theta\sin\theta + d^* r\theta\cos\theta$$

$$\sigma_{rr} = 2b^*\theta + (2c^*/r)\cos\theta - (2d^*/r)\sin\theta$$

$$\sigma_{\theta\theta} = 2b^*\theta$$

$$\tau_{r\theta} = a^*/r^2 - b^* \qquad (3.41)$$

$$u_r = (1/E)[2(1-\nu)b^* r\theta + (1-\nu)c^*\theta \sin\theta + 2c^* \ln r \cos\theta$$

$$+ (1-\nu)d^*\theta\cos\theta - 2d^* \ln r \sin\theta] + \alpha_2\cos\theta + \alpha_3 \sin\theta$$

$$u_\theta = (1/E)[-(1+\nu)a_*/r + (3-\nu)b^* r - 4\,b^* r \ln r - (1+\nu)c^*\sin\theta - 2c^*\ln r \sin\theta$$

$$+ (1-\nu)c^*\theta\cos\theta - (1+\nu)d^*\cos\theta - 2d^*\ln r\cos\theta - (1-\nu)d^*\theta\sin\theta]$$

$$-\alpha_1 r - \alpha_2\sin\theta + \alpha_3\cos\theta$$

In the examples that follow, the stress functions previously listed will be employed to determine the stresses and displacements for problems, which lend themselves to polar coordinates. As the selection of the stress function is often the most difficult phase of the problem, particular emphasis will be placed on the reasoning behind the selection.

3.13 STRESS DISTRIBUTION IN A THIN, INFINITE PLATE WITH A CIRCULAR HOLE SUBJECTED TO UNIAXIAL TENSILE LOADS

A thin plate of infinite length and width with a circular hole is shown in Fig. 3.4. The plate is subjected to a uniform tensile-type load that produces a uniform stress σ_o in the y direction at $r = \infty$. The distribution of the stresses about the hole, along the x axis, and along the y axis can be determined by using the Airy's-stress-function approach.

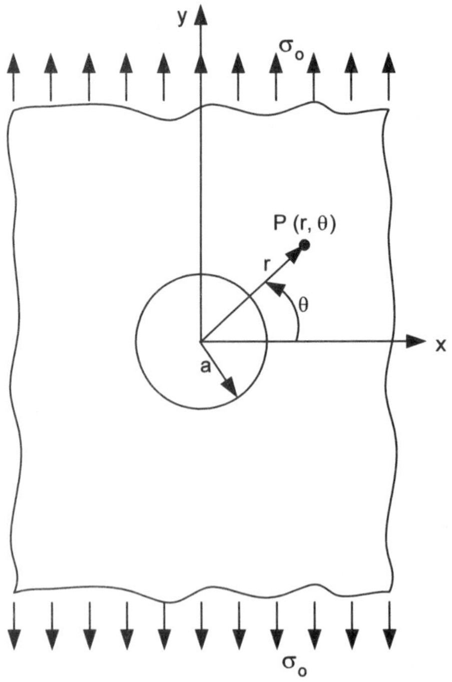

:

Figure 3.4 Thin, infinite plate with a circular hole subjected to a uniaxial tensile stress σ_o.

The boundary conditions that must be satisfied are:

$$\sigma_{rr} = \tau_{r\theta} = 0 \qquad \text{at } r = a \qquad (a)$$

and

$$\sigma_{yy} = \sigma_o \qquad \text{at} \Rightarrow \infty$$
$$\sigma_{xx} = \tau_{xy} = 0 \qquad \text{at} \Rightarrow \infty \qquad \text{(b)}$$

Selection of a stress function for this particular problem is difficult because none of the four functions previously tabulated is satisfactory. In order to overcome this difficulty, a method of superposition is commonly used that employs two different stress functions. The first function is selected such that the stresses associated with it satisfy the boundary conditions at $r \Rightarrow \infty$ but in general violate the conditions on the boundary of the hole. The second stress function must have associated stresses, which cancel the stresses on the boundary of the hole without influencing the stresses at $r \Rightarrow \infty$. An illustration of this superposition process is presented in Fig. 3.5.

The boundary conditions at $r \Rightarrow \infty$ can be satisfied by the uniform stress field associated with the stress function ϕ_2 in Eqs. (3.25). For the case of uniaxial tension in the y direction, ϕ_2 reduces to:

$$\phi_2 = a_2 x^2 = \tfrac{1}{2}\sigma_o x^2 \qquad \text{(c)}$$

The stresses throughout a plate without a hole are

$$\sigma_{yy} = \sigma_o \qquad \sigma_{xx} = \tau_{xy} = 0 \qquad \text{(d)}$$

If an imaginary hole of radius a is cut into the plate, the stresses σ_{rr}, $\sigma_{\theta\theta}$ and $\tau_{r\theta}$ on the boundary of the imaginary hole can be computed from Eqs. (3.31) as follows:

$$\sigma_{rr}{}^{I} = \sigma_o \sin^2\theta = \tfrac{1}{2}\sigma_o (1 - \cos 2\theta)$$
$$\sigma_{\theta\theta}{}^{I} = \sigma_o \cos^2\theta = \tfrac{1}{2}\sigma_o (1 + \cos 2\theta) \qquad \text{(e)}$$
$$\tau_{r\theta}{}^{I} = \sigma_o \sin\theta \cos\theta = \tfrac{1}{2}\sigma_o \sin 2\theta$$

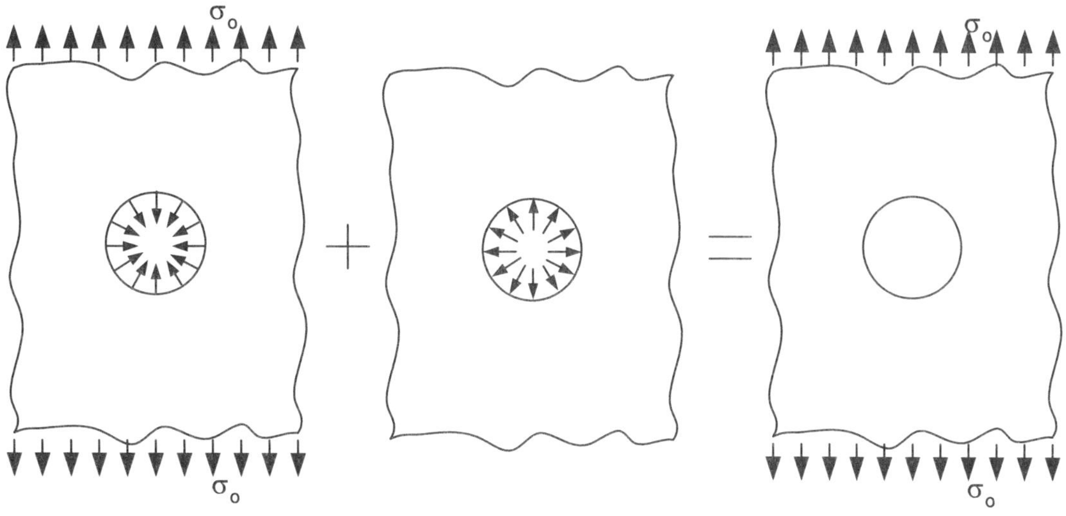

Figure 3.5 Method of superposition.

In the original problem, the boundary conditions at r = a were:

$$\sigma_{rr} = \tau_{r\theta} = 0$$

The boundary conditions to be satisfied by the stresses associated with the second stress function are therefore

$$\sigma_{rr} = -\sigma_o \sin^2\theta = -\tfrac{1}{2}\sigma_o(1-\cos 2\theta) \qquad \text{at } r = a$$

$$\sigma_{rr} = \tau_{r\theta} = \sigma_{\theta\theta} = 0 \qquad \text{at } r \Rightarrow \infty \qquad \text{(f)}$$

$$\tau_{r\theta} = -\sigma_o \sin\theta\cos\theta = -\tfrac{1}{2}\sigma_o \sin 2\theta \qquad \text{at } r = a$$

From Eqs. (f) it is apparent that the stresses σ_{rr} and $\tau_{r\theta}$ are functions of sin 2θ and cos 2θ, which suggests $\phi^{(2)}$ given by Eqs. (3.40) as a possible stress function. Inspection of Eqs. (3.40) indicates, however, that this function can satisfy the boundary conditions only for $\tau_{r\theta}$. From Eqs. (3.38), however, it can be seen that the stresses associated with $\phi^{(0)}$ can satisfy the boundary conditions for σ_{rr} without influencing $\tau_{r\theta}$. Thus, the stress function $\phi^{(0)} + \phi^{(2)}$ may be applicable. From Eqs. (3.38) and (3.40),

$$\phi^{(0)} + \phi^{(2)} = a_0 + b_0 \ln r + c_0 r^2 + d_0 r^2 \ln r + (a_2 r^2 + b_2 r^{-2} + c_2 r^4 + d_2)\cos 2\theta \qquad \text{(g)}$$

$$\sigma_{rr} = (b_0/r^2) + 2c_0 + d_0(1 + 2\ln r) - [2a_2 + 6b_2/r^4 + 4d_2/r^2]\cos 2\theta \qquad \text{(h)}$$

$$\sigma_{\theta\theta} = -(b_0/r^2) + 2c_0 + d_0(3 + 2\ln r) + [2a_2 + 6b_2/r^4 + 10c_2 r^2]\cos 2\theta \qquad \text{(i)}$$

$$\tau_{r\theta} = (2a_2 - 6b_2/r^4 + 6c_2 r^2 - 2d_2/r^2]\sin 2\theta \qquad \text{(j)}$$

Eqs. (h) to (j) contain seven unknowns: b_o, c_o, d_o, a_2, b_2, c_2 and d_2. Because $\sigma_{\theta\theta} = \sigma_{rr} = \tau_{r\theta} = 0$ as $r \Rightarrow \infty$,

$$c_o = d_o = a_2 = c_2 = 0 \qquad \text{(k)}$$

and Eqs. (h) to (j) reduce to

$$\sigma_{rr} = (1/r^2)[b_0 - (6b_2/r^2 + 4d_2)\cos 2\theta] \qquad \text{(l)}$$

$$\sigma_{\theta\theta} = (1/r^2)[-b_0 + (6b_2/r^2)\cos 2\theta] \qquad \text{(m)}$$

$$\tau_{r\theta} = -(1/r^2)[(6b_2/r^2 + 2d_2)\sin 2\theta] \qquad \text{(n)}$$

From the boundary conditions at r = a,

$$\tau_{r\theta} = -(1/a^2)[(6b_2/a^2 + 2d_2)\sin 2\theta] = -\tfrac{1}{2}\sigma_o \sin 2\theta \qquad \text{(o)}$$

$$\sigma_{rr} = (1/a^2)[b_0 - (6b_2/a^2 + 4d_2)\cos 2\theta] = -\tfrac{1}{2}\sigma_o(1-\cos 2\theta) \qquad \text{(p)}$$

Solving Eqs. (o) and (p) for the coefficients gives

$$b_o = -\tfrac{1}{2}\sigma_o a^2 \qquad b_2 = \tfrac{1}{4}\sigma_o a^4 \qquad d_2 = -\tfrac{1}{2}\sigma_o a^2 \qquad \text{(q)}$$

Substituting Eqs. (q) into Eqs. (l) to (n) gives

$$\sigma_{rr}^{II} = -\frac{\sigma_o a^2}{2r^2}\left[1 + \left(\frac{3a^2}{r^2} - 4\right)\cos 2\theta\right]$$

$$\sigma_{\theta\theta}^{II} = \frac{\sigma_o a^2}{2r^2}\left(1 + \frac{3a^2}{r^2}\cos 2\theta\right)$$

$$\tau_{r\theta}^{II} = -\frac{\sigma_o a^2}{2r^2}\left[\left(\frac{3a^2}{r^2} - 2\right)\sin 2\theta\right]$$

The required solution for the original problem is obtained by superposition as follows:

$$\sigma_{rr} = \sigma_{rr}^{I} + \sigma_{rr}^{II} = \frac{\sigma_o}{2}\left\{\left(1 - \frac{a^2}{r^2}\right)\left[1 + \left(\frac{3a^2}{r^2} - 1\right)\cos 2\theta\right]\right\}$$

$$\sigma_{\theta\theta} = \sigma_{\theta\theta}^{I} + \sigma_{\theta\theta}^{II} = \frac{\sigma_o}{2}\left[\left(1 + \frac{a^2}{r^2}\right) + \left(1 + \frac{3a^4}{r^4}\right)\cos 2\theta\right] \qquad (3.42)$$

$$\tau_{r\theta} = \tau_{r\theta}^{I} + \tau_{r\theta}^{II} = \frac{\sigma_o}{2}\left[\left(1 + \frac{3a^2}{r^2}\right)\left(1 - \frac{a^2}{r^2}\right)\sin 2\theta\right]$$

Equations (3.42) give the polar components of stress at any point in the body defined by r, θ. By employing Eqs. (3.42) the stresses along the x axis, along the y axis, and around the boundary of the hole can easily be computed.

The stresses along the x-axis can be obtained by setting θ = 0 and r = x in Eqs. (3.42). Thus:

$$\sigma_{rr} = \sigma_{xx} = \frac{\sigma_o}{2}\left(1 - \frac{a^2}{x^2}\right)\frac{3a^3}{x^2}$$

$$\sigma_{\theta\theta} = \sigma_{yy} = \frac{\sigma_o}{2}\left(2 + \frac{a^2}{x^2} + \frac{3a^4}{x^4}\right) \qquad (3.43)$$

$$\tau_{r\theta} = \tau_{xy} = 0$$

The distribution of σ_{xx}/σ_o and σ_{yy}/σ_o is plotted as a function of position along the x-axis in Fig. 3.6. An examination of this figure clearly indicates that the presence of the hole in the infinite plate under uniaxial tension increases the stress σ_{yy} by a factor of 3. This factor is often called a **stress concentration factor**. In a later chapter it will be shown how photoelasticity can be effectively employed to determine stress-concentration factors.

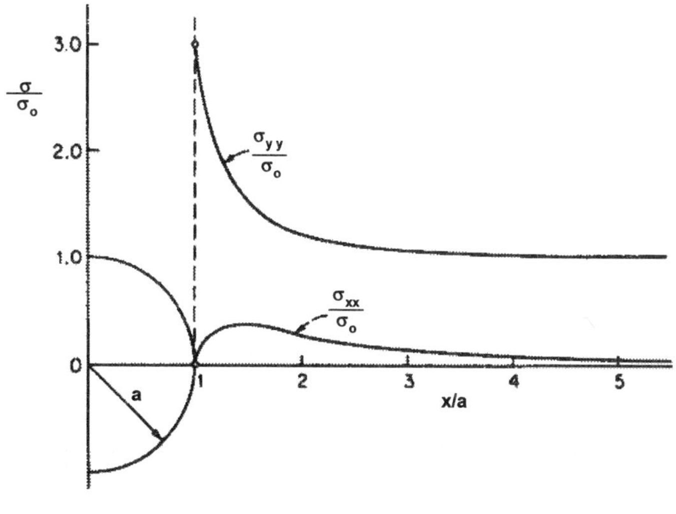

Figure 3.6 Distribution of σ_{xx}/σ_o and σ_{yy}/σ_o along the x-axis.

In a similar manner, the stresses along the y-axis can be obtained by setting $\theta = \pi/2$ and $r = y$ in Eqs. (3.42).

$$\sigma_{rr} = \sigma_{yy} = \frac{\sigma_o}{2}\left(2 - \frac{5a^2}{y^2} + \frac{3a^4}{y^4}\right)$$

$$\sigma_{\theta\theta} = \sigma_{xx} = \frac{\sigma_o}{2}\left(\frac{a^2}{y^2} - \frac{3a^4}{y^4}\right) \qquad (3.44)$$

$$\tau_{r\theta} = \tau_{xy} = 0$$

The distribution of σ_{xx}/σ_o and σ_{yy}/σ_o is plotted as a function of position along the y-axis in Fig. 3.7. In this figure it can be noted that $\sigma_{xx}/\sigma_o = -1$ at the boundary of the hole; thus the influence of the hole not only produces a concentration of the stresses but in this case it also produces a change in the sign of the stresses.

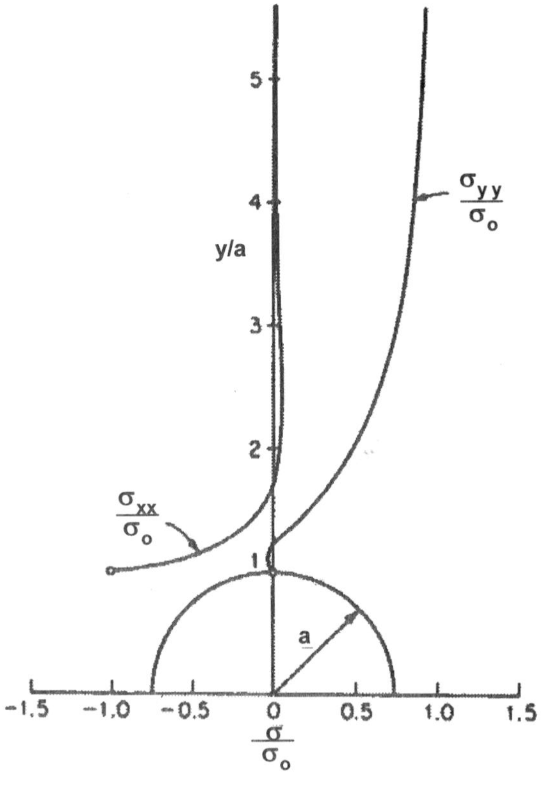

Figure 3.7 Distribution of σ_{xx}/σ_o and σ_{yy}/σ_o along the y axis.

The distribution of $\sigma_{\theta\theta}$ about the boundary of the hole is obtained by setting $r = a$ in Eqs. (3.42).

$$\sigma_{rr} = \tau_{r\theta} = 0 \qquad \sigma_{\theta\theta} = \sigma_o\,(1 + 2\cos 2\theta) \qquad (3.45)$$

The distribution of $\sigma_{\theta\theta}/\sigma_o$ about the boundary of the hole is shown in Fig. 3.8. The maximum $\sigma_{\theta\theta}/\sigma_o$ occurs at the x axis ($\sigma_{\theta\theta}/\sigma_o = 3$), and the minimum stress occurs at the y axis ($\sigma_{\theta\theta}/\sigma_o = -1$). At the point defined by $\theta = 60°$ on the boundary of the hole, all stresses are zero. This type of point is commonly referred to as a **singular point**.

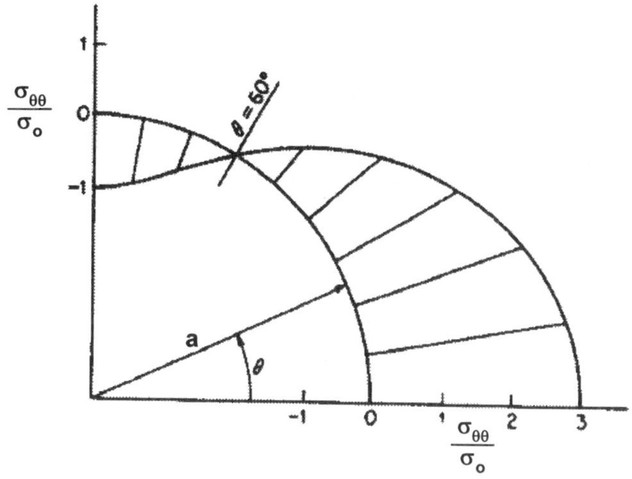

Figure 3.8 Distribution of $\sigma_{\theta\theta}/\sigma_o$ about the boundary of the hole.

EXERCISES

3.1 Give an example of a Type 1 boundary value problem other than the one cited in Section 3.1.

3.2 Give an example of a Type 2 boundary value problem other than the one cited in Section 3.1.

3.3 Give an example of a Type 3 boundary value problem other than the one cited in Section 3.1.

3.4 Give an example of a Type 4 boundary value problem other than the one cited in Section 3.1.

3.5 Name and list all of the field equations. How many field quantities exist in the elastic problem?

3.6 What is the advantage of using the stress equations of compatibility, Eqs. (3.3), in the solution of elasticity problems.

3.7 Discuss the influence of the elastic constants on determining stresses from scale models fabricated from plastics by using Eq. (3.3) as the basis for your discussion.

3.8 Describe the key features that differentiate three dimensional, plane stress and plane strain formulations in elasticity theory.

3.9 Determine the stress-strain and the strain-stress relations from Eqs. (2.16) and (2.17) with the assumptions of the plane-stress problem.

3.10 Determine the stress-strain and the strain-stress relations from Eqs. (2.16) and (2.17) with the assumptions of the plane-strain problem.

3.11 Verify Eq. (3.17c).

3.12 In Eqs. (3.18), the body-force intensities were defined as

$$F_x = -\partial\Omega/\partial x \qquad\qquad F_y = -\partial\Omega/\partial y$$

Determine Ω for a gravitational field of intensity q in the x direction.

3.13 Verify Eqs. (3.26) and Eqs. (3.28).

3.14 Verify Eq. (r) on page xxx.

3.15 Establish Eqs. (3.29).

3.16 Determine the stresses in the uniformly loaded cantilever beam shown in Fig. E3.16, by using the Airy's-stress-function approach. Assume a unit thickness.

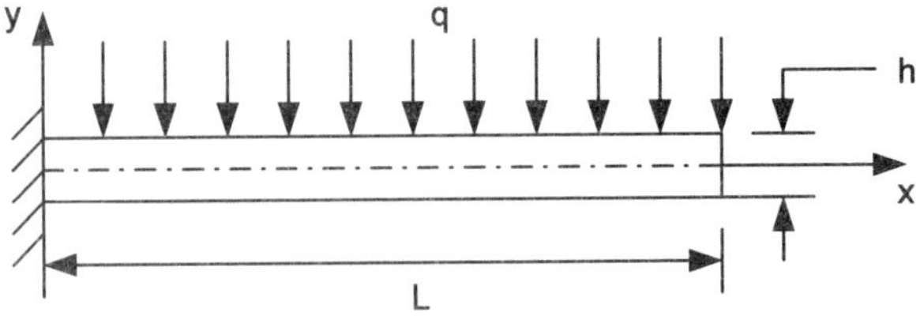

Figure E3.16

3.17 Determine the stresses in the cantilever beam shown in Fig. E3.17 by using the Airy's-stress-function approach.

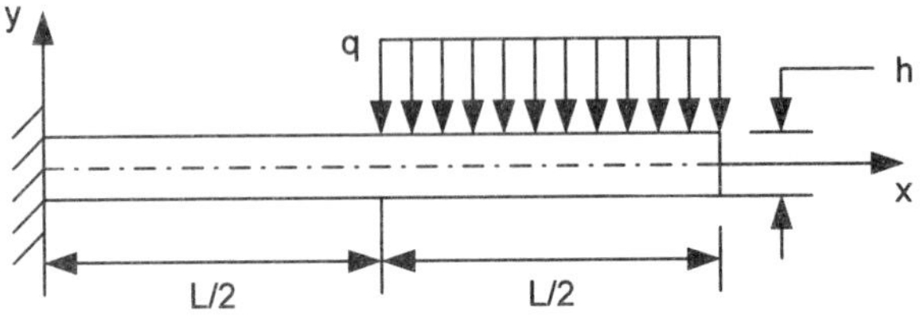

Figure E3.17

3.18 For the built-in triangular plate shown in Fig. E3.18, verify that the applicable stress function is:

$$\phi = \frac{q\cot\alpha}{2(1-\alpha\cot\alpha)}\left[-x^2\tan\alpha + xy + (x^2+y^2)\left(\alpha - \tan^{-1}\frac{x}{y}\right)\right]$$

For the particular case of α = 45° and a plate of unit thickness, determine the normal stress distribution along the line x = L/2 and compare the solution with the results determined by using elementary beam theory.

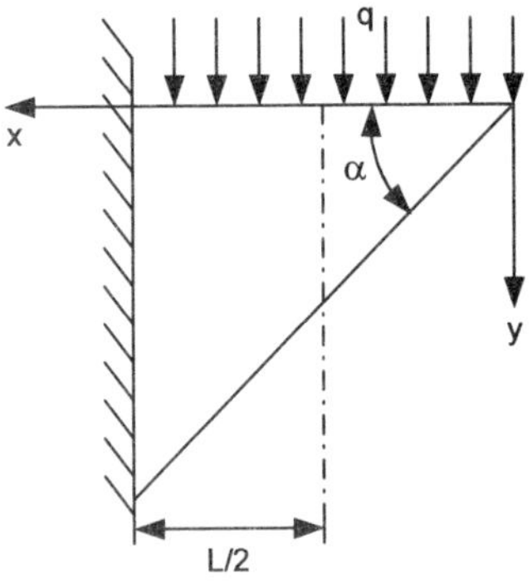

Figure E3.18

3.19 Show that:

$$\frac{\partial^2\phi}{\partial y^2} = \frac{\partial\phi}{\partial r}\left(\frac{\partial^2 r}{\partial y^2}\right) + \frac{\partial^2\phi}{\partial r^2}\left(\frac{\partial r}{\partial y}\right)^2 + 2\frac{\partial^2\phi}{\partial r\partial\theta}\left(\frac{\partial r}{\partial y}\right)\left(\frac{\partial\theta}{\partial y}\right) + \frac{\partial\phi}{\partial\theta}\left(\frac{\partial^2\theta}{\partial y^2}\right) + \frac{\partial^2\phi}{\partial\theta^2}\left(\frac{\partial\theta}{\partial y}\right)^2$$

and verify Eq. (3.32b).

3.20 Show that:

$$\frac{\partial^2\phi}{\partial x\partial y} = \frac{\partial\phi}{\partial r}\left(\frac{\partial^2 r}{\partial x\partial y}\right) + \frac{\partial^2\phi}{\partial r^2}\left(\frac{\partial r}{\partial x}\right)\left(\frac{\partial r}{\partial y}\right) + \frac{\partial^2\phi}{\partial r\partial\theta}\left(\frac{\partial r}{\partial x}\right)\left(\frac{\partial\theta}{\partial y}\right)$$
$$+\frac{\partial\phi}{\partial\theta}\left(\frac{\partial^2\theta}{\partial x\partial y}\right) + \frac{\partial^2\phi}{\partial\theta\partial r}\left(\frac{\partial\theta}{\partial x}\right)\left(\frac{\partial r}{\partial y}\right) + \frac{\partial^2\phi}{\partial\theta^2}\left(\frac{\partial\theta}{\partial x}\right)\left(\frac{\partial\theta}{\partial y}\right)$$

and verify Eq. (3.32c).

3.21 Verify Eqs. (3.37)

3.22 Determine the polar components of the stresses σ_{rr} and $\sigma_{\theta\theta}$ in a thick-walled cylindrical vessel whose inside diameter is 1000 mm and outside diameter is 1400 mm if the vessel is subjected to an internal pressure of 35 MPa. Determine the radial displacement of the outside diameter if the vessel is made from steel (E = 207 GPa and ν = 0.30).

3.23 Show how the radial displacement of the inside and outside surfaces of a cylindrical pressure vessel subjected to internal pressure can be used to determine E and ν.

3.24 Determine the stresses and displacements in the curved beam shown in Fig. E3.24 when subjected to the moment load M.

3.25 Determine the stresses and displacements in the curved beam shown in Fig. E3.25 when subjected to the shear load V.

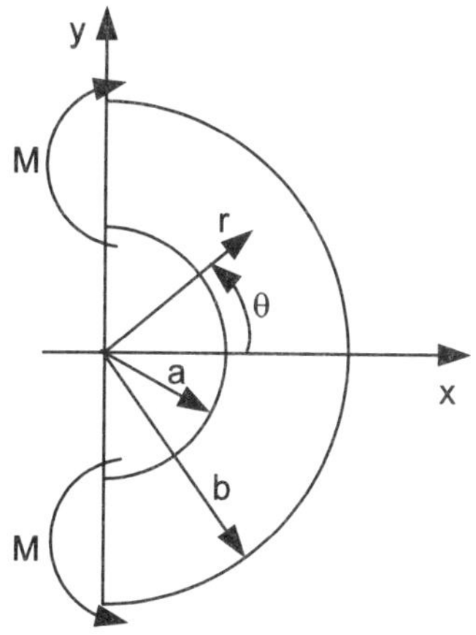

Figure E3.24

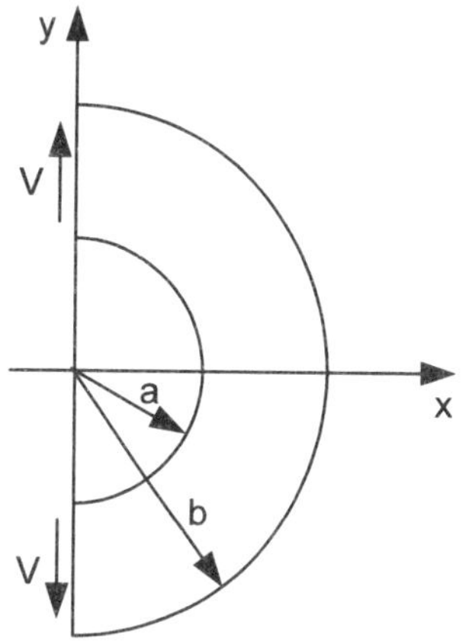

Figure E13.25

3.26 In Exercise 3.22, let a increase while holding (b – a) constant and compare the values of the maximum stress $\sigma_{\theta\theta}$ from curved-beam theory and from straight-beam theory. Draw conclusions regarding the influence of the radius of curvature on $\sigma_{\theta\theta}$.

3.27 Determine the stresses in a semi-infinite plate due to a normal load acting on its edge as shown in Fig. E3.27. Hint: Try the stress function ϕ^*.

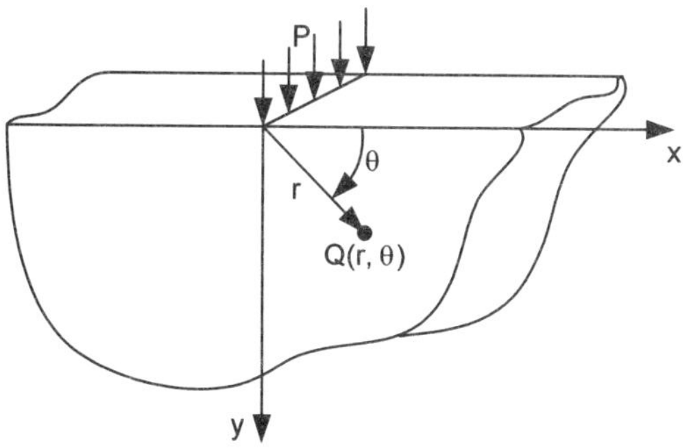

Figure E3.27

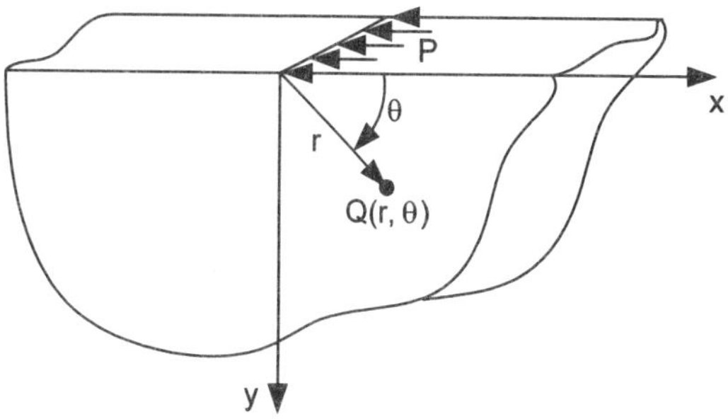

Figure E3.28

3.28 Determine the stresses in a semi-infinite plate due to a shear load acting on its edge as shown in Fig. E3.28. Hint: Use the stress function ϕ^*.

3.29 Determine the stresses in a semi-infinite plate due to an inclined load acting on its edge as shown in Fig. E3.29.

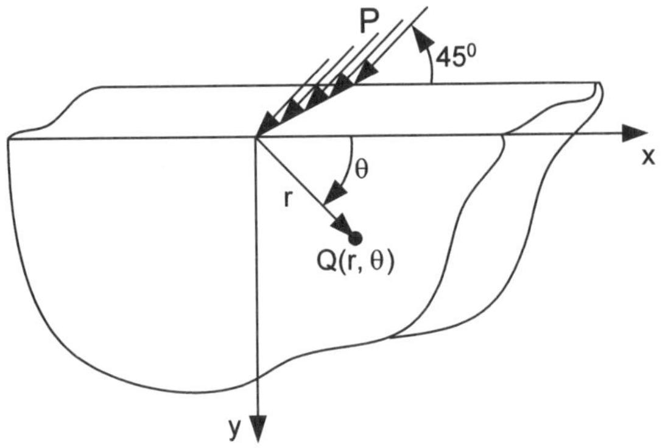

Figure E3.29

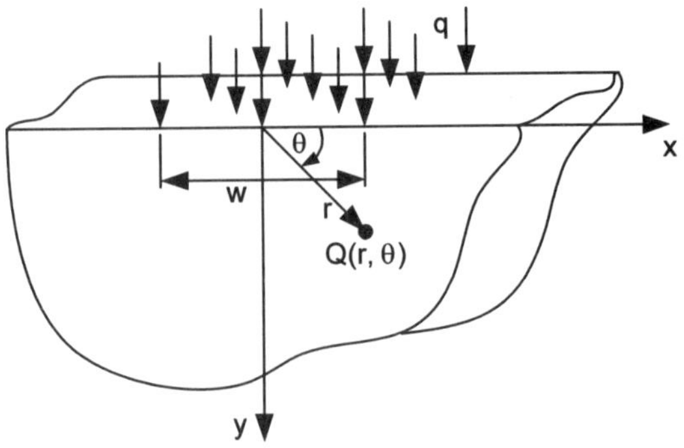

Figure E3.30

3.30 Determine the stresses in a semi-infinite body due to a distributed normal load as shown in Fig. E3.30.

3.31 A steel (E = 207 GPa and ν = 0.30) ring is shrunk onto another steel ring, as shown in Fig. E3.31. Determine the maximum interference possible without yielding one of the rings if the yield strength of both rings is 900 MPa and the dimensions of the rings are as follows:

θ	Case 1	Case 2	Case 3	Case 4
a_i (mm)	100	100	100	0
b_i (mm)	125	115	150	100
b_o (mm)	150	200	175	150

Figure E3.31

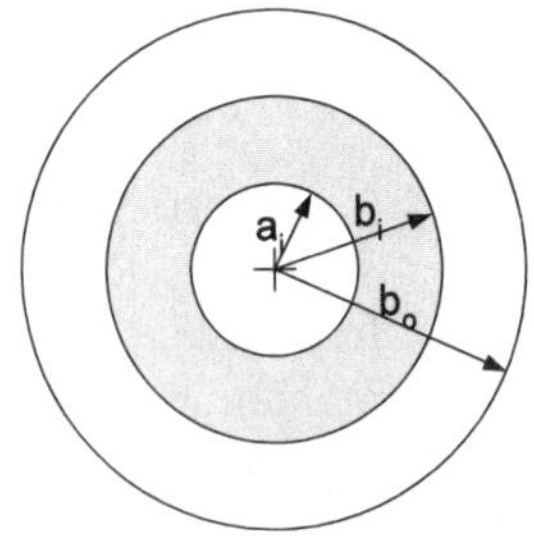

3.32 Discuss the possibility of fabricating gun tubes by shrink fitting two or more long cylinders together to form the tube.

3.33 Using the solution of Exercise 3.30, determine σ_1, σ_2 and τ_{max}. Prepare a drawing illustrating the contour lines for σ_1, σ_2 and τ_{max}. Also, prepare a drawing of the isostatics (stress trajectories) for σ_1 and σ_2.

Bibliography

1. Boresi, A. P., and P. P. Lynn: Elasticity in Engineering Mechanics, Prentice-Hall, Inc., Englewood Cliffs, N.J. 1974.
2. Chou, P. C., and N. J. Pagano: Elasticity, D. Van Nostrand Company, Inc., Princeton, N.J., 1967.
3. Durelli, A. J., E. A. Phillips, and C. H. Tsao: Introduction to the Theoretical and Experimental Analysis of Stress and Strain, McGraw-Hill Book Company, New York, 1958.
4. Sechler, E. E.: Elasticity in Engineering, John Wiley & Sons, Inc., New York, 1952.
5. Timoshenko, S. P., and J. N. Goodier: Theory of Elasticity, 2d ed., McGraw-Hill Book Company, New York, 1951.
6. Kirsch, G.: "Die Theorie der Elasticitat und die Bedurfnisse der Festigkeitlehre," Z. Ver. Dtsch. Ing., vol. 32, pp. 797-807, 1898.
7. Sternberg, E., and M. Sadowsky: "Three-dimensional Solution for the Stress Concentration around a Circular Hole in a Plate of Arbitrary Thickness," J. Appl. Mech., vol. 16, pp. 27-38, 1949.
8. Howland, R. C. J.: "On the Stresses in the Neighborhood of a Circular Hole in a Strip under Tension," Trans. R. Soc., vol. A229, pp. 49-86, 1929.
9. Inglis, C. E.: "Stresses in a Plate Due to the Presence of Cracks and Sharp Corners", Proc. Inst. Nav. Arch., vol. 55, part 1, pp. 219-230, 1913.
10. Greenspan, M.: "Effect of a Small Hole on the Stresses in a Uniformly Loaded Plate," Q. Appl. Math., vol. 2, pp. 60-71, 1944.
11. Jeffrey, G. B.: "Plane Stress and Plane Strain in Bi-polar Co-ordinates," Phil. Trans. R. Soc., vol. A-221, pp. 265-293, 1920.
12. Mindlin, R. D.: "Stress Distribution around a Hole Near the Edge of a Plate under Tension," Proc. SESA, vol. V, no. 2, pp. 56-68, 1948.
13. Ling, C. B.: "Stresses in a Notched Strip under Tension," J. Appl. Mech., vol. 14, pp. 275-280, 1947.
14. Atanackovic and A. Guran: Theory of Elasticity for Scientists and Engineers, Birkhauser, Boston, MA. 2000.
15. Slaughter, W. S.: Linearized Theory of Elasticity, Birkhauser, Boston, MA. 2002.
16. Fung, Y. C. and P. Tang, Classical and Computational Solid Mechanics, World Scientific, Singapore, 2001.
17. Atkin, R. J. and N. Fox: An Introduction to the Theory of Elasticity, Longman, London, 1980.
18. Sadd,M.H.: Elasticity, Theory, Applications and Numerics, Elsevier, 2009.

CHAPTER 4

ELEMENTARY FRACTURE MECHANICS

4.1 INTRODUCTION

In the first three chapters, the theory of elasticity was introduced to show procedures for determining stresses and strains in bodies free of flaws. However, when flaws such as cracks exist in the body, elastic stress-strains are not sufficient to completely predict the onset of failure. The difficulty is due to the geometry of the crack tip. The crack tip is sharp with a radius of curvature approaching zero and this sharpness produces local stresses σ_{xx}, σ_{yy} and τ_{xy} which tend to infinity as one approaches the crack tip. Since the stresses go to infinity for any loading of the body, the theories of failure, such as Tresca or Von Mises, cannot be applied and the load required producing either localized yielding or the onset of crack propagation cannot be predicted.

To treat bodies containing cracks, it is necessary to introduce a method which deals with the singular state of stress at the crack tip. Fracture mechanics, developed by Irwin [1] from the earlier work of Inglis [2], Griffith [3], and Westergaard [4], treats singular stress fields by introducing a quantity known as a stress intensity factor K_I defined as:

$$K_I = \lim_{r \to 0}\left(\sqrt{2\pi r}\,\sigma_{yy}\right) \tag{4.1}$$

where the coordinate system is as shown in Fig. 4.1 and σ_{yy} is evaluated in the limit along the $\theta = 0$ line.

The limit process gives a stress intensity factor K_I that is a linear function of the loads applied to the body. The stress intensity factor remains finite and provides a basis for determining the critical load for failure. The critical condition in fracture mechanics is the onset of crack initiation where the crack extends suddenly at high velocity in cleavage failure or where the crack extends at low velocity by tearing in a shear rupture type failure. In either type of crack extension, the structure has failed. Clearly the introduction of the stress intensity factor provides a means, that is dependent upon the applied state of stress (i.e. σ_{yy}) and the crack length, of judging the criticality of a crack.

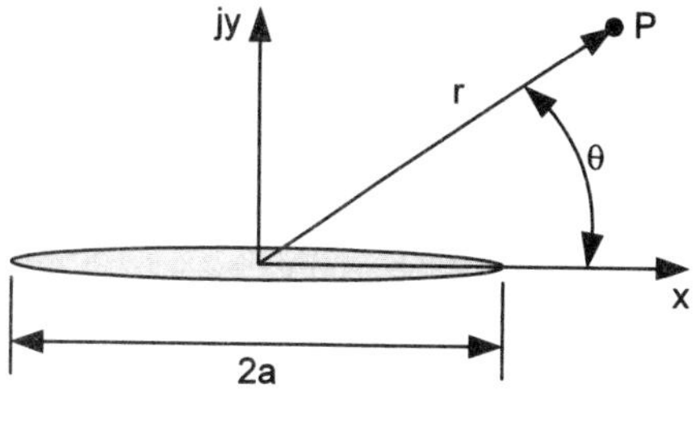

Figure 4.1 Coordinate system defining a double ended crack ($z = x + jy = re^{j\theta}$).

The second and equally important aspect of fracture mechanics is the characterization of the material property that defines the onset of crack initiation. This property is termed fracture toughness and it is determined using well defined testing procedures. For example, the plane strain crack initiation toughness K_{Ic} defines the resistance of the material to crack initiation. When the applied stress intensity factor

$$K_I \geq K_{Ic} \tag{4.2}$$

the crack initiates and extends into the structure. Because K_I is a linear function of the applied load, say P, the failure criterion given in Eq. (4.2) provides an approach for predicting the critical load P_{cr} associated with crack initiation. This approach is illustrated in Fig. 4.2.

Before beginning a study of fracture mechanics, it is important to understand singular stress fields where all of the Cartesian components of stress tend to infinity at a point, while the loads applied to the body remain finite. To show this singular stress state the Inglis solution for a plate under uniaxial load with an elliptical hole will be introduced in the following section (Sec. 4.2).

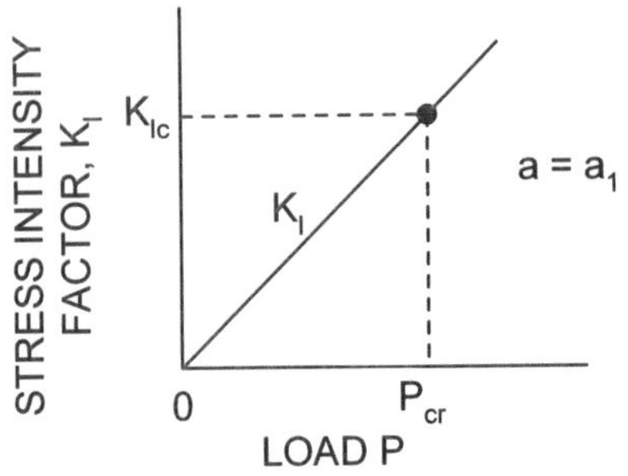

Figure 4.2 Fracture mechanics is used for predicting the critical load P_{cr} associated with crack initiation.

4.2 STRESSES DUE TO AN ELLIPTICAL HOLE IN A UNIFORMLY STRESSED PLATE

Elasticity problems involving elliptic or hyperbolic boundaries are treated using an elliptic coordinate system, defined in Fig. 4.3a, where

$$x = c \cosh \xi \cos \eta \quad \text{and} \quad y = c \sinh \xi \sin \eta \tag{4.3a}$$

with c defined as a constant. Eliminating η from Eq. (4.3a) gives:

$$\frac{x^2}{c^2 \cosh^2 \xi} + \frac{y^2}{c^2 \sinh^2 \xi} = 1 \tag{4.3b}$$

When $\xi = \xi_o$, this is the equation of an ellipse with major and minor axes a and b given by

$$a = c \cosh \xi_o \quad \text{and} \quad b = c \sinh \xi_o \tag{4.4}$$

The foci of the ellipse are at $x = \pm c$. It is clear that the aspect ratio of the ellipse varies as a function of ξ_o. If ξ_o is very large (approaching infinity) the ellipse approaches a circle with a = b. However, if $\xi_o \Rightarrow 0$ the ellipse becomes a double line of length 2c and represents a crack.

Inglis considered the infinite plate with an elliptical hole subjected to uniaxial stresses σ_0 as shown in Fig. 4.3a and found that the stresses σ_η about the hole are given by the equation

$$\sigma_\eta = \sigma_0 e^{2\xi_0} \left[\frac{\sinh 2\xi_0 (1 - e^{-2\xi_0})}{\cosh 2\xi_0 - \cos 2\eta} - 1 \right] \tag{4.5}$$

The boundary stress σ_η is a maximum at the ends of the major axis where $\cos 2\eta = 1$. Substituting this value of η into Eq. (4.5) leads to

$$(\sigma_\eta)_{max} = \sigma_0\left(1+\frac{2a}{b}\right) \tag{4.6}$$

It is instructive to examine the results of Eq. (4.6) for the two limit cases. First, when a = b ($\xi_0 \Rightarrow \infty$), the elliptical hole becomes circular and $(\sigma_\eta)_{max} = 3\sigma_0$. This result confirms the stress concentration for a circular hole in an infinite plate with a uniaxial load as determined by Kirsch [5] and described by Eq. (3.49). Second, when b = 0 ($\xi_0 = 0$), the elliptical hole becomes a flat opening representing a crack. In this case Eq. (4.6) shows that $(\sigma_\eta)_{max} \Rightarrow \infty$ as b ⇒ 0. Note that the maximum stress at the tip of the crack, located at the ends of the major axis of the ellipse, goes to infinity regardless of the magnitude of the applied stress σ_0. This fact indicates that localized yielding will occur at the crack tips for any nonzero load. The commonly employed failure theories such as Tresca and von Mises predict yielding for any level of load and, therefore, do not provide useful information regarding the stability of the crack as the applied stress σ_0 is increased from zero to the value when crack initiation occurs.

When the applied stress σ_0 is parallel to the major axis of the elliptical hole, as shown in Fig. 4.3b, the maximum value of σ_η on the boundary of the hole is at the ends of the minor axes (at point b), and is

$$(\sigma_\eta)_{max} = \sigma_0\left(1+\frac{2b}{a}\right) \tag{4.7}$$

In the limit when b ⇒ 0 and the ellipse represents a crack the stress $(\sigma_\eta)_{max} = \sigma_0$. At the ends of the major axis of the elliptic hole Eq. (4.7) does not apply but $(\sigma_\eta)_{max} = -\sigma_0$ for any value of b/a.

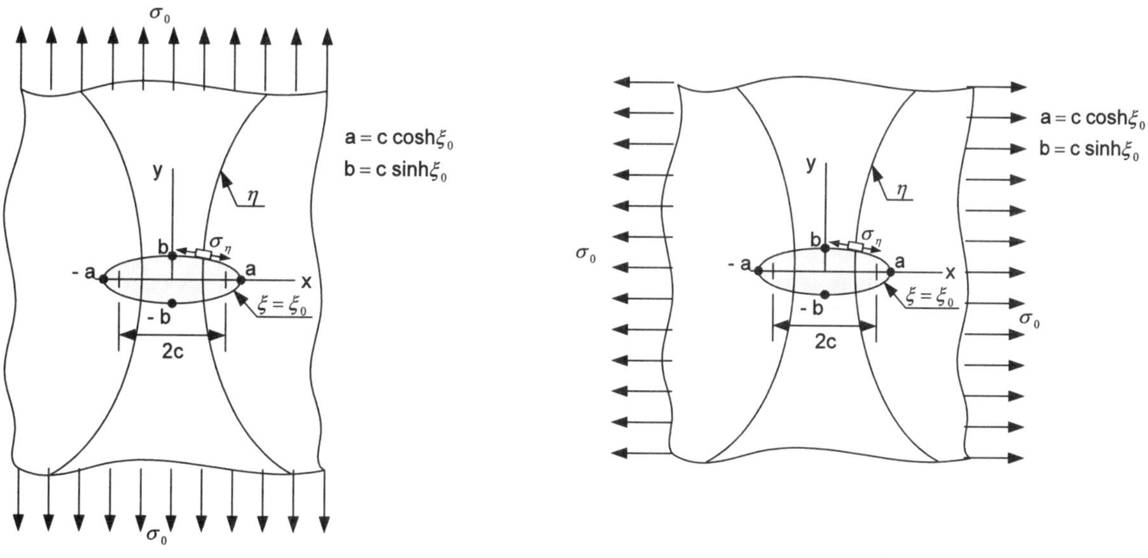

(a) (b)

Figure 4.3 An elliptical hole in an infinite plate: (a) with uniaxial loading σ_0 perpendicular to x; (b) with uniaxial loading σ_0 parallel to x.

The Inglis' solution for the elliptical hole in the plate provides, in the limit as b ⇒ 0, the stress distribution for a crack in a plate. It is evident that the stresses at the tip of the crack are singular when the crack is perpendicular to the applied stress σ_0. The fact that the stresses at the crack tip are singular indicates local yielding will occur for any nonzero stress σ_0 and that methods for predicting structural stability based on Tresca or Von Mises theories for yielding are inadequate. One must introduce concepts of crack stability based on fracture mechanics to predict if the crack will initiate at a specified applied stress σ_0.

4.3 THE WESTERGAARD STRESS FUNCTION

Westergaard introduced a complex stress function Z(z) that is related to the Airy's stress function ϕ by the equation

$$\phi = \mathrm{Re}\bar{\bar{Z}} + y\mathrm{Im}\bar{Z} \tag{4.8}$$

Because Z is a complex function, it is clear that

$$Z(z) = \mathrm{Re}\, Z + j\, \mathrm{Im}\, Z \tag{4.9a}$$

where z is defined as

$$z = x + jy = re^{j\theta} \tag{4.9b}$$

Note that Z is analytic over the region of interest, and the Cauchy-Riemann conditions lead to

$$\nabla^2 \mathrm{Re}\, Z = \nabla^2 \mathrm{Im}\, Z = 0 \tag{4.10}$$

This result shows that the Westergaard stress function automatically satisfies Eq. (3.23b). The bars over the stress function Z in Eq. (4.8) indicate integration. Thus

$$\begin{aligned} \frac{d\bar{\bar{Z}}}{dz} &= \bar{Z} \quad \text{or} \quad \bar{\bar{Z}} = \int \bar{Z}\, dz \\ \frac{d\bar{Z}}{dz} &= Z \quad \text{or} \quad \bar{Z} = \int Z\, dz \\ \frac{dZ}{dz} &= Z' \quad \text{or} \quad Z = \int Z'\, dz \end{aligned} \tag{4.11}$$

where the bar and the prime represents integration and differentiation respectively. Substituting Eq. (4.8) into Eqs. (3.20) gives the Cartesian components of stress in terms of the real and imaginary parts of the Westergaard stress function as

$$\begin{aligned} \sigma_{xx} &= \mathrm{Re}\, Z - y\, \mathrm{Im}\, Z' \\ \sigma_{yy} &= \mathrm{Re}\, Z + y\, \mathrm{Im}\, Z' \\ \tau_{xy} &= -\, y\, \mathrm{Re}\, Z' \end{aligned} \tag{4.12}$$

Equations (4.12) will yield stresses for functions Z(z) that are analytic; however, the stress function must be selected to satisfy boundary conditions corresponding to the problem being investigated. The formulation given by Eqs. (4.12), as originally proposed by Westergaard, correctly accounts for the stress singularity at the crack tip; however, additional terms must be added to adequately represent the stress field in regions adjacent to the crack tip. These additional terms will be introduced in later sections that deal with experimental methods for measuring K_I.

The classic problem in fracture mechanics, shown in Fig 4.4a, is an infinite plate with a central crack of length 2a that is subjected to biaxial tension. The stress function Z applied to solve this problem is:

$$Z = \frac{\sigma_0 z}{\sqrt{z^2 - a^2}} \tag{4.13}$$

Substitution of Eq. (4.13) into Eqs. (4.12) with $z \Rightarrow \infty$ yields $\sigma_{xx} = \sigma_{yy} = \sigma_0$ and $\tau_{xy} = 0$ as required to satisfy the far field boundary conditions. On the crack surface where y = 0 and z = x, for $-a \leq x \leq a$, Re Z = 0 and $\sigma_{yy} = \tau_{xy} = 0$. Clearly, the stress function Z given in Eq. (4.13) satisfies the boundary conditions over the free surfaces of the crack.

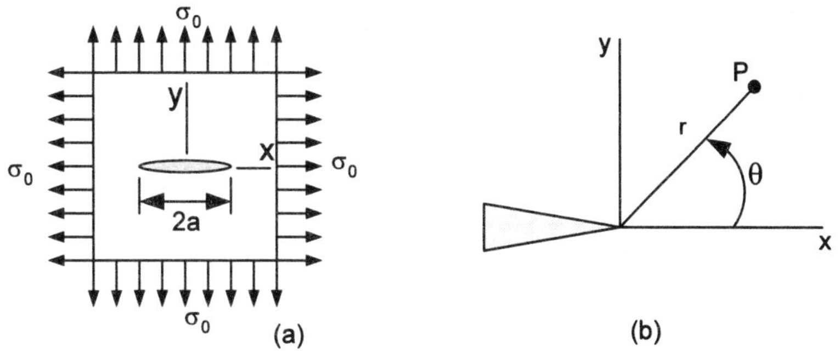

Figure 4.4 (a) A crack of length 2a in an infinite plate subjected to biaxial stress σ_0.
(b) The coordinates are defined with the origin at the crack tip.

It is more convenient to relocate the origin of the coordinate system and place it at the tip of the crack as shown in Fig. 4.4b. The translation of the origin requires that z in Eq. (4.13) be replaced by (z + a) and the new stress function becomes

$$Z = \frac{\sigma_0 (z+a)}{\sqrt{z(z+2a)}} \tag{4.14}$$

Next, consider a small region near the crack tip where z << a. Then, Eq. (4.14) reduces to

$$Z = \sqrt{\frac{a}{2}}\sigma_0 z^{-1/2} \tag{4.15}$$

Substituting Eq. (4.9b) into Eq. (4.15) yields

$$Z = \sqrt{\frac{a}{2r}}\sigma_0 e^{-j\theta/2} \tag{4.16}$$

Recall the identity

$$e^{\pm j\theta} = \cos\theta \pm j\sin\theta \tag{4.17}$$

Substituting Eq. (4.17) into Eq. (4.16) shows that the real part of Z is

$$\mathrm{Re}Z = \sqrt{\frac{a}{2r}}\sigma_0 \cos\frac{\theta}{2} \tag{4.18}$$

Along the crack line where θ and y are both equal to zero Eqs. (4.18) and (4.12) give

$$\sigma_{yy} = \mathrm{Re}Z = \sqrt{\frac{a}{2r}}\sigma_0 \tag{4.19}$$

This result shows that the stress $\sigma_{yy} \Rightarrow \infty$ and is singular with order $1/\sqrt{r}$ as one approaches the crack tip along the x-axis. Finally, Eq. (4.19) can be substituted into Eq. (4.1) to obtain the stress intensity factor K_I as

$$K_I = \sqrt{\pi a}\,\sigma_0 \tag{4.20a}$$

This result shows that the stress intensity factor K_I varies as a linear function of the applied stress σ_0 and increases with the crack length as function of the $\sqrt{a}$ as illustrated in Fig. 4.5. The units of K_I are psi-$\sqrt{\text{in.}}$ or MPa-$\sqrt{\text{m}}$.

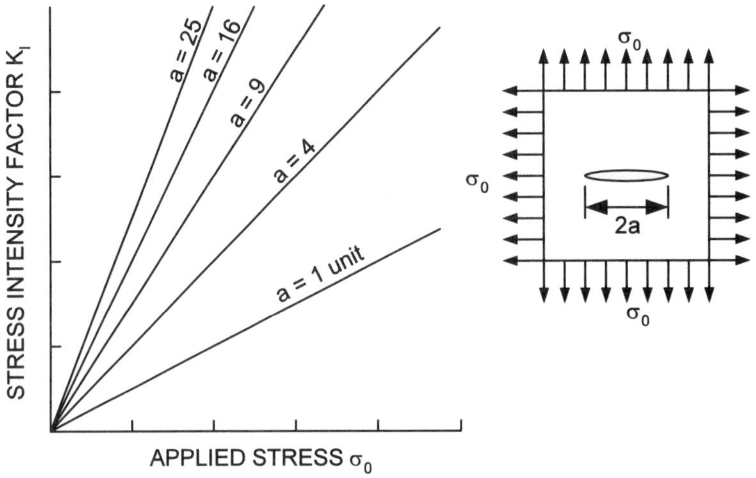

Figure 4.5 Stress intensity factor K_I as a function of the applied stress σ_0 with crack length a as a parameter.

4.4 STRESS INTENSITY FACTORS FOR SELECT GEOMETRIES

Cracks in plane bodies of finite size are important because the crack poses a threat to the stability and safety of the entire structure even though only one plate may be cracked. It is important to determine the stress intensity factor for the specific geometry and loading involved to assess the safety factor for the cracked plate. Many solutions for boundary value problems with cracks exist and a few of the more commonly encountered geometries and loadings will be reviewed in this section. A much more complete listing of solutions is given by Tada, Paris, and Irwin in a handbook [6].

First, consider the tension strip with a centrally located crack as shown in Fig. 4.6. The loading at the ends of the strip, well removed from the crack, is a uniaxial stress σ_0. An approximate solution for K_I, due to Irwin [7], is

$$K_I = \sqrt{\frac{W}{\pi a}\tan\frac{\pi a}{W}}\,\sqrt{\pi a}\,\sigma_0 \tag{4.21}$$

Note that Eq. (4.21) may be written as:

$$K_I = \alpha K_{I\infty} \tag{4.22}$$

where $\alpha = \sqrt{\frac{W}{\pi a}\tan\frac{\pi a}{W}}$ is a multiplier and the term $K_{I\infty} = \sqrt{\pi a}\,\sigma_0$ corresponds to the solution for the infinite plate under biaxial loading. With K_I expressed in the form shown in Eq. (4.22), the effect of the finite boundaries of the body is given in terms of a multiplying factor α that acts on $K_{I\infty}$. The magnitude of α depends on a/W and ranges from 1 for $(a/W) \Rightarrow 0$ to 4 for $(a/W) = 0.487$.

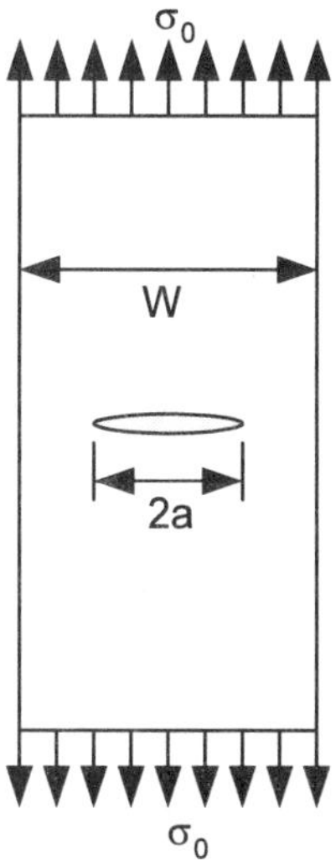

Figure 4.6 Tension strip with a central crack.

Next, consider single and double edge cracks in a tension strip as shown in Figs. 4.7a and 4.7b. Series solutions for these two problems give the multiplier α in terms of (a/W).

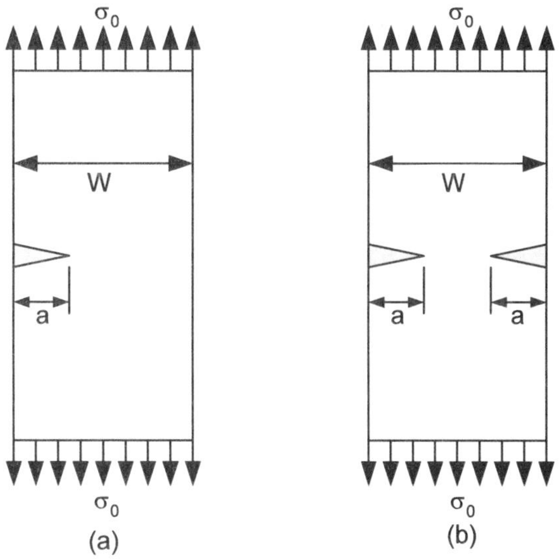

Figure 4.7 Edge-cracked tension strips:
a Single-edge crack (SEC)
b Double-edge-crack (DEC)

For the single edge crack (SEC):

$$\alpha = 1.12 - 0.231(a/W) + 10.55(a/W)^2 - 21.71(a/W)^3 + 30.38(a/W)^4 \qquad (4.23a)$$

For the double edge crack (DEC):

$$\alpha = 1.12 - 0.429(a/W) - 4.78(a/W)^2 + 15.44(a/W)^3 \qquad (4.23b)$$

A listing of multipliers for different a/W ratios is presented in Table 4.1.

Table 4.1
Multiplier α as a function of a/W for single and double edge cracked tension strips.

a/W	SEC	DEC
0	1.12	1.12
0.10	1.18	1.13
0.20	1.37	1.14
0.30	1.66	1.24
0.40	2.10	1.52
0.45	2.42	1.75

As the third example, consider the beam with a crack subjected to three- and four-point bending as shown in Figs. 4.8a and 4.8b. With bending, the form of the solution changes and the multiplying factor α is not used. Instead, the stress intensity factor K_I is expressed in terms of the beam parameters S, W, and B, as defined in Fig. 4.8, and a series in terms of the (a/W) ratio.

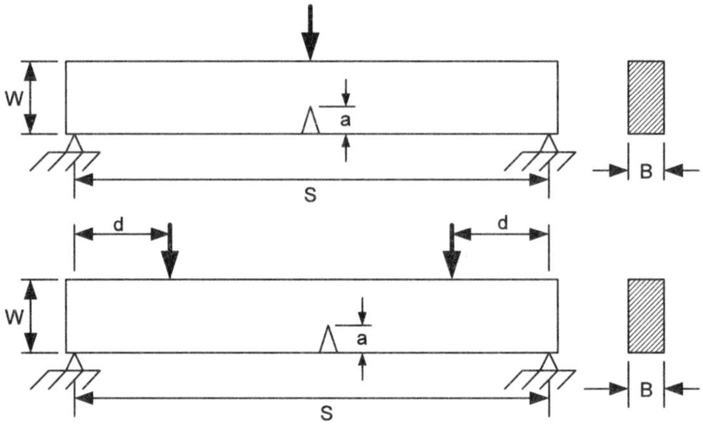

Figure 4.8 Cracked beams in bending.
Top: Symmetric three-point bending.
Bottom: Four-point bending.

For three point bending:

$$K_I = \frac{PS}{BW^{3/2}}\left[2.9\left(\frac{a}{W}\right)^{1/2} - 4.6\left(\frac{a}{W}\right)^{3/2} + 21.8\left(\frac{a}{W}\right)^{5/2} - 37.6\left(\frac{a}{W}\right)^{7/2} + 38.7\left(\frac{a}{W}\right)^{9/2}\right] \quad (4.24a)$$

For the beam subjected to pure bending:

$$K_I = \frac{6M}{BW^2}\sqrt{\pi a}\left[1.122 - 1.4\frac{a}{W} + 7.33\left(\frac{a}{W}\right)^2 - 13.08\left(\frac{a}{W}\right)^3 + 14.0\left(\frac{a}{W}\right)^4\right] \quad (4.24b)$$

As the final example, consider the compact tension specimen, defined in Fig. 4.9 that is loaded through pins positioned over the crack. The stress intensity factor K_I is given as a series in terms of (a/W).

For the compact tension specimen:

$$K_I = \frac{P}{BW^{1/2}}\left[29.6\left(\frac{a}{W}\right)^{1/2} - 185.5\left(\frac{a}{W}\right)^{3/2} + 655.7\left(\frac{a}{W}\right)^{5/2} - 1017\left(\frac{a}{W}\right)^{7/2} + 638.9\left(\frac{a}{W}\right)^{9/2}\right] \quad (4.25)$$

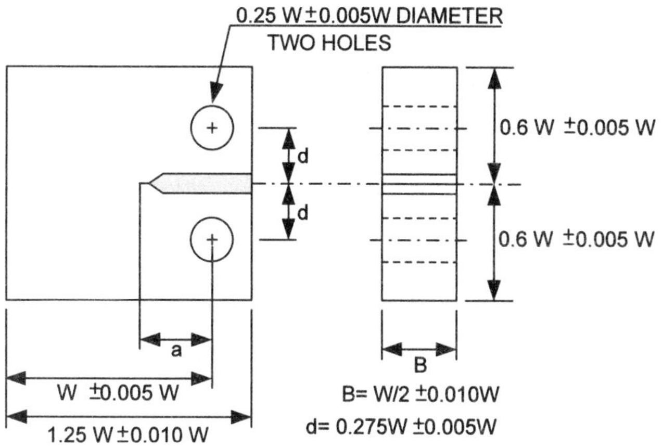

Figure 4.9 Compact tension specimen.

4.5 THE THREE MODES OF LOADING

A crack in a body can be subjected to three different types of loading. The opening mode of loading, illustrated in Fig. 4.10a, involves loads that produce displacements of the crack surfaces perpendicular to the plane of the crack. In this case the crack line or the x-axis, as shown in Fig. 4.4b, is a principal axis. The subscript I on K_I indicates that the stress intensity factor is due to opening mode loading.

The shearing mode of loading, illustrated in Fig. 4.10b, is due to in-plane shear loads that cause the two crack surfaces to slide along one another. The displacement of the crack surfaces is in the plane of the crack and perpendicular to the leading edge of the crack. The subscript II on K_{II} implies that the stress intensity factor is due to shear mode loading. Of course, opening and shear mode loading conditions can occur together. In this case the loading is defined as mixed-mode where both K_I and K_{II} exist in the region near the crack tip.

The tearing mode of loading, shown in Fig. 4.10c, is due to out-of-plane shear loading. The displacements of the crack surfaces are in the plane of the crack and parallel to the leading edge of the crack. The subscript III on K_{III} is used to depict the tearing mode. While the superposition of the three modes of loading gives the most general loading condition, one usually is interested in mode I or mixed-mode loading because they occur more frequently in engineering applications.

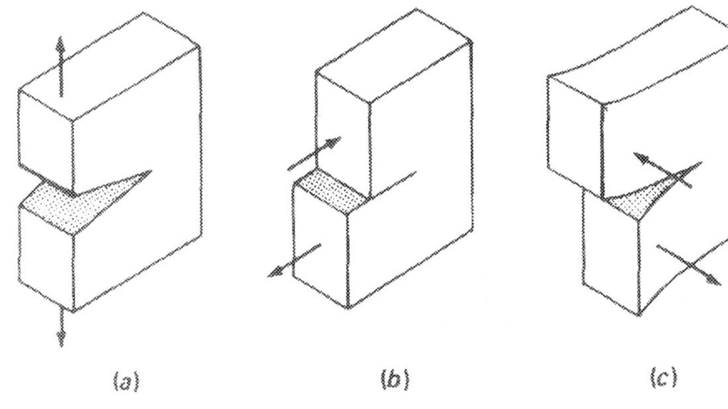

Figure 4.10 Three modes of crack loading: (a) opening; (b) shearing; (c) tearing.

4.6 THE FIELD EQUATIONS FOR THE REGION ADJACENT TO THE CRACK TIP

While it is possible in many cases to determine the stress intensity factors analytically [Section 4.6] or with numerical methods, in other instances it is necessary to measure K_I or K_{II} in carefully controlled experiments. Any experimental method for determining the stress intensity factors depends upon a complete knowledge of the field equations that are valid near the tip of the crack. In this treatment, the region adjacent to the crack tip is divided into three regions as shown in Fig. 4.11. The field quantities (stresses, strains, and displacements) are represented in series form. For example, the stresses are

$$\sigma_{ij} = \sum_{n=0}^{N} A_n r^{(n-1/2)} f_n(\theta) + \sum_{m=0}^{M} B_m r^m g_m(\theta) \tag{4.26}$$

where A_n and B_m are unknown coefficients and $f_n(\theta)$ and $g_m(\theta)$ are trigonometric functions.

Examination of Eq. (4.26) shows that for $r \Rightarrow 0$, the only term in the series that contributes significantly to σ_{ij} is the $n = 0$ term, because all of the other terms vanish. Thus, the very near field, region (1) in Fig. 4.11, is defined as that area adjoining the crack tip where a single term in the series representation is sufficient to determine the field quantity of interest.

As one moves away from the crack tip, the non-singular terms become significant and a single-term representation of the field quantities is not adequate. Additional terms in the series are required to improve the accuracy of the determination of the field quantity. The near-field region, region (2) in Fig. 4.11, is defined as that area beyond region (1) where the field quantities can be accurately represented with a small number of terms, up to say six, of the series.

For still larger values of r, a very large number of terms are required in Eq. (4.26) to accurately describe the stress field. This area is depicted as the far-field region, region (3) in Fig. 4.11. The far-field region is avoided in measuring field quantities in an attempt to determine the stress intensity factor, because the large number of unknown coefficients in the series requires that large amounts of data be taken with very high accuracy.

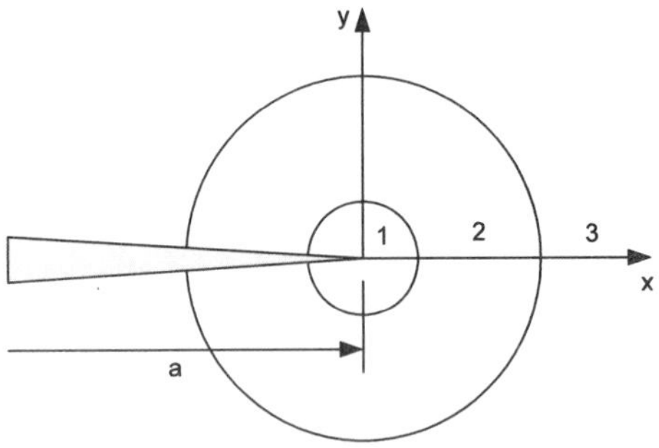

Figure 4.11 The field adjacent to the crack tip is divided into three regions: (1) the very near field, (2) the near field, and (3) the far field.

4.6.1 The Very-Near-Field Equations—Opening Mode

Consider first region (1) where r is small and a single term in the series representation of the field quantities is adequate. In this case the stress function Z may be written as:

$$Z = A_0 z^{-1/2} = A_0 r^{-1/2} e^{-j(\theta/2)} \tag{4.27}$$

where

$$A_0 = \frac{K_I}{\sqrt{2\pi}} \tag{4.28}$$

By taking the real and imaginary parts of Z and Z' and substituting into Eq. (4.12), the relations for the stresses are obtained as:

$$
\begin{aligned}
\sigma_{xx} &= A_0 r^{-1/2} \cos\frac{\theta}{2}\left(1-\sin\frac{\theta}{2}\sin\frac{3\theta}{2}\right)\\
\sigma_{yy} &= A_0 r^{-1/2} \cos\frac{\theta}{2}\left(1+\sin\frac{\theta}{2}\sin\frac{3\theta}{2}\right)\\
\tau_{xy} &= A_0 r^{-1/2} \cos\frac{\theta}{2}\sin\frac{\theta}{2}\cos\frac{3\theta}{2}
\end{aligned}
\tag{4.29}
$$

Next, consider a state of plane-stress where $\sigma_{zz} = 0$ and substitute Eqs. (4.29) into Eqs. (2.19) to determine the strains. Thus,

$$
\begin{aligned}
E\varepsilon_{xx} &= A_0 r^{-1/2} \cos\frac{\theta}{2}\left[(1-\nu)-(1+\nu)\sin\frac{\theta}{2}\sin\frac{3\theta}{2}\right]\\
E\varepsilon_{yy} &= A_0 r^{-1/2} \cos\frac{\theta}{2}\left[(1-\nu)+(1+\nu)\sin\frac{\theta}{2}\sin\frac{3\theta}{2}\right]\\
2\mu\gamma_{xy} &= A_0 r^{-1/2} \sin\theta\cos\frac{3\theta}{2}
\end{aligned}
\tag{4.30}
$$

Finally, the displacements are given by substituting Eqs. (4.30) into Eqs. (2.4) and integrating to obtain:

$$
\begin{aligned}
\mu\, u &= A_0 r^{1/2} \cos\frac{\theta}{2}\left(\frac{1-\nu}{1+\nu}+\sin^2\frac{\theta}{2}\right)\\
\mu\, v &= A_0 r^{1/2} \sin\frac{\theta}{2}\left(\frac{2}{1+\nu}-\cos^2\frac{\theta}{2}\right)\\
E\, w &= -2\nu A_0 r^{-1/2} \cos\frac{\theta}{2}\; B
\end{aligned}
\tag{4.31}
$$

where B is the thickness of the two-dimensional specimen. Note from Eq. (4.31) that only the displacement w is singular as $r \Rightarrow 0$. The displacements u and v go to zero at the crack tip.

The very-near-field equations describe the field quantities in region (1) that is located in the local region of the crack tip. Chona [7] has shown that this region is very small and extends only a distance of 0.02a from the tip of the crack. Because the stress distribution in this region is three-dimensional, neither plane-stress nor plane-strain assumptions are valid. For this reason the very-near-field equations should only be used in a qualitative sense to give first approximations for A_0 or K_I. A more complete set of relations will be provided in Section 4.6.3 that permit improved accuracy in determining A_0 or K_I from the field quantities.

4.6.2 The Very-Near-Field Equations—Shear Mode

Again consider region (1) with a single-term representation of the plane-stress state very near the crack tip. The loading is due to in-plane shear stress τ applied remote to the crack tip. The Airy's stress function for the shear mode is

$$
\phi = -\, y \,\mathrm{Re}\, \overline{Z} \tag{4.32}
$$

Let Z be represented by

$$
Z = C_0\, z^{-1/2} = C_0\, r^{-1/2}\, e^{-j(\theta/2)} \tag{4.33}
$$

where

$$C_0 = \frac{K_{II}}{\sqrt{2\pi}} \tag{4.34}$$

Substituting Eq. (4.32) into Eqs. (3.20) yields

$$\begin{aligned} \sigma_{xx} &= y \operatorname{Re} Z' + 2 \operatorname{Im} Z \\ \sigma_{yy} &= - y \operatorname{Re} Z' \\ \tau_{xy} &= \operatorname{Re} Z - y \operatorname{Im} Z' \end{aligned} \tag{4.35}$$

Substituting Eq. (4.33) into Eqs. (4.35) gives

$$\begin{aligned} \sigma_{xx} &= -C_0 r^{-1/2} \sin\frac{\theta}{2}\left(2 + \cos\frac{\theta}{2}\cos\frac{3\theta}{2}\right) \\ \sigma_{yy} &= C_0 r^{-1/2} \sin\frac{\theta}{2}\cos\frac{\theta}{2}\cos\frac{3\theta}{2} \\ \tau_{xy} &= C_0 r^{-1/2} \cos\frac{\theta}{2}\left(1 - \sin\frac{\theta}{2}\sin\frac{3\theta}{2}\right) \end{aligned} \tag{4.36}$$

Again taking the plane-stress state and substituting Eqs. (4.36) into Eqs. (2.19) gives the strains as:

$$\begin{aligned} E\varepsilon_{xx} &= -C_0 r^{-1/2} \sin\frac{\theta}{2}\left[2 + (1+\nu)\cos\frac{\theta}{2}\cos\frac{3\theta}{2}\right] \\ E\varepsilon_{yy} &= C_0 r^{-1/2} \sin\frac{\theta}{2}\left[2\nu + (1+\nu)\cos\frac{\theta}{2}\cos\frac{3\theta}{2}\right] \\ \mu\gamma_{xy} &= C_0 r^{-1/2} \cos\frac{\theta}{2}\left(1 - \sin\frac{\theta}{2}\sin\frac{3\theta}{2}\right) \end{aligned} \tag{4.37}$$

The displacements are determined by substituting Eqs. (4.37) into Eqs. (2.4) and integrating. Thus,

$$\begin{aligned} \mu\, u &= C_0 r^{1/2} \sin\frac{\theta}{2}\left[\frac{2}{1+\nu} + \cos^2\frac{\theta}{2}\right] \\ \mu\, v &= C_0 r^{1/2} \cos\frac{\theta}{2}\left(-\frac{1-\nu}{1+\nu} + \sin^2\frac{\theta}{2}\right) \\ Ew &= 2\nu\, C_0 r^{-1/2} \sin\frac{\theta}{2} B \end{aligned} \tag{4.38}$$

These relations describe the field quantities in terms of C_0 or K_{II}; however, they are approximate because the plane-stress state in the very-near field (region 1) is not valid.

4.6.3 The Near-Field Equations - Opening Mode

Consider region (2) of Fig. 4.11 where r is sufficiently large to be outside the zone where the stress state is three-dimensional. Of course r is limited so that the field quantities can be expressed with reasonable accuracy (2 to 5%) with a small number of terms (say 3 to 6) in the series expansion. In region (2) the Westergaard equations must be modified, with additional terms added, to account for all of the non-singular terms [8]. In the modified form, one adds to Eq. (4.12) and expresses the stresses in terms of two stress functions Z and Y as:

$$
\begin{aligned}
\sigma_{xx} &= \text{Re}\, Z - y\, \text{Im}\, Z' - y\, \text{Im}\, Y' + 2\, \text{Re}\, Y \\
\sigma_{yy} &= \text{Re}\, Z + y\, \text{Im}\, Z' + y\, \text{Im}\, Y' \\
\tau_{xy} &= -y\, \text{Re}\, Z' - y\, \text{Re}\, Y' - \text{Im}\, Y
\end{aligned}
\tag{4.39}
$$

where the stress functions are given as series relations in terms of z as:

$$
\begin{aligned}
Z(z) &= \sum_{n=0}^{N} A_n z^{(n-1/2)} \\
Y(z) &= \sum_{m=0}^{M} B_m z^m
\end{aligned}
\tag{4.40}
$$

The number of terms used in the series (N + M + 2) will depend on r, the proximity of the boundaries relative to the crack tip, and the influence of the loads applied at finite distances from the crack tip. Substituting Eqs. (4.40) into Eqs. (4.39) gives:

$$
\begin{aligned}
\sigma_{xx} = {} & A_0 r^{-1/2} \cos\frac{\theta}{2}\left(1 - \sin\frac{\theta}{2}\sin\frac{3\theta}{2}\right) + 2B_0 + A_1 r^{1/2}\cos\frac{\theta}{2}\left(1 + \sin^2\frac{\theta}{2}\right) \\
& + 2B_1 r\cos\theta + A_2 r^{3/2}\left(\cos\frac{3\theta}{2} - \frac{3}{2}\sin\theta\sin\frac{\theta}{2}\right) + 2B_2 r^2(\sin^2\theta + \cos 2\theta) \\
\sigma_{yy} = {} & A_0 r^{-1/2}\cos\frac{\theta}{2}\left(1 + \sin\frac{\theta}{2}\sin\frac{3\theta}{2}\right) + A_1 r^{1/2}\cos\frac{\theta}{2}\left(1 - \sin^2\frac{\theta}{2}\right) \\
& + A_2 r^{3/2}\left(\cos\frac{3\theta}{2} + \frac{3}{2}\sin\theta\sin\frac{\theta}{2}\right) + 2B_2 r^2 \sin^2\theta \\
\tau_{xy} = {} & A_0 r^{-1/2}\cos\frac{\theta}{2}\sin\frac{\theta}{2}\cos\frac{3\theta}{2} - A_1 r^{1/2}\sin\frac{\theta}{2}\cos^2\frac{\theta}{2} \\
& - 2B_1 r\sin\theta - 3A_2 r^{3/2}\sin\frac{\theta}{2}\cos^2\frac{\theta}{2} - 2B_2 r^2 \sin 2\theta
\end{aligned}
\tag{4.41}
$$

Substituting Eqs. (4.41) into the stress-strain relations gives

$$E\varepsilon_{xx} = A_0 r^{-1/2}\cos\frac{\theta}{2}\left[(1-\nu)-(1+\nu)\sin\frac{\theta}{2}\sin\frac{3\theta}{2}\right]+2B_0$$
$$+A_1 r^{1/2}\cos\frac{\theta}{2}\left[(1-\nu)+(1+\nu)\sin^2\frac{\theta}{2}\right]+2B_1 r\cos\theta$$
$$+\frac{A_2}{2}r^{3/2}\left[2(1-\nu)\cos\frac{3\theta}{2}-3(1+\nu)\sin\theta\sin\frac{\theta}{2}\right]+2B_2 r^2\left[1-(3+\nu)\sin^2\theta\right]$$

$$E\varepsilon_{yy} = A_0 r^{-1/2}\cos\frac{\theta}{2}\left[(1-\nu)+(1+\nu)\sin\frac{\theta}{2}\sin\frac{3\theta}{2}\right]-2\nu B_0$$
$$+A_1 r^{1/2}\cos\frac{\theta}{2}\left[(1-\nu)-(1+\nu)\sin^2\frac{\theta}{2}\right]-2\nu B_1 r\cos\theta$$
$$+\frac{A_2}{2}r^{3/2}\left[2(1-\nu)\cos\frac{3\theta}{2}+3(1+\nu)\sin\theta\sin\frac{\theta}{2}\right]+2B_2 r^2\left[-\nu+(3\nu+1)\sin^2\theta\right]$$

$$\mu\gamma_{xy} = \frac{A_0}{2}r^{-1/2}\left(\sin\theta\cos\frac{3\theta}{2}\right)-\frac{A_1}{2}r^{1/2}\left(\sin\theta\cos\frac{\theta}{2}\right)-2B_1 r\sin\theta$$
$$-\frac{3A_2}{2}r^{3/2}\left(\sin\theta\cos\frac{\theta}{2}\right)-2B_2 r^2\sin 2\theta \qquad (4.42)$$

4.6.4 The Near-Field Equations - Shearing Mode

The stress field near a single-ended crack tip loaded in the shearing mode in terms of stress functions Z and Y are:

$$\sigma_{xx} = \text{Im } Y + y \text{ Re } Y' + y \text{ Re } Z' + 2 \text{ Im } Z$$
$$\sigma_{yy} = \text{Im } Y - y \text{ Re } Y' - y \text{ Re } Z' \qquad (4.43)$$
$$\tau_{xy} = -y \text{ Im } Y' - y \text{ Im } Z' + \text{Re } Z$$

For the shearing mode, the stress functions are expressed as

$$Z(z) = \sum_{n=0}^{N} C_n z^{(n-1/2)}$$
$$Y(z) = \sum_{m=0}^{M} D_m z^m \qquad (4.44)$$

Note that the stress intensity factor K_{II} is related to C_0 as indicated in Eq. (4.34). From Eqs. (4.43) and (4.44) it is evident that:

$$\sigma_{xx} = \sum_{n=0}^{N} C_n r^{(n-1/2)}\left[\left(n-\frac{1}{2}\right)\sin\theta\cos\left(n-\frac{3}{2}\right)\theta+2\sin\left(n-\frac{1}{2}\right)\theta\right]$$
$$+\sum_{m=0}^{M} D_m r^m\left[\sin m\theta+m\sin\theta\cos(m-1)\theta\right]$$

$$\sigma_{yy} = \sum_{n=0}^{N} C_n r^{(n-1/2)} \left[-\left(n - \frac{1}{2}\right) \sin\theta \cos\left(n - \frac{3}{2}\right)\theta \right] + \sum_{m=0}^{M} D_m r^m \left[\sin m\theta - m \sin\theta \cos(m-1)\theta \right] \tag{4.45}$$

$$\tau_{xy} = \sum_{n=0}^{N} C_n r^{(n-1/2)} \left[\cos\left(n - \frac{1}{2}\right)\theta - \left(n - \frac{1}{2}\right) \sin\theta \sin\left(n - \frac{3}{2}\right)\theta \right] + \sum_{m=0}^{M} D_m r^m \left[-m \sin\theta \sin(m-1)\theta \right]$$

Next, the strains are obtained by using the stress-strain relations. Thus,

$$E\varepsilon_{xx} = -C_0 r^{-1/2} \sin\frac{\theta}{2}\left[(1+\nu)\cos\frac{\theta}{2}\cos\frac{3\theta}{2} + 2\right] + C_1 r^{1/2} \sin\frac{\theta}{2}\left[2 + (1+\nu)\cos^2\frac{\theta}{2}\right] + 2D_1 r \sin\theta + C_2 r^{3/2}\left[2\sin\frac{3\theta}{2} + \frac{3}{2}(1+\nu)\sin\theta\cos\frac{\theta}{2}\right] + 2D_2 r^2 \sin 2\theta$$

$$E\varepsilon_{yy} = C_0 r^{-1/2} \sin\frac{\theta}{2}\left[2\nu + (1+\nu)\cos\frac{\theta}{2}\cos\frac{3\theta}{2}\right] - C_1 r^{1/2} \sin\frac{\theta}{2}\left[2\nu + (1+\nu)\cos^2\frac{\theta}{2}\right] - 2D_1 r \nu \sin\theta - C_2 r^{3/2}\left[2\nu\sin\frac{3\theta}{2} + \frac{3}{2}(1+\nu)\sin\theta\cos\frac{\theta}{2}\right] - 2D_2 r^2 \nu \sin 2\theta \tag{4.46}$$

$$\mu\gamma_{xy} = C_0 r^{-1/2} \cos\frac{\theta}{2}\left(1 - \sin\frac{\theta}{2}\sin\frac{3\theta}{2}\right) + C_1 r^{1/2} \cos\frac{\theta}{2}\left(\sin^2\frac{\theta}{2} + 1\right) + C_2 r^{3/2}\left(\cos\frac{3\theta}{2} - \frac{3}{2}\sin\theta\sin\frac{\theta}{2}\right) - 2D_2 r^2 \sin^2\theta$$

where the series representing Eq. (4.45) were evaluated at N = M = 2. Inspection of Eqs. (4.45) and (4.46) shows that D_0 does not occur in either the stress or strain field for mode II loading.

When the specimen is loaded with tractions or boundary displacements that produce both mode I and II fields, a mixed-mode loading exists. In these cases, the equations representing the stresses or the strains are superimposed. The coefficients in the series A_n, B_m C_n and D_m are determined to satisfy field quantities measured in the near-field region. The stress intensity factors K_I and/or K_{II} are then determined from the coefficients A_0 and C_0 from Eqs. (4.28) and (4.34).

4.7 FRACTURE ENERGY APPROACHES

4.7.1 The Energy Release Rate or Crack Extension Force $\mathcal{G}$

The earliest contribution to the fracture mechanics approach made by Griffith [3] was based on the concept of an energy balance where the energy consumed to extend a crack equals the energy provided by the system. The system includes the plate with a crack and the loading frame. Consider first the case where the loading frame is perfectly rigid and the ends of the stretched plate are fixed. Because the ends are fixed, the frame adds no work and the energy W consumed by the crack during an extension da is provided by a decrease in the strain energy U stored in the plate. The energy equation for the fixed-end condition with a plate of unit thickness is:

$$\frac{d}{da}(U+W)=0 \qquad \text{or} \qquad \frac{dW}{da}=-\frac{dU}{da} \tag{4.47}$$

The right hand side of Eq. (4.47) is defined as:

$$\mathcal{G}=-\frac{dU}{da} \tag{4.48}$$

and is called either the **energy release rate** or the **crack extension force**. The left hand side of Eq. (4.47) is defined as:

$$\mathcal{R}=\frac{dW}{da} \tag{4.49}$$

and is called the **crack growth resistance**. For a stationary crack,

$$\mathcal{G}\le\mathcal{R} \tag{4.50a}$$

for a given material. For an unstable crack that initiates and propagates,

$$\mathcal{G}>\mathcal{R} \tag{4.50b}$$

It is possible to show a relationship between the crack extension force $\mathcal{G}$ and the stress intensity factor K_I by considering the work released near the crack tip as the crack extends some small distance δ. Again take the cracked plate with fixed ends with a crack of length a. Now apply forces to the crack faces over a distance δ that will close the crack as shown in Fig. 4.12. The work due to the closure forces, in the limit as $\delta \Rightarrow 0$, is the same as the crack extension force $\mathcal{G}$. Writing the expression for this work gives:

$$\mathcal{G}=\lim_{\delta\to 0}\frac{1}{\delta}\int_0^{\delta}\sigma_{yy}v\,dx \tag{4.51}$$

With the origin of the coordinate system at the center of the crack Westergaard [4] showed that the crack opening displacement v is given by the expression

$$v=\frac{2\sigma_0}{E}\sqrt{a^2-x^2} \tag{4.52}$$

Substituting $K_I=\sqrt{\pi a}\,\sigma_0$ into Eq. (4.52) gives

$$v = \frac{2K_I}{\sqrt{\pi}\,E}\left(a - \frac{x^2}{a}\right)^{1/2} \qquad (a)$$

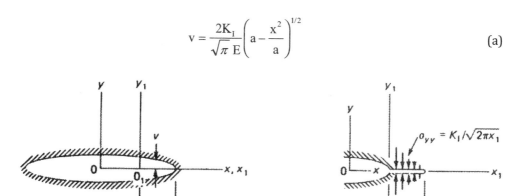

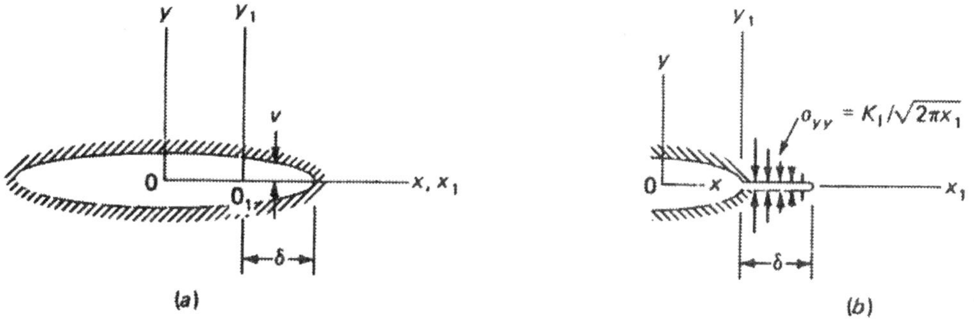

Figure 4.12 Application of forces distributed in proportion to σ_{yy} to close the crack over a length δ.
(a) Coordinates systems Oxy and $O_1x_1y_1$ and the crack opening v.
(b) Closing the crack over an increment length δ.

Consider now a second coordinate system $O_1x_1y_1$, where O_1 is located at the position $x = a - \delta$. It is clear from Fig. 4.12 that

$$x = a - \delta + x_1 \qquad \text{or} \qquad x = r + a - \delta \qquad (b)$$

Now substitute Eq. (b) into Eq. (a) and note that $r << a$ and $\delta << a$. If second order terms are neglected,

$$v = \frac{2K_I}{\sqrt{\pi}E}\sqrt{2(\delta - r)} \qquad (4.53)$$

Next, recall Eqs. (4.28) and (4.29) that give σ_{yy} along the x_1 axis near the crack tip where $r \approx x_1$ as

$$\sigma_{yy} = \frac{K_I}{\sqrt{2\pi r}} \qquad (c)$$

If Eq. (4.53) and Eq. (c) are substituted into Eq. (4.51),

$$\mathcal{G} = \lim_{\delta \to 0} \frac{2K_I^2}{\pi E\delta}\int_0^{\delta}\left[\frac{1 - r/\delta}{r/\delta}\right]^{1/2} dr \qquad (d)$$

Integration of Eq. (d) gives

$$\mathcal{G}_I = \frac{K_I^2}{E} \qquad (4.54)$$

where the subscript I is added to $\mathcal{G}$ to indicate opening mode loading. Recall that plane-stress conditions prevail. In a similar fashion $\mathcal{G}_{II}$ is expressed in terms of K_{II} as

$$\mathcal{G}_{II} = \frac{K_{II}^2}{E} \qquad (4.55)$$

For elastic bodies, including small scale yielding, $\mathcal{G}$ and K are interchangeable by using Eq. (4.54) for mode I and Eq. (4.55) for mode II loadings. For mixed mode loading conditions, the energies from the different modes are added to give

$$\mathcal{G} = \mathcal{G}_I + \mathcal{G}_{II} = \frac{K_I^2 + K_{II}^2}{E} \tag{4.56}$$

4.7.2 Strain Energy Density

The strain energy density W is defined as the strain energy per unit volume that can be expressed as:

$$W = (1/E)[½ (\sigma_{xx}^2 + \sigma_{yy}^2 + \sigma_{zz}^2) - \nu(\sigma_{xx}\sigma_{yy} + \sigma_{yy}\sigma_{zz} + \sigma_{zz}\sigma_{xx}) + (1+\nu)(\tau_{xy}^2 + \tau_{yz}^2 + \tau_{zx}^2)] \tag{4.57}$$

Under plane-stress conditions where $\sigma_{zz} = \tau_{zx} = \tau_{zy} = 0$, Eq. (4.57) reduces to:

$$W = ½ E[\sigma_{xx}^2 + \sigma_{yy}^2 - 2\nu\sigma_{xx}\sigma_{yy} + 2(1+\nu)\tau_{xy}^2] \tag{4.58}$$

The strain energy density in the near-field region about the tip of the crack is of the form:

$$W = S/r + h(r, \theta) \tag{4.59}$$

where $h(r, \theta)$ are non-singular terms and S, the coefficient of 1/r, is called the **strain energy density factor**. Clearly $W \Rightarrow \infty$ as $r \Rightarrow 0$ and the strain energy density is singular at the tip of the crack.

Sih [9] has proposed a critical value of the strain energy density W_c evaluated at a critical distance r_c as a criterion for crack initiation and propagation. If the nonsingular terms in Eq. (4.59) are neglected, it is clear that:

$$S_c = r_c W_c \tag{4.60}$$

If Eq. (4.58) is substituted into Eq. (4.60), the stresses at the critical distance r_c are:

$$\sigma_{xx}^2 + \sigma_{yy}^2 - 2\nu\sigma_{xx}\sigma_{yy} + 2(1+\nu)\tau_{xy}^2 = 2ES_c/r_c \tag{4.61}$$

4.7.3 The J Integral

The quantities K and $\mathcal{G}$ describe the stress state near the crack tip when the field is elastic with a relatively small plastic zone. The plastic enclave is surrounded by elastic material and the plate behaves in an essentially linear elastic manner. However, for tough ductile materials, the plastic zone becomes large and is not small relative to the crack length a. In these cases, K and $\mathcal{G}$ do not provide adequate descriptions of the elastic-plastic behavior of the tough specimens.

To determine an effective energy release rate for specimens where the plasticity effects must be considered, Rice [10] introduced a contour integral taken about the crack tip as shown in Fig. 4.13a. The contour integral J was defined originally by Eshelby [11] as

$$J = \int_\Gamma \left(W dy - \overline{T} \bullet \frac{\partial \overline{u}}{\partial x} ds \right) \tag{4.62}$$

where W is the strain energy density defined in Eq. (4.57); $\overline{u}$ is the displacement vector; ds is an element along the contour Γ; $\overline{T}$ is the tension vector on Γ in the direction of the outer normal **n** to Γ and is given by, $T_{ij} = \sigma_{ij} n_j$.

The J integral vanishes (J = 0) along any closed contour. Also, the integral is path independent and J_1 determined along Γ_1 of Fig. 4.13b is the same as J_2 determined along Γ_2. It is easy to prove that $J_1 = J_2$ because dy = 0 and T = 0 along the crack line. Thus, the crack line may be included in the contour without contributing to the value of J. For this reason points A and B in Fig. 4.13 do not need to coincide.

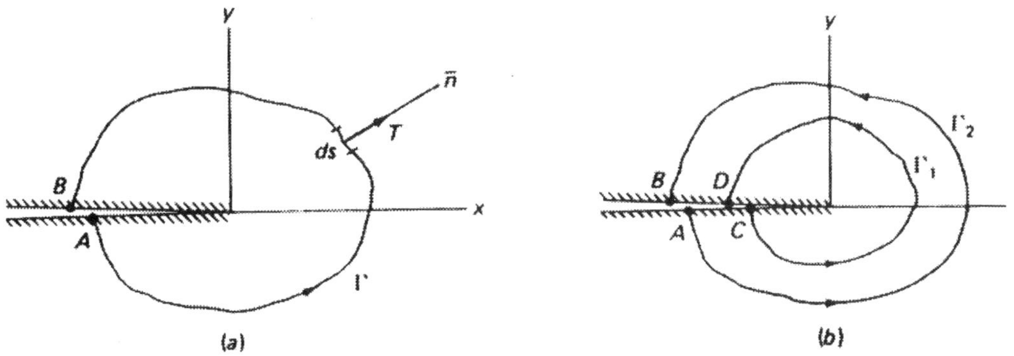

Figure 4.13 Contour integration about the path Γ to obtain J.
(a) Contour about the crack tip. (b) Two contours about the crack tip.

When the contour Γ encircles the crack tip, the J integral represents the change in elastic energy U for a virtual crack extension da and is written as

$$J = -\frac{\partial U}{\partial a} \tag{4.63}$$

For an elastic stress state, Eq. (4.63) is the same as the definition of $\mathcal{G}$ given in Eq. (4.48); hence,

$$\mathcal{G}_I = J_1 = \frac{K_I^2}{2E} \tag{4.64}$$

These relations between $\mathcal{G}_I$, J_1 and K_I hold as long as the yielding at the crack tip is "small scale". When r_p becomes large relative to the crack length a Eq. (4.64) does not hold, the concept of either $\mathcal{G}_I$ or K_I is not valid and elasto-plastic characterization of the field quantities at the crack tip, such as the J integral, are required.

4.8 CRITERIA FOR CRACK INSTABILITY

Griffith first proposed a criterion for crack instability based on the energy consumed in forming the crack extension. For a perfectly elastic and brittle material, that does not exhibit a plastic zone at the crack tip, the incremental energy dW necessary to produce the two new crack surfaces involved in an extension da of the crack is

$$\frac{dW}{da} = 2\gamma = \mathcal{R} \tag{4.65}$$

where γ is the surface energy of the material. Using Eqs. (4.54), (4.50), and (4.20) with Eq. (4.65) leads to the Griffith equation for the critical stress σ_c,

$$\sigma_c = \sqrt{\frac{2\gamma E}{\pi a}} \tag{4.66}$$

Although this relation for σ_c is limited to a narrow class of very brittle materials, it indicates two important concepts in fracture mechanics. First, for a given crack length a, the crack will remain stable if $\sigma_0 < \sigma_c$ in spite of the fact that the stresses at the crack tip are singular. Second, that growth of the crack by any mechanism (i.e. fatigue extension, stress- corrosion cracking, etc.) decreases the critical stress required to initiate the crack.

More than 20 years later Irwin [8] extended the Griffith approach by recasting Eq. (4.65) as:

$$\frac{dW}{da} = 2\gamma + \frac{dW_p}{da} \tag{4.67}$$

where the additional term dW_p/da accounts for the work necessary to form the plastic zone in front of the tip of the crack during an extension. This concept is valid if the plastic zone size remains constant as the crack advances. Experimental observations indicate that the plastic zone size is nearly constant in many different materials when fabricated as thick plates subjected to plane-strain conditions. With the constants imposed by plane-strain

$$\frac{dW}{da} = \mathcal{R} = \mathcal{G}_{Ic} \tag{4.68}$$

where $\mathcal{G}_{Ic}$ is the critical strain-energy release rate that is a material property. Test procedures for establishing $\mathcal{G}_{Ic}$ are described in Ref. 13.

Replacing 2γ with $\mathcal{G}_{Ic}$ in Eq. (4.66) leads to the more modern representation of the critical stress as

$$\sigma_c = \sqrt{\frac{E\mathcal{G}_{Ic}}{\pi a}} \tag{4.69}$$

Before interpreting Eq. (4.69) it is useful to make a final modification of the relation for the critical stress. Note, that the strain-energy release rate $\mathcal{G}_I$ in terms of the stress-intensity factor K_I is:

$$\mathcal{G}_I = \frac{K_I^2}{E} \qquad \text{For plane-stress conditions} \tag{4.54}$$

$$\mathcal{G}_I = \frac{(1-\nu^2)K_I^2}{E} \qquad \text{For plane-strain conditions} \tag{4.70}$$

Finally, substituting Eq. (4.70) into Eq. (4.69) gives

$$\sigma_c = K_{Ic}\sqrt{\frac{(1-\nu^2)}{\pi a}} \tag{4.71}$$

where K_{Ic} is the crack initiation toughness that is a material property determined under plane-strain testing conditions [12].

One may use either $\mathcal{G}_{Ic}$ or K_{Ic} in a crack-stability analysis where the plastic-zone size is considered small with $r_p << a$. For a given material with a toughness property specified in terms of say $\mathcal{G}_{Ic}$, the influence of the applied stress level σ_0 is illustrated in Fig. 4.14. This graph shows $\mathcal{G}_I$ for a wide (thick) plate with a central crack of length 2a, where a is the independent variable. Four values of σ_0 are represented. The intersection of the $\mathcal{G}_I$ – a lines with the material constant line corresponding to $\mathcal{G}_{Ic}$ gives the critical crack length a_{c1}, to a_{c4} for the different applied stress levels σ_0.

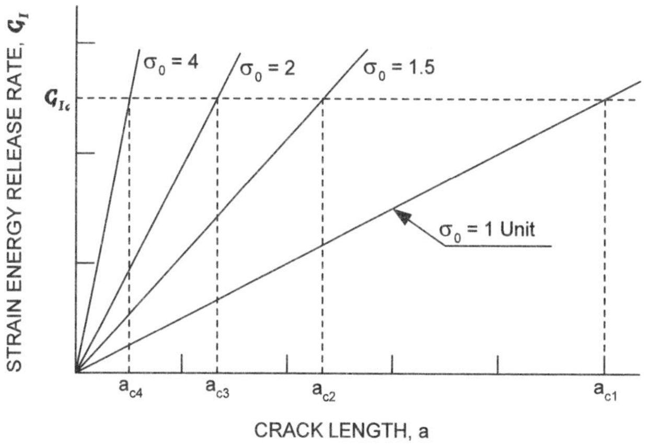

Figure 4.14 Critical crack length for different applied stresses σ_0.

Another graphical display of the stability criteria for small scale yielding is shown in Fig. 4.15. For the same problem, K_I is shown as a function of the far-field stress σ_0 with the crack length a as a parameter. The intersection of the $K_I - \sigma_0$ lines with the material constant line corresponding to K_{Ic} gives the critical stress σ_{c1}, σ_{c2} ..... σ_{c3} necessary to initiate the cracks. It is of interest to note that the line for a = 1 unit does not intersect the K_{Ic} line before $\sigma_0 > \sigma_{ys}$. This fact implies the plate will yield and fail by excessive plastic deformation while the crack remains stationary.

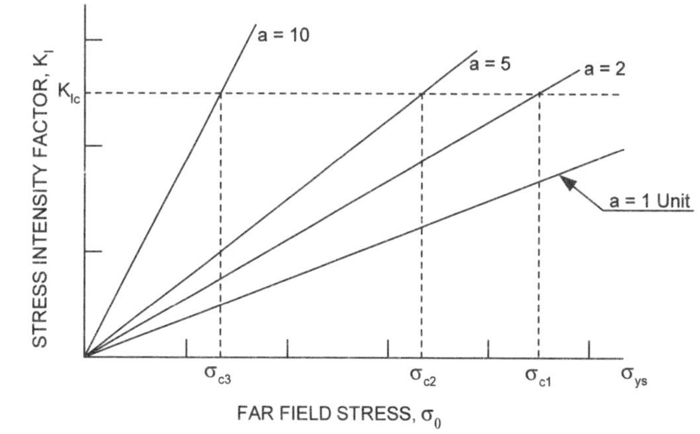

Figure 4.15 The influence of crack length a on the far-field stress σ_0 required to initiate a crack in a material with a crack initiation toughness K_{Ic}.

For large-scale plasticity, the plastic zone size is large and the crack stability analyses using Eqs. (4.69) and (4.71) are not valid. In these cases a critical value, the J integral J_{Ic}, is determined. When the applied stress σ_0 produces J that is

$$J > J_{Ic} \tag{4.72}$$

the crack will begin to grow in a stable manner at very low velocity. Indeed, the stress must be increased to maintain this crack growth in this stable-growth regime of fracture.

4.9 GROWTH OF FATIGUE CRACKS UNDER CYCLIC LOADING

Most brittle fractures are preceded by crack growth due to cyclic loading or stress corrosion cracking. Flaws due to welding or manufacturing defects are essentially small cracks that often cannot be detected in an inspection process. When a structure is subjected to cyclic loading, these flaws act as cracks, which grow slowly until they become sufficient in size to initiate unstable crack propagation. When initiated, these fatigue cracks propagate at high velocity and result in a catastrophic failure often with loss of life and property.

Fortunately, it is possible to use well-established principles of fracture mechanics to predict the behavior of structures containing flaws that are subjected to cyclic loads in service. The approach is based on experimental evidence that shows that the $\log_{10}$ of the crack growth rate, da/dN, is related to the $\log_{10}$ of the alternating component of the stress intensity factor ΔK. Note that the alternating component of the stress intensity factor is $\Delta K = K_{max} - K_{min}$ and that the loading is mode I. In conducting these experiments, researchers noted that the crack growth rate[1] under cyclic loading could be divided into three stages as illustrated in Fig. 4.16.

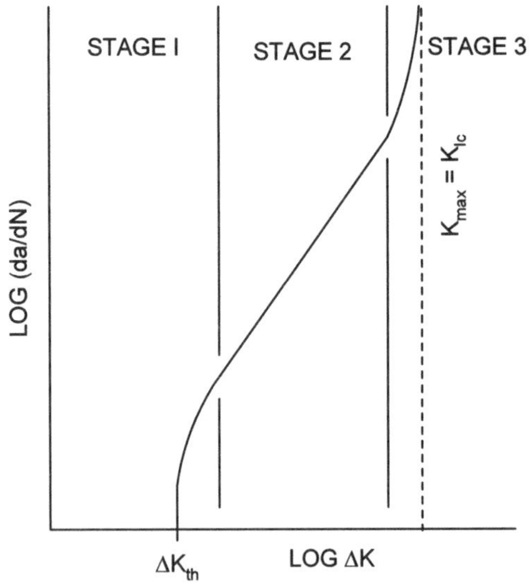

Figure 4.16 An idealized fatigue crack growth rate curve showing the three stages of crack growth with ΔK.

In the first stage, the log of ΔK is small and the crack is either stable or the growth rate is exceedingly small. Indeed, if ΔK is below some threshold value ΔK_{th} the crack is essentially stable. The threshold range of the stress intensity factor ΔK_{th} to insure crack stability is a material property. Some typical values of ΔK_{th} for several different engineering materials are shown in Table 4.2.

When experimental results from fatigue testing programs are plotted on a log-log graph as indicated in Fig. 4.16, a linear relationship is observed in Stage II of the crack growth process, which is given by:

$$\log_{10} (da/dN) = C + n \log_{10} \Delta K \qquad (4.73)$$

where C and n are material constants used in fitting Eq. (4.73) to the experimental data. These constants depend upon the environment, frequency, temperature and the stress ratio $R = \sigma_{min}/\sigma_{max}$. The slope of the crack growth rate curve n is generally in the range $2 < n < 4$.

Rewriting Eq. (4.73) gives:

$$da/dN = C(\Delta K)^n \qquad (4.74)$$

Equation (4.74) is known as the Paris Law to reflect the contribution Paul Paris made in originally formulating this empirical relation in the early 1960s. This equation is important because it enables one to predict the number of cycles to grow a crack to a specified length a_s. To develop the prediction equation, use Eq. (4.20) and Eq. (4.22) to write ΔK as:

$$\Delta K = \alpha\sqrt{\pi a}\,\Delta\sigma \qquad (4.75)$$

where α is a multiplier dependent on the geometry of the body and the ratio a/W and $\Delta\sigma = \sigma_{max} - \sigma_{min}$ is the stress range imposed during cyclic loading.

[1] Crack growth occurs in increments with a small advance occurring during each cycle of loading. Each increment of growth is marked with a discrete striation that appears on the fracture surface. This striation can be observed under a microscope, and the crack growth Δa per cycle is measured as the distance between adjacent striations.

Table 4.2
Threshold ΔK_{th} for common engineering materials*

	ΔK_{th}		σ_{ys}		σ_u	
Material	**MPa-√m**	**ksi-√in.**	**MPa**	**ksi**	**MPa**	**ksi**
Aluminum						
2024-T351	2.8	2.6	372	54	469	68
7075-T651	3.3	3.0	524	79	586	85
Titanium						
Ti-6Al-4V	3.8	3.5	931	135	1000	145
Steel						
4340	6.5	6.0	1069	155	1172	170
A533B	7.1	6.5	483	70	689	100
Stainless Steel						
304	3.8	3.5	276	40	621	90

This data was taken from NASA Report JSC-22267B, March 2000.

Substitute Eq. (4.75) into Eq. (4.74) and rearrange terms to obtain:

$$\frac{da}{a^{n/2}} = C\left[\alpha\sqrt{\pi}\Delta\sigma\right]^n dN \tag{a}$$

Rearrange terms in Eq. (a) and integrate both sides of the resulting relation to write:

$$\int_0^N dN = \frac{1}{C(a\Delta\sigma)^n \pi^{n/2}} \int_{a_i}^{a_s} \frac{da}{a^{n/2}} \tag{b}$$

Performing the integration gives the equation that relates the number of cycles N required to grow a crack from its initial length a_i to its specified length a_s as:

$$N = \frac{2}{(n-2)C(a\Delta\sigma)^n \pi^{n/2}}\left[\frac{1}{a_i^{(n-2)/2}} - \frac{1}{a_s^{(n-2)/2}}\right] \quad \text{for } n \neq 2 \tag{4.76}$$

The number of cycles to failure N_f may be determined from Eq. (4.76) by setting $a_s = a_c$ to obtain:

$$N_f = \frac{2}{(n-2)C(a\Delta\sigma)^n \pi^{n/2}}\left[\frac{1}{a_i^{(n-2)/2}} - \frac{1}{a_c^{(n-2)/2}}\right] \quad \text{for } n \neq 2 \tag{4.77}$$

where a_c is the critical crack length.

To demonstrate the use of Eqs. (4.76) and (4.77), consider a wide plate with an edge crack that is initially 0.1 in. long. The plate is fabricated from material with Paris Law coefficients of n = 4 and C = 6 × 10^{-10} when ΔK is expressed in units of ksi-in$^{1/2}$. The fracture toughness of the plate material K_{Ic} is 50 ksi-in$^{1/2}$. If the plate is subjected to stresses that cycle from 0 to 20 ksi, determine the number of cycles to extend the crack until it is 0.5 in. long.

Substituting these numerical parameters into Eq. (4.87) yields:

$$N = \frac{2}{(4-2)(6\times 10^{-10})(1.12\times 20)^4 \pi^{4/2}}\left[\frac{1}{(0.1)^{(4-2)/2}} - \frac{1}{(0.5)^{(4-2)/2}}\right] = (670.74)(8) = 5366 \text{ cycles} \quad \text{(c)}$$

Next it is prudent to check if the plate will support a stable crack 0.5 in. long. The critical crack length, a_c is determined from Eq. 4.2 and Eq. 4.22 as:

$$a_c = \frac{1}{\pi}\left(\frac{K_{Ic}}{\alpha\sigma_{max}}\right)^2 \quad (4.78)$$

Substituting numerical values into Eq. (4.78) yields:

$$a_c = \frac{1}{\pi}\left(\frac{50}{(1.12)(20)}\right)^2 = 1.586 \text{ in.} \quad \text{(d)}$$

Clearly, the critical crack length for this plate subjected to the prescribed loading is greater that the specified crack length of 0.5 in. Hence, the plate will not fail by unstable crack propagation during the first 5366 cycles as the crack grows from 0.1 to 0.50 in. To determine the number of cycles to failure, substitute numerical values into Eq. (4.77) to obtain:

$$N_f = \frac{2}{(4-2)(6\times 10^{-10})(1.12\times 20)^4 \pi^{4/2}}\left[\frac{1}{(0.1)^{(4-2)/2}} - \frac{1}{(1.586)^{(4-2)/2}}\right] = (670.74)(9.369) = 6281 \text{ cycles}$$

These results show that the crack grew rapidly during the final 915 cycles as it extended from its 0.5 in. length to the critical size of 1.586 in. The fact that n = 4 and the relatively simple expression for ΔK in this example permitted the integration of Eq. (4.74) in closed form. In many practical problems, closed form integration is not possible because the formula for ΔK is more complex. In these cases, the Eq. (4.74) is written as a difference equation and integration is performed numerically.

4.10 FRACTURE CONTROL

Large complex structures usually have inherent flaws of one type or another. These flaws are often introduced in fabrication during the welding process where incomplete welds, embedded slag, holes, inadequate fusion bonding and shrinkage cracks are common. Flaws are sometimes introduced during service. Mechanisms that produce flaws include fatigue, stress-corrosion cracking and impact damage. Regardless of the source of the flaws, they are common and the history of technology contains many examples of dramatic and catastrophic failures that caused significant losses in life and property. Indeed, as recently as 1972 a 584-ft-long tank barge [13] broke almost completely in half while in port with calm seas. The vessel was only one year old indicating that modern methods of design, material procurement, and welding procedures do not always insure safe structures.

Because flaws in structures and components are common, it is essential in the design, fabrication, and maintenance of a structure to establish a fracture-control procedure. If properly implemented, the fracture-control plan can insure the safety of the structure during its service life. Fracture control involves first, the recognition that flaws exist, and second, that the flaws must be maintained in a stable state. To describe the technical aspects of fracture control, consider a simple structure—a very wide plate with a central crack, which exhibits a stress-intensity factor given by:

$$K_I = \sqrt{\pi a}\,\sigma_0 \qquad (4.20a)$$

The crack in this structure will remain stable if

$$K_{Ic} > K_I \qquad (4.2)$$

Combining these two equations gives the stability relation

$$\sqrt{\pi a}\,\sigma_0 < K_{Ic} \qquad (4.79)$$

for this simple structure. This stability relation illustrates the three elements in all fracture-control procedures; namely,

1. Controlling the flaw size, a
2. Knowledge of the applied stress imposed, σ_0
3. Specifying the material toughness, K_{Ic}

Clearly, to insure the integrity of a structure requires that the variables in the stability relation be maintained so that equality in Eq. (4.79) never occurs during the life of a structure.

A typical fracture-control plan usually considers material resistance (i.e. the crack-initiation toughness K_{Ic}) as the first step in maintaining cracks in a stable state. The material selection sets the upper limit for K_{Ic} and it is controlled by the designer in the specification of the material for the structure. Typical values of K_{Ic} for common engineering materials are presented in Table 4.3.

To ensure that the material meets the specification, the supplier provides a certification indicating that tests have been conducted using samples drawn from the "heat" and that the test results exceed the minimum required strength and toughness.

With welded structures, selection and certification of the base plate is only the first step in insuring the specified toughness. Weld materials and weld processes must be selected and then certified. The weld material is often under matched (lower yield strength) relative to the base plate with a higher toughness. The difficulty is in certifying the weld process because this certification requires extensive testing of specimens containing the weld and the heat-affected zone. This testing is expensive but necessary, because experience shows that welding often degrades the toughness of the base plate in the heat-affected zone.

Table 4.3
Initiation toughness for some engineering materials*

	K_{Ic}		σ_{ys}		σ_u	
Material	MPa-√m	ksi-√in.	MPa	ksi	MPa	ksi
Aluminum						
2024-T351	37.0	34	372	54	469	68
7075-T651	30.5	28	524	79	586	85
Titanium						
Ti-6Al-4V	54.5	50	931	135	1000	145
Steel						
4340	147.2	135	1069	155	1172	170
A533B	163.5	150	483	70	689	100
Stainless Steel						
304	218.0	200	276	40	621	90

*This data was taken from NASA Report JSC-22267B, March 2000.

In some applications K_{Ic} decreases with service life. For instance, in reactor pressure vessels, radiation damage due to high neutron flux occurs and the toughness degrades with extended service. Also, some alloys that age harden will exhibit losses in K_{Ic} with service life that is measured in decades.

The second step in a fracture-control plan involves inspection to determine the flaw size a. It is evident from Eq. (4.74) that the critical crack length a_c for a wide plate with a central crack is given by:

$$a_c = \frac{1}{\pi}\left(\frac{K_{Ic}}{\sigma_0}\right)^2 \tag{4.80}$$

Inspection with x-rays, eddy current sensors, and ultrasonic transducers are employed to locate cracks. Large cracks that approach or exceed a_c are repaired. However, small cracks with a length a_d below the minimum detectable length or with a length a_r below the repairable limit remain in the structure. These cracks may grow during service due to fatigue loading or stress-corrosion cracking, and they degrade the strength of the structure as shown in Fig. 4.17. The growth period represents the time interval during which one or more cracks extends from a_r to a_c. This period is usually long and provides the opportunity for periodic inspections during the service life. It is important in a fracture-control plan to specify inspection intervals that are sufficiently short relative to the growth period to provide two or more opportunities to detect the cracks before they achieve critical length. When cracks of length $a \Rightarrow a_c$ are detected, they are repaired (in the field) and the growth period is extended.

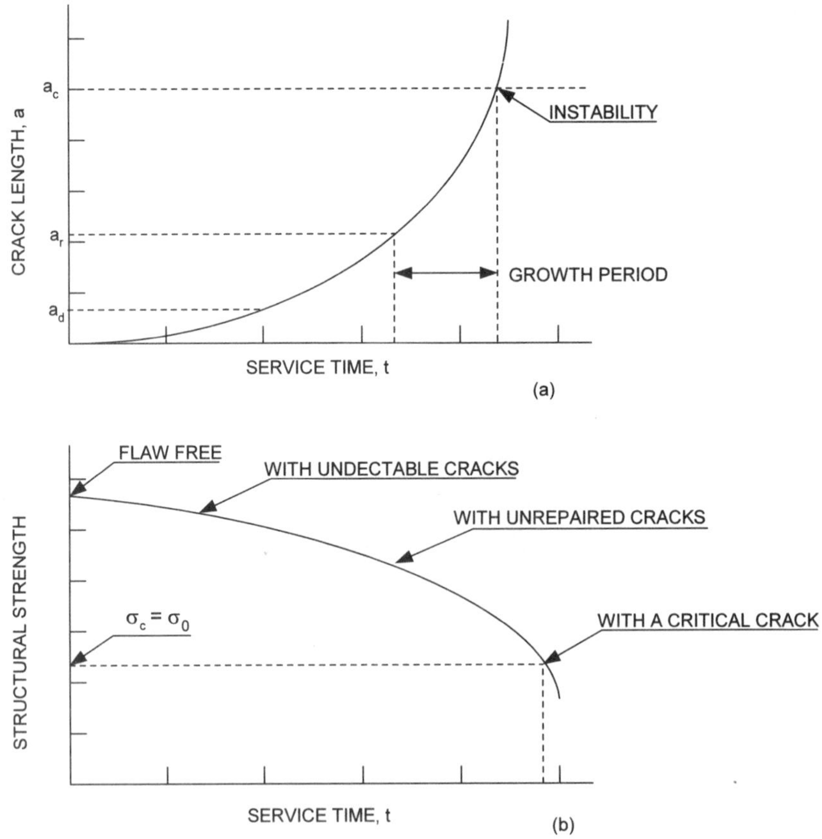

Figure 4.17 (a) Crack growth with time in service.

(b) Loss of structural strength with time and crack growth.

The third step in the fracture-control plan involves the applied load that determines the stress σ_c. From Fig. 4.20, it is evident that the structural strength will exceed σ_c provided $a < a_c$. In many types of structures, periodic inspections and field repair insure that crack growth is controlled and that the cracks do not approach critical length. In these structures, structural stability is achieved simply by maintaining the applied stresses so that $\sigma_0 < \sigma_c$. However, in some structures, accessibility is limited and complete inspection is difficult if not impossible. In these instances, a proof-testing procedure can be applied to the structure to insure safety. In proof testing, loads are applied to the structure that are larger than the normal loading which produces σ_0. These proof loads produce an elevated stress σ_p and if the structure **does not fail**, the cracks in the structure are $a < a_p$ as shown in Fig. 4.18. The single cycle of proof load provides confidence that the cracks in the structure are short and stable. Indeed, depending upon the ratio σ_p/σ_c, a growth period can be predicted where the safety of the structure can be ensured. It is important to divide the growth period into two or more parts to determine the proof-test interval. One repeats proof testing at these intervals to continue to extend the growth period. In this sense, repeated proof testing is similar to repeated inspections. However, with an inspection procedure, the cracks which grow with time in service can be repaired; but with a proof test procedure, the cracks continue to grow and when $a \geq a_p$ the structure will fail during the proof test.

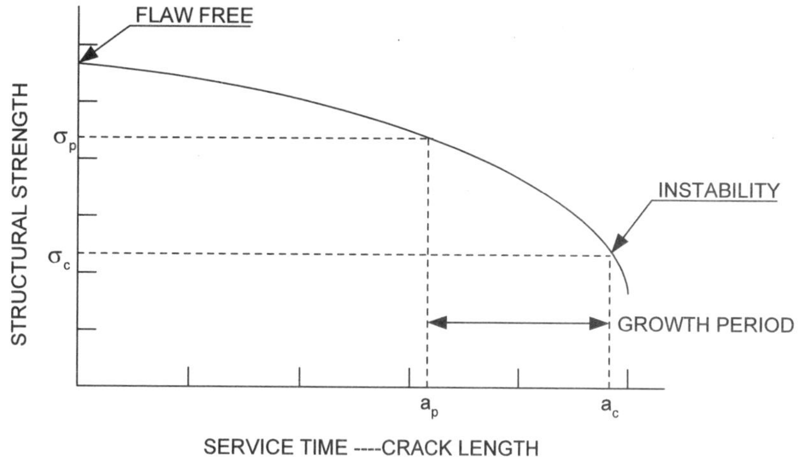

Figure 4.18 Proof testing at a stress σ_p to ensure that $a < a_p$.

EXERCISES

4.1 Your manager directs you to perform a stress analysis of a machine component containing a well-defined crack, and to employ the von-Mises theory of yielding to predict when the structure will fail. How do you respond to this directive?

4.2 If the ratio of the major axis to the minor axis of an ellipse is b/a = 3, determine the coordinate ξ_0 to form an ellipse of these proportions in an elliptical coordinate system. If the dimension b = 10 mm, determine the scaling constant c. Determine the coordinate η that locates a point at the intersection of the x-axis with ξ_0 and another η that locates the intersection of the y-axis with ξ_0.

4.3 Repeat Exercise 4.2 for an ellipse with b/a = 10.

4.4 Write an equation for the stress σ_η about the boundary of an elliptical hole with b/a = 5, if σ_0 = 100. Evaluate this equation for η ranging from 0 to $\pi/2$. Prepare a graph showing the distribution of the stresses about the perimeter of the first quadrant of the elliptical hole.

4.5 Repeat Exercise 4.4 for an elliptical hole with b/a = 20, and b/a = 1. Compare the stress concentration factor $K = (\sigma_\eta/\sigma_0)_{\eta=0}$ for these two cases.

4.6 Determine the stress concentration at the tip of a crack of length 2a when a uniaxial stress σ_0 is applied in a direction parallel to the length of the crack.

4.7 Verify the conversion [from Eq. (4.13) to Eq. (4.14)] for the expression of the stress function Z when the origin is shifted from the center of the crack to the crack tip.

4.8 Beginning with Eq. (4.14) verify Eqs. (4.15) to (4.19).

4.9 A crack 60 mm long develops in a very large plate that is to support a stress σ_0 = 70 MPa. Determine the stress intensity factor K_I. What assumption did you make in this determination?

4.10 If the crack in Exercise 4.9 grows at a rate of 10 mm/month, determine the service life remaining before crack initiation if K_{Ic} = 40 MPa-$\sqrt{m}$.

4.11 Prepare a graph showing α as a function of a/W for a tension strip with a centrally located crack.

4.12 A steel strap 1 mm thick and 20 mm wide with a central crack 4 mm long is loaded to failure. Determine the critical load if K_{Ic} for the strap material is 80 MPa-√m.

4.13 Determine the failure stress σ_0 applied to the strap of Exercise 4.12. Compare this stress to the yield and tensile strengths of 4340 alloy steel.

4.14 Write a program to compute the multiplier α for SEC and DEC cracked tension strips. Evaluate α for $0 < a/W < 0.49$ in increments of 0.01. Display the results in a listing similar to Table 4.1.

4.15 A steel tension bar 8 mm thick and 50 mm wide with a single edge crack 10 mm long is subjected to a uniaxial stress σ_0 = 140 MPa. Determine the stress intensity factor K_I. If K_{Ic} for this steel is 60 MPa-√m, is the crack stable?

4.16 Determine the critical crack length for the steel bar in Exercise 4.15.

4.17 Determine the critical load for the steel bar in Exercise 4.15.

4.18 A steel tension bar 8 mm thick and 50 mm wide with double edge cracks each 5 mm long is subjected to a uniaxial stress σ_0 = 140 MPa. Determine the stress intensity factor K_I. If K_{Ic} for this steel is 160 MPa-√m, is the crack stable?

4.19 Determine the critical crack length for the steel bar in Exercise 4.18.

4.20 Determine the critical load for the steel bar in Exercise 4.18.

4.21 Determine the normalized stress intensity factor $K_IBW^{3/2}/PS$ as a function of a/W for a beam subjected to three point bending.

4.22 Determine the normalized stress intensity factor $K_IBW^2/6M$ as a function of a/W for a beam subjected to pure bending.

4.23 Find the critical load that can be applied to a beam subjected to three point bending if S = 1 m, W = 100 mm, B = 40 mm a = 20 mm and K_{Ic} = 100 MPa-√m.

4.24 Find the critical moment that can be applied to a beam in pure bending if S = 1 m, W = 100 mm, B = 40 mm a = 20 mm and K_{Ic} = 100 MPa-√m.

4.25 Determine the normalized stress intensity factor $K_IBW^{1/2}/P$ as a function of a/W for the compact tension specimen.

4.26 Determine the critical load for a compact tension specimen if B = 50 mm, W = 250 mm and a = 100 mm and K_{Ic} = 120 MPa-√m.

4.27 Give an example of:

a. Opening mode loading.
b. In-plane shear mode loading.
c. Out-of-plane shear mode loading.
d. Mixed mode (I and II) loading.

4.28 Justify the rationale in dividing the field adjacent to the crack tip into 3 regions. Explain how you would describe the stresses in each of these three regions.

4.29 Verify Eqs. (4.29) beginning with Eq. (4.27).

4.30 Verify Eqs. (4.30) beginning with Eq. (4.29).

4.31 Use the strains defined in Eqs. (4.30) and integrate to determine the displacement field given by Eqs. (4.31).

4.32 Verify Eqs. (4.36) beginning with Eq. (4.32).

4.33 Verify Eqs. (4.37) beginning with Eqs. (4.36).

4.34 Use the strains defined in Eqs. (4.37) and integrate to determine the displacement field given in Eqs. (4.38).

4.35 Derive Eqs. (4.41) from Eqs. (4.39) with definitions of Z and Y given by Eqs. (4.40).

4.36 Verify: (a) the first of Eqs. (4.42). (b) the second of Eqs. (4.42). (c) the third of Eqs. (4.42).

4.37 Derive Eqs. (4.45) from Eqs. (4.43) with definitions of Z and Y given by Eqs. (4.44).

4.38 Verify: (a) the first of Eqs. (4.46). (b) the second of Eqs. (4.46). (c) the third of Eqs. (4.46).

4.39 Beginning with Eq. (4.51) derive Eq. (4.54).

4.40 Construct a graph showing the crack opening displacement v as a function of position from the origin to x = a for a central crack if $(K_I / E) = 0.5 \times 10^{-3}\ \sqrt{m}$.

4.41 Derive Eq. (4.58).

4.42 At the point defined by $r = r_c$ and $\theta = \pi/2$, find the relation between K_{Ic} and S_c the critical strain-energy-density factor. Let a = 20 mm.

4.43 Repeat Exercise 4.42 for the point defined by $r = r_c$ and $\theta = 0$. Let $\nu = 1/3$.

4.44 Determine the J integral for the contour a, b, c, d shown in Fig. E4.44. In this determination, find the contribution along each of the four line segments.

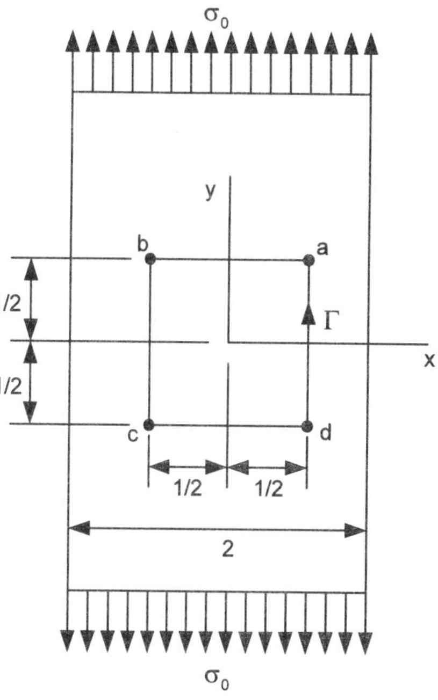

Figure E4.44

4.45 Derive Eq. (4.66).

4.46 The critical value of K_{Ic} for a steel is 70 MPa-$\sqrt{m}$. Determine the critical value of the strain energy release rate under plane-stress conditions and plane-strain conditions.

4.47 The critical value of K_{Ic} for a steel is 70 MPa-$\sqrt{m}$. If the steel is used in a plane-strain application, determine the critical stress for cracks ranging in length from 1 mm to 100 mm.

4.48 For an alloy steel with K_{Ic} = 90 MPa-$\sqrt{m}$ and σ_0 = 900 MPa, determine the crack length 2a where $\sigma_c = \sigma_{ys}$. Assume the crack is centrally located in a thick plate with a uniaxial load giving σ_0.

4.49 Consider a wide plate with an edge crack that is initially 0.2 in. long. The plate is fabricated from material with Paris Law coefficients of n = 4 and $C = 9 \times 10^{-9}$ when ΔK is expressed in units of ksi-in$^{1/2}$. The fracture toughness of the plate material K_{Ic} is 60 ksi-in$^{1/2}$. If the plate is subjected to stresses that cycle from 0 to 25 ksi, determine the number of cycles to extend the crack until it is 0.75 in. long.

4.50 In Exercise 4.49, determine the number of cycles required to grow the crack to its critical length.

4.51 Consider a wide plate with an edge crack that is initially 0.075 in. long. The plate is fabricated from material with Paris Law coefficients of n = 4 and $C = 8 \times 10^{-10}$ when ΔK is expressed in units of ksi-in$^{1/2}$. The fracture toughness of the plate material K_{Ic} is 70 ksi-in$^{1/2}$. If the plate is subjected to stresses that cycle from 0 to 30 ksi, determine the number of cycles to extend the crack until it is 0.6 in. long.

4.52 In Exercise 4.51, determine the number of cycles required to grow the crack to its critical length.

4.53 Describe the three steps involved in fracture control plans.

4.54 Construct a graph showing a_{cr} as a function of K_{Ic}/σ_0 over the range from 0.01 to 0.15 $\sqrt{m}$.

4.55 Write a fracture control plan for a large natural gas storage tank to be located adjacent to a densely populated metropolitan area.

4.56 Why is proof testing used to insure structural integrity?

BIBLIOGRAPHY

1. Irwin, G. R.: Fracture I, Handbuch der Physik VI, Flügge Ed., pp. 558-590, Springer, 1958.
2. Inglis, C. E.: "Stresses in a Plate Due to the Presence of Cracks and Sharp Corners," Transactions Institution of Naval Architects, 55, pp. 219-241, 1913.
3. Griffith, A. A.: "The Phenomena of Rupture and Flow in Solids," Philosophical Transactions, Royal Society of London, A 221, pp. 163-197, 1921.
4. Westergaard, H. M.: "Bearing Pressures and Cracks", Journal of Applied Mechanics, Vol. 61, pp. A49-A53, 1939.
5. Kirsch, G.: "Die Theorie der Elasticität und die Bedürfnisse der Festigkeitlehre, Z. Ver Dtsch Ing. Vol 32, pp. 797-807, 1898.
6. Tada, H., Paris, P. and G. R. Irwin: The Stress Analysis of Cracks Handbook, 2nd Edition, American Society of Mechanical Engineers, New York, 2000.
7. Chona, R.: "Non-Singular Stress Effects in Fracture Test Specimens - A Photoelastic Study," MS Thesis, University of Maryland, Aug. 1985.
8. Irwin, G. R. and J. A. Kies, "Critical Energy Rate Analysis of Fracture Strength of Large Welded Structures," Welding Journal, Vol. 33, Research Supplement, pp. 193s – 198s, 1954
9. Sih, G. C.: "Some Basic Problems in Fracture Mechanics and New Concepts", Engineering Fracture Mechanics, Vol. 5, pp. 365-377, 1973.
10. Rice, J. R.: "A Path Independent Integral and the Approximate Analysis of Strain Concentrations by Notches and Cracks," Journal of Applied Mechanics, pp. 379-386, 1968.
11. Eshelby, J. D.: "Calculation of Energy Release Rate", Prospects of Fracture Mechanics, Editor, Sih, G. C., Von Elst and Broek, D, Noordhoff, pp. 69-84, 1974.
12. Brown, W. F. Jr. and J. E. Srawley: "Fracture Toughness Testing", Fracture Toughness Testing and Its Applications, ASTM - STP # 381, pp. 133-185, 1965.
13. Marine Casualty Report: "Structural Failure of the Tank Barge I.O.S. 3301 Involving the Motor Vessel Martha R. Ingram on 10 January 1972 Without Loss of Life," Report No. SDCG/NTSB, March 1974.
14. Sanford, R. J.: "A Critical Re-examination of the Westergaard Method for Solving Opening-Mode Crack Problems", Mechanics Research Communication, Vol. 6, No.5, 1979.
15. Sanford, R. J.: Principles of Fracture Mechanics, Prentice Hall, Upper Saddle River, 2003.
16. Hertzberg, R. W.: Deformation and Fracture Mechanics of Engineering Materials, 4th Edition, John Wiley & Sons, New York, 1996.
17. Williams, M. L.: "Stress Singularities Resulting from Various Boundary Conditions in Angular Corners of Plates in Extension," Journal of Applied Mechanics, Vol. 19, pp. 526-528, 1952.
18. Williams, M. L.: "On the Stress Distribution at the Base of a Stationary Crack," Journal of Applied Mechanics, Vol. 19, pp. 109-114, 1957.
19. Shukla, A.: "Practical Fracture Mechanics in Design," Marcel Deckker, ISBN 0824758854, 2005.

PART II

DISPLACEMENT AND STRAIN MEASUREMENT METHODS

CHAPTER 5

INTRODUCTION TO STRAIN MEASUREMENTS AND DISPLACEMENT SENSORS

5.1 DEFINITION OF STRAIN AND ITS RELATION TO MEASUREMENTS

A state of strain may be characterized by its six Cartesian strain components or, equally well, by its three principal strain components with the three associated principal directions. The six Cartesian components of strain are defined in terms of the displacement field by the following set of equations when the strains are small (normally the case for elastic analyses):

$$\varepsilon_{xx} = \frac{\partial u}{\partial x} \qquad \varepsilon_{yy} = \frac{\partial v}{\partial x} \qquad \varepsilon_{zz} = \frac{\partial w}{\partial x}$$
$$\gamma_{xy} = \frac{\partial v}{\partial x} + \frac{\partial u}{\partial y} \qquad \gamma_{yz} = \frac{\partial w}{\partial y} + \frac{\partial v}{\partial z} \qquad \gamma_{zx} = \frac{\partial u}{\partial z} + \frac{\partial w}{\partial x} \tag{2.4}$$

The ε symbol represents normal strains and ε_{xx}, for instance, is defined as the change in length of a line segment parallel to the x-axis divided by its original length. The γ symbol represents shearing strains and γ_{xy}, for instance, is defined as the change in the right angle formed by the line segments parallel to the x and y-axes.

Usually strain-gage applications are confined to the free surfaces of a body. The two-dimensional state of stress existing on this surface can be expressed in terms of three Cartesian strains ε_{xx}, ε_{yy} and γ_{xy}. Thus, if the two displacements u and v can be established over the surface of the body, the strains can be determined directly from Eqs. (2.4). In certain cases, the most appropriate approach for establishing the stress and strain field is to determine the displacement field. As an example, consider the very simple problem of a transversely loaded beam. The deflections of the beam w(x) along its longitudinal axis can be accurately determined with relatively simple experimental techniques. The strains and stresses are related to the deflection w(x) of the beam by:

$$\varepsilon_{xx} = \frac{z}{\rho} = z\frac{d^2w}{dx^2} \qquad \sigma_{xx} = E\varepsilon_{xx} = Ez\frac{d^2w}{dx^2} \tag{5.1}$$

where ρ is the radius of curvature of the beam and z is the distance from its neutral axis to a point of interest.

Measurement of the transverse displacements of plates can also be accomplished with relative ease and stresses and strains computed by employing equations similar to Eq. (5.1). In the case of a more general body, however, the displacement field cannot readily be measured. Also, the conversion from displacements to strains requires a determination (by differentiation) of the gradients of experimentally determined displacements at many points on the surface of the specimen. Because the displacements are often difficult to obtain and the differentiation process is subject to large errors, it is advisable to employ a strain gage of one form or another to measure the surface strains directly. Examination of Eqs. (2.4) shows that the strains ε_{xx}, ε_{yy} and γ_{xy} are really the slopes of displacement surfaces u and v. Moreover, these strains are not, in general, uniform; instead, they vary from point to point. The slopes of the displacement surfaces cannot be determine unless the in-plane displacements u and v can be accurately established. Because the in-plane displacements are often exceedingly small in comparison with the transverse (out-of-plane) displacements mentioned previously, their direct measurement over the entire surface of a body is very difficult. To circumvent this difficulty, one displacement component is usually measured over a small region of the body along a short line segment, as illustrated in Fig. 5.1. This displacement measurement is converted to strain by the relationship:

$$\varepsilon_{xx} = \frac{L_x - L_0}{L_0} \frac{\Delta u}{\Delta x} \tag{5.2}$$

where $\Delta u = L_x - L_0$ is the displacement in the x direction over the length of the line segment $L_0 = \Delta x$. Strain measured in this manner is not exact because the determination is made over some finite length L_0 and not at a point, as the definition for the normal strain ε requires. The error involved in this approach depends upon the strain gradient and the length of the line segment L_0. If the strain determination is considered to represent the strain that occurs at the center of the line segment L_0, that is, point x_1 in Fig. 5.1, the error involved for various strain gradients is:

Case 1, strain constant: $\varepsilon_{xx} = k_1$ (no error is induced).
Case 2, strain linear: $\varepsilon_{xx} = k_1x + k_2$ (no error is induced).
Case 3, strain quadratic: $\varepsilon_{xx} = k_1x^2 + k_2x + k_3$.

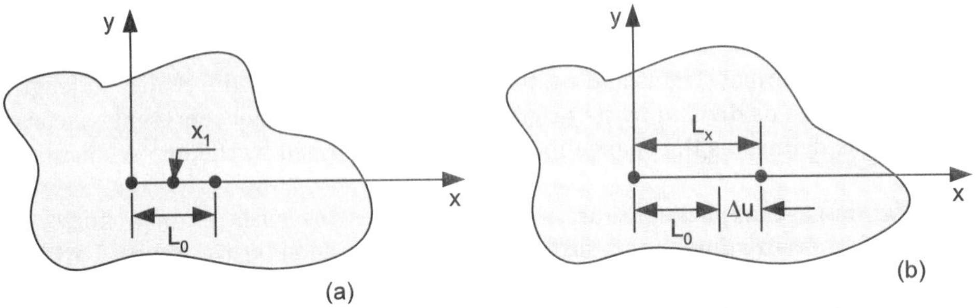

Figure 5.1 Strain measurement over a short line segment of length L_0: (a) before deformation; (b) after deformation.

In case 3, an error is involved because the strain at the midpoint x_1 is not equal to the average strain over the gage length L_0. The average strain ε_{avg} over the gage length L_0 can be determined from:

$$\varepsilon_{avg} = \frac{\int_0^{L_0} (k_1x^2 + k_2x + k_3)dx}{L_0} = \frac{k_1L_0^2}{3} + \frac{k_2L_0}{2} + k_3 \tag{a}$$

and the strain at the midpoint $x_1 = L_0/2$ is given by:

$$\varepsilon_{xx}\big|_{x_1} = \frac{k_1L_0^2}{4} + \frac{k_2L_0}{2} + k_3 \tag{b}$$

The difference between the average and midpoint strains represents the error E that is given by:

$$\mathcal{E} = \frac{k_1 L_0^2}{12} \qquad (5.3)$$

In this example, the error involved depends upon the values of k_1 and L_0. If the gradient is sharp, the value of k_1 will be significant and the error induced will be large unless L_0 is reduced to an absolute minimum. Other examples corresponding to cubic and quartic strain distributions can also be analyzed; however, the fact that an error is induced is established by considering any strain distribution other than a linear one.

In view of the error introduced by the length of the line segment in certain strain fields, great effort has been expended in reducing the gage length L_0. Two factors complicate these efforts. First, mechanical difficulties are encountered when L_0 is reduced. Regardless of the type of gage employed to measure the strain, the gage must have a certain finite size and a certain number of parts. As the size is reduced, the parts become smaller and the dimensional tolerances required on each become prohibitive. Second, the strain to be measured is a very small quantity. Suppose, for example, that strain determinations are to be made with an accuracy of ± 1 $\mu\varepsilon$ over a gage length of 0.1 in. (2.5 mm). The strain gage must measure the corresponding displacement to an accuracy of $\pm 1 \times 10^{-6} \times 0.1 = \pm 1 \text{ x } 10^{-7}$ in. (2.5 nm). These size and accuracy requirements place heavy demands on the talents of investigators in the area of strain-gage development.

The smallest gage developed and sold commercially to date is an electrical-resistance device. This gage is prepared from an ultra thin alloy foil, which is photoetched to produce a grid with a gage length of only 0.008 in (0.2 mm). Smaller strain gages are possible using P or N doped semiconductors as the sensing element; however, these gages are usually produced on small silicon substrates and used in either research applications or in transducers to measure either pressure or acceleration. On the other hand, relatively large mechanical/electrical strain gages , known as extensometers, are still employed in materials testing where an average strain is measured over a gage length of 25 mm during a typical tensile test.

5.2 PROPERTIES OF STRAIN-GAGE SYSTEMS

Historically, the development of strain gages has followed many different paths and gages have been developed which are based on mechanical, optical, electrical, acoustical and pneumatic principles. No single gage system, regardless of the principle upon which it is based, has all the properties required of an optimum gage. Thus there is a need for a wide variety of gage systems to meet the requirements of a wide range of different engineering problems involving strain measurement.

Some of the characteristics commonly used to judge the adequacy of a strain-gage system for a particular application are the following:

1. The calibration constant for the gage should be stable; it should not vary with time, temperature or other environmental factors.
2. The gage should be able to measure strains with an accuracy of ± 1 $\mu\varepsilon$ over a strain range of ± 10%.
3. The gage size, i.e., the gage length L_0 and width w_0, should be small so that strain at a point is approximated with small error.
4. The response of the gage, largely controlled by its inertia, should be sufficient to permit recording of dynamic strains with frequency components exceeding 200 kHz.
5. The gage system should permit on-location or remote readout.
6. The output from the gage during the readout period should be independent of temperature and other environmental parameters.

7. The gage and the associated auxiliary equipment should both be inexpensive to enable wide usage.
8. The gage system should be easy to install and operate.
9. The gage should exhibit a linear response to strain over a wide range.
10. The gage should be suitable for use as the sensing element in other transducer systems where an unknown quantity such as pressure is measured in terms of strain.

No single strain-gage system satisfies all of these characteristics. However, a strain-gage system for a particular application can be selected after proper consideration is given to each of these characteristics in terms of the requirements of the measurement to be made. Over the last 60 years a large number of systems, with wide variations in design, have been conceived, developed and marketed. Each of these systems has four basic characteristics that deserve additional consideration; namely, the gage length L_0, gage sensitivity, range of strain measurement and accuracy.

Strains cannot be measured at a point with any type of gage and, as a consequence, nonlinear strain fields cannot be measured without some degree of error being introduced. In these cases, the error will definitely depend on the gage length L_0 in the manner described in Section 5.1 and may also depend on the gage width w_0. The gage size for a mechanical extensometer is characterized by the distance between the two knife-edges in contact with the specimen (the gage length L_0) and by the width of the movable knife-edge, (the gage width w_0). The gage length and width of the metal-film resistance strain gage is determined by the size of the active area of the grid. In selecting a gage for a given application, gage length is one of the most important conditions.

A second basic characteristic of a strain gage is its sensitivity. Sensitivity is the smallest value of strain that can be read on the scale associated with the strain gage. The term sensitivity should not be mistaken for accuracy or precision, because very large values of magnification can be designed into an instrument system to increase its sensitivity; but friction, wear, noise, drifts etc., introduce large errors which limit its accuracy. In certain applications, gages can be employed with sensitivities of less than 1 $\mu\varepsilon$ if proper procedures are established. In other applications, where sensitivity is not important, 50 to 100 $\mu\varepsilon$ is often sufficient. The choice of a gage is dependent upon the degree of sensitivity required and quite often the selection of a gage with a very high sensitivity, when it is not really necessary, needlessly increases the complexity of the measurement.

A third basic characteristic of a strain gage is its range. Range represents the maximum strain, which can be recorded without re-zeroing or replacing the strain gage. The range and sensitivity are interrelated because very sensitive gages respond to small strains with appreciable response and the range is usually limited to the full-scale deflection or count of an indicator. Often it is necessary to compromise between the range and sensitivity characteristics of a gage to obtain reasonable performance for both these categories.

The final basic consideration is the accuracy or precision. As was previously pointed out, sensitivity does not ensure accuracy. Usually, the very sensitive instruments are quite prone to errors unless they are employed with the utmost care. In a mechanical extensometer, inaccuracies may result from lost motion due to wear or slippage, or deflection of its components. On all strain gages, there is a readout error whether the output of the gage is manually recorded, displayed on a digital multi-meter or saved in computer memory.

5.3 TYPES OF STRAIN GAGE

The problem encountered in measuring strain is to determine the displacement between two points separated by a distance L_0. The physical principles, which have been employed to accomplish this task, are very numerous and a complete survey will not be attempted; however, a few of the more applicable methods will be covered briefly. The principles employed in strain-gage construction can be used as the basis for classifying the gages into the following four groups including, mechanical, optical, electrical and acoustical.

5.4 MECHANICAL STRAIN GAGES

Mechanical strain gages such as the Huggenberger tensometer, or the Berry strain gage are rarely used today because the electrical-resistance strain gages are more accurate, lower in cost and easier to use. Mechanical gages often called extensometers are still widely used today in material test systems. However, these extensometers utilize electrical devices such as displacement transformers or resistance strain gages for sensors. A typical extensometer, shown in Fig. 5.2, is employed in the conventional tensile test where the stress-strain diagram is recorded. The extensometer is equipped with knife-edges and a wire spring that forces the knife-edges into the tension specimen. Elongation or compression of the specimen causes movement of the arms. As these arms move they bend a small cross-flexural element insuring center-point bending over the entire range of the extensometer. The cross-flexural member, which is the sensing element, also provides good lateral stability and requires low actuating forces (about 50 g). Electrical-resistance strain gages, bonded to the cross-flexural element, sense the bending strains and give a signal output that is proportional to the contraction or extension of the tensile specimen. The extensometer provides an accurate response to specimen strain with maximum non-linearity of 0.3% of its range and maximum hysteresis of only 0.1% of its range.

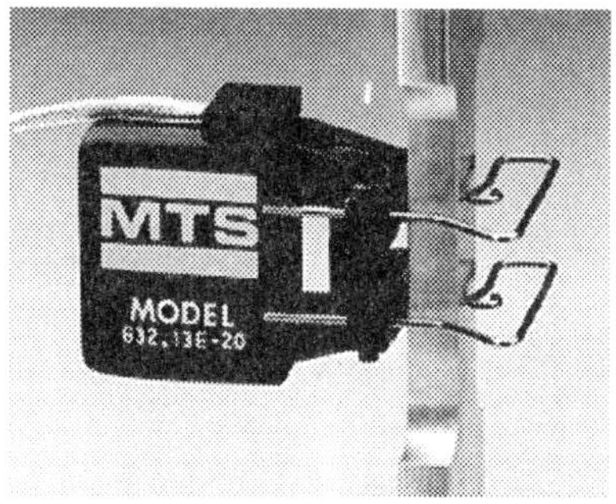

Figure 5.2 An extensometer for measuring average strain on a tensile specimen. (Courtesy of MTS Systems Corporation.)

5.5 OPTICAL STRAIN GAGES

During the past 40 years, considerable research effort has been devoted to the area of optical methods of experimental stress analysis. The availability of gas and ruby lasers as monochromatic, collimated and coherent light sources as well as fiber optics has led to several new developments in strain gages. Three of these developments—the diffraction strain gage, the interferometric strain gage and the fiber optic strain gages—are described in this chapter to indicate the capabilities of optical gages that use coherent light.

5.5.1 The Pryor and North Diffraction Gage

The diffraction strain gage is quite simple in construction; it consists of two blades that are bonded or welded to the specimen, as illustrated in Fig. 5.3. The two blades are separated by a distance b to form a narrow aperture and are fixed to the specimen along its edges to give a gage length L_0. A beam of collimated monochromatic light from a helium-neon laser is directed onto the aperture to produce a diffraction pattern that can be observed as a line of dots on a screen a distance R from the aperture. An example of a diffraction pattern is illustrated in Fig. 5.4.

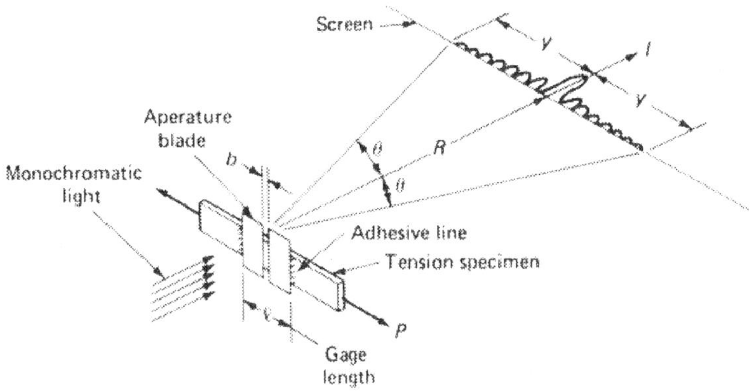

Figure 5.3 Schematic of a diffraction-type strain gage. (After T. R. Pryor and W. P. T. North.)

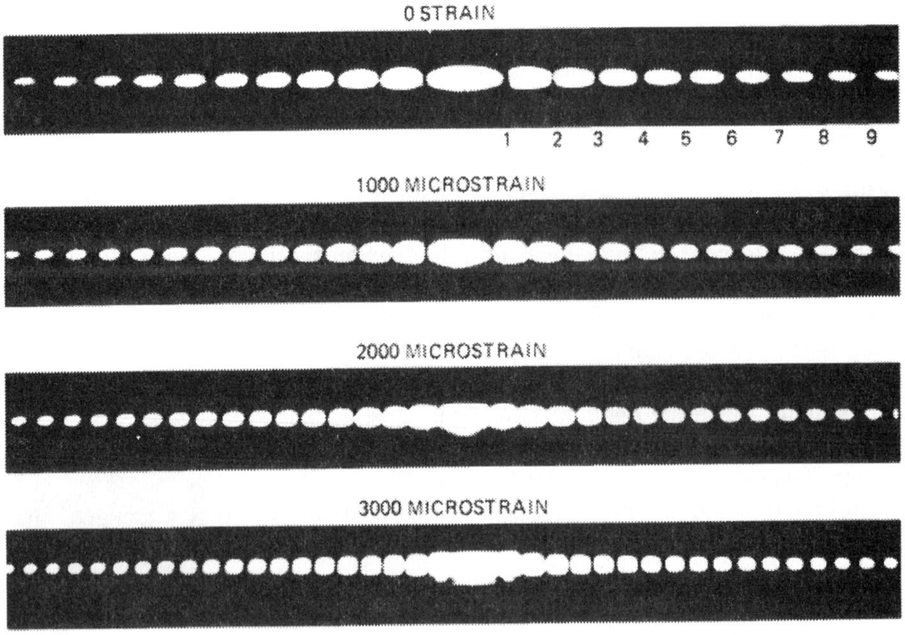

Figure 5.4 Diffractograms showing changes in the diffraction pattern with increasing strain. (Courtesy of T. R. Pryor and W. P. T. North.)

When the distance R to the screen is very large compared with the aperture width b, the distribution of the intensity I of light in the diffraction pattern is given by:

$$I = A_0^2 \frac{\sin^2\beta}{\beta^2} \tag{5.4}$$

where A_0 is the amplitude of the light on the centerline of the pattern ($\theta = 0$) and

$$b = \frac{\pi b}{\lambda}\sin\theta \tag{5.5}$$

where θ is defined in Fig. 5.3 and λ is the wavelength of the light. If the analysis of the diffraction pattern is limited to short distances y from the centerline of the system, sin θ is small enough to be represented by y/R and Eq. (5.5) becomes:

$$\beta = \frac{\pi b}{\lambda}\frac{y}{R} \tag{5.6}$$

The intensity I vanishes according to Eq. (5.4) when sin β = 0 or when $\beta = n\pi$, where n = 1, 2, 3, . By considering those points in the diffraction pattern where I = 0, it is possible to obtain a relationship between their location in the pattern and the aperture width b. Thus:

$$b = \frac{\lambda R n}{y} \tag{5.7}$$

where n is the order of extinction in the diffraction pattern at the position located by y.

As the specimen is strained, the deformation results in a change in the aperture width $\Delta b = \varepsilon L$ and a corresponding change in the diffraction pattern as indicated in Fig. 5.4. The magnitude of this strain ε can be determined from Eq. (5.7) and measurements from the two diffraction patterns. As an example, consider the diffraction pattern after deformation and note

$$b + \Delta b = \frac{\lambda R n^*}{y_1} \tag{a}$$

where n* is a specified order of extinction. Clearly, the diffraction pattern before deformation gives:

$$b = \frac{\lambda R n^*}{y_0} \tag{b}$$

Subtracting Eq. (b) from Eq. (a) and simplifying gives the average strain ε over the gage length L as:

$$\varepsilon = \frac{\Delta b}{L} = \frac{\lambda R n^*}{L}\frac{y_0 - y_1}{y_0 y_1} \tag{5.8}$$

In practice, the order of extinction n* is selected as high as possible consistent with the optical quality of the diffraction pattern. When the higher orders of extinction are used, the distance y can be measured with improved accuracy with a number of different instruments.

In an automated read out system, a linear array of CCD cells replaces the screen in Fig. 5.3. The charge on each cell is proportional to the light intensity and the value of y for each extinction order is determined by locating each cell, which exhibits a minimum intensity. The output from the linear CCD array is monitored with an online computer in real time.

The diffraction strain gage is extremely simple to install and use provided the component can be observed during the test. The method has many advantages for strain measurement at high temperature because it is automatically temperature-compensated if the blades forming the aperture are constructed of the same material as the specimen.

5.5.2 Sharpe's Diffraction Grating Gage

A second optical method of strain measurement utilizes the interference patterns produced when coherent, monochromatic light from a source such as a helium-neon laser is reflected from two shallow V-grooves ruled on a highly polished portion of the specimen surface. The V-grooves are usually cut with a diamond to a depth of approximately 0.000040 in. (0.001 mm) and are spaced approximately 0.005 in. (0.125 mm) apart. Alternatively, indentations produced with a Vickers's diamond indenter can be used to produce reflecting surfaces that make an angle of about 110° with the surface of the specimen.

When the grooves, which serve as reflective surfaces, are small enough to cause the light to diffract and close enough together to permit the diffracted light rays to superimpose, an interference pattern is produced. The intensity of light in the pattern is given by the expression:

$$I = 4A_0^2 \frac{\sin^2\beta}{\beta^2}\cos^2\phi \tag{5.9}$$

where $\beta = (\pi b/\lambda)\sin\theta$; $\phi = (\pi d/\lambda)\sin\theta$; b = width of groove: d = width between grooves and θ = angle from central maximum, as previously defined in Fig. 5.3.

As the light is reflected from the sides of the V grooves, two different interference patterns are formed, as indicated in Fig. 5.5. In an actual experimental situation, the fringe patterns appear as a row of dots on screens located approximately 8 in. (200 mm) from the grooves. The diameter of the dots is consistent with the diameter of the impinging beam of coherent light from a laser.

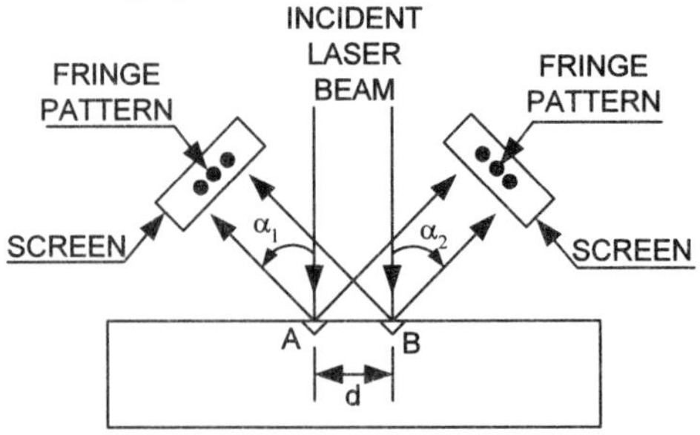

Figure 5.5 Schematic diagram of Sharpe's interferometric strain gage.

The intensity in the interference pattern goes to zero and a dark spot is produced whenever $\beta = n\pi$, with $n = 1, 2, 3,$, or when $\phi = (m + ½)\pi$, with $m = 0, 1, 2,$. When the specimen is strained, both the distance d between grooves and the width b of the grooves change. These effects produce shifts in the fringes of the two interference patterns that can be related to the average strain between the two grooves. The proof is beyond the scope of this presentation; however, it is shown in Reference 4 that:

$$\varepsilon = \frac{(\Delta N_1 - \Delta N_2)\lambda}{2d\sin\alpha} \tag{5.10}$$

where ΔN_1 and ΔN_2 are the changes in fringe order in the two patterns produced by the strain and α is the angle between the incident light beam and the diffracted rays which produce the interference pattern as shown in Fig. 5.5.

More recently, Sharpe has adapted this experimental approach to the measurement of strain in tensile tests of very small samples of polysilicon films 3.5 μm thick. As the films are too thin and brittle, the placement of grooves is not possible. However, gold lines 1 μm wide by 0.5 μm high and 200 μm long are deposited on the polysilicon films during manufacture. The sides of the lines are angled relative to the surface of the specimen; thus, they serve the same function as the grooves or diamond indentations as indicated in Fig. 5.5.

The interferometric gage offers a method for measuring strain without attaching or bonding a device, thus eliminating any reinforcing effects or bonding difficulties. Of course, it is necessary to polish the surface of the specimen and to place grooves, indentations or elevated lines on the specimen's surface. Because no contact is made, the method can be employed on rotating parts or in hostile environments. Temperature compensation is automatic and the method can be employed at very high temperatures. Also, photodiode arrays can be employed to monitor the changes in the fringe patterns with strain eliminating the need to photograph fringe patterns.

5.6 FIBER-OPTIC STRAIN GAGES

While measurements of strain are usually performed with electrical resistance strain gages, described in Chapter 6, there are some inherent advantages of fiber-optic strain gages that should be considered in select applications. These advantages include:

- Lightweight and small size
- Passive with low power requirements
- Free from electromagnetic interference
- High sensitivity
- Environmentally robust
- Stable of extremely long periods of time.

Strain gages constructed from optical fibers are relatively new because their development was dependent on low loss, single mode optical fibers that were not commercially available prior to the 1970s. The first fiber optic strain gage was developed in 1978; since that time a number of different gage designs have evolved.

Optical fiber strain sensors usually based on a change in amplitude (intensity) of light or a change in phase as the light propagates through an optical fiber wave guide. Both amplitude and phase can change as a function of the strain applied to some part of an optical fiber or the sensing element. Fiber optic strain gages are called intrinsic devices because they do not require an external component element in addition to the optical fiber. Fiber-optic strain gages are classified as Mach-Zehnder, Fabry-Perot, Michelson, polarimetric, modal-domain, twin-core, and Bragg grating types. All except the Mach-Zehnder gage can be designed to be insensitive to optical effects in the leads. Accordingly, short gage length measurements are possible. Intrinsic optical fiber strain sensors usually employ a sensing fiber and a reference fiber, either or both of which can be exposed to the strain field. However, the geometric or optical properties of the two fibers must be different if both the reference and sensing fibers are exposed to strain. Each optical arrangement is composed of an input fiber, which carries light from a light source to the strain-sensitive active fiber, followed by an output fiber, which transmits the light from the active fiber to an optical detector. The input and output fibers are also sensitive to the strain; hence, gage designs with insensitive input and output fibers are important and will be described in the following subsections.

The feature common to all intrinsic optical fiber sensors is that they are constructed using only fiber components—fiber couplers replace beam splitters, aluminum-coated fiber ends replace mirrors, optical fiber retarders replace wave plates, etc. The only non-fiber components in a strain sensing system are the light source and detection electronics. These components designed for use with fiber optics are attached directly to the fibers, effectively forming a fully guided optical system. The strain gage becomes a completely closed optical system with the optical fiber laser and the optical phase detection method. While optical fiber strain gages can be produced from suitable combinations of fiber and conventional optics, these hybrid designs are generally more cumbersome to use because the conventional optical components introduce instabilities. The electronic detection schemes employed with optical fiber sensors are as varied as the sensors and are beyond the scope of this textbook.

5.6.1 Mach-Zehnder, Michelson and Fabry-Perot Strain Gage

The Mach-Zehnder Strain Gage

The Mach-Zehnder, Michelson and Fabry-Perot strain sensors use fiber versions of the classical interferometers. The Mach-Zehnder optical fiber sensor, illustrated in Fig. 5.6, is probably the best known, since it was introduced early in the development cycle. The Mach-Zehnder strain interferometer acts in the classic sense—the light propagating in the reference arm is optically interfered with the light propagating in the sensing arm[1]. This resulting coherent interference occurs in the second 2 × 2 coupler that is shown in Fig. 5.6. The intensity and phase shift is modulated by a change in the optical path length due to strain. The change in optical path length is due to a change in the segment length caused by the applied strain and the birefringent effect. The birefringence is a result of a slight dependence of the refractive index on strain. Both these contributions to the change in optical path length are integrated over the active gage length of the fiber. The phase shift is measured by directly counting the fringes, or by using homodyne or heterodyne techniques that will be discussed later.

It is possible to actually form interferometric fringes in space to determine the strain-induced phase change; however, this approach is not recommended, because it introduces instability to the optical arrangement. Instead, the second 2 × 2 coupler is added to provide a stable interference location. A photo-detector records the intensity, which varies as a sinusoid with increasing or decreasing strain as

---

[1] A segment of the sensing fiber is adhesively bonded to a specimen and the strain is transmitted through the bond to increase or decrease the length of the bonded segment.

shown in Fig. 5.7. The results also serve to illustrate the difficulties encountered when attempting to count the fringe orders directly with an interferometer. The intensity-time trace is a sinusoid and unless the strain-time record is monotonic, it is necessary to employ the more advanced detection schemes, such as the active homodyne, to determine the sign of the strain or when the strain changes sign. This constraint applies to all fiber interferometers that undergo a phase shift exceeding 2π, as the intensity monitored by a detector is a periodic function of phase.

Various strain gages, stress gages, and strain rosettes have been developed by using the Mach-Zehnder fiber interferometer by simply making appropriate alterations to the geometry of either the reference or the sensing fiber. Unfortunately, the Mach-Zehnder strain gage has a serious disadvantage, which is the difficulty in isolating the leads from external strain fields.

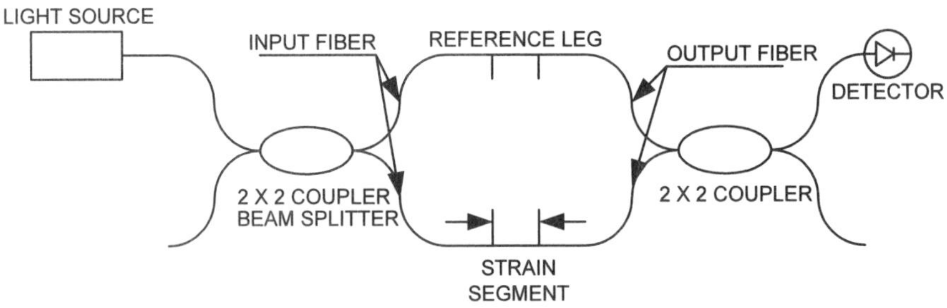

Figure 5.6 The Mach-Zehnder fiber-optic strain gage.

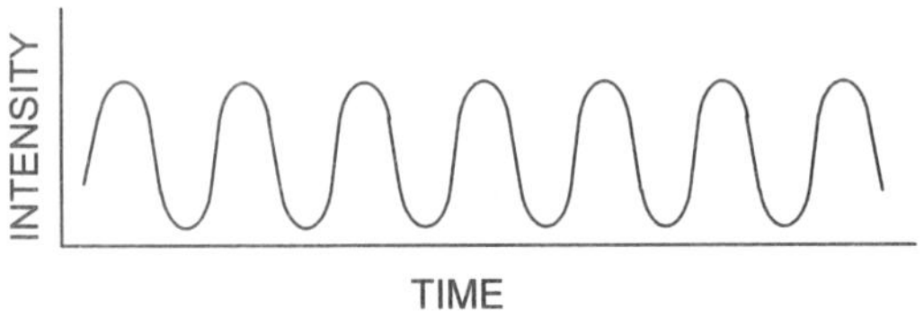

Figure 5.7 Sinusoidal response of a photo-detector illustrating the difficulty of establishing the phase angle when it exceeds 2π.

The Michelson Strain Gage

The Michelson optical fiber strain sensors are Mach-Zehnder strain sensors operating in reflection as illustrated in Fig. 5.8. The optical fibers are cleaved after the first 2 × 2 coupler, and mirrors are formed on the optical fibers by depositing a thin film of aluminum on the cleaved fiber ends. As with Mach-Zehnder strain gages, lead-insensitive Michelson sensors are difficult to design. The most common approach is to subject the input and output fibers to nearly identical strain fields, thus canceling strain or temperature induced effects. This approach to localized measurements works well only when the strain on the surface containing the input and output fibers is small. The sensitivity of the Michelson strain sensor is relatively high, so that gage lengths on the order of a few millimeters are practical.

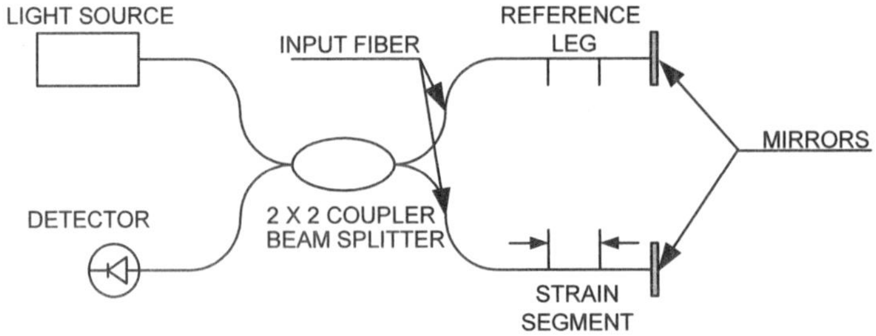

Figure 5.8 The Michelson optical fiber strain gage.

The Fabry-Perot Strain Gage

When the application for embedded optical-fiber strain gages developed, it became important to reduce the number of optical leads required. The Fabry-Perot optical fiber strain sensor is designed so that a single optical fiber carries the optical signal in and out of the strain sensitive cavity. The all-fiber Fabry-Perot strain gage arranged in the reflection mode is shown in Fig. 5.9. The details of the Fabry-Perot cavity, which results in multiple interference of the light entering and leaving it, is shown in Fig. 5.10.

This gage is commercially available[2] in gage length that ranges from 6 to 13 mm. The glass capillary tube that encases the cavity is 250 μm in diameter.

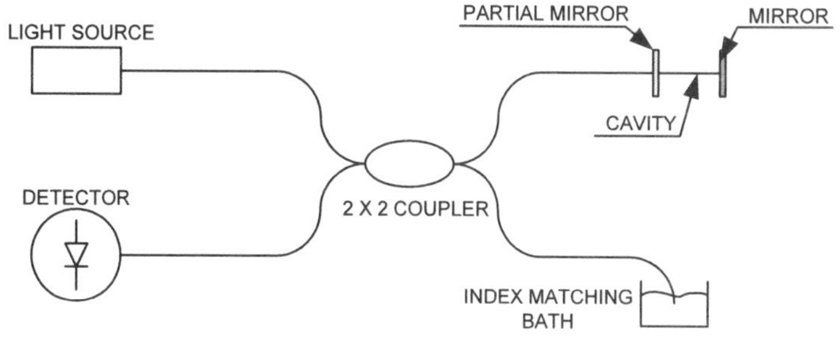

Figure 5.9 The Fabry-Perot strain gage arranged in the reflection mode.

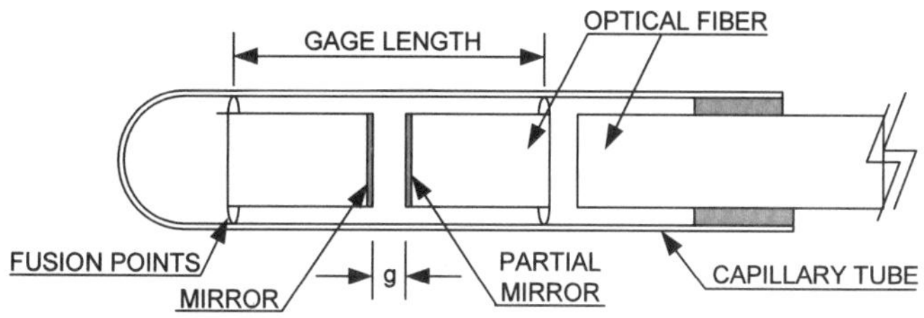

Figure 5.10 Fabry-Perot fiber-optic sensor for measuring strain.

Retardation measurements based on changes in the cavity length with the applied strain produce an optical phase shift, as described previously. The Fabry-Perot strain gage combines the advantages of high sensitivity of the Mach-Zehnder and Michelson strain gage with the advantage of a single optical lead. It has an added advantage in that it can be used with a white light source and with multi-mode fibers. Coupled with a Fizeau interferometer, the Fabry-Perot strain gage provides an absolute measurement of the phase change in the light signal produced by changes in the length of the cavity due to strain.

Another advantage of the Fabry-Perot strain gage is its very small sensitivity to temperature changes (– 0.1 με/°C) prior to bonding the gage to a specimen. The glass capillary tube that encases the fiber has nearly the same coefficient of thermal expansion as the fiber. For this reason, the tube and fiber expand or contract together and the cavity length does not change with temperature fluctuations.

The most significant advantage of the Fabry-Perot strain gage is its stability over extended periods of time when the gage is monitored with a Fizeau interferometer. The readout provides the absolute cavity length and this measurement is not affected by any electronic instabilities. Consequently, the Fabry-Perot strain gage is currently being employed to monitor the state of strain in large structures, such as dams and foundations, over extended periods of time.

Bragg Grating Strain Gage

The Bragg grating sensor, shown in Fig. 5.11, differs markedly from the interferometric sensors described previously. This optical fiber strain gage incorporates a diffraction grating with a pitch that is a function of the strain applied to the active fiber. The grating is written in a germanium-doped optical fiber using high-intensity, ultraviolet, and dual-beam interference that spatially modulates the index of refraction of the fiber core. The pitch of the grating is controlled by adjusting the angle between the two coherent beams forming the interference pattern imposed on the fiber as illustrated in Fig. 5.12. The Bragg grating gage can be monitored with either a narrow-or broadband coherent light source. The narrowband sources offer a high temporal bandwidth signal, while the broadband sources provide the opportunity to multiplex gages that are designed with a single fiber.

[2] FISO Technologies, 500 St-Jean-Baptiste, Suite 195, Quebec, Canada, G2E-5R9.

Multiplexing is accomplished by producing a fiber with a several spatially separated gratings, each with a different pitch. The output of the multiplexed sensor is processed in an optical spectrum analyzer, and the optical signal associated with a specified pitch is centered at the 1/p location in the spectral domain. A strain-induced change in grating pitch is monitored as a perturbation of the resonant frequency of the unstrained grating pitch and provides a direct measure of the strain. The advantages of this sensor include lead insensitivity, very short gage lengths, and the capability for simple multiplexing. The disadvantages include cost of the ultraviolet, high-power laser required to produce the Bragg grating in the fiber core, and the high expense and limited frequency response of the optical spectrum analyzer used in interpreting the output signal. The frequency response of Bragg grating sensors can be improved by using a narrowband source, but this practice requires more complex multiplexing procedures.

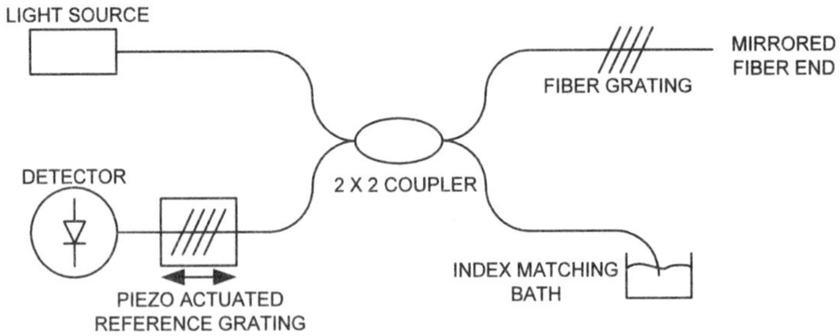

Figure 5.11 A high resolution system for spectral measurements from a fiber grating.

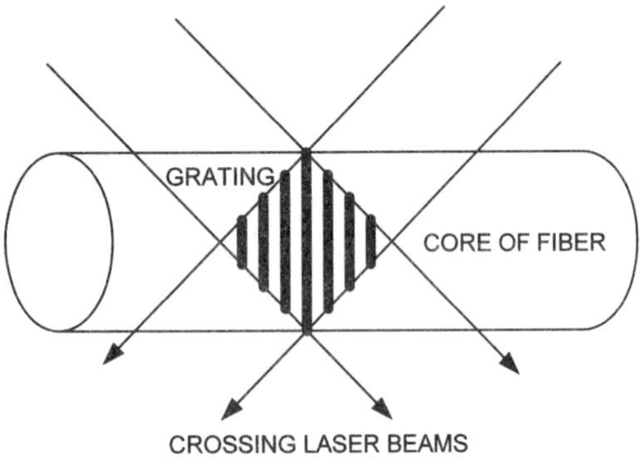

Figure 5.12 Producing a fiber grating with the interference pattern from two crossed laser beams.

Overview

The choice of the type of fiber-optic strain gage is dependent on the application involved in the study, the optical equipment available and the level of expertise of the user. Clearly tradeoffs exist between performance and difficulty in implementation. While optical fiber strain gages have not reached the level of commercialization of resistance strain gages, the intrinsic Fabry-Perot, Bragg grating as well as other sensors, are commercially available from a number of companies. It is anticipated that many of the difficulties encountered in performing difficult measurements with fiber optic strain gages will be resolved in the future as methods and hardware are optimized.

Optical fiber strain gages are finding applications in niche areas not well suited for resistance gages. This is particularly true in electromagnetic noise environments, explosive environments (where electrical sparks cannot be tolerated), and applications requiring monitoring of structures over extended periods of time.

5.6.2 Interpreting Optical-fiber Strain Gage Output

There are a number of methods used to characterize the wavelength and intensity of the light output from fiber-optic sensors, including optical spectrum analyzers, heterodyne methods and spectral interferometric systems. Consider first the optical spectrum analyzers.

Optical Spectrum Analyzers

An optical spectrum analyzer performs power versus wavelength measurements, and is usually employed for characterizing broadband sources such as light emitting diodes (LEDs) and semiconductor lasers. It is also useful in interpreting the signal output from a fiber-optic sensor where either the intensity or the wavelength of the optical signal is a function of the quantity (strain) being measured. The principal of operation of a spectrum analyzer is beyond the scope of this textbook; however, the commercial instruments are equipped with a display that shows the power level or mode spacing of the signal as a function of wavelength. The capability of these instruments is remarkable, as the resolution of the wavelength associated with a spectral peak can be measured to ± 0.030 nm. The mode spacing is particular useful when employing Bragg grating strain gages that exhibit shifts in their spectral peaks with applied strain.

Homodyne Interferometry

For a fiber-optic sensor arranged as an interferometer, the intensity of the signal output will vary as a sinusoid as the strain transmitted to the sensing segment increases or decreases. In this case a photodetector will convert the optical signal to an electrical signal that can be recorded relative to time. It is possible to count the peaks on the voltage-time record to establish the magnitude of the unknown quantity if a calibration constant for the strain gage is known. This is a very simple technique that is effective if a large number of fringes (peaks) are developed during the measurement. If the number of peaks is limited, it is usually necessary to employ a modern electronic phase meter that can resolve phase to about 0.1 degree, which corresponds to resolving a fringe (peak) into 3,600 parts.

Photodetectors are also useful in interpreting signals from fiber-optic sensors that provide an output signal with an intensity variation as a function of the unknown quantity.

Fizeau-CCD-Array Converter

The conversion of the optical signal from a Fabry-Perot strain gage into an electrical signal that can be readily interpreted is often accomplished with a Fizeau interferometer and a linear array of photo diodes or CCDs, as shown in Fig. 5.13. A cylindrical lens that is incorporated into the converter collimates the light signal from the Fabry-Perot strain gage. This light illuminates a Fizeau interferometer that is about 25 mm in diameter. This interferometer consists of a spatially distributed cavity whose thickness varies from almost zero to about some tens of micrometers. The dimensions of the wedge-like cavity correspond to the same values as the minimum and maximum dimensions of the Fabry-Perot cavity of the fiber-optic strain gage. The purpose of the Fizeau wedge is to resolve the spectrum into its wavelength components. The maximum intensity of light is transmitted through the Fizeau cavity at the exact location where its thickness is equal to the length of the cavity in the Fabry-Perot sensor. Thus, the Fizeau interferometer provides an instantaneous measurement of the optical signal for the entire range of gap measurement possible with the Fabry-Perot strain gage. Electronic processing locates the position of the maximum response with the data from the linear CCD array thus establishing the absolute cavity length. This measurement of cavity length is termed absolute because it corresponds to the true cavity length of the Fabry-Perot interferometer at the time of the measurement of the optical signal. The ability to make absolute measurements is critical in applications where long term or static measurements are required. The optical signal is converted in cavity length at a frequency given by the sampling rate of the conversion unit, which can approach 1,000 Hz. The conversion is accurate to ± 1 nm over a working range of 15,000 nm. This corresponds to an accuracy of 0.0067% of full scale for a strain gage

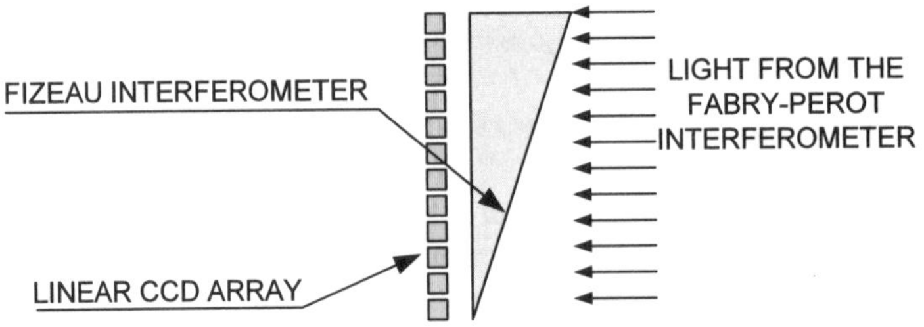

Figure 5.13 A Fizeau interferometer and a CCD array used in optical signal analysis.

5.7 ELECTRICAL STRAIN GAGES

During the past 70 years, electrical strain gages have become so widely accepted that they now dominate the entire strain-gage field except for special applications. The most important electrical strain gage is the resistance type, which will be covered in detail in Chapter 6. Two less commonly employed electrical strain gages, the capacitance type and the inductance type, will be introduced in this section. Although these sensors have limited use in conventional strain analysis, they are often employed in transducer applications and often are employed in measuring displacements.

5.7.1 Capacitance Strain Gages

The capacitance sensor, illustrated in Fig. 5.14, consists of a target plate and a second plate termed as the sensor head. These two plates are separated by an air gap of thickness h and form the two terminals of a capacitor that exhibits a capacitance C given by:

$$C = \frac{kKA}{h} \tag{5.11}$$

where C is the capacitance in picofarad (pF) and A is the area of the sensor head ($\pi D/4$).
K is the relative dielectric constant for the medium in the gap (K = 1 for air).
k = 0.225 is a proportionality constant for dimensions specified in in.
k = 0.00885 for dimensions given in mm.

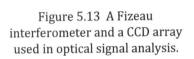

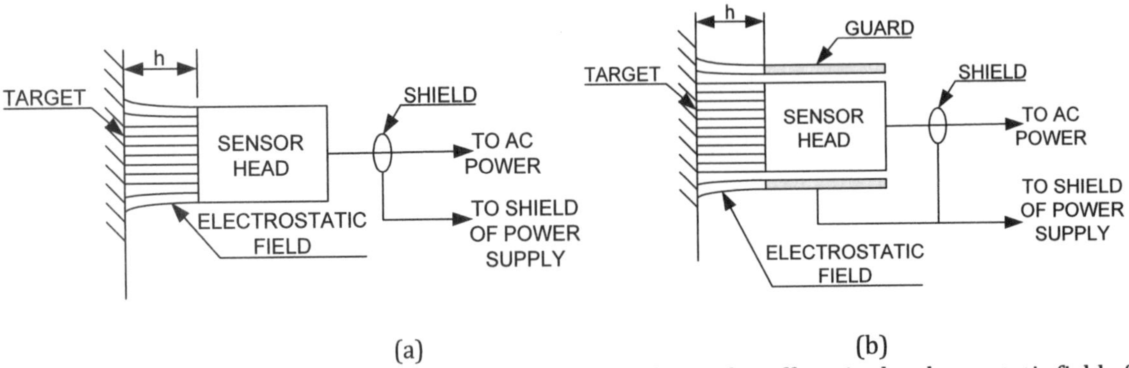

Figure 5.14 Capacitor sensors: (a) without a guard ring, where edge effects in the electrostatic field affect the range of linearity, and (b) with a guard ring to extend the range of linearity.

If the separation between the head and the target is changed by an amount Δh, then the capacitance C becomes:

$$C + \Delta C = \frac{kKA}{h + \Delta h} \qquad (a)$$

Equation (a) can be written as:

$$\frac{\Delta C}{C} = -\frac{\Delta h / h}{1 + (\Delta h / h)} \qquad (5.12)$$

This result indicates that $(\Delta C/C)$ is non-linear because of the presence of Δ/h in the denominator of Eq. (5.12). To avoid the difficulty of employing a capacitance sensor with a non-linear output, the change in the impedance due to the capacitor is measured. Note that the impedance Z_C of a capacitor is given by:

$$Z_C = -j/(\omega C) \qquad (b)$$

With a capacitance change ΔC, it can be shown that:

$$Z_C + \Delta Z_C = -\frac{j}{\omega}\left[\frac{1}{C + \Delta C}\right] \qquad (c)$$

Substituting Eq. (b) into Eq. (c) and solving for $\Delta Z_C/Z_C$ gives:

$$\frac{\Delta Z_C}{Z_C} = -\frac{\Delta C / C}{1 + \Delta C / C} \qquad (5.13)$$

Finally, substituting Eq. (5.12) into Eq. (5.13) yields:

$$\frac{\Delta Z_C}{Z_C} = -\frac{\Delta h}{h} \qquad (5.14)$$

From Eq. (5.14) it is clear that the capacitive impedance Z_C is linear in Δh and that methods of measuring ΔZ_C will permit extremely simple plates (the target as ground and the sensor head as the positive terminal) to act as a sensor to measure the displacement Δh. Cylindrical sensor heads are linear and Eq. (5.14) is valid provided $0 < h < D/4$ where D is the diameter of the sensor head. Fringing in the electric field produces non-linearity if $(h + \Delta h)$ exceeds D/4. The linear range can be extended to $h \approx D/2$ if a guard ring surrounds the sensor as shown in Fig. 5.14b. The guard ring essentially moves the distorted edges of the electric field to the outer edge of the guard, significantly improving the uniformity of the electric field over the sensor area and extending its linear range. An example of a button type head for a capacitance gage, presented in Fig. 5.15, show the guard ring that surround the sensing head.

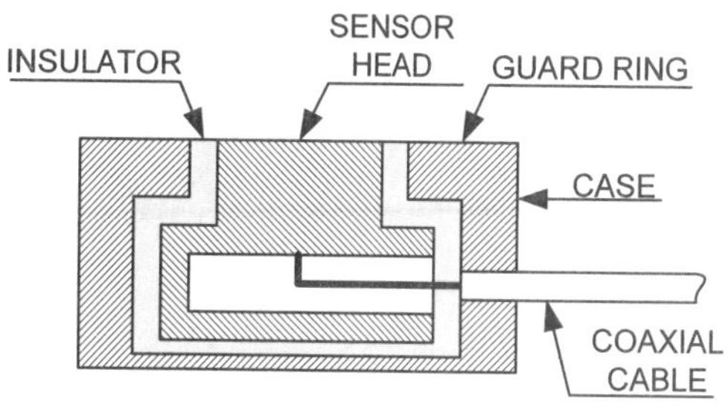

Figure 5.15 A button capacitor gage showing the guard ring and other construction details.

The sensitivity of the capacitance probe is given by Eqs. (b), (5.11), and (5.14) as:

$$S = \frac{\Delta Z_C}{\Delta h} = \left|\frac{Z_C}{h}\right| = \left|\frac{1}{\omega Ch}\right| = \left|\frac{1}{\omega KkA}\right| \tag{5.15}$$

The sensitivity can be improved by reducing the area of the probe; however, as noted previously, the range of the probe is limited by linearity to about D/2. Clearly there is a range-sensitivity trade-off. Of particular importance is the circular frequency ω in Eq. (5.15). Low frequency improves sensitivity but limits frequency response of the instrument, another trade-off. It is also important to note that the frequency of the ac power supply must remain constant to maintain a stable calibration constant.

The capacitance sensor has several advantages. It is non-contacting and can be used with any target material provided the material exhibits a resistivity less than 100 Ω-cm (any metal). The sensor is extremely rugged and can be subjected to high shock loads (5,000 g's) and intense vibratory environments. Their use as a sensor at high temperature is particularly impressive. They can be constructed to withstand temperatures up to 2,000° F and they exhibit a constant sensitivity S over an extremely wide range of temperature (74 to 1,600° F). Examination of the relation for S in Eq. (5.15) shows that the dielectric constant K is the only parameter that can change with temperature. Because K is constant for air over a wide range of temperature, the capacitance sensor has excellent temperature stability.

The change in the capacitive impedance Z_C is usually measured with the circuit illustrated schematically in Fig. 5.16. The probe, its shield, and guard ring are powered with a constant current ac supply. A digital oscillator is used to drive the ac supply and to maintain a constant frequency at 15.6 kHz. This oscillator also provides the reference frequency for the synchronous detector. The voltage drop across the probe is detected with a low capacitance preamplifier. The signal from the preamplifier is then amplified again with a fixed-gain instrument amplifier. The high voltage ac signal from the instrument amplifier is rectified and given a sign in the synchronous detector. The rectified signal is filtered to eliminate high frequency ripple and give a dc output voltage related to Δh. A linearizing circuit is used to extend the range of the sensor by accommodating for the influence of the fringes in the electrostatic field. Finally, the signal is passed through a second instrument amplifier where the gain and zero offset can be varied to adjust the sensitivity and the zero reading on a DVM display. When the gain and zero offset are properly adjusted, the DVM reads Δh directly to a scale selected by the operator.

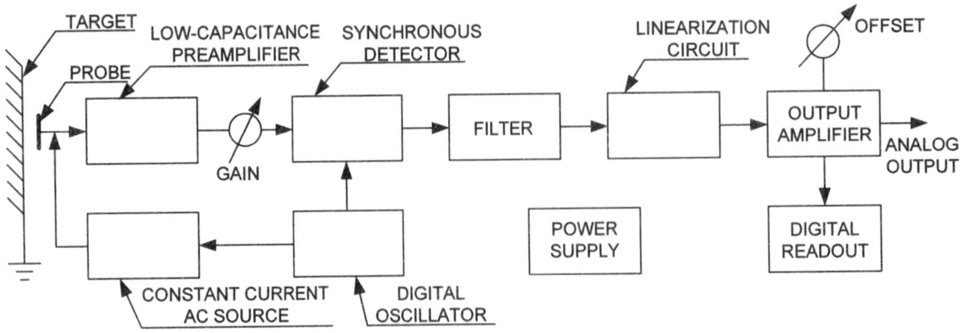

Figure 5.16 Schematic diagram of an instrument system employed with a capacitance sensing system.

5.7.2 The Inductance Strain Gage

Of the many types of inductance measuring systems, which could be employed to measure strain or displacement, the differential-transformer system is the most commonly used. The linear differential transformer is an effective sensor for converting mechanical displacement into an electrical signal. It can be employed in a large variety of transducers including: strain, displacement, pressure, acceleration, force and temperature. A schematic illustration of a linear differential transformer employed as a strain-gage transducer is shown in Fig. 5.17. Mechanical knife-edges are displaced over the gage length L_0 by the strain induced in the specimen. This displacement is transmitted to the core, which moves relative to the coils and an electrical output is produced across the coils.

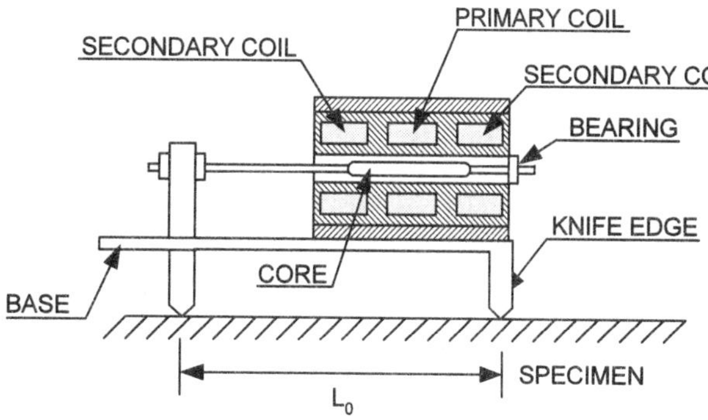

Figure 5.17 Schematic illustration of a linear differential transformer employed as a strain transducer.

The most popular variable-inductance sensor for linear displacement measurements is the linear variable differential transformer (LVDT). A LVDT, illustrated in Fig 5.18a, consists of three symmetrically spaced coils wound onto an insulated bobbin. A magnetic core, which moves through the bobbin without contact, provides a path for magnetic flux linkage between coils. The position of the magnetic core controls the mutual inductance between the center or primary coil and the two outer or secondary coils.

When ac excitation is applied to the primary coil, voltages are induced in the two secondary coils. The secondary coils are wired in a series-opposing circuit, as shown in Fig. 5.18b. When the core is centered between the two secondary coils, the voltages induced in the secondary coils are equal but out of phase by 180 degrees. Because the coils are in a series-opposing circuit, the voltages v_1 and v_2 in the two coils cancel and the output voltage is zero. When the core is moved from the center position, an imbalance in mutual inductance between the primary and secondary coils occurs and an output voltage, $v_o = v_2 - v_1$, develops. The output voltage is a linear function of core position if the motion of the core is within the operating range of the LVDT. The direction of motion can be determined from the phase of the output voltage relative to the input voltage.

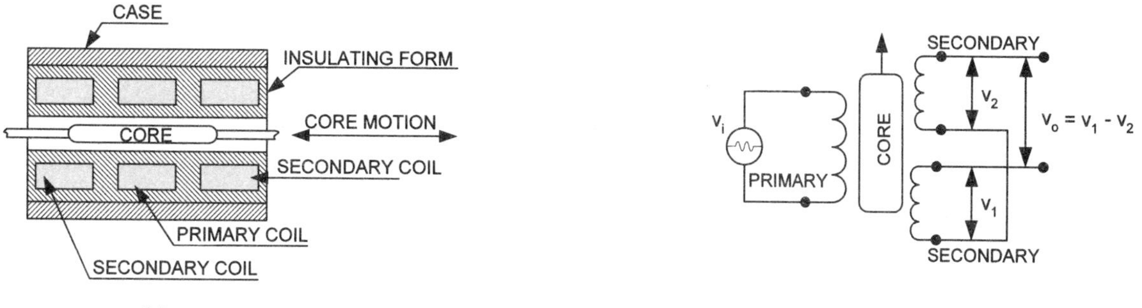

(a) (b)

Figure 5.18(a) Sectional view of a linear variable differential transformer (LVDT).
(b) Schematic circuit diagram for the LVDT coils.

The frequency of the voltage applied to the primary winding can range from 50 to 25,000 Hz. If the LVDT is used to measure transient or periodic displacements, the carrier frequency should be 10 times greater than the highest frequency component in the dynamic signal. The highest sensitivities are attained with excitation frequencies between 1 and 5 kHz. The input voltage can range from 3 to 15 V rms. The power required is usually less than 1 W. Sensitivities of different LVDTs vary from 0.02 to 0.2 V/mm of displacement per volt of excitation applied to the primary coil. At rated excitation voltages, sensitivities vary from about 30 to 350 mV per V_i – mm of displacement. The higher sensitivities are associated with short-stroke LVDTs, with an operating range of ± 0.13 mm; the lower sensitivities are for long-stroke LVDTs, with an operating range of ± 254 mm.

Because the LVDT, a passive sensor, requires ac excitation at a frequency different from common ac supplies, signal-conditioning circuits are needed for its operation. A typical signal conditioner, shown in Fig. 5.19, provides a power supply, a frequency generator to drive the LVDT, and a demodulator to convert the ac output signal from the LVDT to an analog dc output voltage. Finally, a dc amplifier is incorporated in the signal conditioner to increase the magnitude of the output voltage.

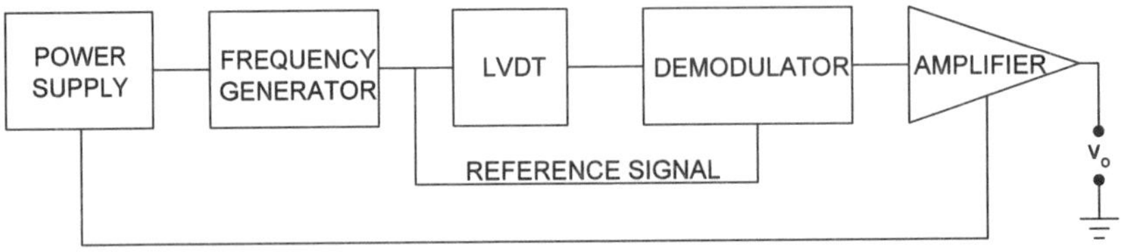

Figure 5.19 Block diagram of the signal conditioning circuit for a LVDT.

Many different types of LVDTs are commercially available today. They are designed for normal applications, for rugged environments, for high temperature (220 °C) and high-pressure measurements and to monitor motion sensitive mechanisms. Most of the LVDTs are packaged in a stainless steel tube that encases the coils and shields the electronics from noise.

With the development of microelectronics, circuits have been developed that permit miniaturization of the signal conditioner shown in Fig. 5.19. These miniaturized circuits are packaged within the case of an LVDT to produce a small self-contained sensor known as a direct current differential transformer (DCDT). A DCDT operates from a battery or a regulated power supply (± 15VDC nominal) and provides an amplified output signal that can be monitored on either a digital voltmeter (DVM) or an oscilloscope. The output impedance of a DCDT is relatively low (about 100 Ω).

The LVDT and the DCDT have many advantages as sensors for measuring displacement. There is no contact between the core and the coils; therefore, friction and hysteresis are eliminated. Because the output is continuously variable with input, resolution is determined by the characteristics of the voltage recorder. Non-contact also ensures that life will be very long with no significant deterioration of performance over this period. The small core mass and freedom from friction give the sensor a limited capability for dynamic measurements. Finally, the sensors are not damaged by over travel; therefore, they can be employed as feedback transducers in servo-controlled systems where over travel may occur due to accidental deviations beyond the control band.

In more recent years, the DCDT has been equipped with an analog to digital converter (A/D) to provide a digitized output. Their input voltage may be varied from 8.5 to 30 V, and their output signal has a resolution of at least 15 bits or 1 part in 32,768.

5.8 ACOUSTICAL STRAIN GAGES

Acoustical strain gages have been employed in a variety of forms since the late 1920s; however, they have been largely supplemented by the electrical-resistance strain gage with one important exception. They are unique among all forms of strain gages in view of their long-term stability and freedom from drift over extended time periods. The acoustical gage described here was developed in 1944 and today is still representative of the devices currently being employed. The strain-measuring system is based on the use of two identical gages identified as a test gage and a reference gage. The significant parts of a gage are shown schematically in Fig. 5.20.

In this illustration it is evident that the gage has the common knife-edges for mounting the device. One knife-edge is mounted to the main body, which is fixed, while the other knife-edge is mounted in a bearing suspension and is free to elongate with the specimen. The gage length L_0 is 3 in. (76 mm). One end of a steel wire is attached to the movable knife-edge while the other end of the wire passes through a small hole in the fixed knife-edge and is attached to a tension screw. The movable knife-edge is connected to a second tension screw by a leaf spring. This arrangement permits the initial tension in the wire to be applied without transmitting the wire preload to the knife-edges.

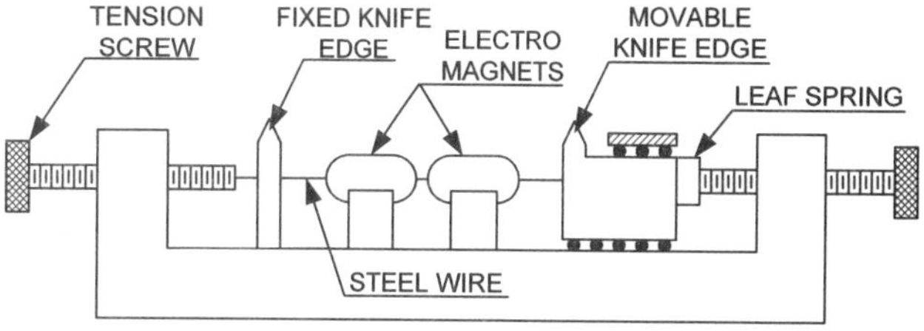

Figure 5.20 Schematic drawing illustrating design features of an acoustical strain gage.

The wire passes between the pole pieces of two small electromagnets. One of these magnets is used to keep the wire vibrating at its natural frequency; the other is employed to monitor the frequency of the system. Electrically both magnets operate together—the signal from the monitoring magnet is amplified and fed back into the driving magnet to keep the string excited in its natural frequency.

The reference gage is identical to the test gage except that the knife-edges are removed and a micrometer is used to tension the wire. A helical spring is employed in series with the wire to permit larger rotations of the micrometer head for small changes of stress in the wire.

To operate the system, the test gage is mounted and adjusted and the reference gage is placed in a location suitable for compensating for temperature effects. Both gages are energized and each wire emits a musical note. If the frequency of vibrations from the two gages is not the same, beats will occur. The micrometer setting is varied on the reference gage until the beat frequency decreases to zero. The reading of the micrometer is then taken and the strain is applied to the test gage. The change in tension in the wire of the test gage produces a change in frequencies and it is necessary to adjust the reference gage with the micrometer until the beats are eliminated. This new micrometer reading is proportional to the strain.

If the test gage is located in a remote position and the beat signals from the test and reference gages cannot be by the operator, it is possible to balance the two gages by using an oscilloscope. The voltage output from the pickup coils of each gage is displayed while operating the oscilloscope in the xy mode. The resulting Lissajous figure provides the display enabling the adjustment of the micrometer on the reference gage to match the frequency of the test gage.

The natural frequency f of a wire held between two fixed points is given by the expression:

$$f = \frac{1}{2L}\sqrt{\frac{\sigma}{\rho}} \tag{5.16}$$

where L is the length of wire between supports; σ is the stress and ρ is the density of the wire, respectively. In terms of strain in the wire, the frequency is governed by the following equation, which comes directly from Eq. (5.16):

$$f = \frac{1}{2L}\sqrt{\frac{E\varepsilon}{\rho}} \tag{5.17}$$

where E is the modulus of elasticity.

The sensitivity of this instrument is very high, with possible determinations of displacements of the order of 0.1 μin. (2.5 μm). The range is limited to about one-thousandth of the wire length before over or under stressing of the sensing wire becomes critical. The gage is temperature-sensitive unless the thermal coefficients of expansion of the base and wire are closely matched over the temperature range encountered during a test. Finally, the force required to drive the transducer is relatively large and it should not be employed in high compliance systems where the large driving force will affect its accuracy.

5.9 SEMICONDUCTOR STRAIN GAGES

The development of semiconductor strain gages was an outgrowth of research at the Bell Telephone Laboratories, which led to the introduction of the transistor in the early 1950s. Smith determined the piezoresistive properties of semiconducting silicon and germanium in 1954. Further development of semiconductor transducers by Mason and Thurston in 1957 eventually led to the commercial marketing of piezoresistive strain gages in 1960.

Basically, the semiconductor strain gage consists of a small, thin rectangular filament of single crystal silicon. The semiconducting materials exhibit a very high strain sensitivity S_A, with values ranging from 50 to 175 depending upon the type and amount of impurity diffused into the pure silicon crystal. The resistivity ρ of a single-crystal semiconductor with impurity concentrations of the order of 10^{16} to 10^{20} atoms/cm^3 is given by:

$$\rho = \frac{1}{\mathrm{eN}\mu} \tag{5.18}$$

N is number of charge carriers, which depends on the concentration of the impurity, and μ is the mobility of the charge carriers, which depends on the strain and its direction relative to the crystal axes.

Piezoresistive strain gages occupy a niche in the strain-gage market. Their advantage of a high sensitivity is balanced by several disadvantages, which include higher cost, limited range, and large temperature effects.

The importance of Eq. (5.18) can be better understood if it is considered in terms of the sensitivity S_A of a semiconductor to strain, which can be written as:

$$S_A = 1 + 2\nu + \frac{d\rho / \rho}{\varepsilon} \tag{5.19}$$

For metallic conductors, $1 + 2\nu \approx 1.6$ and ranges from 0.4 to 2.0 for the common strain-gage alloys. For semiconductor materials $(d\rho/\rho)/\varepsilon$ can be varied between −125 and +175 by selecting the type and concentration of the impurity. Thus, very high conductor sensitivities are possible where the resistance change is about 100 times larger than that obtained for the same strain with metallic-alloy gages. Also, negative gage factors are possible which permit large electrical outputs from Wheatstone-bridge circuits where multiple strain gages are employed.

The semiconductor materials have another advantage over metallic alloys for strain-gage applications. The resistivity of P-type silicon is of the order of 5,000 μΩ-cm, which is 1,000 times greater than the resistivity of Advance or Constantan, which is 49 μΩ-cm. Because of this very high resistivity, semiconductor strain gages often do not utilize grid geometries. They are usually very short single elements with leads as shown in Fig. 5.21.

In producing semiconductor strain gages, ultra pure single-crystal silicon is employed. Boron is used as the trace impurity in producing the P-type (positive gage factor) piezoresistive material. Arsenic is used to produce the N-type (negative gage factor) material. The very high sensitivity of semiconductor gages to strain and their high resistivity have led to their application in measuring extremely small strains, in miniaturized transducers and in very-high-signal-output transducers.

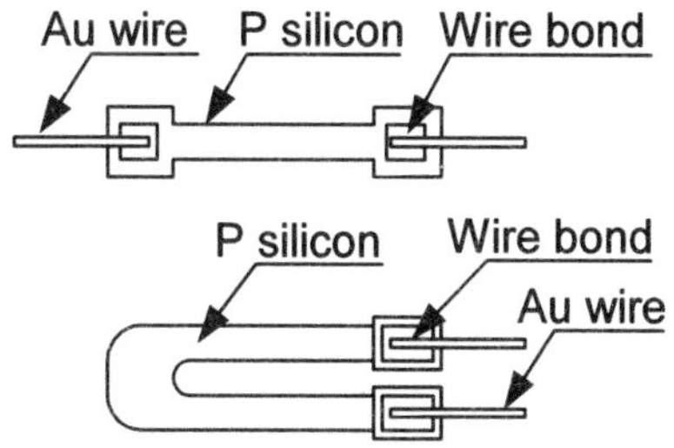

Figure 5.21 Construction details for semiconductor strain gages.

5.9.1. Piezoresistive Properties of Semiconductors

A crystal of a semiconducting material is electrically anisotropic; consequently, the relation between the potential gradient $\boldsymbol{E}$ and the current density $\boldsymbol{I}$ is formulated relative to the directions of the crystal axes to give components $\boldsymbol{E_1}$, $\boldsymbol{E_2}$, and $\boldsymbol{E_3}$ of the vector $\boldsymbol{E}$. Thus

$$\begin{aligned} \boldsymbol{E_1} &= \rho_{11}\boldsymbol{I_1} + \rho_{12}\boldsymbol{I_2} + \rho_{13}\boldsymbol{I_3} \\ \boldsymbol{E_2} &= \rho_{21}\boldsymbol{I_1} + \rho_{22}\boldsymbol{I_2} + \rho_{23}\boldsymbol{I_3} \\ \boldsymbol{E_3} &= \rho_{31}\boldsymbol{I_1} + \rho_{32}\boldsymbol{I_2} + \rho_{33}\boldsymbol{I_3} \end{aligned} \qquad (5.20)$$

where the first subscript of each resistivity coefficient indicates the component of the voltage field to which it contributes and the second identifies the component of current. The single crystal will permit isotropic conduction only if:

$$\rho_{11} = \rho_{22} = \rho_{33} = \rho$$

$$\rho_{12} = \rho_{13} = \rho_{21} = \rho_{23} = \rho_{31} = \rho_{32} = 0$$

Because this situation exists for unstressed cubic crystals, Eqs. (5.20) reduce to

$$\boldsymbol{E_1} = \rho\boldsymbol{I_1} \qquad \boldsymbol{E_2} = \rho\boldsymbol{I_2} \qquad \boldsymbol{E_3} = \rho\boldsymbol{I_3} \qquad (5.21)$$

When a state of stress is imposed on the crystal, it responds by exhibiting a piezoresistive effect that can be described by the expression:

$$\rho_{ij} = \delta_{ij}\,\rho + \pi_{ijkl}\,\tau_{kl} \qquad (5.22)$$

where the subscripts i,j,k, and l range from 1 to 3 and π_{ijkl} is a fourth-rank piezoresistivity tensor, which is a function of the crystal and the level and type of the impurities. A complete discussion of the results obtained from Eq. (5.22) is beyond the scope of this text. Fortunately, several reductions in the complexity of Eq. (5.22) are possible for the P and N doped single crystal silicon used in the fabrication of semiconductor strain gages. It can be shown that the sensitivity π_g of the gage to stress can be varied by changing the orientation of the gage axis relative to the crystal axes or by changing the doping (N or P) and the doping concentration.

The difference in voltage gradient $\Delta\boldsymbol{E}$ across the semiconductor element before and after stressing can be written as:

$$\Delta\boldsymbol{E}/\rho = \boldsymbol{I_g}(1 + \pi_g\sigma_g) - \boldsymbol{I_g} = \boldsymbol{I_g}\pi_g\sigma_g \qquad (5.23)$$

Normalizing this relationship with respect to the voltage gradient in the unstressed state gives:

$$\frac{\Delta \boldsymbol{E_g}}{\boldsymbol{E_g}} = \frac{\Delta R_g}{R_g} = \pi_g \sigma_g \tag{5.24}$$

Because a uniaxial state of stress exists in the gage element,

$$\sigma_g = E\,\varepsilon \tag{a}$$

where E is the modulus of elasticity of the semiconducting silicon and ε is the strain transmitted to the element from the specimen. Substituting Eq. (a) into Eq. (5.24) gives:

$$\Delta R_g/R_g = \pi_g E\varepsilon = S_{sc}\,\varepsilon \tag{5.25}$$

where $S_{sc} = \pi_g E$ is the strain sensitivity of the piezoresistive material due to strain-induced changes in resistivity. It should be noted that S_{sc} is sufficiently large (≈ 100) such that the net effect of changing length and cross sectional area on the sensitivity of the conducting element is very small (≈ 1.5).

5.9.2 Temperature Effects on Semiconductor Strain Gages

The results of the previous discussion indicated that the response of a semiconductor strain gage is linear with respect to strain. Unfortunately, this is a simplification that is not true in the general case. For lightly doped semiconducting materials (10^{19} atoms/cm^3 or less), the sensitivity S_A is markedly dependent upon both strain and temperature with

$$S_A = \frac{T_0}{T}(S_A)_0 + C_1\left(\frac{T_0}{T}\right)^2 \varepsilon + C_2\left(\frac{T_0}{T}\right)^3 \varepsilon^2 + \cdots \tag{5.26}$$

where $(S_A)_0$ is the room-temperature zero-strain sensitivity, as defined in Eq. (5.19).
T is the temperature, with T_0 = 294 °K.
C_1, C_2 are constants, which depend on the type of impurity, the level of doping and the orientation of the element with respect to the crystal axes.

The variation of the semiconductor sensitivity S_A as a function of doping level for P-type silicon is shown in Fig. 5.22. As the impurity concentration is increased from 10^{16} to 10^{20} atoms/cm^3, the sensitivity S_A decreases from 155 to 50. Two significant advantages related to the temperature effect compensate for this loss of sensitivity:

1. The effect of temperature on the sensitivity S_A is greatly diminished (because the temperature coefficient of sensitivity approaches zero) when the impurity concentration approaches 10^{20} atoms/cm^3, as shown in Fig. 5.22.
2. The temperature coefficient of resistance decreases from 0.009/°C for impurity concentrations of 10^{16} atoms/cm^3 to 0.00036/°C for impurity concentrations of 10^{19} atoms/cm^3.

Experiments with P-type strain gages indicate that the effects of temperature are minimized with impurity concentrations of 10^{19} atom/cm^3 that results in a sensitivity S_A = 109. However, temperature compensation of single element P-type semiconductor strain gages is not possible because the gage responds to temperature and the gage factor changes. As a result, a temperature induced apparent strain occurs.

Figure 5.22 Alloy sensitivity and temperature coefficient of sensitivity as a function of dopant concentration for P-type silicon.

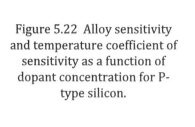

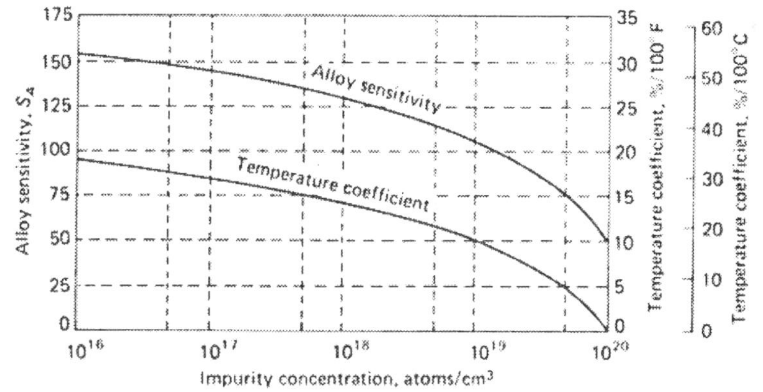

5.9.3 Linearity and Strain Limits

It is clear from Eq. (5.26) that the response of semiconductor strain gages may be nonlinear with respect to strain. For piezoresistive materials with a low impurity concentration, the non-linearity is significant. Increasing the impurity level to 10^{19} to 10^{20} atoms/cm^3 markedly improves the linearity. Unfortunately, most commercial gages are P-type with a gage factor of about 140, which corresponds to an impurity level of 10^{17} atoms/cm^3 that is below the optimum for linearity. Typical linearity specifications are $\pm$ 0.25% to 600 $\mu\varepsilon \pm$ 1.5% to 1,500 $\mu\varepsilon$.

The small elements used in fabricating semiconductor strain gages are removed from a single crystal of silicon by a slicing. The single crystal is sectioned and then sliced with a diamond saw to produce thin plates of P-type silicon with their plane in the [111] crystal axis. The thin plates are lapped and etched to eliminate flaws induced by sawing and to improve the surface finish. The small sensing elements are then etched from the plate by using photoresist images to protect the sensing elements.

While silicon exhibits a very high strength, it is glasslike and accordingly semiconductor strain gages must be treated with care during mounting. On flat surfaces, installation presents no problems, and the gages can be subjected to approximately 3,000 $\mu\varepsilon$ before the gages begin to fail by brittle fracture. When semiconductor gages are installed in a fillet and bending stresses are imposed during installation, extreme care must be exercised to avoid rupture of the element. Gages can be mounted on fillet radii from 0.1 to 0.25 in. (2.5 to 6 mm); however, the strains induced in the element during installation reduce its strain range in subsequent experiments.

The fatigue life of semiconductor gages is rated at 10^7 cycles for cyclic strains of $\pm$ 500 $\mu\varepsilon$. This is considerably less than the capability of metallic-foil-type gages; however, semiconductor strain gages are usually employed in low-magnitude strain fields, so that the relatively low strain limits placed on single-cycle and multiple-cycle measurements usually are not a significant concern.

5.9.4 Semiconductor Strain Gage Overview

Semiconductor strain gages have the advantage of a high gage factor and a high resistivity, which leads to miniaturization of the sensing element. They have the disadvantage of exhibiting significant temperature effects. The primary application of semiconductor gages is as sensors for very rigid miniature transducers which provide high output signals and very wide bandwidth. In these applications, the mechanical element in the transducer is fabricated from single crystal silicon. The semiconductor gages are usually implanted by a diffusion process into this silicon element. Semiconductor gages are less suited as sensors in the more common general purpose high-accuracy transducers that incorporate mechanical elements fabricated from metal. The use of semiconductor gages in experimental stress or fracture analysis is rare. They are applied only to measure very small strains with high accuracy under stable thermal conditions.

5.10 GRID METHOD OF STRAIN ANALYSIS

The grid method of strain analysis is one of the oldest techniques known to experimentally determine in plane displacements. The method requires placement of a grid (a series of well-defined parallel lines) on the surface of the specimen. Next, the grid is carefully photographed before and after loading the specimen to obtain records, which will show the distortion of the grid. Measurements of the distance between the grid lines before and after deformation give lengths L_i and L_f, respectively. These lengths may be interpreted in terms of strain in several different ways:

Lagrangian strain: $\varepsilon = (L_f - L_i)/L_i$ (5.27)

Eulerian strain: $\varepsilon = (L_f - L_i)/L_f$ (5.28)

Natural strain: $\varepsilon = \ln (L_f / L_i)$ (5.29)

The exact form used to determine the strain will depend upon the purpose of the analysis and the amount of deformation experienced by the model.

If the grid consists of two series of orthogonal parallel lines over the entire surface of the specimen (see Fig 5.23), the grid method will give strain components over the entire field. The strain component ε_{xx} in the horizontal direction is obtained by comparing the distances between the vertical grid lines before and after deformation. The vertical or ε_{yy} component of strain can be obtained from the horizontal array of grid lines, and the $\varepsilon_{45°}$ component of strain can be obtained from measurements across the diagonals if the orthogonal grid array is square.

In general, two main difficulties are associated with the utilization of the grid method for measuring strain. First, the strains being measured are usually quite small, and in many cases the displacement readings cannot be made with sufficient precision to keep the accuracy of the strain determinations within reasonable limits. Also, the definition of the grid lines is often poor when the image is magnified, so that appreciable errors are introduced in the displacement readings.

In order to effectively employ the grid method and avoid these two difficulties, the deformations applied to the model must be large enough to impose strain levels of the order of 5%. Usually, model deformations this large are to be avoided in an elastic stress analysis; hence other experimental methods such as the moiré method or electrical-resistance strain gages are preferred. However, when large deformations are associated with the stress-analysis problem, e.g., plastic deformations or deformations in rubber like materials, the grid method is one of the most effective methods known. In fact, if the strains exceed 10 to 30 percent, the range of the commonly employed strain gages is exceeded, and only the moiré method or the grid method is adequate for measuring these large strains.

5.10.1 Circular Spot Arrays

With significant advancements in video cameras, low cost memory and image processing methods and software, new approaches to the grid method have been developed. As an example, consider the grid produced with a periodic array of black spots on a white background as illustrated in Fig. 5.24. Note that the pattern includes a single T mark that is used to indicate rigid-body rotations.

The array of circular dots is placed on the surface of a plane specimen and digital images of the pattern are made and stored during the loading process. Comparison of the digital images from two different loads, known as pattern matching, is simply the modern adaptation of the well-established grid method described previously. The new features are the digital-image acquisition hardware and the image-analysis procedures used in tracking the displacement fields.

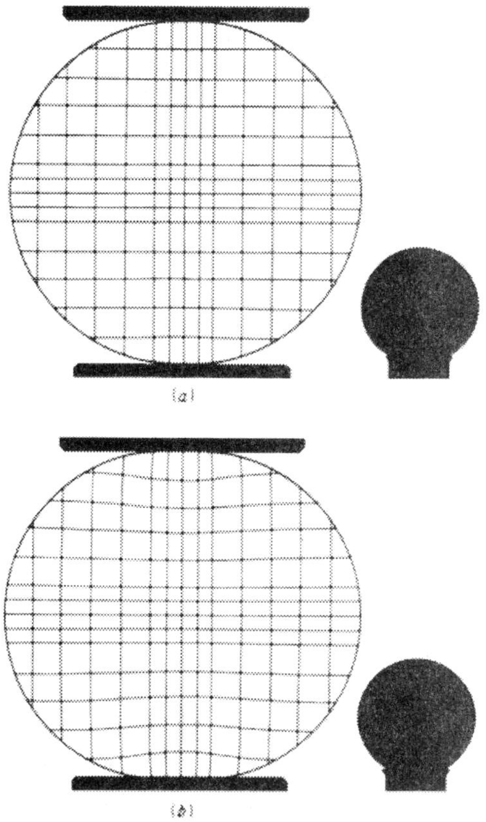

Figure 5.23 An embedded rubber-thread grid in a transparent urethane model of a circular disk subjected to diametrical compression: (a) Before deformation; (b) After deformation.

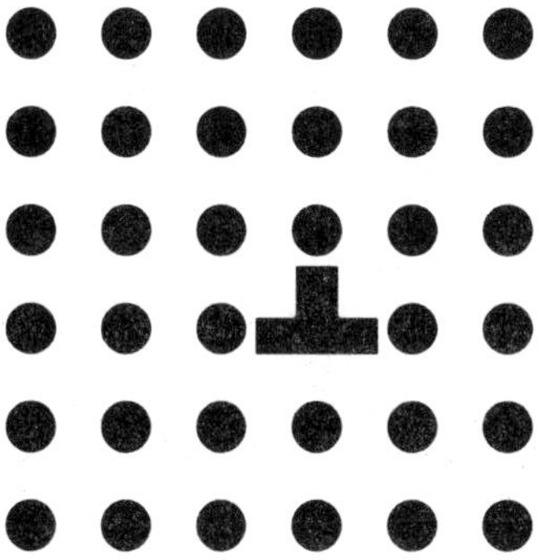

Figure 5.24 An array of black circular dots on a white background.

The digital-image acquisition hardware includes a video camera and a digitizing (A to D) circuit board that converts the camera signals into digital format and stores the intensity level for each picture element (pixel) in computer memory. The size and shape of the pixels depend upon the video camera. For this discussion consider that the pixels are rectangular in shape and arranged in an imaging matrix as illustrated in Fig. 5.25. The photosensitive portion of the pixel is shown as the crosshatched area A_s.

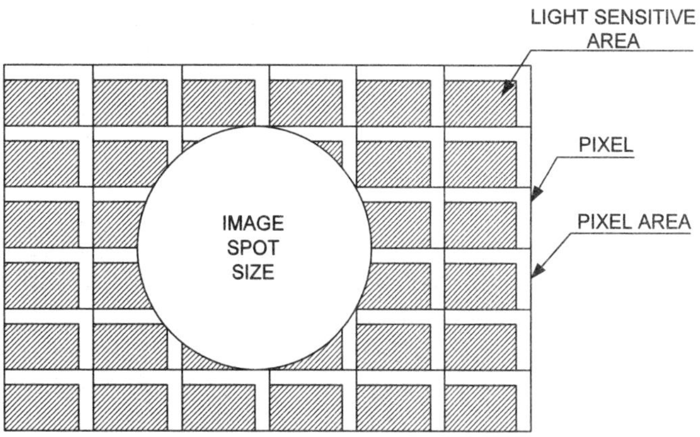

Figure 5.25 Imaging matrix showing the pixel area A_p, the light sensitive area A_s and a circular spot.

The intensity I_{kl} registered at each pixel location is given by:

$$I_{kl} = \frac{1}{A_s}\int I(x, y) \tag{5.30}$$

where k and l locate the pixel in the image matrix and I(x, y) is the intensity of light on the pixel.

The intensity I_{kl} is an analog signal, which is converted to a digital value called a gray level. The gray level G_{kl} is an integer number ranging from 0 to 255 for an 8-bit A/D converter. The gray level G_{kl} is related to a registered intensity at each pixel location by the equation:

$$G_{kl} = INT(I_{kl}) \tag{5.31}$$

where INT is an operator, which rounds I_{kl} downward to the nearest integer between 0 and 259.

Now consider the intensity of light I(x,y) for the grid spot projected over a region of pixels as indicated in Fig. 5.25. For this configuration, it is clear that

$$I(x, y) = \begin{cases} G_D & \text{for } r \le R \\ 0 & \text{for } r > R \end{cases} \tag{5.32}$$

where R is the radius of the spot and $r = \sqrt{x^2 + y^2}$

G_D is the gray level difference between the black spot and the white (0) background.

If the spot completely covers the pixel, $I_{kl} = G_D$. If the pixel is located outside the spot, $I_{kl} = 0$. However, if the spot partially covers the pixel, I_{kl} is obtained by integrating Eq. (5.36). The gray levels G_D are stored for each pixel location. With typical 1392 × 1024 arrays, an 8-bit word corresponding to 256 gray levels is stored for each of the 1.425×10^6 pixel locations.

The next step in the image analysis is to analyze this image to determine the position of the centroid for each circular spot. The centroid for a typical spot is determined from:

$$\bar{y} = \frac{\sum_{k=k_0}^{k_f}\sum_{l=l_0}^{l_f} k(T - G_{kl})}{\sum_{k=k_0}^{k_f}\sum_{l=l_0}^{l_f} (T - G_{kl})}$$

$$\bar{x} = \frac{\sum_{k=k}^{k_f}\sum_{l=l_0}^{l_f} l(T - G_{kl})}{\sum_{k=k_0}^{k_f}\sum_{l=l_0}^{l_f} (T - G_{kl})} \tag{5.33}$$

where k_0, k_f, l_0, and l_f define a window of pixels including the grid point and T is a threshold level. Only those pixels with $G_{kl} < T$ are included in computing the centroid locations given by Eq. (5.33).

For a given circular spot, analyses of the two images yield

$$u = \bar{x}_2 - \bar{x}_1$$
$$v = \bar{y}_2 - \bar{y}_1 \qquad (5.34)$$

where subscripts 1 and 2 refer to two load levels. Because a large number of points can be employed over the surface of the body, the method can be classified as full field.

The use of Eq. (5.34) assumes that rigid body rotations do not occur between the two load levels. If rigid body rotations do occur it is necessary to analyze the T mark. In this image analysis, the directions of the principal inertia axes are determined for each load level to give θ_{y1}, and θ_{y2}. The rigid body rotation ω, defined in Fig. 5.26, is given by:

$$\omega = \theta_{y2} - \theta_{y1} \qquad (5.35)$$

If ω is not zero, it is necessary to adjust both $\bar{x}_2$ and $\bar{y}_2$ for each circular spot to account for the displacements due to rigid-body rotation.

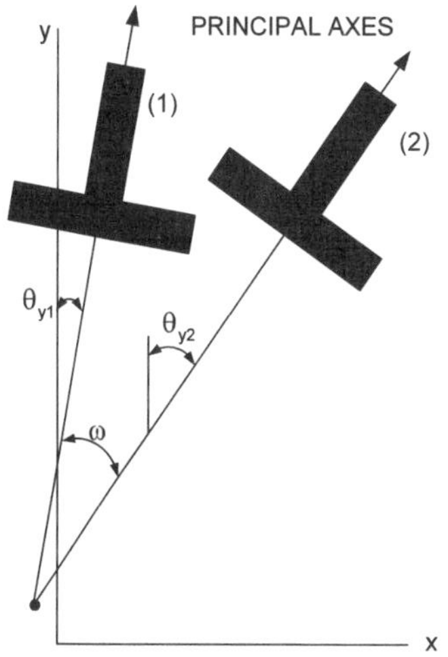

Figure 5.26 Image of the T pattern used to determine rigid body rotations.

Sirkis has examined the parameters which affect the accuracy of this method, which include the fill factor, the size of the circular spot T, the gray level difference G_D, and signal to noise ratio. The fill factor is defined as the ratio of the sensitive area A_s to the entire area A_p of the pixel. Clearly, as the sensing arrays in video cameras improve [1], the ratio A_s/A_p will increase and the error due to the fill factor will decrease.

The spot radius markedly affects the error as indicated in Fig. 5.27. Large spot radii, 10 to 20 times the pixel size, are necessary to limit error. As the fill factor increases, the spot size required for a given error decreases. The gray level difference G_D should be maximized. This is controlled by the operator in forming the grid (contrast), lighting, adjusting the gain and offset of the A/D converter and the number of bits for gray level distinction.

Noise, like fill factor, is dependent upon equipment. The noise can be reduced with image averaging, electromagnetic isolation and cooling the CCD array in the camera. With careful techniques and an imaging system with high signal to noise ratio (200), it is possible to limit error to $\pm$ 0.05 pixels. The corresponding surface displacement depends upon the magnification of the optical system used to project the image of the circular spots onto the sensing array of the camera. The time required to

[1] To improve the sensitivity and fill factors for CCD arrays, Eastman Kodak Co. has recently introduced an array that incorporates a stacked micro-lenticular array. This is an array of cylindrical lens that collect the incoming light and focus it on the sensitive areas of the pixel thereby increasing the effective fill ratio of the CCD array.

process an array of 1,000 circular spots is negligible when a modern PC, with 3 GHz processor speed and several GB of RAM memory, is used to process the image.

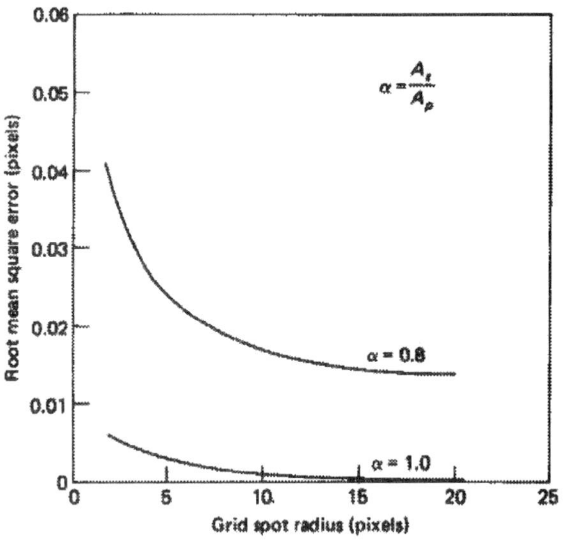

Figure 5.27 Error as a function of spot size. (After Sirkis)

5.11 EDDY CURRENT SENSORS

An eddy current sensor measures distance between the sensor head and an electrically conducting surface, as illustrated in Fig. 5.28. Sensor operation is based on eddy currents that are induced at the conducting surface as magnetic flux lines from the sensor intersect with the surface of the conducting material. The magnetic flux lines are generated by the active coil in the sensor, which is driven at a very high frequency (1 MHz). The magnitude of the eddy current produced at the surface of the conducting material is a function of the distance between the active coil and the surface. The eddy currents increase as the distance decreases.

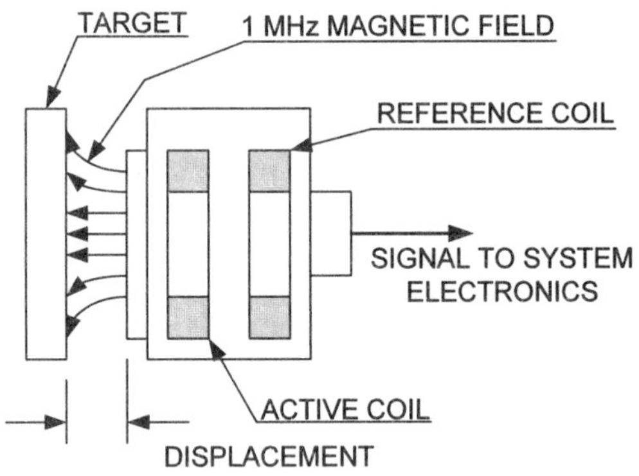

Figure 5.28 Schematic diagram for an eddy current sensor..

Changes in the eddy currents are sensed with an impedance (inductance) bridge. Two coils in the sensor are used for two arms of the bridge. The other two arms are housed in an associated electronic package illustrated in Fig. 5.29. The first coil in the sensor (active coil), which changes inductance with target movement, is wired into the active arm of the bridge. The second coil is wired into an opposing arm of the same bridge, where it serves as a compensating coil to balance and cancel the effects of temperature change. The output from the impedance bridge is demodulated and becomes the analog signal, which is linearly proportional to distance between the sensor and the target. This signal is then amplified prior to recording on some analog recorder or conversion to a digital signal.

The sensitivity of the sensor is dependent upon the target material with higher sensitivity associated with higher conductivity materials. For aluminum targets, the sensitivity is typically 100 mV/mil (4 V/mm). Thus, it is apparent that eddy current sensors are high-output devices if the specimen material is non-magnetic; however, the sensitivity decreases significantly if the specimen material is magnetic. While the sensitivity is dependent on the resistivity (conductivity) and the magnetism of the target material, this fact does not affect the accuracy of the sensor because it is easily calibrated for the specific target material.

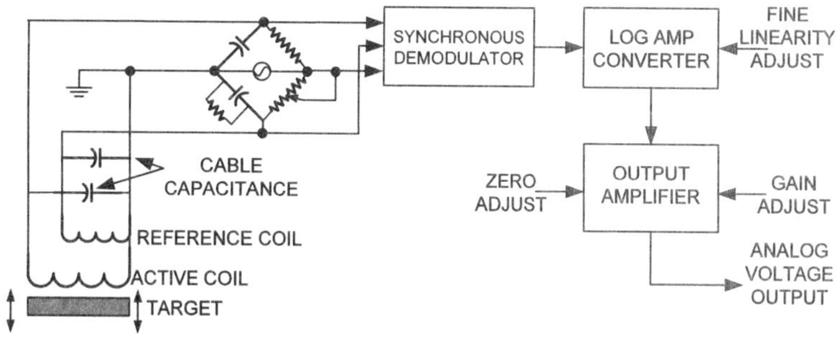

Figure 5.29 Block diagram of the signal condition circuits used in converting impedance changes due to target displacement to an analog output voltage.

For non-conducting, poorly conducting or magnetic materials, it is possible to bond a thin film of aluminum foil to the surface of the target at the location of the sensor to improve the sensitivity. Because the penetration of the eddy currents into the material is minimal, the thickness of the foil can be as small as 0.2 mm (ordinary kitchen type aluminum foil).

The effect of temperature on the output of the eddy current sensor is small. The sensing head with dual coils is temperature compensated; however, a small error can be produced by temperature changes in the target material because the resistivity of the target material is a function of temperature. For instance, if the temperature of an aluminum target is increased by 500 °F, its resistivity increases from 3 to 6 μΩ-cm. and the bridge output is reduced by about 2 percent for this change in resistivity. For aluminum, the temperature sensitivity of the eddy current sensor is 0.004% °F.

The range of the sensor is controlled by the coil diameter with the larger sensors exhibiting the larger ranges. The range to diameter ratio is usually about 0.25. Linearity is typically better than ± 0.5 percent and resolution is better than 0.05 percent of full scale. The frequency response is typically 20 kHz, although small-diameter coils can be used to increase this response to 50 kHz.

The fact that eddy current sensors do not require contact for measuring displacement is quite important. As a result of this feature, they are often used in transducer systems for automatic control of dimensions in fabrication processes. They are also applied extensively to determine thickness of organic coatings that are non-conducting.

5.12 PIEZOELECTRIC SENSORS

A piezoelectric material, as its name implies, produces an electric charge when it is subjected to a force or pressure. Piezoelectric materials, such as single-crystal quartz or polycrystalline barium titanate, contain molecules with asymmetrical charge distributions. When pressure is applied, the crystal deforms and there is a relative displacement of the positive and negative charges within the crystal. This displacement of internal charges produces external charges of opposite sign on the external surfaces of the crystal. If these surfaces are coated with metallic electrodes, as illustrated in Fig. 5.30, the charge q that develops can be determined from the output voltage v_o because:

$$q = v_o C \tag{5.36}$$

where C is the capacitance of the piezoelectric crystal.

The surface charge q is related to the applied pressure p by:

$$q = S_q A p \tag{5.37}$$

where S_q is the charge sensitivity of the piezoelectric crystal and A is the area of the electrode.

The charge sensitivity S_q is a function of the orientation of the sensor (usually a cylinder) relative to the axes of the piezoelectric crystal. Typical values of S_q for common piezoelectric materials are given in Table 5.1.

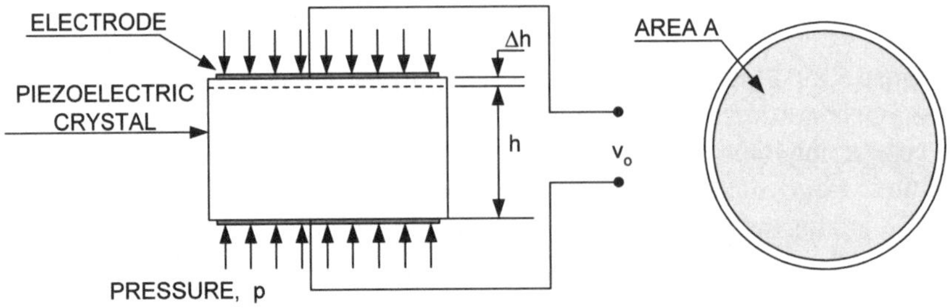

Figure 5.30 Piezoelectric crystal deforming under the action of an applied pressure.

The output voltage v_o developed by the piezoelectric sensor is obtained by substituting Eqs. (5.11) and (5.37) into Eq. (5.36). Thus:

$$v_0 = \left(\frac{S_q}{kK}\right) h\, p \qquad \text{(a)}$$

The voltage sensitivity S_v of the sensor can be expressed as:

$$S_v = \left(\frac{S_q}{kK}\right) \qquad (5.38)$$

The output voltage v_o of the sensor is then:

$$v_0 = S_v h\, p \qquad (5.39)$$

Again, voltage sensitivity S_v of the sensor is a function of the orientation of the axis of the cylinder relative to the crystallographic axes. Typical values of S_v are also presented in Table 5.1.

Table 5.1
Typical charge S_q and voltage S_v sensitivities of piezoelectric materials

Material	Orientation	S_q (pC/N)	S_v (V-m/N)
Quartz SiO_2 Single Crystal	X-cut, length longitudinal	2.2	0.055
	X-cut, thickness longitudinal	– 2.0	– 0.05
	Y-cut, thickness shear	4.4	0.11
Barium Titanate $BaTiO_3$ Ceramic Poled	Parallel to polarization	130	0.011
	Perpendicular to polarization	– 56	– 0.004

Most piezoelectric transducers are fabricated from single-crystal quartz because it is the most stable of the piezoelectric materials, and is nearly loss free both mechanically and electrically. Its properties are: modulus of elasticity 86 GPa, resistivity 10^{14} Ω-cm, and dielectric constant 40.6 pF/m. It exhibits excellent high-temperature properties and can be operated up to 550 °C. The charge sensitivity of quartz is very low when compared to barium titanate; however, with high-gain charge amplifiers available for processing the output signal, the lower sensitivity is not a serious disadvantage.

Barium titanate is a polycrystalline material that can be polarized by applying a high voltage to the electrodes while the material is at a temperature above the Curie point (125 °C). The electric field aligns the ferroelectric domains in the barium titanate and it becomes piezoelectric. If the polarization voltage is maintained while the material is cooled well below the Curie point, the piezoelectric characteristics become permanent and stable after a short aging period.

The mechanical stability of barium titanate is excellent; it exhibits high mechanical strength and has a high modulus of elasticity (120 GPa). It is more economical than quartz and can be fabricated in a wide variety of sizes and shapes. While its application in transducers is second to quartz, it is frequently used in ultrasonic applications as a driver. In this application, a voltage is applied to the electrodes and the barium titanate deforms and delivers energy to a work piece or test specimen.

Most sensors exhibit relatively low output impedance (in the range of 100 to 1,000 Ω). However, when piezoelectric crystals are used as the sensing elements in transducers, the output impedance is usually extremely high. The output impedance of a small cylinder of quartz depends the geometry of the crystal and on the frequency ω associated with the applied pressure. Because the sensor acts like a capacitor, the output impedance is given by:

$$Z_C = \frac{1}{j\omega C} = -\frac{j}{\omega C} \qquad \text{(a)}$$

Clearly, the impedance ranges from infinity for static applications to about 10 kΩ for very high-frequency applications (100 kHz). With this high output impedance, care must be exercised in monitoring the output voltage; otherwise, very serious errors can occur.

A circuit diagram of a typical system used to measure a voltage produced by a piezoelectric sensor is shown in Fig. 5.31. The piezoelectric sensor acts as a charge generator. In addition to the charge generator, the sensor is represented by parallel components including a capacitor C_p (about 10 pF) and a leakage resistor R_p (about 10^{14} Ω). The capacitance of the lead wires C_L must also be considered, because even relatively short lead wires have a capacitance larger than the sensor. The amplifier is either a cathode follower or a charge amplifier with sufficient input impedance to isolate the piezoelectric sensor in spite of its very large output impedance. If a pressure is applied to the sensor and maintained for a long period of time, the charge q developed by the piezoelectric material leaks because a small current flows through both R_p and the amplifier resistance R_A. The time available for readout of the signal depends upon the effective time constant τ_e of the circuit that is given by:

$$\tau_e = R_e C_e = \frac{R_p R_A}{R_p + R_A}(C_p + C_L + C_A) \qquad (5.40)$$

where R_e is the equivalent resistance of the circuit and C_e is the equivalent capacitance of the circuit.

Time constants ranging from 1,000 to 100,000 s can be achieved with quartz sensors and commercially available charge amplifiers. These time constants are sufficient to permit measurement of quantities that vary slowly with time or measurement of static quantities for short periods of time. Problems associated with measuring the output voltage diminish as the frequency of the mechanical excitation increases and the output impedance Z_C of the sensor decreases.

The inherent dynamic response of the piezoelectric sensor is very high, because the resonant frequency of the small cylindrical piezoelectric element is so large. The resonant frequency of the transducer depends upon its mechanical design as well as the mass and stiffness of the sensor. However, the most significant advantage of the piezoelectric sensor is its very high-frequency response.

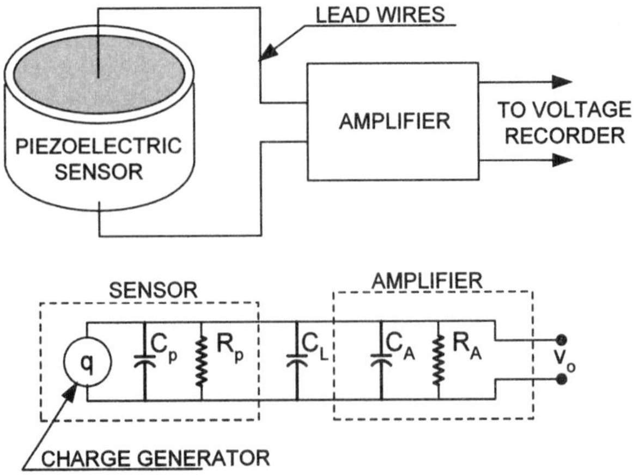

Figure 5.31 Schematic diagram of a measuring system with a piezoelectric sensor.

5.13 PHOTOELECTRIC SENSORS

In many applications, where contact cannot be made with the object being examined, a photoelectric sensor can often be employed to monitor changes in the intensity of light, that sense the quantity being measured. When light impinges on a photoelectric sensor, it either creates or modulates an electrical signal. Most photoelectric devices employ semiconductor materials and operate by either generating a current or by changing resistivity. These devices are photo detectors that respond quickly to changes in light intensity.

5.13.1 Photoconducting Sensors

Photoconductive cells, illustrated in Fig. 5.32, are fabricated from semiconductor materials, such as cadmium sulfide (CdS) or cadmium selenide (CdSe), which exhibit a strong photoconductive response. When a photon with sufficient energy strikes a molecule of, say CdS, an electron is driven from the valence band to the conduction band and a hole or vacancy remains in the valence band. This hole and the electron both serve as charge carriers, and with continuous exposure to light, the concentration of charge carriers increases and the resistivity decreases. A circuit used to detect the resistance change ΔR of the photoconductor is also shown in Fig. 5.32. The resistance of a typical photoconducting detector changes over about 3 orders of magnitude as the incident radiation varies from dark to very bright.

When a photoconductor is placed in a dark environment, its resistance is high and only a small dark current flows. If the sensor is exposed to light, the resistance decreases significantly (the ratio of maximum to minimum resistance for R_d in Fig. 5.32 ranges from 100 to 10,000 in common commercial sensors); therefore, the output current can be quite large. The sensitivity depends on cell area, type of cell (CdS or CdSe), and the power limit for the cell. If the maximum voltage v_i is used to supply the cell at its maximum power limit, then sensitivities of up to 0.2 mA/lx result for CdS type cells. A typical cadmium sulfide photoconductor exhibits a maximum dark resistance of about 1 to 2 MΩ. When subjected to a light intensity of about 2 foot-candles the resistance decreases to about 1 to 5 kΩ. The change of about three orders of magnitude is large enabling many applications with very simple control circuits.

Photoconductors respond to radiation ranging from long thermal radiation through the infrared, visible, and ultraviolet regions of the electromagnetic spectrum. The sensitivity S, changes significantly with wavelength and drops sharply at both short and long wavelengths; consequently, photoconductive cells exhibit the same disadvantage as most other photodetectors because its calibration constant depends on the wave length of the impinging light.

The photocurrent requires some time to develop after the excitation is applied and some time to decay after the excitation is removed. The rise and fall time for commercially available photoconductors is usually about a second. Because of these delays, the CdS and CdSe photoconductors are not suitable for dynamic measurements. Instead, their simplicity, high sensitivity and low cost lend them to applications involving counting and switching based on a slowly varying light intensity.

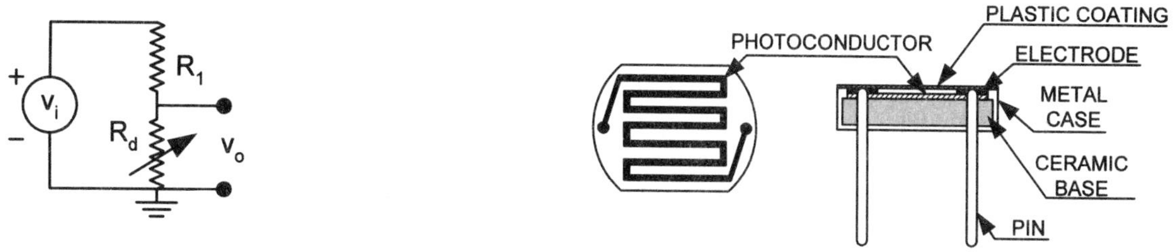

Figure 5.32 When a photoconductor is used as R_d, the output voltage will vary with the light intensity.

5.13.2 Photodiode Sensors

Photodiodes are semiconductor devices that are responsive to high-energy particles and photons. Photodiodes operate by absorption of photons or charged particles and generate a current that flows in an external circuit, which is proportional to the incident power. Photodiodes can be used to detect the presence or absence of minute quantities of light and can be calibrated for extremely accurate measurements from intensities below 1 pW/cm^2 to intensities above 100 mW/cm^2. Silicon is the most common semiconductor material used in fabricating planar diffused photodiodes. These photodiodes are employed in such diverse applications as spectroscopy, photography, analytical instrumentation, optical position sensors, beam alignment, surface characterization, laser range finders, optical communications and medical imaging instruments.

Planar diffused silicon photodiodes are P-N junction diodes. A P-N junction can be formed by diffusing either a P-type impurity (anode), such as boron, into a N-type bulk silicon wafer, or a N-type impurity, such as phosphorous, into a P-type bulk silicon wafer. The diffused area defines the photodiode active area. To form an ohmic contact, it is necessary to diffuse another impurity into the backside of the wafer. The impurity is an N-type for P-type active area and P-type for an N-type active area. Then contact pads are deposited on the front active area on defined areas, and on the backside, completely covering the device. The active area is then covered with an anti-reflection coating to reduce the reflection of the light for a specific predefined wavelength. The non-active area on the top is covered with a thick layer of silicon oxide. A schematic illustration of a planar diffused photodiode fabricated from N-type silicon is presented in Fig. 5.33.

Figure 5.33 The diffusion areas used to create a P-N junction in a N-type silicon photodiode.

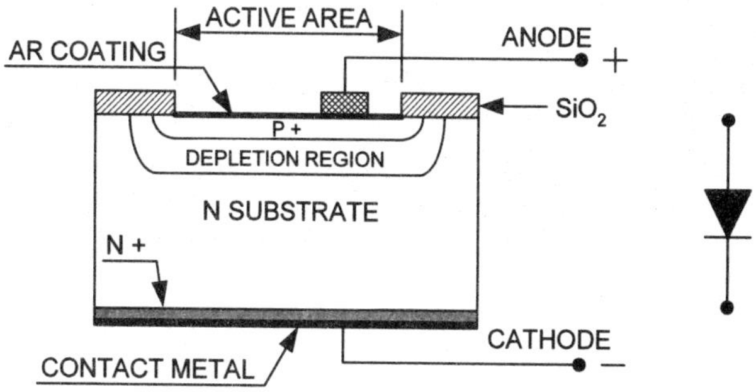

By controlling the thickness of bulk substrate, the speed and responsivity of the photodiode can be controlled. Note that the photodiodes, when biased, are operated in the reverse bias mode, i.e. a negative voltage applied to anode and positive volt-age to cathode.

Electrical Characteristics of Photodiodes

A silicon photodiode can be represented by a current source in parallel with an ideal diode as shown in Fig. 5.34. The current source represents the current generated by the incident light, and the diode represents the P-N junction. In addition, a junction capacitance (C_J) and a shunt resistance (R_{SH}) are in parallel with the other components. A resistance (R_S) is connected in series with all components in this model.

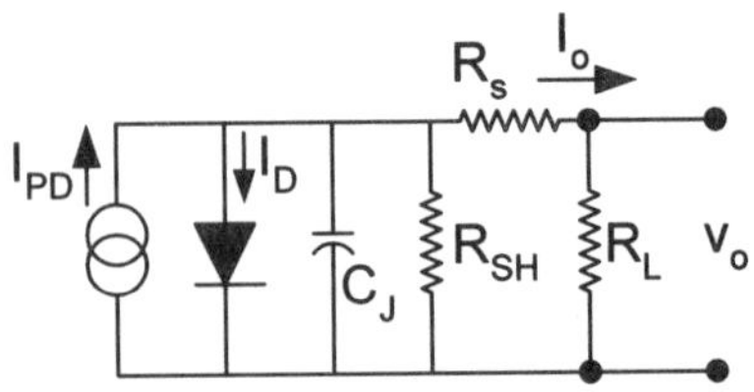

Figure 5.34 Representative circuit for a photodiode.

Physically the shunt resistance is the slope of the current-voltage curve of the photodiode at the origin, i.e. V=0. Although an ideal photodiode should have an infinite shunt resistance, actual values range from 10s to several 1,000 MΩ. Experimentally the shunt resistance is determined by applying ± 10 mV across the diode, measuring the resulting current and calculating the resistance from Ohm's law. Shunt resistance is used to determine the noise current in the photodiode with no bias (photovoltaic mode). For superior photodiode performance, a very high shunt resistance should be specified.

The series resistance of a photodiode arises from the resistance of the contacts and the resistance of the undepleted silicon shown in Fig. 5.33. The series resistance is given by:

$$R_s = \frac{(h_s + h_d)\rho}{A} + R_C \tag{5.41}$$

where h_s is the thickness of the substrate and h_d is the thickness of the depleted region
A is the diffused area of the junction
ρ is the resistivity of the substrate
R_C is the contact resistance.

The series resistance is used to determine the linearity of the photodiode in photovoltaic mode[2]. Although an ideal photodiode should have zero series resistance, typical values ranging from 10 Ω to as much as 1,000 Ω are measured.

The junction capacitance C_J is due to the geometry of the planar structure of the photodiode. The boundaries of the depletion region, shown in Fig. 5.33, act as the plates of a parallel plate capacitor. The junction capacitance is directly proportional to the diffused area and inversely proportional to the thickness of the depletion region. Higher resistivity substrates have lower junction capacitance because of the larger voltage drop across the substrate. Furthermore, the capacitance is dependent on the magnitude of the reverse bias voltage. As an example, a typical p-i-n diode exhibits C_J = 260 pF with a reverse bias voltage of 1 V; however, this capacitance decreases to 30 pF when the reverse bias voltage is increased to 100 V. The junction capacitance is very important in dynamic measurements of light intensity because it markedly affects the response time of the photodiode.

The rise time t_r of a photodiode is defined as the time for the signal to rise or fall from 10% to 90% or from 90% to 10% of the final value, respectively. This parameter can be expressed in term of the frequency response f_r (the frequency at which the photodiode output decreases by 3dB). The rise time may be approximated by:

$$t_r = 0.35/f_r \tag{5.42}$$

Circuits showing photodiodes operating in the photoconduction mode and in the photovoltaic mode are presented in Fig. 5.35.

---

[2] Photodiodes act as photovoltaic deices when no bias voltage is applied.

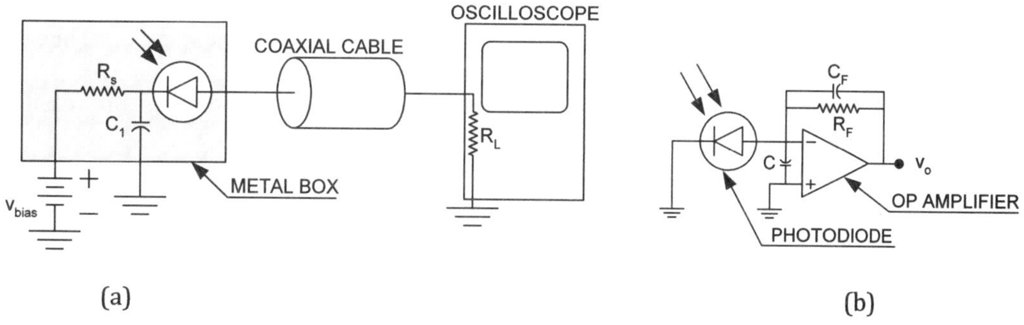

Figure 5.35 Circuits used with photodiodes. (a) High intensity light at high frequency. (b) Low intensity light with low frequency.

Continued improvements in semiconductor photodiodes expand the sensors that are commercially available. New products include a line of a large numbers of photodiodes to permit measurement of the variation of light at discrete points along a line. Also area arrays of photodiodes are used in digital cameras. These new developments have greatly expanded the number of applications for photodiodes in the measurement of full field light patterns.

EXERCISES

5.1 The stress distribution in a thin wide steel plate (E = 200 GPa and ν = 0.30) with a central circular hole is given by Eqs. (3.42) when the plate is subjected a uniaxial tensile or compressive load. Determine the error made in the determination of the maximum stress σ_{max} on the boundary of the hole if a strain gage having a gage length L_0 = 3 mm is used. The diameter of the hole in the plate is 10 mm.

5.2 Determine the error associated with measurements of ε_{yy} along the longitudinal axis of the plate of Exercise 5.1 if gages with $L_0/a = 0.1$ are located at y/a = 1.5, 2.0, and 3.0.

5.3 Determine the error associated with measurements of ε_{yy} along the transverse axis of the plate of Exercise 5.1 if gages with $w_0/a = 0.1$ are located at x/a = 1.5, 2.0, and 3.0.

5.4 For the plate of Exercise 5.1, determine the error associated with measurements of ε_{yy} along the transverse axis of the plate due to both gage width and gage length. Take $w_0/a = 0.1$ and $L_0/a = 0.1$ and consider gages located at x/a = 1.5, 2.0, and 3.0.

5.5 Consider a strain field given by the expression:

$$\varepsilon_{xx} = b \sin(\pi x/a)$$

Determine the error made in the measurement of ε_{xx} at x = 0, a/4, and a/2 as a function of L_0/a. Plot this error for $0.01 \leq L_0/a \leq 1$.

5.6 Consider a strain field given by the expression:

$$\varepsilon_{xx} = b + b \sin(\pi x/a)$$

Determine the error made in the measurement of ε_{xx} at x = a/2 when $L_0/a = 1/4$. Compare with the results from Exercise 5.5.

5.7 Consider a strain field given by the expression:

$$\varepsilon_{xx} = b + bx + b \sin(\pi x/a)$$

Determine the error made in the measurement of ε_{xx} at x = a/2 when $L_0/a = 1/4$. Compare with the results from Exercises 5.5 and 5.6. What is the effect of adding constant and linear contributions to the strain field on the overall error?

5.8 With the advent of digital data processing where varying calibration constants are easily programmed, why is it still important that a strain gage should respond linearly.

5.9 Justify in a single paragraph each of the ten characteristics that are used to judge the adequacy of a strain gage system.

5.10 Describe the design features of the extensometer shown in Fig. 5.2 and indicate the importance of each of these features to an individual using the extensometer to determine stress-strain curves in a tension test of round bar specimens.

5.11 A helium-neon laser (λ = 632.8 nm) is used to illuminate a diffraction type strain gage with a gage length of 20 mm. The diffraction pattern is displayed on a screen, which is located 3 m from the aperture. The initial aperture width b is 0.1 mm.

(a) Determine the density n/y of the diffraction pattern.
(b) If the gage is subjected to a strain of 1,000 $\mu\varepsilon$, what is the new aperture width and the new density n/y_1?
(c) Suppose the diffraction pattern is sufficiently well defined for the +8 and −8 orders of extinction to be clearly observed before and after subjecting the gage to the strain. If distances y_0 and y_1 can be determined from scale measurements on the screen to ± 1 mm, estimate the percent error in the measurement of strain.

5.12 Repeat Exercise 5.11 if the initial aperture width b = 0.05 mm.
5.13 Derive Eq. (5.9).
5.14 Derive Eq. (5.10).
5.15 Cite the advantages of measuring strain with fiber-optic strain gages.
5.16 Cite the disadvantages of measuring strain with fiber-optic strain gages.
5.17 Sketch the optical arrangement for the Mach-Zehnder fiber-optic sensor.
5.18 Sketch the optical arrangement for the Michelson fiber-optic sensor.
5.19 Sketch the optical arrangement for the Fabry-Perot fiber-optic sensor.
5.20 Describe the method for monitoring the output of a fiber-optic sensor incorporating a Bragg diffraction grating.
5.21 Describe the methods for interpreting the output from a Fabry-Perot fiber optic sensor.
5.22 Compare the electrical resistance strain gage with a Fabry-Perot fiber optic strain gage. Cite an application where you would specify a Fabry-Perot fiber optic sensor and recommend against using an electrical resistance strain gage.
5.23 A capacitance strain gage (see Fig. 5.14) has a gage length L_0 = 20 m and h = 0.10 mm. Construct a graph of capacitance C versus strain ε as the strain is varied from 0 to 100%.

(a) Is the output from the gage linear over the entire range?
(b) If not, what is the linear range?
(c) Why was the maximum value in part (b) selected as the limit of the linear range?

5.24 Repeat Exercise 5.23 but construct a graph of change in impedance $\Delta Z_c \backslash Z_c$ with strain over the same range. Comment on the effect of measuring reactance Z_c instead of capacitance C.
5.25 Prepare a graph of the sensitivity S of a capacitance sensor as a function of frequency ω. Assume the dielectric in the gap is air and consider probe diameters of 1, 2, 5 and 10 mm.
5.26 Write an engineering brief describing the advantages and disadvantages of capacitance sensors.
5.27 Describe the operating principles of the instrument, shown in Fig. 5.16, which is used to monitor the output of a capacitance sensor.
5.28 Design a strain extensometer, incorporating a commercially available DCDT, which can be utilized with a computer controlled data logging system to generate stress-strain curves automatically during tensile tests of standard ASTM specimens of 1020 carbon steel in a universal testing machine. Select the range of the DCDT, specify the input voltage, decide on the gage length of the

extensometer, design the linkage (the specimen may fracture with the extensometer in place), determine the output voltage as a function of stress on the specimen, and specify the characteristics of the data logging system to be used to plot the stress-strain curves automatically.

5.29 Prepare a block diagram representing the electronic components in a LVDT. Describe the function of each component.

5.30 Prepare a sketch of the output signal as a function of time for an LVDT with its core located in a fixed off-center position if:

(a) The demodulator is functioning.
(b) The demodulator is removed from the circuit.

5.31 Prepare a sketch of the output signal as a function of time for an LVDT with its core moving at constant velocity from one end of the LVDT through the center to the other end if:

(a) The demodulator is functioning.
(b) The demodulator is removed from the circuit.

5.32 Describe the basic differences between an LVDT and a DCDT.

5.33 Design an acoustical strain gage, which can be installed in a concrete dam and monitored over the life (estimated to exceed two centuries), of the dam. The gage will be monitored periodically to record the change in strain with rising and falling head, to record any change in effective modulus of the concrete due to cracking or other deterioration of the concrete, and to estimate damage to the structure after any natural occurrence such as an earthquake. Items to be considered in the design include selection of materials for the components of the gage, gage length, wire size, wire type, and frequency range during operation. Estimate the accuracy of the strain measurement and comment on the ability of the gage to detect structural damage or deterioration of the concrete.

5.34 Show the matrix, which represents the 81 terms in the fourth-rank piezoresistivity tensor π_{ijkl} given in Eq. (5.22).

5.35 Show that Eq. (5.22) reduces Eq. (5.23) when the tensors ρ_{ij} and τ_{ij} are symmetrical and that the piezoresistive coefficients are reduced to the three terms defined in the simplified π matrix.

5.36 One of the advantages of semiconductor strain gages is related to the fact that both positive and negative gage factors can be achieved with P- and N-type silicon. Show the gain in sensitivity, based on gage factor, achieved in the design of a beam type deflection transducer when four gages (two of the P-type and two of the N-type) are employed.

5.37 Outline the two methods that can be used to compensate for temperature induced apparent strains in semiconductor strain gages.

5.38 Describe the precautions that must be taken in installing and cycling semiconductor strain gages due to their inherently brittle characteristics.

5.39 Determine the power that can be dissipated by a semiconductor strain gage 1.5 mm long. If the gage resistance is 500 Ω, determine the current passing through the gage and the voltage drop across the gage when the maximum power is being dissipated.

5.40 Describe the difference among Lagrangian strain, Eulerian strain and natural strain.

5.41 Suppose you had a high quality CCD camera and were required to measure strain in a large flat panel subjected to axial loading. How would you use the circular spot arrays to determine the strain field at critical locations on the plate?

5.42 Suppose you are to purchase a CCD camera to be used in strain measurement made on flat plates. Describe the features that would be important in your selection other than price.

5.43 Design a 50-mm strain extensometer to be used for a simple tension test of mild steel. If the strain extensometer is to be used only in the elastic region and to detect the onset of yielding, specify the maximum range. What is the advantage of limiting the range?

5.44 Can eddy current sensors be employed with the following target material?

(a) Magnetic materials
(b) Polymers
(c) Non-magnetic metallic foils

Indicate procedures that permit usage of the sensor in these three cases.

5.45 Determine the charge q developed when a piezoelectric crystal having A = 15 mm^2 and h = 8 mm is subjected to a pressure p = 2 MPa if the crystal is:

(a) X-cut, length-longitudinal quartz
(b) Parallel to polarization barium titanate

5.46 Determine the output voltages for the piezoelectric crystals described in Exercise 5.28.

5.47 Compare the use of quartz and barium titanate as materials for:

(a) Piezoelectric sensors
(b) Ultrasonic signal sources

5.48 The equivalent circuit (see Fig. 5.31) for a measuring system incorporating a piezoelectric crystal consists of the following: R_p = 10 TΩ, R_A = 1 GΩ, C_p = 20 pF, C_A = 15 pF. Let C_L be a variable ranging from 10 pF to 1000 pF. Determine the effective time constant τ_e for the circuit. If the error must be limited to 5 percent, determine the time available for measurement of the magnitude of a step pulse of unit magnitude.

5.49 Compare the characteristics of a piezoresistive sensor with those of a piezoelectric sensor.

5.50 What advantages does the piezoresistive sensor have over the common (metal) electrical resistance strain gage? What are some of the disadvantages?

5.51 For the silicon photodetector with the output current-light intensity specification shown in Fig. 5.33, determine the responsivity R. Assume the output voltage is due to the voltage drop across resistances of 10, 100, 1,000 and 10,000 Ω.

5.52 Design a circuit to turn on outside lights at your home as it begins to get dark. Use a photoconduction cell in the circuit and provide for an adjustment to control the intensity level for switch activation.

5.53 Sketch the circuit for a photovoltaic cell and write a paragraph explaining its operation. Write another paragraph stating the advantages and disadvantages of this light sensor.

5.54 Sketch the circuit for a photodiode used in the photoconduction mode and write a paragraph explaining its operation. Write another paragraph stating the advantages and disadvantages of this light sensor.

5.55 Write your own summary of the important topics in this chapter.

BIBLIOGRAPHY

1. Pryor, T. R., and W. P. T North: "The Diffractographic Strain Gage," Experimental Mechanics, Vol. 11, No. 12, pp. 565-578, 1971.
2. Pryor, T. R., O. L.: Hageniers, and W. P. T. North: "Displacement Measurement along a Line by the Diffractographic Method," Experimental Mechanics, Vol. 12, No. 8, pp. 384-386, 1972.
3. Sharpe, W. N. Jr.: "The Interferometric Strain Gage," Experimental Mechanics, Vol. 8, No. 4, pp. 164-170, 1968.
4. Sharpe, W. N. Jr. et al: "Measurements of Young's Modulus, Poisson's Ratio and Tensile Strength of Polysilicon," Proceedings of the 10th International Workshop on Micromechanical Systems, Nagoya Japan, 1997, pp. 424-429
5. Harting, D. R.: "Evaluation of a Capacitive Strain Measuring System for use to 1500 °F", Instrument Society of America, ASI Publication No. 75251, pp. 289-297, 1975.
6. Foster R. L. and S. P. Wnuk Jr.: "High Temperature Capacitive Displacement Sensing", Instrument Society of America, Paper # 85-0123, 0096-7238, pp. 245-252, 1985.
7. Herceg, E. E.: Handbook of Measurement and Control, Schaevitz Engineering, Pennsauken, N.J., 1976.
8. Shepherd, R.: "Strain Measurement Using Vibrating-Wire Gages", Experimental Mechanics, Vol. 4, No. 8, pp. 244-248, 1964.
9. Potocki, F. P.: "Vibrating-Wire Strain Gauge for Long-Term Internal Measurements in Concrete," Engineer, Vol. 206, pp. 964-967, 1958.
10. Mason, W. P., and R. N. Thurston: "Piezoresistive Materials in Measuring Displacement, Force, and Torque," Journal Acoustical Society America, Vol. 29, No. 10, pp. 1096-1101, 1957.
11. Geyling, F. T., and J. J. Forst: "Semiconductor Strain Transducers," Bell System Technical Journal., Vol. 39, 1960.
12. Mason, W. P.: "Semiconductors in Strain Gages," Bell Laboratory Record., Vol. 37, No.1, pp. 7-9, 1959.
13. Smith, C. S.: "Piezoresistive Effect in Germanium and Silicon," Physical Review, Vol.94, pp. 42-49, 1954.
14. Padgett, E. D., and W. V. Wright: "Silicon Piezoresistive Devices," pp. 1-20, in M. Dean and R. D. Douglas (eds.), Semiconductor and Conventional Strain Gages, Academic Press, Inc., New York, 1962.
15. O'Regan, R.: "Development of the Semiconductor Strain Gage and Some of Its Applications," pp. 245-257, in M. Dean and R. D. Douglas (eds.), Semiconductor and Conventional Strain Gages, Academic Press, Inc., New York, 1962.
16. Mason, W. P., J. J. Forst, and L. M. Tornillo: "Recent Developments in Semiconductor Strain Transducers," pp. 109-120, in M. Dean and R. D. Douglas (eds.), Semiconductor and Conventional Strain Gages. Academic Press, Inc., New York, 1962.
17. Kurtz, A. D.: "Adjusting Crystal Characteristics to Minimize Temperature Dependence," pp. 259-272, in M. Dean and R. D. Douglas (eds.), Semiconductor and Conventional Strain Gages, Academic Press, Inc., New York, 1962.
18. Sanchez, J. C., and W. V. Wright: "Recent Developments in Flexible Silicon Strain Gages," pp. 307-346, in M. Dean and R. D. Douglas (eds.), Semiconductor and Conventional Strain Gages, Academic Press, Inc., New York, 1962.
19. Durelli, A. J., J. W. Dally, and W. F. Riley: "Developments in the Application of the Grid Method to Dynamic Problems," Journal Applied Mechanics, Vol.26, No. 4, pp. 629-634, 1959.
20. Durelli, A. J., and W. F. Riley: "Developments in the Grid Method of Experimental Stress Analysis," Proceedings. SESA, Vol. XIV, No. 2, pp. 91-100, 1957.
21. Parks, V. J., and A. J. Durelli: "Various Forms of Strain Displacement Relations Applied to Experimental Strain Analysis," Experimental Mechanics, Vol.4, No. 2, pp. 37-47, 1964.
22. Fail, W. F. and Taylor, C. E.: "An Application of Pattern Mapping to Plane Motion", Experimental Mechanics, Vol. 30, No. 4, pp.404-410, 1990.
23. Pratt, W. K.: Digital Image Processing, John Wiley and Sons, Inc., New York, 1978.
24. Sirkis, J. S.: "Improved Grid Methods Through Displacement Pattern Matching", Proceedings 1989 SEM Spring Conference on Experimental Mechanics., pp. 439-444.

25. Butter, C. D. and G. B. Hocker: "Fiber Optics Strain Gage," Applied Optics, Vol. 17, No. 18, pp. 2868-2869, 1978.
26. Sirkis, J. S. and C. E. Taylor: "Interferometric Fiber Optic Strain Sensor," Experimental Mechanics, Vol. 28, No. 2, pp. 170-176, 1988.
27. Meltz, G. R., W. W. Morey, and W. H. Glen: "Formation of Bragg Gratings in Optical Fibers by a Transverse Holographic Method," Optical Letters, Vol. 14, pp823-825, 1989.
28. Udd, E. et al: "Progress on Multidimensional Strain Field Measurements Using Fiber Optic Grating Sensors," Proceedings SPIE Smart Structures Symposia Conference, Newport Beach, March 6-9, 2000.
29. Schultz, W. L. et al: "Single and Multiaxis Fiber Grating Based Strain Sensors for Civil Structure Applications," Proceedings SPIE, Vol. 3586, p. 41, 1999.
30. Choquet, P., F. Juneau and J. Bessette: "New Generation of Fabry-Perot Fiber Optic Sensors for Monitoring of Structures," Proceedings SPIE Smart Structures Symposia Conference, Newport Beach, March 6-9, 2000.
31. Lee, C. E. et al: "Optical-fiber Fabry-Perot Embedded Sensors, Optical Letters, Vol. 13, No. 21, pp. 1225-1227, 1989.
32. Rashleigh, S. C.: "Origins and Control of Polarization in Single Mode Optical Fibers, Journal Lightwave Technology, Vol. 1, No. 2, pp. 312-331, 1983.
33. Stetson, K. A.: "Optical Heterodyning," Chapter 10, Handbook on Experimental Mechanics, 2nd Ed., ed. A. S. Kobayashi, VCH Publishers, New York, pp. 477-490, 1993.
34. Narendran, N.; Shukla, A.; Letcher, S. Application of fiber optic sensor to a fracture mechanics problem. Eng. Fracture Mech. 1991, 38, 491-498.
35. Narendran, N.; Shukla, A.; Letcher, S. Determination of fracture parameters using embedded fiber optic sensors. Experimental Mechanics, 1991, 31, 360-365.

CHAPTER 6

ELECTRICAL RESISTANCE STRAIN GAGES

6.1 INTRODUCTION

In the preceding chapter, several different strain-measuring systems were introduced, and their performance characteristics such as range, sensitivity, gage length and precision of measurement were discussed. None of these different systems, regardless of the principle upon which the gage was based, exhibits all the properties required for an optimum device; however, the electrical-resistance strain gage approaches the requirements for an optimum system. As such, the electrical-resistance strain gage is the most frequently used device in stress-analysis work throughout the world today. Electrical-resistance strain gages are also widely used as sensors in transducers designed to measure such quantities as load, torque, pressure and acceleration.

In 1856 Lord Kelvin discovered the principle upon which the electrical-resistance strain gage is based. He loaded copper and iron wires in tension and noted that their resistance increased with the strain applied to the wire. Furthermore, he observed that the iron wire showed a greater increase in resistance than the copper wire when they were both subjected to the same strain. Finally, Lord Kelvin employed a Wheatstone bridge to measure the resistance change. In this classic experiment, he established three vital facts, which have greatly aided in the development of the electrical-resistance strain gage:

1. Resistance of the wire changes as a function of strain.
2. Different materials have different sensitivities.
3. The Wheatstone bridge can be used to measure these resistance changes accurately.

It is remarkable that 80 years passed before strain gages based on Lord Kelvin's experiments became commercially available.

Today, after about 75 years of commercial development and extensive utilization by industrial and academic laboratories throughout the world, the bonded-foil strain gage monitored with a Wheatstone bridge has become a highly perfected measuring system. Precise results for surface strains can be obtained quickly using relatively simple methods and inexpensive gages and instrumentation systems. In spite of the relative ease in employing strain gages, there are many features of the gages that must be thoroughly understood to obtain optimum performance from a strain measurement for applied stress analysis. In this chapter, the electrical-resistance strain gage will be examined in detail to illustrate each feature affecting its performance. Strain-gage circuits and recording instruments used in measuring the strain-related resistance changes are also reviewed in Chapter 7.

6.2 STRAIN SENSITIVITY IN METALLIC ALLOYS

Lord Kelvin noted that the resistance of a wire increases with increasing strain and decreases with decreasing strain. The question then arises whether this change in resistance is due to the dimensional change in the wire under strain or to the change in resistivity of the wire with strain. It is possible to answer this question by performing a very simple analysis and comparing the results with experimental data that have been compiled on the characteristics of certain metallic alloys.

The resistance R of a uniform conductor with a length L, cross-sectional area A, and specific resistance ρ is given by:

$$R = \rho \frac{L}{A} \tag{6.1}$$

Differentiating Eq. (6.1) and dividing by the total resistance R leads to:

$$\frac{dR}{R} = \frac{d\rho}{\rho} + \frac{dL}{L} - \frac{dA}{A} \tag{a}$$

The term dA represents the change in cross-sectional area of the conductor due to the transverse strains, which are equal to $-\nu(dL/L)$. If the diameter of the conductor[1] before the application of the axial strain is noted as d_0, then the diameter after the strain is applied is given by:

$$d_f = d_0\left(1 - \nu \frac{dL}{L}\right) \tag{b}$$

and from Eq. (b) it is clear that:

$$\frac{dA}{A} = -2\nu \frac{dL}{L} + \nu^2\left(\frac{dL}{L}\right)^2 \approx -2\nu \frac{dL}{L} \tag{c}$$

Substituting Eq. (c) into Eq. (a) gives:

$$\frac{dR}{R} = \frac{d\rho}{\rho} + \frac{dL}{L}(1 + 2\nu) \tag{6.2}$$

this can be rewritten as:

$$S_A = \frac{dR/R}{\varepsilon} = 1 + 2\nu + \frac{d\rho/\rho}{\varepsilon} \tag{6.3}$$

where S_A is the sensitivity of the metallic alloy used in the conductor and is defined as the resistance change per unit of initial resistance divided by the applied strain.

Examination of Eq. (6.3) shows that the strain sensitivity of any alloy is due to two factors—the change in the dimensions of the conductor, as expressed by the $(1 + 2\nu)$ term, and the change in specific resistance, as represented by the $(d\rho/\rho)/\varepsilon$ term. Experimental results show that S_A varies from about 2 to 4 for most metallic alloys. For pure metals, the range is from −12.1 (nickel) to +10.1 (platinum). This fact implies that the change in specific resistance can be quite large for certain metals because $1 + 2\nu$ usually ranges between 1.4 and 1.7. The change in specific resistance is due to the number of free electrons and the variation of their mobility with applied strain.

A list of some metallic alloys commonly employed in commercial strain gages, together with their sensitivities, is presented in Table 6.1. It should be noted that the sensitivity depends upon the particular alloy being considered. Moreover, the values assigned to S_A in Table 6.1 are not necessarily

---

[1] A conductor with a circular cross section is used in this derivation.

constants. The value of the strain sensitivity S_A depends upon the degree of cold working imparted to the conductor in its formation, the impurities in the alloy and the range of strain over which the measurement of S_A is made.

Table 6.1
Strain sensitivity S_A for common strain-gage alloys

Material	Composition (%)	S_A
Advance or Constantan	45 Ni, 55 Cu	2.1
Nichrome V	80 Ni, 20 Cu	2.1
Isoelastic	36 Ni, 8 Cr, 0.5 Mo, 55.5 Fe	3.6
Karma	74 Ni, 20 Cr, 3 Al, 3 Fe	2.0
Armour D	70 Fe, 20 Cr, 10 Al	2.0
Platinum-Tungsten	92 Pt, 8 W	4.0

Most electrical-resistance strain gages produced today are fabricated from the copper-nickel alloy known as Advance or Constantan. A typical curve showing the percent change in resistance $\Delta R/R$ as a function of percent strain for this alloy is given in Fig. 6.1. This alloy is well suited for strain-gages for the following reasons:

1. The value of the strain sensitivity S_A is linear over a wide range of strain, and the hysteresis of a bonded conductor is extremely small.
2. The value of S_A does not change significantly as the material transitions from elastic behavior to plastic flow.
3. The alloy has a high specific resistance (ρ = 49 $\mu\Omega$-cm).
4. The alloy has excellent thermal stability and is not influenced appreciably by temperature changes when bonded to common structural materials.
5. The small temperature-induced changes in resistance of the alloy can be controlled with trace impurities or by heat treatment.

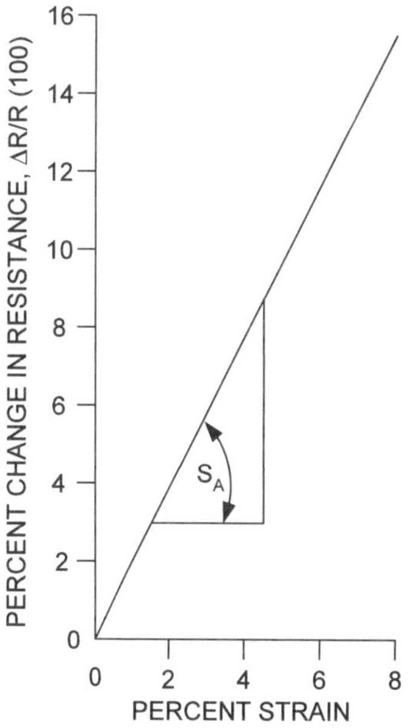

Figure 6.1 Change in resistance as a function of strain for an Advance alloy (45% Ni, 55% Cu).

The first advantage of the Advance-type alloy over other alloys implies that the calibration constant for the gage will not vary with strain level; therefore, a single calibration constant is adequate for all levels of strain. The wide range of linearity with strain (even into the alloy's plastic region) indicates that it can be employed for measurements of both elastic and plastic strains in most structural materials. The high specific resistance of the alloy is useful when constructing a small gage with a relatively high resistance. Finally, the temperature characteristics of selected melts of the alloy permit the fabrication of temperature-compensating strain gages for each structural material. With temperature-compensating strain gages, the temperature-induced $\Delta R/R$ on a given material can be maintained at less than 10^{-6}/°C.

The Isoelastic alloy (36% Ni, 8% Cr, 0.5% Mo, 55.5% Fe) is also employed in commercial gages because of its high sensitivity (3.6 for Isoelastic compared with 2.1 for Advance) and its high fatigue strength. The increased sensitivity is advantageous in dynamic applications where the strain-gage output must be amplified to a considerable degree before recording. The high fatigue strength is useful when the gage is to operate in a cyclic strain field where the alternating strains exceed 1500 $\mu\varepsilon$. In spite of these two advantages, use of the Isoelastic alloy is limited because it is extremely sensitive to temperature changes. When used in a strain gage bonded to a steel specimen, a change in temperature of 1 °C will produce an apparent strain indication of 300 to 400 $\mu\varepsilon$ as shown in Fig. 6.2. Isoelastic gages can be used in dynamic applications only when the temperature is stable over the time required for the dynamic measurement.

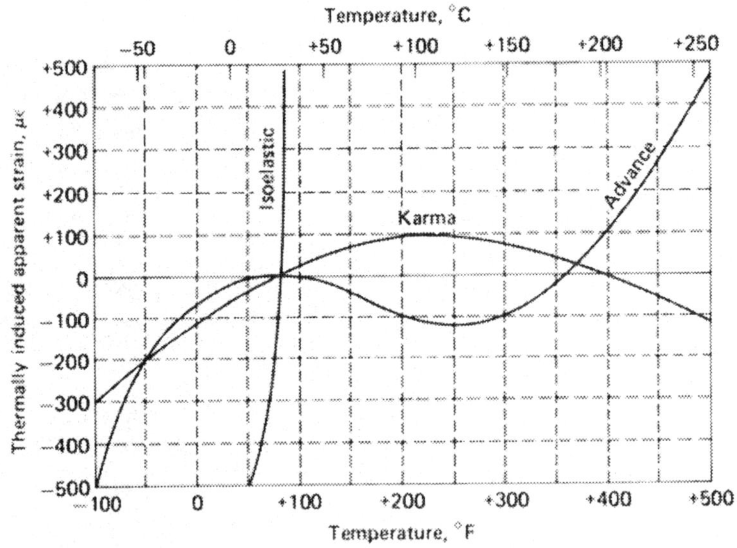

Figure 6.2 Thermally induced apparent strain with temperature for three common strain-gage alloys.

The Karma alloy (74% Ni, 20% Cr, 3% Al, 3% Fe) has properties that are similar to Advance alloy. Indeed, its fatigue limit is higher than Advance but lower than Isoelastic. In addition, Karma exhibits excellent stability with time and is always used when strain measurements are made over extended periods (weeks or months). Another advantage is that the temperature compensation that can be achieved with Karma is better over a wider range of temperature than the compensation that can be achieved with the Advance alloy. The thermally induced apparent strain with temperature for Advance, Isoelastic and Karma, shown in Fig. 6.2, clearly indicates more accurate compensation at the temperature extremes for the Karma alloy. Finally, Karma can be used to 260 °C in static strain measurements whereas Advance is limited to 200 °C. The primary disadvantage of Karma is the difficulty encountered in soldering lead wires to the tabs.

The other alloys, Nichrome V, Armour D, and the platinum-tungsten alloy are more stable and oxidation resistant at higher temperatures. These alloys are used for special-purpose gages that permit measurements of strain to be made at temperatures in excess of 500 °F (260 °C).

6.3 GAGE CONSTRUCTION

It is theoretically possible to measure strain with a single length of wire as the sensing element of a strain gage. However, circuit requirements, which are needed to prevent overloading of the power supply and to minimize heat generated by the gage current, place a lower limit of approximately 100-Ω on the gage resistance. As a result, a 100-Ω strain gage fabricated from the finest standard wire is about 100 mm long.

The very earliest electrical-resistance strain gages were of the unbonded type, where the conductors were straight wires strung between a movable frame and a fixed frame. This gage was large and required knife-edges for mounting, which greatly limited its applicability. The problem of conductor length and gage mounting was solved in the mid-1930s by Ruge and Simmons. They formed a grid pattern containing the required length of wire and solved the conductor-length problem. Bonding the wire grid directly to the specimen with suitable adhesives solved the attachment problem. Bonded-wire strain gages were employed for strain measurements almost exclusively, from the mid-1930's to the mid 1950's. They are still used on rare occasions today when long gage lengths are necessary; but in most instances, they have been replaced by the bonded-foil strain gage.

Saunders and Roe produced the first metal-foil strain gages in England in 1952. With this type of gage, the grid configuration is produced from metal foil by a photoetching process. Because the process is quite versatile, a wide variety of gage sizes and grid shapes can be produced. Typical examples of the variety of gages marketed commercially are shown in Fig. 6.3. The shortest gage length available in a metal-foil gage is 0.008 in. (0.20 mm). The longest gage length is 4.00 in. (102 mm). Standard gage resistances are 120 and 350 Ω; however, high gage resistances (500, 1,000, and 3,000 Ω) are commercially available in select sizes for transducer applications.

Figure 6.3 A collage of strain gages showing the various sizes and types of gages that is commercially available. Courtesy of Micro Measurements Group.

Multiple-element gages, shown in Fig. 6.4, are available with 10 gages arranged along a line. These strip gages are usually installed in fillets occurring at reentrant corners where high strain gradients exist and it is difficult to locate the point where the strain is a maximum. Grid spacing is also closer than what can usually be achieved with individual gages, thus yielding better resolution of strain fields with significant non-linear gradients.

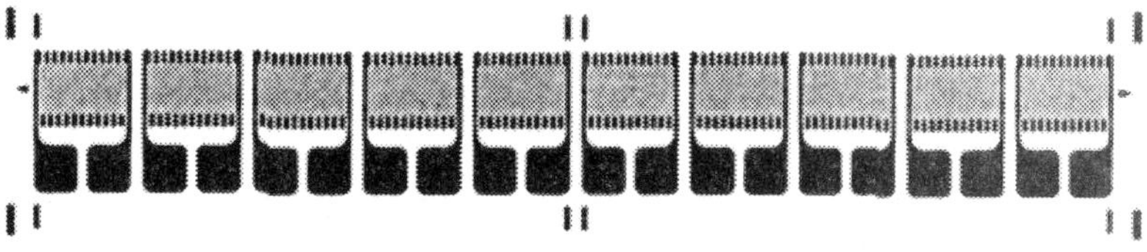

Figure 6.4 Linear array of 10 strain gages on a single carrier.

Overall dimensions for the complete gage patterns vary with the grid and solder-tab configurations. When preferred, some types of the gages can be cut to produce smaller strips with fewer grids. Most strip gages are commercially available in two different versions—with all the grids oriented either parallel or perpendicular to the long axis of the carrier. Some of the strip gages are designed with a common lead, connected to one side of all the grids. Others provide two leads for each of the 10 strain gages deployed along the strip.

Two-and three-element rosettes, shown in Fig. 6.5, are available in either the in-plane or stacked configuration in a wide range of sizes. Rosettes are employed for strain measurements in biaxial stress fields. The biaxial rosette pattern has two grids perpendicular to one another as illustrated in Fig. 6.5a. Planar rosettes, like the one shown here, are constructed with all grids on the same plane. Stacked rosettes are also available with separate grids layered on top of one another. With two independent measurements of strain made in perpendicular directions about a point, the principal strains can be determined when their directions are known.

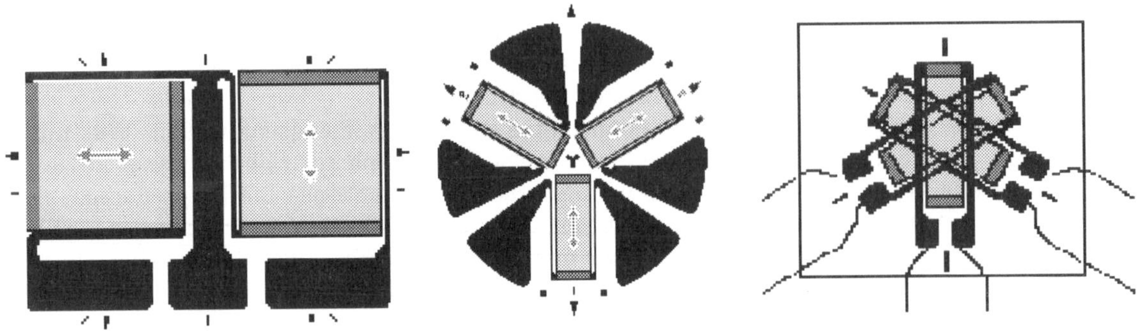

(a) (b) (c)

Figure 6.5 Two and three element rosettes. (Measurements Group)

Three-element rosettes are used when the principal directions for the strain field are not known. The rectangular rosette pattern (not shown here) has its three independent grids oriented at 0, 45, and 90 degrees. A planar delta rosette, with grids at 30, 150 and 270 degrees, is shown in Fig. 6.5b. A stacked delta rosette is shown in Fig. 6.5c. With three independent measurements of strain at a point, the principal strains and their directions can be determined.

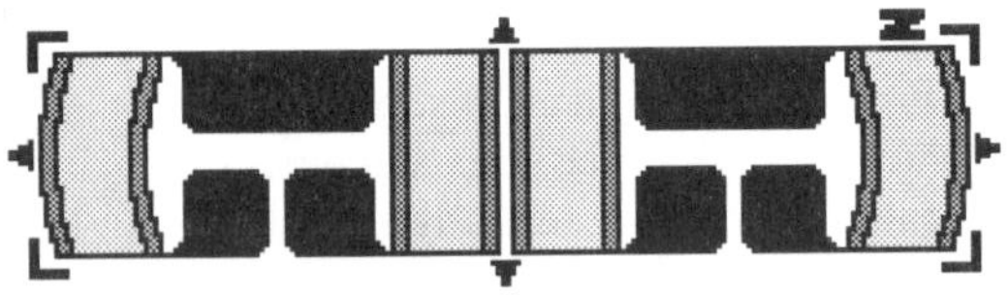

Figure 6.6 A diaphragm gage designed for a circular diaphragm pressure transducer.

Special gage configurations are also available for use in transducers. A typical example is shown in Fig. 6.6 for the diaphragm-type pressure transducer. Two of the four gage elements are designed to sense circumferential strains near the outside edge of the circular diaphragm and the remaining two elements sense the radial strains near the center of the diaphragm. Strain gages produced for transducers are usually fabricated with alloys and carrier materials that are designed to minimize creep of the gage over extended periods of time.

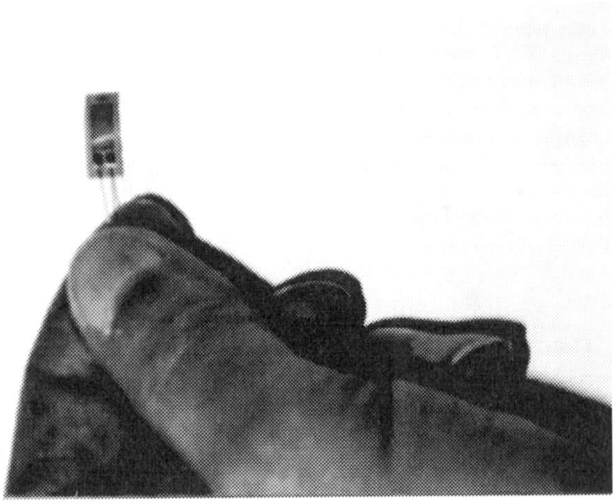

Figure 6.7 Thin plastic carriers are necessary to handle the fragile foil strain gages.

The etched metal-film grids are very fragile and easy to distort, wrinkle or tear. For this reason, the metal film is usually bonded to a thin plastic sheet, which serves as a backing or carrier before photoetching. The carrier material also provides electrical insulation between the gage and the component after the gage is mounted. The use of a backing sheet to serve as a carrier for the grid is illustrated in Fig. 6.7, which shows a gage being handled. Markings for the centerline of the gage length and width are also displayed on the carrier as indicated in Fig. 6.7.

Very thin paper was the first carrier material employed in the production of wire-type gages. However, paper has been replaced with a thin (0.001-in. or 0.025-mm) film of polyimide that is a tough and flexible plastic. For transducer applications, where precision and linearity are very important, a very thin high-modulus epoxy film is used for the carrier. The epoxy backing is not suitable for general-purpose strain gages because it is brittle and can easily be broken during gage installation. Glass-fiber reinforced epoxy-phenolic polymer is employed as a carrier when the strain gage will be exposed to high-level cyclic strains and fatigue life of the gage system is important. In this application, the carrier encapsulates the grid, as shown in Fig. 6.8. Glass-reinforced epoxy-phenolic carriers are also used for moderate temperature applications up to 750 °F (400 °C). For very high temperature applications, a strippable carrier is used. This carrier is removed during application of the gage, and a ceramic adhesive serves to maintain the grid configuration and to insulate the gage.

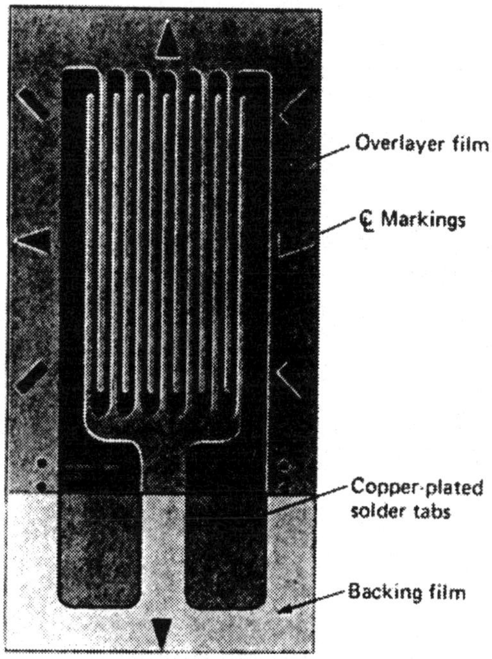

Figure 6.8 Encapsulation of the grid with a thin plastic film enhances fatigue life of the strain gage.

Another type of gage, originally developed for high temperature strain measurement, is the free filament strain gage shown in Fig. 6.9. It consists of a grid that is supported over only a part of its area by a fiberglass carrier. The grid is partially bonded to the specimen with a ceramic adhesive and the carrier is removed. The remainder of the gage is then bonded to the specimen to complete the installation. More detail on high temperature strain gages will be presented later in this chapter.

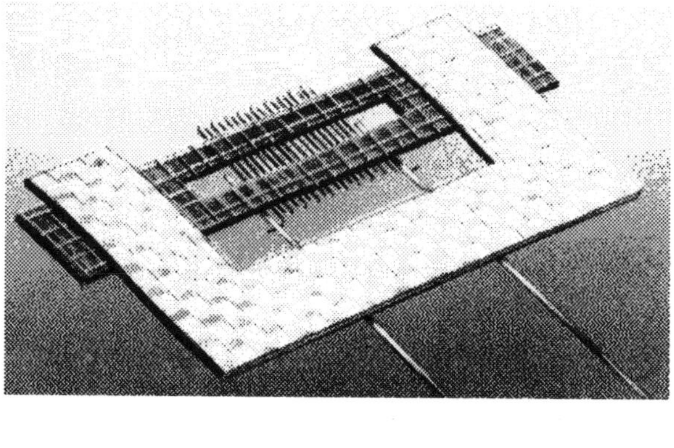

Figure 6.9 A free filament strain gage for high temperature measurements.

Currently, weldable gages are available with strain gages that have been bonded to a shim fabricated from a high temperature alloy. Two different approaches are employed in producing these gages. For applications at temperatures ranging from −195 to +260 °C, a foil grid etched from a Karma alloy and supported with a glass fiber reinforced epoxy-phenolic carrier is bonded to the metal shim with polymeric adhesives. A second approach utilizes the free filament gage that is bonded to the metal shim with ceramic adhesives. This approach provides a gage that is suitable for measuring transient or dynamic strain signals up to temperatures of about 1,100 °C. Both of these gages can be welded onto a number of different metals with a capacitor discharge welder and can be used immediately. This simplicity in application represents a significant advantage over other high-temperature strain-gage systems that require rather elaborate installation techniques. Indeed, the simplicity of the welding installation is attractive whenever gages must be mounted in the field under adverse conditions regardless of the temperature anticipated. The weldable gage is extremely rugged and waterproof.

Of the various gages described, the metal-foil strain gage is the most frequently employed for both general-purpose stress analysis and transducer applications. There are occasional special-purpose applications for which bonded-wire, weldable, semiconductor or fiber-optic strain gages are more suitable, but these applications are not common.

6.4 STRAIN-GAGE ADHESIVES AND MOUNTING METHODS

The bonded resistance strain gage is a high-quality precision resistor that must be attached to the specimen with a suitable adhesive. For precise strain measurements both the correct adhesive and proper bonding procedures must be employed.

The adhesive serves a vital function in the strain-measuring system; it must transmit the strain from the specimen to the gage's sensing element without distortion. It may appear that this role can be easily accomplished if the adhesive is suitably strong; however, the characteristics of the polymeric adhesives used to bond strain gages are such that the adhesive can influence gage resistance, apparent gage factor, hysteresis characteristics, resistance to stress relaxation, temperature-induced zero drift and insulation resistance.

The singularly unimpressive feat of bonding a strain gage to a specimen is perhaps one of the most critical steps in the entire process of measuring strain with a bonded resistance strain gage. The improper use of an adhesive costing a few dollars per test can seriously degrade the validity of an experimental stress analysis, which may cost many thousands of dollars.

When mounting a strain gage, it is important to carefully prepare the surface of the component where the gage is to be located. This preparation consists of sanding away any paint or rust to obtain a smooth but not highly polished surface. Next, solvents are employed to remove all traces of oil or grease and the surface is etched with an appropriate acid. Finally, the clean, sanded, degreased, and etched surface is neutralized (treated with a basic solution) to give it the proper chemical affinity for the adhesive.

The gage location is then marked on the specimen and the gage is positioned by using a rigid transparent tape in the manner illustrated in Fig. 6.10. The tape maintains the position and orientation of the gage as the adhesive is applied and as the gage is pressed into place by squeezing out the excess adhesive.

After the gage is installed, the adhesive must be exposed to a proper combination of pressure and temperature for a suitable length of time to ensure a complete cure. This curing process is quite critical because the adhesive will expand because of temperature, experience a volume reduction due to polymerization, exhibit a contraction upon cooling, and on occasion experience post cure shrinkage. Because the adhesive is sufficiently rigid to control deformation of the strain-sensitive element in the gage, residual stresses developed in the adhesive will influence the output of the strain gage. Of particular importance is the post cure shrinkage, which may influence the gage output long after the adhesive is supposedly completely cured. If a long-term strain measurement is attempted with an incompletely cured adhesive, the stability of the gage will be seriously impaired and the accuracy of the measurements compromised.

A wide variety of adhesives are available for bonding strain gages. Factors influencing the selection of a specific adhesive include the carrier material, the operating temperature, the curing temperature and the maximum strain to be measured. A discussion of the characteristics of several different adhesive systems in common use follows.

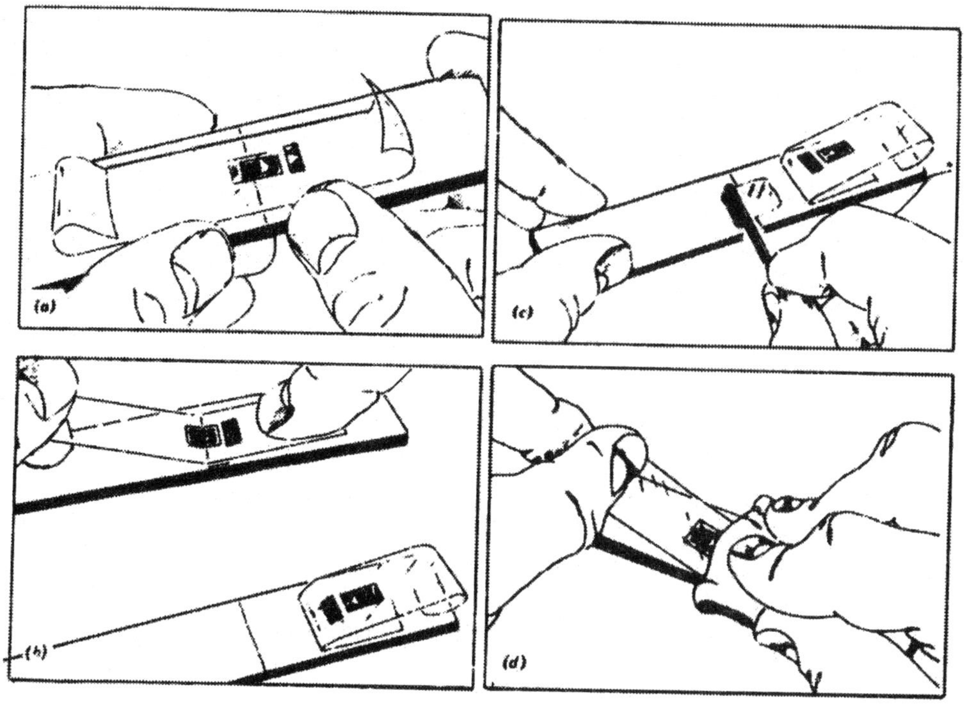

Figure 6.10 The tape method of bonding foil strain gages with flexible carriers: (a) position the gage and overlay it with transparent tape; (b) peel the tape back to expose the gage bonding area; (c) apply thin layer of adhesive over the bonding area; (d) replace the tape in the overlay position with a wiping action to clear the excessive adhesive. (Courtesy of the Measurements Group).

6.4.1 Strain Gage Adhesives

Epoxy Cements

Epoxies are a class of thermosetting plastics that exhibit higher bond strength and higher failure strain than other types of adhesives used to mount strain gages. Epoxy systems are usually composed of two constituents—a monomer and a hardening agent. The monomer, or base epoxy, is a light amber fluid that is usually quite viscous. A hardening agent mixed with the monomer will induce polymerization. Amine-type curing agents produce an exothermic reaction, which releases sufficient heat to accomplish curing at room temperature or at relatively low curing temperatures. Anhydride-type curing agents require the application of heat to promote polymerization. Temperatures in excess of 250 °F (120 °C) must be applied for several hours to complete polymerization of the base monomer. In some cases solvents and/or diluents are added to reduce viscosity. So many epoxies and curing agents are commercially available today that it is impossible to be specific in the coverage of their properties or their behavior. However, the following remarks will be valid and useful for any system of epoxies used in strain-gage applications.

With both types of curing agent, particularly the amine type, the amount of hardener added to the monomer is extremely important. The adhesive curing temperatures and the residual stresses produced during polymerization can be significantly influenced by small deviations from the specified values. For this reason, the quantities of both the monomer and the curing agent should be carefully weighed before they are mixed together.

In general, pure epoxies do not liberate volatiles during cure; therefore, post-cure heat cycling is not necessary to evaporate chemical by-products released during polymerization. Solvents, which are volatile, should not be added to the epoxies to improve their viscosity for general-purpose applications. A filler material with micron-size particles of pure silica can be added in moderate quantities (5 to 10 percent by weight) to improve the bond strength and to reduce the temperature coefficient of expansion of the epoxy.

A modest clamping pressure of 5 to 20 psi (35 to 140 kPa) is recommended for the epoxies during the cure period to ensure as thin a bond line (adhesive layer) as possible. In transducer applications, diluent-thinned epoxies are frequently specified to reduce viscosity so that extremely thin (less than 200 μin., or 0.005 mm) void-free bond lines can be obtained. Thin bond lines tend to minimize creep, hysteresis, and linearity problems. For these transducer applications, clamping pressures of approximately 50 psi (350 kPa) are recommended.

Many different epoxy systems are available from strain-gage manufacturers in kit form with the components pre-weighed. These systems are recommended because the epoxies are especially formulated for strain-gage applications. They are easy and convenient to use, have an adequate pot life and the time-temperature curve for the curing cycle is specified. The use of hardware-store variety two-tube epoxy systems is discouraged because these systems usually incorporate modifiers or plasticizers to improve the toughness of the adhesive. The modifiers cause large amounts of creep and hysteresis that is undesirable in strain-gage applications.

The best indication of a properly cured adhesive system can be obtained by measuring the resistance between the gage grid and the specimen (the resistance through the adhesive layer). A properly cured installation will exhibit a resistance to ground exceeding 10,000 MΩ. Minute traces of either solvent or water in the adhesive will lower the resistance of the adhesive layer and influence gage performance.

Cyanoacrylate Cement

A modified form of a pressure-curing adhesive consisting of a methyl-2-cyanoacrylate compound is commonly employed as a strain-gage adhesive. This adhesive system is simple to use, and the strain gage can be employed approximately 10 min after bonding. Chemically, this adhesive is quite unusual in that it requires neither heat nor a catalyst to induce polymerization. Apparently, when this adhesive is spread in a thin film between two components to be bonded, minute traces of water or other weak bases on the surfaces of the components are sufficient to trigger the polymerization process. A catalyst can be applied to the bonding surfaces to decrease the reaction times, but it is not essential.

In strain-gage applications, a thin film of the adhesive is placed between the gage and the specimen and a gentle pressure is applied for about 1 or 2 min to induce polymerization. Once initiated, the polymerization will continue at room temperature without maintaining the pressure.

The rapid room-temperature cure of the cyanoacrylate adhesive makes it ideal for general-purpose strain-gage applications. The performance of this adhesive system, however, will deteriorate markedly with time, moisture absorption or elevated temperature. It should not be used where extended life of the gage system is important. Coatings such as microcrystalline wax, silicone rubber, polyurethane, etc., can be used to protect the adhesive from moisture in the air and extend the life of an installation.

Polyester Adhesives

Polyesters, like epoxies, are two-component adhesives. The polyesters exhibit a high shear strength and modulus; however their peel strength is low and they are less resistant to solvents than epoxies. Their primary advantage is the ability to polymerize at a relatively low temperature 40 °F (5 °C) which permits gage installations at temperatures only modestly above freezing.

Ceramic Cements

Two different approaches are used in bonding strain gages with ceramic adhesives. The first utilizes a blend of finely ground ceramic powders such as alumina and silica combined with a phosphoric acid. Usually this blend of powders is mixed with a solvent such as isopropyl alcohol and an organic binder to form a liquid mixture that facilitates handling. A pre-coat of the ceramic cement is applied and fired to form a thin layer of insulation between the gage grid and the component. A second layer of ceramic cement is then applied to bond the gage. In this application, the carrier for the gage is removed, and the grid is totally encased in the ceramic adhesive.

Because many of the ceramic cements marketed commercially have proprietary compositions, it is often difficult to select the most suitable product. A cement developed by the National Institute for Standards and Technology (NIST) for high temperature strain measurements is described in Table 6.2. It is recommended for high-temperature applications because it exhibits a very high resistivity at temperatures up to 1,800 °F (980 °C). Gage resistance to ground with this cement will normally exceed 6 MΩ. The ceramic cements are used primarily for high-temperature application or in radiation environments, where organic adhesives cannot be employed.

Table 6.2
Composition of a high temperature ceramic cement*

Constituent	Parts by weight
Alumina, Al_2O_3	100
Silica, SiO_2	100
Chrome anhydride, CrO_3	2.5
Colloidal silica solution	200
Orthophosphoric acid, H_3PO_4	30

*Developed by the National Institute for Standards and Technology (NIST).

A second method for bonding strain gages with a ceramic material utilizes a flame-spraying Rokide process. A special gun is used to apply the ceramic particles to the gage. For this type of application, gages are constructed with a grid fabricated from wire, which is mounted on slotted carriers. The carrier, which holds the gage and lead wires in position during attachment, is made from a glass-reinforced Teflon tape. The tape is resistant to the molten flame-sprayed ceramic particles; however, after the gage grid is partially secured to the component, the tape is removed and the grid is completely encased with flame-sprayed ceramic particles.

The flame-spray gun utilizes an oxyacetylene gas mixture in a combustion chamber to produce very high temperatures. Ceramic material in rod form is fed into the combustion chamber; there the rod decomposes into softened partially melted particles that are forced from the chamber by the burning oxyacetylene gas. The particles impinge on the surface of the component and form a continuous coating. Because the particles (even in their softened state) are somewhat abrasive, foil grids are not normally used because they are subject to damage by erosion due to particle impact.

Most gage installations are made with pure alumina, or with a composition of 98% alumina and 2% silica. For applications involving temperatures less than 800 °F (425 °C), the 98 alumina-2 silica rod is employed because it is easier to melt and apply. The pure alumina rod is intended for higher-temperature applications.

6.4.2 Testing a Strain-Gage Adhesive System

After a strain gage has been bonded to the surface of a specimen, it must be inspected to determine the adequacy of the bond. The inspection procedure attempts to establish if any voids exist between the gage and the specimen and whether the adhesive is totally cured. Voids can result from bubbles originally in the cement or from the release of volatiles during the curing process. Tapping or pressing the gage installation with a soft rubber eraser and observing the effect on a strain indicator can detect these voids. If the strain indicator shows a variation in its output, the gage bond is not satisfactory and voids exist between the sensing element and the specimen.

The stage of the cure of the adhesive is much more difficult to establish. Two procedures are frequently employed which give some indication of the relative completeness of the cure. The first procedure utilizes measurement of the resistance between the gage grid and the specimen as an indication of the stage of the cure. The resistance of the adhesive layer increases as the adhesive cures. Typical gage installations should exhibit a resistance across the adhesive layer of the order of 10,000 MΩ. Resistance values of 100 to 1000 MΩ normally indicate a need for additional time or temperature to cure the adhesive.

The second procedure for determining the completeness of the cure involves subjecting the gage installation to a strain cycle while measuring the resistance change. The change in the zero-load strain reading (zero-shift) after a strain cycle or the area enclosed by a strain-resistance change curve is a measure of the degree of cure. Strain-gage installations with completely cured adhesives when cycled to 1000 $\mu\varepsilon$ will exhibit zero shifts of less than 2 $\mu\varepsilon$ of apparent strain. If a larger zero shift is observed, the adhesive should be subjected to a post cure temperature cycle.

In the event that the gages are mounted on a component or a structure that cannot be strain-cycled before testing, a temperature cycle can be substituted for the strain cycle. If the adhesive in a gage installation has been thoroughly cured, no change in the zero strain will be observed when the gage is returned to its original temperature. If the adhesive cure is incomplete, the temperature cycle will result in additional polymerization with associated shrinkage of the adhesive, which causes a shift in the zero reading of the gage.

After the gages have been bonded to the structure or component, it is necessary to attach lead wires so that the change in resistance can be monitored on a suitable instrumentation system. Because the metal-foil strain gages are relatively fragile, care must be exercised in attaching the lead wires to the soldering tabs. Rugged anchor terminals that are bonded to the component are usually employed, as illustrated in Fig. 6.11. A small-diameter wire (32- to 36-gage) is used to connect the gage tab to the anchor terminal. Three lead wires are soldered to the anchor terminals. The use of three lead wires to ensure temperature compensation in the Wheatstone-bridge measuring circuit will be discussed in Chapter 7. The size of lead wire employed will depend upon the distance between the gage and the instrumentation system. For normal laboratory installations, where lengths rarely exceed 10 to 14 ft (3 to 5 m), wire sizes between 26 and 30 gages are frequently used. Stranded wire is usually preferred to solid wire because it is more flexible and suffers less breakage due to improper stripping or lead-wire movement during the test.

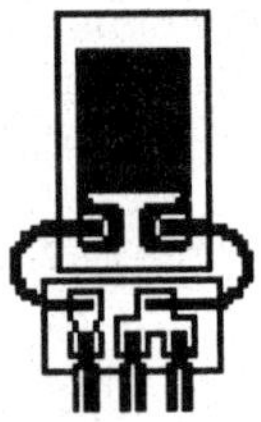

Figure 6.11 Three-lead wire connection to a strain gage using a bonded terminal to anchor the wires.

6.5 GAGE SENSITIVITIES AND GAGE FACTOR

The strain sensitivity of a single, uniform length of a conductor was previously defined as:

$$S_A = \frac{dR/R}{\varepsilon} \approx \frac{\Delta R/R}{\varepsilon}$$

where ε is a uniform strain along the conductor and in the direction of the axis of the conductor. This sensitivity S_A is a function of the alloy employed to fabricate the conductor and its metallurgical condition. Whenever the conductor is formed into a grid to yield a short gage length that is required for measuring strain, the gage exhibits sensitivity to both axial and transverse strain.

With the older flat-grid wire gages, the transverse sensitivity was due primarily to the end loops in the grid pattern, which placed part of the conductor in the transverse direction. In the foil gages, the end loops are enlarged and desensitized to a large degree. The axial segments of the grid pattern, however, have a large width-to-thickness ratio; thus, some amount of transverse strain will be transmitted through the adhesive and carrier material to the axial segments of the grid pattern to produce a response in addition to the axial-strain response. The magnitude of the transverse strain transmitted to the grid segments depends upon the thickness and elastic modulus of the adhesive, the carrier material, the grid material and the width-to-thickness ratio of the axial segments of the grid.

The response of a bonded strain gage to a biaxial strain field can be expressed as:

$$\frac{\Delta R}{R} = S_a\varepsilon_a + S_t\varepsilon_t + S_s\gamma_{at} \tag{6.4}$$

where ε_a is the normal strain along axial direction of gage.
ε_t is the normal strain along transverse direction of gage.
γ_{at} is the shearing strain.
S_a is the sensitivity of gage to axial strain.
S_t is the sensitivity of gage to transverse strain.
S_s is the sensitivity of gage to shearing strain.

In general, the gage sensitivity S_s is small and can be neglected. By setting $S_s = 0$, the response of the gage can then be expressed as:

$$\frac{\Delta R}{R} = S_a(\varepsilon_a + K_t\varepsilon_t) \tag{6.5}$$

where $K_t = S_t/S_a$ is defined as the transverse sensitivity factor for the gage.

Strain-gage manufacturers provide a calibration constant known as the **gage factor** S_g for each gage. The gage factor S_g relates the resistance change to the axial strain as:

$$\frac{\Delta R}{R} = S_g\varepsilon_a \tag{6.6}$$

The gage factor for each gage type produced from a roll of foil is determined by mounting sample gages from this roll onto a specially designed calibration beam. The beam is then deflected to produce a known strain ε_a. The resistance change ΔR is then measured and the gage factor S_g is determined from Eq. (6.6).

With the beam-in-bending method of calibration, the strain field experienced by the gage is biaxial, with:

$$\varepsilon_t = -\nu_0\,\varepsilon_a \tag{a}$$

where ν_0 = 0.285 is Poisson's ratio of the beam material. If Eq. (a) is substituted into Eq. (6.5), the resistance change occurring in the calibration process is:

$$\frac{\Delta R}{R} = S_a \varepsilon_a (1 - \nu_0 K_t) \tag{6.7}$$

Because the resistance changes given by Eqs. (6.6) and (6.7) are identical, the gage factor S_g is related to S_a and K_t by the expression:

$$S_g = S_a (1 - \nu_0 K_t) \tag{6.8}$$

Typical values of S_g, S_a and K_t for several different gage configurations are shown in Table 6.3.

It is important to recognize that error will occur in a strain-gage measurement when Eq. (6.6) is employed except for the two special cases—where either the stress field is uniaxial or the transverse sensitivity factor K_t for the gage is zero. The magnitude of the error can be determined by considering the response of a gage in a general biaxial field with strains ε_a and ε_t. Substituting Eq. (6.8) into Eq. (6.5) gives:

$$\frac{\Delta R}{R} = \frac{S_g \varepsilon_a}{1 - \nu_0 K_t}\left(1 + K_t \frac{\varepsilon_t}{\varepsilon_a}\right) \tag{b}$$

from Eq. (b), the true value of the strain ε_a can be expressed as:

$$\varepsilon_a = \frac{\Delta R/R}{S_g}\,\frac{1 - \nu_0 K_t}{1 + K_t(\varepsilon_t/\varepsilon_a)} \tag{c}$$

Table 6.3

Gage factor S_g, axial sensitivity S_a, transverse sensitivity S_t, and transverse-sensitivity factor K_t for several different foil-type strain gages.

Gage Designation	S_g	S_a	S_t	K_t (%)
EA-06-015CK-120	2.13	2.14	0.0385	1.8
EA-06-030TU-120	2.02	2.03	0.0244	1.2
WK-06-030TU-350	1.98	1.98	0.0040	0.2
EA-06-062DY-120	2.03	2.04	0.0286	1.4
WK-06-062DY-350	1.96	1.96	−0.0098	−0.5
EA-06-125RA-120	2.06	2.07	0.0228	1.1
WK-06-125RA-350	1.99	1.98	−0.0297	−1.5
EA-06-250BG-120	2.11	2.11	0.0084	0.4
WA-06-250BG-120	2.10	2.10	−0.0063	−0.3
WK-06-250BG-350	2.05	2.03	−0.0690	−3.4
WK-06-250BF-1000	2.07	2.06	−0.0453	−2.2
EA-06-500AF-120	2.09	2.09	0.0	0
WK-06-500AF-350	2.04	1.99	−0.1831	−9.2
WK-06-500BH-350	2.05	2.01	−0.1347	−6.7
WK-06-500BL-1000	2.06	2.03	−0.0893	−4.4

*This data is approximate as the values depend on the lot of foil used in gage fabrication.

The apparent strain ε_a', which is obtained if only the gage factor is considered, can be determined from Eq. (6.6) as:

$$\varepsilon_a' = \frac{\Delta R/R}{S_a} \tag{d}$$

Comparing Eqs. (c) and (d) yields:

$$\varepsilon_a = \varepsilon_a' \frac{1-\nu_0 K_t}{1+K_t(\varepsilon_t/\varepsilon_a)} \tag{6.9}$$

The percent error E involved in neglecting the transverse sensitivity of the gage is given by:

$$\mathcal{E} = \frac{\varepsilon_a' - \varepsilon_a}{\varepsilon_a}(100) \tag{6.10}$$

Substituting Eq. (6.9) into Eq. (6.10) yields:

$$\mathcal{E} = \frac{K_t(\varepsilon_t/\varepsilon_a + \nu_0)}{1-\nu_0 K_t}(100) \tag{6.11}$$

The results of Eq. (6.11), shown graphically in Fig. 6.12, indicate that the error is a function of K_t and the strain biaxial ratio $\varepsilon_t/\varepsilon_a$. Because the errors can be significant when both K_t and $\varepsilon_t/\varepsilon_a$ are large, it is important that corrections be made to account for the transverse sensitivity of the gage. Methods to correct for these errors are presented in Chapter 8.

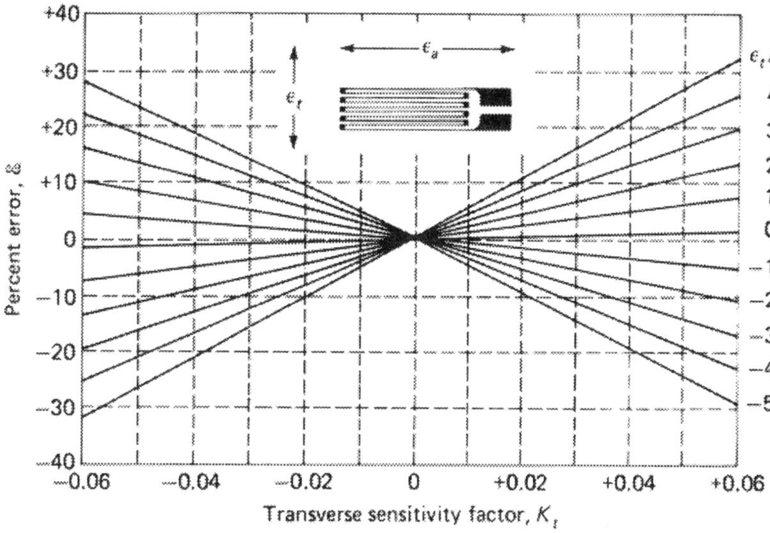

Figure 6.12 Error as a function of the transverse sensitivity factor with biaxial strain ratio as a parameter.

6.6 PERFORMANCE CHARACTERISTICS OF FOIL STRAIN GAGES

Foil strain gages are small precision resistors mounted on a carrier that is bonded to a component part in a typical application. The gage resistance is accurate to ± 0.3%, and the gage factor, based on a lot calibration, is certified to ± 0.5%. These specifications indicate that foil-type gages provide a means for making precise measurements of strain. The results actually obtained, however, are a function of the installation procedures, the state of strain being measured and environmental conditions during the test. All these factors affect the performance of a strain-gage system.

6.6.1 Strain-gage Linearity, Hysteresis and Zero Shift

One measure of the performance of a strain-gage system (system here implies gage, adhesive, lead wires, switches and instrumentation) involves considerations of linearity, hysteresis and zero shift. If gage output, in terms of measured strain, is plotted as a function of applied strain as the load on the component is cycled, results similar to those shown in Fig. 6.13 will be obtained. A slight deviation from linearity is typically observed, and the unloading curve falls below the loading curve to form a hysteresis loop. Also, when the applied strain is reduced to zero, the gage output indicates a small negative strain—termed zero shift. The magnitudes of the deviation from linearity, hysteresis, and zero shift depend on the strain level, the adequacy of the bond, the degree of cold work of the foil material, and the viscoelastic characteristics of the carrier material.

For properly installed gages, deviations from linearity should be approximately 0.1% of the maximum strain for polyimide carriers and 0.05% for epoxy carriers. First-cycle hysteresis and zero shift depend strongly on the strain range as shown in Fig. 6.14. Metallurgical changes in the gage alloy produce small but permanent changes in the resistance that accumulates with the number of cycles. It should be noted that the zero shift per cycle is much larger during the first 5 to 10 cycles. For this reason it is recommended that a strain gage installation be cycled to 125 % of the maximum test strain for at least five cycles prior to establishing the zero strain readings for all of the gages in the installation.

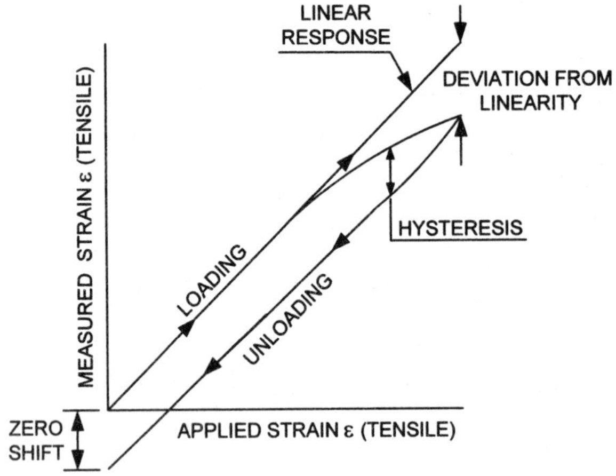

Figure 6.13 A typical strain cycle showing non-linearity, hysteresis and zero shift (scale exaggerated).

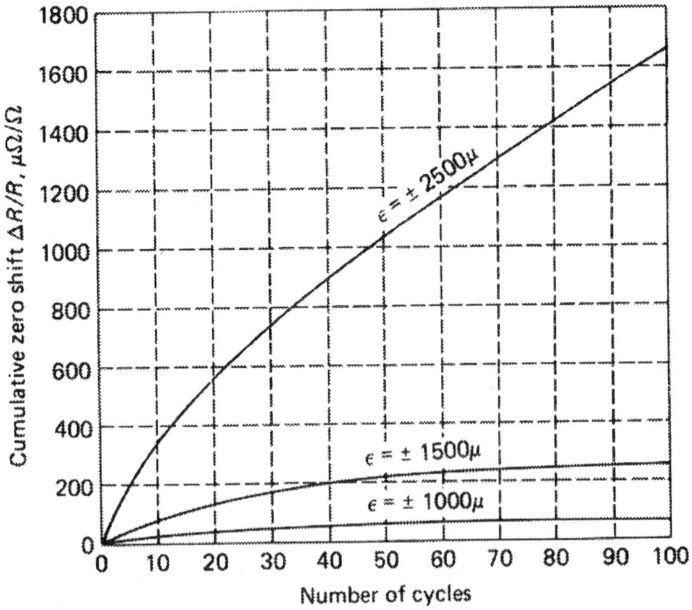

Figure 6.14 Cumulative zero shift with the number of strain cycles for Advance type strain gages.

6.6.2 Temperature Compensation

In many experiments, the strain-gage installation is subjected to temperature changes during the test period, and careful consideration must be given to determining whether the change in resistance is due to applied strain or temperature change. When the ambient temperature changes, four effects occur, which may alter the performance characteristics of the strain gage:

1. The strain sensitivity S_A of the metal alloy used for the grid changes.
2. The gage grid either elongates or contracts ($\Delta L/L = \alpha\Delta T$).
3. The base material upon which the gage is mounted either elongates or contracts ($\Delta L/L = \beta \Delta T$).
4. The resistance of the gage changes because of the influence of the temperature coefficient of resistivity of the gage material ($\Delta R/R = \gamma\Delta T$).

The change in the strain sensitivity S_A of Advance, Karma and other strain-gage alloys with variations in temperature is shown in Fig. 6.15. These data indicate that $\Delta S_A/\Delta T$ equals 0.735 and –0.975%/100 °C for Advance and Karma alloys, respectively. As a consequence, the variations of S_A with temperature are neglected for room-temperature experiments, where the temperature fluctuations rarely exceed ± 10 °C. In thermal-stress problems, however, larger temperature variations are possible; therefore, the change in S_A should be taken into account by adjusting the gage factor as the temperature changes during the experiment.

The effects of gage-grid elongation, base-material elongation and increase in gage resistance with increases in temperature combine to produce a temperature-induced change in resistance of the gage $(\Delta R/R)_{\Delta T}$ which can be expressed as:

$$\left(\frac{\Delta R}{R}\right)_{\Delta T} = (\beta - \alpha)S_g\Delta T + \gamma\Delta T \qquad (6.12)$$

where α and β are the thermal coefficient of expansion of gage and base materials, respectively.
γ is the temperature coefficient of resistivity of gage material.
S_g is the gage factor.

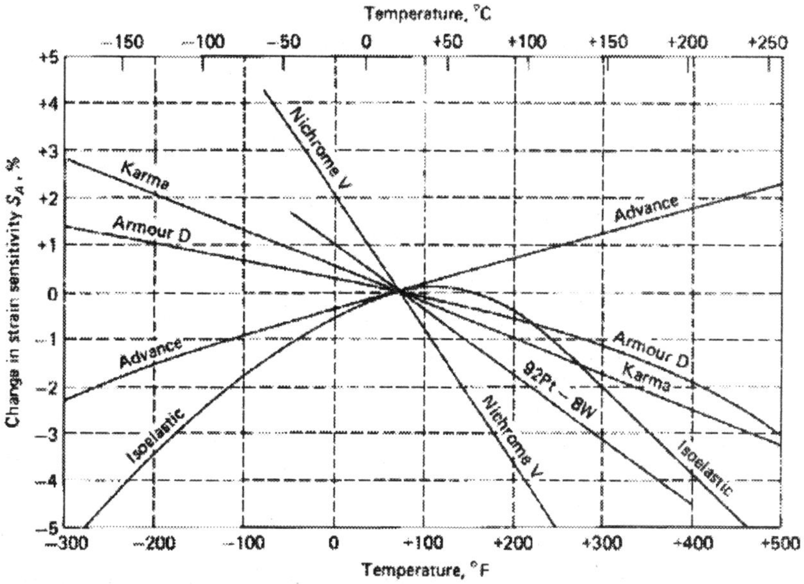

Figure 6.15 Change in alloy sensitivity S_A of several strain-gage alloys as a function of temperature.

If there is a differential expansion between the gage and the base material due to temperature change (that is, $\sigma \neq \beta$), the gage will be subjected to a mechanical strain $\varepsilon = (\beta - \alpha)\Delta T$, which does not occur in the specimen. The gage reacts to this strain by indicating a change in resistance in the same manner that it indicates a change for a strain due to the load applied to the specimen. Unfortunately, it is impossible to separate the apparent strain due to the change in temperature from the strain due to the applied load. If the gage alloy and the base material have identical coefficients of expansion this component of the thermally induced $(\Delta R/R)_{\Delta T}$ vanishes. The gage may still register a change of resistance with temperature, however, if the coefficient of resistivity γ is not zero. This component of $(\Delta R/R)_{\Delta T}$ indicates an apparent strain, which does not exist in the specimen.

Two approaches can be employed to affect temperature compensation in a gage system. The first involves compensation in the gage so that the net effect of the three factors in Eq. (6.12) is canceled out. The second involves compensation for the effects of the temperature change in the signal conditioner required to convert $\Delta R/R$ to voltage output. The second method will be discussed in Chapter 7 when instrument systems are considered in detail.

In producing temperature-compensated gages it is possible to obtain compensation by perfectly matching the coefficients of expansion of the base material and the gage alloy while holding the temperature coefficient of resistivity at zero. Compensation can also be obtained with a mismatch in the coefficients of expansion if the effect of a finite temperature coefficient of resistivity cancels out the effect of the mismatch in temperature coefficients of expansion.

The values of the factors α and γ influencing the temperature response of the strain gage mounted on a specimen with thermal characteristics specified by the value of β are quite sensitive to the composition of the strain-gage alloy, its impurities and the degree of cold working used in its manufacture. It is common practice for strain-gage manufacturers to determine the thermal response characteristics of sample gages from each lot of alloy material that they employ in their production. Because of variations in α and γ between each melt and each roll of foil, it is possible to select foils of Advance and Karma alloys, which are suitable for use with almost any type of base material. The gages produced by using this selection technique are known as **selected-melt** or **temperature-compensated** gages and are commercially available with the self-temperature-compensating numbers listed in Table 6.4. Some widely used materials and approximate values for their temperature coefficients of expansion are also listed in this table.

Unfortunately, these gages are not perfectly compensated over a wide range in temperature because of the nonlinear character of both the expansion coefficients and the resistivity coefficients with temperature. A typical curve showing the apparent strain (temperature-induced) as a function of temperature for a temperature-compensated strain gage fabricated from Advance alloy is shown in Fig. 6.16. These results show that the errors introduced by small changes in temperature in the neighborhood of 75 °F (24 °C) are quite small with apparent strains of less than 1 $\mu\varepsilon$/°F or ½ $\mu\varepsilon$/°C. However, when the change in temperature is large, the apparent strains can become significant, and corrections to account for the thermally induced apparent strains are necessary. These corrections involve measurement of the test temperature at the gage site with a thermocouple and use of a calibration curve similar to the one shown in Fig. 6.16. Note that a calibration curve is provided by the strain-gage manufacturer for each lot of gages produced.

The range of temperature over which an Advanced alloy strain gage can be employed is approximately from – 20 to 380 °F (–30 to 193 °C). For temperatures above and below these respective limits, the gage will function; however, very small changes in temperature will produce large apparent strains that can be difficult to account for properly in the analysis of the data.

An extended range of test temperatures is obtained with gages fabricated with Karma alloy. The thermally induced apparent strains as a function of temperature for the Karma alloy are also shown in Fig. 6.16. These data indicate that the strain-temperature curve has a modest slope over the entire range of temperature from – 100 to 500 °F (– 73 to 260 °C). Temperature measurements are still required whenever temperature variations are large; however, the small slope of the calibration curve makes it possible to accurately account for the thermally induced apparent strains.

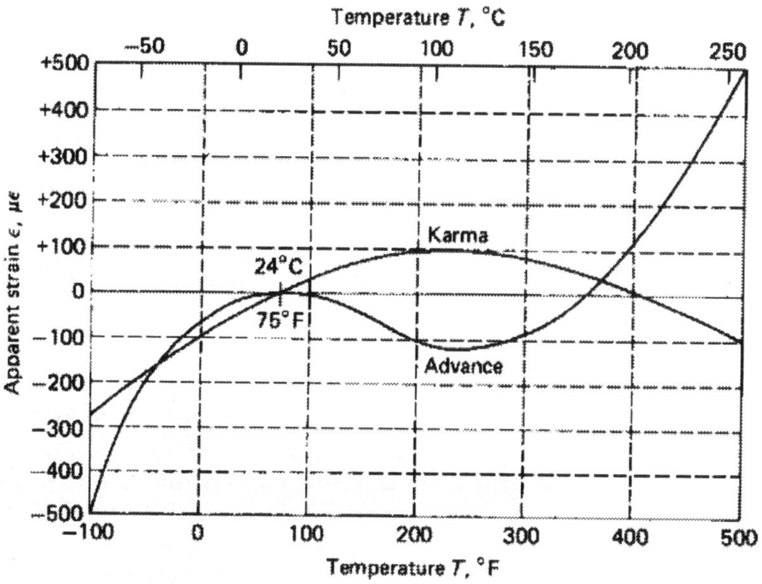

Figure 6.16 Apparent strain with temperature for Advance and Karma alloy temperature-compensated strain gages mounted onto a specimen having a matching temperature coefficient of expansion.

Table 6.4
Expansion coefficients available in temperature-compensated gages

Specimen Material	Expansion Coefficient $10^{-6}/°F$	Expansion Coefficient $10^{-6}/°C$	Self-temperature compensation #
Invar	0.8	1.4	00
Quartz	0.3	0.5	00
Titanium Silicate	0.0	0.0	00
Alumina	3.0	5.4	03
Molybdenum	2.7	4.9	03
Tungsten	2.4	4.3	03
Zirconium	3.1	5.6	03
Glass	5.1	9.2	05
410 Stainless Steel	5.5	9.9	05
Titanium Alloy 6Al-4V	4.9	8.8	05
Beryllium	6.4	11.5	06
Cast iron	6.0	10.8	06
Inconel X	6.7	12.1	06
Monel	7.5	13.5	06
Nickel A	6.6	11.9	06
4340 Steel	6.3	11.3	06
1018 Steel	6.7	12.1	06
17-7PH Stainless Steel	5.7	10.3	06
Beryllium Copper	9.3	16.7	09
Phosphor Bronze	10.2	18.4	09
304 Stainless Steel	9.6	17.3	09
310 stainless Steel	8.0	14.4	09
2024-T4 Aluminum	12.9	23.2	13
Cu 70, 30 Zn Brass	11.1	20.0	13
Tin	13.0	23.4	13
AZ-31B Magnesium	14.5	26.1	15

6.6.3 Elongation Limits

The maximum strain that can be measured with a foil strain gage depends on the gage length, the foil alloy, the carrier material and the adhesive. The Advance and Karma alloys with polyimide carriers, used for general-purpose strain gages, can be employed to strain limits of ± 5 and ± 1.5% strain, respectively. This strain range is adequate for elastic analyses on metallic and ceramic components, where yield or fracture strains rarely exceed 1%; however, these limits can easily be exceeded in plastic analyses, where strains in the post yield range can become large. In these instances, a special post yield gage is normally employed; it is fabricated using a double annealed Advance foil grid with a high-elongation polyimide carrier. Urethane-modified epoxy adhesives are generally used to bond post yield gages to the structure. If proper care is exercised in preparing the surface of the specimen, roughening the back of the gage, formulating a high-elongation plasticized adhesive system and attaching the lead wires without significant stress raisers, it is possible to approach strain levels of 20% before cracks begin to occur in the solder tabs or at the ends of the grid loops.

Special-purpose strain-gage alloys are not applicable for measurements of large strains. The Isoelastic alloy will withstand ± 2% strain; however, it undergoes a change of sensitivity at strains larger than 0.75%. Armour D and Nichrome V are primarily used for high-temperature measurements and are limited to maximum strain levels of approximately ± 1%.

For very large strains, where specimen elongations of 100% may be encountered, liquid-metal strain gages can be used. The liquid-metal strain gage is simply a micro-bore Tygon tube filled with mercury or a gallium-indium-tin alloy, as indicated in Fig. 6.17. When the specimen to which the gage is attached is strained, the volume of the tube cavity remains constant because Poisson's ratio of Tygon is approximately 0.5. Thus the length of the tube increases ($\Delta L = \varepsilon L$) while the diameter D of the tube decreases ($\Delta D = -\nu\varepsilon D$). The resistance of such a gage increases with strain, and it can be shown that the gage factor is given by:

$$S_g = 2 + \varepsilon \tag{6.13}$$

The resistance of a liquid-metal gage is very small (less than 1 Ω) because the capillary tubes used in their construction have a relatively large inside diameter 0.007 in. (0.18 mm). As a consequence, the gages are usually used in series with a large fixed resistor to form a total resistance of 120 Ω so that the gage can be monitored with a conventional Wheatstone bridge. The response of a liquid-metal gage, as shown in Fig. 6.18, is slightly nonlinear with increasing strain due to the increase in gage factor with strain.

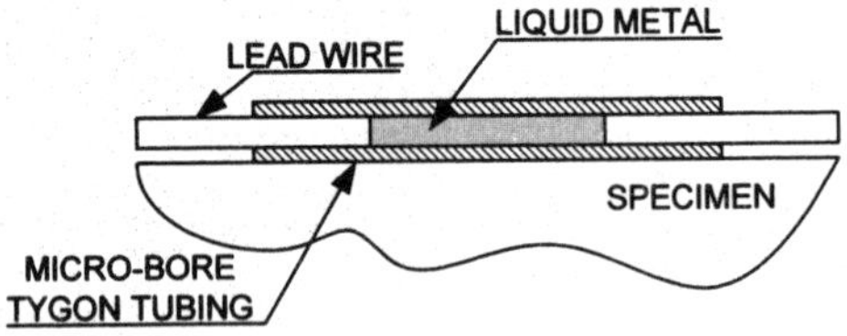

Figure 6.17 A liquid-metal, electrical resistance strain gage.

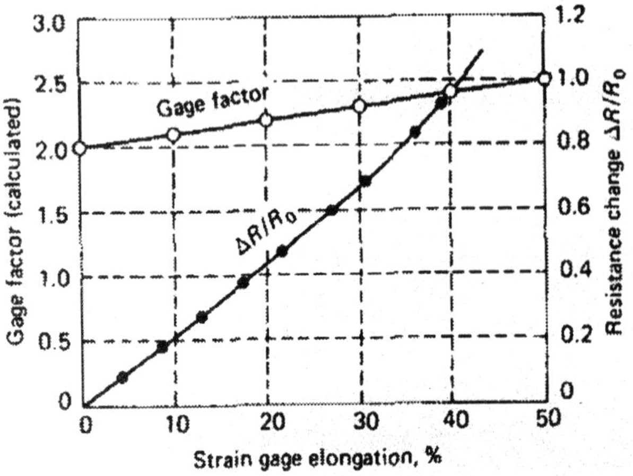

Figure 6.18 Resistance change and gage factor as a function of strain for a liquid metal strain gage. (After Harting).

6.6.4 Dynamic Response of Strain Gages

In dynamic applications of strain gages, the question of their frequency response often arises. This question can be resolved into two parts—the response of the gage in its thickness direction i.e., how long it takes for an element of the gage to respond to the strain in the specimen beneath it, and the response of the gage due to its length. It is possible to estimate the time required to transmit the strain from the specimen through the adhesive and carrier to the strain-sensing element by considering a gage mounted on a specimen as shown in Fig. 6.19. A strain wave is propagating through the specimen with velocity c_1. This specimen strain wave induces a shear-strain wave in the adhesive and carrier, which propagates with a velocity c_2. The transit time, which is given by $t = h/c_2$ equals 50 ns for typical carrier and adhesive combinations, where c_2 = 40,000 in./s (1000 m/s) and h = 0.002 in. (0.05 mm). The time required for the conductor to respond may exceed this transit time by a factor of 3 to 5; therefore, the response time should be approximately 200 ns. Experiments conducted by Oi and an analysis by Bickle indicate that transit times are approximately 100 ns.

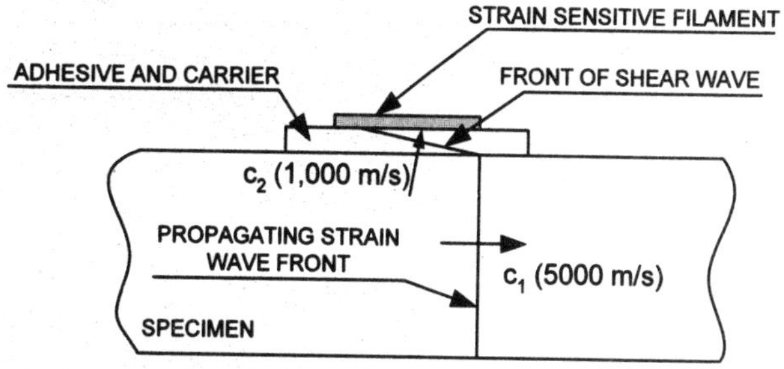

Figure 6.19 Dynamic strain transmission between the specimen and the gage.

The rise time in nanoseconds for a strain gage responding to a step pulse is given by:

$$t_r = (L_0/c_1) + 100 \qquad (6.14)$$

where L_0 is the length of the gage and the 100 ns term is added to account for the transmission time through the carrier and adhesive.

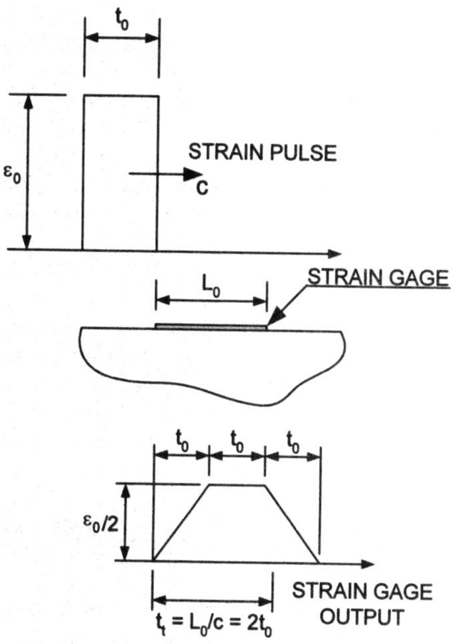

Figure 6.20 Dynamic response of a gage to a step-pulse.

A typical rise time for a 0.125-in. (3.17-mm) gage mounted on a steel bar (c_2 = 200,000 in./s or 5000 m/s) is 600 + 100 = 700 ns. Such short rise times are often neglected when long strain pulses are encountered and the rate of change of strain with time is small. When short, steep-fronted strain pulses are encountered, however, the response time of the gage should be considered because the measured strain pulse can be distorted, as shown in Fig. 6.20. In this instance, the gage is mounted on a specimen in which a strain pulse is propagating with amplitude ε_0, a time duration t_0 and a velocity c. The front of the pulse will reach and just pass over a gage of length L_0 in a transit time $t_t = L_0/c$. In Fig. 6.20 the gage length L_0 is selected so that the transit time t_t equals $2t_0$. If the gage records average strain over its length, its output will rise linearly to a value of $\varepsilon_0/2$ over a time period of t_0. The output will then remain constant at $\varepsilon_0/2$ for a period of time equal to t_0 and then decrease linearly to zero over a final time period of t_0. The effect of the gage length in this example was to decrease the amplitude of the output by a factor of 2 and to increase the total time duration of the pulse by a factor of 3. The distortion of the pulse as indicated by the gage will depend upon the ratio t_t/t_0; and as this ratio goes to zero, the distortion vanishes.

It is possible to correct for this distortion, by noting that the indicated strain ε at time t is given by:

$$\bar{\varepsilon}(t) = \frac{1}{L_0}\int_{c_1t-L_0}^{c_1t} \varepsilon(x)dx \tag{6.15}$$

Differentiating Eq. (6.15) gives:

$$\frac{d\bar{\varepsilon}}{dt} = \frac{1}{L_0}\left\{\int_{c_1t-L_0}^{c_1t} \frac{\partial}{\partial t}[\varepsilon(x)]dx + \varepsilon(a)c_1 - \varepsilon(b)c_1\right\} \tag{6.16}$$

where ε(a) and ε(b) are the strains at the two ends of the gage. If the pulse shape remains constant during propagation,

$$\frac{\partial}{\partial t}[\varepsilon(x)]dx = 0$$

and Eq. (6.16) reduces to:

$$\frac{d\bar{\varepsilon}}{dt} = \frac{c_1}{L_0}\{\varepsilon(a) - \varepsilon(b)\} \tag{6.17}$$

Because

$$\varepsilon(b,t) = \varepsilon\left(a, t - \frac{L_0}{c_1}\right)$$

then

$$\varepsilon(a,t) = \frac{L_0}{c_1}\frac{d\bar{\varepsilon}}{dt} + \varepsilon\left(a, t - \frac{L_0}{c_1}\right) \tag{6.18}$$

Numerical methods are used to solve Eq. (6.18).

6.6.5 Heat Dissipation

It is well recognized that temperature variations can significantly influence the output of strain gages, particularly those that are not properly temperature compensated. The temperature of the gage is of course influenced by ambient-temperature variations and by the power dissipated in the gage when it is connected into a Wheatstone bridge or a potentiometer circuit. The power P is dissipated in the form of heat and the temperature of the gage must increase above the ambient temperature to dissipate the heat. The exact temperature increase required is very difficult to specify because many factors influence the heat balance for the gage. The heat to be dissipated depends upon the voltage applied to the gage and the gage resistance. Thus:

$$P = \frac{V^2}{R} = I^2R \tag{6.19}$$

where P = power, W; I = gage current, A; R = gage resistance, Ω; and V = voltage across the gage, V.

Factors that govern the heat dissipation include:

1. Gage size, w_0 and L_0.
2. Grid configuration, spacing and size of conducting elements.
3. Carrier, type of polymer and thickness.
4. Adhesive, type of polymer and thickness.
5. Specimen material, thermal diffusivity.
6. Specimen volume in the local area of the gage.
7. Type and thickness of overcoat used to waterproof the gage.
8. Velocity of the air flowing over the gage installation.

A parameter often used to characterize the heat-dissipation characteristics of a strain-gage installation is the power density P_D, which is defined as:

$$P_D = P/A \tag{6.20}$$

where P is the power that must be dissipated by the gage and A is the area of the grid of the gage. Power densities that can be tolerated by a gage are strongly related to the specimen which serves as the heat sink because conduction to the sink is much more significant than convection to the air. Recommended values of P_D for different materials and conditions are listed in Table 6.5.

Table 6.5
Allowable power densities

Power Density P_D		**Specimen Conditions**
W/in^2	mW/mm^2	
5-10	8-16	Heavy aluminum or copper sections
2-5	3-8	Heavy steel sections
1-2	1.5-3	Thin steel sections
0.2-0.5	0.3-0.8	Fiberglass, glass and ceramics
0.02-0.05	0.03-0.08	Unfilled plastics

When a Wheatstone bridge with four equal arms is employed, the bridge excitation voltage V_s is related to the power density on the strain gage by:

$$V_s^2 = 4AP_D\,R \tag{6.21}$$

Allowable bridge voltages for specified grid areas and power densities are shown for the 120-Ω gages in Fig. 6.21. Typical grid configurations are identified along the abscissa in the figure to illustrate the effect of gage size on allowable bridge excitation. It should be noted that small gages mounted on a poor heat sink ($P_D < 1$ W/in² or 1.5 mW/mm²) result in lower allowable bridge voltages than those employed in many commercial strain indicators (3 to 5 V). In these instances, it is necessary; to use a higher-resistance gage (350 Ω or 1,000 Ω in place of 120 Ω) or a gage with a larger grid area.

6.6.6 Stability

In some strain-gage applications it is necessary to record strains over a period of months or years without having the opportunity to unload the specimen and recheck the zero resistance. The duration of the readout period is important and makes this application of strain gages one of the most difficult. All the factors, which can influence the behavior of the gage, have an opportunity to do so; moreover, there is enough time for the individual contribution to the error from each of the factors to become quite significant. For this reason, it is imperative that every precaution be taken in employing the resistance-type gage if meaningful data are to be obtained.

Drift in the zero reading from an electrical-resistance strain-gage installation is due to the effects of moisture or humidity variations on the carrier and the adhesive, the effects of long-term stress relaxation of the adhesive, the carrier, and the strain-gage alloy, and instabilities in the resistors in the inactive arms of the Wheatstone bridge.

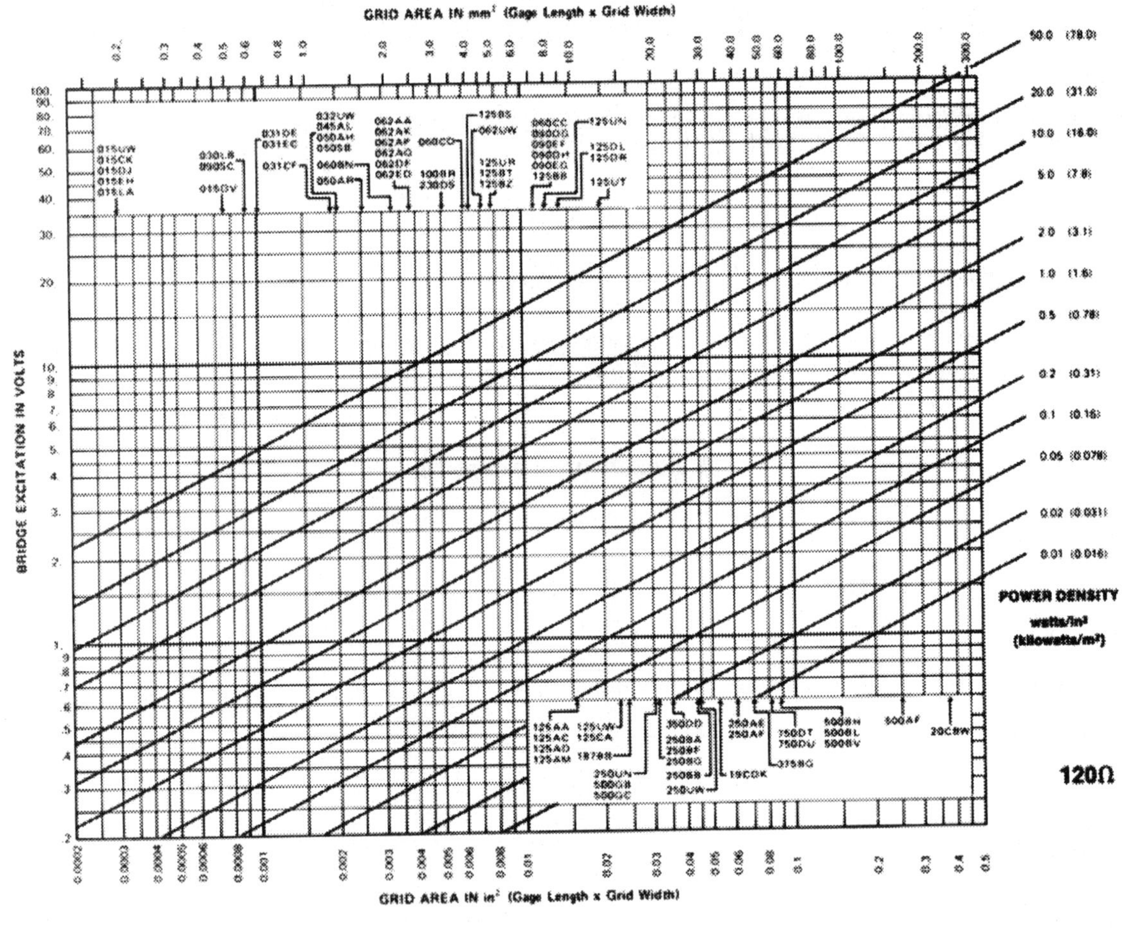

Figure 6.21 Allowable bridge voltage as a function of grid area with power density as a parameter for 120-Ω gages.

Results of an interesting series of stability tests by Freynik are presented in Fig. 6.22. In evaluating a typical general-purpose strain gage with a grid fabricated from Advance alloy and a polyimide carrier, zero shifts of 270 με were observed after 30 days. Because this installation was carefully waterproofed, the large drifts were attributed to stress relaxation in the polyimide carrier over the period of observation.

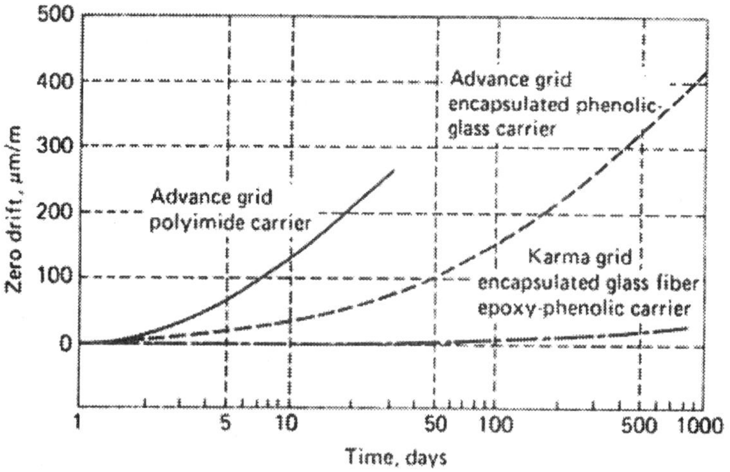

Figure 6.22 Zero drift as a function of time for several gage types at 167 °F (65 °C). (Data from Freynik)

Results from a second strain-gage installation with a grid fabricated from Advance alloy and a carrier from glass-fiber-reinforced phenolic were more satisfactory with zero drift of approximately 100 με after 50 days. The presence of the glass fibers essentially eliminated drift due to stress relaxation in the carrier material; the drift measured was attributed to instabilities in the Advance alloy grid at the test temperature of 167 °F (75 °C). The final and most satisfactory results were obtained with a Karma grid and an encapsulating glass-reinforced epoxy-phenolic carrier. In this case, the zero shift averaged only 30 με after an observation period of 900 days. In similar tests, with this gage installation at room temperature, the zero drift was only – 25 με. These results show that electrical-resistance strain gages can be used for long-term measurements provided Karma grids with glass-fiber-reinforced epoxy-phenolic carriers are employed with a well-cured epoxy adhesive system. The gage installation should be waterproofed to minimize the effects of moisture penetration. It is also important to specify hermetically sealed bridge-completion resistors to ensure stability of the Wheatstone bridge over the long observation periods.

6.7 ENVIRONMENTAL EFFECTS

The environment markedly affects the performance of resistance strain gages. Moisture, temperature extremes, hydrostatic pressure, nuclear radiation and cyclic loading produce changes in gage behavior which must be accounted for in the installation of the gage and in the analysis of the data to obtain meaningful results. Each of these parameters is discussed in the following subsections.

6.7.1 Effects of Moisture and Humidity

A strain-gage installation can be detrimentally affected by direct contact with water or by the water vapor normally present in the air. Both the adhesive and carrier absorb water, and the gage performance is affected in several ways. First, the moisture decreases the gage-to-ground resistance. If this value of resistance is reduced sufficiently, the effect is the same as that of placing a shunt resistor across the active gage. The water also degrades the strength and rigidity of the bond and reduces the effectiveness of the adhesive in transmitting the strain from the specimen to the gage. If this loss in adhesive strength or rigidity is sufficient, the gage will not develop its stated calibration factors and measuring errors are introduced.

Plastics also expand when they absorb water and contract when they release it; thus, any change in the moisture concentration in the adhesive will produce strains in the adhesive, which will in turn be transmitted to the strain gage. These moisture-induced adhesive strains will produce a strain-gage response that cannot be separated from the response due to the applied mechanical strain. Finally, the presence of water in the adhesive will cause electrolysis when current passes through the gage. During the electrolysis process, the gage filament will erode and a significant increase in resistance will occur. Again, the strain gage will indicate a tensile strain due to this electrolysis that cannot be differentiated from the applied mechanical strain.

Many methods for waterproofing strain gages have been developed; however, the extent of the measures taken to protect the gage from moisture depends to a large degree on the application and the extent of the gage exposure to water. For normal laboratory work, where the readout time is relatively short, a thin layer of microcrystalline wax or an air-drying polyurethane coating is usually sufficient to protect the gage installation from moisture in the air. For much more severe applications, e.g., prolonged exposure to seawater, it is necessary to build up a seal out of soft wax, synthetic rubber, metal foil and a final coat of rubber. A cross section of a well-protected gage installation is shown in Fig. 6.23. Care should be exercised in forming the seal at the lead-wire terminal because the seal usually fails at this location. Also, the lead wire insulation should be rubber, and splices in the cable must be avoided. If water gains entry to the lead wires it will be transmitted for significant distances by capillary action to the gage installation.

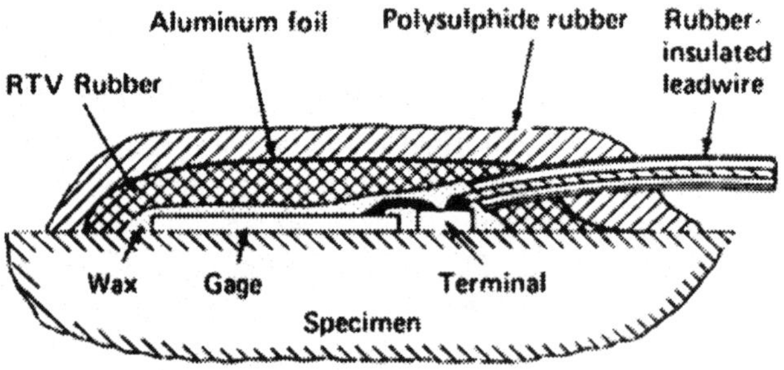

Figure 6.23 Waterproofing a strain gage for severe seawater exposure.

6.7.2 Effects of Hydrostatic Pressure

In the stress analysis of pressure vessels and piping systems, strain gages are frequently employed on interior surfaces where they are exposed to a gas or fluid pressure that acts directly on the sensing element of the gage. Under such conditions, pressure-induced resistance changes occur which must be accounted for in the analysis of the strain-gage data.

Milligan and Brace independently studied this effect of pressure by mounting a gage on a small specimen, placing the specimen in a special high-pressure vessel, and monitoring the strain as the pressure was increased to 140,000 psi (965 MPa). In this type of experiment, the hydrostatic pressure p produces a strain in the specimen, which is given by Eq. (2.16) as:

$$\varepsilon = \frac{-p}{E} - \frac{\nu}{E}[(-p)+(-p)] = -\frac{1-2\nu}{E}p = K_T p \qquad (6.22)$$

where $K_T = -(1-2\nu)/E$ is often referred to as the compressibility constant for a material. The strain gages were monitored during the pressure cycle, and it was observed that the indicated strains were less than the true strains as predicted by Eq. (6.22).

The pressure effect can be characterized by defining the slope of the pressure-strain curve as an apparent compressibility constant for the material. Thus:

$$K_1 = -\frac{\Delta\varepsilon}{\Delta p} \qquad (6.23)$$

The difference D_p between the observed and predicted effects due to pressure can then be written as:

$$D_p = \frac{K_T - K_1}{K_T} \qquad (6.24)$$

Experimental results by Milligan, shown in Fig. 6.24, indicate that D_p depends upon the compressibility constant for the specimen material, the curvature of the specimen where the gage is mounted, and the type of strain-sensing alloy used in fabricating the gage. In spite of relatively large values for D_p, however, it was observed that the strain response of the gage remained linear with pressure.

The true strain ε_t can be expressed in terms of the indicated strain ε_i and a correction term ε_{cp} as:

$$\varepsilon_t = \varepsilon_i - \varepsilon_{cp} \qquad (6.25)$$

The magnitude of the correction term ε_{cp} is expressed in terms of the experimentally determined values of D_p as:

$$\varepsilon_{cp} = D_p K_T p \qquad (6.26)$$

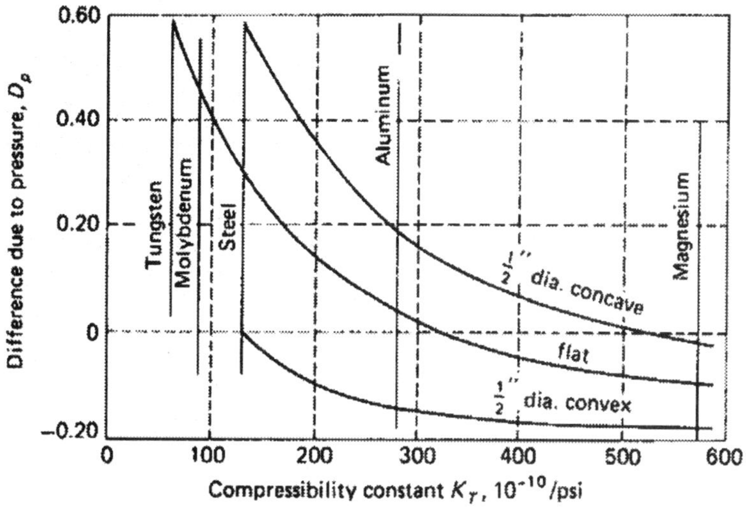

Figure 6.24 Difference D_p due to pressure as a function of the compressibility constant K_T for Advance foil gages on specimens with different curvature. (From data by Milligan.)

For a flat steel specimen, $K_T = 133 \times 10^{-10}$ in²/lb (1.93×10^{-12} m²/N) and $D_p = 0.3$; hence, the correction ε_{cp} is approximately 4 $\mu\varepsilon$ for a pressure of 1,000 psi (7 MPa). Because this is a very small correction, it is usually possible to neglect pressure effects for pressures less than approximately 3,000 psi (20 MPa).

For hydrostatic pressure applications, foil strain gages with the thinnest possible carriers should be employed. The gage should be mounted on a smooth surface with a thinned adhesive to obtain the thinnest possible bond line. Bubbles in the adhesive layer cannot be tolerated because the pressure normal to the surface of the gage will force the sensing element into any void beneath the gage and additional erroneous resistance changes will result.

6.7.3 Effects of Nuclear Radiation

Several difficult problems are encountered when electrical-resistance strain gages are employed in nuclear-radiation fields. The most serious difficulty involves the change in electrical resistivity of the strain gage and lead wires as a result of the fast-neutron dose. This effect is significant; changes of 2 to 3 percent in $\Delta R/R$ have been observed with a neutron dose of 10^{18} nvt. These changes in resistivity produce zero drift with time that can be as large as an apparent strain of 10,000 to 15,000 $\mu\varepsilon$. The exact rate of change of resistivity is a function of the strain-gage and lead-wire materials, the state of strain in the gage and the temperature. Typical changes in resistance with integrated fast-neutron flux are shown in Fig. 6.25. Because the changes in resistivity are a function of the magnitude and sign of the strain, the use of dummy gages to cancel the effects of radiation exposure is not effective. Because the change in electrical resistivity appears to be a linear function of the logarithm of the dose, the most satisfactory solution to neutron-induced zero drift is to employ preexposed strain-gage installations and to reduce test times to a minimum. Frequent unloading and reestablishing the zero resistance of the gage is essential.

The neutrons also produce a change in the sensitivity of the strain-gage alloys. Variations in S_A for Advance alloy, shown in Fig. 6.26, are large and unpredictable when the integrated neutron exposure exceeds 10^{16} nvt.

The fast neutrons also produce mechanical effects which are detrimental to strain-gage installations. With exposure to the fast neutrons, the strain-gage alloy exhibits an increase in its yield strength and modulus of elasticity and a decrease in elongation capability. Radiation-induced cross-linking in polymers also destroys the original organic structure of the bond. For this reason, ceramic adhesives are normally employed in any long-term tests where exposure will accumulate.

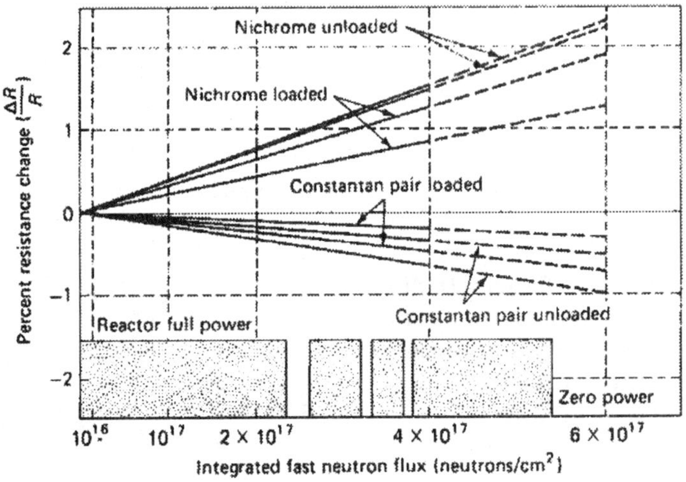

Figure 6.25 Percent change in resistance due to of exposure to neutron flux. (From data by Tallman.)

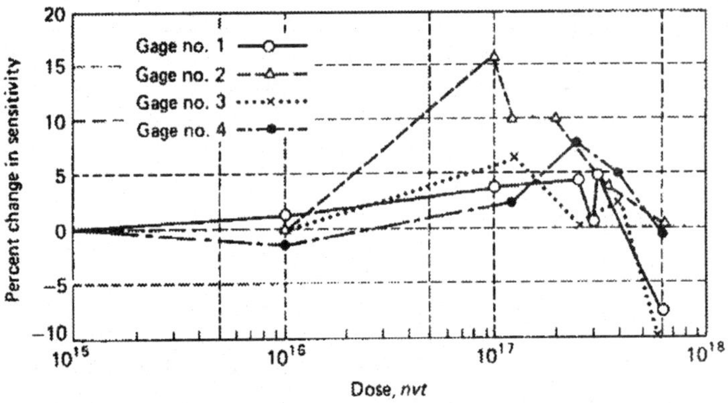

Figure 6.26 Percent change in sensitivity of an Advance alloy strain gage with exposure to fast neutrons. (From data by Tallman.)

In nuclear-radiation fields with high gamma flux, considerable energy is transferred to the gage and specimen; therefore, temperature changes can be large. For precise measurements, the temperature change must be predicted or determined so that the strain-gage results can be corrected for the effects of temperature. It is possible to measure strains in strong radiation fields; however, the special precautions must be taken or serious errors will occur.

6.7.4 Effects of High Temperature

Resistance-type strain gages can be employed at elevated temperatures for both static and dynamic stress analyses; however, the measurements require many special precautions, which depend primarily on the temperature and the time of observation. At elevated temperatures, the resistance R of a strain gage must be considered to be a function of temperature T and time t in addition to strain ε. Thus:

$$R = f(\varepsilon, T, t) \tag{6.27}$$

The resistance change ΔR/R is then given by:

$$\frac{\Delta R}{R} = \frac{1}{R}\frac{\partial f}{\partial \varepsilon}\Delta\varepsilon + \frac{1}{R}\frac{\partial f}{\partial T}\Delta T + \frac{1}{R}\frac{\partial f}{\partial t}\Delta t \tag{6.28}$$

where $\frac{1}{R}\frac{\partial f}{\partial \varepsilon} = S_g$ is the gage sensitivity to strain (gage factor); $\frac{1}{R}\frac{\partial f}{\partial T} = S_T$ is the gage sensitivity to temperature and $\frac{1}{R}\frac{\partial f}{\partial t} = S_t$ is the gage sensitivity to time.

Equation (6.28) can then be expressed in terms of the three sensitivity factors as:

$$\Delta R/R = S_g\,\Delta\varepsilon + S_T\,\Delta T + S_t\,\Delta t \tag{6.29}$$

In the previous discussion of performance characteristics of foil strain gages, it was shown that the sensitivity of the gages to temperature and time was made negligibly small at normal operating temperatures of 0 to 150 °F (–18 to 65 °C) by proper section of the strain-gage alloy and carrier materials. As the test temperature increases above this level, however, the performance of the gage deteriorates, and S_T and S_t are not usually negligible.

As the temperature increases, temperature compensation is less effective, and corrections must be made to account for the apparent strain, as shown in Fig. 6.16. Comparisons of these results indicate that the Karma strain-gage alloy is more suitable for the higher-temperature applications than Advance. The Karma gages can be employed to temperatures up to about 500 °F (260 °C) with temperature-induced apparent strains of less than ± 100 με.

The stability of a strain-gage installation is also affected by temperature; and strain-gage drift becomes a more serious problem as the temperature and the time of observation are increased. Stability is affected by stress relaxation in the adhesive bond and in the carrier material and by metallurgical changes (phase transformations and annealing) in the strain-gage alloy. The carrier material controls the upper temperature limit on commercially available Karma gages. Glass-reinforced epoxy-phenolic carriers are rated at 550 °F (288 °C); however, Karma gages with this type of carrier drift with time as shown in Fig. 6.27. If the time of loading and observation is long, corrections must be made for the zero drift. Drift rates will depend upon both the strain level and the temperature. For high-temperature strain analyses, it is recommended that a series of strain-time calibration curves, similar to the one shown in Fig. 6.27, be developed to cover the range of strains and temperatures to be encountered. Zero-drift corrections can then be taken from the appropriate curve.

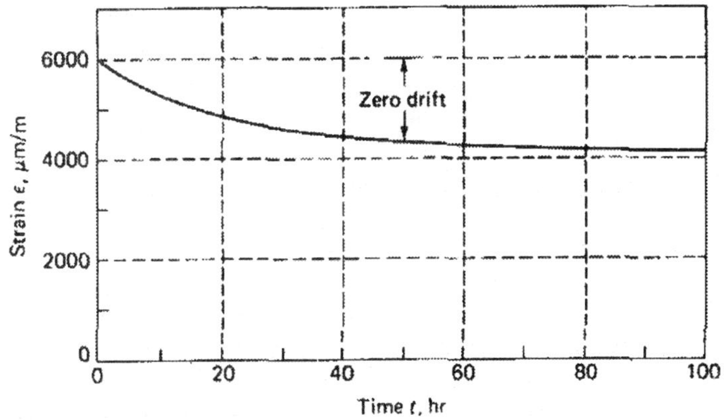

Figure 6.27 Zero drift with time for a Karma alloy strain gage with a glass fiber epoxy phenolic carrier at 560 °F (293 °C). (Data from Hayes.)

Changes in the gage factor with temperature are relatively small as indicated in Fig. 6.15. Also, because the change is linear with temperature, corrections can be made to accurately account for this effect.

The problem of gage stability and apparent strain due to temperature changes is greatly reduced if the period of observation is short. For dynamic analyses, where relatively short strain-time records are used (times usually less the 1 s), the temperature does not have sufficient time to change and temperature effects are therefore small. For this reason, dynamic strain-gage analyses can be made at very high temperatures with good precision if the proper alloy and adhesives are employed.

Resistance-strain-gage measurements at temperature higher than 550 °F (288 °C) require special gages, special techniques for mounting the gage, and special procedures for monitoring the strain-gage signal. At these higher temperatures, polymeric materials can no longer be used for the carrier or the adhesive. The gages must be mounted to the specimen with the ceramic cements described previously. The carrier is either removed entirely or replaced with a thin stainless-steel shim. Strain gages for the very high temperature applications are fabricated using materials such as Nichrome V, Armour D, or alloy 479 (See Table 6.1). Of these different materials, the platinum-tungsten alloy is often preferred because of its inherent oxidation resistance and metallurgical stability. Performance characteristics of this alloy, shown in Fig. 6.28, indicate that the gage factor drops about 30 percent as the temperature increases from 70 to 1,600 °F (21 to 871 °C). Unfortunately, the temperature coefficient of resistance of this alloy is very high and large apparent strains (40 to 80 με/°F or 70 to 140 με/°C) are indicated by the gage when the temperature changes. The material is stable, and zero drift is negligible to temperatures of 800 °F (427 °C). At temperatures above 800 °F (427 °C), drift will occur with the rate of drift increasing with temperature.

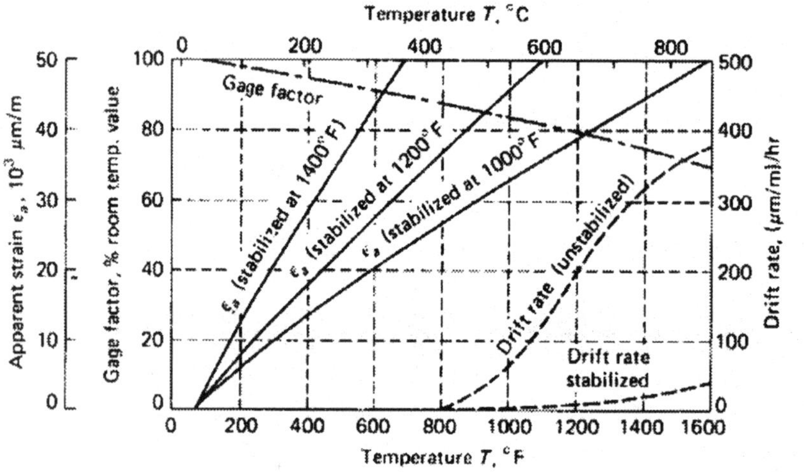

Figure 6.28 performance characteristics of a platinum-tungsten alloy strain gage as a function of temperature. (From data by Wnuk.)

In the early 1990s, NASA Glenn introduced an improved alloy for high temperature strain measurements. The alloy, palladium with 13% chromium by weight (hereafter, PdCr) is structurally stable and oxidation resistant up to at least 1,100 °C (2,000 °F). Its temperature-induced resistance change is linear and repeatable. The alloy is not sensitive to the rates of heating and cooling. Initial strain gages fabricated from 25-μm diameter PdCr wire provided reliable static strain measurements to 800 °C. By further improving the purity of the alloy and by developing gage fabrication techniques using sputter-deposition, photolithography patterning and chemical etching, a PdCr thin-film strain gage was developed. This thin film gage was capable of measuring dynamic and static strain to at least 1,100 °C. For static strain measurements, a thin film Pt element gage served as a temperature compensator to further minimize the temperature effect of the gage. These thin-film gages provide the advantage of minimally intrusive surface strain measurements and give highly repeatable readings with low drift at temperatures from ambient to 1,100 °C. This development represents a 300 °C advance in operating temperature over the PdCr wire gage and a 500 °C advance over commercially available gages made of other materials.

Stabilizing the alloy can minimize the effects of drift rate. The stabilization process consists of annealing the alloy at the test temperature for 12 to 16 h. The stabilized drift is relatively low, with rates of 20 με/h reported at 1,400 °F (760 °C). The effect of apparent strain with temperature is much more difficult to treat and usually requires the use of dual-element or four-element gages, which compensate for the effects of temperature by signal subtraction in the Wheatstone bridge. An example of a four-element (complete-bridge) gage is shown in Fig. 6.29. A reasonable degree of temperature compensation has been achieved (± 300 με) for temperatures up to 1400 °F (760 °C). The gage is useful over a wide range of temperature if the apparent strain is accounted for by correcting the gage output.

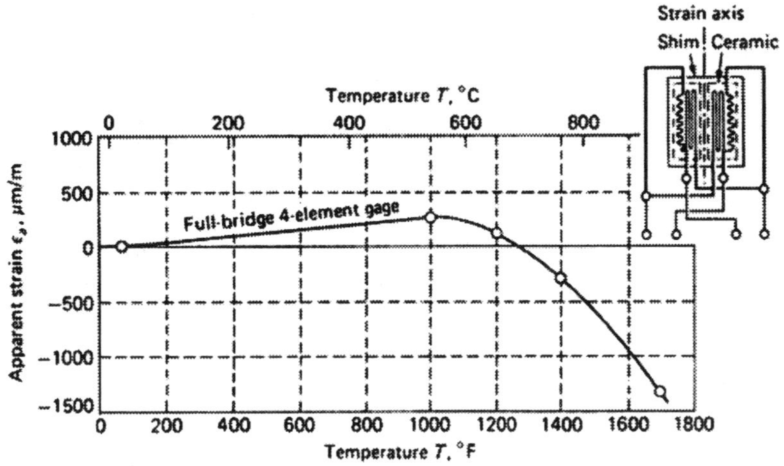

Figure 6.29 Apparent strain as a function of temperature for a four-element gage. (Data from Wnuk).

6.7.5 High Temperature Strain Measurements with Capacitive Gages

Measurement of static strains at high temperatures is difficult with resistance strain gages. The gage factor changes, the apparent strains are large, zero drift with time is significant, and even the ceramic adhesives which serves as the gage insulation begin to conduct at high temperature. Strain gages based on capacitive sensors are often more suitable than gages using resistive elements as sensors.

The CERL (Central Electricity Research Laboratory) gage, illustrated in Fig. 6.30, utilizes a simple parallel-plate capacitor mounted between two curved beam elements. The two curved beam elements are attached at their ends and the assembly is spot welded to the specimen. When the specimen is strained, the length L_0 of the gage changes, the curved beams deflect and the air gap between the plates change in proportion to the strain. The curved beam elements can be fabricated from the same material as the test specimen to provide automatic temperature compensation. Because the dielectric constant of air is nearly constant with temperature, the calibration constant is stable up to 650 °C.

The capacitance varies non-linearly with strain changing from about 1.3 to 0.4 pF as the strain increase from 0 to 1%. The use of a transformer ratio arm bridge operating at 50 V with a carrier frequency of 1.6 kHz permits the strain to be resolved to ± 1με.

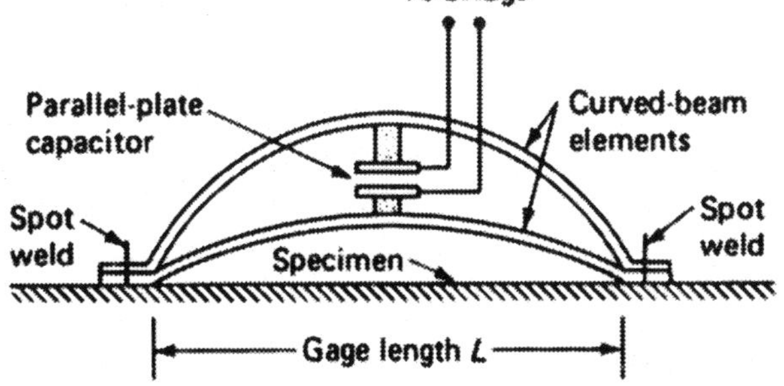

Figure 6.30 Schematic illustration of the CERL capacitance strain gage.

6.7.6 Effects of Cryogenic Temperatures

Strain can be measured with electrical-resistance strain gages to liquid-nitrogen temperatures (−320 °F or −196 °C) and below. Two factors must be carefully accounted for in using strain gages at this temperature extreme. The first effect is the change in gage factor with temperature, illustrated in Fig. 6.31. These results indicate that this correction is small—only −1.8 and +3.8 percent, respectively, for Advance and the selected-melt Karma alloy SK-15 at – 320 °F (– 196 °C).

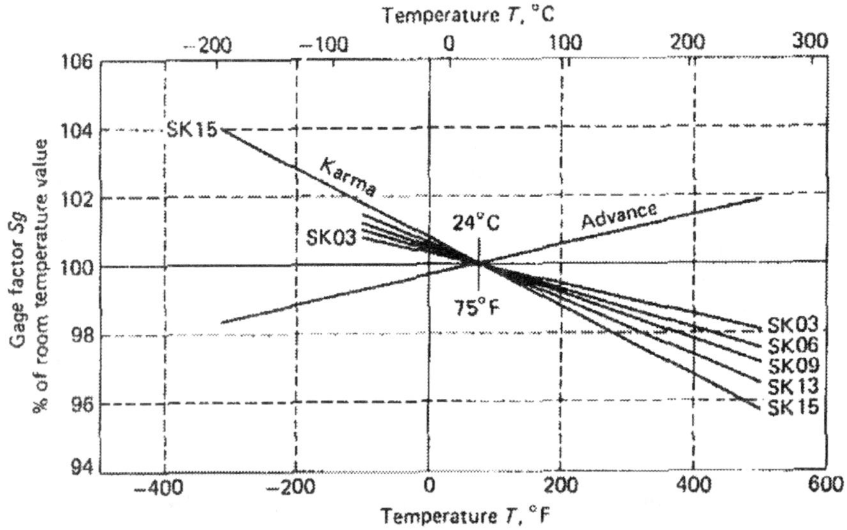

Figure 6.31 Changes in gage factor with temperature. (Data from Telinde.)

The second effect, which is extremely important to account for when measuring strains at cryogenic temperatures, is the very large apparent strains introduced by small changes in temperature. Karma alloys are preferred to Advance alloys because of their better stability at the temperature extremes, as shown in Fig. 6.16. Temperature-induced apparent strains for Karma strain gages mounted on a variety of materials are shown in Fig. 6.32.

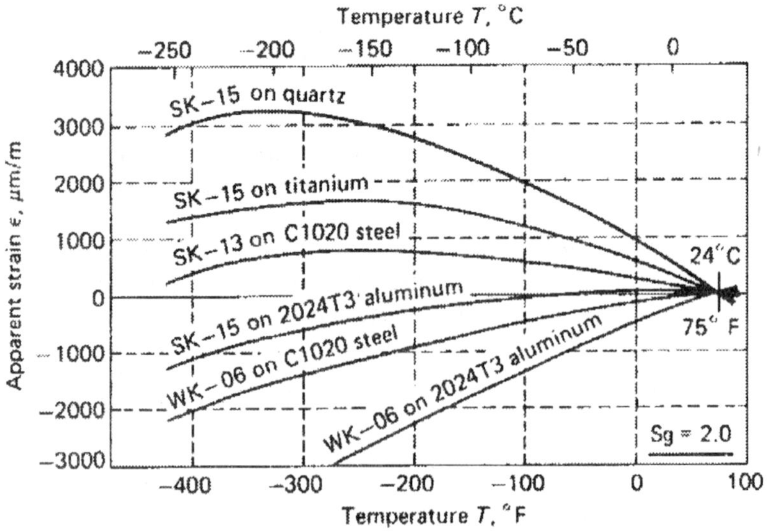

Figure 6.32 Apparent strain as a function of temperature for several selected-melt Karma alloys on different materials. (Data from Telinde.)

Although both Advance and Karma strain gages are temperature-compensated, the compensation is limited to the temperature range from – 100 to 500 °F (– 73 to 260 °C). As the test temperature is reduced below – 100 °F (– 73 °C), the strain gages become more sensitive to temperature changes and large apparent strains are produced. It is possible to minimize the effects of temperature changes at cryogenic temperatures by selecting compensating gages that are mismatched relative to the component material. For instance, the Karma gage compensated for a material with a temperature coefficient of expansion equal to 13 ppm/°F and 23.4 ppm/°C exhibits a relatively stable response to temperature variations from – 100 to – 400 °F (– 73 to – 240 °C) when mounted on a steel specimen having a temperature coefficient of expansion of 6 ppm/°F (10.8 ppm/°C) as shown in Fig. 6.32.

If temperature sensors are mounted with the strain gages, the measured temperatures can be used with curves similar to those shown in Fig. 6.32 to correct for the apparent strains. It is also possible to use a four-element gage, which is connected to produce signal cancellation in the Wheatstone bridge that compensates for the temperature-induced response of the strain gages.

Cryogenic temperatures are usually achieved in experiments by using liquid nitrogen, liquid hydrogen or liquid helium. All three are excellent insulating materials, and it is not necessary or advisable to use electrical insulation compounds between the gage grid and the cryogenic fluid. The gage should be coated with a silicone grease to provide a heat shield and to eliminate the possibility of liquid boiling over the gage grid during the experiment.

Special consideration must also be given to changes in the mechanical properties of component materials at cryogenic temperatures. The effect of cryogenic temperatures is to increase the elastic modulus from 5 to 20%. The higher values of the elastic modulus must be employed in the stress-strain equations when converting strain to stress.

6.7.7 Effects of Strain Cycling

Strain gages are frequently mounted on components subjected to cyclic loading, and the fatigue life of the component during which the gage must be monitored can exceed several million cycles. Three factors, which must be considered in a fatigue application, are zero shift, change in gage factor and failure of the gage in fatigue.

When the strain gage is subjected to repeated cyclic strain, the gage grid work hardens and its specific resistance changes. The specific resistance change produces a zero shift. The amount of this zero shift depends on the magnitude of the strain, the number of cycles, the grid alloy, the original state of cold work of the grid alloy, and the type of carrier employed in the gage construction. Typical examples of zero shift as a function of number of strain cycles are shown in Fig. 6.33 for four different gages. The poorest gage for this type of application is the open-faced Advance grid (EA type gage), which begins to exhibit noticeable zero shift at 10^3 cycles. Encapsulating the grid in a glass-reinforced epoxy-phenolic resin (WA type gage) improves the life, and exposures of 10^5 cycles at $\pm$ 2100 $\mu\varepsilon$ occur before zero shift becomes apparent. Further improvement can be achieved by using Karma (K) or Isoelastic (D) alloys fully encapsulated in the glass-reinforced epoxy-phenolic carrier material with factory-installed lead wires.

The changes in gage factor are quite small due to strain cycling and in general can be neglected if the zero shift is less than 200 $\mu\varepsilon$.

A very important factor to consider when using strain gages under fatigue conditions is the increase in sensitivity that occurs once a fatigue crack develops in the grid of the gage. The crack will usually develop in the grid near the lead wire on the tab. The presence of the crack produces a small increase in the resistance of the gage, which is monitored as an apparent strain. Thus, the apparent gage factor has increased with the development and subsequent propagation of the fatigue crack. Typical results showing this increase in gage factor are shown in Fig. 6.34. It is important to note that the gage factor increased from 10 to 32% for the three gages which failed in fatigue. Also, the error in "calibration" existed over an appreciable fraction of the life of the gage.

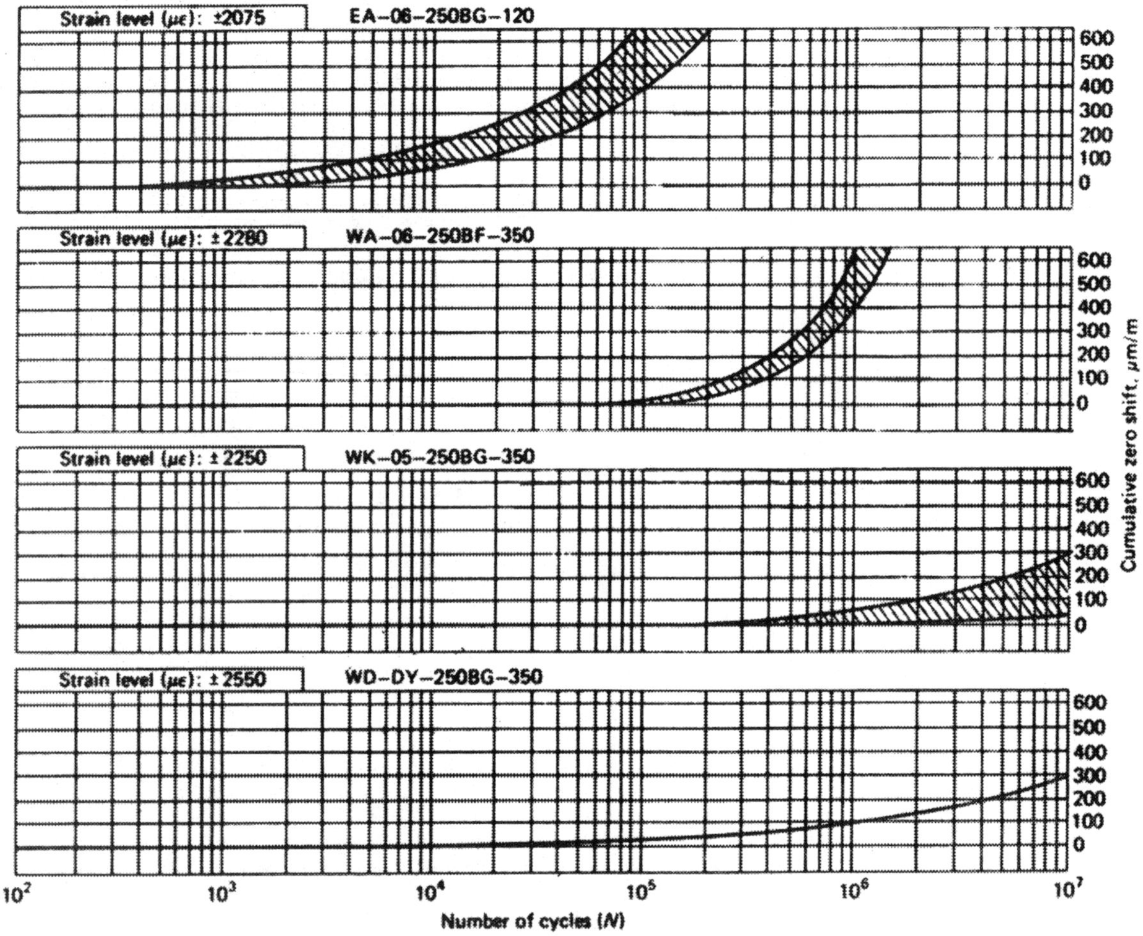

Figure 6.33 Zero shift as a function of fatigue exposure for (a) Advanced, open faced; (b) Advance, encapsulated grid; (c) Karma, encapsulated grid; (d) Isoelastic, encapsulated grid.

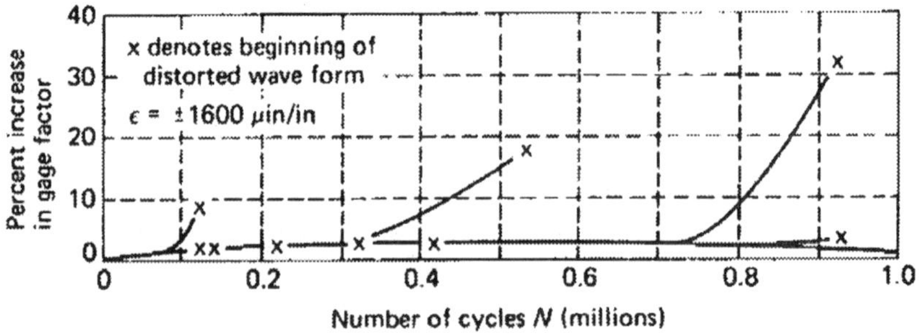

Figure 6.34 Increase in gage factor as a function of fatigue exposure for an annealed Advanced foil gage. (Measurements Group, Inc.)

6.8 SUMMARY AND FUTURE TRENDS

Since the mid-1950s and the introduction of the foil-grid construction and selected-melt gages, there has been no major developments in metallic alloy electrical-resistance strain gages. Continued research and development by the user community has led to an improved awareness of the range of application[2] of the gages and an appreciation of the degree of precision that can be obtained. Continued development and investment by strain gage manufacturers has led to a broad product line with very tightly controlled specifications that ensure a high-quality gage that will perform in a reliable and stable manner. As the electrical-resistance strain gage begins its seventh decade of development, it has become part of a mature technology, where advances are made in very small increments through the contributions of many users and suppliers.

More significant developments have occurred in strain gage instrumentation as improvements in semiconductor technology have led to high-gain stable amplifiers, more highly regulated power supplies, precision digital voltmeters, computer-controlled, multi-channel data acquisition systems and effective software for on-line computer processing of strain gage data in real time. These developments have reduced both the time and the costs and improved the accuracies achieved in experiments where strain gages are used in large numbers to characterize the stress field in complex structures.

The capability of high-gain amplifiers has inhibited the application of semiconductor strain gages for general-purpose stress analysis. The cost of the semiconductor gages is justified only in those cases where measurements of extremely low strains with great precision are required. Piezoresistive technology is of growing importance today in developing transducers to measure load and pressure. In these transducers, silicon is usually used as the mechanical element and the wiring for a complete Wheatstone bridge is incorporated directly into the silicon by utilizing integrated circuit manufacturing procedures. The result is a very high output voltage with excellent linearity and essentially negligible hysteresis.

New developments are occurring in the strain measurement field, but in an area where several relatively new technologies including fiber optics, laser diodes, photodetectors and CCD arrays are being combined to develop a wide variety of optical fiber sensors. Laboratory work by several investigators has lead to new and novel methods for measuring strain that extend the measurement capabilities of the experimentalist.

EXERCISES

6.1 The strain sensitivity of most metallic alloys is about 2.0. What portion of this sensitivity is due to dimension changes in the conductor? What portion is due to changes in the number of free electrons and their mobility?

6.2 Advance or Constantan alloy exhibits a linear response of $\Delta R/R$ to strain ε for strains as large as 8%. Discuss why this is a remarkably large range of linearity with respect to strain when the elastic limit of the material may be as low as 145 MPa and the modulus of elasticity is 152 GPa. Base your discussion on Eq. (6.3) and include the fact that Poisson's ratio for the material will increase from 0.3 to 0.5 as the material transforms from the elastic to the plastic state.

6.3 Describe the advantages and disadvantages of a strain gage fabricated using an Isoelastic alloy.

6.4 Describe the advantages and disadvantages of a strain gage fabricated using a Karma alloy.

---

[2] The development of the new strain gage alloy PdCr is an example of a development that has markedly extended the high temperature capabilities of the electrical resistance strain gage.

6.5 Table 5.2 shows S_A = 4.0 for a Platinum-Tungsten alloy. This sensitivity is higher than that of the Isoelastic alloy. Why isn't it used instead of:

the Isoelastic alloy.
the Karma alloy.
the Advance alloy.

6.6 List four different carrier materials used in strain-gage construction and give the reasons for their use.

6.7 Estimate the shear stress induced in the adhesive layer of a bonded strain gage subjected to 4000 $\mu\varepsilon$. The gage has an L/w ratio of 2 and is fabricated from advance alloy 30 μm thick. The carrier is polyimide with a thickness of 50 μm.

6.8 Why is post cure shrinkage of an adhesive used to bond a strain gage detrimental?

6.9 Briefly discuss the conditions which would dictate using the following cements in a strain-gage installation:

(a) Polyester (b) Epoxy cement
(c) Cyanoacrylate (d) Ceramic

6.10 Describe two procedures used to evaluate the completeness of the cure of an adhesive being used to bond a strain gage to a specimen.

6.11 Describe a procedure for determining if bubbles exist under a strain gage.

6.12 If you observe a bubble under the center of the grid will it affect the gage output if the gage is located in a field of: (a) uniform tensile strain. (b) uniform compressive strain.

6.13 Determine the transverse sensitivity factor for a strain gage fabricated using Constantan alloy with S_A = 2.11 if it has been calibrated and found to exhibit a gage factor S_g = 2.08.

6.14 The transverse sensitivity factor K_t for a new grid configuration must be determined. It has been suggested that a simple tension specimen can be used with gages oriented in the axial and transverse directions to make the determination. Develop an expression for K_t in terms of Poisson's ratio ν of the specimen material and the ratio of the outputs from the two strain gages.

6.15 A value of K_t = 0.03 was obtained for the gage in Exercise 6.14 by assuming Poisson's ratio ν = 0.30 for the specimen material. The true value of Poisson's ratio was 0.25. Determine the percent error introduced in the determination of K_t.

6.16 Determine the percent error introduced by neglecting transverse sensitivity if a WK-06-062DY-350 type strain gage is used to measure the longitudinal strain in a steel (ν = 0.30) beam.

6.17 Determine the percent error introduced by neglecting transverse sensitivity if a WK-06-250BG-350 type strain gage is used to measure the maximum tensile strain on the surface of a circular steel (ν = 0.30) shaft subjected to a pure torque.

6.18 Determine the percent error introduced by neglecting transverse sensitivity if an EA-06-125RA-120 type strain gage is used to measure the maximum tensile strain on the outside surface of a steel (ν = 0.29) thin-walled cylindrical pressure vessel which is being subjected to an internal pressure.

6.19 Specify a test procedure that will determine the zero shift associated with a strain-gage installation. Specify a second test procedure that will minimize this zero shift.

6.20 Determine the change in alloy sensitivity S_A of Constantan as the temperature is increased from −100 to +200 °C.

6.21 Determine the change in alloy sensitivity S_A of Karma alloy as the temperature is increased from −100 to +200 °C.

6.22 A strain-gage analysis is to be conducted on a component that will be maintained at (65 ± 20 °C) throughout the test period. What alloy should be specified for the strain-gage grid? If temperature-compensated gages were specified, what errors would occur due to the temperature fluctuations?

6.23 Repeat Exercise 6.22 by changing (65 ± 20 °C) to (200 ± 30 °C).

6.24 Repeat Exercise 6.22 by changing (65 ± 20 °C) to (24 ± 8 °C).

6.25 Specify the special procedures required for a strain-gage analysis where plastic strains of the order of 15 percent are anticipated.

6.26 For the liquid-metal strain gage illustrated in Fig. 6.17, show that the gage factor S_g is given by $S_g = 2 + \varepsilon$.

6.27 A strain wave is propagating in the x direction through a specimen with a characteristic velocity $c_1 = 5000$ m/s. The strain wave may be represented by the equation $\varepsilon = A \sin (2\pi/\lambda)(x - c_1 t)$, where A is the amplitude and λ is the wavelength of the wave. If $\lambda = nL_0$, determine the amplitude of strain recorded by a strain gage having a length L_0. Prepare a graph of the output as n varies from 0.1 to 10.

6.28 Determine the allowable bridge voltage for the grid configurations 015LA, 060BN, 125DL, 250BA, 375BG, and 500GH if 120-Ω strain gages are mounted on:

(a) An aluminum engine head. (b) A large steel beam.
(c) A fiberglass tank. (d) A steel ship hull.

6.29 Write a test specification outlining special procedures to be used in measuring strain over a 2-year period on a large steel smokestack subjected to wind loads.

6.30 Write a test specification to be followed by a laboratory technician in waterproofing a strain gage for:

(a) a short-duration test in the laboratory.
(b) a long-term test under seawater exposure.

6.31 A strain gage is mounted on the inside surface of a large-diameter steel pressure vessel which is subjected to an internal pressure of 150 MPa. Determine the true strain in the hoop direction if the inside diameter is 450 mm and the outside diameter is 800 mm. Compute the error involved in neglecting pressure effects on the output of the strain gages.

6.32 Describe a procedure to minimize the effect of neutron-induced zero drift in a strain-gage installation subjected to intense radiation.

6.33 Describe a procedure for measuring the strain on a mechanical component at 500 °F (260 °C) if the loading cycle requires:

(a) 1 s (b) 1 min (c) 1 h (d) 1 day (e) 1 yr

6.34 Describe a procedure for measuring the strain on a component at a temperature of 900 °F (482 °C) if the strain cycle occurs in 1 h.

6.35 Describe a procedure for measuring the strain on a mechanical component at a temperature of 1600 °F (872 °C) if the strain is transient and a loading cycle requires only 0.01 s.

6.36 A strain-gage test is to be conducted with a Karma strain gage at – 200 F (– 129 °C). The gage factor is specified as 2.05 at room temperature. What gage factor should be used in reducing the value of $\Delta R/R$ to strain?

6.37 Specify the self-temperature-compensating number for a Karma strain gage to be employed on a stainless-steel tank at liquid-nitrogen temperatures.

6.38 Select a gage to operate under cyclic strain conditions where unloading of the structure is not practical. The strain is:

(a) ± 1200 με for 10^5 cycles. (b) ± 2500 με for 10^6 cycles.

6.39 Will a continuity check on the gage resistance indicate that a fatigue crack has initiated in the grid and that the gage will respond to strain with a supersensitive gage factor? Why?

BIBLIOGRAPHY

1. Thomson, W. (Lord Kelvin): "On the Electrodynamic Qualities of Metals," Proceedings Royal Society, 1856.
2. Tomlinson, H.: "The Influence of Stress and Strain on the Action of Physical Forces," Philosophical Transactions, Royal Society, London, vol. 174, pp. 1-172, 1883.
3. Simmons, E. E., Jr.: Material Testing Apparatus, U.S. Patent 2,292,549, Feb. 23, 1940.
4. Kammer, E. W., and T. E. Pardue: "Electric Resistance Changes of Fine Wires during Elastic and Plastic Strains," Proceedings SESA, Vol VII, no. 1, pp. 7-20, 1949.
5. Maslen, K. R.: "Resistance Strain Characteristics of Fine Wires," Royal Aircraft Establishment, Technical Note, Instruments, 127, 1952.
6. Meyer, R. D.: "Application of Unbonded-Type Resistance Gages," Instruments, vol. 19, no. 3. pp. 136-139, 1946.
7. Campbell, W. R.: "Performance of Wire Strain Gages: I, Calibration Factors in Tension," NACA Tech. Note 954, 1944: II, Calibration Factors in Compression, NACA Technical Note 978, 1945.
8. DeForest, A. V.: "Characteristics and Aircraft Applications of Wire Wound Resistance Strain Gages," Instruments, vol. 15, no. 4, pp. 112, 1942.
9. Jones, E., and K. R. Maslen: "The Physical Characteristics of Wire Resistance Strain Gages," Royal Aircraft Establishment, R and M 2661, November 1948.
10. Maslen, K. R., and I. G. Scott: "Some Characteristics of Foil Strain Gauges," Royal Aircraft Establishment, Technical Note, Instruments, p. 134, 1953.
11. Wnuk, S. P.: "Recent Developments in Flame Sprayed Strain Gages," Proceedings Western Regional. Strain Gage Committee, 1965, pp. 1-6.
12. Stein, P. K.: Adhesives: "How They Determine and Limit Strain Gage Performance," Semiconductor and Conventional Strain Gages, pp. 45-72 in M. Dean and R. D. Douglas (eds.), Academic Press, Inc., New York, 1962.
13. Pitts, J. W., and D. G. Moore: Development of High Temperature Strain Gages, National. Bureau Standards, Monograph, 26, 1966.
14. Bodnar, M. J., and W. H. Schroder: "Bonding Properties of a Solventless Cyanoacrylate Adhesive," Modern Plastics, September 1958.
15. Campbell, W. R.: "Performance Tests of Wire Strain Gages: IV, Axial and Transverse Sensitivities," NACA Technical Note, 1042, June 1946.
16. Meier, J. H.: "On The Transverse Strain Sensitivity of Foil Gages," Experimental Mechanics, vol. 1, no. 7, pp. 39-40, 1961.
17. Wu, C. T.: "Transverse Sensitivity of Bonded Strain Gages," Experimental Mechanics, Vol 2, no. 11, pp. 338-344, 1962.
18. "Errors Due to Transverse Sensitivity in Strain Gages," Micro-Measurements Tech. Note 509, 1993.
19. Matlock, H., and S. A. Thompson: "Creep in Bonded Electrical Strain Gages," Proceedings SESA, vol. XII, no. 2, pp. 181-188, 1955.
20. Bloss, R. L.: "Characteristics of Resistance Strain Gages," pp. 123-142 in M. Dean and R. D. Douglas (eds.), Semiconductor and Conventional Strain Gages, Academic Press, Inc., New York, 1962.
21. Barker, R. S.: "Self-Temperature Compensating SR-4 Strain Gages," Proceedings SESA, Vol XI, no. 1, pp. 119-128, 1953.
22. Hines, F. F., and L. J. Weymouth: "Practical Aspects of Temperature Effects on Resistance Strain Gages," Strain Gage Readings, vol. IV, no. 1, 1961.
23. Harding, D.: "High Elongation Measurements with Foil and Liquid Metal Strain Gages," Proceedings Western Regional. Strain Gage Committee, 1965, pp. 23-31.
24. Stone, J. E., Madsen, N. H., Milton, J. L., Swinson, W. F. and J. L. Turner, "Developments in the Design and Use of Liquid-Metal Strain Gages", Strain-Gage and Transducer Techniques, no. 1, Experimental Mechanics, pp. 45-55, 1984.
25. Stein, P. K.: "Advanced Strain Gage Techniques," chap. 2, Stein Engineering Services, Phoenix, Ariz., 1962.

26. Bickle, L. W.: "The Response of Strain Gages to Longitudinally Sweeping Strain Pulses," Experimental Mechanics, vol. 10, no. 8, pp. 333-337, 1970.
27. Oi, K.: "Transient Response of Bonded Strain Gages," Experimental Mechanics, vol.6, no. 9, pp. 463-469, 1966.
28. Freynik, H. S., Jr.: "Investigation of Current Carrying Capacity of Bonded Resistance Strain Gages," M.S. Thesis, Massachusetts Institute of Technology, Cambridge, MA., 1961.
29. "Optimizing Strain Gage Excitation Levels," Micro-Measurements Tech. Note TN-502, 1979.
30. Beyer, F. R., and M. J. Lebow: "Long-Time Strain Measurements in Reinforced Concrete," Proceedings SESA, vol. XI, no. 2, pp. 141-152, 1954.
31. Freynik, H. S., and G. R. Dittbenner: "Strain Gage Stability Measurements for a Year at 75 °C in Air," Univ. Calif. Radiation Lab. Rep. 76039, 1975.
32. Dean, M.: "Strain Gage Waterproofing Methods and Installation of Gages in Propeller Strut of U.S.S. Saratoga," Proceedings SESA, vol. XVI, no. 1, pp. 137-150, 1958.
33. Wells, F. E.: "A Rapid Method of Waterproofing Bonded Wire Strain Gages," Proceedings SESA, Vol. XV, no. 2, pp. 107-110, 1958.
34. Milligan, R. V.: "The Effects of High Pressure on Foil Strain Gages," Experimental Mechanics, vol. 4, no. 2, pp. 25-36, 1964.
35. Brace, W. F.: "Effect of Pressure on Electrical-Resistance Strain Gages," Experimental Mechanics, vol. 4, no. 7, pp. 212-216, 1964.
36. Milligan, R. V.: "Effects of High Pressure on Foil Strain Gages on Convex and Concave Surfaces," Experimental Mechanics, vol. 5, no. 2, pp. 59-64, 1965.
37. Anderson, S. D., and R. C. Strahm: "Nuclear Radiation Effects on Strain Gages," Proceedings Western Regional. Strain Gage Committee, 1968, pp. 9-16.
38. Tallman, C. R.: "Nuclear Radiation Effects on Strain Gages," Proceedings Western Regional. Strain Gage Committee, 1968, pp. 17-25.
39. Wnuk, S. P.: "Progress in High Temperature and Radiation Resistant Strain Gage Development," Proceedings Western Regional. Strain Gage Committee, 1964, pp. 41-47.
40. Day, E. E.: "Characteristics of Electric Strain Gages at Elevated Temperatures," Proceedings SESA, vol. IX, no. 1, pp. 141-150, 1951.
41. Wnuk, S. P.: "New Strain Gage Developments," Proceedings Western Regional. Strain Gage Committee, 1964, pp. 1-7.
42. Hayes, J. K., and G. Roberts: "Measurements of Stresses under Elastic-Plastic Strain Conditions at Elevated Temperatures," Proceedings Technical. Committee for Strain Gages 1969, pp. 20-39.
43. Denyssen, I. P.: "Platinum-Tungsten Foil Strain Gage for High-Temperature Applications," Proceedings Western Regional. Strain Gage Committee,1964, pp. 19-23.
44. Weymouth, L. J.: "Strain Measurement in Hostile Environment," Applied Mechanics Reviews, vol. 18, no. 1, pp. 1-4, 1965.
45. Noltingk, B. E.: "Measuring Static Strains at High Temperatures," Experimental Mechanics 15, no. 10, pp. 420-423, 1975.
46. Harting, D. R.: "Evaluation of a Capacitive Strain Measuring System for Use to 1500 °F," Instrument Society of America, ASI 75251, pp. 289-297, 1975.
47. Telinde, J.C.: "Strain Gages in Cryogenic Environment," Experimental Mechanics vol. 10, no. 9, pp. 394-400, 1970.
48. Telinde, J. C.: "Strain Gages in Cryogenics and Hard Vacuum," Proceedings Western Regional. Strain Gage Committee, 1968, pp. 45-54.
49. Dorsey, J.: "New Developments and Strain Gage Progress," Proceedings Western Regional. Strain Gage Committee, 1965, pp. 1-10.
50. "Fatigue Characteristics of Micro-Measurement Strain Gages," Micro-Measurements Tech. Note 508-1, 1991.
51. Lei, J.-F.: "High Temperature Thin Film Strain Gages," HITEMP Review 1994, NASA CP-10146, Vol. I, 1994, pp. 25-1 to 13.

52. Castelli, M. G.; and Lei, J.-F.: "A Comparison between High Temperature Extensometry and PdCr Based Resistance Strain Gages With Multiple Application Techniques," HITEMP Review 1994, NASA CP-10146, Vol. II, Oct. 1994, pp. 36-1 to 12.
53. Lei, J.-F.: "A Resistance Strain Gage with Repeatable and Cancelable Apparent Strain for Use to 800 °C," NASA CR-185256, 1990.
54. Lei, J.-F.: "Palladium-Chromium Static Strain Gages for High Temperatures," NASA Langley Measurement Technology Conference Proceedings, NASA CP-3161, 1992, pp. 189-209.

CHAPTER 7

STRAIN GAGE CIRCUITS AND INSTRUMENTATION

7.1 INTRODUCTION

An electrical-resistance strain gage changes its resistance with applied strain according to Eq. (6.6), which indicates that:

$$\frac{\Delta R}{R} = S_g \varepsilon_a \tag{6.6}$$

where the gage axis coincides with the x axis and $\varepsilon_{yy} = -0.285\ \varepsilon_{xx}$. In order to employ the electrical-resistance strain gage in any experimental stress analysis, the quantity $\Delta R/R$ must be measured and converted to the strain that produced the resistance change. The Wheatstone-bridge circuit is commonly employed to convert the value of $\Delta R/R$ to a voltage signal (denoted here as Δv) that can be measured with a suitable recording instrument.

This chapter presents the basic theory for the Wheatstone bridge and includes equations for the circuit sensitivities and effective range for each circuit. Temperature compensation and signal addition are discussed in detail. Also covered are the constant-current circuits recommended for gages that exhibit large changes in $\Delta R/R$. Methods of calibrating the Wheatstone-bridge circuits are described, and the effects of lead wires on noise, calibration and temperature compensation are discussed.

Insofar as possible, this discussion is based on fundamental principles without reference to commercially available circuits. However, the material presented here is applicable to commercial instruments because their design is based on these fundamental principles.

7.2 THE WHEATSTONE BRIDGE

The Wheatstone bridge is an electrical circuit employed to determine the change in resistance that a gage undergoes when it is subjected to a strain. The Wheatstone bridge can be used to determine both dynamic and static strain-gage readings. The bridge may be used as a direct-readout device, where the output voltage Δv_0 is measured and related to strain. Also, the bridge may be used as a null-balance system, where the output voltage Δv_0 is adjusted to a zero value by adjusting the resistive balance of the bridge. In either of these modes of operation, the bridge can be effectively employed in a wide variety of strain-gage applications.

The Wheatstone bridge may be powered with a constant voltage or with a constant current. While these circuits appear to be similar, there is a difference in the circuit characteristics. Let's first consider the constant voltage Wheatstone bridge.

7.2.1 Constant Voltage Wheatstone Bridge

The electrical circuit for the Wheatstone bridge is shown in Fig 7.1. The output voltage v_0 of the bridge can be determined by treating the top and bottom parts of the bridge as individual voltage dividers. Thus,

$$v_{AB} = \frac{R_1}{R_1 + R_2} v_s \quad \text{and} \quad v_{AD} = \frac{R_4}{R_3 + R_4} v_s \qquad \text{(a)}$$

The output voltage v_0 of the bridge is given by:

$$v_0 = v_{BD} = v_{AB} - v_{AD} \qquad \text{(b)}$$

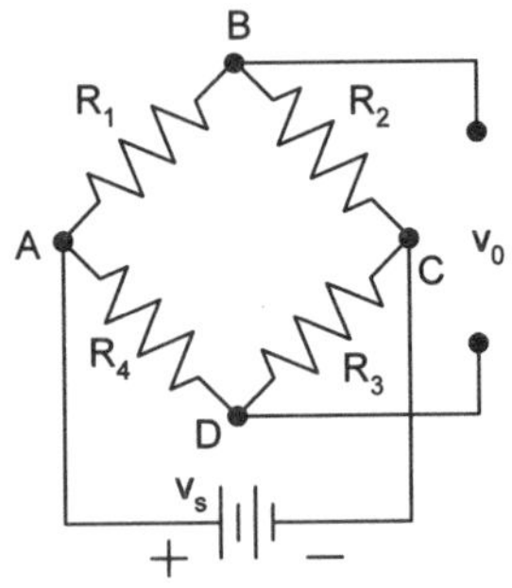

Figure 7.1 The constant voltage Wheatstone bridge circuit.

Substituting Eqs. (a) into (b) yields:

$$v_0 = \frac{R_1R_3 - R_2R_4}{(R_1 + R_2)(R_3 + R_4)} v_s \qquad (7.1)$$

Equation (7.1) indicates that the initial output voltage will vanish ($v_0 = 0$) if:

$$R_1R_3 = R_2R_4 \qquad (7.2)$$

When Eq. (7.2) is satisfied, the bridge is in **balance**. The ability to balance the bridge (and zero v_0) represents a significant advantage, because it is much easier to measure small changes in voltage Δv_0 from a null voltage than from an elevated voltage v_0, which may be as much as 1,000 times greater than Δv_0.

With an initially balanced bridge, an output voltage Δv_0 develops when resistances R_1, R_2, R_3 and R_4 are varied by amounts ΔR_1, ΔR_2, ΔR_3 and ΔR_4, respectively. From Eq. (7.1), with these new values of resistance,

$$\Delta v_0 = \frac{(R_1 + \Delta R_1)(R_3 + \Delta R_3) - (R_2 + \Delta R_2)(R_4 + \Delta R_4)}{(R_1 + \Delta R_1 + R_2 + \Delta R_2)(R_3 + \Delta R_3 + R_4 + \Delta R_4)} v_s \qquad \text{(c)}$$

Expanding, neglecting higher-order terms, and substituting Eq. (7.1) into Eq. (c) yields:

$$\Delta v_0 = \frac{R_1R_2}{(R_1 + R_2)^2}\left(\frac{\Delta R_1}{R_1} - \frac{\Delta R_2}{R_2} + \frac{\Delta R_3}{R_3} - \frac{\Delta R_4}{R_4}\right) v_s \qquad (7.3)$$

An equivalent form of this equation is obtained by substituting $r = R_2/R_1$ in Eq. (7.3) to give:

$$\Delta v_0 = \frac{r}{(1+r)^2}\left(\frac{\Delta R_1}{R_1} - \frac{\Delta R_2}{R_2} + \frac{\Delta R_3}{R_3} - \frac{\Delta R_4}{R_4}\right) v_s \tag{7.4}$$

Equations (7.3) and (7.4) indicate that the output voltage from the bridge is a linear function of the resistance changes. This apparent linearity results from the fact that the higher-order terms in Eq. (d) were neglected. If the higher-order terms are retained, the output voltage Δv_0 is a nonlinear function of the $\Delta R/R$'s, which can be expressed as

$$\Delta v_0 = \frac{1}{(1+r)^2}\left(\frac{\Delta R_1}{R_1} - \frac{\Delta R_2}{R_2} + \frac{\Delta R_3}{R_3} - \frac{\Delta R_4}{R_4}\right)(1-\eta) v_s \tag{7.5}$$

where

$$\eta = \frac{1}{1 + \dfrac{r+1}{\dfrac{\Delta R_1}{R_1} + \dfrac{\Delta R_4}{R_4} + r\left(\dfrac{\Delta R_2}{R_2} + \dfrac{\Delta R_3}{R_3}\right)}} \tag{7.6}$$

In a commonly used arrangement for the bridge, $R_1 = R_2 = R_3 = R_4$. In this case, Eq. (7.6) reduces to:

$$\eta = \frac{\sum_{i=1}^{4} \dfrac{\Delta R_i}{R_i}}{\sum_{i=1}^{4} \dfrac{\Delta R_i}{R_i} + 2} \tag{7.7}$$

The error due to the nonlinear effect is a function of $\Delta R_1/R_1$ and r. The results from Eq. (7.6) with r = 1, show that $\Delta R_1/R_1$ must be less than 0.02, if the error due to the nonlinear effect is not to exceed 1%. While this limit may appear quite restrictive, the Wheatstone bridge is usually employed with transducers that exhibit very small changes in $\Delta R_1/R_1$.

The sensitivity S_c of a Wheatstone bridge with a constant-voltage power supply and a single active arm is determined from Eq. (7.4) as:

$$S_c = \frac{\Delta v_0}{\Delta R_1/R_1} = \frac{r}{(1+r)^2} v_s \tag{7.8}$$

Again it is clear that increasing the supply voltage produces an increase in sensitivity; however, the power p_T that can be dissipated by the transducer limits the supply voltage v_s to:

$$v_s = I_T(R_1 + R_2) = I_T R_T(1+r) = (1+r)\sqrt{p_T R_T} \tag{7.9}$$

Substituting Eq. (7.9) into Eq. (7.8) gives:

$$S_{cv} = \frac{r}{1+r}\sqrt{p_T R_T} \tag{7.10}$$

Equation (7.10) indicates that the circuit sensitivity of the constant-voltage Wheatstone bridge is due to two factors; (1) the circuit efficiency r/(1 + r) and (2) the characteristics of the transducer as indicated by p_T and R_T. Increasing r increases circuit efficiency; however, r should not be so high as to require unusually large supply voltages. For example, a 500 Ω sensor capable of dissipating 0.2 W in a bridge with r = 4 (80% circuit efficiency) will require a supply voltage v_s = 50 V which is higher than the capacity of most highly regulated power supplies.

The selection of a sensor with a high resistance and a high heat dissipating capability is much more effective in maximizing circuit sensitivity than increasing the circuit efficiency beyond 70 or 80%. The product $p_T R_T$ for commercially available sensors can range from about 1 W-Ω to 1,000 W-Ω; therefore, much more latitude exists for increasing circuit sensitivity S_{cv} by transducer selection than by increasing circuit efficiency.

Circuit sensitivity S_{cv} can also be increased, as indicated by Eq. (7.4), by using multiple sensors (one in each arm of the bridge). In most cases, however, the cost of the additional sensors is not warranted. Instead, it is usually more economical to use a high-gain differential amplifier to increase the output signal Δv_0 from the Wheatstone bridge.

Load effects in a Wheatstone bridge are negligible if a high-impedance voltage-measuring instrument (such as a DVM for static signals or a computer equipped with a signal conditioning circuit board for dynamic signals) is used with the bridge. The output impedance Z_B of the bridge can be determined by using Thevenin's theorem. Thus,

$$Z_B = R_B = \frac{R_1 R_2}{R_1 + R_2} + \frac{R_3 R_4}{R_3 + R_4} \tag{7.11}$$

In most bridge arrangements, R_B is less than 10^4 Ω. Because the input impedance Z_B of modern voltage recording devices exceeds 10^6 Ω, the ratio $Z_B/Z_M < 0.01$ and loading errors are usually less than 1%.

7.2.2 Constant Current Wheatstone Bridge

There are advantages of using a constant-current power supply with the Wheatstone bridge that are determined by replacing the voltage source in Figure 7.1 with a constant current supply. The current from the source I_s divides at point A into currents I_1 and I_2 where:

$$I_s = I_1 + I_2 \tag{a}$$

The voltage drops across resistances R_1 and R_4 are given by:

$$v_{AB} = I_1 R_1 \qquad v_{AD} = I_2 R_4 \tag{b}$$

Thus, the output voltage v_0 from the bridge is given by:

$$v_0 = v_{BD} = v_{AB} - v_{AD} = I_1 R_1 - I_2 R_4 \tag{7.12}$$

From Eq. (7.12), it is clear that the bridge will be balanced ($v_0 = 0$) if,

$$I_1 R_1 = I_2 R_4 \tag{c}$$

This balance equation is not in a useful form, because the currents I_1 and I_2 are unknowns. However, the magnitudes of these currents can be determined by observing that the voltage v_{AC} can be expressed in terms of I_1 and I_2 as

$$v_{AC} = I_1(R_1 + R_2) = I_2(R_3 + R_4) \qquad \text{(d)}$$

From Eqs. (a), (c), and (d),

$$I_1 = \frac{R_3 + R_4}{R_1 + R_2 + R_3 + R_4} I_s \quad \text{and} \quad I_2 = \frac{R_1 + R_2}{R_1 + R_2 + R_3 + R_4} I_s \qquad \text{(e)}$$

Substituting Eqs. (e) into Eq. (7.12) gives:

$$v_0 = \frac{I_s}{R_1 + R_2 + R_3 + R_4}(R_1R_3 - R_2R_4) \qquad (7.13)$$

This equation shows that the balance requirement for the constant-current Wheatstone bridge is:

$$(R_1R_3 = R_2R_4) \qquad (7.2)$$

which is the same as that for the constant-voltage Wheatstone bridge.

The open-circuit output voltage v_0, from an initially balanced bridge, due to resistance changes ΔR_1, ΔR_2, ΔR_3 and ΔR_4 is determined from Eq. (7.13) as:

$$\begin{aligned} \Delta v_0 &= \frac{I_s}{\sum R + \sum \Delta R}\left[(R_1 + \Delta R_1)(R_3 + \Delta R_3) - (R_2 + \Delta R_2)(R_4 + \Delta R_4)\right] \\ &= \frac{I_s R_1 R_3}{\sum R + \sum \Delta R}\left(\frac{\Delta R_1}{R_1} - \frac{\Delta R_2}{R_2} + \frac{\Delta R_3}{R_3} - \frac{\Delta R_4}{R_4} + \frac{\Delta R_1}{R_1}\frac{\Delta R_3}{R_3} - \frac{\Delta R_2}{R_2}\frac{\Delta R_4}{R_4}\right) \end{aligned} \qquad (7.14)$$

where $\Sigma R = R_1 + R_2 + R_3 + R_4 = \Sigma\Delta R = \Delta R_1 + \Delta R_2 + \Delta R_3 + \Delta R_4$

Equation (7.14) shows that the constant-current Wheatstone bridge exhibits a nonlinear output voltage Δv_0. The non-linearity is due to the $\Sigma\Delta R$ term in the denominator and to the two second-order terms within the bracketed quantity. Consider a typical application with a single strain gage transducer in arm R_1 and fixed-value resistors in the other three arms of the bridge such that

$$R_1 = R_4 = R_T, \qquad R_2 = R_3 = rR_T, \qquad \Delta R_2 = \Delta R_3 = \Delta R_4 = 0 \qquad \text{(f)}$$

For this example, Eq. (7.14) reduces to:

$$\Delta v_0 = \frac{I_s R_T r}{2(1+r) + (\Delta R_T/R_T)}\left(\frac{\Delta R_T}{R_T}\right) \qquad (7.15)$$

this can also be expressed as:

$$\Delta v_0 = \frac{I_s R_T r}{2(1+r)}\frac{\Delta R_T}{R_T}(1+\eta) \qquad (7.16)$$

where

$$\eta = \frac{\Delta R_T/R_T}{2(1+r) + (\Delta R_T/R_T)} \qquad (7.17)$$

It is evident from Eq. (7.17) that increasing r can reduce the nonlinear effect. The error due to nonlinearity for a constant current Wheatstone bridge is significantly less than the error for a bridge powered with a constant voltage source. Comparing the results of Eq. (7.6) and Eq. (7.17) clearly shows the advantage of the constant-current power supply in extending the range of the Wheatstone bridge circuit.

The circuit sensitivity S_c, for the conditions specified in Eq. (f), obtained from Eq. (7.16) is:

$$S_{cc} = \frac{\Delta v_0}{\Delta R_T / R_T} = \frac{I_s R_T r}{2(1+r)} \tag{7.18}$$

For the example being considered, the bridge is symmetric; therefore, the current $I_T = I_s/2$. The power dissipated by the transducer is:

$$p_T = I_T^2 R_T = \tfrac{1}{4} I_s^2 R_T \tag{g}$$

Substituting Eq. (g) into Eq. (7.18) yields a relation that is identical with Eq. (7.10), which indicates that the circuit sensitivity is the same for constant-voltage and constant-current Wheatstone bridges.

The significant advantage of the Wheatstone bridge is the ability to initially balance the bridge to produce a zero output voltage ($v_0 = 0$). Another advantage is the ability to use the bridge in a null-balance mode[1] that eliminates the need for a precise voltage-measuring instrument.

7.2.3 The Wheatstone Bridge as a Direct Readout Device

To show the principle of operation of the Wheatstone bridge as a direct-readout device (where Δv_0 is measured to determine the strain), consider the circuit shown in Fig. 7.1. The output voltage Δv_0 from the Wheatstone bridge is given by Eq. (7.5) with the nonlinear term given by Eq. (7.6)

The voltage Δv_0 will go to zero and the bridge will be considered in balance when:

$$R_1 R_3 = R_2 R_4 \tag{7.2}$$

It is this zeroing feature, which permits the Wheatstone bridge to be employed for static strain measurements. The bridge is initially balanced ($\Delta v_0 = 0$) before strains are applied to the gages in the bridge. Then the strain-induced voltage Δv_0 can be measured relative to a zero voltage for both static and dynamic applications.

7.2.4 Wheatstone-Bridge Sensitivity to Strain

The sensitivity of the Wheatstone bridge must be considered from two points of view:

1. With a fixed voltage applied to the bridge regardless of gage current (a condition which exists in many commercially available instruments).
2. With a variable voltage whose upper limit is determined by the power dissipated by the arm of the bridge that contains the strain gage.

---

[1] The null balance method is less important today because of the availability of accurate digital voltmeters (DVM) with high input impedance at reasonable costs.

By recalling the definition for the circuit sensitivity, and using the basic bridge relationship given in Eq. (7.5) for Δv_0, it is clear that:

$$S_c = \frac{\Delta v_0}{\varepsilon} = \frac{v_s}{\varepsilon}\frac{r}{(1+r)^2}\left(\frac{\Delta R_1}{R_1} - \frac{\Delta R_2}{R_2} + \frac{\Delta R_3}{R_3} - \frac{\Delta R_4}{R_4}\right) \qquad \text{(a)}$$

Note that the nonlinear term $(1 - \eta)$ in Eq. (7.5) has been neglected, because it is quite small if the strains being measured are less than 5%.

If a multiple-gage circuit is considered with n gages (where n = 1, 2, 3, or 4) whose outputs sum when placed in the bridge circuit, it is possible to write:

$$\sum_{m=1}^{m=n} \frac{\Delta R_m}{R_m} = n\frac{\Delta R}{R} = nS_g\varepsilon \qquad \text{(b)}$$

Substituting Eq. (b) into Eq. (a) gives the circuit sensitivity as:

$$S_c = v_s \frac{r}{(1+r)^2} nS_g \qquad (7.19)$$

This sensitivity equation is applicable in those cases where the bridge supply voltage v_s is fixed and independent of the bridge resistance. The equation shows that the sensitivity of the bridge depends upon the number n of active arms employed, the gage factor S_g, the supply voltage v_s and the ratio of the resistances $r = R_2/R_1$. A graph of r versus $r/(1 + r)^2$ (the circuit efficiency) in Fig. 7.2 shows that maximum circuit sensitivity occurs when r = 1. With four active arms in this bridge, a circuit sensitivity of $S_g v_s$ can be achieved, whereas with one active gage and r = 1 a circuit sensitivity of only $S_g v_s/4$ can be obtained.

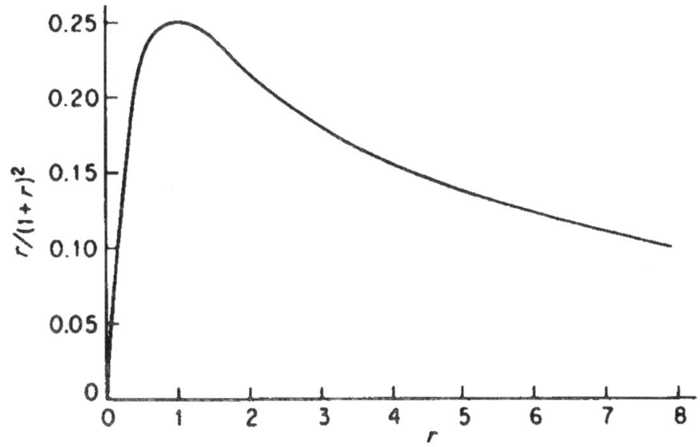

Figure 7.2 Circuit efficiency as a function of r for a Wheatstone bridge with a fixed-supply voltage.

When the bridge supply voltage v_s is selected to drive the gages in the bridge so that they dissipate the maximum allowable power, a different sensitivity equation must be employed. Because the gage current is a limiting factor in this approach, the number of gages used in the bridge and their relative position are important. To show this fact, consider the following four cases, illustrated in Fig. 7.3, which represent the most common bridge arrangements.

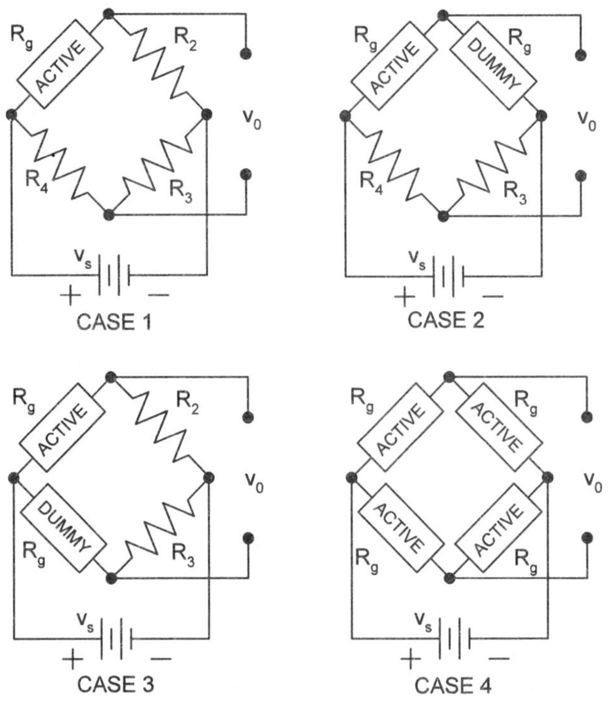

Figure 7.3 Four common arrangements of the Wheatstone bridge circuit.

Case 1: This bridge arrangement consists of a single active gage in position R_1 and is employed for many dynamic and some static strain measurements where temperature compensation in the circuit is not critical. The value of $R_1 = R_g$, but the value of the other three resistors may be selected to maximize circuit sensitivity, provided the initial balance condition $R_1R_3 = R_2R_4$ is maintained. The power dissipated by the gage can be determined from:

$$v_s = I_g(R_1 + R_2) = I_gR_g(1+r) = (1+r)\sqrt{P_gR_g} \qquad \text{(c)}$$

By combining Eqs. (7.19) and (c) and recalling that $r = R_2/R_1$, it is possible to write the circuit sensitivity in the following form:

$$S_c = \frac{r}{1+r} S_g \sqrt{P_gR_g} \qquad (7.20)$$

Here it is evident that the circuit sensitivity of the bridge is due to two factors, namely, the circuit efficiency, which is given by $r/(1 + r)$, and the gage selection, represented by the term $S_g\sqrt{P_gR_g}$. For this bridge arrangement, r should be selected as high as possible to increase circuit efficiency but not high enough to increase the supply voltage beyond reasonable limits. A value of $r = 9$ gives a circuit efficiency of 90% and, with a 120-Ω dissipating 0.15 W, requires a supply voltage of 42.4 V.

Case 2: This bridge arrangement employs one active gage in arm R_1 and one dummy gage in arm R_2 that is utilized for temperature compensation. The value of the gage current that passes through both gages (note $R_1 = R_2 = R_g$) is given by:

$$v_O = 2\, I_gR_g \qquad \text{(d)}$$

Substituting Eq. (d) into Eq. (7.19) with $n = 1$ and $r = 1$ gives S_c as:

$$S_c = \frac{1}{2} S_g \sqrt{P_g R_g} \qquad (7.21)$$

In this instance the circuit efficiency is fixed at 50% because the condition $R_1 = R_2 = R_g$ requires that r = 1. Thus, it is clear that the placement of a dummy gage in position R_2 to affect temperature compensation reduces circuit efficiency significantly. The gage selection maintains its importance in this arrangement because the term $S_g \sqrt{P_g R_g}$ again appears in the circuit-sensitivity equation. In fact, this term will appear in the same form in all four bridge arrangements identified in Fig. 7.3.

Case 3: The bridge arrangement in this case incorporates an active gage in the R_1 position and a dummy gage in the R_4 position. Fixed resistors of any value are placed in positions R_2 and R_3. With these gage positions, the bridge is temperature-compensated because the temperature-introduced resistance changes are canceled out in the Wheatstone-bridge circuit. The current through both the active gage and the dummy gage is given by:

$$v_s = I_g (R_1 + R_2) = (1 + r)\sqrt{P_g R_g} \qquad (e)$$

Substituting Eq. (e) into Eq. (7.19) and recalling for this case that n = 1 gives the circuit sensitivity as:

$$S_c = \frac{r}{1 + r} S_g \sqrt{P_g R_g} \qquad (7.22)$$

The circuit sensitivity for this bridge arrangement (Case 3) is identical to that which can be achieved with the bridge circuit shown in Case 1. Thus, temperature compensation can be obtained without any loss in circuit sensitivity only if the dummy gage is placed in position R_4.

Case 4: In this bridge arrangement, four active gages are placed in the bridge, with one gage in each of the four arms. If the gages are placed on, say, a beam in bending, as shown in Fig.7.4, the signals from each of the four gages will add and the value of n given in Eq. (7.19) will be equal to 4. The power dissipated by each of the four gages is given by:

$$v_s = 2 I_g R_g = \sqrt{P_g R_g} \qquad (f)$$

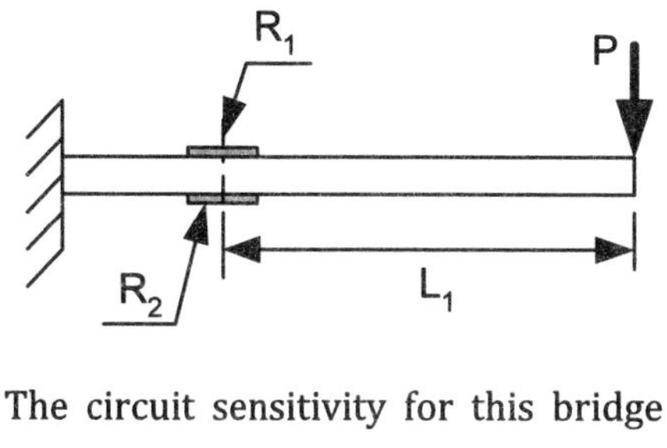

Figure 7.4 Strain gages attached to the top and bottom surfaces of a cantilever beam in bending.

Because the resistance is the same for all four gages, r = 1. The circuit sensitivity for this bridge arrangement is obtained by substituting Eq. (f) into Eq. (7.19):

$$S_c = 2 I_g R_g S_g = 2 S_g \sqrt{P_g R_g} \qquad (7.23)$$

A four-active-arm bridge is slightly more than twice as sensitive as the single-active-arm bridges (Case 1 or Case 3). Also, this bridge arrangement is temperature compensated. However, the four gages employed are a relatively high price to pay for this increased sensitivity. The two-active-arm bridge presented in Exercise 7.17 exhibits a sensitivity, which approaches that given in Eq. (7.23).

Examination of these four bridge arrangements plus the fifth arrangement presented in Exercise 7.17 shows that the circuit sensitivity can be varied between 0.5 and 2 times $S_g\sqrt{P_gR_g}$. Temperature compensation for a single active gage in position R_1 can be effected without loss in sensitivity by placing the dummy gage in position R_4. If the dummy gage is placed in position R_2, the circuit sensitivity is reduced by a factor of 2. Circuit sensitivities can be improved by the use of multiple-active-arm bridges, as illustrated in cases 4 and 5. However, the cost of installing additional gages for a given strain measurement is rarely warranted except for transducer applications. For experimental stress analyses, single-active-arm bridges are normally employed, and the signal from the bridge is usually amplified from 10 to 1000 times before readings are taken.

7.2.5 Commercial Strain Indicators

Advances in voltage regulators, stable amplifiers and analog-to-digital converters have permitted the replacement of strain indicators incorporating null-balance bridges with instruments containing direct reading liquid crystal LCC displays that present the recorded strain directly in με. An example of a commercially available strain gage indicator is presented in Fig. 7.5. This unit is powered with line voltage or current drawn from a USB port on a PC. Alternatively it is powered with batteries to provide portability. The batteries, two D cells, provide power for up to 600 hours.

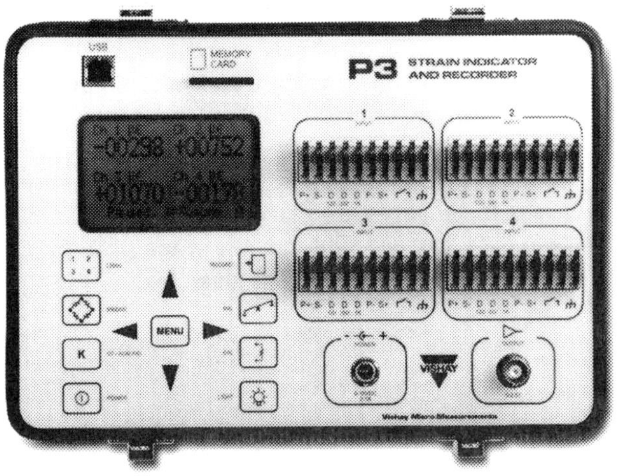

Figure 7.5 A portable strain indicator. (Vishay Micro-Measurements)

This strain indicator is capable of simultaneously accepting four inputs from quarter, half, and full-bridge strain gage circuits, including strain gage based transducers. It can be used in a wide variety of measurement applications, because it functions as a bridge amplifier, static strain indicator and digital data logger. The bridge is powered by a battery supply equipped with a voltage regulator that applies a fixed 1.5 V dc across its input terminals. The negative power terminal is grounded with the negative of the battery supply. The bridge is equipped with internal resistors and may be used in ¼, ½ or full bridge configurations. The ¼ or ½ bridge arrangement accommodates 120, 350 and 1,000 Ω strain gages and the full and ½ bridge configuration accommodates 60 to 2,000 Ω gages. Shunt resistors are provided across the dummy resistor gages that produce a calibration reading of 5,000 με on the display if the gage factor is set at 2.000. A stable measurement circuit, regulated bridge excitation supply, and a precisely settable gage factor enable measurements of ±0.1% accuracy and 1 με resolution (± 3 counts). The range of the instrument is ± 31,000 με. Also, input connections and switches are provided for remote shunt calibration of transducers and full-bridge circuits.

The strain indicator, shown in Fig. 7.5, incorporates a large LCD display, with a 128 × 64 dot matrix structure, for readout of setup information and acquired data. A menu-driven user interface is accessed through a front-panel keypad to configure the instrument for a strain gage test. User selections include input and output channels, bridge configuration, measurement units, bridge balance, calibration method and recording options.

Strain gages are connected with eccentric-lever release terminal blocks. Data can be recorded at a rate of up to 1 reading per channel per second. This data is stored on a removable flash card and is transferred by USB to a host computer for subsequent storage, reduction and presentation. The instrument can also be configured and operated directly from a PC with suitable software.

The gage factor is adjusted with the front panel key pad that controls the reference voltage of the analog-to-digital converter. The LCC display is used in monitoring the adjustment of the reference voltage and in this manner the indicator can be adjusted for a specific gage factor from 0.500 to 9.900 with a resolution of 0.001. The output from the bridge is amplified and filtered and the analog signal (2.5 V maximum) is available on a BNC terminal to provide for an oscilloscope connection. The analog signal is converted to a digital signal with a D/A converter capable of 480 samples per second.

7.2.6 Summary of the Wheatstone Bridge Circuit

The Wheatstone-bridge circuit can be employed for both static and dynamic strain measurements because it can be initially balanced to yield a zero output voltage. The output voltage Δv_0 from the Wheatstone bridge is nonlinear with respect to resistance change ΔR. No emphasis was placed on nonlinearity or on loading effects because they are not usually troublesome in the typical experimental stress analysis. Nonlinear effects can become significant when semiconductor gages are used to measure relatively large strains because of the large ΔR's involved. In these applications, the constant-current bridge circuits should be used. Loading effects should never be a problem because high-quality, low-cost measuring instruments with high input impedances are readily available.

The Wheatstone bridge can be used to add or subtract signals from multiple-strain-gage installations. A four-active-arm bridge is slightly more than twice as sensitive as an optimum single-active-arm bridge. For transducer applications, the gain in sensitivity as well as cancellation of gage response from components of load which are not being measured are important factors; therefore, the Wheatstone bridge is used almost exclusively for transducers using strain gages as sensors.

Temperature compensation can be employed without a loss of sensitivity provided a separate dummy gage or another active gage is employed in arm R_4 of the bridge. Grounding of the Wheatstone bridge can be accomplished only at point C (see Fig. 7.1); consequently, noise can become a problem with the Wheatstone bridge. For further details on the treatment of noise in strain-gage circuits, see Section 7.5.

7.3 CALIBRATING STRAIN-GAGE CIRCUITS

A strain-measuring system usually consists of a strain gage, a power supply (either constant voltage or constant current), circuit-completion resistors, an amplifier and a recording instrument of some type. A schematic illustration of a typical strain measuring system is shown in Fig. 7.5. It is possible to calibrate each component of the system and determine the voltage-strain relationship from the equations developed previously in this chapter. However, this procedure is time-consuming and subject to calibration errors associated with each of the components involved in the complete system. A more precise and direct procedure is to obtain a single calibration for the complete system so that readings from the recording instrument can be directly related to the strains that produced them.

Direct system calibration can be achieved by shunting a fixed resistor R_{sh} across one arm (say R_2) of the Wheatstone bridge, as shown in Fig. 7.5. If the bridge is initially balanced and the switch is closed to place R_{sh} in parallel with R_2; the effective resistance R_{2e} of this arm of the bridge is given by:

$$R_{2e} = \frac{R_2 R_{sh}}{R_2 + R_{sh}} \tag{a}$$

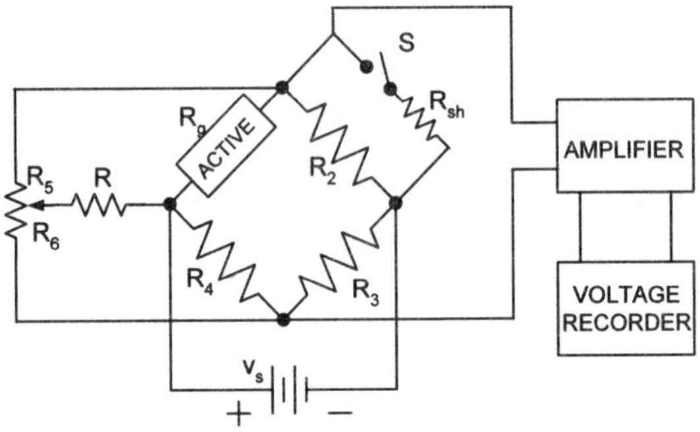

Figure 7.5 Typical strain instrument system with a calibration resistor.

Because of the shunt resistance R_{sh}, the ratio of the change in resistance to the original resistance in arm R_2 of the bridge is:

$$\frac{\Delta R_2}{R_2} = \frac{R_{2e} - R_2}{R_2} \tag{b}$$

Combining Eqs. (a) and (b) gives:

$$\frac{\Delta R_2}{R_2} = -\frac{R_2}{R_2 + R_{sh}} \tag{c}$$

Substituting Eq. (c) into Eq. (7.3) gives the output voltage of the bridge produced by R_{sh}. Thus:

$$\Delta v_0 = \frac{R_1 R_2}{(R_1 + R_2)^2} \frac{R_2}{R_2 + R_{sh}} v_s \tag{d}$$

Note also that a single active gage in position R_1 of the bridge would produce an output due to a strain ε of:

$$\Delta v_0 = \frac{R_1 R_2}{(R_1 + R_2)^2} (S_g \varepsilon) v_s \tag{e}$$

Equating Eqs. (d) and (e) gives:

$$\varepsilon_{sh} = \frac{R_2}{S_g (R_2 + R_{sh})} \tag{7.24}$$

where ε_{sh} is the calibration strain that would produce the same voltage output from the bridge as the calibration resistor R_{sh}.

For example, consider a bridge with $R_2 = R_g = 350\ \Omega$ and $S_g = 2.05$. If $R_{sh} = 100\ k\Omega$, the calibration strain $\varepsilon_{sh} = 1{,}700\mu\varepsilon$. If the recording instrument is operated during the period of time when the switch S is closed, an instrument deflection $d_{sh} = 1{,}700\ \mu\varepsilon$ will be recorded as shown in Fig. 7.6. The switch can then be opened and the load-induced strain can be recorded in the normal manner to give a strain-time pulse similar to the one illustrated in Fig. 7.6. The peak strain associated with this strain-time pulse produces an instrument deflection dp which can be numerically evaluated as:

$$\varepsilon = \frac{d_p}{d_{sh}} \varepsilon_{sh} \tag{7.25}$$

This method of shunt calibration is accurate and easy to employ. It provides a means of calibrating the complete system regardless of the number of components in it. The calibration strain produces a reading on the recording instrument; all other readings are linearly related to this calibration value.

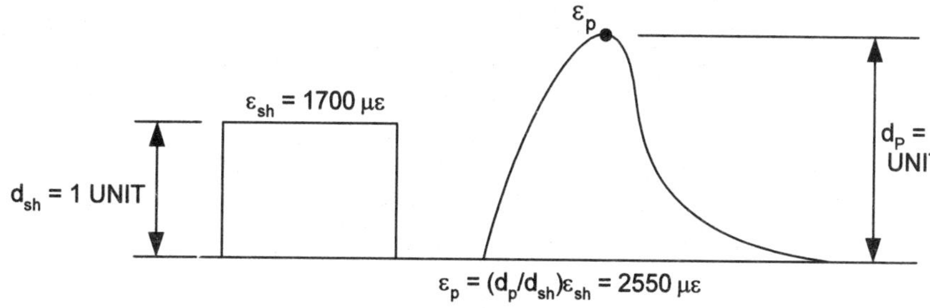

Figure 7.6 Shunt calibration used to determine peak value on a strain time trace.

7.4 EFFECTS OF LEAD WIRES, SWITCHES, AND SLIP RINGS

The resistance change for a metallic-foil strain gage is quite small [0.7 mΩ/με for a 350-Ω gage]. As a consequence, anything that produces even a very small resistance change within the Wheatstone bridge is extremely important. The components within a Wheatstone bridge almost always include lead wires, soldered joints, terminals and binding posts. Frequently, switches and slip rings are also included. The effects of each of these components on the output of the Wheatstone-bridge circuit are discussed in the following subsections.

7.4.1 Effect of Lead Wires

Consider first a two-lead wire system, illustrated in Fig. 7.7, where a single active gage is positioned on a test structure at a location remote from the bridge and recording system. If the length of the two-lead wires is long, three detrimental effects occur; signal attenuation, loss of balancing capability and loss of temperature compensation.

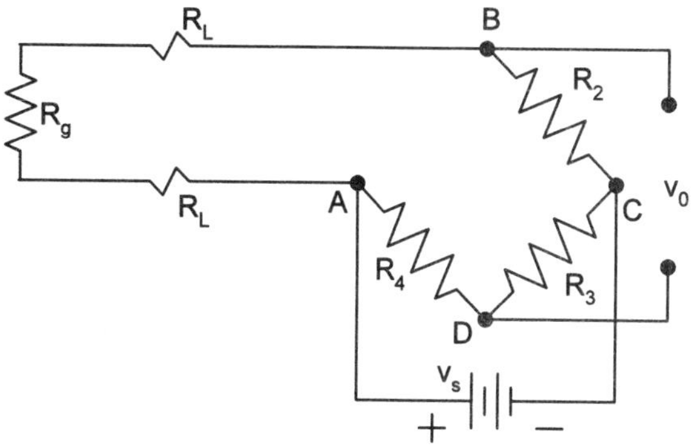

Figure 7.7 A strain gage connected to a Wheatstone bridge with a two-lead-wire system.

To show that signal attenuation may occur note that:

$$R_1 = R_g + 2R_L \tag{a}$$

where R_L is the resistance of a single lead wire. Note also that:

$$\frac{\Delta R_1}{R_1} = \frac{\Delta R_g}{R_g + 2R_L} = \Delta\frac{DR_g/R_g}{1 + 2R_L/R_g} \qquad \text{(b)}$$

Equation (b) may be expressed in terms of a signal loss factor $\mathcal{L}$. Thus

$$\frac{\Delta R_1}{R_1} = \frac{\Delta R_g}{R_g}(1 - \mathcal{L}) \qquad \text{(c)}$$

From Eqs. (b) and (c), the signal loss factor $\mathcal{L}$ for the two-lead wire system can be expressed as:

$$\mathcal{L} = \frac{2R_L/R_g}{1 + 2R_L/R_g} \approx \frac{2R_L}{R_g} \qquad \text{if } 2R_L/R_g \leq 0.1 \qquad (7.26)$$

The signal loss factor $\mathcal{L}$ is presented as a function of the ratio of lead resistance to gage resistance in Fig. 7.8. This graph clearly shows that $\mathcal{L}$ increases rapidly as R_L becomes a significant fraction of R_g. In order to limit lead-wire losses to less than 2%, R_L/R_g must be less than 0.01. If test conditions dictate long leads, then large-gage wire must be employed to limit R_L. It is also advantageous to use 350 or 1,000-Ω gages instead of 120-Ω gages. The resistance of a 100-ft (30.5-m) length of lead wire as a function of wire gage size is listed in Table 7.1 for copper wire.

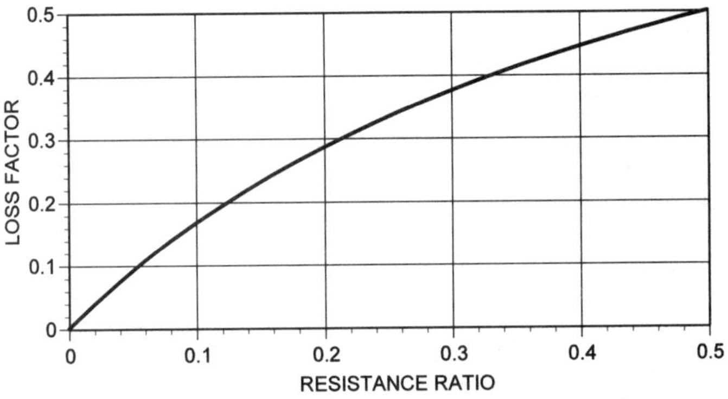

Figure 7.8 Signal loss factor as a function of resistance ratio R_L/R_g.

Table 7.1
Resistance of solid-conductor copper wire

Gage No.	Resistance Ω per 100 ft	Gage No.	Resistance Ω per 100 ft
12	0.159	28	6.49
14	0.253	30	10.31
16	0.402	32	16.41
18	0.639	34	26.09
20	1.015	36	41.48
22	1.614	38	65.96
24	2.567	40	104.90
26	4.081		

The second detrimental effect of the two-lead wire system is loss of the ability to initially balance the bridge. For the bridge shown in Fig. 7.7,

$R_1 = R_g + 2R_L$ $\quad$ $R_2 = R_3 = rR_g$ $\quad$ (fixed resistors)

$R_4 = R_g$ $\quad$ $\Delta R_2 = \Delta R_3 = 0$

With the addition of $2R_L$ in arm R_1 of the bridge, it is obvious that the initial balance condition $R_1R_3 = R_2R_4$ is not satisfied. Of course, a parallel-balance resistor similar to the one shown in Fig. 7.5 is available in most commercial bridges to obtain initial balance; however, if $R_L/R_g > 0.02$, the range of the balance potentiometer is exceeded and initial balance of the bridge cannot be achieved.

The third detrimental effect of the two-lead wire system is that temperature compensation of the measuring circuit is lost. First, consider a temperature-compensating dummy gage with relatively short leads in arm R_4 of the bridge shown in Fig. 7.3. The output voltage due to resistance changes in arms R_1 and R_4, as obtained from Eq. (7.4), is

$$\Delta v_0 = v_s \frac{r}{(1+r)^2}\left(\frac{\Delta R_1}{R_1} - \frac{\Delta R_4}{R_4}\right) \tag{d}$$

If the gages are subjected to a temperature difference ΔT at the same time that the active gage is subjected to a strain ε, then Eq. (d) becomes:

$$\Delta v_0 = v_s\left[\left(\frac{\Delta R_g}{R_g + 2R_L}\right)_\varepsilon + \left(\frac{\Delta R_g}{R_g + 2R_L}\right)_{\Delta T} + \left(\frac{2\Delta R_L}{R_g + 2R_L}\right)_{\Delta T} - \left(\frac{\Delta R_g}{R_g}\right)_{\Delta T}\right] \tag{7.27}$$

This relation shows that temperature compensation is not achieved in the Wheatstone bridge because the second and fourth terms in the bracketed quantity are not equal. Also, the lead wires can suffer significant resistance changes due to temperature; therefore, the third term in the bracketed quantity can produce significant errors in the measurement of strain with the two-lead-wire system.

Employing the three-lead-wire system shown in Fig. 7.9 can minimize the detrimental effects of long lead wires. In this circuit, both the active and dummy gages are placed at the remote location. One of the three wires is used to transfer terminal A of the bridge to the remote location. It is not considered a lead wire because it is not within either arm R_1 or arm R_4 of the bridge. The active and dummy gages each have one long lead wire with resistance R_L and one short lead wire with negligible resistance (see Fig. 7.9). With the three-lead-wire system, the signal loss factor $\mathcal{L}$ is reduced to:

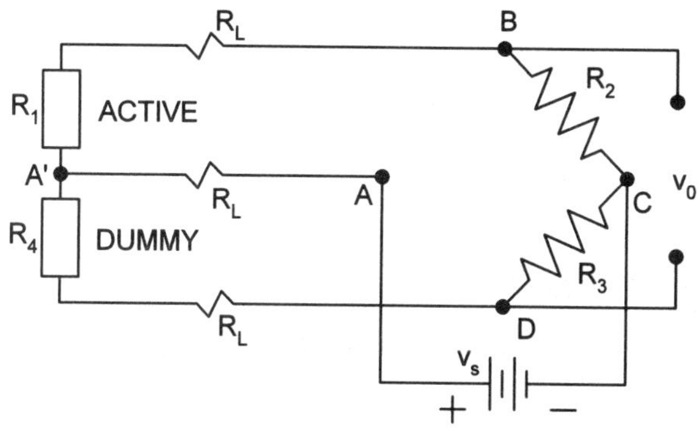

Figure 7.9 A Wheatstone bridge with a three-lead-wire arrangement.

$$\mathcal{L} = \frac{R_L/R_g}{1+R_L/R_g} = \frac{R_L}{R_g} \tag{7.28}$$

The bridge retains its initial balance capability because the resistance of both arms R_1 and R_4 is increased by R_L. With the three-lead-wire system, Eq. (7.27) becomes:

$$\Delta v_0 = v_s \frac{r}{(1+r)^2}\left[\left(\frac{\Delta R_g}{R_g + R_L}\right)_{\varepsilon} + \left(\frac{\Delta R_g}{R_g + R_L}\right)_{\Delta T} + \left(\frac{\Delta R_L}{R_g + R_L}\right)_{\Delta T} - \left(\frac{\Delta R_g}{R_g + R_L}\right)_{\Delta T} - \left(\frac{\Delta R_L}{R_g + R_L}\right)_{\Delta T}\right] \tag{7.29}$$

Examination of the relation above shows that temperature compensation is achieved because all the temperature-related terms in the bracketed quantity cancel.

7.4.2 Effect of Switches

In many strain-gage applications, a large number of strain gages are installed and monitored several times during the test. When several strain gages are involved in an experiment, it is not economically feasible to employ a separate recording instrument for each gage. Instead, a single recording instrument is used, and the gages are switched in and out of this instrument. Two different methods of switching are commonly used in multiple-gage installations.

The first method, illustrated in Fig. 7.10, involves switching one side of each active gage, in turn, into arm R_1 of the bridge. The other side of each of the active gages is connected to terminal A of the bridge by a common lead wire. A single dummy gage or fixed resistor is used in arm R_4 of the bridge. With this arrangement, the switch is located within arm R_1 of the bridge; therefore, an extremely high quality switch with negligible resistance (less than 1 mΩ) must be used. If the switching resistance is not negligible, the switch resistance adds to ΔR_g to produce an error in the strain measurement. The quality of a switch can be checked quite easily, because excessive switch resistance produces variations in the zero strain readings.

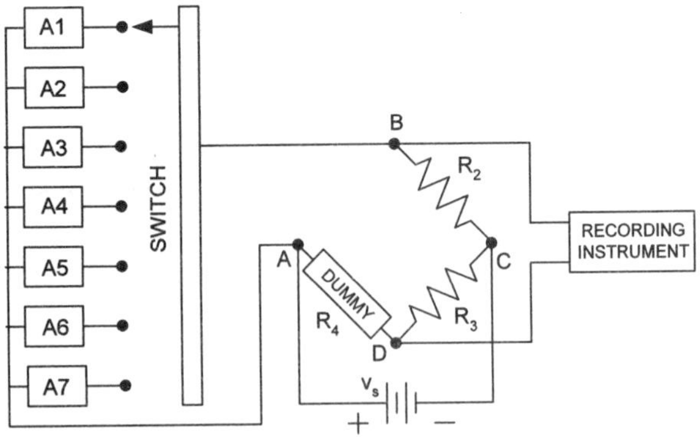

Figure 7.10 A method for switching single active gages into arm R_1 of a Wheatstone bridge.

The second method, illustrated in Fig. 7.11, involves switching the complete bridge. In this method, a three-pole switch is employed to connect the leads between the bridge and the power supply and the recording instrument. Because the switch is not located in the arms of the bridge, switching resistance is not as important. Switching the complete bridge is more expensive, however, because a separate dummy gage and two bridge-completion resistors are required for each active gage.

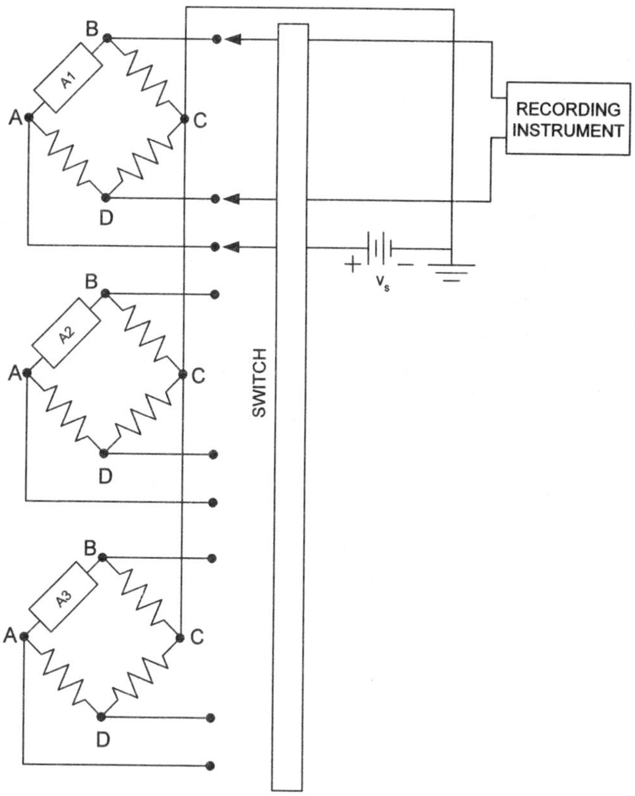

Figure 7.11 Schematic drawing showing a switching circuit for a complete bridge.

7.4.3 Effect of Slip Rings

Strain gages are frequently used on rotating machinery, where it is impossible to use ordinary lead wires to connect the active gages to the recording instrument. Slip rings are employed in these applications to provide lead-wire connections. The rings are mounted on a shaft that is attached to the end of the rotating member so that the rings rotate with the member. The shell of the slip-ring assembly is stationary and it usually carries several brushes for each slip ring. Lead wires from the strain-gage bridge, which rotate with the member, are connected to the slip rings. Lead wires from the power supply and recording instrument, which are stationary, are connected through the brushes to the appropriate slip ring. Depending on the design of the slip-ring assembly, satisfactory operation at rotary speeds up to 24,000 rpm can be achieved.

Wear debris collecting on the slip rings and brush tend to generate electrical noise in this type of strain-gage system. The use of multiple brushes wired in parallel helps to minimize the noise; however, the resistance changes between the rings and the brushes are usually so large that slip rings are not recommended for use within an arm of the bridge. Instead, a complete bridge should be assembled on the rotating member, as shown in Fig. 7.12, and the slip rings should be used to connect the bridge to the power supply and recording instrument. In this way, the effects of resistance changes due to the slip rings are minimized.

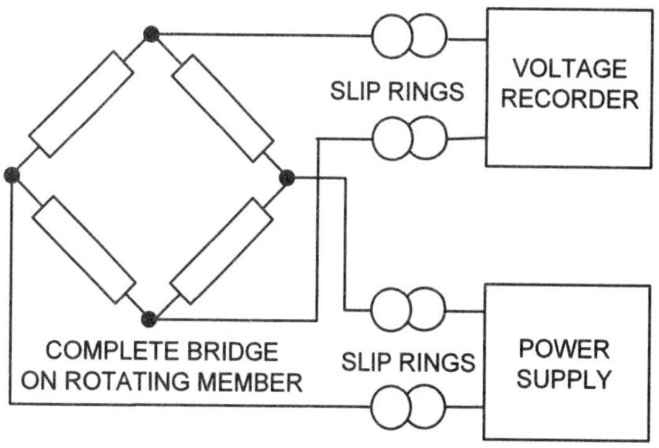

Figure 7.12 Slip rings connecting a complete bridge to a recording instrument and a power supply.

7.5 ELECTRICAL NOISE

The voltage Δv_o from a typical strain gage circuit is quite small (usually less than 10 μV/με). As a consequence, electrical noise is an important consideration in strain-gage circuit design. Electrical noise in strain-gage circuits is produced by the magnetic fields generated when currents flow through supply wires located in close proximity to the strain-gage lead wires, as shown in Fig. 7.13. When an alternating current flows in the supply wire, a time-varying magnetic field is produced, which cuts both wires of the signal circuit and induces a voltage in the signal loop. The induced voltage is proportional to the current I and the area enclosed by the signal loop but inversely proportional to the distance from the supply wire to the signal circuit.

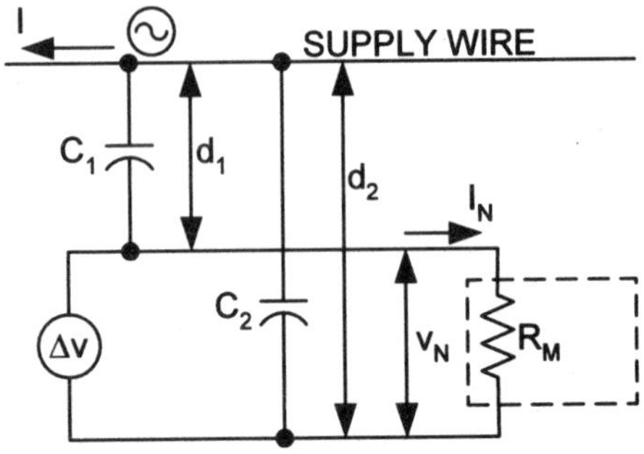

Figure 7.13 Electrical noise generation from a supply wire.

Because the distances d_1 and d_2 in Fig. 7.13 are not equal, the difference in magnetic fields at the two signal leads induces a noise voltage vN that is superimposed on the strain-gage signal voltage (v0. In certain instances, where the magnetic fields are large, the noise voltage becomes significant and makes separation of the true strain-gage signal from the noise signal difficult.

There are three procedures that should normally be employed to reduce noise to a minimum.

Procedure 1: All lead wires should be tightly twisted together to minimize the area in the signal loop and make the distances d_1 and d_2 equal. In this way, the noise voltage is minimized.

Procedure 2: Shielded cables should be used for the strain gage leads and the shields should be connected only to the signal ground as indicated in Fig. 7.14. If the shield is connected to both the signal ground and the system ground, as shown in Fig. 7.15, a ground loop is formed. Because two different grounds are seldom at the same absolute voltage, a noise signal can be generated by the potential difference that exists between the two grounding points. A second ground loop, from the signal source through the cable shield to the amplifier, also occurs with the grounding method shown in Fig. 7.15. Alternating currents in the shield, due to this second ground loop, are coupled to the pair through the distributed capacitance in the signal cable (see Fig 7.15). Either of these ground loops is capable of generating a noise signal 100 times larger than the strain gage signal.

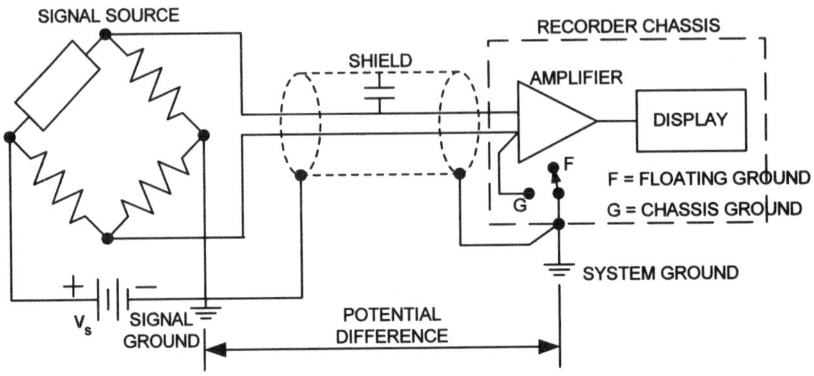

Figure 7.14 Proper method for grounding a shielded cable.

In most recording instruments, the third conductor in the power cord is used to provide the system ground. Because this ground is connected to the enclosure of the recording instrument, care should be exercised to insulate the enclosure from any other building ground. In most modern recording instruments, the amplifier can be operated in either a floating mode or a grounded mode, as shown in Figs. 7.14 and 7.15, respectively. In the floating mode, the amplifier is isolated from the system ground; and with the signal source also isolated from the system ground, the arrangement is correct for minimizing the noise signal. The proper point of attachment for the signal ground on the Wheatstone-bridge circuit is at the negative terminal of the power supply. The power supply itself should be floated relative to the system ground to avoid a ground loop at the supply.

Figure 7.15 Incorrect method of grounding a shielded cable that results in a ground loop.

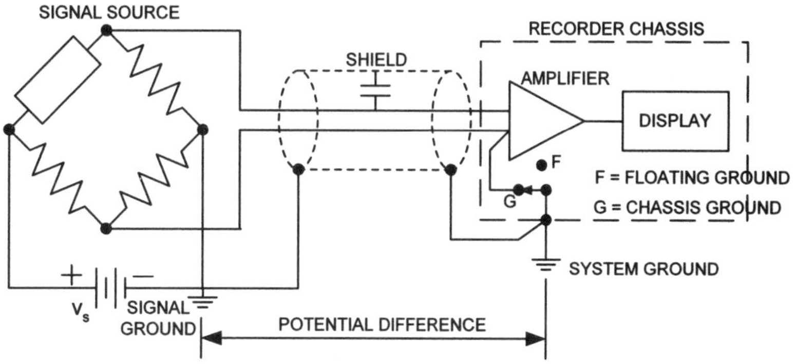

Procedure 3: The third way to eliminate noise is by common-mode rejection. Here the lead wires are arranged so that any noise signals will be equal in magnitude and phase on both the lead wires. If a differential amplifier is employed in the recording system, the noise signals are rejected and the strain signals are amplified. Unfortunately, the common-mode rejection of the very best differential amplifiers is not perfect, and the amplifier transmits a very small portion of the common-mode voltage. The common-mode rejection for good low-level data amplifiers is about 10^5 to 1 at 60 Hz; thus, it is evident that significant noise suppression can be achieved in this manner.

7.6 TRANSDUCER APPLICATIONS

Because the electrical-resistance strain gage is such a remarkable measuring device—small, lightweight, linear, precise and inexpensive—it is used as the sensor in a wide variety of transducers. In a transducer such as a load cell, the unknown quantity is measured by sensing the strain developed in a mechanical member. Because the load is linearly related to the strain, providing the mechanical member remains elastic, the load cell can be calibrated so that the output signal is proportional to load.

Transducers of many different types and models are commercially available. Included are load cells, torque meters, pressure gages, displacement gages and accelerometers. In addition, many different special-purpose transducers are custom-designed, with strain gages as the sensing device, to measure other quantities.

7.6.1 Load Cells

The design of strain-gage transducers for general-purpose measurement is relatively easy if accuracy is limited to ± 2%. Consider, for instance, a tensile load cell fabricated from a tension specimen, as shown in Fig. 7.16. To convert the simple tension bar into a load cell, four strain gages are mounted on the central region of the bar with two opposing gages in the axial direction and two opposing gages in the transverse direction. When a load P is applied to the tension member, the axial and transverse strains produced are:

$$\varepsilon_a = \frac{P}{AE} \qquad \varepsilon_t = -\frac{\nu P}{AE} \tag{7.30}$$

where A is the cross-sectional area, E is the modulus of elasticity and ν is the Poisson's ratio.

If the four gages are positioned in the Wheatstone bridge as shown in Fig. 7.16, the ratio of output voltage to supply voltage $\Delta v_0 / v_s$ is given by:

$$\frac{\Delta v_0}{v_s} = \frac{1}{4}\left(\frac{\Delta R_1}{R_1} - \frac{\Delta R_2}{R_2} + \frac{\Delta R_3}{R_3} - \frac{\Delta R_4}{R_4}\right) \tag{7.31}$$

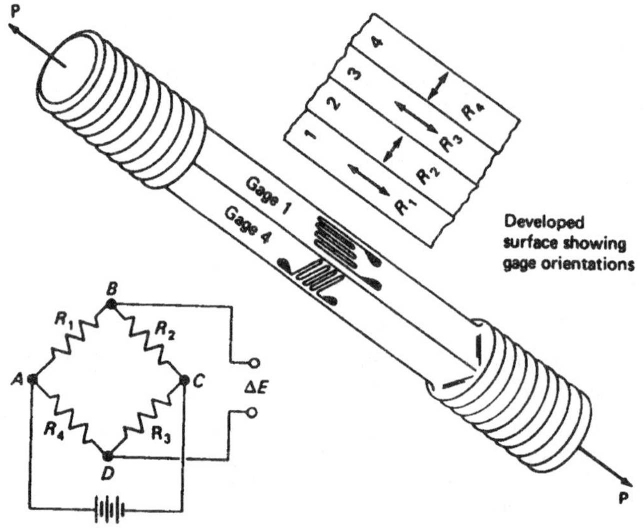

Figure 7.16 Strain gages mounted on a simple tension specimen to produce a load cell.

The changes in resistance of the four gages on the tension member are obtained from Eqs. (6.6) and (7.30) as:

$$\frac{\Delta R_1}{R_1} = \frac{\Delta R_3}{R_3} = S_g \varepsilon_a = \frac{S_g P}{AE}$$

$$\frac{\Delta R_2}{R_2} = \frac{\Delta R_4}{R_4} = S_g \varepsilon_t = -\frac{\nu S_g P}{AE} \tag{a}$$

Substituting into Eq. (7.31) gives:

$$\frac{\Delta v_0}{v_s} = \frac{S_g P}{2AE}(1+\nu) \approx \frac{P}{AE}(1+\nu) \tag{7.32}$$

when $S_g \approx 2.00$.

From Eq. (7.32), it is evident that the output signal $\Delta v_0/v_s$ is linearly related to the load P. The magnitude of $\Delta v_0/v_s$ will depend upon the design of the tension member, i.e., its cross-sectional area A and the material constants E and ν. In most commercial load cells $\Delta v_0/v_s$ varies between 0.001 and 0.003. Steel with E = 30 x 10^6 psi (207 GPa) and ν = 0.30 is usually used to fabricate the tension member. The range of the load cell P_R (the maximum load) is then given by:

$$P_R = A\frac{\Delta v_0}{v_s}\frac{E}{1+\nu} \tag{7.33}$$

The upper limit on the output signal $\Delta v_0/v_s$ is determined by the strength of the tensile member and the fatigue limit of the strain gages. The maximum stress in the tension member is obtained from Eq. (7.33) as:

$$\sigma = \frac{P}{A} = \frac{\Delta v_0}{v_s}\frac{E}{1+\nu} \tag{7.34}$$

With steel tension members and $\Delta v_0/v_s = 0.003$, the stress σ = 69,000 psi (476 MPa) developed in the tension member is well within the fatigue limit of heat-treated alloy steel such as 4340. However, the axial strain $\varepsilon_a = 2300\ \mu\varepsilon$ is near the fatigue limit of most strain gages.

Placement of the strain gages on the four sides of the tension member, as shown in Fig. 7.16, provides a load cell which is essentially independent of either bending or torsional loads. Consider a bending moment M applied to the tension member either by a transverse load or by an eccentrically applied axial load. The moment M may have any direction relative to the axes of symmetry of the cross section as shown in Fig. 7.17. The components M_1 and M_2 of the moment M will produce resistance changes in the gages as follows:

$$\left.\frac{\Delta R_2}{R_2}\right|_{M1} = -\left.\frac{\Delta R_4}{R_4}\right|_{M1} \quad \text{and} \quad \left.\frac{\Delta R_1}{R_1}\right|_{M1} = -\left.\frac{\Delta R_3}{R_3}\right|_{M1} = 0$$

$$\left.\frac{\Delta R_3}{R_3}\right|_{M1} = -\left.\frac{\Delta R_1}{R_1}\right|_{M2} \quad \text{and} \quad \left.\frac{\Delta R_2}{R_2}\right|_{M2} = -\left.\frac{\Delta R_4}{R_4}\right|_{M2} = 0 \tag{a}$$

Substituting Eq. (a) into Eq. (7.31) shows that the effects of bending moments applied to the load cell are canceled in the Wheatstone bridge because $\Delta v_0/v_s$ vanishes for both M_1 and M_2.

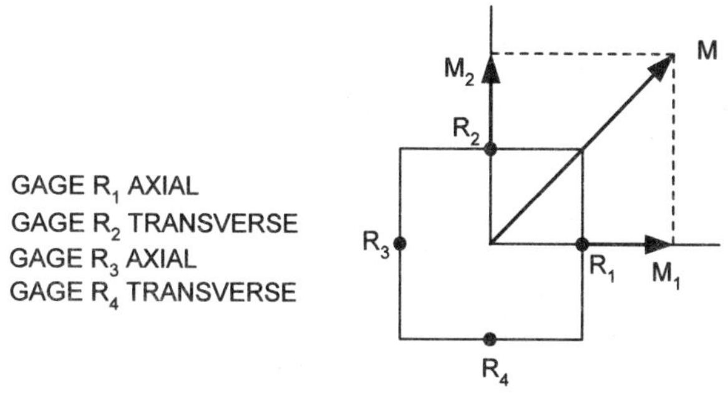

Figure 7.17 Resolution of moment M into its components M_1 and M_2.

Consider next the tension member subjected to a torque T, as shown in Fig. 7.18. The state of stress in the tension member for this form of loading has been shown to be:

$$\tau_{max} = \frac{4.81T}{a^3} \qquad \sigma_{xx} = \sigma_{yy} = \sigma_{zz} = 0$$

Thus

$$\varepsilon_{xx} = \varepsilon_{yy} = \varepsilon_{zz} = 0 \tag{b}$$

When Eqs. (b) are substituted into Eq. (6.6),

$$\frac{\Delta R_1}{R_1} = \frac{\Delta R_3}{R_3} = S_g \varepsilon_{xx} = 0$$
$$\frac{\Delta R_2}{R_2} = \frac{\Delta R_4}{R_4} = S_g \varepsilon_{xx} = 0 \qquad \text{(c)}$$

Substitution of Eqs. (c) into Eq. (7.31) indicates that the output of the tensile load cell is independent of the applied torque because $\Delta v_0/v_s$ is again equal to zero. Temperature compensation is also achieved with the four active strain gages in the bridge.

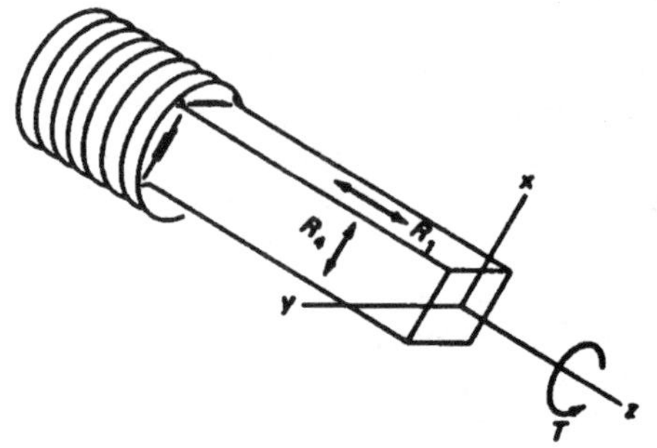

Figure 7.18 Torque T applied to the tension member of an axial load cell.

7.6.2 Diaphragm Pressure Transducers

A second type of transducer that utilizes a strain gage as the sensing element is the diaphragm type of pressure transducer. Here a special-purpose strain gage is mounted on one side of the diaphragm while the other side is exposed to the pressure. The diaphragm pressure transducer is small, easy to fabricate, inexpensive and has a reasonable natural frequency.

The special-purpose diaphragm strain gage, shown in Fig. 7.19, has been designed to maximize the output voltage from the transducer. The strain distribution in the diaphragm is given by:

$$\varepsilon_{rr} = \frac{3p(1-\nu^2)}{8Et^2}\left(R_0^2 - 3r^2\right)$$
$$\varepsilon_{\theta\theta} = \frac{3p(1-\nu^2)}{8Et^2}\left(R_0^2 - r^2\right) \qquad (7.35)$$

where p is the pressure, t is the thickness of the diaphragm, R_0 is its outside radius and r is the position parameter.

Examination of this strain distribution indicates that the circumferential strain $\varepsilon_{\theta\theta}$ is always positive and assumes its maximum value at r = 0. The radial strain ε_{rr} is positive in some regions but negative in others and assumes its maximum negative value at r = R_0. Both these strain distributions are shown in Fig. 7.20. The special-purpose diaphragm strain gage has been designed to take advantage of this distribution. Circumferential grids are employed in the central region of the diaphragm, where $\varepsilon_{\theta\theta}$ is maximum. Similarly, radial grids are employed near the edge of the diaphragm, where ε_{rr} is a maximum. It should so be noted that the circumferential and radial grids are each divided into two parts so that the special-purpose gage actually consists of four separate strain sensors. Terminals are provided which permit the individual gages to be connected into a bridge with the circumferential elements in arms R_1 and R_3 and the radial elements in arms R_2 and R_4

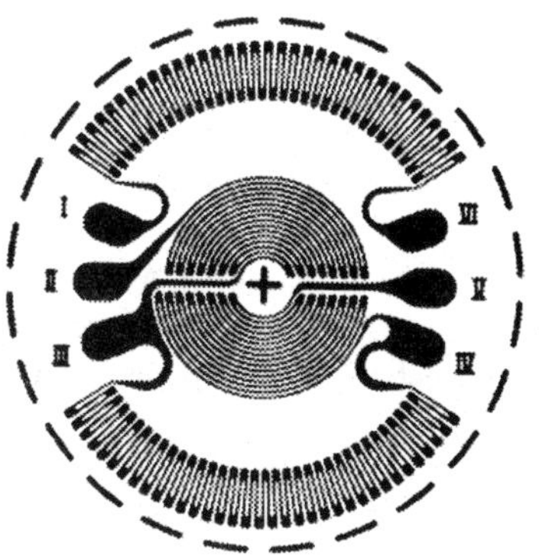

Figure 7.19 Diaphragm type pressure transducer.

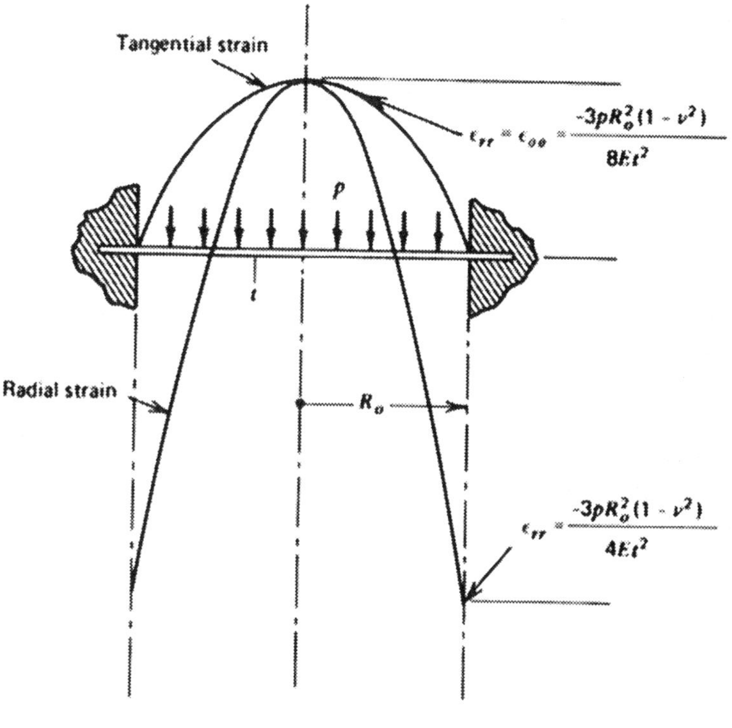

Figure 7.20 Distribution of radial and tangential strain over a diaphragm.

If the strains are averaged over the areas of the circumferential and radial grids, and if the average values of ΔR/R obtained are substituted into Eq. (7.31), the signal output can be approximated by:

$$\frac{\Delta v_0}{v_s} = 0.82\frac{pR_0^2\left(1-\nu^2\right)}{t^2E} \tag{7.36}$$

Special-purpose diaphragm strain gages are commercially available in four sizes ranging from 0.182 to 0.455 in. (4.62 to 7.56 mm) in diameter with resistances of 350 and 1,000 Ω.

Under the action of pressure, the diaphragm deflects and changes from a flat circular plate to a segment of a large-radius shell. As a consequence, the strain in the diaphragm is nonlinear with respect to the applied pressure. Acceptable linearity can be maintained by limiting the deflection of the diaphragm. The center deflection w_c of the diaphragm can be expressed as:

$$w_c = \frac{3pR_0^4(1-\nu^2)}{16t^3E} \tag{7.37}$$

If $w_c \leq t/4$, $\Delta v_0/v_s$ will be linear to within 0.3% over the pressure range of the transducer.

Frequently, diaphragm pressure transducers are employed to measure pressure transients. In these dynamic applications, the natural frequency of the diaphragm should be considerably higher than the highest frequency present in the pressure pulse. Depending upon the degree of damping incorporated in the design of the transducer, an undamped natural frequency of 5 to 10 times greater than the highest applied frequency should be sufficient to avoid resonance effects. The natural frequency f_n of the diaphragm can be expressed as:

$$\omega_n = 2p\, f_n = \frac{10.21t}{R_0^2}\sqrt{\frac{gE}{12(1-\nu^2)\gamma}} \tag{7.38}$$

where γ is the density of the diaphragm material and g is the gravitational constant. If the thickness of the diaphragm cannot be determined accurately, the natural frequency can be determined experimentally by tapping the transducer at the center of the diaphragm to excite the fundamental mode and recording the vibratory response on an oscilloscope. The peak-to-peak period is the reciprocal of the natural frequency.

7.7 AMPLIFIERS

An amplifier is one of the most important components in an instrumentation system. Amplifiers are used in nearly every system to increase low-level signals from a transducer to a higher level sufficient for recording with a voltage measuring instrument or for conversion to a digital code using an analog to digital converter (ADC). An amplifier is represented in schematic diagrams of instrumentation systems by the triangular symbol shown in Fig. 7.21. The voltage input to the amplifier is v_i; the voltage output is v_0. The ratio v_0/v_i is the gain G of the amplifier. As the input voltage is increased, the output voltage increases in the linear range of the amplifier according to the relationship:

$$v_0 = Gv_i \tag{7.39}$$

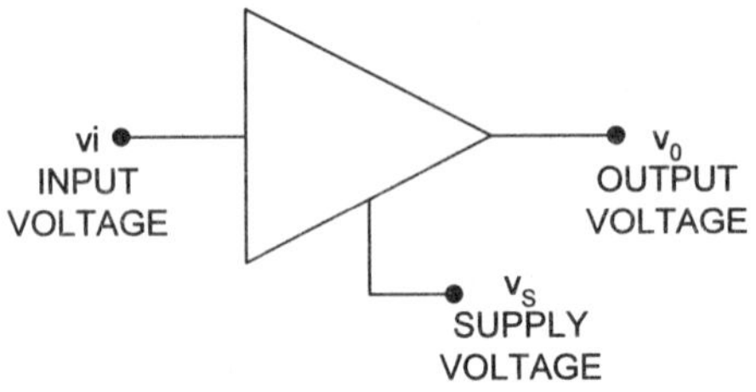

Figure 7.21 Symbol for an amplifier.

The linear range of an amplifier is finite because the supply voltage and the characteristics of the amplifier components limit the output voltage. A typical input-output graph for an amplifier is shown in Fig. 7.22. If the amplifier is driven beyond the linear range, serious errors can result if the gain G is treated as a constant.

The frequency response or bandwidth of an amplifier must also be given careful consideration during design of an instrumentation system. The gain is a function of the frequency of the input signal and there is always a high frequency where the gain of the amplifier will be less than its value at the lower frequencies. This frequency effect on amplifier gain is similar to inertia effects in a mechanical system. A finite time (transit time) is required for the input current to pass through all of the components in the amplifier and reach the output terminal. Also, time is required for the output voltage to develop because some capacitance is always present in the circuits of the amplifier and the recording instrument.

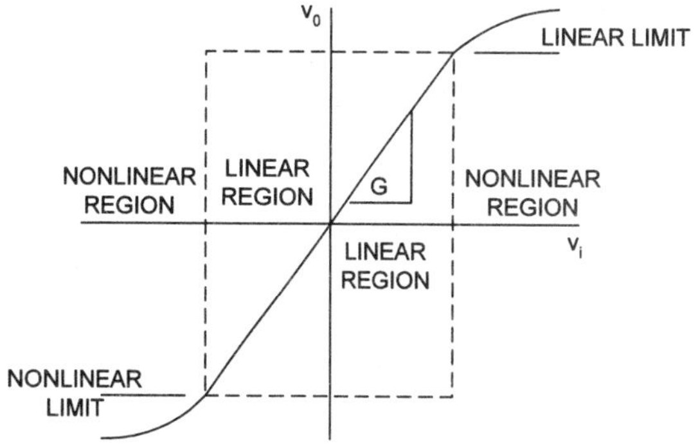

Figure 7.22 A typical input versus output voltage for an amplifier.

The frequency response of an amplifier-recorder system can be illustrated in two different ways. First, the output voltage can be described as a function of time for a step input as shown in Fig. 7.23a. The rise in output voltage for this representation can be approximated by an exponential function of the form:

$$v_0 = \left(1 - e^{-t/\tau}\right) \tag{7.40}$$

where τ is the time constant for the amplifier.

The second method of illustrating frequency effects utilizes a graph showing gain plotted as a function of frequency as shown in Fig. 7.23b. The output of the amplifier is flat between the lower and upper frequency limits f_L and f_U. Thus, a dynamic signal with all frequency components within the band between f_L and f_U will be amplified with a constant gain. DC or dc-coupled amplifiers are designed with input circuits that maintain a constant gain down to zero frequency. However, if a capacitor is placed in series with the input to the amplifier to block the dc components of the input signal, the gain G goes to zero as the frequency of the input signal decreases toward zero. The addition of the series capacitor at the input terminal produces an ac-coupled amplifier with a variable gain for frequencies between zero and f_L.

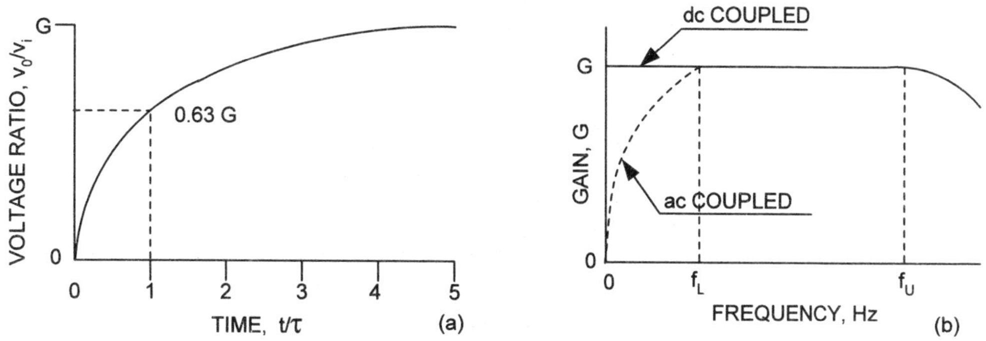

Figure 7.23 (a) Frequency response of a typical amplifier to a step function input voltage. (b) Gain of a typical amplifier as a function of frequency of the input voltage.

Amplifiers are classified as either single ended or dual input. With single-ended types both the input and output voltages are referenced to ground as indicated in Fig. 7.24a. Single-ended amplifiers can be used only when the output from the signal conditioning circuit is referenced to ground. The output from a

Wheatstone bridge is not referenced to ground; therefore, single-ended amplifiers cannot be used with this signal conditioning circuit. Dual input or differential amplifiers must be employed (see Fig. 7.24b) where two separate voltages, each referenced to ground, are connected to the inputs. The output from a differential amplifier is single-ended and referenced to ground. The ideal output voltage from a differential amplifier is given by:

$$v_o = G(v_{i1} - v_{i2}) \tag{7.41}$$

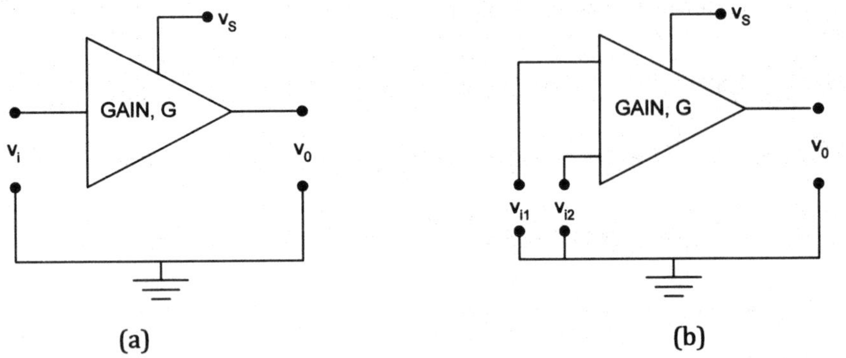

Figure 7.24 Amplifier with single ended input voltage. (b) Amplifier with double ended input voltage.

Generally, the input voltages to a differential amplifier are expressed as:

$$v_{i1} = v + \Delta v \qquad \text{and} \qquad v_{i2} = v \tag{a}$$

where v is the common mode voltage and Δv is the small difference voltage that is to be amplified. Unfortunately, due to slight differences in the amplifier's components, the output voltage v_o is not zero as indicated by Eq. (7.41) when Δv is zero. It is more accurate to write Eq. (7.41) as:

$$v_o = G_d\,\Delta v + G_c\,v \tag{7.42}$$

where G_d is the gain for the difference voltage Δv and G_c is the gain for the common mode voltage v.

One measure of the quality of a differential amplifier is the common mode rejection ratio (CMRR) where:

$$\mathrm{CMRR} = \frac{G_d}{G_c} \tag{7.43}$$

To minimize G_c relative to G_d, a very high value of the CMRR is preferred. Values of CMRR ranging from 1,000 to 20,000 are typical for differential amplifiers with the lower values occurring at the higher frequencies. With CMRR > 1000, Eq. (7.42) is closely approximated by Eq. (7.41).

A high value for the CMRR in the difference mode is important because it implies that spurious signals common to both inputs v_{i1} and v_{i2} such as noise, power supply ripple, and temperature-induced drift are canceled. The ability of the differential amplifier to eliminate these undesirable components of the input signal is a significant advantage. Differential amplifiers can be used with all types of signal conditioning circuits.

Another measure of the quality of an amplifier is related to the signal-to-noise ratio S/N, which is written as:

$$(S/N)_i = \left(\frac{v_i}{v_{ni}}\right)^2 \tag{7.44}$$

where v_{ni} is voltage superimposed on the input signal due to electronic noise.

Note that the voltage ratio v_i/v_{ni} in Eq. (7.44) is squared because the signal-to-noise ratio is defined as the ratio of the signal power to the noise power.

To evaluate the quality of an amplifier in limiting noise on the output signal we consider the noise parameter F_n, which is defined by:

$$F_n = 10\ \log\left[\frac{(S/N)_i}{(S/N)_0}\right] = 10\log\ (NF) \tag{7.45}$$

where NF is the noise factor given by:

$$NF = \left[\frac{(S/N)_i}{(S/N)_0}\right] > 1 \tag{7.46}$$

Note the signal-to-noise ratio on the output is:

$$(S/N)_0 = \left(\frac{v_0}{v_{n0}}\right)^2 = \frac{G_p v_i^2}{G_p v_{ni}^2 + v_{nA}^2} \tag{7.47}$$

where v_{nA} is the noise introduced by the amplifier and G_p is the power gain of the amplifier.

Substituting Eqs. (7.44) and (7.47) into Eq. (7.46) gives:

$$NF = 1 + \frac{v_{nA}^2}{G_p v_{ni}^2} \tag{7.48}$$

It is important to minimize the term $[v_{nA}^2/(G_p v_{ni}^2)]$ so that the noise factor NF is approaches one. If NF = 2, it is evident from Eq. (7.48) that the amplifier and input source is adding a noise signal equal to the noise in the input signal. Clearly, the addition of electronic noise on the signal at any point in the instrumentation system is objectionable.

Another measure of quality of an amplifier is its dynamic range R_d that is defined as:

$$R_d = 20\ \log\left(\frac{v_m}{v_n}\right) \tag{7.49}$$

where v_m is the maximum input signal before the amplifier becomes nonlinear and $v_n = v_{nA}/G$. A high dynamic range is sought to extended usage in the linear range of the amplifier.

7.8 OPERATIONAL AMPLIFIERS

An operational amplifier (op-amp) is an integrated circuit where miniaturized transistors, diodes, resistors and capacitors have been placed on a small silicon chip to form a complete amplifier circuit. Operational amplifiers serve many functions because they can easily be adapted to perform several mathematical operations by adding a small number of external passive components, such as resistors or capacitors. Operational amplifiers have an extremely high gain ($G = 10^5$ is a typical value). Consequently, G is usually considered infinite in the analysis and design of circuits containing an op-amp. The input impedance (typically R = 4 MΩ and C = 8 pF) is so high that circuit loading usually is not a consideration. Output resistance (of the order of 100 Ω) is sufficiently low to be considered negligible in most applications.

Figure 7.25 shows the symbols used to represent the internal op-amp circuit in schematic diagrams. The two input terminals are identified as the inverting (–) terminal and the non-inverting (+) terminal. The output voltage v_0 of an op-amp is given by the expression:

$$v_0 = G(v_{i1} - v_{i2}) \tag{7.41}$$

It is evident from Eq. (7.41) that the op-amp is a differential amplifier; however, it is not used as an instrument differential amplifier because of its extremely high gain and **poor stability**. The op-amp can be used effectively, however, as a part of a larger circuit (with more accurate and more stable passive elements) for many applications. Several applications of the op-amp, including inverting amplifiers, voltage followers, summing amplifiers, integrating amplifiers and differentiating amplifiers are summarized in the next subsection.

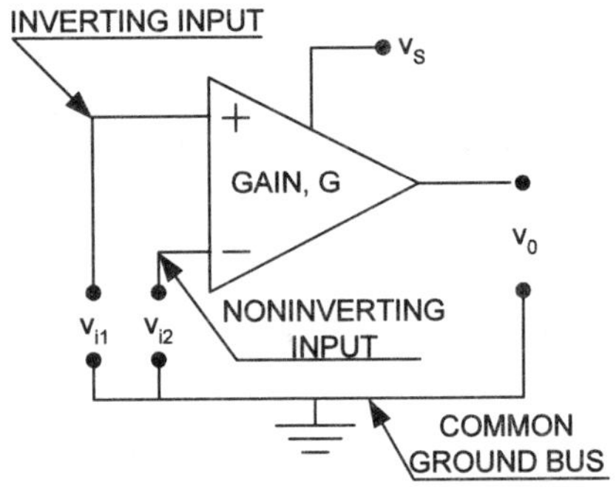

Figure 7.25 Circuit diagram of an operational amplifier

7.8.1 Op-Amp Applications

Several applications of op-amps are described in this section. However, the circuit analysis is presented for only one of these many applications—an inverting amplifier. An inverting amplifier with single-ended input and output can be assembled from an op-amp and resistors, as shown in Fig. 7.26. In this circuit, the input voltage v_i is applied to the negative input terminal of the op-amp through an input resistor R_1. The positive input terminal of the op-amp is connected to a common ground wire. The output voltage v_0 is fed back to the negative terminal of the op-amp through a feedback resistor R_f.

The gain of the inverting amplifier can be determined by considering the sum of the currents at point A in Fig 7.26. Thus,

$$I_1 + I_f = I_a \tag{a}$$

If v_a is the voltage drop across the op-amp is given by:

$$I_1 = \frac{v_i - v_a}{R_1}, \qquad I_f = \frac{v_0 - v_a}{R_f}, \qquad I_a = \frac{v_a}{R_a} \tag{b}$$

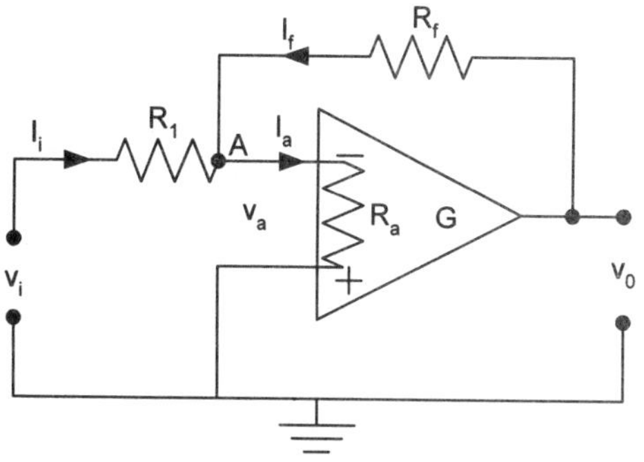

Figure 7.26 An inverting amplifier with single ended input and output.

The voltage drop across the op-amp v_a is related to the output voltage v_0 by the gain. Therefore,

$$v_a = -\frac{v_0}{G} \tag{c}$$

From Eqs. (a), (b), and (c),

$$\frac{v_0}{v_i} = -\frac{R_f}{R_1}\left[\frac{1}{1+\frac{1}{G}\left(1+\frac{R_f}{R_1}+\frac{R_f}{R_a}\right)}\right] \tag{7.50}$$

As an example, consider an op-amp with a gain G = 200,000 and R_a = 4 MΩ, with R_1 = 0.1 MΩ and R_f = 1 MΩ. Substituting these values into Eq. (7.50) yields:

$$\frac{v_0}{v_i} = -10\left[\frac{1}{1+5.6\times 10^{-5}}\right] = -9.9994400 \approx -10 \tag{d}$$

Thus, it is obvious that the term containing the gain G in Eq.(7.50) can be neglected without introducing appreciable error (0.0056% in this example), and the gain of the circuit G can be accurately approximated by:

$$G_c = -\frac{v_0}{v_i} \approx -\frac{R_f}{R_1} \tag{7.51}$$

Op-amps can also be used in non-inverting amplifiers and differential amplifiers in addition to the inverting amplifiers. The circuits for each of these amplifiers are shown in Fig. 7.27, and the governing equations for each of these circuits are given by:

For the non-inverting amplifier:

$$G_c = \frac{v_0}{v_i} = \frac{G}{1+\frac{GR_1}{R_1+R_f}} \approx 1+\frac{R_f}{R_1} \tag{7.52}$$

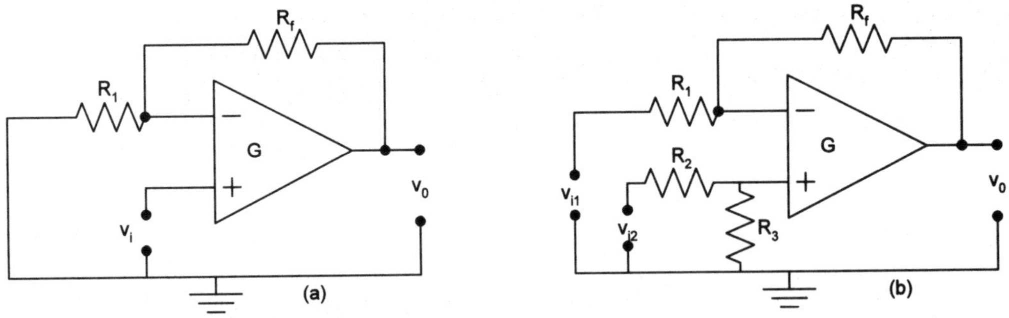

Figure 7.27 Amplifiers designed using op-amps. (a) Non-inverting amplifier. (b) Differential amplifier.

For the differential amplifier:

$$v_0 \approx \frac{R_3}{R_2}\left(\frac{1+\dfrac{R_f}{R_1}}{1+\dfrac{R_3}{R_2}}\right)v_{i2} - \left(\frac{R_f}{R_1}\right)v_{i1} \qquad (7.53)$$

If $R_f/R_1 = R_3/R_2$, then Eq. (7.53) reduces to:

$$G_c = \frac{v_0}{v_{i2} - v_{i1}} \approx \frac{R_f}{R_1} \qquad (7.54)$$

The circuits shown in Fig. 7.27 have been simplified to illustrate the concept of developing an amplifier with a gain G_c that is essentially independent of the op-amp gain G. In practice, these circuits must be modified to account for zero-offset voltages because, ideally, the output voltage v_0 of the amplifier should be zero when the both of the inputs (+) and (–) of the op-amp are connected to the common ground buss. In typical circuits, this zero voltage is not achieved automatically, because the op-amps exhibit a zero-offset voltage. It is necessary to add a biasing circuit to the amplifier that can be adjusted to restore the output voltage to zero; otherwise, serious measurement errors can occur. Because the magnitude of the offset voltage changes (drifts) as a result of temperature, time and power-supply voltage variations, it is advisable to adjust the bias circuit periodically to restore the zero output conditions.

A biasing circuit for an inverting amplifier with single-ended input and output is shown in Fig. 7.28. Common values of resistances, R_2, R_3 and R_4 are $R_3 = R_1$, $R_2 = 10\ \Omega$ and $R_4 = 25\ k\Omega$. Voltages $v = \pm 15$ V are often used because the zero-offset voltage of the op-amp can be either positive or negative. The magnitude of the bias voltage that must be applied to the op-amp seldom exceeds a few millivolts.

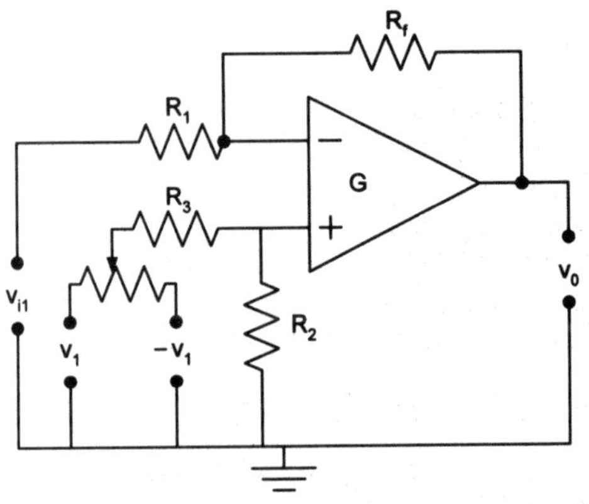

Figure 7.28 Biasing circuit for an amplifier with a single ended input and output.

7.8.2 Additional Circuits Containing Op-amps

Voltage Follower

An op-amp can also be used to construct an instrument with very high input impedance for use with transducers that incorporate piezoelectric sensors. The high-impedance circuit, shown in Fig. 7.29, is known as a voltage follower and has a circuit gain of unity ($G_c = 1$). The purpose of a voltage follower is to serve to adjust the impedance between the transducer and the voltage-recording instrument.
The gain G_c of the voltage follower is given by:

$$G_c = \frac{v_0}{v_i} = \frac{G}{1+G} \tag{7.55}$$

When the gain G of the op-amp is very large, the gain G_c of the circuit approaches unity. The input resistance R_{ci} of the voltage follower circuit is given by:

$$R_{ci} = (1 + G)R_a \tag{7.56}$$

Because both G and R_a are very large for op-amps, the input impedance R_{ci} of the voltage follower circuit can be made quite large. In the design of a voltage follower, electrometer op-amps are specified because they exhibit $G \approx 5 \times 10^4$ and $R_a \approx 10^{13}\ \Omega$. This circuit gives an input impedance of $R_{ci} \approx 5 \times 10^{17}\ \Omega$ that is sufficiently large to minimize the effect of charge drain on measurements of short duration events.

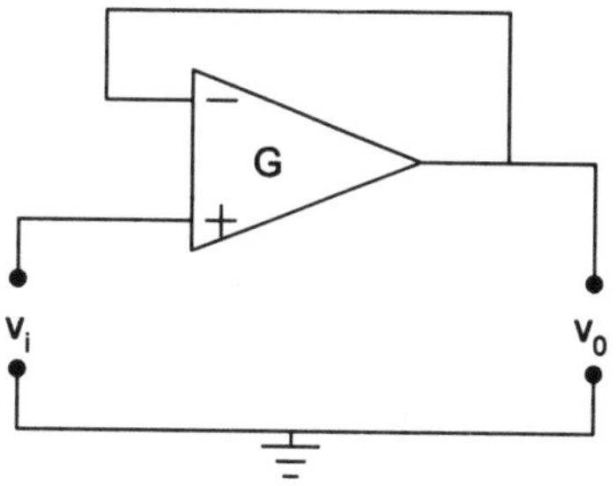

Figure 7.29 A high input impedance voltage follower circuit.

Summing Amplifier

In some data analysis applications, signals from two or more sources are added to obtain an output signal that is proportional to the sum of the input signals. Adding can be accomplished with an op-amp circuit, known as a summing amplifier, shown in Fig. 7.30.

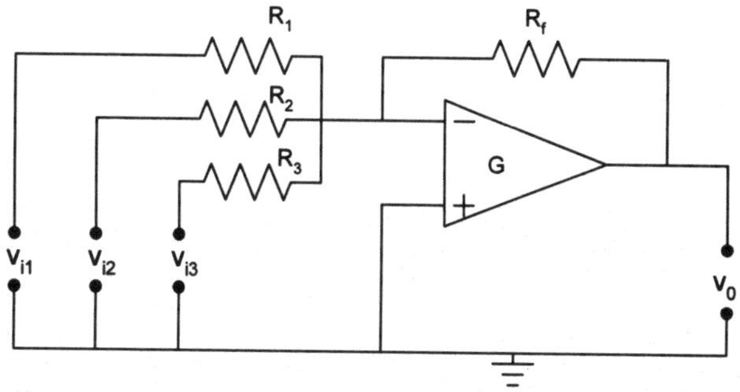

Figure 7.30 A summing amplifier circuit designed with an op-amp.

The output voltage from this summing amplifier is given by:

$$v_0 = -R_f\left(\frac{v_{i1}}{R_1}+\frac{v_{i2}}{R_2}+\frac{v_{i3}}{R_3}\right) \qquad (7.57)$$

Equation (7.57) indicates that the input signals v_{i1} , v_{i2} and v_{i3} are scaled by ratios R_f/R_1, R_f/R_2 and R_f/R_3, respectively, and then summed. If $R_1 = R_2 = R_3 = R_f$, the inputs sum without scaling and Eq. (7.57) reduces to:

$$v_0 = -(v_{i1} + v_{i2} + v_{i3}) \qquad (7.58)$$

Integrating Amplifier

An integrating amplifier utilizes a capacitor in the feedback loop as shown in Fig 7.31. An expression for the output voltage from the integrating amplifier can be established by following the procedure used for the summing amplifier. The output voltage of an integrating amplifier is related to the input voltage by:

$$v_0 = -\frac{1}{R_1 C_f}\int_0^t v_i dt \qquad (7.59)$$

It is clear from Eq. (7.59) that the output voltage v_0 from the circuit of Fig. 7.31 is the integral of the input voltage v_0 with respect to time multiplied by the constant $-R_1C_f$.

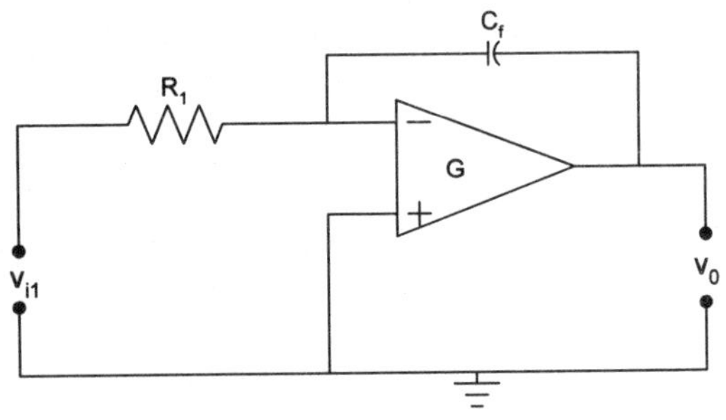

Figure 7.31 An integrating amplifier circuit designed with an op-amp.

Differentiating Amplifier

The differentiating amplifier is similar to the integrating amplifier except that the positions of the resistor and capacitor of Fig. 7.31 are interchanged. An expression for the output voltage v_0 of the differentiating amplifier is given by:

$$v_0 = -R_f C_1 \frac{dv_i}{dt} \qquad (7.60)$$

Considerable care must be exercised to minimize noise on the input signal when the differentiation amplifier is used. Noise superimposed on the input voltage is differentiated and contributes significantly to the output voltage producing large error. The effects of high-frequency noise can be suppressed by placing a capacitor across resistance R_f, however, the presence of this capacitor affects the differentiating process, and Eq. (7.60) must be modified to account for its effects.

7.9 FILTERS

In many instrumentation applications, the signal from the transducer is combined with noise or some other parasitic signal. These parasitic voltages can often be eliminated with a filter that is designed to attenuate the undesirable noise signals, but transmit the transducer signal without significant attenuation or distortion. Filtering of the signal is possible if the frequencies of the parasitic and transducer signals are sufficiently different. Two filters utilizing passive components that are commonly employed in signal conditioning include: the RC high-pass filter and the RC low-pass filter. Schematic diagrams of these filters are shown in Fig 7.32.

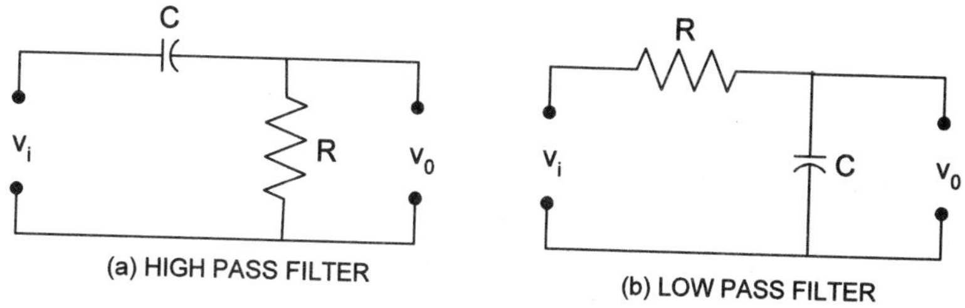

Figure 7.32 Filter circuit that utilize passive components R and C.

High Pass Filter

A simple yet effective high-pass resistance-capacitance (RC) filter is illustrated in Fig. 7.32a. The behavior of this filter in response to a sinusoidal input voltage of the form:

$$v_i = v_a e^{j\omega t} \qquad \text{(a)}$$

can be determined by summing the voltage drops around the loop of Fig. 7.32a. Thus,

$$v_i - \frac{q}{C} - RI = 0 \qquad \text{(b)}$$

where q is the charge on the capacitor. Equation (b) can be expressed in a more useful form by differentiating with respect to time to obtain:

$$RC\frac{dI}{dt} + I = j\omega C v_a e^{j\omega t} \qquad \text{(c)}$$

Solve Eq. (c) by letting $I = I_a e^{j\omega t}$ to obtain:

$$I = \frac{j\omega C v_a}{1 + j\omega RC} e^{j\omega t} \qquad \text{(d)}$$

The output voltage v_0 is the voltage drop across the resistance R; therefore, from Eq. (d):

$$v_0 = IR = \frac{\omega RC v_a}{\sqrt{1 + (\omega RC)^2}} e^{(j\omega t + \phi)} \qquad \text{(e)}$$

where the phase angle $\phi = (\pi/2) - \tan^{-1}\omega RC$

The ratio of the amplitudes of the output and input voltages v_0/v_i obtained from Eqs. (a) and (e), gives the frequency response function for the high-pass filter as:

$$\frac{v_0}{v_i} = H(\omega) = \frac{j\omega RC}{1 + j\omega RC} = |H(\omega)| e^{j\phi} \qquad \text{(f)}$$

where

$$|H(\omega)| = \frac{\omega RC}{\sqrt{1 + (\omega RC)^2}} \qquad (7.61)$$

Equation (7.61) indicates that $v_0/v_i \Rightarrow 1$ as the frequency becomes large; thus, this filter is known as a high-pass filter. At zero frequency (dc), the voltage ratio v_0/v_i vanishes, which indicates that the filter completely blocks any dc component of the output voltage. This dc-blocking capability of the high-pass RC filter can be used to great advantage when a low-amplitude transducer signal is superimposed on a large dc output voltage. Because the RC filter eliminates the dc voltage, a low-magnitude but a frequency dependent signal from a transducer can be amplified to produce a satisfactory display. When a high-pass RC filter is used, $\omega RC > 5$ is necessary to ensure that the input signal is transmitted through the filter with an attenuation that is less than 2%.

Low Pass Filter

Interchanging the position of the resistor and capacitor of the high-pass RC filter produces a low-pass RC filter. This modified RC circuit, shown in Fig. 7.32b, has transmission characteristics that are opposite to those of the high-pass RC filter; namely, it transmits low-frequency signals and attenuates high-frequency signals.

Using the methods of circuit analysis described in the previous, it is easy to show that the ratio of the output and input voltages for this filter is given by:

$$\frac{v_0}{v_i} = H(\omega) = \frac{1}{1+j\omega RC} = |H(\omega)|e^{-j\phi} \tag{7.62}$$

where
$$|H(\omega)| = \frac{1}{\sqrt{1+(\omega RC)^2}} \quad \text{and} \quad \phi = \tan^{-1}(\omega RC)$$

These results indicate that $H(\omega)$ varies from 1 to 0 as ωRC changes from 0.1 to 100. Both limits of this response function are important. The low frequency portion is important because it controls the attenuation of the input signal. Note that a 2% attenuation occurs when $\omega RC = 0.203$. To avoid errors greater than 2%, values for R and C must be selected so that $\omega RC < 0.203$ when designing the low pass filter.

The high-frequency response of the filter is also important because it controls the attenuation of the parasitic or noise signal. A reduction of 90% in the noise signal can be achieved if $\omega_p RC = 10$ (ω_p is the circular frequency of the parasitic signal). It is not always possible with this passive filter to simultaneously limit the attenuation of the input signal to 2% while reducing the parasitic voltages by 90%, because this requires that $\omega_p/\omega_i > 20$. If $\omega_p/\omega_i < 20$, it will be necessary to accept a higher ratio of parasitic signal, or to accept a higher loss of the input signal.

Active Filters

Operational amplifiers are employed to construct active filters where select frequencies can be attenuated and the signal amplified during the filtering process. Several of these active filters are illustrated in Fig. 7.33. The circuit shown in Fig. 7.33a is a low-pass filter similar to that presented in Fig. 7.32b. The addition of the operational amplifier permits the gain of the filter to be adjusted independent of the critical frequency ω_c.

A two-pole Bessel filter is presented in Fig. 7.33b. The addition of the second pole (the use of a second capacitor) increases the rate of roll off in attenuating the higher frequency signals and improves filter performance. Exercises concerning filters are given at the end of the chapter to enable one to compare performances of different types of filters.

Noise at 60 Hz is extremely common and annoying in most laboratories. This noise is due to the presence of motors and lights operating at line frequency. If proper shielding does not eliminate these noise signals, a notch filter, which attenuates signals at 60 Hz, can be employed. A twin-tee notched filter, shown in Fig. 7.33c, has an attenuating notch that depends on the critical frequency f_c. Selecting appropriate values for R and C, positions f_c at 60 Hz. The frequencies of the input signal should be considerably lower or higher than f_c because the notch is broad.

Many other filters such as the fourth-order low-pass Butterworth filter and the Chebyshev filter are often used in signal processing applications.

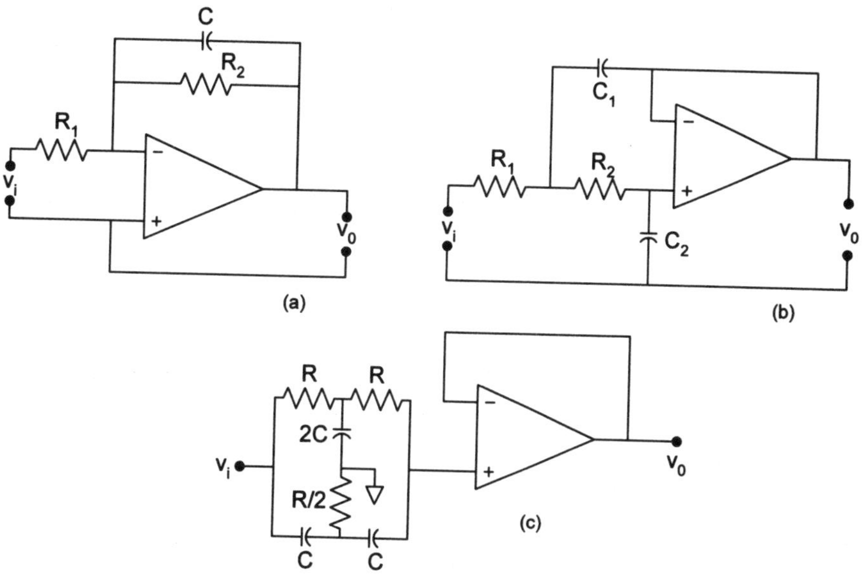

Figure 7.33 Active filter circuits that employ operational amplifiers.

Filters are used to eliminate undesirable signals such as noise, a dc signal, or a high frequency carrier signal. Filters must be selected very carefully; otherwise, the filter may attenuate both the noise signal and the input signal (if the frequencies are similar) and produce serious error. Active filters that employ operational amplifiers combine amplification and filtering functions. They can also be employed to produce notched filters, which are useful in eliminating 60-Hz noise signals.

EXERCISES

7.1 A constant-voltage Wheatstone-bridge circuit is employed with a displacement transducer (potentiometer type) to convert resistance change to output voltage. If the displacement transducer has a total resistance of 2,000 Ω, then $\Delta R = \pm$ 1,000 Ω if the wiper is moved from the center position to either end. If the transducer is placed in arm R_1 of the bridge and, if $R_1 = R_2 = R_3 = R_4$ = 1,000 Ω, determine the magnitude of the nonlinear term η as a function of ΔR. Prepare a graph of η versus ΔR as ΔR varies from – 1,000 Ω to +1,000 Ω.

7.2 Determine the output voltage v_O as a function of ΔR for the displacement transducer and Wheatstone bridge described in Exercise 7.1 if v_s = 8V.

7.3 The nonlinear output voltage of Exercise 7.2 makes data interpretation difficult. How can the Wheatstone-bridge circuit be modified to improve the linearity of the output voltage v_O?

7.4 A strain gage with $R_g = 350\ \Omega$, $p_T = 0.25$ W, and $S_g = 2.05$ is used in arm R_1 of a constant-voltage Wheatstone bridge. Determine the values of R_2, R_3, and R_4 needed to maximize v_0 if the available power supply is limited to 15 V. Also determine the circuit sensitivity of the bridge of with these resistors.

7.5 If the strain gage of Exercise 7.4 is subjected to strain of 1,200 με, determine the output voltage v_0.

7.6 Four strain gages are installed on a cantilever beam as shown in Fig. E7.6 to produce a displacement transducer:

(a) Indicate how the gages should be wired into a Wheatstone bridge to maximize signal output.
(b) Determine the circuit sensitivity if $R_g = 350\ \Omega$, $p_T = 0.15$ W and S = 2.00.
(c) Determine the calibration constant $C = \delta/v_0$ for the transducer.

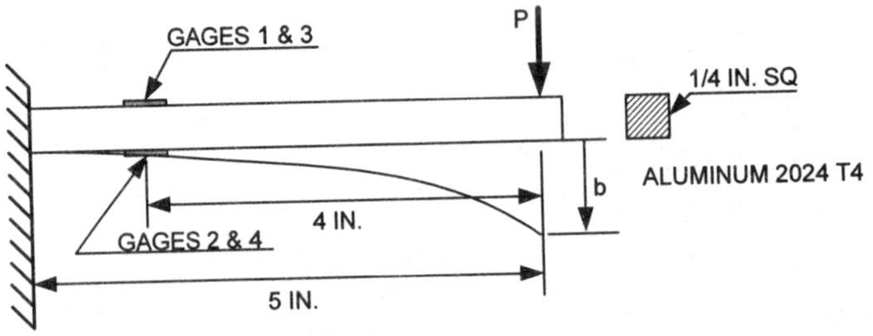

Figure E7.6

7.7 If the cantilever beam of Exercise 7.6 is used as a load transducer, determine the calibration constant $C = P/v_0$.

7.8 A strain gage with $R_g = 350\ \Omega$, $p_T = 0.10$ W, and $S_g = 2.00$ is used in arm R_1 of a constant-current Wheatstone bridge. Determine:

(a) Values of R_2, R_3 and R_4 needed to maximize v_0 if the available power supply can deliver a maximum of 10 mA.
(b) The circuit sensitivity of the bridge of part (a)
(c) The output voltage v_0 if the gage is subjected to a strain of 1,500 με.

7.9 If the displacement transducer of Exercise 6.1 is used with a constant-current Wheatstone bridge, determine the magnitude of the nonlinear term η as a function of ΔR. Prepare a graph of η versus ΔR as ΔR varies from – 1,000 Ω to +1,000 Ω.

7.10 Determine the output voltage v_0 as a function of ΔR for the displacement transducer and Wheatstone bridge described in Exercise 7.6 if I = 20 mA.

7.11 Select the following gages from an on-line strain-gage catalog:

(a) The gage having the smallest gage length.
(b) The gage having the largest gage length.
(c) A gage having an area of approximately 38 mm², a resistance of 350 Ω and a gage factor of about 2.
(d) A gage having an area of approximately 19 mm², a resistance of 1,000 Ω and a gage factor of about 2.
(e) A gage having an area of approximately 19 mm², a resistance of 350 Ω and a gage factor of about 3.5.

If the allowable power density is (0.78 mW/mm²) for each of these gages, determine S_c, v_s, and R_b for a Wheatstone bridge with r = 3. Discuss the results.

7.12 For the best and worst cases in Exercise 7.11 construct a graph of output voltage Δv_0 as a function of strain ε. For strain levels associated with the yield strength of low-carbon steels (σ_y = 250 MPa), determine the magnitude of Δv_0. Is this a high- or low-level signal?

7.13 Three strain gages are placed in series in arm R_1 of a fixed-voltage Wheatstone bridge, as shown in Fig. E7.13. If the strain experienced by all gages is the same, determine the increase in circuit sensitivity over that obtained with a single gage.

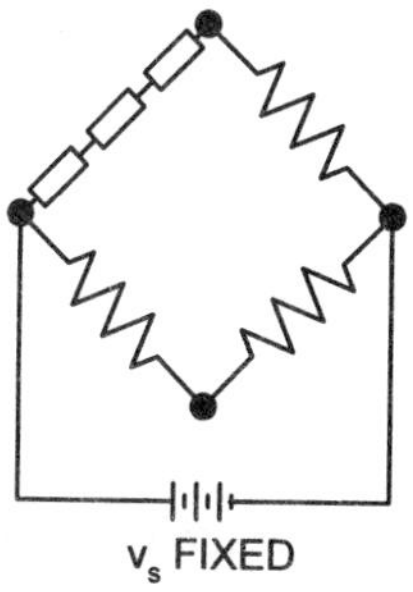

Figure E7.13

7.14 If the voltage can be varied in the Wheatstone bridge of Fig. E7.13, does the insertion of the additional gages improve the sensitivity? If r = 4, determine the percent change in sensitivity for the three gages versus a single gage. Compute the required voltage v_s for the three series connected gages and for a single gage if R_g = 350 Ω and P_g = 0.02 W.

7.15 Compare the circuit sensitivities of the three bridges shown in Fig. E7.15. Compute the required voltages for bridges b and c.

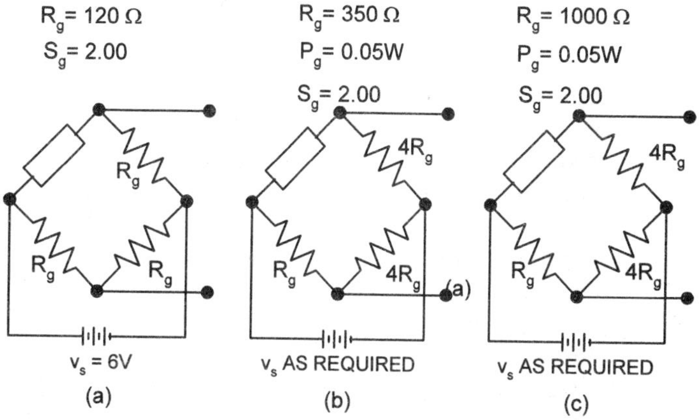

Figure E7.15

7.16 Show that the bridge arrangements represented in cases 2 to 4 of Fig. 7.3 are all temperature compensating.

7.17 Determine the circuit sensitivity for the bridge arrangement shown in Fig. E7.17. Compare this sensitivity with that given by Eq. (7.23) for case 4 of Fig. 7.5. Discuss the results obtained.

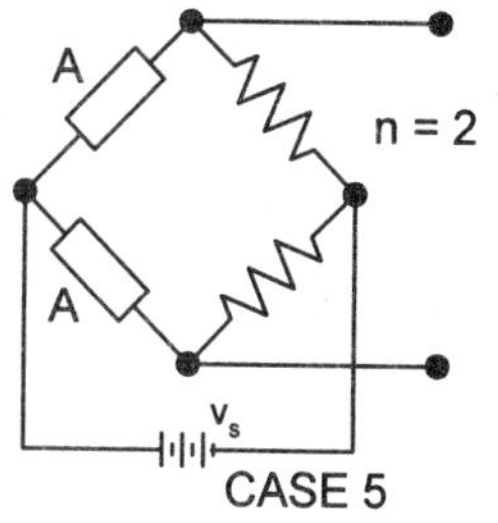

Figure E7.17

7.18 For a Wheatstone bridge powered with a constant current supply, show that:

$$\Delta v_0 = \frac{I R_g r}{2(1+r) + \Delta R_g / R_g} \frac{\Delta R_g}{R_g}$$

Note in deriving this relation assume that $R_1 = R_4 = R_g$ $R_2 = R_3 = rR_g$ $\Delta R_2 = \Delta R_3 = 0$.

7.19 Beginning with the equation shown in Exercise 7.18, verify that the nonlinear term for the Wheatstone bridge powered with a constant current supply is given by Eq. (7.17).

7.20 Prepare a graph showing the calibration strain ε_c as a function of R_c for R_2 = 120, 350, 500, and 1,000 Ω. Assume S_g = 2.00.

7.21 Could the calibration resistance be placed across arm R_1 of the bridge that contains the active gage? When would this procedure be recommended?

7.22 Determine the signal loss factor for a two-lead-wire system if 18-gage wire is employed and the recording instrument and strain gage are separated by 200 m. Determine the signal loss factor if a three-lead-wire system is employed.

7.23 Determine the signal loss factor for a three-lead-wire system as a function of lead length ranging from 1 to 1000 m for wire sizes:

(a) #12 (b) #18 (c) #22 (d) #26

7.24 For the two-lead-wire system of Exercise 7.23, determine the apparent strain introduced by an average temperature change of 18 °F (10 °C) over the length of the lead wires. Assume S_g = 2.00.

7.25 Specify a switch that would be adequate to use in the switching arrangement shown in Fig. 7.10.

7.26 Specify a switch that would be adequate to use in the switching arrangement shown in Fig. 7.11.

7.27 Briefly describe electrical noise and indicate why it occurs.

7.28 Outline three procedures that should be followed to minimize electrical noise in a strain-gage recording system.

7.29 Describe signal ground, system ground and building ground. Describe a ground loop. How are ground loops avoided? How many grounds are usually employed in a recording system? How is a ground loop avoided if more than one ground is employed?

7.30 Design a torque cell to measure torques which range from 0 to 100 kN-m. Specify the size and shape of the mechanical member, the gage layout, the circuit design and the voltage required. The torque cell must be insensitive to axial loads and bending moments.

7.31 Prove that the torque cell designed in Exercise 7.30 is insensitive to axial loads and bending moments.

7.32 Design a load cell to measure a 100-MN axial tensile load. Specify the size and shape of the mechanical member, the gage layout, the circuit design and the voltage required.

7.33 Design a displacement transducer having a cantilever beam as the mechanical member. The transducer must be capable of measuring the displacement to an accuracy of ± 0.004 mm over a range of 5 mm. Specify the size of the mechanical member, the gage layout, the circuit design, and the voltage required.

7.34 Design a line of diaphragm type pressure transducers to measure pressures of 100, 200, 500, 1,000, 2,000, 5,000 and 10,000 kPa. If the diameter of this line of transducers is 15 mm and the diaphragms are fabricated from stainless steel, specify the diaphragm thickness t for each pressure if $\Delta v_o/v_s$ = 0.003. Determine the displacement at the center of the diaphragm at each maximum pressure. Determine the natural frequency of the diaphragm for each case.

7.35 Design a load-measuring transducer having a cantilever beam as the mechanical member and four electrical-resistance strain gages as the sensing elements. Position the gages on the beam and connect them into a fixed-voltage Wheatstone bridge to achieve maximum sensitivity. The output from the system must be proportional to the load P, which can be positioned anywhere on the right half of the beam, as shown in Fig. E7.35.

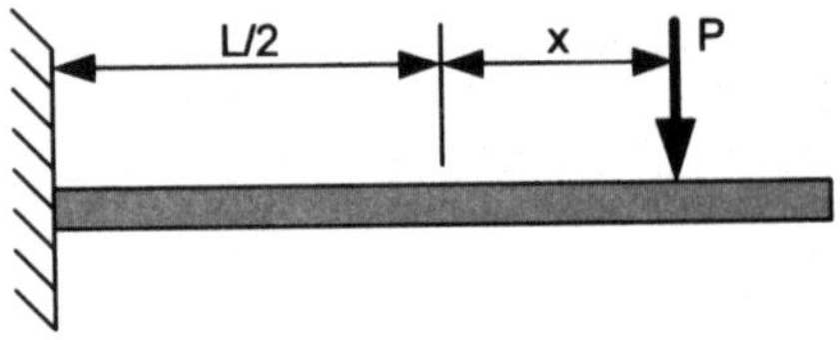

Figure E7.35

7.36 Design a scale to be used to measure the weight of first class mail. The scale is a subsystem in a larger automatic system used to place the correct postage on a large volume of mail each day.

7.37 Design a probe to be used to measure the viscosity of oil flowing in a pipe with a diameter of 200 mm. The range of the absolute viscosity is from 1 to 10^4 MPa-s.

7.38 Prepare a graph showing v_o/v_i for an amplifier responding to a step input voltage. Let τ = 10 μs and consider gains of 10, 100, and 1000.

7.39 Sketch simple circuits showing the difference between single ended and differential amplifiers.

7.40 Suppose a common voltage of 0.1 V is imposed on the input to a differential amplifier with a gain G_d = 500. When measuring a voltage difference Δv = 10 mV, determine the output voltage v_o if the common mode rejection ratio is:

(a) 1,000 (c) 10,000 (b) 5,000 (d) 20,000

7.41 Prepare a graph showing the dynamic range R as a function of gain G for an amplifier with a maximum input voltage v_i = 500 mV. Assume the amplifier noise v_{nA} is 5 μA.

7.42 Use an op-amp with a gain of 100 dB and R_a = 7 MΩ to design an inverting amplifier with a gain of:

(a) 10 (c) 50 (b) 20 (d) 100

7.43 Use an op-amp with a gain of 100 dB and R_a = 7 MΩ to design a non-inverting amplifier with a gain of:

(a) 10 (c) 50 (b) 20 (d) 100

7.44 Use an op-amp with a gain of 100 dB and R_a = 7 MΩ to design a differential amplifier with a gain of

(a) 10 (c) 50 (b) 20 (d) 100

7.45 Determine the input and output impedances for a voltage follower that incorporates an op-amp having a gain of 120 dB and R_a = 10 MΩ.

7.46 Verify Eq. (7.56).

7.47 Verify Eq. (7.59).

7.48 Three signals v_{i1}, v_{i2} and v_{i3} are to be summed so that the output voltage v_o is proportional to $v_{i1} + 3v_{i2} + (1/3)\,v_{i3}$. Select resistances R_1, R_2, R_3 and R_f to accomplish this operation.

7.49 Show that the op-amp circuit shown in Fig. E7.49 is a combined adding/scaling and subtracting/scaling amplifier by deriving the following equation for the output voltage v_o:

$$v_0 = \frac{R_f^*}{R_4} v_{i4} + \frac{R_f^*}{R_5} v_{i5} - \frac{R_f}{R_1} v_{i1} - \frac{R_f}{R_2} v_{i2} - \frac{R_f}{R_3} v_{i3}$$

where

$$R_f^* = R_f \left(\frac{\dfrac{1}{R_1} + \dfrac{1}{R_2} + \dfrac{1}{R_3} + \dfrac{1}{R_f}}{\dfrac{1}{R_4} + \dfrac{1}{R_5} + \dfrac{1}{R}} \right)$$

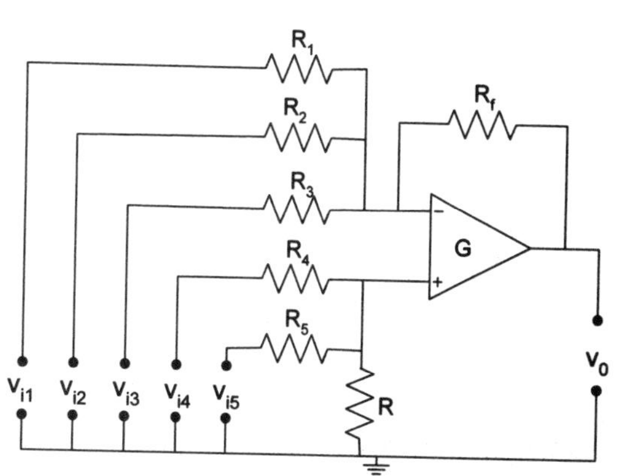

Figure E7.49

7.50 The signals shown in Fig. E7.50 are to be used as input to an integrating amplifier having R_1 = 1 MΩ and C_f = 0.5 μF. Sketch the output signal corresponding to each of the input signals.

7.51 Discuss potential problem areas associated with the output voltages from signals (a) and (c) of Exercise 7.50.

7.52 Repeat Exercise 7.50 with a differentiating amplifier in place of the integrating amplifier.

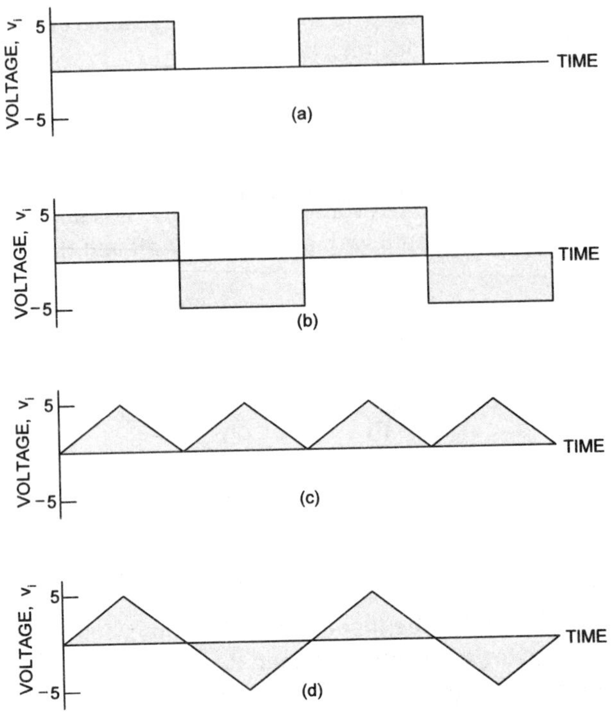

Figure E7.50

7.53 The sawtooth voltage pulse shown in Fig. E7.53 is the input to the filter. Resolve this voltage pulse into its Fourier components and consider the pulse distortion as the first five components pass through the filter. Discuss the results obtained.

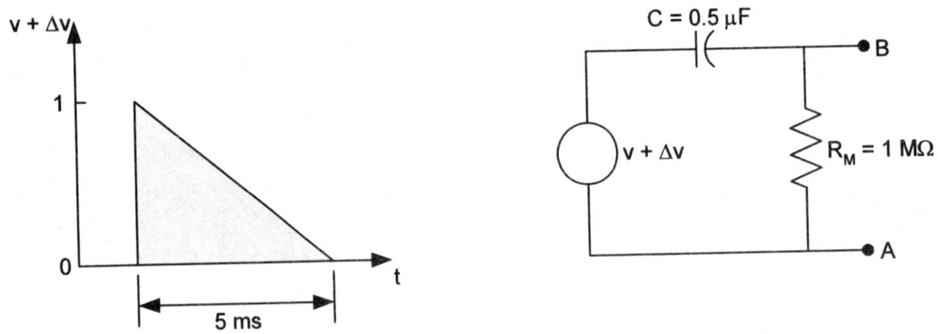

Figure E7.53

7.54 How could the simple filter shown in Fig. E7.53 be changed to improve the fidelity of the signal?

7.55 The triangular voltage pulse shown in Fig. E7.55 is the input to the filter. Resolve this voltage pulse into its Fourier components and consider the pulse distortion as the first five components pass through the filter. Discuss the results obtained and indicate how this filter design can be improved.

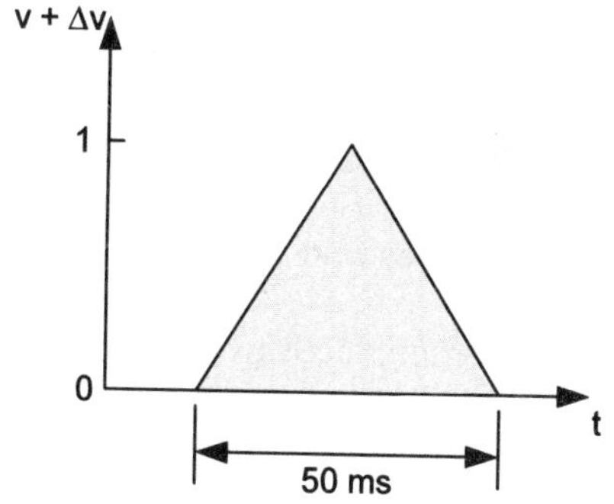

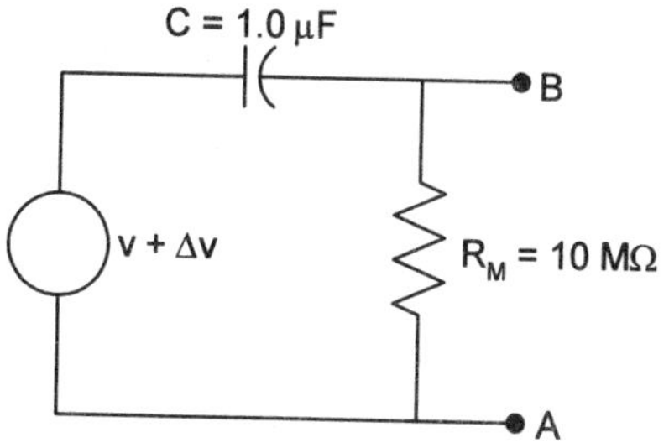

Figure E7.55

7.56 Draw circuits for the simple high-pass and low-pass RC filters. Sketch response curves for these filters.

7.57 Verify Eq. (7.61).

7.58 Verify Eq. (7.62).

7.59 Select R and C for a low-pass filter that will block 60 Hz noise but transmit the following low frequency signals with less than 1% loss:

(a) 5-Hz (b) 10-Hz (c) 20 Hz.

7.60 Select R and C in Fig. 7.33 to give a notch filter with a critical frequency f_c of

(a) 60-Hz (b) 1200-Hz (c) 10,000-Hz.

BIBLIOGRAPHY

1. Stein, P. K.: "Strain Gage Circuits for Semiconductor Gages," in M. Dean and R. D. Douglas (eds.), Semiconductor and Conventional Strain Gages, Academic Press, Inc., New York, pp. 273-282,1962.
2. Frank, E.: "Strain Indicator for Semiconductor Gages," in M. Dean and R. D. Douglas (eds.), Semiconductor and Conventional Strain Gages, Academic Press, Inc., New York, pp. 283-306, 1962.
3. Murray, W. M., and P. K. Stein: Strain Gage Techniques, Lectures and laboratory exercises by authors at M.I.T. Cambridge, MA, 1960.
4. Perry, C. C., and H. R. Lissner: The Strain Gage Primer, 2nd Edition, McGraw-Hill Book Company, New York, pp. 200-217, 1962.
5. "Noise Control in Strain Gage Measurements," TN-501-2, Measurements Group Inc., 1992.
6. "Strain Gage Based Transducers," Measurements Group, Inc, 1988.
7. "Design Considerations for Diaphragm Pressure Transducers," Measurements Group Tech. Note 510-1, 1992.
8. Timoshenko, S.: Strength of Materials, Part II, "Advanced Theory and Problems," 3rd Edition, D. Van Nostrand Company, Inc., Princeton, N.J. pp. 96-97, 1956.
9. Timoshenko, S.: Vibration Problems in Engineering, 3rd Edition, D. Van Nostrand Company, Inc., Princeton, N.J., pp. 449-451, 1955.
10. Ahmed, H. and P. J. Spreadbury: Analogue and Digital Electronics for Engineers, 2nd Edition, Cambridge University Press, New York, 1984.
11. Barna, A. and D. I. Porat: Operational Amplifiers, 2nd Edition, John Wiley & Sons, New York, 1989.
12. Brophy, J. J.: Basic Electronics for Scientists, 5th Edition, McGraw-Hill, New York, 1990.
13. Doebelin, E. O.: Measurement Systems, 5th Edition, McGraw-Hill, New York, 2007.
14. Hilburn, J. L. and D. E. Johnson: Manual of Active Filter Design, 2nd Edition, McGraw-Hill, New York, 1983.
15. Hughes, F. W.: Op-Amp Handbook, 2nd Edition, Prentice Hall, Englewood Cliffs, NJ, 1986.
16. Irvine, R. G.: Operational Amplifier Characteristics and Applications, 2nd edition, Prentice Hall, Englewood Cliffs, NJ, 1987.
17. Lenk, J. D.: Handbook of Modern Solid-State Amplifiers, Prentice-Hall, Englewood Cliffs, NJ, 1974.
18. Malmstadt, H. V., C. G. Enke, and S. R. Crouch: Electronics and Instrumentation for Scientists, Benjamin/Cummings, Menlo Park, CA., 1981.
19. Meiksin, Z. H.: Complete Guide to Active Filter Design, Op Amps, and Passive Components, Prentice-Hall, Englewood Cliffs, NJ, 1989.
20. Parks, T. W. and C. S. Burrus: Digital Filter Design, John Wiley & Sons, New York, 1987.
21. Stephenson, F. W.: RC Active Filter Design Handbook, John Wiley & Sons, New York, 1985.
22. Van Valkenburg, M. E.: Analog Filter Design, Holt, Rinehart, and Winston, New York, 1982.

CHAPTER 8

STRAIN ANALYSIS METHODS

8.1 INTRODUCTION

Electrical-resistance strain gages are normally employed on the free surface of a specimen to establish the stress at a particular point on this surface. In general, it is necessary to measure three strains at a point to completely define either the stress or the strain field. In terms of principal strains, it is necessary to measure ε_1, ε_2, and the direction of ε_1 relative to the x-axis as given by the principal angle φ. Conversion of the strains into stresses requires, in addition, a knowledge of the elastic constants E and ν of the specimen material.

In certain special cases the state of stress can be established with a single strain gage. Consider first a uniaxial state of stress where $\sigma_{yy} = \tau_{xy} = 0$ and the direction of σ_{xx} is known. In this case, a single-element strain gage is mounted with its axis coincident with the x axis. The stress σ_{xx} is given by:

$$\sigma_{xx} = E\,\varepsilon_{xx} \tag{8.1}$$

Next, consider an isotropic state of stress where $\sigma_{xx} = \sigma_{yy} = \sigma_1 = \sigma_2$ and $\tau_{xy} = 0$. In this case a strain gage may be mounted with any orientation since all directions are principal, and the magnitude of the stresses can be established from:

$$\sigma_{xx} = \sigma_{yy} = \sigma_1 = \sigma_2 = \frac{E}{1-\nu}\varepsilon_\theta \tag{8.2}$$

where ε_θ is the strain measured in any direction in the isotropic stress field.

When less is known beforehand regarding the state of stress in the specimen, it is necessary to employ multiple-element strain gages to establish the magnitude of the stress field. If the specimen being investigated has an axis of symmetry, or if a brittle-coating analysis has been conducted to establish the principal-stress directions, this knowledge can be used to reduce the number of gage elements required from three to two. A two-element rectangular rosette similar to the one illustrated in Figs. 6.5a is mounted on the specimen with its axes coincident with the principal directions. The two principal strains, ε_1 and ε_2, obtained from the gages can be employed to give the principal stresses σ_1 and σ_2:

$$\sigma_1 = \frac{E}{1-\nu}(\varepsilon_1 + \nu\varepsilon_2) \quad \text{and} \quad \sigma_2 = \frac{E}{1-\nu}(\varepsilon_2 + \nu\varepsilon_1) \tag{8.3}$$

These relations give the complete state of stress since the principal directions are known a priori. The stresses on any plane can be established by employing Eqs. (1.16) with the results obtained from Eq. (8.3).

In the most general case, no knowledge of the stress field or its directions is available before the experimental analysis is conducted. Three-element rosettes are required in these instances to completely establish the stress field. To show that three strain measurements are sufficient, consider three strain gages aligned along axes A, B, and C, as shown in Fig. 8.1.

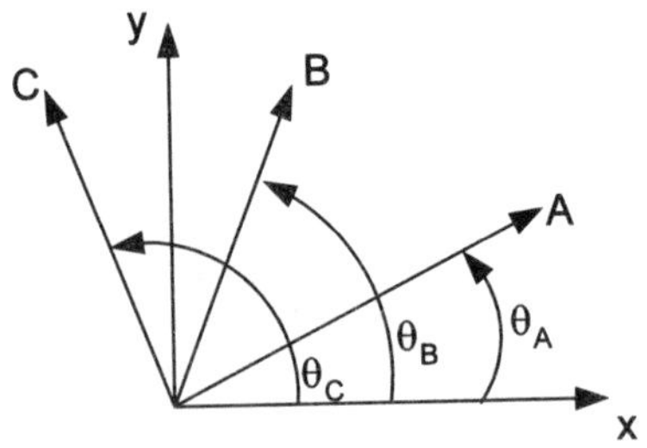

Figure 8.1 Three strain gage elements placed at arbitrary angles relative to the x and y axes.

From Eqs. (2.18) it is evident that:

$$\varepsilon_A = \varepsilon_{xx} \cos^2 \theta_A + \varepsilon_{yy} \sin^2 \theta_A + \gamma_{xy} \sin \theta_A \cos \theta_A$$

$$\varepsilon_B = \varepsilon_{xx} \cos^2 \theta_B + \varepsilon_{yy} \sin^2 \theta_B + \gamma_{xy} \sin \theta_B \cos \theta_B \qquad (8.4)$$

$$\varepsilon_C = \varepsilon_{xx} \cos^2 \theta_C + \varepsilon_{yy} \sin^2 \theta_C + \gamma_{xy} \sin \theta_C \cos \theta_C$$

The Cartesian components of strain ε_{xx}, ε_{yy} and γ_{xy} can be determined from a simultaneous solution of Eqs. (8.4). The principal strains and the principal directions can then be established by employing Eqs. (2.7), (1.12), and (1.14). The results are:

$$\varepsilon_1 = \frac{1}{2}\left(\varepsilon_{xx} + \varepsilon_{yy}\right) + \frac{1}{2}\sqrt{\left(\varepsilon_{xx} - \varepsilon_{yy}\right)^2 + \gamma_{xy}^2}$$

$$\varepsilon_2 = \frac{1}{2}\left(\varepsilon_{xx} + \varepsilon_{yy}\right) - \frac{1}{2}\sqrt{\left(\varepsilon_{xx} - \varepsilon_{yy}\right)^2 + \gamma_{xy}^2} \qquad (8.5)$$

$$\tan\, 2\phi = \frac{\gamma_{xy}}{\varepsilon_{xx} - \varepsilon_{yy}}$$

where φ is the angle between the principal axis σ_1 and the x-axis. The principal stresses can then be computed from the principal strains by utilizing Eqs. (2.20).

In practice, three-element rosettes with defined angles (that is θ_A, θ_B and θ_C fixed at specified values) are employed to provide sufficient data to completely define the stress field. These rosettes are described by the fixed angles as the rectangular rosette, the delta rosette, and the tee-delta rosette. Examples of commercially available three-element rosettes are presented in Figs. 6.5b and 6.5c. The three-element rosette gage will be discussed in detail in the Section 8.2.

8.2 THE THREE-ELEMENT RECTANGULAR ROSETTE

The three-element rectangular rosette employs gages placed at the 0, 45°, and 90° positions, as indicated in Fig. 8.2. For this particular rosette, it is clear from Eqs. (8.4) that:

$$\varepsilon_A = \varepsilon_{xx} \qquad \varepsilon_B = ½\left(\varepsilon_{xx} + \varepsilon_{yy} + \gamma_{xy}\right) \qquad \varepsilon_C = \varepsilon_{yy} \qquad (8.6)$$

$$\gamma_{xy} = 2\varepsilon_B - \varepsilon_A - \varepsilon_C$$

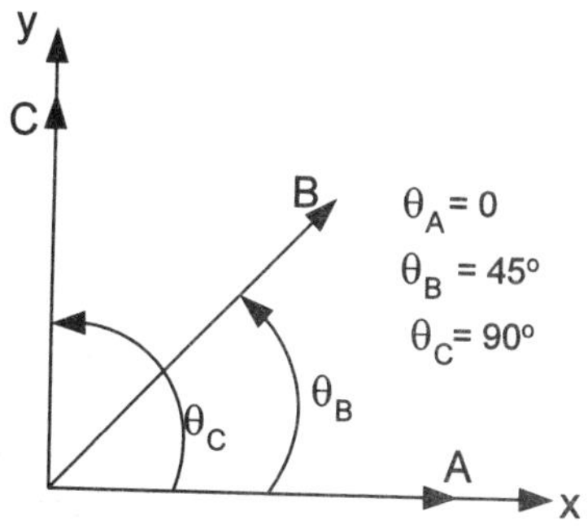

Figure 8.2 Strain gage positions in a three element rectangular rosette.

Thus, by measuring the strains ε_A, ε_B and ε_C, the Cartesian components of strain ε_{xx}, ε_{yy} and ε_{xy} can be quickly and simply established through the use of Eqs. (8.6). Next, by utilizing Eqs. (8.5), the principal strains ε_1 and ε_2 can be established as:

$$\varepsilon_1 = \frac{1}{2}(\varepsilon_A + \varepsilon_C) + \frac{1}{2}\sqrt{(\varepsilon_A - \varepsilon_C)^2 + (2\varepsilon_B - \varepsilon_A - \varepsilon_C)^2}$$
$$\varepsilon_2 = \frac{1}{2}(\varepsilon_A + \varepsilon_C) - \frac{1}{2}\sqrt{(\varepsilon_A - \varepsilon_C)^2 + (2\varepsilon_B - \varepsilon_A - \varepsilon_C)^2} \qquad (8.7a)$$

The principal angle ϕ is given by:

$$\tan 2\phi = \frac{2\varepsilon_B - \varepsilon_A - \varepsilon_C}{\varepsilon_A - \varepsilon_C} \qquad (8.7b)$$

The solution of Eq. (8.7b) gives two values for the angle ϕ. The first ϕ_1 refers to the angle between the x-axis and the axis of the maximum principal strain ε_1. The second ϕ_2 refers to the angle between the x-axis and the axis of the minimum principal strain ε_2. These angles are illustrated in the Mohr's strain circle shown in Fig. 8.3. It is possible to show (see Exercise 8.5) that applying the following rules can identify the principal axes:

$$\begin{aligned}
&0 < \phi_1 < 90° && \text{when } \varepsilon_B > \tfrac{1}{2}(\varepsilon_A + \varepsilon_C)\\
&-90 < \phi_1 < 0° && \text{when } \varepsilon_B < \tfrac{1}{2}(\varepsilon_A + \varepsilon_C)\\
&\phi_1 = 0° && \text{when } \varepsilon_A > \varepsilon_C \text{ and } \varepsilon_A = \varepsilon_1\\
&\phi_1 = \pm 90° && \text{when } \varepsilon_A < \varepsilon_C \text{ and } \varepsilon_A = \varepsilon_2
\end{aligned} \qquad (8.8)$$

Finally, the principal stresses occurring in the component can be established by employing Eqs. (8.7) together with Eqs. (8.3) to obtain:

$$\sigma_1 = E\left[\frac{\varepsilon_A + \varepsilon_C}{2(1-\nu)} + \frac{1}{2(1+\nu)}\sqrt{(\varepsilon_A - \varepsilon_C)^2 + (2\varepsilon_B - \varepsilon_A - \varepsilon_C)^2}\right]$$
$$\sigma_2 = E\left[\frac{\varepsilon_A + \varepsilon_C}{2(1-\nu)} - \frac{1}{2(1+\nu)}\sqrt{(\varepsilon_A - \varepsilon_C)^2 + (2\varepsilon_B - \varepsilon_A - \varepsilon_C)^2}\right] \qquad (8.9)$$

The use of Eqs. (8.6) to (8.9) permits a determination of the Cartesian components of strain, the principal strains and their directions, and the principal stresses by using a totally analytical approach. However, it is also possible to determine these quantities with a graphical approach, as illustrated in Fig. 8.3. A Mohr's strain circle is initiated by laying out the ε (abscissa) and the ½ γ (ordinate) axes. The three strains ε_A, ε_B and ε_C are plotted as points on the abscissa. Vertical lines are then drawn through these three points. The shearing strain γB_{xyB} is computed from Eqs. (8.6), and ½ γ_{xy} is plotted positive downward or negative upward along the vertical line drawn through ε_A to establish point A. This shearing strain may also be plotted as positive upward or negative downward along the vertical line through ε_C to establish point C. The diameter of the circle is then determined by drawing a line between A and C that intersects the abscissa and defines the center of the circle at a distance ½ ($\varepsilon_A + \varepsilon_C$) from the origin. A circle is then drawn from this center passing through points A and C. The circle will intersect the vertical line drawn through ε_B, and this point of intersection is labeled as B. A straight line through the center of the circle and point B should be a perpendicular bisector of the diameter AC. The values of the principal strains ε_1 and ε_2 are given by the intersections of the circle with the abscissa. The principal angle $2\phi_1$ is given by the angle $AO\varepsilon_1$ and is negative if point A lies above the ε axis. The principal angle $2\phi_2$ is given by the angle $AO\varepsilon_2$ and is positive if point A lies above the ε axis. The maximum shearing strain is established by a vertical line drawn through the center of the circle to give point D at the intersection. The projection of point D onto the ½ γ axis determines the value of ½ γ_{max}. The principal stresses can be determined directly from the principal strains by employing Eqs. (8.3).

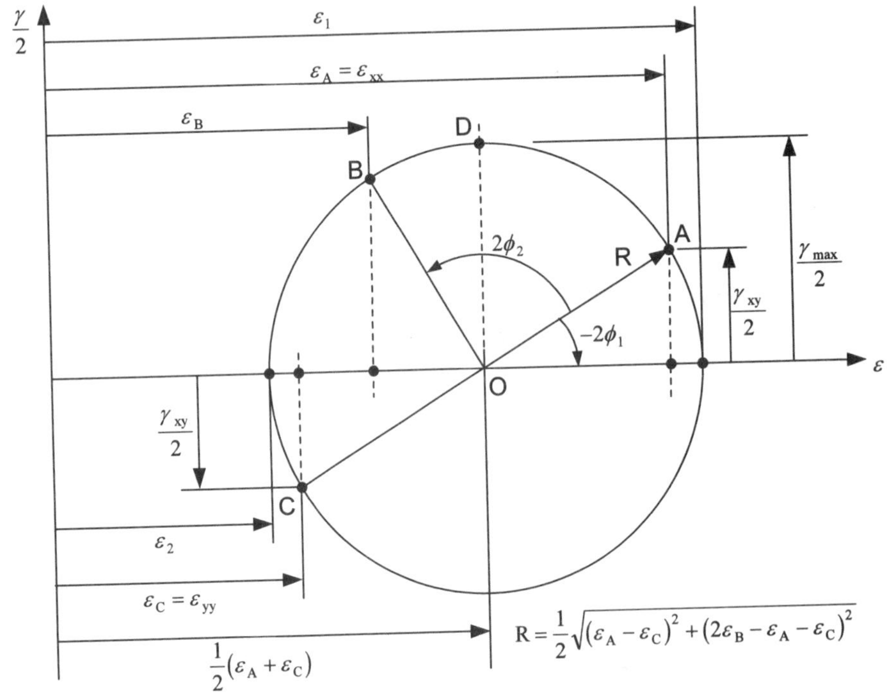

Figure 8.3 Graphical solution for the principal strains and their directions from a rectangular rosette.

This graphical approach for reducing the data obtained from a rectangular rosette is quite applicable when the data to be reduced are limited to a few gages. However, if large amounts of data must be reduced, the analytical approach is normally preferred since it requires less time.

Nomographs are available for use when data from a very large number of rosettes must be reduced. Also, the analytical approach for obtaining the principal strains and stresses and their directions can be programmed in LabVIEW as described in Appendix A.

8.3 CORRECTIONS FOR TRANSVERSE STRAIN EFFECTS

In Section 6.5, it was noted that foil-type resistance strain gages exhibit a sensitivity S_t to transverse strains. Reference to Fig. 6.12 shows that in certain instances this transverse sensitivity can lead to large errors, and it is important to correct the data to eliminate this effect. Two different procedures for correcting data have been developed.

The first procedure requires a priori knowledge of the ratio $\varepsilon_t/\varepsilon_a$ of the strain field. The correction factor is evident in Eq. (6.9), where:

$$\varepsilon_a = \varepsilon_a' \frac{1-\nu_0 K_t}{1+K_t(\varepsilon_t/\varepsilon_a)} \tag{6.9}$$

The term ε_a' is the apparent strain, and the correction factor CF is given by:

$$CF = \frac{1-\nu_0 K_t}{1+K_t(\varepsilon_t/\varepsilon_a)} \tag{8.10}$$

It is possible to correct the strain gage for this transverse sensitivity by adjusting its gage factor. The corrected gage factor S_g^* that should be used with a strain indicator is given by:

$$S_g^* = S_g \frac{1+K_t(\varepsilon_t/\varepsilon_a)}{1-\nu_0 K_t} \tag{8.11}$$

Correction for the cross-sensitivity effect when the strain field is unknown is more involved and requires the experimental determination of strain in both the x and y directions. If ε'_{xx} and ε'_{yy} are the apparent strains recorded in the x and y directions, respectively, then from Eq. (6.9) it is evident that:

$$\begin{aligned} \varepsilon'_{xx} &= \frac{1}{1-\nu_0 K_t}\left(\varepsilon_{xx} + K_t \varepsilon_{yy}\right) \\ \varepsilon'_{yy} &= \frac{1}{1-\nu_0 K_t}\left(\varepsilon_{yy} + K_t \varepsilon_{xx}\right) \end{aligned} \tag{8.12}$$

where the unprimed quantities ε_{xx} and ε_{yy} are the true strains. Solving Eqs. (8.12) for ε_{xx} and ε_{yy} gives:

$$\begin{aligned} \varepsilon_{xx} &= \frac{1-\nu_0 K_t}{1-K_t^2}\left(\varepsilon'_{xx} - K_t \varepsilon'_{yy}\right) \\ \varepsilon_{yy} &= \frac{1-\nu_0 K_t}{1-K_t^2}\left(\varepsilon'_{yy} - K_t \varepsilon'_{xx}\right) \end{aligned} \tag{8.13}$$

Equations (8.13) give the true strains ε_{xx} and ε_{yy} in terms of the apparent strains ε'_{xx} and ε'_{yy}. Correction equations for transverse strains in two- and three-element rosettes are cited in Reference [1].

8.4 THE STRESS GAGE

The transverse sensitivity which was shown in the previous section to result in errors in strain measurements can be employed to produce a special- purpose transducer known as a stress gage. The stress gage looks very much like a strain gage except that its grid is designed to give a select value of K_t so that the output ΔR/R is proportional to the stress along the axis of the gage. The stress gage serves a very useful purpose when a stress determination in a particular direction is the ultimate objective of the analysis, for it can be obtained with a stress gage rather than a three-element rosette.

The principle of a stress gage is shown in the following derivation. The output ΔR/R of any gage is expressed by Eq. (6.7) as:

$$\frac{\Delta R}{R} = S_a \varepsilon_a (1 - \nu_0 K_t) \tag{6.7}$$

The relationship between stress and strain for plane stress is given by Eqs. (2.19) as:

$$\begin{aligned} \varepsilon_a &= (1/E)[\sigma_a - \nu\sigma_t] \\ \varepsilon_t &= (1/E)[\sigma_t - \nu\sigma_a] \end{aligned} \tag{2.19}$$

Substituting Eqs. (2.19) into Eq. (6.7) yields:

$$\frac{\Delta R}{R} = \frac{\sigma_a S_a}{E}(1 - \nu K_t) + \frac{\sigma_t S_a}{E}(K_t - \nu) \tag{8.14}$$

Examination of Eq. (8.14) indicates that the output of the gage ΔR/R will be independent of σ_t if $K_t = \nu$. It can also be shown that the axial sensitivity S_a of a gage is related to the alloy sensitivity S_A by the expression given by:

$$S_a = \frac{S_A}{1 + K_t} \tag{8.15}$$

Substituting Eq. (8.15) into Eq. (8.14) and letting $K_t = \nu$ leads to:

$$\sigma_a = \frac{E}{S_A(1 - \nu)} \frac{\Delta R}{R} \tag{8.16}$$

Because the factor $E/[S_A(1 - \nu)]$ is a constant for a given gage alloy and specimen material, the gage output in terms of ΔR/R is linearly proportional to stress. In practice, the stress gage is made with a V-type grid configuration. Further analysis of the stress gage is necessary to understand its operation in a strain field which is unknown and in which the strain gage is placed in an arbitrary direction. Consider the placement of the gage, as shown in Fig. 8.4, along an arbitrary x-axis which is at some unknown angle ϕ with the principal axis corresponding to σ_1. The grid elements are at a known angle θ relative to the x-axis.

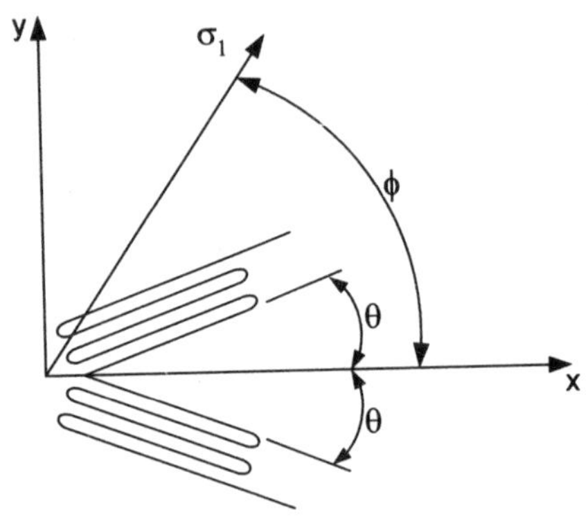

Figure 8.4 A stress gage positioned relative to the x-axis and the principal axis associated with σ_1.

The strain along the top grid element is given by a modified form of Eq. (8.4) as:

$$\varepsilon_{\phi-\theta} = ½\,(\varepsilon_1 + \varepsilon_2) + ½\,(\varepsilon_1 - \varepsilon_2)\cos 2(\phi - \theta) \qquad \text{(a)}$$

The strain along the lower grid element is:

$$\varepsilon_{\phi+\theta} = ½\,(\varepsilon_1 + \varepsilon_2) + ½\,(\varepsilon_1 - \varepsilon_2)\cos 2(\phi + \theta) \qquad \text{(b)}$$

Summing Eqs. (a) and (b) and expanding the cosine terms yields:

$$\varepsilon_{\phi-\theta} + \varepsilon_{\phi+\theta} = (\varepsilon_1 + \varepsilon_2) + (\varepsilon_1 - \varepsilon_2)\cos 2\phi \cos 2\theta \qquad \text{(c)}$$

Note from the Mohr's strain circles presented earlier in this chapter that:

$$\varepsilon_{xx} + \varepsilon_{yy} = \varepsilon_1 + \varepsilon_2 \qquad \text{(d)}$$

$$\varepsilon_{xx} - \varepsilon_{yy} = (\varepsilon_1 - \varepsilon_2)\cos 2\phi \qquad \text{(e)}$$

Substituting Eqs (d) and (e) into Eq. (c) gives:

$$\varepsilon_{\phi-\theta} + \varepsilon_{\phi+\theta} = (\varepsilon_{xx} + \varepsilon_{yy}) + (\varepsilon_{xx} - \varepsilon_{yy})\cos 2\theta = 2(\varepsilon_{xx}\cos^2\theta + \varepsilon_{yy}\sin^2\theta)$$

$$= 2\cos^2\theta\,(\varepsilon_{xx} + \varepsilon_{yy}\tan^2\theta) \qquad \text{(f)}$$

If the gage is manufactured so that θ is equal to $\tan^{-1}\sqrt{\nu}$, then:

$$\tan^2\theta = \nu \qquad \cos^2\theta = \frac{1}{1+\nu}$$

and Eq. (f) becomes:

$$\varepsilon_{\phi-\theta} + \varepsilon_{\phi+\theta} = \frac{2}{1+\nu}\left(\varepsilon_{xx} + \nu\varepsilon_{yy}\right) \qquad \text{(g)}$$

Substituting Eq. (g) into Eqs. (2.20) yields:

$$\sigma_{xx} = \frac{E}{2(1-\nu)}\left(\varepsilon_{\phi-\theta} + \varepsilon_{\phi+\theta}\right) \qquad (8.17)$$

where $½\,(\varepsilon_{\phi-\theta} + \varepsilon_{\phi+\theta})$ is the average strain from the two gage elements and is equal to $(\Delta R/R)/S_g$.

The gage reading will give $½\,(\varepsilon_{\phi-\theta} + \varepsilon_{\phi+\theta})$, and it is only necessary to multiply this reading by $E/(1-\nu)$ to obtain σ_{xx}. The stress gage will thus give σ_{xx} directly with a single gage. However, it does not give any data regarding σ_{yy} or the principal angle ϕ_1. Moreover, σ_{xx} may not be the most important stress since it may differ appreciably from σ_1. If the directions of the principal stresses are known, the stress gage may be used more effectively by choosing the x axis to coincide with the principal axis corresponding to σ_1 so that $\sigma_{xx} = \sigma_1$. In fact, when the principal directions are known, a conventional single-element strain gage can be employed as a stress gage.

This adaptation is possible if the gage is located along a line which makes an angle θ with respect to the principal axis, as shown in Fig. 8.5. In this case, the strains will be symmetrical about the principal axis; hence, it is clear that:

$$\varepsilon_{\phi-\theta} = \varepsilon_{\phi+\theta} = \varepsilon_\theta$$

and Eq. (8.17) reduces to:

$$\sigma_1 = \frac{E}{1-\nu}\varepsilon_\theta \tag{8.18}$$

The value ε_θ is recorded with the strain gage and converted to σ_1 directly by multiplying by E/(1 – ν). This procedure reduces the number of gages necessary if only the value of σ_1 is to be determined. The saving of a gage is of particular importance in dynamic strain measurements when the instrumentation required becomes complex and the number of available channels of recording equipment may be limited.

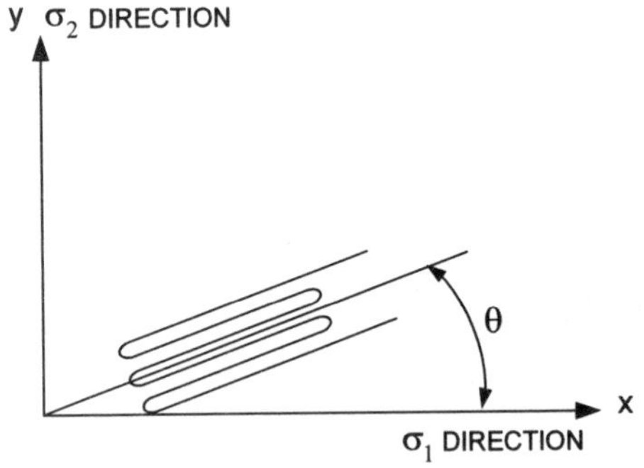

Figure 8.5 A single element strain gage employed as a stress gage when the principal directions are known.

8.5 PLANE-SHEAR OR TORQUE GAGE

Consider two strain gages, A and B, oriented at angles $\theta_A = -\theta_B$ with respect to the x axis, as shown in Fig. 8.6. The strains along the gage axes are given by a modified form of Eqs. (2.18) as:

$$\begin{aligned}\varepsilon_A &= \frac{\varepsilon_{xx}+\varepsilon_{yy}}{2} + \frac{\varepsilon_{xx}-\varepsilon_{yy}}{2}\cos 2\theta_A + \frac{\gamma_{xy}}{2}\sin 2\theta_A \\ \varepsilon_B &= \frac{\varepsilon_{xx}+\varepsilon_{yy}}{2} + \frac{\varepsilon_{xx}-\varepsilon_{yy}}{2}\cos 2\theta_B + \frac{\gamma_{xy}}{2}\sin 2\theta_B\end{aligned} \tag{8.19}$$

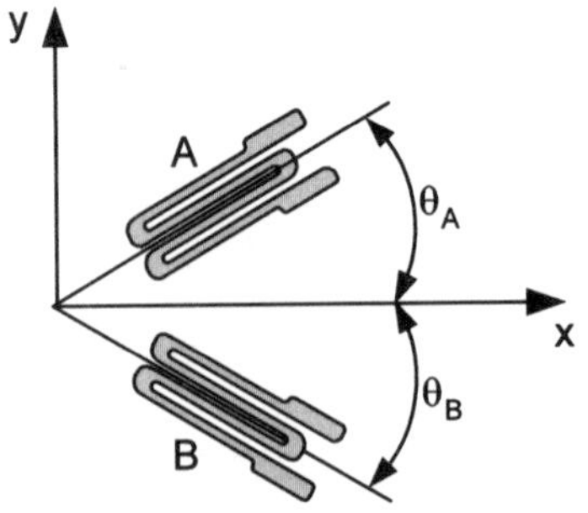

Figure 8.6 Positions of gages A and B for measuring γBxyB.

From Eqs. (8.19), the shear strain γ_{xy} is given by:

$$\gamma_{xy} = \frac{2(\varepsilon_A - \varepsilon_B) - (\varepsilon_{xx} - \varepsilon_{yy})(\cos 2\theta_A - \cos 2\theta_B)}{\sin 2\theta_A - \sin 2\theta_B} \tag{a}$$

Note that gages A and B are oriented such that $\theta_A = -\theta_B$; hence:

$$\cos 2\theta_A = \cos(-\theta_B) = \cos 2\theta_B \tag{b}$$

because the cosine is an even function. As a result, Eq. (a) reduces to:

$$\gamma_{xy} = \frac{2(\varepsilon_A - \varepsilon_B)}{\sin 2\theta_A - \sin 2\theta_B} = \frac{\varepsilon_A - \varepsilon_B}{\sin 2\theta_A} \tag{8.20}$$

because the sine is an odd function. Thus, the shearing strain γB_{xyB} is proportional to the difference between the normal strains experience by gages A and B when they are oriented with respect to the x axis as shown in Fig. (8.6). The angle $\theta_A = -\theta_B$ can be arbitrary; however, for the angle $\theta_A = \pi/4$, Eq. (8.20) reduces to:

$$\gamma_{xy} = \varepsilon_A - \varepsilon_B \tag{8.21}$$

Equation (8.21) indicates that the shearing strain γ_{xy} can be measured with a two-element rectangular rosette by orienting the gages at 45° and – 45° with respect to the x axis and connecting one gage into arm R_1 and the other into arm R_4 of a bridge. The subtraction $\varepsilon_A - \varepsilon_B$ will be performed automatically in the bridge, and the output will give $2\gamma_{xy}$ directly.

Both two- and four-element shear gages are marketed commercially for this measurement. The four-element gages provide a complete four-arm bridge with twice the output of the two-element gages. Typical shear gages are illustrated in Fig. 8.7.

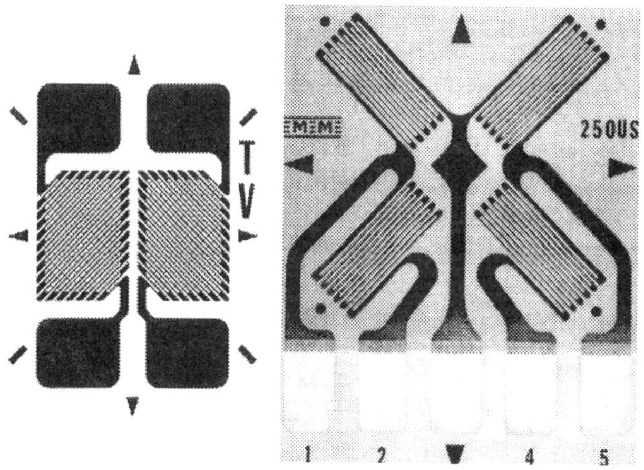

Figure 8.7 Two and four element rosettes that are used for shear strain measurement.

8.6 THE STRESS INTENSITY FACTOR GAGE K_I

Consider a two dimensional body with a single-ended through crack as shown in Fig. 4.7a. The stability of this crack is determined by the opening mode stress intensity factor K_I. If the specimen is fabricated from a brittle material, the crack will be initiated when:

$$K_I \geq K_{Ic} \tag{4.2}$$

It is possible to determine K_I as a function of loading on a structure by placing one or more strain gages near the crack tip. To show an effective approach to this measurement consider a series representation, using three terms, of Eqs. 4.42. The three-term representation of the strain field is:

$$
\begin{aligned}
E\varepsilon_{xx} &= A_0 r^{-1/2}\cos\frac{\theta}{2}\left[(1-\nu)-(1+\nu)\sin\frac{\theta}{2}\sin\frac{3\theta}{2}\right]+2B_0 \\
&\quad + A_1 r^{1/2}\cos\frac{\theta}{2}\left[(1-\nu)+(1+\nu)\sin^2\frac{\theta}{2}\right] \\
E\varepsilon_{yy} &= A_0 r^{-1/2}\cos\frac{\theta}{2}\left[(1-\nu)+(1+\nu)\sin\frac{\theta}{2}\sin\frac{3\theta}{2}\right]-2\nu B_0 \\
&\quad + A_1 r^{1/2}\cos\frac{\theta}{2}\left[(1-\nu)-(1+\nu)\sin^2\frac{\theta}{2}\right] \\
\mu\gamma_{xy} &= \frac{A_0}{2} r^{-1/2}\left(\sin\theta\cos\frac{3\theta}{2}\right)-\frac{A_1}{2} r^{1/2}\left(\sin\theta\cos\frac{\theta}{2}\right)
\end{aligned}
\tag{8.22}
$$

where A_0, B_0 and A_1 are unknown coefficients that depend on the geometry of the specimen and the loading. Recall that A_0 and K_I are related by:

$$A_0 = \frac{K_I}{\sqrt{2\pi}} \tag{4.28}$$

Equations (8.22) could be used to determine the unknown coefficients A_0, B_0 and A_1 if three or more strain gages are placed at appropriate positions in the near field region. However, the number of gages required for the determination of A_0 or K_I can be reduced to one by considering a gage oriented at an angle α and positioned along the Px' axis as shown in Fig. 8.8. For the rotated coordinates shown in Fig. 8.8, the strain $\varepsilon_{x'x'}$ is obtained from Eqs. (8.22) and Eqs. (2.18) as:

$$
\begin{aligned}
2\mu\varepsilon_{x'x'} &= A_0 r^{-1/2}\left[k\cos\frac{\theta}{2}-\frac{1}{2}\sin\theta\sin\frac{3\theta}{2}\cos 2\alpha+\frac{1}{2}\sin\theta\cos\frac{3\theta}{2}\sin 2\alpha\right] \\
&+B_0(k+\cos 2\alpha)+A_1 r^{1/2}\cos\frac{\theta}{2}\left[k+\sin^2\frac{\theta}{2}\cos 2\alpha-\frac{1}{2}\sin\theta\sin 2\alpha\right]
\end{aligned}
\tag{8.23}
$$

where

$$k = \frac{1-\nu}{1+\nu} \tag{8.24}$$

Figure 8.8 Definition of a rotated coordinate system positioned at point P.

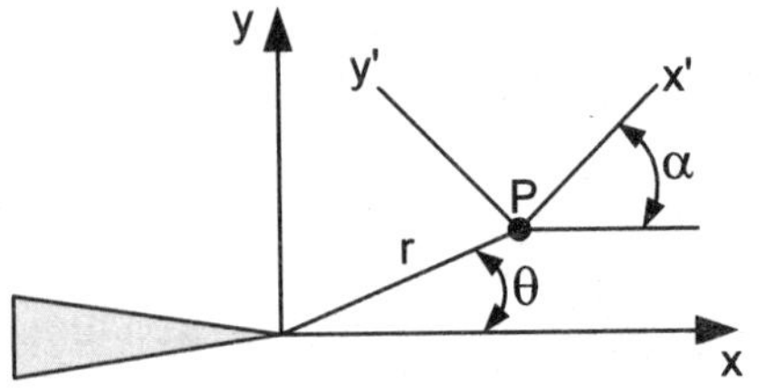

The coefficient of the B_0 term is eliminated by selecting the angle α as:

$$\cos 2\alpha = -k = -\frac{1-\nu}{1+\nu} \tag{8.25}$$

Next, the coefficient of A_1 vanishes if the angle θ is selected as:

$$\tan(\theta/2) = -\cot 2\alpha \tag{8.26}$$

By the proper placement of a single strain gage with angles α and θ determined to satisfy Eqs. (8.25) and (8.26), the strain $\varepsilon_{x'x'}$ is related directly to the stress intensity factor K_I by:

$$2\mu\varepsilon_{x'x'} = \frac{K_I}{\sqrt{2\pi r}}\left[k\cos\frac{\theta}{2} - \frac{1}{2}\sin\theta\sin\frac{3\theta}{2}\cos 2\alpha + \frac{1}{2}\sin\theta\cos\frac{3\theta}{2}\sin 2\alpha\right] \tag{8.27}$$

The choice of the angles α and θ depends only on Poisson's ratio as indicated in Table 8.1.

Table 8.1
Angles α and θ as a Function of Poisson's Ratio ν

ν	θ, degrees	α, degrees
0.2350	73.74	63.43
0.300	65.16	61.29
0.333	60.00	60.00
0.400	50.76	57.69
0.500	38.97	54.74

Consider, for example, an aluminum plate (ν = 1/3) with a through crack. In this case α = θ = 60° (see Table 8.1) and a single strain gage is placed at any point located by r in the near field region along a 60° radial line drawn from the crack tip. For this example with ν = 1/3 and α = θ = 60° Eq. (8.27) reduces to:

$$K_I = E\varepsilon_g\sqrt{\frac{8\pi r}{3}} \tag{8.28}$$

where $\varepsilon_g = \varepsilon_{x'x'}$ is the strain indicated by the single strain gage.

8.6.1 Strain Gradient Errors

The use of strain gages to measure stress intensity factors was delayed for nearly 30 years because of an unfounded concern about error introduced by the strain gradient effect. To explore the magnitude of the error due to strain gradients, consider a single-element gage positioned in the near field region with ν = 1/3, k = 1/2 and α = θ = 60°. The strain along this radial line is determined from Eq. (8.27) as:

$$\varepsilon_{x'x'} = q r^{-1/2} \tag{8.29}$$

where

$$q = \frac{K_I}{E\sqrt{8\pi/3}} \tag{8.30}$$

The gage senses the strain $\varepsilon_{x'x'}$ and the gage signal represents the average strain over the gage length which is given by:

$$(\varepsilon_{x'x'})_{avg} = \frac{q}{r_0 - r_i}\int_{r_i}^{r_0} r^{-1/2}dr \tag{8.31}$$

where r_0 and r_i are positions defining the location of the gage grid as shown in Fig. 8.9. Integrating Eq. (8.31) yields

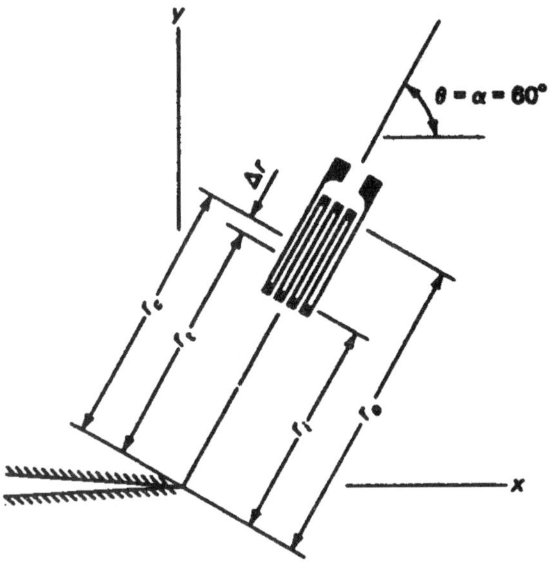

Figure 8.9 Definition of radii associated with placement of a strain gage near a crack tip.

$$(\varepsilon_{x'x'})_{avg} = \frac{2q}{\sqrt{r_0} - \sqrt{r_i}} \quad \text{(a)}$$

The strain gage response $(\varepsilon_{x'x'})_{ave}$ corresponds to the true strain $\varepsilon_{x'x'}$ at a specific point r_t along the gage length. By equating Eq (a) with Eq. (8.29) it is clear that r_t is:

$$r_t = \frac{1}{4}\left[\sqrt{r_0} + \sqrt{r_i}\right]^2 \quad (8.32)$$

Note that the position of the center of the gage r_c is given by:

$$r_c = ½ (r_0 + r_i) \quad \text{(b)}$$

Defining Δr as the distance between the center of the gage and the true strain location r_t gives:

$$\Delta r = r_c - r_t = \frac{1}{4}\left[r_0 - 2\sqrt{r_0}\sqrt{r_i} + r_i\right] \quad \text{(c)}$$

Next write the gage length L as:

$$L = r_0 - r_i \quad \text{(d)}$$

Finally, combine Eqs. (b), (c), and (d) to obtain:

$$\frac{\Delta r}{r_c} = \frac{1}{2}\left\{1 - \left[1 - \left(\frac{L}{2r_c}\right)^2\right]^{1/2}\right\} \quad (8.33)$$

where $r_c > L/2$ to avoid placing any portion of the gage over the crack tip. The results of Eq. (8.33) are shown in Fig. 8.10. The strain-gradient effect is a maximum with $\Delta r/r_c = 0.5$ when the gage is placed as close as possible to the crack tip with $r_c = L/2$. This is a serious error; however, placement at this location should be avoided in any event since it locates the gage in region 1 (see Fig. 4.11), where the plane-stress analysis of the strain field is not valid. For gages placed in the near field region, r_c/L will probably exceed 2 and $\Delta r/r_c < 0.016$. If a correction is required to account for Δr, the radius r_t is determined from Eq. (8.32) and this value is substituted into Eq. (8.28) to determine K_I. In many applications, the difference between r_c and r_t is negligible. For example, consider a gage with L = 1 mm positioned at r_c = 5 L and note from Fig. 8.10 that $\Delta r/r_c = 0.0025$. The correction $\Delta r = 0.0125$ mm is much less than the accuracy that can be achieved in measuring r_c.

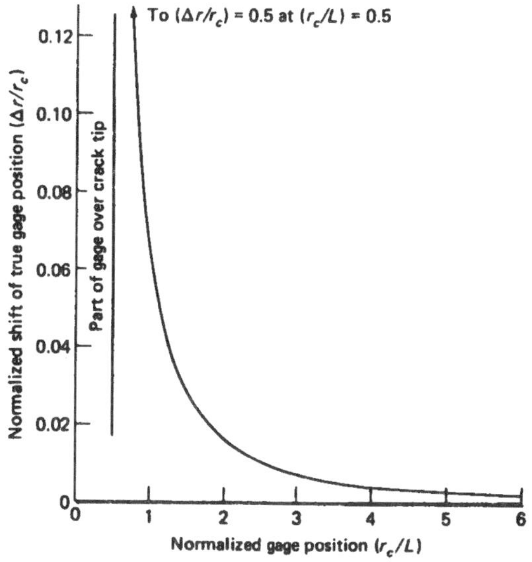

Figure 8.10 Normalized shift of the true gage position $\Delta r/r_c$ as a function of normalized gage position r_c/L.

8.7 DETERMINING MIXED MODE STRESS INTENSITY FACTORS

Frequently, the loading on a body containing a crack produces a mixed mode state of stress. The mixed mode indicates that the opening mode and the shearing mode occur together. The strain field in the region near the crack tip, for a mixed mode state of stress, is given by superposition of Eqs. (4.42) and (4.46). Consider, again, a strain gage positioned at point P with an orientation α, as shown in Fig. 8.8. The strain in the x' direction may be written by using Eqs. (4.42), (4.46) and the strain transformation equation Eq. (2.18) to give:

$$
\begin{aligned}
E\varepsilon_{x'x'} = {} & A_0 r^{-1/2}\left\{\begin{array}{l}\cos\dfrac{\theta}{2}\left[(1-\nu)-(1+\nu)\sin\dfrac{\theta}{2}\sin\dfrac{3\theta}{2}\right]\cos^2\alpha+\cos\dfrac{\theta}{2}\left[(1-\nu)+(1+\nu)\sin\dfrac{\theta}{2}\sin\dfrac{3\theta}{2}\right]\sin^2\alpha \\ +(1+\nu)\sin\theta\cos\dfrac{3\theta}{2}\sin\alpha\cos\alpha\end{array}\right\} \\
& + 2B_0\left(\cos^2\alpha-\nu\sin^2\alpha\right) \\
& + A_1 r^{1/2}\left\{\begin{array}{l}\cos\dfrac{\theta}{2}\left[(1-\nu)+(1+\nu)\sin^2\dfrac{\theta}{2}\right]\cos^2\alpha+\cos\dfrac{\theta}{2}\left[(1-\nu)+(1+\nu)\sin^2\dfrac{\theta}{2}\right]\sin^2\alpha \\ -(1+\nu)\sin\theta\cos\dfrac{\theta}{2}\sin\alpha\cos\alpha\end{array}\right\} \\
& + 2B_1 r\left[\cos\theta\cos^2\alpha-\nu\cos\theta\sin^2\alpha-2(1+\nu)\sin\theta\sin\alpha\cos\alpha\right]
\end{aligned}
\tag{8.34}
$$

$$+C_0 r^{-1/2}\begin{Bmatrix} \sin\frac{\theta}{2}\left[(1+\nu)\cos\frac{\theta}{2}\cos\frac{3\theta}{2}+2\nu\right]\sin^2\alpha - \sin\frac{\theta}{2}\left[(1+\nu)\cos\frac{\theta}{2}\cos\frac{3\theta}{2}+2\right]\cos^2\alpha \\ +2(1+\nu)\cos\frac{\theta}{2}\left[1-\sin\frac{\theta}{2}\sin\frac{3\theta}{2}\right]\sin\alpha\cos\alpha \end{Bmatrix}$$

$$+C_1 r^{1/2}\begin{Bmatrix} \sin\frac{\theta}{2}\left[(1+\nu)\cos^2\frac{\theta}{2}+2\right]\cos^2\alpha - \sin\frac{\theta}{2}\left[(1+\nu)\cos^2\frac{\theta}{2}+2\nu\right]\sin^2\alpha \\ +2(1+\nu)\cos\frac{\theta}{2}\left[\sin^2\frac{\theta}{2}+1\right]\sin\alpha\cos\alpha \end{Bmatrix}$$

$$+2D_1 r\left[\sin\theta\left(\cos^2 - \nu\sin^2\alpha\right)\right]$$

where an eight-term series is used in Eq. (8.34) to represent the mixed mode strain field.

Again, if cos 2α = – k, the coefficients of the B_0 and D_1 terms vanish. Also, if tan (θ/2) = – cot 2α, the coefficient of the A_1 term vanishes. The angles α and θ that permit these simplifications are exactly the same as those listed in Table 8.1. The D_0 term is not required. The reduction of Eq. (8.34) to a four-term series containing A_0, B_1, C_0 and C_1, is given as exercise.

To show an application of strain gages to the independent measurement of K_I and K_{II} for a mixed-mode state of stress, consider an aluminum specimen with ν = 1/3 and α = θ = ± 60°. For strain gages deployed along the radial line defined by θ = +60°, Eq (8.34) reduces to:

$$\varepsilon_{x'x'+} = \frac{\sqrt{3}}{2}A_0 r^{-1/2} - 2B_1 r + \frac{1}{2}C_0 r^{-1/2} + C_1 r^{1/2} \tag{8.35}$$

and for strain gages deployed along the radial line defined by θ = – 60°,

$$\varepsilon_{x'x'-} = \frac{\sqrt{3}}{2}A_0 r^{-1/2} - 2B_1 r - \frac{1}{2}C_0 r^{-1/2} - C_1 r^{1/2} \tag{8.36}$$

If Eqs. (8.35) and (8.36) are added:

$$E(\varepsilon_{x'x'+} + \varepsilon_{x'x'-}) = \frac{\sqrt{3}}{2}A_0 r^{-1/2} - 4B_1 r \tag{8.37}$$

The results of Eq. (8.37) shows that the sum of the two strains measured at the same position r along the +60° and – 60° radial lines gives a quantity that is independent of the shearing mode. If ($\varepsilon_{x'x'+} + \varepsilon_{x'x'-}$) is determined at two or more positions of r, then the unknown coefficient can be calculated. With two values of ($\varepsilon_{x'x'+} + \varepsilon_{x'x'-}$), deterministic methods of computation are employed. However, if more than two values of ($\varepsilon_{x'x'+} + \varepsilon_{x'x'-}$) are measured, then overdeterministic methods of solution of the equations are useful.

Subtracting Eq. (8.36) for Eq. (8.35) yields:

$$E(\varepsilon_{x'x'+} - \varepsilon_{x'x'-}) = C_0 r^{-1/2} + 2C_1 r^{1/2} \tag{8.38}$$

This result shows that the difference of two strains measured at the same position r along the +60° and – 60° radial lines gives a quantity that is independent of the opening mode. Solution of Eq. (8.38) by either deterministic or overdeterministic methods yields C_0 that is related to the shearing mode stress intensity factor by the equation:

$$K_{II} = \sqrt{2\pi}\, C_0 \tag{8.39}$$

8.8 OVERDETERMINISTIC METHODS OF STRAIN ANALYSIS

In fracture mechanics, one knows the form of the solution for the stresses or strains in the near-field region. This solution is given by series relations such as those given in Eqs. (4.41) and (4.42). As the size of the near-field region is enlarged, the number of terms in the series representation must be increased to accurately describe the field quantities. The increase in the number of terms in the series increases the number of unknown coefficients A_n and B_m and eventually the use of deterministic methods becomes inappropriate. With a large number of unknown coefficients, small errors in the measurement of strain, $\varepsilon_{x'x'}$, gage position (r, θ), and gage orientation α combine to produce large errors in the determination of A_n and B_m. To avoid these errors, additional strain measurements are made so that the amount of data available exceeds the number of unknowns by a factor of about 3 or 4. With this additional data, overdeterministic solutions for the unknown coefficients A_n and B_m can be employed and the results are improved by averaging in a least-squares sense.

To illustrate the overdeterministic method, consider that n strain gages are deployed in a radial pattern (α = θ) in the near-field region with an arbitrary selection of r and θ. These gages record a radial strain ε_{rr} which is described by:

$$\begin{aligned}
2\mu\varepsilon_{rr} &= A_0 r^{-1/2}\left\{k\cos\frac{\theta}{2} - \frac{1}{2}\sin\theta\sin\frac{3\theta}{2}\cos 2\theta + \sin^2\theta\cos\theta\cos\frac{3\theta}{2}\right\} \\
&+ B_0(k+\cos 2\theta) + A_1 r^{1/2}\cos\frac{\theta}{2}\left\{k + \sin^2\frac{\theta}{2}\cos 2\theta - \sin^2\theta\cos\theta\right\} \\
&+ B_1 r\cos\theta\left[k+1-6\sin^2\theta\right] + A_2 r^{3/2}\left\{k\cos\frac{3\theta}{2} - \frac{3}{2}\sin\frac{\theta}{2}\cos 2\theta - 3\sin^2\theta\cos\theta\cos\frac{\theta}{2}\right\} \\
&+ B_2\frac{1}{1+\nu} r^2\left[1-(3+2\nu)\sin^2\theta\cos^2\theta + \nu\sin^2\theta\left(1+\sin^2\theta\right)\right]
\end{aligned} \tag{8.40}$$

For n strain readings for a given load, Eq. (8.40) is used repeatedly to form a system of equations in terms of the unknown coefficients A_0, A_1, A_2, B_0, B_1 and B_2. Thus:

$$\begin{aligned}
2\mu\varepsilon_{rr1} &= A_0 r_1^{-1/2} g_{01} + B_0 r_1^0 h_{01} + A_1 r_1^{1/2} g_{11} + B_1 r_1^1 h_{11} + A_2 r_1^{3/2} g_{21} + B_2 r_1^2 h_{21} \\
2\mu\varepsilon_{rr2} &= A_0 r_2^{-1/2} g_{02} + B_0 r_2^0 h_{02} + A_1 r_2^{1/2} g_{12} + B_1 r_2^1 h_{12} + A_2 r_2^{3/2} g_{22} + B_2 r_1^2 h_{22} \\
&\cdots\cdots\cdots\cdots\cdots\cdots\cdots\cdots\cdots\cdots \\
&\cdots\cdots\cdots\cdots\cdots\cdots\cdots\cdots\cdots\cdots \\
&\cdots\cdots\cdots\cdots\cdots\cdots\cdots\cdots\cdots\cdots \\
2\mu\varepsilon_{rrn} &= A_0 r_n^{-1/2} g_{0n} + B_0 r_n^0 h_{0n} + A_1 r_n^{1/2} g_{1n} + B_1 r_n^1 h_{1n} + A_2 r_n^{3/2} g_{2n} + B_2 r_n^2 h_{2n}
\end{aligned} \tag{8.41}$$

where g and h are the functions of k and θ given in Eq. (8.40). It is more convenient to write Eq. (8.41) in matrix form as:

$$\{\varepsilon\} = [fg]\{AB\} \tag{8.42}$$

where {ε} is a column matrix containing n data points.

[fg] is a 6 x n matrix containing the known coefficients.
{AB} is a column matrix containing the six unknown coefficients.

Equation (8.42) is overdeterministic with six unknown coefficients. A least-squares solution for these 6 unknowns requires only two matrix manipulations. First, multiply both sides by the transpose $[fg]^T$

$$[fg]^T \{\varepsilon\} = [fg]^T \{AB\} \tag{a}$$

to obtain a 6 x 6 system of linear equations. Second, solve for {AB} by noting that:

$$\{AB\} = [[fg]^T [fg]]^{-1} [fg]\{\varepsilon\} \tag{8.43}$$

where $[[fg]^T [fg]]^{-1}$ is the inverse matrix.

The matrix solutions for the unknown coefficients A_n, B_m are easy to program. The method provides an effective technique for using statistical procedures (least squares) to minimize the errors in the determination of A_0 or K_I.

8.9 RESIDUAL STRESS DETERMINATION

8.9.1 Uniaxial Residual Stresses

Residual stresses, are self-equilibrated, and often occur in a component or structure without the application of an external load. Residual stresses are usually the result of fabrication processes or assembly processes, although they can be produced if the structure is loaded in the plastic range and then unloaded.

Residual stresses are difficult to measure since they are independent of the load and they are imposed on the component before strain gages can be installed. To sense the presence of residual stresses with strain gages, it is necessary to unload (relieve) the residual stresses after the strain gage is mounted at some point P on the component. Residual stresses are relieved by cutting away select regions of the component and in this process the component is destroyed. For this reason, the method of residual stress determination with strain gages is classified as destructive.

There are several different methods for relieving residual stresses but the hole drilling technique is the simplest and the most widely used. To show the application of the hole drilling technique, consider a thin plate subjected to a uniaxial residual stress σ_0 which is in the y direction and uniformly distributed through the thickness h of the plate and over the region of measurement. The state of stress in polar coordinates prior to drilling the hole at any point P (r, θ) is obtained from Eq. (1.11) as:

$$\begin{aligned} \sigma^{I}_{rr} &= \frac{\sigma_0}{2}(1-\cos 2\theta) \\ \sigma^{I}_{\theta\theta} &= \frac{\sigma_0}{2}(1+\cos 2\theta) \\ \tau^{I}_{r\theta} &= \frac{\sigma_0}{2}\sin 2\theta \end{aligned} \tag{8.44}$$

A hole of radius a is now drilled completely through the thin plate as shown in Fig. 8.11. Drilling the hole markedly changes the stress distribution in the plate since a cylinder of stressed material of radius a has been removed. To determine the stress distribution in the post-drilled state consider the stress state in a uniaxially loaded infinite plate with a circular hole obtained by the Kirsch and discussed in Chapter 3. The stresses are given by Eqs. (3.42b) as:

$$\sigma_{rr} = \frac{\sigma_0}{2}\left\{\left(1-\frac{a^2}{r^2}\right)\left[1+\left(\frac{3a^2}{r^2}-1\right)\cos 2\theta\right]\right\}$$
$$\sigma_{\theta\theta} = \frac{\sigma_0}{2}\left[\left(1+\frac{a^2}{r^2}\right)+\left(1+\frac{3a^4}{r^4}\right)\cos 2\theta\right] \qquad (3.42)$$
$$\tau_{r\theta} = \frac{\sigma_0}{2}\left[\left(1+\frac{3a^2}{r^2}\right)\left(1-\frac{a^2}{r^2}\right)\sin 2\theta\right]$$

where r > a.

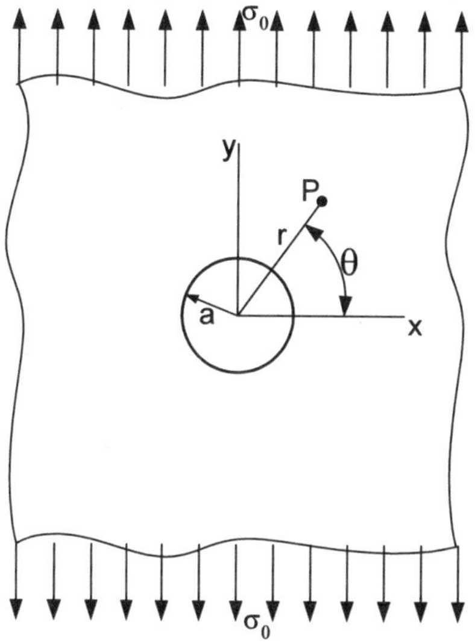

Figure 8.11 Coordinates of point P located in a region of a plate adjacent to a through hole of radius a.

Subtracting the initial stresses given by Eqs. (8.44) from the stresses in the plate after drilling the hole that are described by Eq. (3.42a) gives:

$$\sigma_{rr}^{II} = -\frac{\sigma_0 a^2}{2r^2}\left[1+\left(\frac{3a^2}{r^2}-4\right)\cos 2\theta\right]$$
$$\sigma_{\theta\theta}^{II} = \frac{\sigma_0 a^2}{2r^2}\left(1+\frac{3a^2}{r^2}\cos 2\theta\right) \qquad (8.45)$$
$$\tau_{r\theta}^{II} = -\frac{\sigma_0 a^2}{2r^2}\left[\left(\frac{3a^2}{r^2}-2\right)\sin 2\theta\right]$$

Substituting Eqs. (8.45) into Eqs. (2.19) gives:

$$\varepsilon_{rr} = -\frac{\sigma_0 a^2(1+\nu)}{2Er^2}\left[1+\frac{3a^2}{r^2}\cos 2\theta-\frac{4\cos 2\theta}{1+\nu}\right]$$
$$\varepsilon_{\theta\theta} = \frac{\sigma_0 a^2(1+\nu)}{2Er^2}\left[1+\frac{3a^2}{r^2}\cos 2\theta-\frac{4\nu\cos 2\theta}{1+\nu}\right] \qquad (8.46)$$

These relations give the strain distribution in the region about the hole that is due to relief of the uniaxial residual stress σ_0. It is possible to simplify this analysis by considering the strain along the x axis where $\theta = 0°$, $\varepsilon_{\theta\theta} = \varepsilon_{yy}$ and $\varepsilon_{rr} = \varepsilon_{xx}$:

$$\frac{2E\varepsilon_{rr}}{\sigma_0(1+\nu)} = -\left(\frac{a}{r}\right)^2\left[1+3\left(\frac{a}{r}\right)^2 - \frac{4}{1+\nu}\right]$$
$$\frac{2E\varepsilon_{\theta\theta}}{\sigma_0(1+\nu)} = \left(\frac{a}{r}\right)^2\left[1+3\left(\frac{a}{r}\right)^2 - \frac{4\nu}{1+\nu}\right] \tag{8.47}$$

The normalized strains $(3E\varepsilon)/(2\sigma_0)$ corresponding to Eq. (8.47) with $\nu = 1/3$, are shown as a function of position r/a in Fig. 8.12. Inspection of these curves shows that both strains exhibit sharp gradients with respect to r at locations close to the edge of the hole. This implies that precise positioning of a strain gage relative to the hole is necessary for gage locations with $1 < r/a < 1.5$. If a single gage is to be used to determine σ_0, it should be positioned at $r/a = 1.75$ and oriented along the x-axis to measure ε_{rr}. With this placement of the gage, the residual stress for a material with $\nu = 1/3$ is given by:

$$\sigma_0 = 4.502\ E\varepsilon_g \tag{8.48}$$

The placement of the gage at $r/a = 1.75$ maximizes the signal due to ε_{rr} in the region where strain gradients are minimized.

This analysis of uniaxial residual stress illustrates the approach used to measure σ_0 in the simplest possible case. However, residual stresses are usually biaxial and it is necessary to determine the magnitudes of the principal residual stresses σ_1^R and σ_2^R as well as the principal directions. The strain-gage method used for the determination of biaxial residual stresses is covered in the following section.

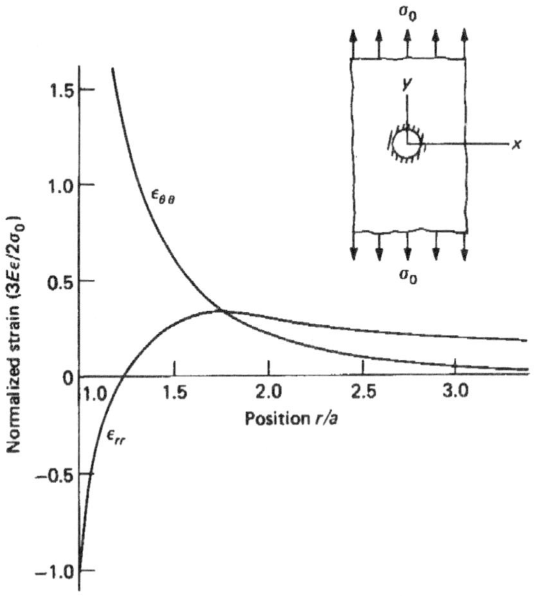

Figure 8.12 Distribution of strains along the x-axis due to residual stress relief by hole drilling (ν = 1/3).

8.9.2 Biaxial Residual Stresses

Consider a biaxial state of residual stress described with σ_1^R and σ_2^R as illustrated in Fig. 8.13. Drilling a through hole of radius a in the plate relieves the residual stresses and produces a change in the strain in the local region near the hole. The change in strain can be determined by using Eqs. (8.46) and the principle of superposition. Before applying the principle of superposition, rewrite Eq. (8.46) for ε_{rr} in the following concise form:

$$\varepsilon_{rr} = \sigma_1^R\,[C_1 - C_2 \cos 2\theta_2] \tag{8.49}$$

where

$$C_1 = -\frac{1+\nu}{2E}\left(\frac{a}{r}\right)^2 \qquad C_2 = -\frac{1+\nu}{2E}\left(\frac{a}{r}\right)^2\left[-3\left(\frac{a}{r}\right)^2 + \frac{4}{1+\nu}\right]$$

and θ_2 is the angle between the principal direction σ_2^R and the radial line through point P. Note, in writing Eq. (8.49) that σ_1^R was used in place of σ_0. In a similar manner, the strain ε_{rr} due to relief of σ_2^R can be written as:

$$\varepsilon_{rr} = \sigma_{2P}{}^{RP}\,[C_1 - C_2 \cos 2\theta_1] \qquad \text{(a)}$$

Figure 8.13 A biaxial state of residual stress σ_1^R and σ_2^R uniformly distribute through the thickness of the plate.

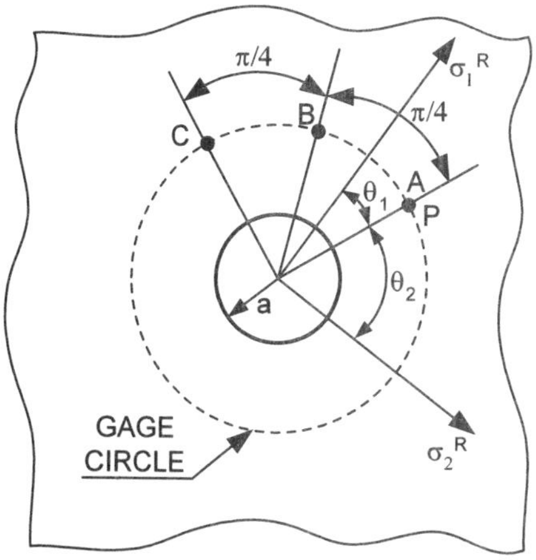

Because,

$$\cos 2\theta_1 = \cos 2[\theta_2 + (\pi/2)] = -\cos 2\theta_2 \qquad \text{(b)}$$

it is clear that:

$$\varepsilon_{rr} = \sigma_2^R\,[C_1 + C_2 \cos 2\theta_2] \qquad (8.50)$$

Superimposing the radial strains given by Eqs. (8.49) and (8.50) gives:

$$\varepsilon_{rr} = C_1(\sigma_1^R + \sigma_2^R) + C_2\,(\sigma_2^R - \sigma_1^R)\cos 2\theta_2 \qquad (8.51)$$

At this stage in the analysis, it is clear that Eq. (8.51) is the basic relation for ε_{rr} needed to solve for the unknowns σ_1^R, σ_2^R and the angle defining the orientation of this orthogonal pair of principal stresses. A three gage rectangular rosette with the gages deployed as shown in Fig. 8.14 is used to measure three strains ε_A, ε_B and ε_C. For the strains measured at points A, B, and C in Fig. 8.13, Eq. (8.51) yields:

$$\begin{aligned}
\varepsilon_A &= C_1\left(\sigma_1^R + \sigma_2^R\right) + C_2\left(\sigma_2^R - \sigma_1^R\right)\cos 2\theta_2 \\
\varepsilon_B &= C_1\left(\sigma_1^R + \sigma_2^R\right) + C_2\left(\sigma_2^R - \sigma_1^R\right)\cos 2\left(\theta_2 + \frac{\pi}{4}\right) \\
\varepsilon_C &= C_1\left(\sigma_1^R + \sigma_2^R\right) + C_2\left(\sigma_2^R - \sigma_1^R\right)\cos 2\left(\theta_2 + \frac{\pi}{2}\right)
\end{aligned} \qquad (8.52)$$

Solving Eqs. (8.52) for σ_1^R and σ_2^R gives:

$$
\begin{aligned}
\sigma_1^R &= \frac{\varepsilon_A + \varepsilon_C}{4C_1} + \frac{\sqrt{2}}{4C_2}\sqrt{(\varepsilon_A - \varepsilon_B)^2 + (\varepsilon_B - \varepsilon_C)^2} \\
\sigma_2^R &= \frac{\varepsilon_A + \varepsilon_C}{4C_1} - \frac{\sqrt{2}}{4C_2}\sqrt{(\varepsilon_A - \varepsilon_B)^2 + (\varepsilon_B - \varepsilon_C)^2}
\end{aligned}
\tag{8.53}
$$

or an alternative form of Eq. (8.53), often found in the literature, can be written as:

$$
\begin{aligned}
\sigma_1^R &= \frac{\varepsilon_A + \varepsilon_C}{4C_1} + \frac{1}{4C_3}\sqrt{(\varepsilon_A - \varepsilon_C)^2 + (\varepsilon_A + \varepsilon_C - 2\varepsilon_B)^2} \\
\sigma_2^R &= \frac{\varepsilon_A + \varepsilon_C}{4C_1} - \frac{1}{4C_3}\sqrt{(\varepsilon_A - \varepsilon_C)^2 + (\varepsilon_A + \varepsilon_C - 2\varepsilon_B)^2}
\end{aligned}
\tag{8.54}
$$

where $C_3 = C_2/\sqrt{2}$ and

$$
\tan 2\theta = \frac{\varepsilon_A - 2\varepsilon_B + \varepsilon_C}{\varepsilon_C - \varepsilon_A} \tag{8.55}
$$

where θ is the angle from gage A to the nearer of the two principal axes. If $\varepsilon_C > \varepsilon_A$, θ refers to σ_1, but if $\varepsilon_C < \varepsilon_A$, θ refers to σ_2.

The rosettes used in residual stress investigations are illustrated in Fig. 8.14. The strain gages are all positioned about a circle of radius r_0. The position of the hole and its radius a are located with patterns etched in foil. All the gages and the hole centering aids are carried on a single film carrier and a covering film is often used to encapsulate the installation. This covering film protects the fragile gage grids from the chips produced as the hole is drilled.

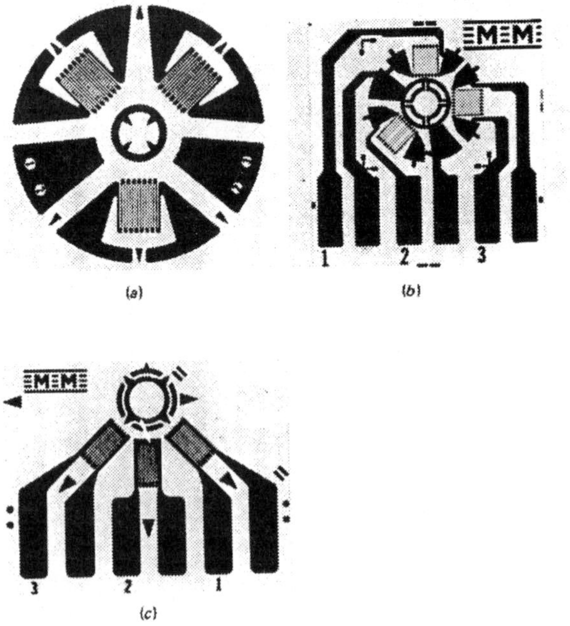

Figure 8.14 Rectangular rosettes adapted for the hole drilling method for determining biaxial residual stresses.

The constants C_1 and C_2 are defined in Eq. (8.49); however, it should be recognized that this equation was developed by considering the strain gage as a point sensor. In reality, the strain gage has a grid area and the size and orientation of the grids affect the values of C_1 and C_2. To avoid errors due to the finite size and orientation of the strain gage grid, the gage installation is calibrated and the constants determined experimentally.

Calibration is usually performed with a wide plate that is loaded in uniaxial tension. The rosette is mounted so that gage C is in the direction of loading and gage A is in the transverse direction. For thin specimens, the calibration plate (stress free) is the same thickness as the part being investigated. The rosette is then calibrated by following the procedure described below:

1. Load the calibration specimen to produce a uniaxial calibration stress σ_c.
2. Measure the strains ε_C' and ε_A' developed in the calibration specimen prior to drilling the through hole.
3. Unload the specimen and carefully drill the hole at the exact center of the rosette.
4. Reload the calibration specimen to establish the same calibration stress σ_c.
5. Measure the strains ε_C'' and ε_A'' with the hole.

The calibration strains corresponding to the calibration stress imposed on the specimen is given by:

$$\begin{aligned} \varepsilon_{Cc} &= \varepsilon_C'' - \varepsilon_C' \\ \varepsilon_{Ac} &= \varepsilon_A'' - \varepsilon_A' \end{aligned} \qquad \text{(a)}$$

From Eq. (8.49) and the geometry defined in Fig. 8.13, it is evident that:

$$\begin{aligned} \varepsilon_{Cc} &= \sigma_c\left[\overline{C}_1 - \overline{C}_2\cos(0)\right] = \sigma_c\left[\overline{C}_1 - \overline{C}_2\right] \\ \varepsilon_{Ac} &= \sigma_c\left[\overline{C}_1 - \overline{C}_2\cos(\pi)\right] = \sigma_c\left[\overline{C}_1 + \overline{C}_2\right] \end{aligned} \qquad (8.56)$$

where $\overline{C}_1$ and $\overline{C}_2$ are calibration constants used to replace the coefficients C_1 and C_2 in Eqs. (8.53) and (8.54).

Solving Eq. (8.56) for $\overline{C}_1$ and $\overline{C}_2$ yields:

$$\overline{C}_1 = \frac{\varepsilon_{Cc} + \varepsilon_{Ac}}{2\sigma_c} \quad \text{and} \quad \overline{C}_2 = \frac{\varepsilon_{Cc} - \varepsilon_{Ac}}{2\sigma_c} \qquad (8.57)$$

The coefficients $\overline{C}_1$ and $\overline{C}_2$ are functions of the geometry of the hole (diameter and depth) and the geometry of the grid of the strain gages for a specified set of elastic constants E and ν. Consequently, different size rosettes can be characterized by the calibration constants $\overline{C}_1$ and $\overline{C}_2$ providing the hole diameter and depth are scaled accordingly.

Two new coefficients $\overline{\overline{C}}_1$ and $\overline{\overline{C}}_2$ are introduced to remove material dependence from $\overline{C}_1$ and $\overline{C}_2$. The coefficients $\overline{\overline{C}}_1$ and $\overline{\overline{C}}_2$ are defined as:

$$\overline{\overline{C}}_1 = -\frac{2E\overline{C}_1}{1+\nu} \quad \text{and} \quad \overline{\overline{C}}_2 = -2E\overline{C}_2 \qquad (8.58)$$

A comparison of Eq. (8.58) with the expressions for C_1 and C_2 in Eq. (8.49), shows that the material dependence of $\overline{\overline{C}}_1$ is accommodated. The coefficient $\overline{\overline{C}}_2$ shows a slight dependence on Poisson's ratio; however, the dependence is sufficiently small (less than 2%) to be neglected if $0.25 < \nu < 0.35$.

Table 8.2

Calibration coefficients $\overline{\overline{C}}_1$ and $\overline{\overline{C}}_2$ for the ASTM E837 hole drilling rosette

Hole Depth	Coefficient $\overline{\overline{C}}_1$					Coefficient $\overline{\overline{C}}_2$				
	Hole Diameter, D/D_m					Hole Diameter, D/D_m				
h/D_m	0.30	0.35	0.40	0.45	0.50	0.30	0.35	0.40	0.45	0.50
0.000	0.000	0.000	0.000	0.000	0.000	0.000	0.000	0.000	0.000	0.000
0.025	−0.011	−0.015	−0.021	−0.028	−0.033	−0.021	−0.028	−0.037	−0.048	−0.058
0.050	−0.027	−0.036	−0.049	−0.063	−0.081	−0.050	−0.067	−0.088	−0.111	−0.140
0.075	−0.043	−0.059	−0.079	−0.102	−0.132	−0.082	−0.111	−0.145	−0.181	−0.229
0.100	−0.058	−0.080	−0.107	−0.137	−0.175	−0.115	−0.155	−0.210	−0.250	−0.311
0.125	−0.072	−0.099	−0.131	−0.167	−0.210	−0.147	−0.196	−0.252	−0.312	−0.382
0.150	−0.084	−0.114	−0.150	−0.190	−0.236	−0.175	−0.232	−0.297	−0.366	−0.443
0.175	−0.094	−0.127	−0.165	−0.208	−0.256	−0.200	−0.264	−0.335	−0.410	−0.492
0.200	−0.100	−0.135	−0.176	−0.221	−0.270	−0.220	−0.290	−0.367	−0.447	−0.531
0.225	−0.105	−0.142	−0.184	−0.230	−0.280	−0.237	−0.312	−0.392	−0.476	−0.562
0.250	−0.108	−0.146	−0.189	−0.236	−0.286	−0.251	−0.329	−0.413	−0.499	−0.587
0.300	−0.112	−0.149	−0.193	−0.241	−0.290	−0.271	−0.352	−0.4414	−0.530	−0.619
0.350	−0.112	−0.150	−0.190	−0.240	−0.289	−0.282	−0.367	−0.457	−0.548	−0.639
0.400	−0.111	−0.150	−0.190	−0.236	−0.285	−0.288	−0.376	−0.466	−0.558	−0.648
0.450	−0.108	−0.145	−0.187	−0.233	−0.282	−0.290	−0.378	−0.471	−0.563	−0.653
0.500	−0.106	−0.141	−0.183	−0.228	−0.276	−0.291	−0.379	−0.472	−0.564	−0.654

8.9.3 Residual Stresses in Thick Plates

In the preceding discussion, it was assumed that the residual stresses developed in a relatively thin flat plate, and they were uniform through the thickness of this plate. This assumption is very limiting—in most cases, residual stresses occur in bodies that are neither thin nor flat. To determine the magnitude of residual stresses in the more general case, a blind hole drilling technique is employed. With this approach an end mill is used to drill the hole of diameter D in increments into the thickness of the body as illustrated in Fig. 8.15. Surface mounted strain gages surrounding this hole with a mean diameter D_m provide readings of several strains as the depth h of the hole is increased.

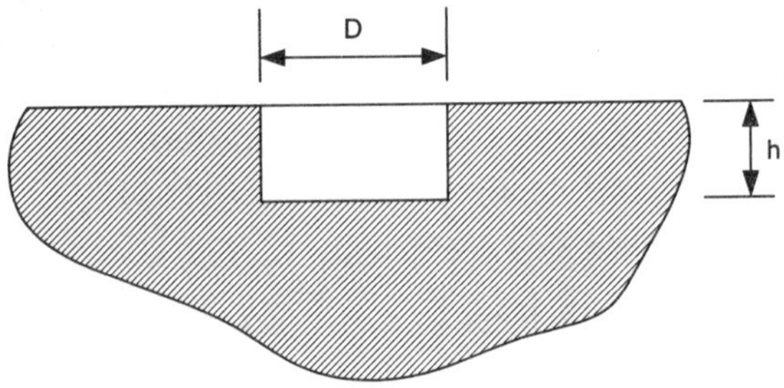

Figure 8.15 Blind hole drilled into a thick specimen to determine residual stresses.

For thick specimens, it is recommended that the blind hole be drilled in increments and the calibration process repeated for each increment of hole depth Δh. The analysis is performed for each depth to provide data to determine coefficients $\overline{C}_1$ and $\overline{C}_2$ for each h/D ratio. These constants can be used with Eq. (8.53) or Eq. (8.54) to determine the residual stresses as a function of depth. Another approach is to use the coefficients $\overline{\overline{C}}_1$ and $\overline{\overline{C}}_2$ for the ASTM E837 hole-drilling rosette. If the elastic constants are known, the coefficients $\overline{C}_1$ and $\overline{C}_2$ can be determined from Eq. (8.56) by using the values of $\overline{\overline{C}}_1$ and $\overline{\overline{C}}_2$ that are listed in Table 8.2

EXERCISES

8.1 Suppose a state of pure shear stress occurs (say on a circular shaft under pure torsion). Show how a single-element strain gage can be employed to determine the principal stresses σ_1 and σ_2.

8.2 In a thin-walled cylinder loaded with internal pressure ($\sigma_h = 2\ \sigma_a$), show how a single-element strain gage in a hoop direction can be used to establish the hoop and axial stresses σ_h and σ_a.

8.3 A two-element rectangular rosette is being used to determine the two principal stresses at the point shown in Fig. E8.3. If ε_1 = 860 $\mu\varepsilon$ and ε_2 = −390 $\mu\varepsilon$, determine σ_1 and σ_2. Use E = 207 GPa and ν = 0.30.

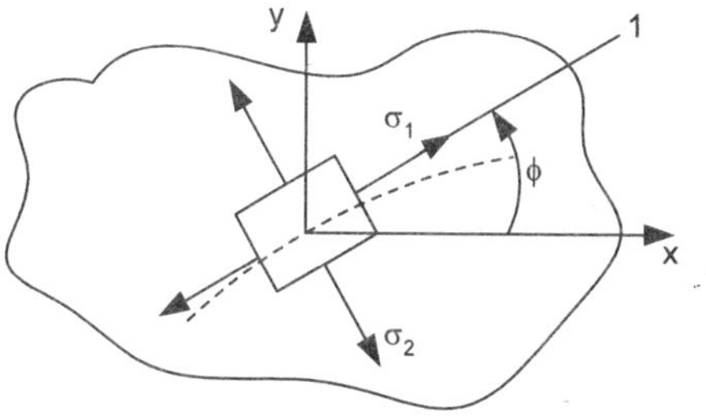

Figure E8.3

8.4 If strain data from the two-element rectangular rosette shown in Fig. E8.3 yields σ_1 = 180 MPa and σ_2 = 60 MPa, find σ_{xx}, σ_{yy} and τ_{xy} when ϕ is (a) 30°, (b) 45°, (c) 60° and (d) 90°.

8.5 Prove the validity of Eqs. (8.8) by using the Mohr's strain diagram presented in Fig. 8.3.

8.6 Derive an expression for the maximum shear stress τ_{max} at a point in terms of the strains obtained from a three-element rectangular rosette by using Eqs. (8.9)

8.7 The following observations are made with a rectangular rosette mounted on a steel (E = 207 GPa and ν = 0.30) specimen. Determine the principal strains, the principal stresses, and the principal angles ϕ_1 and ϕ_2.

Case No.	ε_A, $\mu\varepsilon$	ε_B, $\mu\varepsilon$	ε_B, $\mu\varepsilon$
1	1000	−500	0
2	1800	600	−400
3	−1000	400	400
4	1600	−200	−1800
5	−400	0	400

8.8 Write a computer program to determine the principal strains ε_1 and ε_2, the principal angle ϕ_1, and the principal stresses σ_1 and σ_2 using the strains ε_A, ε_B and ε_C measured with a three-element rectangular rosette.

8.9 For the three-element delta rosette with θ_A = 0°, θ_B = 120° and θ_C = 240°, determine the equation for the principal strains ε_1 and ε_2 in terms of ε_a, ε_B and ε_C measured with a three-element rectangular rosette.

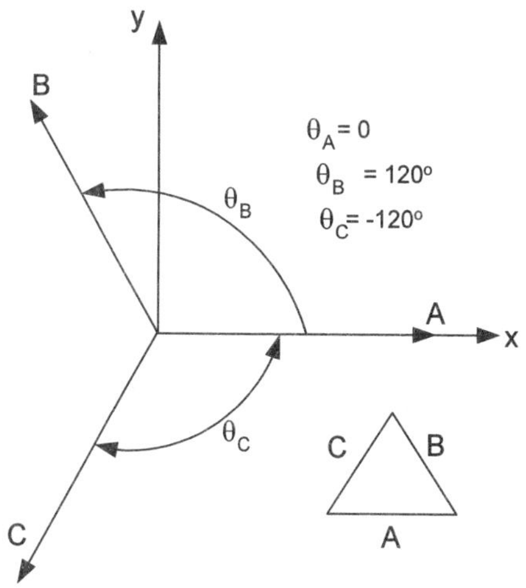

Fig. E8.9

8.10 For the three-element delta rosette shown in Fig. E8.9 determine the equation for the principal angle ϕ in terms of ε_a, ε_B and ε_C.

8.11 Construct a Mohr's circle for strains using the strains ε_a, ε_B and ε_C measured with a three-element delta rosette. Show that the radius of the circle is:

$$R = \{[\varepsilon_A - (1/3)(\varepsilon_A + \varepsilon_B + \varepsilon_C)]^2 + (1/3)(\varepsilon_C - \varepsilon_C)^2\}^{1/2}.$$

8.12 Using the result of Exercise 8.11 show that the principal angles can be identified by applying the following rules:

$0° < \phi_1 < 90°$	when $\varepsilon_C > \varepsilon_B$
$-90° < \phi_1 < 0°$	when $\varepsilon_C < \varepsilon_B$
$\phi_1 = 0°$	when $\varepsilon_B = \varepsilon_c$ and $\varepsilon_A > \varepsilon_B = \varepsilon_C$
$\phi_1 = \pm 90°$	when $\varepsilon_B = \varepsilon_c$ and $\varepsilon_A < \varepsilon_B = \varepsilon_C$

8.13 Using the results of Exercise 8.9, derive the equations for the principal stresses σ_1 and σ_2 in terms of the strains ε_a, ε_B and ε_C measured with a three-element delta rosette.

8.14 The following observations were made with a three-element delta rosette mounted on a steel (E = 207 GPa and ν = 0.30) specimen. Determine the principal strains ε_1 and ε_2, the principal stresses σ_1 and σ_2, and the principal angles ϕ_1 and ϕ_2.

Case No.	ε_A, με	ε_B, με	ε_C, με
1	800	−400	400
2	1600	800	0
3	−1200	600	800
4	1400	0	−1400
5	−600	200	800

8.15 Verify the expressions listed in Table 8.3 (shown on the next page) for the rectangular four-element rosette.

8.16 Verify the expressions listed in Table 8.3 (shown on the next page) for the tee-delta rosette.

8.17 Determine the error due to cross sensitivity when a WK-06-125RA-350 type strain gage is used on aluminum (E = 70 GPa and ν = 0.33) to measure (a) a state of hydrostatic compression, (b) a state of pure shear, and (c) Poisson's ratio.

8.18 Solve problem 8.17 if the gage is used on steel (E = 207 GPa and ν = 0.30).

8.19 The following apparent strain data were obtained with two-element rectangular rosettes:

Rosette No.	ε'_{xx}, (με)	ε'_{yy}, (με)
1	1200	600
2	−400	1400
3	2400	800
4	1200	−600

Determine the true strains ε_{xx} and ε_{yy} if $K_t = 0.02$. In each case, determine the error which would have occurred if the cross sensitivity of the gage had been neglected.

8.20 Solve Exercise 8.19 if $K_t = 0.015$.

8.21 Solve Exercise 8.19 if $K_t = -0.015$.

8.22 Determine the included angle between elements in a V-type stress gage designed for use on (a) glass (ν = 0.25), (b) steel (ν = 0.30), and (c) aluminum (ν = 0.33).

Table 8.3

A summary of equations for principal strains, principal stresses and their directions from four types of rosettes.

Type of Rosette	Gage arrangement 	Principal strain and principal stress	Principal angle	Identification $0 < \phi_1 < 90°$
Three-element rectangular	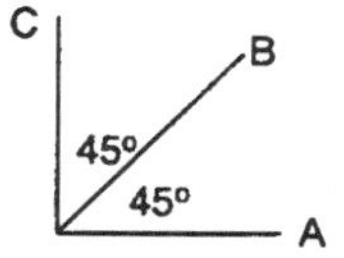	$\varepsilon_{1,2} = \frac{\varepsilon_A + \varepsilon_C}{2} \pm \frac{1}{2}\sqrt{(\varepsilon_A - \varepsilon_C)^2 + (2\varepsilon_B - \varepsilon_A - \varepsilon_C)^2}$ $\sigma_{1,2} = \frac{E}{2}\left[\frac{\varepsilon_A + \varepsilon_C}{1-\nu} \pm \frac{1}{1+\nu}\sqrt{(\varepsilon_A - \varepsilon_C)^2 + (2\varepsilon_B - \varepsilon_A - \varepsilon_C)^2}\right]$	$\tan 2\phi_1 = \frac{2\varepsilon_B - \varepsilon_A - \varepsilon_C}{\varepsilon_A - \varepsilon_C}$	$\varepsilon_B > \frac{\varepsilon_A + \varepsilon_C}{2}$
Delta	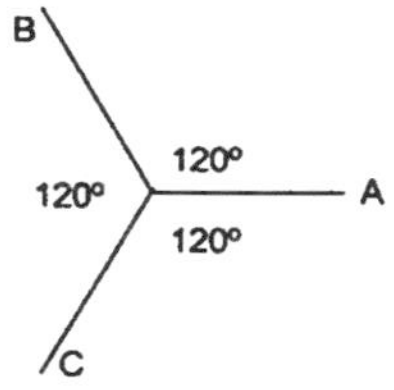	$\varepsilon_{1,2} = \frac{\varepsilon_A + \varepsilon_B + \varepsilon_C}{3} \pm \frac{\sqrt{2}}{3}\sqrt{(\varepsilon_A - \varepsilon_B)^2 + (\varepsilon_B - \varepsilon_C)^2 + (\varepsilon_C - \varepsilon_A)^2}$ $\sigma_{1,2} = \frac{E}{3}\left[\frac{\varepsilon_A + \varepsilon_B + \varepsilon_C}{1-\nu} \pm \frac{\sqrt{2}}{1+\nu}\sqrt{(\varepsilon_A - \varepsilon_B)^2 + (\varepsilon_B - \varepsilon_C)^2 + (\varepsilon_C - \varepsilon_A)^2}\right]$	$\tan 2\phi_1 = \frac{\sqrt{3}\,(\varepsilon_C - \varepsilon_B)}{2\varepsilon_A - (\varepsilon_B + \varepsilon_C)}$	$\varepsilon_C > \varepsilon_B$
Four-element rectangular	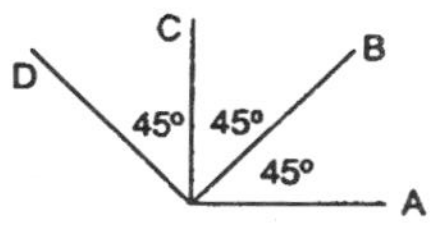	$\varepsilon_{1,2} = \frac{\varepsilon_A + \varepsilon_B + \varepsilon_C + \varepsilon_D}{4} \pm \frac{1}{2}\sqrt{(\varepsilon_A - \varepsilon_C)^2 + (\varepsilon_B - \varepsilon_D)^2}$ $\sigma_{1,2} = \frac{E}{2}\left[\frac{\varepsilon_A + \varepsilon_B + \varepsilon_C + \varepsilon_D}{2(1-\nu)} \pm \frac{1}{1+\nu}\sqrt{(\varepsilon_A - \varepsilon_C)^2 + (\varepsilon_B - \varepsilon_D)^2}\right]$	$\tan 2\phi_1 = \frac{\varepsilon_B - \varepsilon_D}{\varepsilon_A - \varepsilon_C}$	$\varepsilon_B > \varepsilon_D$
Tee-delta	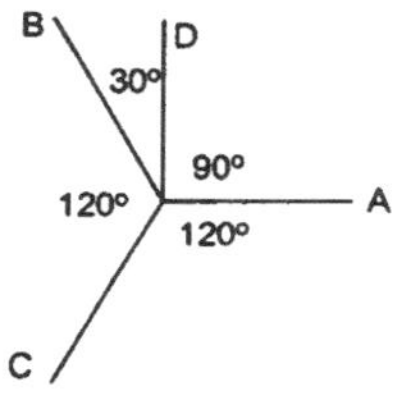	$\varepsilon_{1,2} = \frac{\varepsilon_A + \varepsilon_D}{2} \pm \frac{1}{2}\sqrt{(\varepsilon_A - \varepsilon_D)^2 + \frac{4}{3}(\varepsilon_C - \varepsilon_B)^2}$ $\sigma_{1,2} = \frac{E}{2}\left[\frac{\varepsilon_A + \varepsilon_D}{1-\nu} \pm \frac{1}{1+\nu}\sqrt{(\varepsilon_A - \varepsilon_D)^2 + \frac{4}{3}(\varepsilon_C - \varepsilon_B)^2}\right]$	$\tan 2\phi_1 = \frac{2(\varepsilon_C - \varepsilon_B)}{\sqrt{3}(\varepsilon_A - \varepsilon_D)}$	$\varepsilon_C > \varepsilon_B$

8.23 Determine $\Delta R/R$ for the circular-arc gage element shown in Fig. E8.23. Neglect transverse-sensitivity effects.

Figure E8.23

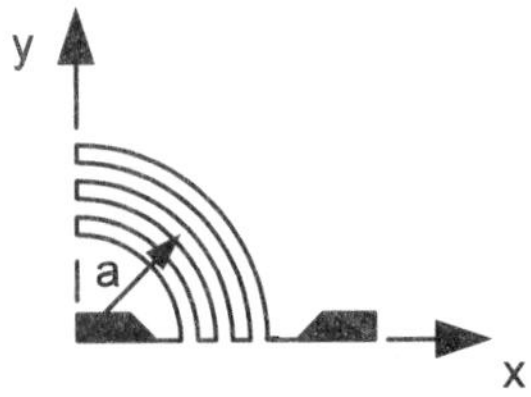

8.24 Show that two circular-arc gage elements (see Fig. E8.23) located in adjacent quadrants and positioned in arms R_1 and R_4 of a Wheatstone bridge have an output $\Delta v_0/v_s$ proportional to $2\gamma_{xy}/\pi$.

8.25 Determine $\Delta v_0/v_s$ for a shear gage consisting of four circular-arc elements arranged in all four quadrants and connected into the four arms of a Wheatstone bridge.

8.26 Beginning with Eq. (8.29) verify Eq. (8.33).

8.27 If a single strain gage with a gage length of 2 mm is placed a distance r_c = 6mm from the crack tip, determine the distance Δr between the center of the gage and the true strain location r_t.

8.28 Repeat Exercise 8.31 if the gage length is reduced to 1 mm.

8.29 Repeat Exercise 8.31 if the gage is placed at a position r_c = 9 mm from the crack tip.

8.30 A strain gage with a gage length of 1 mm is mounted on aluminum with $\theta = \alpha = 60°$ at a distance of 6 mm from the crack tip. Determine the stress intensity factor K_I if the strain is 900 $\mu\varepsilon$.

8.31 Verify Eq. (8.34).

8.32 If $\cos 2\alpha = -k$ and if $\tan(\theta/2) = -\cot 2\alpha$, show that Eq. (8.34) reduces to a four-term series containing A_0, B_1, C_0 and C_1.

8.33 Verify Eq. (8.37) and Eq. (8.38) beginning with the results of Exercise 8.36 subjected to $\nu = 1/3$ and $\theta = \alpha = 60°$.

8.34 Verify Eq. (8.38).

8.35 Verify Eq. (8.40).

8.36 Write a program in a suitable computer language that implements the theory described in Section 8.8. The program should accept as input ε_{rr}, r, θ, ν and k and give as output the unknown coefficients A_0, A_1, A_2, B_0, B_1 and B_2.

8.37 Write a computer program to determine the effectiveness of the solution provided by the results of Exercise 8.36. Explain the strategy used as the basis for this program.

8.38 An aluminum panel with a uniaxial residual stress σ_0 = 100 MPa in the y direction was investigated using the hole-drilling technique. If a strain gage is placed at a location 12 mm from the center of the hole and oriented in the x direction, find the strain developed as an 8-mm diameter hole is drilled through the panel.

8.39 If the hole in Exercise 8.38 is enlarged from a diameter of 8 mm to diameters of 12, 16, and then 20 mm, determine the strain developed at the gage as the hole is enlarged. Does this exercise suggest an overdeterministic method for determining residual stresses?

8.40 Verify Eq. (8.51).

8.41 Beginning with Eq. (8.51) verify Eq. (8.53).

8.42 A residual stress rosette like the one shown in Fig. 8.14 is designed with r/a = 3. This type of rosette was mounted on three steel panels and strain measurements were taken for each plate after drilling the center hole. Determine the biaxial residual stresses in the three plates.

Plate	ε_A, $\mu\varepsilon$	ε_B, $\mu\varepsilon$	ε_C, $\mu\varepsilon$
1	700	− 700	0
2	300	600	900
3	500	500	500

BIBLOGRAPHY

1. Ades, C. S.: "Reduction of Strain Rosettes in the Plastic Range," Experimental Mechanics, Vol.2, no. 11, pp. 345-249, 1962.
2. Bossart, K. J., and G. A. Brewer: "A Graphical Method of Rosette Analysis," Proceedings SESA, Vol. IV, no. 1, pp. 1-8, 1946.
3. McClintock, F. A.: "On Determining Principal Strains from Strain Rosettes with Arbitrary Angles," Proceedings SESA, Vol. IX, no. 1, pp. 209-210, 1951.
4. Murray, W. M.: "Some Simplifications in Rosette Analysis," Proceedings SESA, Vol. XV, no. 2, pp. 39-52, 1958.
5. Stein, P. K.: "A Simplified Method of Obtaining Principal Stress Information from Strain Gage Rosettes," Proceedings SESA, Vol. XV, no.2, pp. 21-38, 1958.
6. Meier, J. H.: "On the Transverse-Strain Sensitivity of Foil Gages," Experimental Mechanics, Vol. 1, no. 7, pp. 39-40, 1961.
7. Meyer, M. L.: "A Simple Estimate for the Effect of Cross Sensitivity on Evaluated Strain-Gage Measurements," Experimental Mechanics, Vol. 7, no. 11, pp. 476-480, 1967.
8. Wu, C. T.: "Transverse Sensitivity of Bonded Strain Gages," Experimental Mechanics, Vol. 2, no. 11, pp. 338-344, 1962.
9. "Errors Due to Transverse Sensitivity in Strain Gages," Micro-Measurements TN-509, 1993.
10. Williams, S. B.: "Geometry in the Design of Stress Measuring Circuits," Proceedings SESA, Vol. XVII, no. 2, pp. 161-178, 1960.
11. Hines, F. F.: "The Stress-Strain Gage," Proceedings 1st International Congress Experimental Mechanics 1963, pp 237-253.
12. Lissner, H. R., and C. C. Perry: "Conventional Wire Strain Gage Used as a Principal Stress Gage," Proceedings SESA, Vol. XIII, no. 1, pp. 25-32, 1955.
13. Perry, C. C.: "Plane-Shear Measurement with Strain Gages," Experimental Mechanics, Vol. 9, no. 1, 1969.
14. Dally, J. W. and R. J. Sanford: "Strain Gage Methods for Measuring the Opening Mode Stress Intensity Factor," Experimental Mechanics, Vol. 27, no. 4, pp. 381-388, 1987.
15. Dally, J. W. and J. R. Berger: "Strain Gage Method for Determining K_I and K_{II} in a Mixed Mode Field", Proceedings of 1986 Spring Conf. SEM, June 1986.
16. Berger, J. R. and J. W. Dally: "An Overdeterministic Approach for Measuring the Opening Mode Stress Intensity Factor Using Strain Gages," Experimental Mechanics, Vol. 28, no. 2, pp. 142-145, 1988.
17. "Determining Residual Stresses by the Hole-drilling Strain-Gage Method." ASTM Standard E837-85.
18. Rendler, J. J. and I. Vigness: "Hole-drilling Strain-gage Method of Measuring Residual Stresses," Proceedings, SESA Vol. XXIII, no. 2, pp. 577-586, (1966).
19. Bynum, J. E.: "Modifications to the Hole-drilling Technique of Measuring Residual Stresses for Improved Accuracy and Reproducibility," Experimental Mechanics, Vol. 21, no. 1, pp. 21-33, 1981.
20. Niku-Lari, A. J. Lu, and J. F. Flavenot,: "Measurement of Residual Stress Distribution by the Incremental Hole-Drilling Method," Experimental Mechanics, Vol. 25, pp. 175-185, 1985.
21. Measurement of Residual Stresses by the Hole Drilling Strain Gage Method, Measurements Group, TN-503-5, 1993.
22. Schajer, G. S., "Application of Finite Element Calculations to Residual Stress Measurement," Journal of Engineering Materials and Technology, Vol 130, pp. 157-163, 1981.
23. Schajer, G. S., G. Roy, M. T. Flaman and J. Lu, "Hole Drilling and Ring Core Methods," Chapter 2 in Handbook of Measurement of Residual Stresses, Society for Experimental Mechanics, Fairmont Press, Lilburn, GA, pp. 5-34, 1996.
24. "Errors Due to Transverse Sensitivity in Strain Gages," Micro-Measurements TN-509, 1993.

PART III

OPTICAL METHODS OF STRESS, STRAIN AND DISPLACEMENT ANALYSIS

CHAPTER 9

BASIC OPTICS

9.1 THE NATURE OF LIGHT

A typical technical description of **basic optics** begins with a section on the "nature of light" where the history of the contributions of prominent scientists in developing the "theory of light" is presented. In this description, the conflict between two competing theories—wave theory and particle theory—is discussed. Surprisingly, the generation of light and its physical characteristics that depend on its method of production are not covered. This chapter begins with the classical description of the nature of light. The physics involved in the production of light is then introduced in Section 9.2.

The phenomenon of light has attracted the attention of men and women from the earliest times. The ancient Greeks considered light to be an emission of small particles by a luminous body which entered the eye and returned to the body. Empedocles (484-424 B.C.) suggested that light takes time to travel from one point to another; however, Aristotle (384-322 B.C.) later rejected this idea as being too much to assume. The ideas of Aristotle concerning the nature of light persisted for approximately 2000 years.

In the seventeenth century, considerable effort was devoted to a study of the optical effects associated with thin films, lenses and prisms. Huygens (1629-1695) and Hooke (1635-1703) attempted to explain some of these effects with a wave theory. In this wave theory, a hypothetical substance of zero mass, called the ether, was assumed to occupy all space. Initially, light propagation was assumed to be a longitudinal vibratory disturbance moving through the ether. The idea of secondary wavelets, in which each point on a wavefront can be regarded as a new source of waves, was proposed by Huygens to explain refraction. Huygens' concept of secondary wavelets is widely used today to explain, in a simple way, other optical effects such as diffraction and some types of interference. At about the same time, Newton (1642-1727) proposed his corpuscular theory, in which light is visualized as a stream of small but swift particles emanating from shining bodies. The theory was able to explain many of the optical effects observed at the time.

A revival of interest in the wave theory of light began with the work of Young (1773-1829), who demonstrated that the presence of a refracted ray at an interface between two materials was to be expected from a wave theory, while the corpuscular theory of Newton could explain the effect only with difficulty. His two-pinhole experiment, which demonstrated the interference of light, together with the work of Fresnel (1788-1827) on polarized light, which required transverse rather than longitudinal vibrations, firmly established the transverse ether wave theory of light.

The next major step in the evolution of the theory of light was due to Maxwell (1831-1879). His electromagnetic theory predicts the presence of two vector fields in light waves, an electric field and a magnetic field. Because these fields can propagate through space unsupported by any known matter, the need for the hypothetical ether of the previous wave theory was eliminated. The electromagnetic wave theory also unites visible light with all the other invisible entities of the electromagnetic spectrum, e.g., cosmic rays, gamma rays, x-rays, ultraviolet rays, infrared rays, microwaves, radio waves and electric-power-transmission waves. The wide range of wavelengths and frequencies available for study in the electromagnetic spectrum has led to rapid development of additional theory and understanding.

Although Maxwell's electromagnetic wave theory was sufficient to describe most of the observed behavior of light, it could not explain the **photoelectric effect**. The photoelectric effect is the ejection of electrons from the surface of a metal when it is exposed to radiation. Experiments that showed that the kinetic energy of the electrons released from the metal's surface are independent of the amplitude of the light wave. Of course, the wave theory predicted the electrons ejected would be more energetic when the metal was exposed to a more intense beam of light. Einstein expanded Plank's quantum model that assumes the energy is released in discrete (not continuous) steps. The photoelectric effect is due to the energy transferred from a photon when it impacts an electron in a metal. Clearly the photon is acting like a particle in this transfer of energy. Einstein showed that the energy $\mathcal{E}$ of a photon is proportional to the frequency of the electromagnetic wave:

$$\mathcal{E} = h\,f \qquad (9.1)$$

where $h = 6.63 \times 10^{-34}$ J-s and f is the frequency of the electromagnetic wave.

Einstein's theory retains features of both the wave and particle theories. In practice, wave theory is used to describe most of the observed behavior of light. When wave theory is not adequate, particle theory is employed. For most of the effects to be described in later sections of this text, the wave properties of light are important and the particle characteristics of individual photons have limited application.

9.2 THE PHYSICS OF LIGHT

The physics of light pertains to the way light (a photon) is produced. Unfortunately, this topic is relegated to one of the final chapters in the typical college physics book and often neglected in the classroom. However, it is important to recognize how photons are produced before one can understand the properties of coherent and incoherent (ordinary) light. As indicated in the previous section, photons interact with some of the electrons in atoms. The production of a single photon occurs when an electron changes its energy state.

To understand energy states, consider Bohr's model of a hydrogen atom with its single orbiting electron as illustrated in Fig. 9.1. In Bohr's model, the electrons are permitted to orbit in only specific orbits with discrete radii. The radii of the permissible orbits are given by:

$$r_n = \frac{n^2 h^2}{mke^2} \qquad n = 1, 2, 3, \ldots. \qquad (9.2)$$

where m is the mass of the electron, h is Planck's constant (6.326×10^{-3}), k is Coulomb's constant and e is the elementary charge.

The radius of the first permissible orbit in the hydrogen atom is obtained for n = 1 as $r_1 = 0.0529$ nm. The second and third permissible orbits are at $r_2 = 0.1058$ nm and $r_3 = 0.1587$ nm.

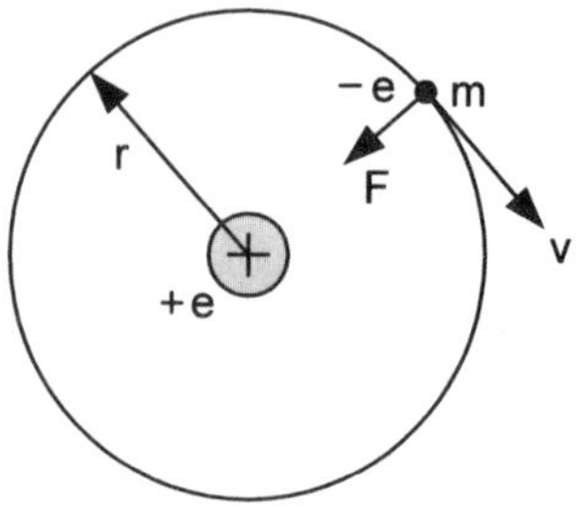

Figure 9.1 Model of a hydrogen atom.

The energy associated with each permissible orbit (quantum) states is given by:

$$\mathcal{E}_n = -\frac{mk^2e^4}{2h^2}\left(\frac{1}{n^2}\right) \qquad (9.3)$$

The lowest energy state, known as the ground state, occurs when n = 1 with $\mathcal{E}_1 = -13.6$ eV. The energy states for the second and third orbits (quantum numbers) are $\mathcal{E}_2 = -3.40$ eV and $\mathcal{E}_3 = 1.51$ eV. An energy level diagram for the hydrogen atom is presented in Fig. 9.2.

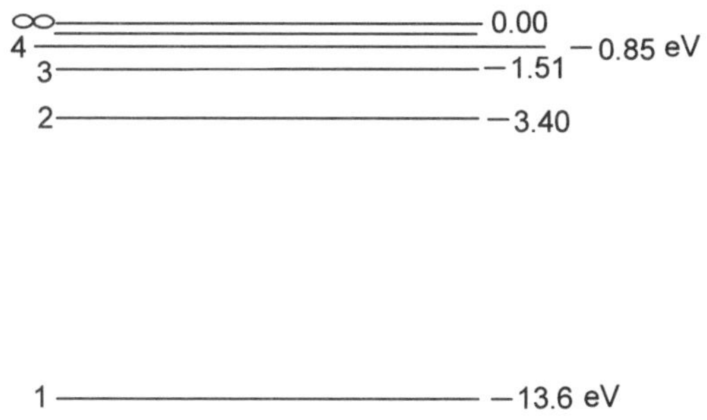

Figure 9.2 The energy level diagram for hydrogen. Quantum numbers are shown for each orbit on the left and the energy levels in eV on the right.

When an electron jumps from one orbit to another it reduces its energy state and in the process a photon is emitted. The frequency f of this photon is given by:

$$f = \frac{\mathcal{E}_{nf} - \mathcal{E}_{ni}}{h} \qquad (9.4)$$

This understanding of energy states is essential in describing absorption of light, production of spontaneous light and stimulated emission. Consider first absorption of light by a gas containing atoms with many permissible energy states. Before the atoms in the gas are exposed to light, they are in the ground state and are stable. However, when the gas is exposed to light, only those photons with energies that match the separation between energy levels $\Delta\mathcal{E} = h\,f$ will be absorbed by the atoms in the gas. The energy levels of those atoms that absorb photons are elevated to a higher state and considered to be excited. The atoms in the excited state are unstable and transition back to the ground (stable) state in about 10^{-8} s.

When an atom is in an excited state, its higher orbit electron may jump back to ground state orbit and in the process releases a photon that propagates into space. A graphical illustration of the photon production process is presented in Fig. 9.3 where the electron transitions from an excited state to a ground state. This process is known as spontaneous emission and occurs within about 10^{-8} s after the atom is excited.

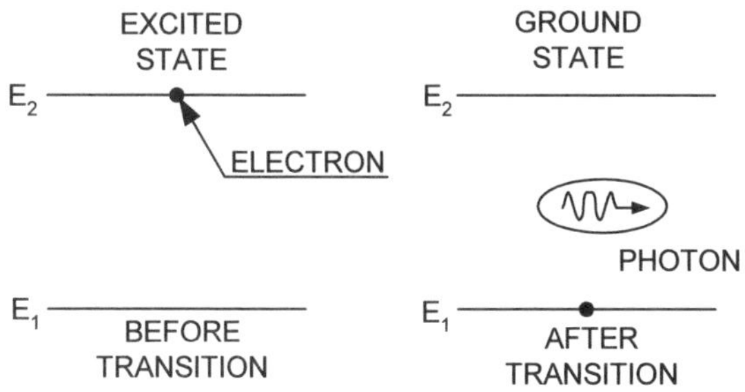

Figure 9.3 Spontaneous emission of a photon as an atom transitions from an excited state to a ground state.

Incoherent (ordinary) light is produced by heating a solid or a vapor until many of its atoms are raised to an excited state. The tungsten filament in an incandescent bulb and the vapor in a sodium lamp are examples of this process. As the temperature is increased, more and more atoms are excited and these unstable atoms undergo a spontaneous transition emitting photons to produce light. This light consists of numerous short pulses originating form a large number of different atoms. Each pulse consists of a finite number of oscillations known as a wave train. Each wave train is thought to be a few meters long with a duration of approximately 10^{-8} s. Because the light emissions occur in many different atoms which do not act together in a cooperative manner, the wave trains differ from each other with respect to their planes of vibration, frequency, amplitude and phase. Thus, radiation produced by spontaneous emission is referred to as incoherent light.

Another process for producing photons (light) is by stimulated emission, which is illustrated in Fig. 9.4. Consider an atom existing in an excited state as shown on the left side of Fig. 9.4. While the atom is in an excited state, a photon with energy of $\Delta\mathcal{E}$ is inserted into it. The presence of the photon stimulates the transition from its excited state to the ground state and a second photon is emitted. This photon is identical (same frequency and phase) as the inserted photon. These photons can stimulate other nearby atoms existing in the excited state to emit photons creating a chain reaction. The result is the beam of coherent light produced in a laser.

Figure 9.4 An illustration showing stimulated emission of a photon from an excited atom when stimulated by inserting a photon into an atom.

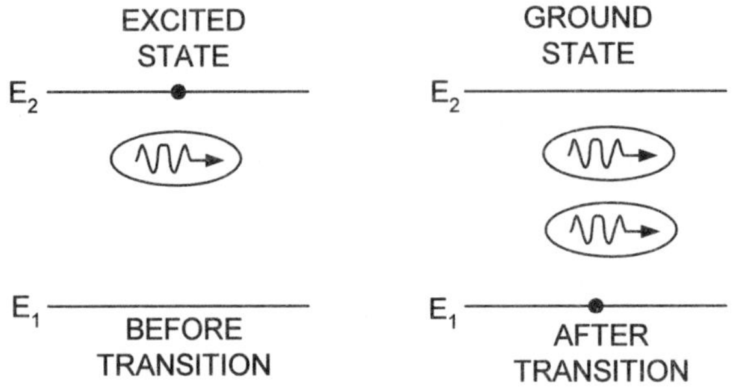

The laser[1] has become an extremely important light source that is essential in conducting many of the experiments described in later chapters. At the core of any laser is the lasing medium, typically a crystal such as a ruby or a gas filled tube containing argon or helium-neon. An example of a laser cavity is illustrated in Fig. 9.5. The tube that contains the lasing medium is excited by a pumping source, creating an inverted population. Pumping is accomplished with an external radiation source. At each end of the laser cavity are two reflective mirrors. On one end, the mirror is 100% reflective and the other mirror is 99% reflective, permitting about 1% of the beam to escape from the cavity. This stream of photons generates the output beam of monochromatic light that can be observed. The two mirrors are parallel so that the photons created reflect back and forth between the mirrors stimulating further emission of photons. The parallel orientation of the two mirrors at either end of the cavity also produces a highly collimated light beam.

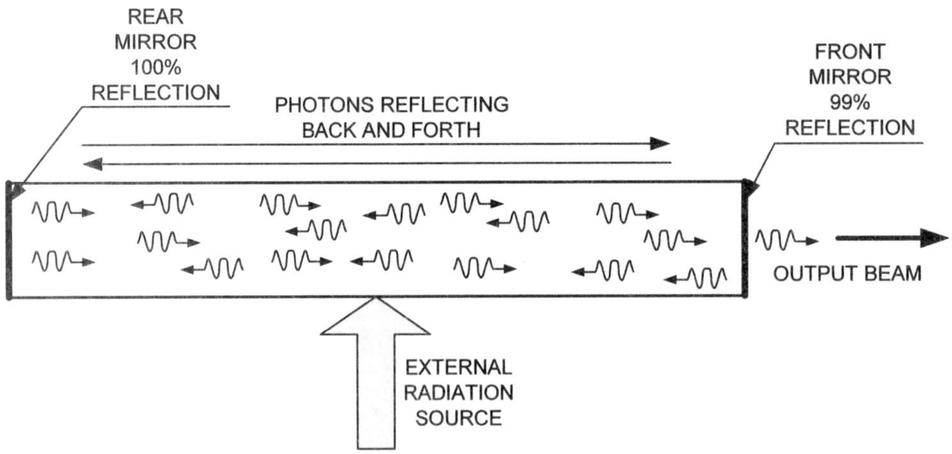

Figure 9.5 Illustration of the pumping action to produce simulated emission in a laser cavity.

[1] The word laser is an acronym for Light Amplification by Simulated Emission of Radiation.

Light sources such as the laser, in which the atoms act cooperatively in emitting light, produce coherent light. The wave trains in coherent light are monochromatic, in phase, linearly polarized, collimated and extremely intense. For the many of the interference effects that will be discussed in later chapters, coherent wave trains are required.

9.3 WAVE THEORY OF LIGHT

Electromagnetic radiation is predicted by Maxwell's theory to be a transverse wave motion which propagates with an extremely high velocity. Associated with this wave are oscillating electric and magnetic fields which can be described with electric and magnetic vectors **E** and **H**. These vectors are in phase, perpendicular to each other and at right angles to the direction of propagation. A simple representation of the electric and magnetic vectors associated with an electromagnetic wave at a given instant of time is illustrated in Fig. 9.6. For simplicity and convenience of representation, the wave has been given sinusoidal form.

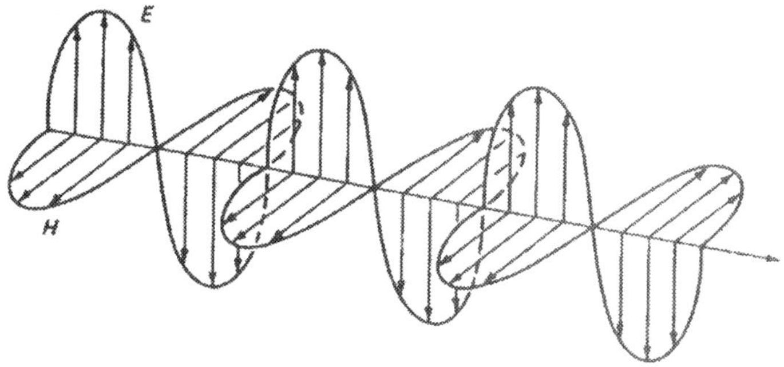

Figure 9.6 The electric and magnetic vectors associated with a plane electromagnetic wave.

All types of electromagnetic radiation propagate with the same velocity in free space, approximately 3×10^8 m/s (186,000 mi/s). Two parameters used to differentiate between the various radiations are wavelength and frequency. These two quantities are related to the velocity of light by the relationship:

$$\lambda f = c \tag{9.5}$$

where λ is the wavelength, f is the frequency and c is the velocity of propagation.

The electromagnetic spectrum has no upper or lower limits. The radiations commonly observed have been classified in the broad general categories shown in Fig. 9.7.

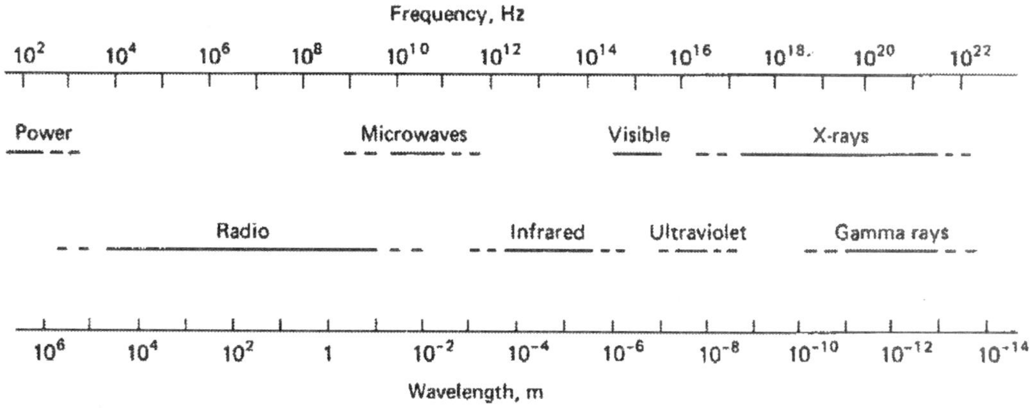

Figure 9.7 The electromagnetic spectrum.

Light is usually defined as radiation that can affect the human eye. From Fig. 9.7, it is evident that the visible range of the spectrum is a small band centered about a wavelength of approximately 550 nm. The limits of the visible spectrum are not well defined because the eye ceases to be sensitive at both long and short wavelengths; however, normal vision is usually in the range from 400 to 700 nm. Within this range, the eye interprets the wavelengths as the different colors listed in Table 9.1. Light from a source that emits a continuous spectrum with nearly equal energy for every wavelength is known as **white light**. Light of a single wavelength or frequency is called **monochromatic light**.

Table 9.1
The Visible Spectrum

Wavelength Range (nm)	Color
400-450	Violet
450-480	Blue
480-510	Blue-green
510-550	Green
550-570	Yellow-green
570-590	Yellow
590-630	Orange
630-670	Red

Electromagnetic waves can be classified as one-, two-, or three-dimensional according to the number of dimensions in which they propagate. Light waves which emanate radially from a small source are three-dimensional. Two quantities associated with a propagating wave which will be useful in discussions involving geometrical and physical optics are wavefronts and rays. For a three-dimensional pulse of light emanating from a source, both the electric vector and the magnetic vector exhibit the periodic oscillation in magnitude along any radial line as shown in Fig. 9.6. The locus of points on different radial lines from the source exhibiting the same disturbance at a given instant of time, e.g., maximum or minimum values, is a surface known as a **wavefront**. The wavefront surface moves as the pulse propagates. If the medium is optically homogeneous and isotropic, the direction of propagation is at right angles to the wavefront. A line normal to the wavefront, indicating the direction of propagation of the waves, is called a **ray**. When waves propagate outward in all directions from a point source, the wavefronts are spherical and the rays are radial lines. At large distances from the source, the spherical wavefronts have very little curvature, and over a limited region they can be treated as plane. Plane wavefronts can also be produced by using a lens or mirror to direct a cone of light from a point source into a parallel (collimated) beam.

In ordinary (incoherent) light, which is emitted from, say, a tungsten-filament light bulb, the light vector is not restricted in any sense and may be considered to be composed of a number of arbitrary transverse vibrations. Each of the components has a different wavelength, different amplitude, a different orientation (plane of vibration) and a different phase. The vector used to represent the light wave can be either the electric vector or the magnetic vector. Both exist simultaneously, as shown in Fig. 9.6, and either can be used to describe the optical effects associated with photoelasticity, moiré, interferometry and holography.

9.3.1 The Wave Equation

Because the disturbance producing light can be represented by a transverse wave, it is possible to express the magnitude of the light (electric) vector in terms of the solution of the one-dimensional wave equation:

$$E = f(z - ct) + g(z + ct) \tag{9.6}$$

where E is the magnitude of light vector, z is the position along axis of propagation and t is the time.

The term f(z - ct) represents a wave propagating in the positive z direction and the term g(z + ct) represents a wave propagating in the negative z direction. Optical effects of interest in experimental mechanics can be described with a simple sinusoidal or harmonic waveform. Thus, light propagating in the positive z direction away from the source can be represented as:

$$E = f(z - ct) = \frac{K}{z} \cos \frac{2\pi}{\lambda}(z - ct) \tag{9.7}$$

where K is related to the strength of the source and K/z is an attenuation coefficient associated with the expanding spherical wavefront. At distances far from the source, the attenuation is small over short observation distances, and therefore attenuation is usually neglected. For plane waves, the attenuation is not a factor because the beam of light maintains a constant cross section. Equation (9.7) can then be written as:

$$E = a \cos \frac{2\pi}{\lambda}(z - ct) \tag{9.8}$$

where a is the amplitude of the wave. A graphical representation of the magnitude of the light vector as a function of position along the positive z axis, at two different times, is shown for a plane light wave in Fig. 9.8. The length from peak to peak on the magnitude curve for the light vector is the wave length λ. The time required for passage of two successive peaks at some fixed value of z is defined as the period T of the wave and is given by:

$$T = \lambda/c \tag{9.9}$$

The frequency of a light vector is the number of its oscillations per second. Clearly, the frequency is the reciprocal of the period, or:

$$f = 1/T = c/\lambda \tag{9.10}$$

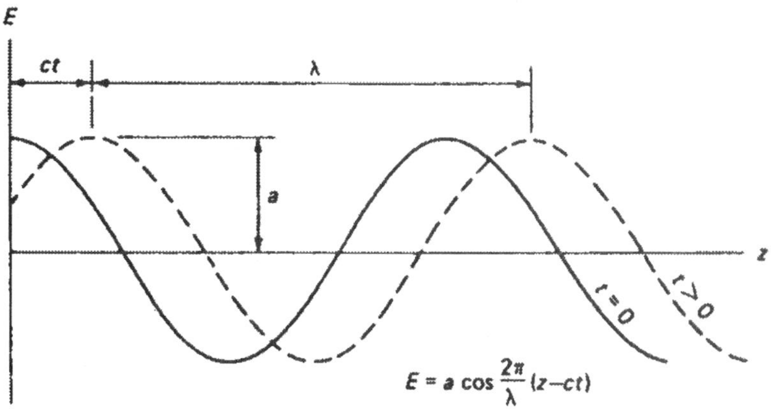

Figure 9.8 Magnitude of the light vector as a function of position along the axis of propagation at two different times.

The terms angular frequency and wave number are frequently used to simplify the argument in a sinusoidal representation of a light wave. The angular frequency ω and the wave number ξ are given by:

$$\omega = 2\pi/T = 2\pi f \tag{9.11}$$

$$\xi = 2\pi/\lambda \tag{9.12}$$

Substituting Eqs. (9.11) and (9.12) into Eq. (9.8) yields:

$$E = a \cos(\xi z - \omega t) \qquad \text{(a)}$$

Two waves having the same wavelength and amplitude but a different phase are shown in Fig. 9.9. The two waves can be expressed by:

$$\begin{aligned} E_1 &= a \cos\frac{2\pi}{\lambda}(z + \delta_1 - ct) \\ E_2 &= a \cos\frac{2\pi}{\lambda}(z + \delta_2 - ct) \end{aligned} \qquad (9.13)$$

where δ_1 is the initial phase of wave E_1 and δ_2 is the initial phase of E_2.
$\delta = \delta_2 - \delta_1$ is the linear phase difference between the two waves.

The linear phase difference δ is often referred to as **retardation** because the second wave trails the first.

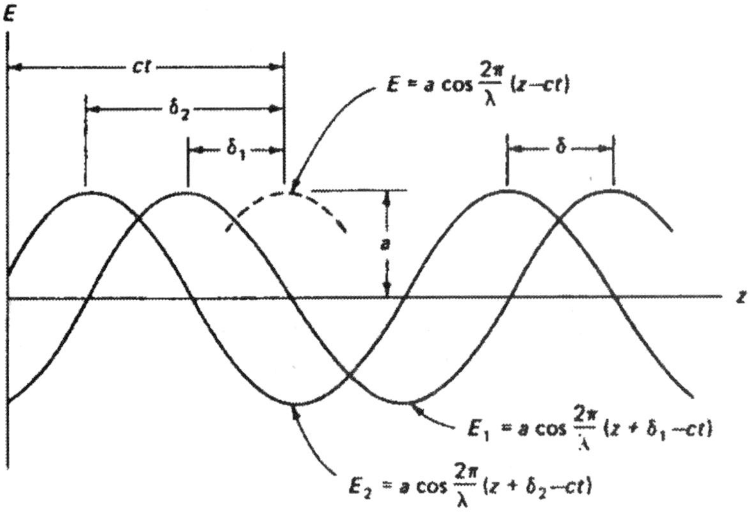

Figure 9.9 Magnitude of the light vector components as a function of position along the axis of propagation for two waves with different initial phases.

The magnitude of the light vector can also be plotted as a function of time at a fixed position along its ray. This representation is useful for many applications because the eye, photographic films, digital cameras and other light-detecting devices are located at fixed positions along the z axis for observations.

9.3.2 Superposition of Waves

In later chapters on photoelasticity, interferometry and moiré, the phenomena associated with the superposition of two waves having the same wave length or frequency but different amplitude and phase will be encountered. At a fixed position z_0 along a light ray, where the observations will be made, the equations for the two waves can be expressed as:

$$\begin{aligned} E_1 &= a_1 \cos\frac{2\pi}{\lambda}(z_0 + \delta_1 - ct) = a_1 \cos(\phi_1 - \omega t) \\ E_2 &= a_2 \cos\frac{2\pi}{\lambda}(z_0 + \delta_2 - ct) = a_2 \cos(\phi_2 - \omega t) \end{aligned} \qquad (9.14)$$

where ϕ_1 and ϕ_2 are the phase angles of wave E_1 and wave E_2 at position z_0, respectively.
a_1 and a_2 are the amplitudes of waves E_1 and E_2, respectively.

Consider, first, the case where the light vectors associated with the two waves oscillate in the same plane. The magnitude E of the resulting light vector is the sum:

$$E = E_1 + E_2 \tag{a}$$

Substituting Eqs. (9.14) into Eq. (a) yields:

$$E = a_1 (\cos\omega t \cos\phi_1 + \sin\omega t \sin\phi_1) + a_2 (\cos\omega t \cos\phi_2 + \sin\omega t \sin\phi_2) = a \cos(\phi - \omega t) \tag{b}$$

where

$$a^2 = a_1^2 + a_2^2 + 2a_1a_2 \cos(\phi_2 - \phi_1) \tag{9.15}$$

and

$$\tan\phi = \frac{a_1 \sin\phi_1 + a_2 \sin\phi_2}{a_1 \cos\phi_1 + a_2 \cos\phi_2} \tag{9.16}$$

Equation (b) indicates that the resulting wave has the same frequency as the original waves but a different amplitude and a different phase angle. This procedure can easily be extended to the addition of three or more waves.

A special case often arises where the amplitudes of the two waves are equal ($a_1 = a_2$). In this case, the amplitude of the resulting wave is given by Eq. (9.15) as:

$$a = \sqrt{2a_1^2\left(1 + \cos\frac{2\pi\delta}{\lambda}\right)} = \sqrt{4a_1^2 \cos^2\frac{\pi\delta}{\lambda}} \tag{c}$$

In applications of optical methods, the amplitude of the resulting wave is important, **but the time variation is not**. This conclusion is due to the fact that the eye and all light sensing instruments respond to the intensity of light (intensity is proportional to the square of the amplitude) but cannot detect the wave's rapid time fluctuations (for sodium light the frequency is 5.1×10^{14} Hz). For the special case of two waves of equal amplitude, the combined intensity is given by:

$$I \propto a^2 = 4a_1^2 \cos^2\frac{\pi\delta}{\lambda} \tag{9.17}$$

Equation (9.17) indicates that the intensity of the light wave resulting from the superposition of two waves of equal amplitude is a function of the linear phase difference δ between the waves. The intensity of the resultant wave assumes its maximum value when $\delta = n\lambda$, n = 0, 1, 2, 3,..... When the linear phase difference is an integral number of wavelengths, then:

$$I = 4a_1^2 \tag{d}$$

and the intensity of the resultant wave is four times the intensity of one of the individual waves. The intensity of the resultant wave assumes its minimum value when $\delta = [(2n + 1)/2]\lambda$, n = 0, 1, 2, 3,... When the linear phase difference is an odd number of half wavelengths, then:

$$I = 0 \tag{e}$$

The modification of intensity by superposition of light waves is referred to as an **interference effect**. The effect represented by Eq. (d) is **constructive interference**. The effect represented by Eq. (e) is **destructive interference**. Interference effects have important application in photoelasticity, moiré, interferometry and holography.

When the electric vector used to describe the light wave is restricted to a single plane, the condition is known as **plane** or **linearly polarized light**. Two other important forms of polarized light arise as a result of the superposition of two linearly polarized light waves having the same frequency but mutually perpendicular planes of vibration, as shown in Fig. 9.10. At a fixed position z_0 along the light ray, the equations for the two waves can be expressed as:

$$E_x = a_x \cos\frac{2\pi}{\lambda}(z_0 + \delta_x - ct) = a_x \cos(\phi_x - \omega t)$$

$$E_y = a_y \cos\frac{2\pi}{\lambda}(z_0 + \delta_y - ct) = a_y \cos(\phi_y - \omega t) \qquad (9.18)$$

$$\mathbf{E} = a\cos(\phi - \omega t)$$

where ϕ_x and ϕ_y are phase angles associated with waves in the x-z and y-z planes.
a_x, a_y are the amplitudes of waves in the x-z and y-z planes.

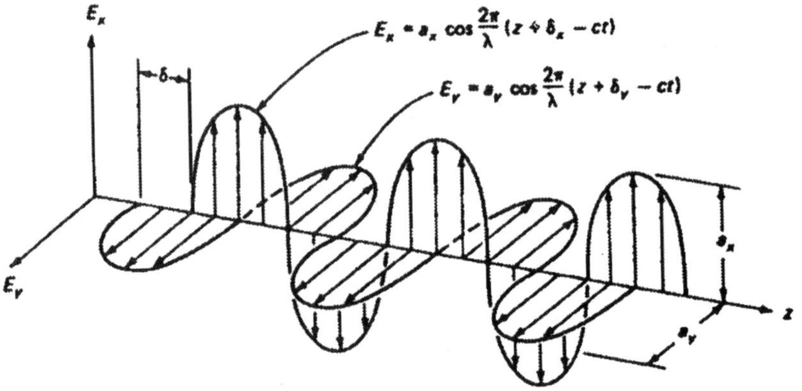

Figure 9.10 Two linearly polarized light waves having the same frequency but mutually perpendicular planes of vibration.

The magnitude of the resulting light vector is given by vector addition as:

$$E = \sqrt{E_x^2 + E_y^2} \qquad (f)$$

Considerable insight into the nature of the light resulting from the super-position of two mutually perpendicular waves is provided by a study of the trace of the tip of the resulting electric vector on a plane perpendicular to the axis of propagation at the point z_0. An expression for this trace can be obtained by eliminating time from Eqs. (9.18) to give:

$$\frac{E_x^2}{a_x^2} - 2\frac{E_x E_y}{a_x a_y}\cos(\phi_y - \phi_x) + \frac{E_y^2}{a_y^2} = \sin^2(\phi_y - \phi_x) \qquad (g)$$

because:

$$\phi_y - \phi_x = \frac{2\pi}{\lambda}(\delta_y - \delta_x) = \frac{2\pi\delta}{\lambda}$$

$$\frac{E_x^2}{a_x^2} - 2\frac{E_x E_y}{a_x a_y}\cos\frac{2\pi\delta}{\lambda} + \frac{E_y^2}{a_y^2} = \sin^2\frac{2\pi\delta}{\lambda} \qquad (9.19)$$

Equation (9.19) is the equation of an ellipse and light exhibiting this behavior is known as **elliptically polarized light**. The tips of the electric vectors at different positions along the z axis form an elliptical helix, as shown in Fig. 9.11. During an interval of time t, the helix will translate a distance z = ct in the positive direction. As a result, the electric vector at position z_0 will rotate in a counterclockwise direction as the translating helix is observed in the positive z direction. The locus of points representing the trace of the tip of the light vector on the perpendicular plane is the ellipse described by Eq. (9.19) and illustrated in Fig. 9.12.

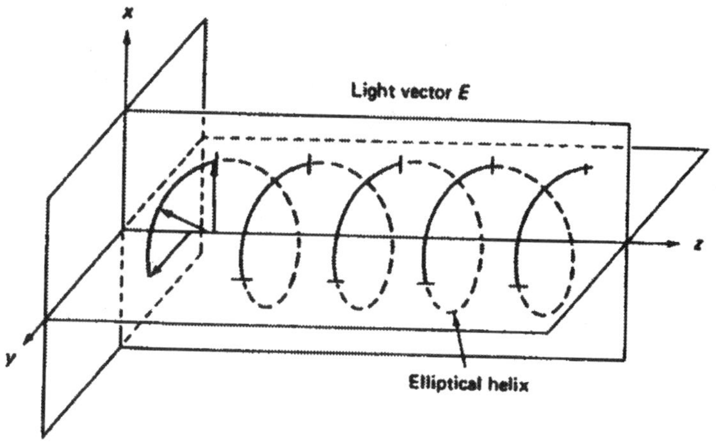

Figure 9.11 An elliptical helix is formed by the tip of the light vector as it propagates along the z axis.

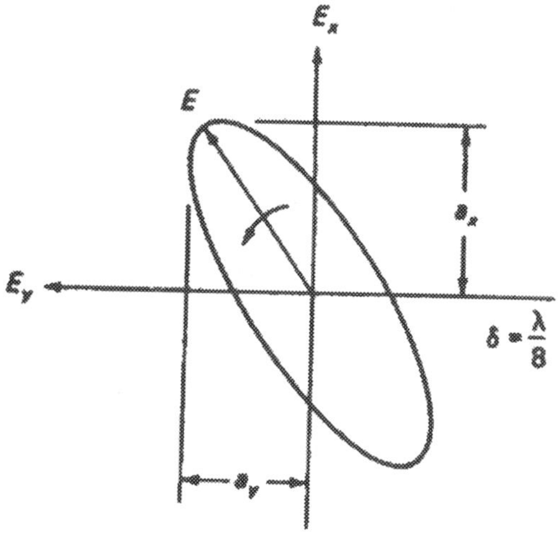

Figure 9.12 Trace of the light vector's tip on a plane perpendicular to the z axis at position z_0.

A special case of elliptically polarized light occurs when the amplitudes of the two waves E_x and E_y are equal and $\delta = [(2n + 1)/4]\lambda$, n = 0, 1, 2,..., so that Eq. (9.19) reduces to the equation of a circle.

$$E_x^2 + E_y^2 = a^2 \tag{h}$$

Light exhibiting this behavior is known as **circularly polarized light**, and the tips of the light vectors form a circular helix along the z axis. For $\delta = \lambda/4, 5\lambda/4,...$ the helix is a left circular helix, and the light vector rotates counterclockwise with time as the translating helix is observed in the positive z direction. For $\delta = 3\lambda/4, 7\lambda/4,...,$ the helix is a right circular helix, and the electric vector rotates in the clockwise direction.

A second special case of elliptically polarized light occurs when the linear phase difference δ between the two waves E_x and E_y is an integral number of half wavelengths ($\delta = n\lambda/2$, n = 0, 1, 2,...). For this case, Eq. (9.19) reduces to:

$$E_y = (a_y^2/a_x^2)\, E_x \tag{i}$$

Equation (i) describes a straight line and light exhibiting this behavior is known as **plane-** or **linearly polarized light**. The amplitude of the plane wave depends upon the amplitudes of the two original waves because:

$$a = \sqrt{a_x^2 + a_y^2} \tag{k}$$

In this discussion, light is treated as a wave motion without beginning or end. The light emitted by a conventional light source, e.g., a tungsten-filament light bulb, consists of numerous short pulses originating from many different atoms. Each pulse consists of a finite number of oscillations known as a wave train—each a few meters long with a duration of approximately 10^{-8} s. Because the light emissions occur in many different atoms which do not act together in a cooperative manner, the wave trains differ from each other in plane of vibration, frequency, amplitude and phase. Radiation produced in this by spontaneous emission is referred to as incoherent light. Light sources such as the laser, in which the atoms act cooperatively in emitting light, produce coherent light, in which the wave trains are monochromatic, in phase, linearly polarized, collimated and intense.

9.3.3 The Wave Equation in Complex Notation

A convenient way to represent both the amplitude and phase of a light wave, such as the one represented by Eq. (9.8), for calculations involving a number of optical elements is through the use of complex or exponential notation. Recall the Euler identity:

$$e^{j\theta} = \cos\theta + j\sin\theta \tag{a}$$

where $j^2 = -1$. The sinusoidal wave of Eq. (9.8) is represented by the real part of the complex expression with:

$$\overline{E} = ae^{j(2\pi/\lambda)(z-ct)} = ae^{j(\phi-\omega t)} \tag{b}$$

The imaginary part of Eq. (a) could also be used to represent the wave; however, it is normally assumed that the real part of a complex quantity is the one having physical significance. If the amplitude of the wave is also considered to be a complex quantity, then:

$$\overline{a} = a_r + ja_i = ae^{j[(2\pi/\lambda)\delta]} \tag{c}$$

where:

$$a = \sqrt{a_r^2 + a_i^2} \tag{d}$$

and

$$\tan\frac{2\pi}{\lambda}\delta = \frac{a_i}{a_r} \tag{e}$$

A wave with an initial phase δ can be expressed in exponential notation as:

$$\overline{E} = \overline{a}e^{j(2\pi/\lambda)(z-ct)} = ae^{j(2\pi/\lambda)(z+\delta-ct)} \tag{9.20}$$

The physical waves previously represented by Eqs. (9.13) are simply the real part of Eq. (9.20) when written in exponential notation. Superposition of two or more waves, having the same frequency but different amplitude and phase, is easily performed with the exponential representation. The real and imaginary parts of the amplitudes of the individual waves are added separately in an algebraic manner. The resultant complex amplitude gives the amplitude and phase of a single wave equivalent to the sum of the individual waves. Extensive use will be made of this representation in Chapter 10, where the theory of photoelasticity is discussed.

The real and imaginary parts of a complex quantity such as the amplitude of a wave may also be written as:

$$a_r = \frac{1}{2}(\bar{a} + \bar{a}^*) \qquad a_i = \frac{1}{2}(\bar{a} - \bar{a}^*) \qquad \text{(f)}$$

where $\bar{a}^* = a_r - ia_i$ (g)

is the complex conjugate of the original complex amplitude.

From Eq. (d) it is clear that:

$$a^2 = a_r^2 + a_i^2 \qquad \text{(h)}$$

From Eq. (h) and Eq. (f):

$$a^2 = \bar{a}\,\bar{a}^* \qquad (9.21)$$

This representation for the square of the amplitude of a complex quantity will be used in developments of optical methods based on the intensity of light.

9.4 REFLECTION AND REFRACTION

In the previous section, the electromagnetic wave nature of light was discussed and wavefronts and rays were defined. The discussions were limited to light propagating in free space. Most optical effects of interest, however, occur as a result of the interaction between a ray of light and some physical material. In free space, light propagates with a velocity c, which is approximately 3 x 10^8 m/s. In any other medium, the light's velocity is reduced. The ratio of the velocity in free space to the velocity in a medium depends on the index of refraction n of the medium. The index of refraction for most gases is only slightly greater than unity (for air, n = 1.0003). Values for liquids range from 1.3 to 1.5 (for water, n = 1.33) and for solids it ranges from 1.4 to 1.8 (for glass, n = 1.5). The index of refraction for a material is not constant but varies slightly with wavelength of the light being transmitted. This dependence of index of refraction on wavelength is referred to as **dispersion.**

Because the frequency of a light wave is independent of material through which it propagates, the wavelength is shorter in a solid or a liquid than in free space. For this reason, a wave propagating in any material will develop a linear phase shift δ with respect to a similar wave propagating in the free space. The magnitude of the phase shift, in terms of the index of refraction of the material, can be developed as follows. The time required for passage through a material of thickness h is given by:

$$t = h/v \qquad \text{(a)}$$

where h is the thickness of the material along the path of light propagation and v is the velocity of light in the material. The distance s traveled during the same time by a wave in free space is:

$$s = ct = (ch)/v = nh \qquad \text{(b)}$$

The distance δ by which the wave in the material trails the wave in free space is given by the difference as:

$$\delta = s - h = nh - h = h(n - 1) \qquad (9.22)$$

The retardation δ is a positive quantity because the index of refraction of a material is always greater than unity. The relative position of one wave with respect to another can be described by including the retardations in the phase of the appropriate wave equation.

When a ray of light strikes a surface between two transparent materials with different indices of refraction, it is divided into a reflected ray and a refracted ray, as shown in Fig. 9.13. The reflected and refracted rays lie in the plane formed by the incident ray and the normal to the surface and is known as the **plane of incidence**. The angle of incidence α, the angle of reflection β, and the angle of refraction γ are related as indicated below:

For reflection: $$\alpha = \beta \tag{9.23}$$

For refraction $$\frac{\sin\alpha}{\sin\gamma} = \frac{n_2}{n_1} = n_{21} \tag{9.24}$$

where n_1 and n_2 are the indices of refraction of materials 1 and 2, respectively and n_{21} is the index of refraction of material 2 with respect to material 1.

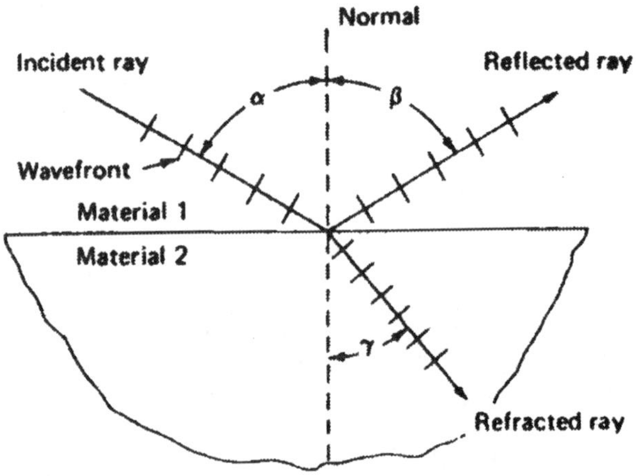

Figure 9.13 Reflected and refraction of a plane light wave at an interface between two transparent materials.

If the incident light ray is propagating in the material having the higher index of refraction, n_{21} will be a number less than unity. Under these conditions, some critical angle of incidence α_c is reached for which the angle of refraction is 90°. For angles of incidence greater than the critical angle, there is no refracted ray and total internal reflection occurs. Total internal reflection cannot occur when the incident ray is in the medium with the lower index of refraction.

The laws of reflection and refraction give the direction of reflected and refracted rays but do not describe their intensity. Intensity relationships, which are derived from Maxwell's equations, indicate that the intensity of a reflected ray depends upon both the angle of incidence and the direction of polarization of the incident ray. Consider a completely unpolarized ray of light impinging on a surface between two transparent materials, as shown in Fig. 9.14. The electric vector for each wavetrain in the ray can be resolved into two components, one perpendicular to the plane of incidence (the perpendicular component) and the other parallel to the plane of incidence (the parallel component). For completely unpolarized incident light, the two components have equal intensity. The intensity of the reflected ray can be expressed as:

$$I_r = RI_i \tag{9.25}$$

where I_i is the intensity of incident ray, I_r is the intensity of reflected ray and R is the reflection coefficient.

The reflection coefficient is different for the perpendicular and parallel components of the reflected wave. For the perpendicular component:

$$R_{90^\circ} = \frac{\sin^2(\alpha-\gamma)}{\sin^2(\alpha+\gamma)} \tag{9.26}$$

For the parallel component:

$$R_{0^\circ} = \frac{\tan^2(\alpha-\gamma)}{\tan^2(\alpha+\gamma)} \tag{9.27}$$

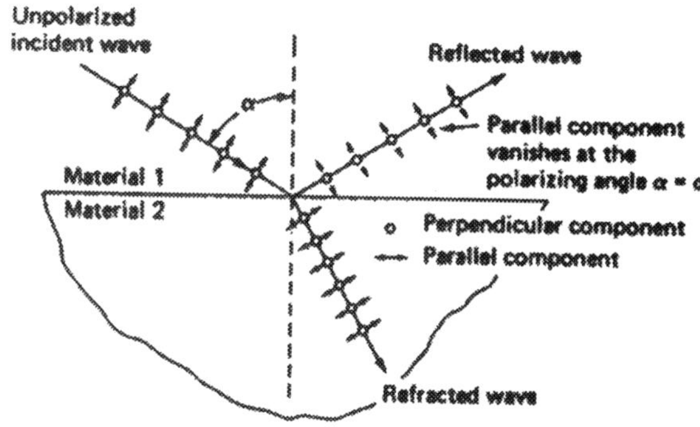

Figure 9.14 Reflection and refraction at the polarizing angle.

Reflection coefficients for an air-glass interface (n_{21} = 1.5) are shown in Fig. 9.15. These data indicate that there is a particular angle of incidence for which the reflection coefficient for the parallel component is zero. This angle is referred to as the **polarizing angle** α_p. Because the parallel component vanishes when the angle of incidence is equal to the polarizing angle, the ray reflected from the surface is plane-polarized with the plane of vibration perpendicular to the plane of incidence. It is also observed that a plane phase change of $\delta = \lambda/2$ occurs when light is incident from the medium with the lower index of refraction. When light is incident from the medium with the higher index of refraction, no phase change occurs upon reflection.

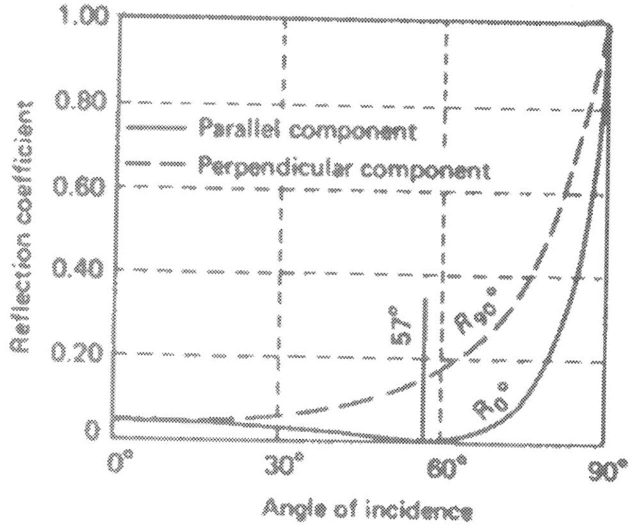

Figure 9.15 Reflection coefficients for an air-glass interface (n_{21} = 1.5)

Metal surfaces exhibit relatively large reflection coefficients, as shown in Fig. 9.16. At oblique incidence, the coefficients for light polarized parallel to the plane of incidence are less than the coefficients for the perpendicular component, and a change of phase also occurs. Unfortunately, the phase change varies with both the angle of incidence and direction of polarization. As a result, plane-polarized light is converted by oblique reflection to elliptically polarized light.

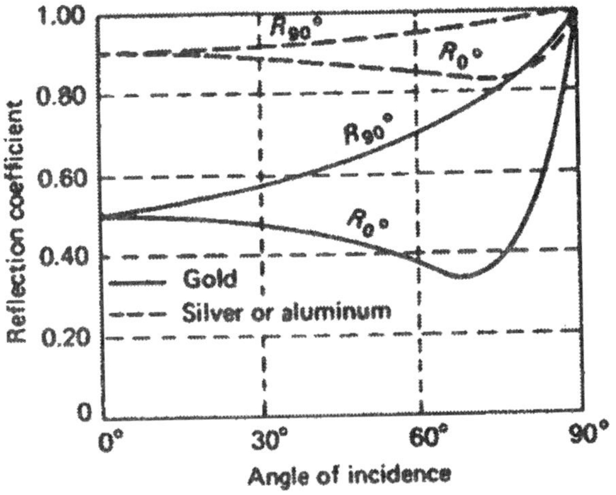

Figure 9.16 Reflection coefficients for several air-metal interfaces.

9.5 IMAGE FORMATION BY LENSES AND MIRRORS

In the previous section, reflection and refraction of a plane light wave at a plane interface between two materials was considered. More complicated situations frequently arise in the optical systems used for studies in experimental mechanics. Because lenses and mirrors are widely used in many of these systems, a brief discussion of the significant features of these elements is provided here for future reference.

9.5.1 Plane Mirrors

An object O placed at a distance u in front of a plane mirror is shown in Fig.9.17. The light from each point on the object (such as point A) is a spherical wave which reflects from the mirror. When the eye or other light-sensing instrument intercepts the reflected rays, an image I of the object O at a distance v behind the mirror is perceived. In this instance, the image I is a virtual image because light rays do not pass through image points such as A'. From the geometry of Fig. 9.17, it is obvious that the magnitudes of u and v are the same. The image is erect and has the same height as the object. One difference between the object and the image, not apparent from Fig. 9.17, is that left and right are interchanged (an image of a left hand appears as a right hand in the mirror).

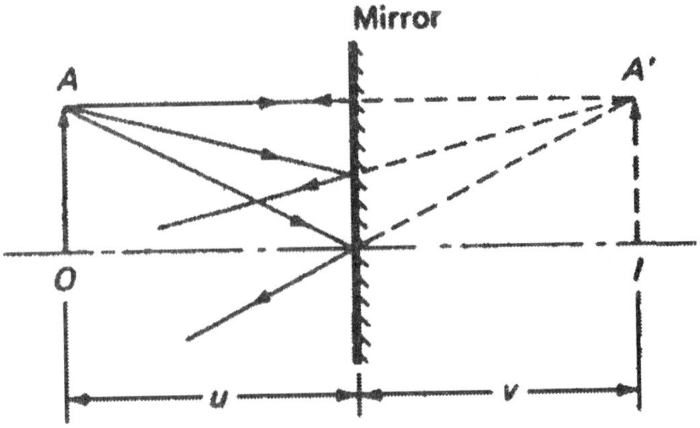

Figure 9.17 Image formation by a plane mirror.

9.5.2 Spherical Mirrors

An object O placed at a distance u in front of a concave spherical mirror is shown in Fig. 9.18. The center of curvature of the mirror is located at C and the focal point is at F. The light from each point on the object reflects as shown in Fig. 9.18 for a typical point A. In this instance, the image I is a real image because light rays pass through the image point A'. From the geometry of Fig. 9.18 it can be shown that if all rays from the object make a small angle with respect to the axis of the mirror, then the distance to the image satisfies the expression:

$$\frac{1}{u}+\frac{1}{v}=\frac{2}{r}=\frac{1}{f} \tag{9.28}$$

where r is the radius of curvature of the mirror. For the case illustrated, both u and v are positive because both the object and the image are real. The distance v will be negative when it defines the position of a virtual image. With this sign convention, Eq. (9.28) applies to all concave, plane and convex mirrors. The ratio of the size of the image to the size of the object is known as the **magnification** M and is given by the expression:

$$M=-\frac{v}{u} \tag{9.29}$$

where the minus sign is used to indicate an inverted image. Equation (9.29) also applies to all concave, plane and convex mirrors. For example, when Eqs. (9.28) and (9.29) are applied to the plane mirror of Fig. 9.17, v = – u and M = 1. These results indicate that the image, which is imaginary (virtual), will be erect and identical in size to the object.

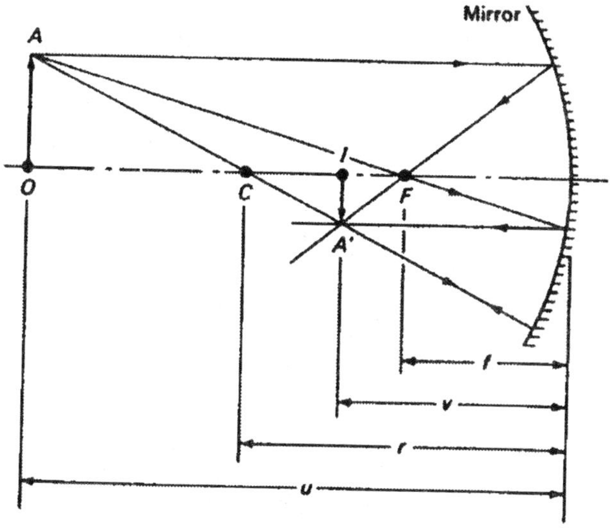

Figure 9.18 Image formation by a concave spherical mirror.

9.5.3 Thin Lenses

At least one and often a series combination of lenses is employed in an optical arrangement to magnify and focus the image of an object on a photographic plate or a CCD image sensor. For this reason, the passage of light through a single convex lens and a pair of convex lenses arranged in series will be examined in detail. The treatment will be limited to instances where the thin-lens approximation can be applied; i.e., the thickness of the lens can be neglected with respect to other distances such as the focal length f of the lens and the object and image distances u and v.

Single-Lens System

The optical representation of a single-lens system is shown in Fig. 9.19. The light from each point on the object can be considered as a spherical wave which is reflected and refracted at the air-glass interface, as indicated in Section 9.4. With mirrors, the reflected rays are of interest; however, with lenses, the refracted rays produce the desired optical effects. From the geometry of Fig. 9.19 and the small angle assumption, it is clear that the object distance u, the image distance v and the focal length f of the lens are related by the expression:

$$\frac{1}{u}+\frac{1}{v}=\frac{1}{f} \qquad \text{(a)}$$

and that the magnification is given by the expression:

$$M=-\frac{v}{u} \qquad \text{(b)}$$

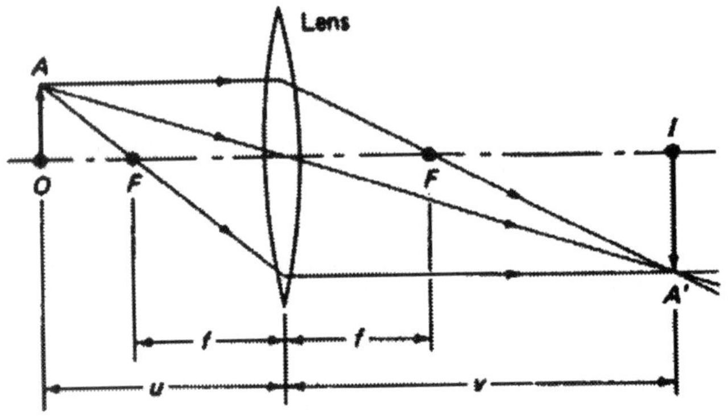

Figure 9.19 Image formation by a single convex lens with the object outside the focal point.

Equations (a) and (b) are identical to Eqs. (9.28) and (9.29) for mirrors. The image in Fig. 9.19 is a real image; therefore, v is positive even though the image is on the opposite side of the lens from the object. With v positive, Eq. (b) indicates that the magnification is negative (thus the image is inverted, as shown).

The situation illustrated in Fig. 9.19 occurs when the object is located beyond the focal point of the lens. However, the object may also be placed between the focal point and the lens surface, as illustrated in Fig. 9.20. In this case, the distance u is positive and v is negative because a virtual image is formed. Equation (b) then yields a positive magnification, which indicates that the image is erect (as shown).

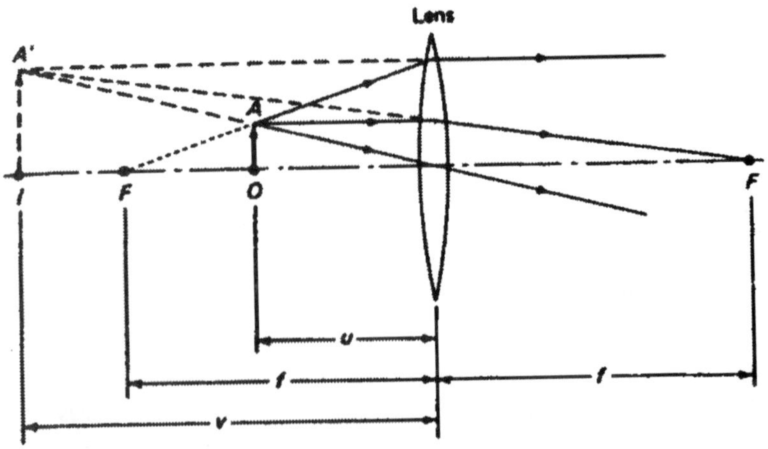

Figure 9.20 Image formation by a single convex lens with the object inside the focal point.

A similar analysis for a concave lens indicates that Eqs. (9.28) and (9.29) apply so long as v is considered negative for virtual images and positive for real images. The proof is left as an exercise.

If Eqs. (9.28) and (9.29) are combined and solved for u and v in terms of M and f,

$$u = \frac{M-1}{M} f \qquad \text{and} \qquad v = (1-M)f \tag{9.30}$$

The total length L from the object to the focusing plane of the image in terms of M and f is given by:

$$L = u + v = \left(2 - M - \frac{1}{M}\right) f \tag{9.31}$$

These results show that the distances u, v and L are directly related to the focal length of the lens. Because the focal length of the camera lens on many optical instruments may be relatively large, the length of an optical bench may exceed 1 or 2 m. This length, of course, depends upon the magnification capabilities of the camera, as shown in Table 9.2.

Table 9.2
Influence of the magnification M on the dimensions u, v and L in a camera assembly

– M	L/f	u/f	v/f
¼	6.25	5.00	1.25
½	4.50	3.00	1.50
1	4.00	2.00	2.00
2	4.50	1.50	3.00
4	6.25	1.25	5.00

Two Lenses in Series

The classical optical representation of a series combination of two convex lenses is shown in Fig. 9.21. This series arrangement of convex lenses is often employed in optical arrangements to decrease their total length. The object is positioned inside the focal point of the long-focal-length lens F_1. The slowly diverging rays from lens F_1 are then converged by a shorter-focal-length lens F_2. By applying Eq. (9.28) to both lenses, it is clear that:

For the lens F_1:

$$\frac{1}{u_1} + \frac{1}{v_1} = \frac{1}{f_1} \tag{a}$$

For the lens F_2:

$$\frac{1}{u_2} + \frac{1}{v_2} = \frac{1}{f_2} \tag{b}$$

Because the virtual image from the field lens serves as the object for the condenser lens,

$$u_2 = -v_1 + s \tag{c}$$

The magnification M_c for the combined lens system is given by:

$$M_c = M_1 M_2 = \left(-\frac{v_1}{u_1}\right)\left(-\frac{v_2}{u_2}\right) = \frac{v_1 v_2}{u_1 u_2} \tag{d}$$

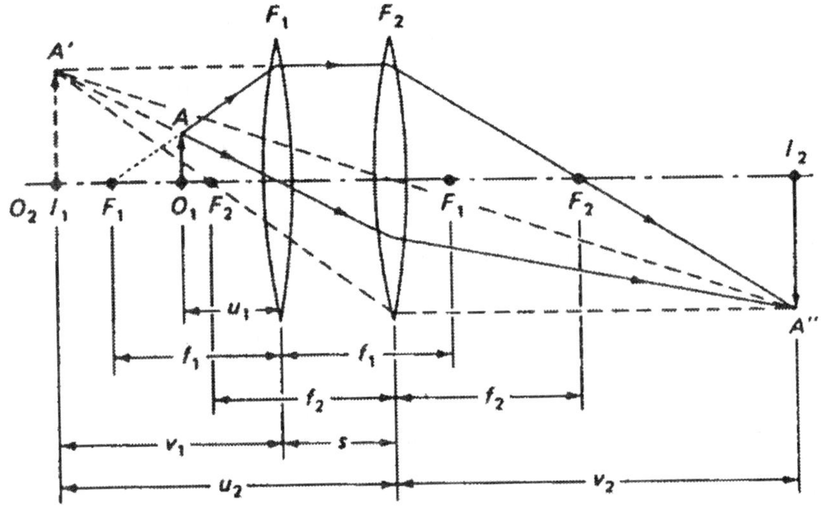

Figure 9.21 Image formation by a series combination of two convex lenses.

By combining Eqs. (a) to (d) and solving for M, u_1, v_2 and s it is possible to show that:

$$M = \frac{f_1 f_2}{(u_1 - f_1)(s - f_2) - u_1 f_1} \tag{9.32}$$

$$s = \frac{f_2(f_1 - u_1) - u_1 f_1 - f_1 f_2 / M}{f_1 - u_1} \tag{9.33}$$

$$u_1 = \frac{f_2 f_2[(1 - M)/M] + s f_1}{s - f_1 - f_2} \tag{9.34}$$

$$v_2 = M\left[s\left(\frac{u_1}{f_1} - 1\right) - u_1\right] \tag{9.35}$$

These results can be employed to design optical systems, i.e., to choose f_1 and f_2 to give a range of magnification M within a certain length $L = u_1 + v_2$. Or if the optical system already exists, where f_1 and f_2 are fixed, u_1 and s can be varied within limits. These equations can be employed to determine M within these limits. It should be noted that M is a function of f_1, f_2, u_1 and s; thus, a given magnification M can be achieved in many different ways.

All the relationships for the series lens arrangement, where the object O_1 is placed inside the focal point of lens F_1, are applicable for the case where the object is placed outside the focal point of the lens. This fact can be illustrated by assuming in Fig. 9.21 that the image I_2 is the object O_1. If the rays are then traced through the lens combination in a reverse direction from the ones shown, the object O_1 in the figure becomes the image I_2. The magnification in the two cases is different. In practice, the position of the object and the location of the viewing screen will depend on the object being studied and the characteristics of the lenses available for the series combination.

The previous discussion for lenses and mirrors was based on the assumption that all light rays from an object were confined to a region near the axis of the element and made small angles with respect to the axis. In systems where the rays are not confined to a region near the axis, all rays from an object point do not focus at exactly the same image point. This effect is known as a **spherical aberration**. Similarly, when an object point is located such that rays from the point make an appreciable angle with respect to the axis of the element, the image tends to become two mutually perpendicular line segments rather than a point image. This defect in optical elements is known as **astigmatism**. Other lens defects which must be considered in the design of optical systems include distortion, coma, curvature of field and chromatic aberration. The interested reader should consult an optics textbook for a more complete discussion of defects in images produced by mirrors and lenses.

9.6 OPTICAL DIFFRACTION AND INTERFERENCE

Consider light issued by a point source that is being used to illuminate a viewing screen. If an opaque plate with a large hole is placed between the source and the screen, the area illuminated on the screen can be defined by rays drawn from the source to the boundary of the hole and extended to the screen. As the hole in the plate is made smaller and smaller, the area illuminated on the screen decreases until some minimum area of illumination is achieved. Further decreases in the size of the hole cause the area of illumination to increase. This phenomenon, associated with an apparent bending of the light rays, is known as **diffraction**. Diffraction plays an important role in several methods of experimental stress analysis.

Diffraction of waves at a small aperture is explained by Huygens' principle—each point on a wavefront may be regarded as a new source of waves. For example, consider a series of plane wavefronts impinging on a rectangular slit as shown in Fig. 9.22.

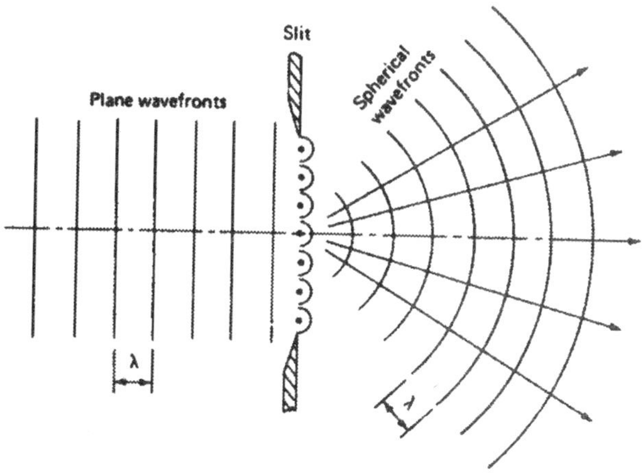

Figure 9.22 Diffraction of plane wave fronts at a narrow rectangular slit.

When the width of the slit is small (less than 0.10 mm), diffraction becomes important. In accordance with Huygens' principle, the light emerging from the slit can be considered to consist of a series of spherical wavefronts which expand and ultimately illuminate a viewing screen placed in the path of the propagating waves. Equation (9.7) can be modified to describe these spherical wavefronts as:

$$E = \frac{Kb}{r}\cos\frac{2\pi}{\lambda}(r - ct) \tag{9.7a}$$

The amplitude of the wave Kb/r depends on the strength of the source K, the width of the slit b and the radial position r of the expanding spherical wavefront.

If the viewing screen is placed at a large distance from the slit, or if a lens is used to focus parallel rays from the slit to a point P on the screen, the intensity distribution I as a function of the angle of inclination θ of the rays may be determined. For example, consider the emerging parallel rays from each spherical source shown in Fig. 9.23. The contribution to the total amount of light reaching P from an increment of width ds of the slit above and below the optical axis can be expressed as:

$$dE_p(s) = \frac{Kds}{r}\cos\frac{2\pi}{\lambda}(r - ct - s\sin\theta) \tag{a}$$

$$dE_p(-s) = \frac{Kds}{r}\cos\frac{2\pi}{\lambda}(r - ct + s\sin\theta) \qquad \text{(b)}$$

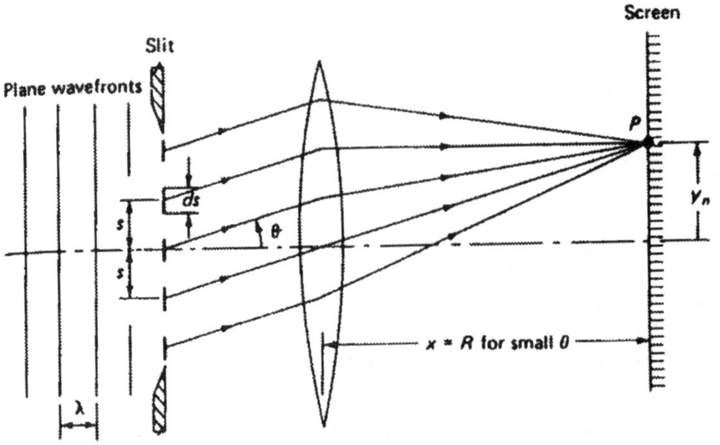

Figure 9.23 Light rays from a rectangular slit of width b which contribute to the light intensity at P.

The total amount of light reaching the point P can be determined by summing the contributions from each increment of width of the slit. This is most easily accomplished by adding Eqs. (a) and (b) and integrating over half the width of the slit. Thus:

$$\begin{aligned} E_p &= \int_0^{b/2} dE_p(s) + dE_p(-s) \\ &= \int_0^{b/2} \frac{2K}{r}\left[\cos\frac{2\pi}{\lambda}(r - ct)\cos\frac{2\pi}{\lambda}(s\sin\theta)\right]ds \qquad (9.36) \\ &= \frac{Kb}{r}\frac{\sin\beta}{\beta}\cos\frac{2\pi}{\lambda}(r - ct) \end{aligned}$$

where

$$\beta = \frac{\pi b \sin\theta}{\lambda} \qquad (9.36a)$$

Because the intensity at P is proportional to the square of the amplitude,

$$I = \left(\frac{Kb}{r}\right)^2 \frac{\sin^2\beta}{\beta^2} \qquad (9.37)$$

For maximum or minimum intensity $dI/d\beta = 0$, and

$$\frac{2\sin\beta}{\beta^3}(\beta\cos\beta - \sin\beta) = 0$$

Thus either

$$\beta = n\pi \qquad \text{or} \qquad \beta = \tan\beta \qquad \text{(c)}$$

The condition $\beta = n\pi$ yields zero intensity except at $\beta = 0$. The roots of $\beta = \tan\beta$ define positions of maximum intensity. The maximums are located at $\beta = 0, 1.430\pi, 2.459\pi, 3.471\pi, 4.477\pi, ...$ If the intensity at $\beta = 0$ ($\theta = 0$) is arbitrarily taken as 1, the intensities of successive maxima are 1/21, 1/61, 1/120, ...,

The dark bands which form in regions of zero intensity are called **diffraction** fringes. For a rectangular slit, the fringes will be straight parallel lines. For a circular pinhole, the fringes will be an array of concentric circles about the axis of the optical system. If the incident light is parallel and a lens is used to produce a focused image on a screen, the phenomenon is referred to as **Fraunhofer diffraction**. If the incident light comes from a point source at a finite distance from the slit or pinhole, the wavefronts are divergent and the phenomenon is known as **Fresnel diffraction**.

For a narrow slit, the fringe positions can be determined by substituting Eq. (c) into Eq. (9.36a) to give:

$$\frac{nb\sin\theta}{\lambda} = n\pi$$

For small values θ, sin θ = y_n/R, where R is the distance between the lens and the screen and y_n is the distance from the axis of the optical system to the diffraction fringe of order n. Thus:

$$y_n = \frac{n\lambda R}{b} \qquad n = 1, 2, 3, ... \qquad (9.38)$$

The spacing of diffraction fringes is used as the basis for several strain-measurement methods.

EXERCISES

9.1 Verify Eqs. (9.15) and (9.16).

9.2 The wavelength of light from a helium-neon laser is 632.8 mm. Determine:
  (a) The frequency of this light.
  (b) The wavelength of this light in a glass plate (n = 1.522).
  (c) The velocity of propagation in the glass plate.
  (d) The linear phase shift (in terms of wavelength in free space) after the light has passed through a 25-mm-thick glass plate.

9.3 Verify Eq. (9.19).

9.4 Show that the waves described by Eqs. (9.13) are the real part of the expression for E as given in Eq. (9.20).

9.5 Verify Eq. (9.21).

9.6 A plate of glass having an index of refraction of 1.57 with respect to air is to be used as a polarizer. Determine the polarizing angle and the angle of refraction of the transmitted ray.

9.7 Unpolarized light is directed onto a plane glass surface (n = 1.57) at an angle of incidence of 60°. Determine reflection coefficients associated with the parallel and perpendicular components of the reflected ray.

9.8 A light source is located 15 m below the surface of a body of water (n = 1.33) Determine the maximum distance (measured from a point directly above the source) at which the source will be visible from the air side of the air-water interface.

9.9 A ray of monochromatic light is directed at oblique incidence onto the surface of a glass plate. The ray emerges from the opposite side of the plate in a direction parallel to its initial direction but with a transverse displacement. Develop an expression for this transverse displacement in terms of the plate thickness h, the index of refraction of the glass n, and the angle of the incidence α of the light ray.

9.10 The radius of a concave spherical mirror is 500 mm. An object is located 1000 mm from the mirror. Determine the image location and the magnification. Show the results on a sketch similar to Fig. 9.18.

9.11 Solve Exercise 9.10 if the object is located 750 mm from the mirror.

9.12 Solve Exercise 9.10 if the object is located 100 mm from the mirror.

9.13 A concave mirror will be used to focus the image of an object onto a screen 1.50 m from the object. If a magnification of – 3 is required, what radius of curvature must the mirror have?

9.14 Determine the image location and the magnification for an object located 800 mm from a mirror if the mirror is (a) a plane mirror and (b) a convex mirror with a radius of curvature of 1000 mm.

9.15 A convex mirror has a radius of curvature of 2500 mm. Determine the magnification and the image location for an object located 1000 mm from the mirror. Show the results on a sketch similar to Fig. 9.18.

9.16 A thin convex lens has a focal length of 600 mm. An object is located 900 mm to the left of the lens. Determine the image location and the magnification. Show the results on a sketch similar to Fig. 9.19.

9.17 Solve Exercise 9.16 if the object is located 300 mm from the lens.

9.18 An object is placed 500 mm to the left of a thin convex lens having a focal length of 250 mm. A second lens having a focal length of 300 mm is placed 600 mm to the right of the first lens. Determine the location and magnification of the resulting image. Show the results on a sketch similar to Fig. 9.21.

9.19 An object is located 225 mm to the left of a thin convex lens having a focal length of 300 mm. A second lens having a focal length of 250 mm is located 150 mm to the right of the first lens. Determine the location and magnification of the resulting image. Show the results on a sketch similar to Fig. 9.21.

9.20 An object is located 125 mm to the left of a thin convex lens having a focal length of 250 mm. A spherical concave mirror having a radius of curvature of 500 mm is located 1000 mm to the right of the lens. Determine the location and magnification of the image resulting from this lens-mirror combination. Show the results on a sketch similar to Fig. 9.21.

9.21 Two thin lenses having focal lengths f_1 and f_2 are placed in contact with one another. Develop an expression for the equivalent focal length of this series combination.

9.22 Verify Eqs. (9.32) through (9.35).

9.23 Parallel light of wavelength 546.1 mm is directed at normal incidence onto a slit having a width of 0.050 mm. A lens having a focal length of 1000 mm is positioned behind the slit and is used to focus the light passing through the slit onto a screen. Determine the distance (a) from the center of the diffraction pattern to the first diffraction fringe (minimum intensity) and (b) between the third and fourth diffraction fringes.

9.24 Solve Exercise 9.23 if light from a helium-neon laser is used which has a wavelength of 632.8 nm.

9.25 In a single-slit diffraction pattern, the distance from the first minimum on the left to the first minimum on the right is 25 mm. The screen on which the pattern is displayed is 5 m from the slit. The wavelength of the light is 589.3 nm. Determine the slit width.

9.26 Verify Eq. (9.36).

BIBLIOGRAPHY

1. Born, M., and E. Wolf: Principles of Optics, 5th Edition, Pergamon Press, New York, 1975.
2. Jenkins, F. A., and H. E. White: Fundamentals of Optics, 4th Edition, McGraw-Hill Book Company, New York, 1976.
3. Garbuny, M.: Optical Physics, Academic Press, New York, 1965.
4. Serway, R. A., and J. S. Faughn: College Physics, 3rd Edition, Harcourt Brace College Publishers, New York, 1992.
5. Cloud, G. L.: Optical Methods of Engineering Analysis, Cambridge University Press, New York, 1998.

CHAPTER 10

PHOTOELASTICITY

10.1 INTRODUCTION

Of the many optical methods of stress or strain analysis that have been developed, photoelasticity is the easiest to use. The optical instruments needed to conduct a photoelastic analysis are simple and relatively inexpensive. A table isolated from vibrations, required for interferometric methods, is not needed for a polariscope. Ordinary incoherent light, even white light, may be employed in forming photoelastic images (fringe patterns). Models are relatively easy to construct from transparent birefringent polymers. The data provides a full field map of the differences in the principal stresses. Both two- and three-dimensional analyses are possible. Finally, photoelastic theory is relatively easy for instructors to teach and students to learn. For this reason, photoelasticity will be the first of the optical methods of experimental stress analysis to be described in this part of the book.

Many transparent non-crystalline materials that are optically isotropic when free of stress become optically anisotropic and display characteristics similar to crystals when they are stressed. These characteristics persist while loads on the material are maintained but disappear when the loads are removed. This behavior, known as **temporary double refraction**, was first observed by Sir David Brewster in 1816. The method of photoelasticity is based on this physical behavior of transparent non-crystalline materials.

The optical anisotropy (temporary double refraction) which develops in a material as a result of stress can be represented by an ellipsoid, known in this case as the index ellipsoid. The semi-axes of the **index ellipsoid** represent the principal indices of refraction of the material at the point, as shown in Fig. 10.1. Any radius of the ellipsoid represents a direction of light propagation through the point. A plane through the origin, which is perpendicular to the radius, intersects the ellipsoid as an ellipse. The semi-axes of the ellipse represent the indices of refraction associated with light waves having planes of vibration which contain the radius vector and an axis of the ellipse. For a material which is optically isotropic, the three principal indices of refraction are equal, the index ellipsoid becomes a sphere, and the index of refraction is the same for all directions of light propagation through the material.

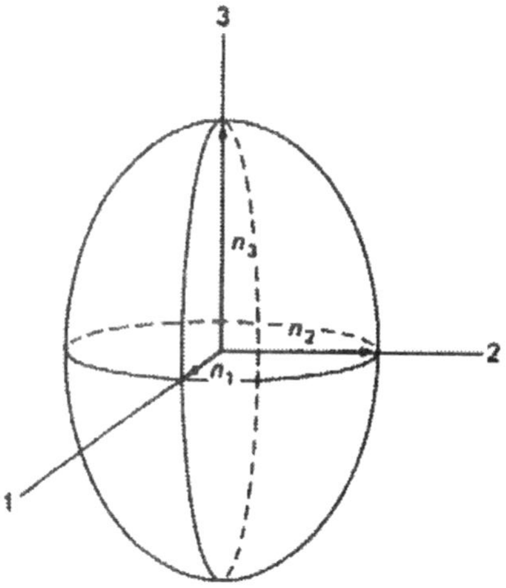

Figure 10.1 The index ellipsoid.

The similarities which exist between the stress ellipsoid for the state of stress at a point and the index ellipsoid for the optical properties of a material exhibiting temporary double refraction suggest the presence of a relationship between the two quantities. This relationship, which forms the basis for an experimental determination of stresses (or strains), is known as the **stress-optic law**.

10.2 THE POLARISCOPE

The polariscope is the optical instrument utilized in photoelasticity. The polariscope takes advantage of the properties of polarized light in its operation. For experimental stress-analysis work, two types are frequently employed, the plane polariscope and the circular polariscope[1]. The names follow from the type of polarized light used in their operation.

In practice, plane-polarized light is produced with an optical element known as a **plane** or **linear polarizer**. Production of circularly polarized light or the more general elliptically polarized light requires the use of a linear polarizer together with an optical element known as a **wave plate**. A brief discussion of linear polarizers, wave plates and their series combination follows.

10.2.1 Linear or Plane Polarizers

When a light wave strikes a plane polarizer, this optical element resolves the wave into two mutually perpendicular components, as shown in Fig. 10.2. The component parallel to the axis of polarization is transmitted while the component perpendicular to the axis of polarization is absorbed.

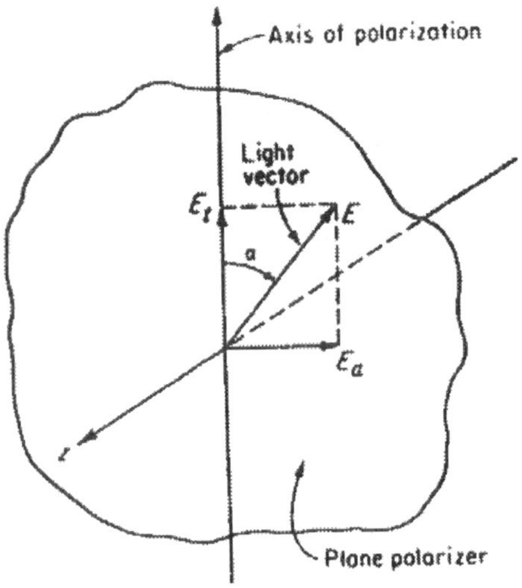

Figure 10.2 Absorbing and transmitting characteristics of a plane polarizer.

For a plane polarizer fixed at z_0, the equation for the light vector is:

$$\mathbf{E} = a \cos \frac{2\pi}{\lambda}(z_0 - ct) \qquad \text{(a)}$$

Because the initial phase of the wave is not important in photoelasticity, by using Eqs. (13.10) and (13.11), Eq. (a) can be reduced to:

$$\mathbf{E} = a \cos 2\pi ft = a \cos \omega t \qquad (10.1)$$

[1] A grey field polariscope, which utilizes circularly polarized light, has been introduced in recent years. This new version of the classical circular polariscope is described in Section 10.8.

The quantity $\omega = 2\pi f$ is the circular frequency of the wave. The absorbed and transmitted components of the light vector are:

$$E_a = a \cos \omega t \sin \alpha$$
$$E_t = a \cos \omega t \cos \alpha \qquad \text{(b)}$$

where α is the angle between the axis of polarization and the incident light vector as shown in Fig. 10.2.

Polaroid filters are almost always used for producing polarized light in polariscopes. They have the advantage of providing a large field of very well polarized light at a relatively low cost. Most modern polariscopes containing linear polarizers employ Polaroid H sheet, a transparent material with strained and oriented molecules. In the manufacturing of Polaroid films, a thin sheet of polyvinyl alcohol is heated, stretched, and immediately bonded to a supporting sheet of cellulose acetate butyrate. The polyvinyl face of the assembly is then stained by a liquid rich in iodine. The amount of iodine diffused into the sheet determines its quality. Filters are available in a number of grades, according to the amount of light they transmit. Because the quality of a polarizer is judged by its transmission ratio, grade HN-22, (with the best transmission ratio) is recommended for photoelastic purposes.

10.2.2 Wave Plates

A wave plate is an optical element which has the ability to resolve a light vector into two orthogonal components and to transmit the components with different velocities. Such a material is called **doubly refracting** or **birefringent**. The doubly refracting plate illustrated in Fig. 10.3 has two principal axes labeled 1 and 2. The transmission of light along axis 1 proceeds at velocity c_1 and along axis 2 at velocity c_2. Because c_1 is greater than c_2, axis 1 is called the **fast axis** and axis 2 the **slow axis**.

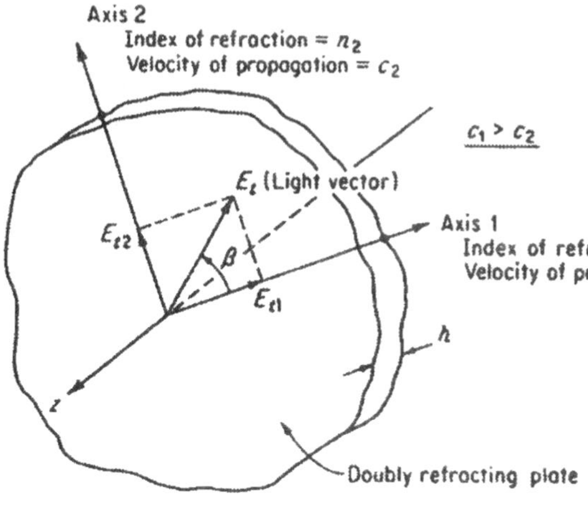

Figure 10.3 A plane polarized light vector entering a doubly refracting plate.

If this doubly refracting plate is placed in a field of plane-polarized light so that the light vector $\mathbf{E_t}$ makes an angle β with axis 1 (the fast axis), then on entering the plate the light vector is resolved into two components E_{t1} and E_{t2} along axes 1 and 2 with magnitudes given by:

$$E_{t1} = \mathbf{E_t} \cos \beta = a \cos \alpha \cos \omega t \cos \beta = k \cos \omega t \cos \beta$$
$$E_{t2} = \mathbf{E_t} \sin \beta = a \cos \alpha \cos \omega t \sin \beta = k \cos \omega t \sin \beta \qquad \text{(a)}$$

where $k = a \cos \alpha$. The light components E_{t1} and E_{t2} travel through the plate with different velocities c_1 and c_2; therefore, the two components emerge from the plate at different times. In other words, one component is retarded in time relative to the other component. This retardation produces a relative phase shift between the two components. From Eq. (9.22), the linear phase shifts for components E_{t1} and E_{t2} with respect to a wave in air can be expressed as:

$$\delta_1 = h(n_1 - n)$$
$$\delta_2 = h(n_2 - n) \quad \text{(b)}$$

where n is the index of refraction of air.

The relative linear phase shift is then computed simply as:

$$\delta = \delta_2 - \delta_1 = h(n_2 - n_1) \quad (10.2)$$

The relative angular phase shift Δ between the two components as they emerge from the plate is given by:

$$\Delta = \frac{2\pi}{\lambda}\delta = \frac{2\pi h}{\lambda}(n_2 - n_1) \quad (10.3)$$

The relative angular phase shift Δ produced by a doubly refracting plate is dependent upon its thickness h, the wavelength of the light λ and the properties of the plate as described by $(n_2 - n_1)$. When the doubly refracting plate is designed to give $\Delta = \pi/2$, it is called a **quarter-wave plate**. Doubly refracting plates designed to give angular retardations of π and 2π are known as **half-** and **full-wave plates**, respectively. Upon emergence from a wave plate exhibiting retardation Δ, the two components of light are described by:

$$E'_{t1} = k \cos\beta \cos \omega t$$
$$E'_{t2} = k \sin\beta \cos (\omega t - \Delta) \quad (10.4)$$

With this representation, only the relative phase shift between components has been considered. The additional phase shift suffered by both components, as a result of passage through the wave-plate material (as opposed to free space), has been neglected because it has no effect on the phenomenon being considered. The amplitude of the light vector which is produced by these two components can be expressed as:

$$E'_t = \sqrt{(E'_{t1})^2 + (E'_{t2})^2} = k\sqrt{\cos^2\beta\cos^2\omega t + \sin^2\beta\cos^2(\omega t - \Delta)} \quad (10.5)$$

The angle that the emerging light vector makes with axis 1 is given by:

$$\tan\gamma = \frac{E'_{t2}}{E'_{t1}} = \frac{\cos(\omega t - \Delta)}{\cos\omega t}\tan\beta \quad (10.6)$$

It is clear that both the amplitude and the rotation of the emerging light vector can be controlled by the wave plate. Controlling factors are the relative phase difference Δ and the orientation angle β. Various combinations of Δ and β and their influence on the type of polarized light produced will be discussed in Section 10.2.3.

Wave plates employed in a photoelastic polariscope usually consist of a single plate of quartz or calcite cut parallel to the optic axis, or a sheet of oriented polyvinyl alcohol. In recent years, as the design of the modern polariscope has tended toward a relatively large diameter field, most wave plates employed have been fabricated from oriented sheets of polyvinyl alcohol. These wave plates, called **retarders**, are manufactured by warming and unidirectionally stretching the sheet. Because the oriented polyvinyl alcohol sheet is only about 20 μm thick (for a quarter-wave plate), the commercial retarders are usually laminated between two sheets of cellulose acetate butyrate.

10.2.3 Conditioning Light with a Linear Polarizer and Wave Plate in Series

The magnitude and direction of the light vector emerging from a series combination of a linear polarizer and a wave plate is always polarized; however, the type of polarization may be plane, circular, or elliptical. The factors which control the type of polarized light produced by this combination are the relative phase difference Δ imposed by the wave plate and the orientation angle β. Three well-defined cases exist.

Case 1: Plane Polarized Light

If the angle β is set equal to zero and the relative retardation Δ is not restricted in any sense, the magnitude and direction of the emerging light vector are given by Eqs. (10.5) and (10.6) as:

$$E'_t = k \cos \omega t \qquad \text{and} \qquad \gamma = 0 \tag{a}$$

Because $\gamma = 0$, the light vector is not rotated as it passes through the wave plate and the light remains plane-polarized. The wave plate in this instance does not influence the light except to produce retardation with respect to a wave in free space which depends on the plate thickness and the index of refraction n_1. Similar results are obtained by letting $\beta = \pi/2$ with:

$$E'_t = k \cos(\omega t - \Delta) \qquad \text{and} \qquad \gamma = \pi/2 \tag{b}$$

Case 2: Circularly Polarized Light

If a wave plate is selected so that $\Delta = \pi/2$, that is, a quarter-wave plate, and $\beta = \pi/4$, the magnitude and direction of the light vector as it emerges from the plate are given by Eqs. (10.5) and (10.6) as:

$$E'_t = \frac{\sqrt{2}}{2} k \sqrt{\cos^2 \omega t + \sin^2 \omega t} = \frac{\sqrt{2}}{2} k \qquad \text{and} \qquad \gamma = \omega t \tag{c}$$

The light vector described here has a constant magnitude and the tip of the light vector traces out a circle as it rotates. The vector rotates with a constant angular velocity in a counterclockwise direction when viewed in the positive z direction. Such light is known as **left circularly polarized light**. Right circularly polarized light could be obtained by setting $\beta = 3\pi/4$. The light vector would then rotate with a constant angular velocity in the clockwise direction.

Case 3: Elliptically Polarized Light

If a quarter-wave plate ($\Delta = \pi/2$) is selected and $\beta \neq n\pi/4$ (n = 0, 1, 2, 3, ...) then, by Eqs. (10.5) and (10.6), the magnitude and direction of the emerging light vector are:

$$E'_t = k\sqrt{\cos^2 \beta \cos^2 \omega t + \sin^2 \beta \sin^2 \omega t}$$

$$\tan \gamma = \tan \beta \tan \omega t \tag{d}$$

The amplitude of the light vector in this case varies with angular position in such a way that the tip of the light vector traces out an ellipse as it rotates. The shape and orientation of the ellipse and the direction of rotation of the light vector depend on β.

Consider the significance of Eq. (10.3) in the production of circularly polarized light, and note that a quarter-wave plate is required; therefore, the phase difference $\Delta = \pi/2$. It is clear that the thickness h can be determined to give $\Delta = \pi/2$ after the plate material, $(n_2 - n_1)$ is selected, and the wavelength λ of the light is fixed. However, a quarter-wave plate suitable for one wavelength of monochromatic light, i.e., a constant wavelength, will not be suitable for a different wavelength. Also, quarter-wave plates cannot be designed for white light because white light is comprised of different wavelengths.

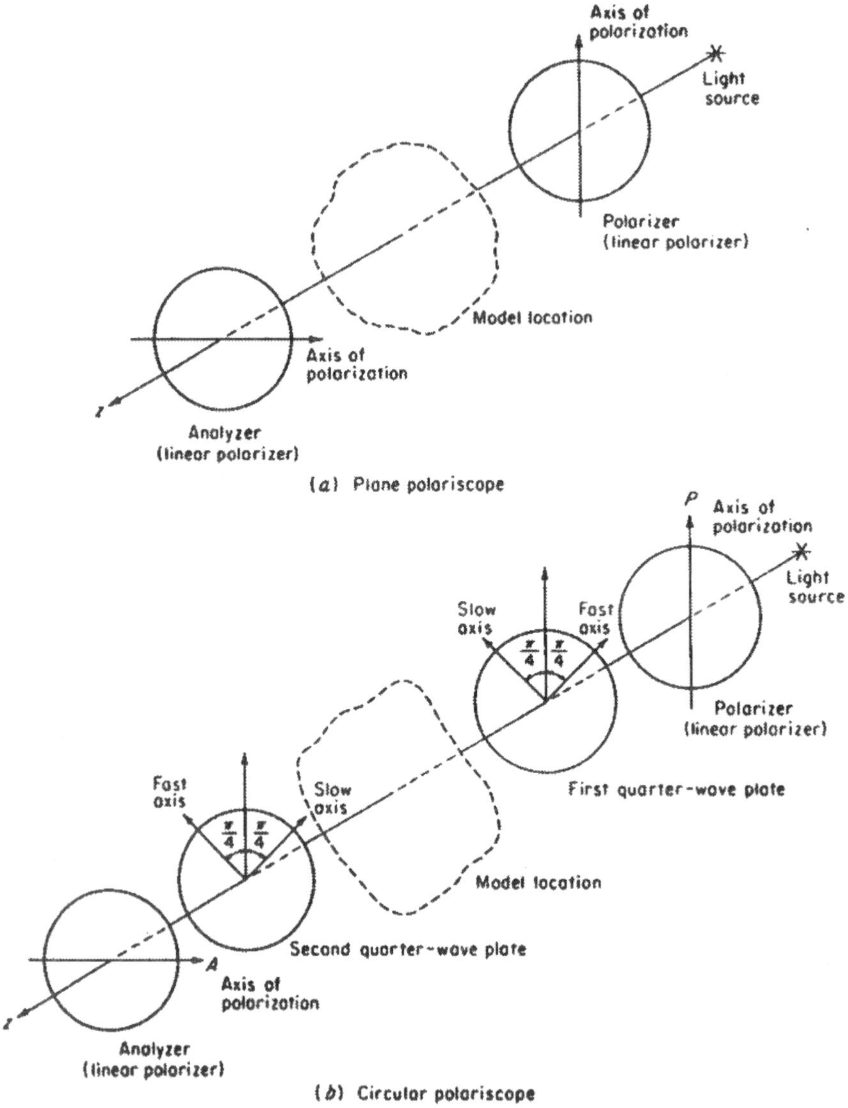

Figure 10.4
Arrangement of the optical elements in plane and circular polariscopes.

10.2.4 Arrangement of the Optical Elements in a Polariscope

The Plane Polariscope

The plane polariscope is the simplest optical system used in photoelasticity; it consists of two linear polarizers and a light source arranged as illustrated in Fig. 10.4a. The linear polarizer nearest the light source is called the polarizer, while the second linear polarizer is known as the analyzer. In the plane polariscope, the two axes of polarization are always crossed—no light is transmitted through the analyzer— and this optical system produces a dark field. In operation, a photoelastic model is inserted between the two crossed elements and viewed through the analyzer. The behavior of the photoelastic model in a plane polariscope will be covered in Section 10.5.

The Circular Polariscope

As the name implies, the circular polariscope employs circularly polarized light. The photoelastic apparatus contains four optical elements and a light source as illustrated in Fig. 10.4b. The first element following the light source is a linear polarizer, which converts ordinary light into plane-polarized light. The second element is a quarter-wave plate set at an angle $\beta = \pi/4$ to the plane of polarization. This quarter-wave plate converts the plane-polarized light into circularly polarized light. The second quarter-wave plate is set with its fast axis parallel to the slow axis of the first quarter-wave plate. The purpose of this element is to convert the circularly polarized light into plane-polarized light vibrating in the vertical plane. The last element is the analyzer, with its axis of polarization horizontal, and its purpose is to extinguish the light. This series of optical elements constitutes the standard arrangement for a circular polariscope, and it produces a dark field. Four arrangements of the optical elements in the circular polariscope are possible, depending upon whether the polarizers and quarter-wave plates are crossed or parallel. These optical arrangements are described in Table 10.1.

Table 10.1
Four arrangements of the optical elements in a circular polariscope

Arrangement	Quarter Wave Plates	Polarizer and Analyzer	Field
A	Crossed	Crossed	Dark
B	Crossed	Parallel	Light
C	Parallel	Crossed	Dark
D	Parallel	Parallel	Light

Arrangements A and B are normally recommended for light- and dark-field use of the polariscope because the error introduced by imperfect quarter-wave plates, (i.e., both quarter-wave plates differ from $\pi/2$ by a small amount) is minimized. Because quarter-wave plates are often of poor quality, this fact is important in selecting the optical arrangement.

10.2.5 Construction Details of Diffused-Light Polariscopes

The arrangement of the optical elements discussed previously is not sufficiently complete or detailed for the visualization of a working polariscope. The degree of complexity of a polariscope varies widely with the investigator and ranges from highly complex lens systems with servomotor drives on the four optical elements to very simple arrangements with no lenses and no provision for rotation of any element.

The diffused-light polariscope described here is one of the simplest and least expensive polariscopes available; yet it can be employed to produce very high quality photoelastic results. This polariscope requires only one lens; however, its field can be made very large because its diameter is dependent only upon the size of the available linear polarizers and quarter- wave plates. Diffused-light polariscopes with field diameters up to 450 mm can readily be constructed. A schematic illustration of the construction details of a diffused-light polariscope is shown in Fig. 10.5.

Figure 10.5 Construction details of a diffused-light polariscope. (1) Light house; (2) Monochromatic light (low pressure sodium bulb); (3) White light, 300 W tungsten-filament lamps; (4) diffusing plate; (5) First polarizer; (6) First quarter- wave plate; (7) Loading frame; (8) Second quarter-wave plate; (9) Second polarizer; (10) Long focal length lens; (11) Digital camera.

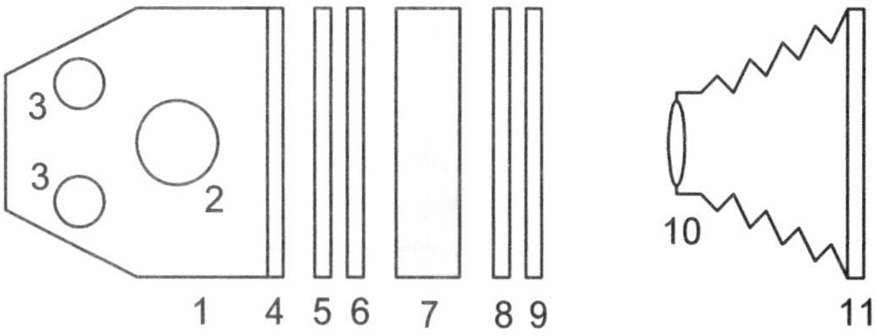

10.3 THE STRESS-OPTIC LAW

The theory which relates changes in the indices of refraction of a material exhibiting temporary double refraction to the state of stress in the material is due to Maxwell, who reported the phenomenon in 1853. Maxwell noted that the changes in the indices of refraction were linearly proportional to the loads and thus to stresses or strains for a linearly elastic material. The relationships can be expressed in equation form as:

$$n_1 - n_0 = c_1 \sigma_1 + c_2 (\sigma_2 + \sigma_3)$$

$$n_2 - n_0 = c_1 \sigma_2 + c_2 (\sigma_3 + \sigma_1) \qquad (10.7)$$

$$n_3 - n_0 = c_1 \sigma_3 + c_2 (\sigma_1 + \sigma_2)$$

where σ_1, σ_2 and σ_3 are the principal stresses at point.

n_0 is the index of refraction of material in unstressed state.

n_1, n_2 and n_3 are the principal indices of refraction which coincide with the principal stress directions.

c_1 and c_2 are constants known as **stress-optic coefficients.**

Equations (10.7) are the fundamental relationships between stress and optical effect and are known as the **stress-optic law**. These equations indicate that the complete state of stress at a point can be determined by measuring the three principal indices of refraction and establishing the directions of the three principal optical axes. Because the measurements are extremely difficult to make in a three-dimensional model, most applications have been limited to cases of plane stress ($\sigma_3 = 0$). For plane-stress states, Eqs. (10.7) reduce to:

$$n_1 - n_0 = c_1 \sigma_1 + c_2 \sigma_2$$

$$\qquad (10.8)$$

$$n_2 - n_0 = c_1 \sigma_2 + c_2 \sigma_1$$

Three-dimensional stress states, where $\sigma_3 \neq 0$, can be studied with photoelasticity using the stress freezing process that will be described later in Chapter 11.

10.4 THE STRESS-OPTIC LAW—IN RELATIVE RETARDATION

Equations (10.7) describe the changes in index of refraction due to applied stress experienced by a material exhibiting temporary double refraction. The method of photoelasticity makes use of relative changes in index of refraction which can be written by eliminating n_0 from Eqs. (10.7) as:

$$n_2 - n_1 = (c_2 - c_1)(\sigma_1 - \sigma_2) = c(\sigma_1 - \sigma_2)$$

$$n_3 - n_2 = (c_2 - c_1)(\sigma_2 - \sigma_3) = c(\sigma_2 - \sigma_3) \qquad (10.9)$$

$$n_1 - n_3 = (c_2 - c_1)(\sigma_3 - \sigma_1) = c(\sigma_3 - \sigma_1)$$

where $c = c_1 - c_2$ is the relative stress-optic coefficient first expressed in terms of brewsters (1 Brewster = 10^{-13} cm^2/dyne = 10^{-12} m^2/N = 6.895 x 10^{-9} 1/psi). Photoelastic materials are considered to exhibit positive birefringence when the velocity of propagation of the light wave associated with the principal stress σ_1 is greater than the velocity of the wave associated with the principal stress σ_2. Because the principal stresses are ordered such that $\sigma_1 \geq \sigma_2 \geq \sigma_3$, the principal indices of refraction of a positive doubly refracting material can be ordered such that $n_3 \geq n_2 \geq n_1$. The form of Eqs. (10.9) has been selected to make the relative stress-optic coefficient c a positive constant.

Because a stressed photoelastic model behaves like a temporary wave plate, Eq. (10.3) can be used to relate the relative angular phase shift Δ (or relative retardation) to changes in the indices of refraction in the material resulting from the stresses. For example, consider a slice of material (thickness h) oriented perpendicular to one of the principal-stress directions at the point of interest in the model. If a beam of plane-polarized light is passed through the slice at normal incidence, the relative retardation Δ accumulated along each of the principal-stress directions can be obtained by substituting Eq. (10.3) into each of Eqs. (10.9) to yield:

$$\Delta_{12} = \frac{2\pi \,\mathrm{hc}}{\lambda}(\sigma_1 - \sigma_2)$$

$$\Delta_{23} = \frac{2\pi \,\mathrm{hc}}{\lambda}(\sigma_2 - \sigma_3) \tag{10.10}$$

$$\Delta_{31} = \frac{2\pi \,\mathrm{hc}}{\lambda}(\sigma_3 - \sigma_1)$$

where Δ_{12} is the magnitude of the relative angular phase shift (relative retardation) developed between components of a light beam propagating in the σ_3 direction. Similar interpretations can be applied to the retardations Δ_{23} and Δ_{31}.

The relative retardation Δ is linearly proportional to the difference between the two principal stresses having directions perpendicular to the path of propagation of the light beam. The third principal stress, having a direction parallel to the path of propagation of the light beam, has no effect on the relative retardation. Also, the relative retardation Δ is linearly proportional to the model or slice thickness h and inversely proportional to the wavelength λ of the light being used.

The relative stress-optic coefficient c is usually assumed to be a material constant that is independent of the wavelength of the light being used. A study by Vandaele-Dossche and van Geen [10] has shown, however, that this coefficient may depend on wavelength in some cases as the model material passes from the elastic to the plastic state. The dependence of the relative stress-optic coefficient c on the wavelength of the light being used is referred to as **photoelastic dispersion** or **dispersion of birefringence**.

From an analysis of the general three-dimensional state of stress at a point, and by considering the change in index of refraction with direction of light propagation in the stressed material, it can be shown that Eqs. (10.10) apply not only for principal stresses but also for secondary principal stresses σ'_1 and σ'_2. Thus:

$$\Delta' = \frac{2\pi \,\mathrm{hc}}{\lambda}(\sigma'_1 - \sigma'_2) \tag{10.11}$$

The secondary principal stresses ($\sigma'_1 \geq \sigma'_2$) at the point of interest lie in the plane whose normal vector is coincident with the path of the light beam. The stress-optic law in terms of secondary principal stresses is widely used in application of three-dimensional photoelasticity.

For two-dimensional plane-stress bodies where $\sigma_3 = 0$, the stress-optic law for light at normal incidence to the plane of the model can be written without the subscripts on the retardation simply as:

$$\Delta = \frac{2\pi \,\mathrm{hc}}{\lambda}(\sigma_1 - \sigma_2) \tag{10.12}$$

In Eq. (10.12) it is understood that σ_1 and σ_2 are in-plane principal stresses and that σ_1 is greater than σ_2, but not greater than $\sigma_3 = 0$ if both in-plane stresses are compressive. The form of Eq. (10.12) is usually simplified to:

$$\sigma_1 - \sigma_2 = \frac{Nf_\sigma}{h} \tag{10.13}$$

$$N = \frac{\Delta}{2\pi} \tag{10.14}$$

Where N is the relative retardation in terms of cycles of retardation, and counted as the fringe order. The term f_σ is:

$$f_\sigma = \frac{\lambda}{c} \tag{10.15}$$

The material fringe value f_σ is a property of the model material for a given wavelength λ and h is the model thickness.

It is immediately apparent from Eq. (10.13) that the stress difference $\sigma_1 - \sigma_2$ in a two-dimensional model can be determined if the relative retardation N can be measured and if the material fringe value f_σ can be established by means of calibration. Actually, the function of the polariscope is to determine the value of N at each point in the model.

If a photoelastic model exhibits a perfectly linear elastic behavior, the difference in the principal strains $\varepsilon_1 - \varepsilon_2$ can also be measured by establishing the fringe order N. The stress-strain relations for a two-dimensional or plane state of stress are given by Eqs. (2.19) as:

$$\varepsilon_1 = \frac{1}{E}(\sigma_1 - \nu\sigma_2)$$

$$\varepsilon_2 = \frac{1}{E}(\sigma_2 - \nu\sigma_1)$$

or;

$$\varepsilon_1 - \varepsilon_2 = \frac{1+\nu}{E}(\sigma_1 - \sigma_2)$$

Substituting these results into Eq. (10.13) yields:

$$\frac{Nf_\sigma}{h} = \frac{E}{1+\nu}(\varepsilon_1 - \varepsilon_2) \tag{10.16}$$

This is rewritten as:

$$\frac{Nf_\varepsilon}{h} = (\varepsilon_1 - \varepsilon_2) \tag{10.17}$$

It is clear that:

$$f_\varepsilon = \frac{1+\nu}{E} f_\sigma \tag{10.18}$$

where f_ε is the material fringe value in terms of strain.

For a perfectly linear elastic photoelastic model, the determination of N is sufficient to establish both $\sigma_1 - \sigma_2$ and $\varepsilon_1 - \varepsilon_2$ if any three of the material properties E, ν, f_σ or f_ε are known. However, many photoelastic materials exhibit viscoelastic properties, and Eqs. (2.19) and (10.18) are not valid. The viscoelastic behavior of photoelastic materials is discussed later in Chapter 11.

10.5 EFFECTS OF A STRESSED MODEL IN A PLANE POLARISCOPE

It is clear that the principal-stress difference $\sigma_1 - \sigma_2$ can be determined in a two-dimensional model if the fringe order N is measured at each point in the model. Moreover, the optical axes of the model (a temporary wave plate) coincide with the principal-stress directions. These two facts can be effectively utilized to determine $\sigma_1 - \sigma_2$ if a method to measure the optical properties of a stressed model has been established.

Consider, first, the case of a model subjected to plane-stress inserted into the field of a plane polariscope with its normal coincident with the axis of the polariscope, as illustrated in Fig. 10.6. Note that the principal-stress direction at the point under consideration in the model makes an angle α with the axis of polarization of the polarizer.

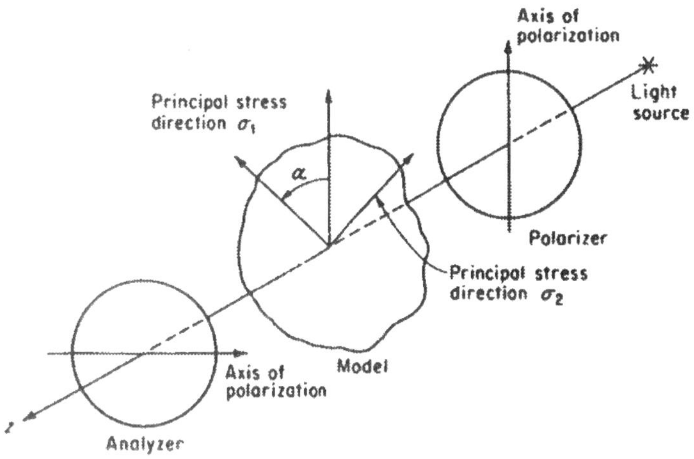

Figure 10.6 A stressed photoelastic model in a plane polariscope.

It is clear that a plane polarizer resolves an incident light wave into components which vibrate parallel and perpendicular to the axis of the polarizer. The component parallel to the axis is transmitted and the component perpendicular to the axis is internally absorbed. Because the initial phase of the wave is not important in the development which follows, the plane-polarized light beam emerging from the polarizer can be represented by the simple expression:

$$E_{py} = k \cos \omega t \tag{a}$$

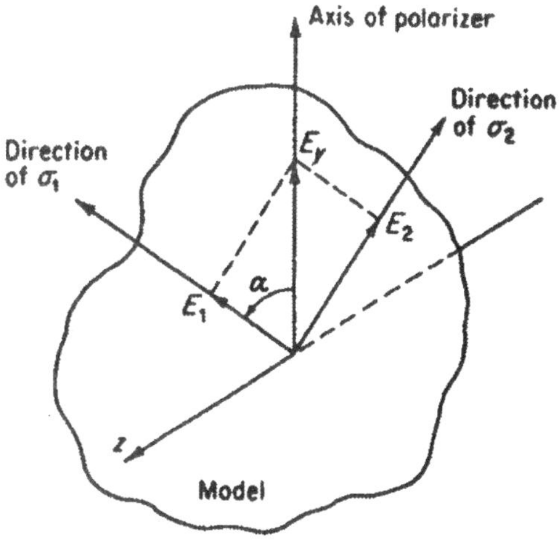

Figure 10.7 Resolution of the light vector as it enters a stressed model in a plane polariscope.

After leaving the polarizer, this plane-polarized light wave enters the model, as shown in Fig. 10.7. Because the stressed model exhibits the optical properties of a wave plate, the incident light vector is resolved into two components E_1 and E_2 with vibrations parallel to the principal stress directions at the entry point. Thus:

$$E_1 = k \cos \alpha \cos \omega t$$
$$E_2 = k \sin \alpha \cos \omega t \qquad \text{(b)}$$

Because the two components propagate through the model with different velocities ($c > v_1 > v_2$), they develop phase shifts Δ_1 and Δ_2 with respect to a wave in air. The waves upon emerging from the model can be expressed as:

$$E'_1 = k \cos \alpha \cos (\omega t - \Delta_1)$$
$$E'_2 = k \sin \alpha \cos (\omega t - \Delta_2) \qquad \text{(c)}$$

where $\Delta_1 = \frac{2\pi h}{\lambda}(n_1 - 1)$ and $\Delta_2 = \frac{2\pi h}{\lambda}(n_2 - 1)$

After leaving the model, the two components continue to propagate without further change and enter the analyzer in the manner shown in Fig. 10.8. The light components E'_1 and E'_2 are resolved when they enter the analyzer into horizontal components E''_1 and E''_2 into vertical components. Because the vertical components are internally absorbed in the analyzer, they have not been shown in Fig. 10.8.

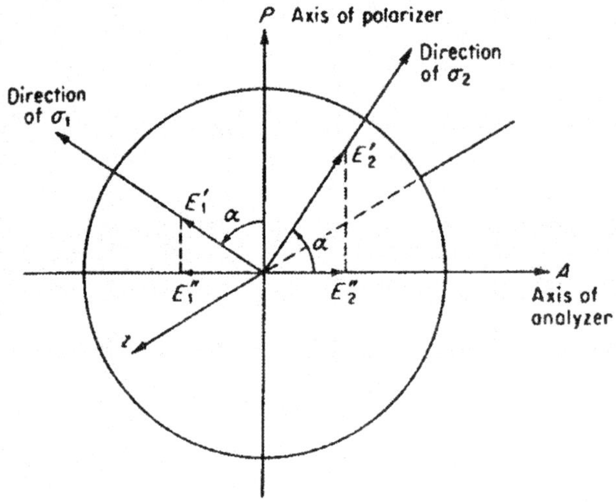

Figure 10.8 Components of the light vectors which are transmitted through the analyzer of a plane polariscope.

The horizontal components transmitted by the analyzer combine to produce an emerging light vector E_{ax}, which is given by:

$$E_{ax} = E''_2 - E''_1 = E'_2 \cos \alpha - E'_1 \sin \alpha \qquad \text{(d)}$$

Substituting Eqs. (c) into Eq. (d) yields:

$$E_{ax} = k \sin \alpha \cos \alpha \left[\cos (\omega t - \Delta_2) - \cos (\omega t - \Delta_1)\right]$$

$$E_{ax} = k \sin 2\alpha \, \sin \frac{\Delta_2 - \Delta_1}{2} \, \sin\left(\omega t - \frac{\Delta_2 + \Delta_1}{2}\right) \qquad (10.19)$$

It is interesting to note in Eq. (10.19) that the average angular phase shift $(\Delta_2 + \Delta_1)/2$ affects the phase of the light wave emerging from the analyzer but not the amplitude (coefficient of the time-dependent term). It has no influence on the intensity (intensity is proportional to the square of the amplitude) of the light emerging from the analyzer. The relative retardation $\Delta_2 - \Delta_1$, appears in the amplitude of the wave; therefore, it is one of the two important parameters that control the intensity of light emerging from the analyzer. Because the average angular phase shift $(\Delta_2 + \Delta_1)/2$ had no effect on the intensity, it does not contribute to the optical patterns observed in a photoelastic model. In future developments of photoelasticity theory only relative retardations will be considered.

Because the intensity of light is proportional to the square of the amplitude of the light wave, the intensity of the light emerging from the analyzer of a plane polariscope is determined from Eq. (10.19) as:

$$I = K \sin^2 2\alpha \sin^2 \frac{\Delta}{2} \tag{10.20}$$

where

$$\Delta = \Delta_2 - \Delta_1 = \frac{2\pi h}{\lambda}(n_2 - n_1) = \frac{2\pi hc}{\lambda}(\sigma_1 - \sigma_2)$$

Examination of Eq. (10.20) indicates that extinction (I = 0) occurs either when $\sin^2 2\alpha = 0$ or when $\sin^2 (\Delta/2) = 0$. Clearly, one condition for extinction is related to the principal-stress directions and the other is related to the principal-stress difference.

10.5.1 Effect of Principal-Stress Directions

When $2\alpha = n\pi$, where n = 0, 1, 2, ..., $\sin^2 2\alpha = 0$ and extinction occurs. This relation indicates that, when one of the principal-stress directions coincides with the axis of the polarizer ($\alpha = 0$, $\pi/2$, or any exact multiple of $\pi/2$), the intensity of the light is zero. Because the analysis of the optical effects produced by a stressed model in a plane polariscope was conducted for an arbitrary point in the model, the analysis is valid for all points of the model. When the entire model is viewed in the polariscope, a fringe pattern is observed; the fringes are loci of points where the principal-stress directions (either σ_1 or σ_2) coincide with the axis of the polarizer. The fringe pattern produced by the $\sin^2 2\alpha$ term in Eq.(10.20) is the isoclinic fringe pattern. Isoclinic fringe patterns are used to determine the principal-stress directions at all points of a photoelastic model. Because isoclinics represent a very important segment of the data obtained from a photoelastic model, the topic of isoclinic-fringe-pattern interpretation will be treated in more detail later.

10.5.2 Effect of Principal-Stress Difference

When $\Delta/2 = n\pi$, where n = 0 1, 2, 3, ..., $\sin^2 (\Delta/2) = 0$ and extinction occurs. When the principal-stress difference is either zero (n = 0) or sufficient to produce an integral number of wavelengths of retardation (n = 1, 2, 3, ...), the intensity of light emerging from the analyzer is zero. When a model is viewed in the polariscope, this condition for extinction yields a second fringe pattern where the fringes are loci of points exhibiting the same order of extinction (n = 0, 1, 2, 3, ...). The fringe pattern produced by the $\sin^2 (\Delta/2)$ term in Eq. (10.20) is the isochromatic fringe pattern. The nature of the optical effect producing the isochromatic fringe pattern requires some additional discussion.

Recall from Eq. (10.12.) that the relative retardation Δ may be expressed as:

$$\Delta = \frac{2\pi hc}{\lambda}(\sigma_1 - \sigma_2)$$

and

$$n = \frac{\Delta}{2\pi} = \frac{hc}{\lambda}(\sigma_1 - \sigma_2) \tag{e}$$

Examination of Eq. (e) indicates that the order of extinction n depends on both the principal-stress difference $\sigma_1 - \sigma_2$ and the wavelength λ of the light. When a model is viewed in monochromatic light, the isochromatic fringe pattern appears as a series of dark bands because the intensity of light is zero when n = 0, 1, 2, 3, ... However, when a model is viewed with white light (all wavelengths of the visible spectrum present), the isochromatic fringe pattern appears as a series of colored bands. The intensity of light is zero, and a black fringe appears only when the principal-stress difference is zero and a zero order of extinction occurs for all wavelengths of light. No other region of zero intensity is possible because the value of $\sigma_1 - \sigma_2$ required to produce a given order of extinction is different for each of the wavelengths. For non-zero values of $\sigma_1 - \sigma_2$ only one wavelength can be extinguished from the white light. The colored bands form in regions where $\sigma_1 - \sigma_2$ produces extinction of a particular wavelength of the white light. For example, when $\sigma_1 - \sigma_2$ produces extinction of the green wavelengths, the complementary color, red, appears as the isochromatic fringe. At the higher levels of principal-stress difference, where several wavelengths of light can be extinguished simultaneously, e. g., second order red and third order violet, the isochromatic fringes become pale and very difficult to identify; and should not be used for analysis.

With monochromatic light, the individual fringes in an isochromatic fringe pattern remain sharp and clear to very high orders of extinction. Because the wavelength of the light is fixed, Eq. (e) can be written in terms of the material fringe value f_σ and the isochromatic fringe order N as:

$$n = N = \frac{h}{f_\sigma}(\sigma_1 - \sigma_2) \tag{f}$$

The number of fringes appearing in an isochromatic fringe pattern is controlled by the magnitude of the principal-stress difference $\sigma_1 - \sigma_2$, the thickness h of the model and the sensitivity of the photoelastic material, as denoted by the material fringe value f_σ.

In general, the principal-stress difference $\sigma_1 - \sigma_2$ and the principal stress directions vary from point to point in a photoelastic model. As a result, the isoclinic fringe pattern and the isochromatic fringe pattern are superimposed, as shown in Fig. 10.9. Separation of the patterns requires special techniques which will be discussed later.

Figure 10.9 Superimposed isochromatic and isoclinic fringe patterns for a ring loaded in diametric compression.

Theoretically, the isoclinic and isochromatic fringes should be lines of zero width; however, the photograph in Fig. 10.9 shows the fringes as bands with considerable width. The width of the fringes is due to the characteristics of the eye and the recording media and not to inaccuracies in the analytical development. If the intensity of light emerging from the analyzer is measured with a digital camera, a minimum intensity is recorded at some point near the center of the fringe that coincides with the extinction line.

10.5.3 Frequency Response of a Polariscope

The frequency f for light in the visible spectrum is of the order of 10^{14} Hz (see Fig. 9.7). As a result of this very high frequency, neither the eye nor any type of existing high-speed photo detectors can monitor the periodic extinction associated with the ωt term in the expression for the light wave. The light wave represents the carrier of the information (signal) in a polariscope. Because the carrier frequency should be about 10 times larger than the signal frequency for fidelity in dynamic recording, the frequency response of a polariscope is of the order of 10^{14} Hz.

10.6 EFFECTS OF A STRESSED MODEL IN A CIRCULAR POLARISCOPE (DARK FIELD, ARRANGEMENT A)

When a stressed photoelastic model is placed in the field of a circular polariscope with its normal coincident with the z axis, the optical effects differ significantly from those obtained in a plane polariscope. The use of a circular polariscope eliminates the isoclinic fringe pattern while it maintains the isochromatic fringe pattern. To illustrate this effect, consider the stressed model in the circular polariscope (arrangement A) shown in Fig. 10.10.

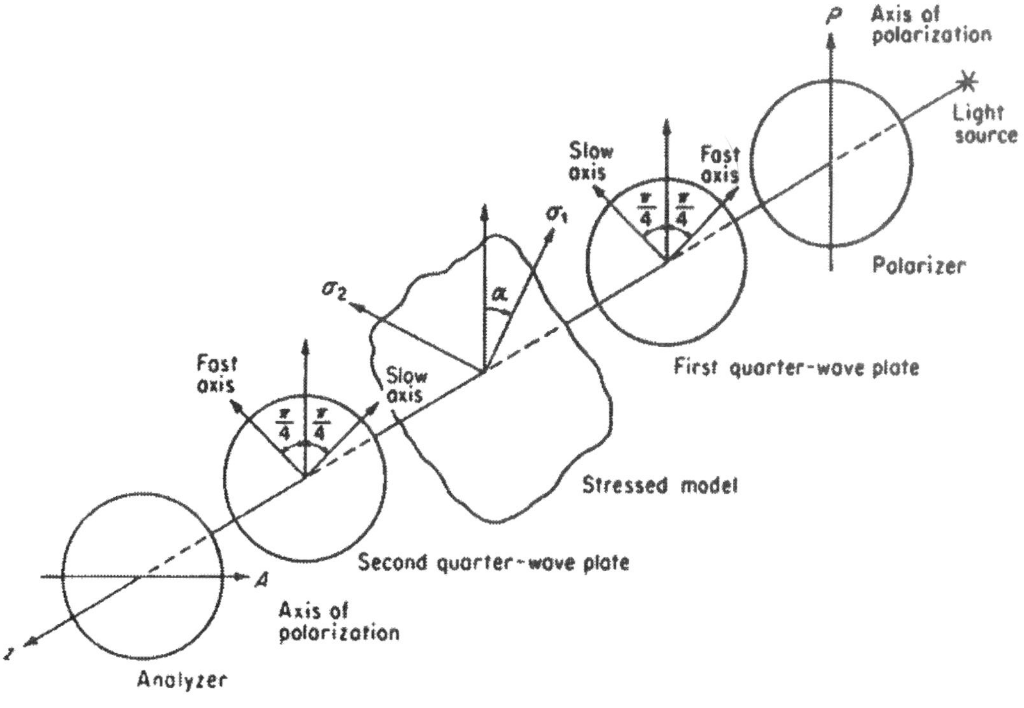

Figure 10.10 A stressed photoelastic model in a circular polariscope (arrangement A, crossed polarizer and analyzer, crossed quarter-wave plates).

The plane-polarized light beam emerging from the polarizer can be represented by the same simple expression used for the plane polariscope, namely,

$$E_{py} = k \cos \omega t \qquad \text{(a)}$$

When the light enters the first quarter-wave plate, it is resolved into components E_f and E_s with vibrations parallel to the fast and slow axes, respectively. Because the axes of the quarter-wave plate are oriented at 45° with respect to the axis of the polarizer, we can write:

$$E_f = \frac{\sqrt{2}}{2} k \cos\omega t \quad \text{and} \quad E_s = \frac{\sqrt{2}}{2} k \cos\omega t$$

As the components propagate through the plate, they develop a relative angular phase shift $\Delta = \pi/2$; and the components emerging from the plate are out of phase. Thus:

$$E'_f = \frac{\sqrt{2}}{2} k \cos\omega t$$
$$E'_s = \frac{\sqrt{2}}{2} k \cos\left(\omega t - \frac{\pi}{2}\right) = \frac{\sqrt{2}}{2} k \sin\omega t \qquad \text{(b)}$$

Recall it was shown previously that these two plane-polarized beams represent circularly polarized light with the light vector rotating counterclockwise as it propagates between the quarter-wave plate and the model.

After leaving the first quarter-wave plate, the components of the light vector enter the model in the manner illustrated in Fig. 10.11. Because the stressed model exhibits the characteristics of a temporary wave plate, the components E'_f and E'_s are resolved into components E_1 and E_2, which have directions coincident with principal-stress directions in the model. Thus:

$$E_1 = E'_f \cos\left(\frac{\pi}{4} - \alpha\right) + E'_s \sin\left(\frac{\pi}{4} - \alpha\right)$$
$$E_2 = E'_s \cos\left(\frac{\pi}{4} - \alpha\right) - E'_f \sin\left(\frac{\pi}{4} - \alpha\right) \qquad \text{(c)}$$

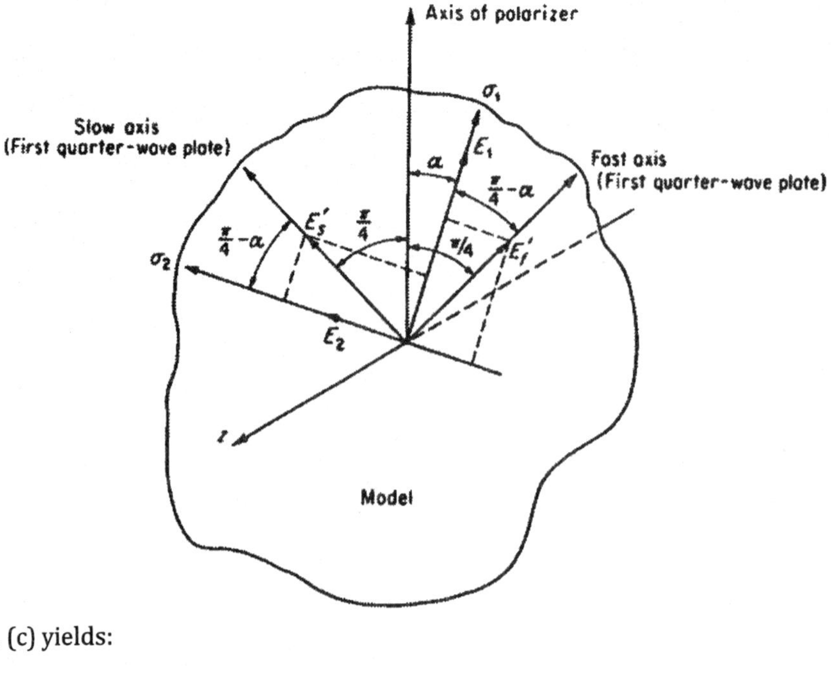

Figure 10.11 Resolution of the light components as they enter the stressed model.

Substituting Eqs. (b) into Eqs. (c) yields:

$$E_1 = \frac{\sqrt{2}}{2} k \cos\left(\omega t + \alpha - \frac{\pi}{4}\right)$$
$$E_2 = \frac{\sqrt{2}}{2} k \sin\left(\omega t + \alpha - \frac{\pi}{4}\right)$$

The two components E_1 and E_2 propagate through the model with different velocities. The additional relative retardation Δ accumulated during passage through the model is given by Eq. (10.12) and the waves upon emerging from the model can be expressed as:

$$E_1' = \frac{\sqrt{2}}{2} k \cos\left(\omega t + \alpha - \frac{\pi}{4}\right)$$
$$E_2' = \frac{\sqrt{2}}{2} k \sin\left(\omega t + \alpha - \frac{\pi}{4} - \Delta\right) \qquad \text{(d)}$$

The light emerging from the model propagates to the second quarter-wave plate and enters it according to the diagram shown in Fig. 10.12. The components associated with the fast and slow axes of the second quarter-wave plate are:

$$E_f = E_1' \sin\left(\frac{\pi}{4} - \alpha\right) + E_2' \cos\left(\frac{\pi}{4} - \alpha\right)$$
$$E_s = E_1' \cos\left(\frac{\pi}{4} - \alpha\right) - E_2' \sin\left(\frac{\pi}{4} - \alpha\right) \qquad \text{(e)}$$

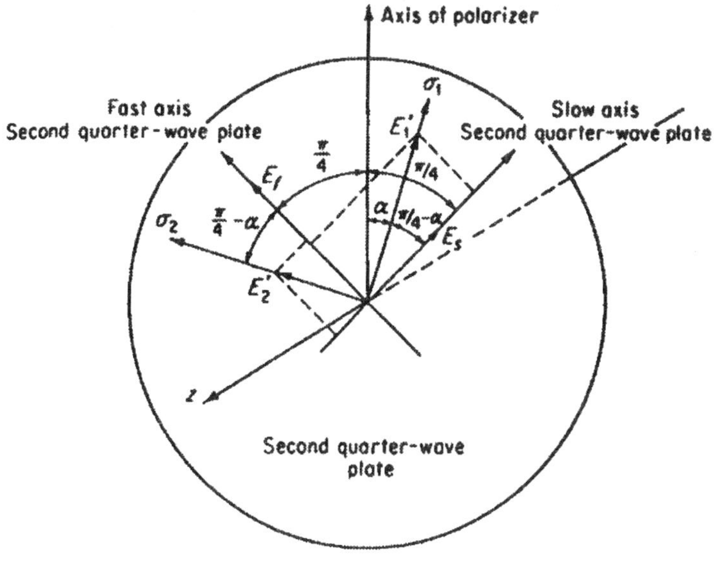

Figure 10.12 Resolution of the light components as they enter the second quarter-wave plate.

Substituting Eqs. (d) into Eqs. (e) yields:

$$E_f = \frac{\sqrt{2}}{2} k \left[\cos\left(\omega t + \alpha - \frac{\pi}{4}\right)\sin\left(\frac{\pi}{4} - \alpha\right) + \sin\left(\omega t + \alpha - \frac{\pi}{4} - \Delta\right)\cos\left(\frac{\pi}{4} - \alpha\right)\right]$$
$$E_s = \frac{\sqrt{2}}{2} k \left[\cos\left(\omega t + \alpha - \frac{\pi}{4}\right)\cos\left(\frac{\pi}{4} - \alpha\right) - \sin\left(\omega t + \alpha - \frac{\pi}{4} - \Delta\right)\sin\left(\frac{\pi}{4} - \alpha\right)\right]$$

As the light passes through the second quarter-wave plate, a relative phase shift of $\Delta = \pi/2$ develops between the fast and slow components. The waves emerging from the plate can be expressed as:

$$E'_f = \frac{\sqrt{2}}{2}k\left[\cos\left(\omega t + \alpha - \frac{\pi}{4}\right)\sin\left(\frac{\pi}{4} - \alpha\right) + \sin\left(\omega t + \alpha - \frac{\pi}{4} - \Delta\right)\cos\left(\frac{\pi}{4} - \alpha\right)\right]$$
$$E'_s = \frac{\sqrt{2}}{2}k\left[\sin\left(\omega t + \alpha - \frac{\pi}{4}\right)\cos\left(\frac{\pi}{4} - \alpha\right) + \cos\left(\omega t + \alpha - \frac{\pi}{4} - \Delta\right)\sin\left(\frac{\pi}{4} - \alpha\right)\right] \quad (10.21)$$

Finally, the light enters the analyzer, as shown in Fig. 10.13. The vertical components of E'_f and E'_s are absorbed in the analyzer while the horizontal components are transmitted to give:

$$E_{ax} = \frac{\sqrt{2}}{2}(E'_s - E'_f) \quad (f)$$

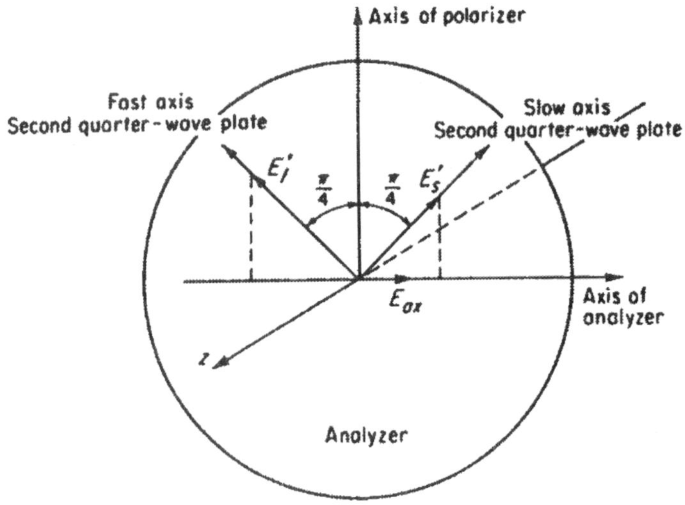

Figure 10.13 Components of the light vectors which are transmitted through the analyzer (dark field).

Substituting Eqs. (10.21) into Eq. (f) gives an expression for the light emerging from the analyzer of a circular polariscope (arrangement A). Thus:

$$E_{ax} = k\sin\frac{\Delta}{2}\sin\left(\omega t + 2\alpha - \frac{\Delta}{2}\right) \quad (10.22)$$

Because the intensity of light is proportional to the square of the amplitude of the light wave, the light emerging from the analyzer of a circular polariscope (arrangement A) is given by:

$$I = K\sin^2\frac{\Delta}{2} \quad (10.23)$$

This result indicates that the intensity of the light beam emerging from the circular polariscope is a function only of the principal-stress difference because the angle α does not appear in the expression for the amplitude of the wave. This fact indicates that the isoclinics have been eliminated from the fringe pattern observed with the circular polariscope. From the $\sin^2(\Delta/2)$ term in Eq. (10.23) it is clear that extinction will occur when $\Delta/2 = n\pi$, where n = 0, 1, 2, 3, ... This type of extinction is identical with that previously described for the plane polariscope for the isochromatic fringe pattern. An example of a dark field fringe pattern is shown in Fig. 10.14.

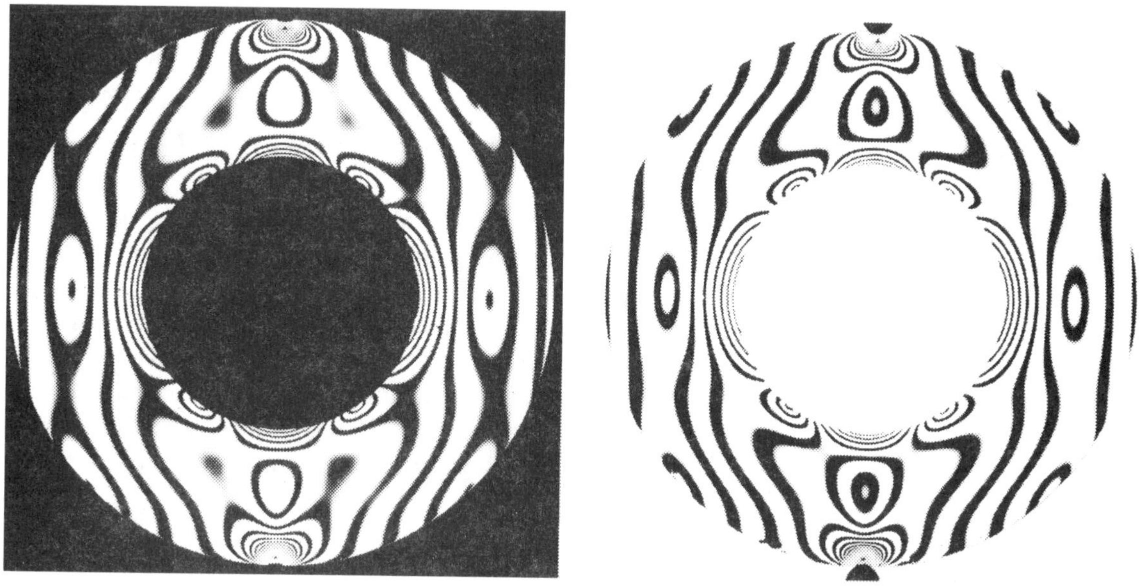

Figure 10.14 Dark and light field fringe patterns of a ring loaded in diametric compression.

10.6.1 Effects of a Stressed Model in a Circular Polariscope (Light Field)

A circular polariscope is usually employed with both the dark- and light-field arrangements (A and B). The circular polariscope can be converted from dark field (arrangement A) to light field (arrangement B) simply by rotating the analyzer through 90°, The advantage of employing both light- and dark-field arrangements is that twice as many fringes are obtained for the whole-field determination of $\sigma_1 - \sigma_2$. Recall that the order of the fringes N coincides with n for the plane polariscope and for the dark-field circular polariscope; therefore, the fringes are counted in the sequence 0, 1, 2, 3, ... However, with the light-field arrangement of the circular polariscope, N and n do not coincide. Instead, it can be shown that the intensity I is given by:

$$I = K\cos^2\frac{\Delta}{2} \tag{10.24}$$

Equation (10.24) shows that extinction (I = 0) will occur when

$$\frac{\Delta}{2} = \frac{1+2n}{2}\pi \qquad \text{for } n = 0,\ 1,\ 2,\ 3,\ \ldots\ldots$$

and

$$N = \frac{\Delta}{2\pi} = \frac{1}{2} + n$$

This relation implies that the order of the first fringe observed in a light-field polariscope is ½, which corresponds to n = 0. An example of a light-field isochromatic fringe pattern is shown in Fig. 10.14.

By using the circular polariscope with both light- and dark-field arrangements, it is possible to obtain two photographs or digital recordings of the resulting isochromatic fringe patterns. The data from the light and dark fields give a whole-field representation of the fringes to the nearest ½ order.

Interpolation between fringes often permits an estimate of the order of fringes to ± 0.1, giving accuracies for $\sigma_1 - \sigma_2$ of $\pm\ 0.1f_\sigma/h$. If more accurate determinations are necessary, the Tardy method of compensation described in the following section can be used.

10.7 EFFECTS OF A STRESSED MODEL IN A CIRCULAR POLARISCOPE (ARBITRARY ANALYZER POSITION, TARDY COMPENSATION)

The analysis for the dark- and light-field arrangements of the circular polariscope can be carried one step further to include rotation of the analyzer through some arbitrary angle γ. The purpose of such a rotation is to provide a means for determining fractional fringe orders.

In the previous derivations, the optical effects produced by a stressed model in both plane and circular polariscopes were studied by using a trigonometric representation of the light wave. For more complicated situations, the derivations with this representation quickly become an exercise in the manipulation of trigonometric identities and very little of the physical significance of the problem is retained. Under such circumstances, an exponential representation of the light wave can be used to simplify the derivations; therefore, it will be used in this development.

Consider, first, the passage of light through the optical elements of a circular polariscope (arrangement A). With the exponential representation, the light wave emerging from the polarizer can be expressed as:

$$E_{py} = ke^{j\omega t} \tag{a}$$

When the light enters the first quarter-wave plate, it is resolved into components parallel to the fast and slow axes of the plate. The components develop a phase shift $\Delta = \pi/2$ as they propagate through the plate and emerge as:

$$E'_f = \frac{\sqrt{2}}{2}ke^{j\omega t} \quad \text{and} \quad E'_s = -j\frac{\sqrt{2}}{2}ke^{j\omega t} \tag{b}$$

As the light enters the model, it is resolved into components parallel to the principal-stress directions. The components develop an additional phase shift Δ which depends on the principal-stress difference and emerge from the model as:

$$\begin{aligned} E'_1 &= \frac{\sqrt{2}}{2}ke^{j(\omega t+\alpha-\pi/4)} \\ E'_2 &= -j\frac{\sqrt{2}}{2}ke^{j(\omega t+\alpha-\pi/4-\Delta)} \end{aligned} \tag{c}$$

When the light enters the second quarter-wave plate, it is again resolved into components parallel to the fast and slow axes of the plate. The components experience a phase shift of $\Delta = \pi/2$ as they propagate through the plate and emerge as:

$$\begin{aligned} E'_f &= \frac{\sqrt{2}}{2}k\left[\sin\left(\frac{\pi}{4}-\alpha\right) - je^{-j\Delta}\cos\left(\frac{\pi}{4}-\alpha\right)\right]e^{j(\omega t+\alpha-\pi/4)} \\ E'_s &= \frac{\sqrt{2}}{2}k\left[e^{-j\Delta}\sin\left(\frac{\pi}{4}-\alpha\right) - j\cos\left(\frac{\pi}{4}-\alpha\right)\right]e^{j(\omega t+\alpha-\pi/4)} \end{aligned} \tag{10.25}$$

Finally, as the light passes through the analyzer, the vertical components of E'_f and E'_s are absorbed while the horizontal components are transmitted. Thus:

$$\begin{aligned} E_{ax} &= \frac{k}{2}\left[\left(e^{-j\Delta} - 1\right)\sin\left(\frac{\pi}{4} - \alpha\right) + j\left(e^{-j\Delta} - 1\right)\cos\left(\frac{\pi}{4} - \alpha\right)\right] e^{j(\omega t + \alpha - \pi/4)} \\ &= \frac{k}{2}\left(e^{-j\Delta} - 1\right) e^{j(\omega t + 2\alpha)} \end{aligned} \tag{10.26}$$

Recall from Eq. (9.21) that the square of the amplitude of a wave in exponential notation is the product of the amplitude and its complex conjugate. Thus:

$$I \approx E_{ax} E^*_{ax} = K \sin^2 \frac{\Delta}{2} \tag{10.27}$$

This expression for the intensity of the light emerging from the analyzer of a circular polariscope (arrangement A) is identical with that previously determined using a trigonometric representation of the light wave and presented in Eq. (10.23).

To establish the effect of a stressed model in a circular polariscope with the analyzer oriented at an arbitrary angle γ with respect to its dark-field position, it is only necessary to consider the light components emerging from the second quarter-wave plate, as represented by Eqs. (10.25), and their transmission through the analyzer, when it is positioned as shown in Fig. 10.15. Thus:

$$E_{ay} = E'_s \cos\left(\frac{\pi}{4} + \gamma\right) - E'_f \sin\left(\frac{\pi}{4} + \gamma\right) \tag{d}$$

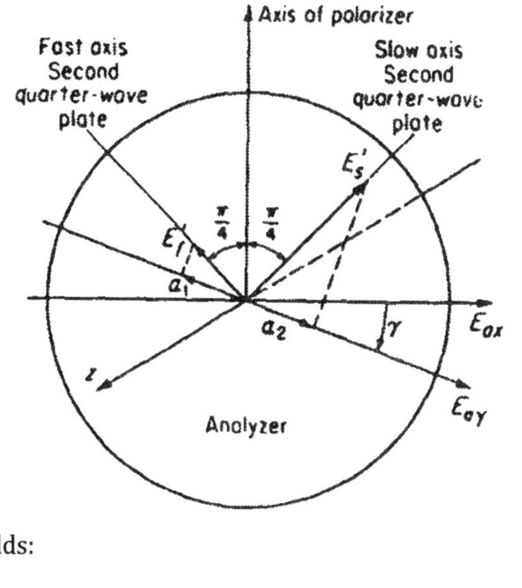

Figure 10.15 Rotation of the analyzer to obtain extinction in the Tardy method of compensation.

Substituting Eqs. (10.25) into Eq. (d) and combining terms yields:

$$E_{ay} = \frac{\sqrt{2}}{2} k \left\{ \begin{aligned} &\sin\left(\frac{\pi}{4} - \alpha\right)\left[e^{-j\Delta}\cos\left(\frac{\pi}{4} + \gamma\right) - \sin\left(\frac{\pi}{4} + \gamma\right)\right] \\ &+ j\cos\left(\frac{\pi}{4} - \alpha\right)\left[e^{-j\Delta}\sin\left(\frac{\pi}{4} + \gamma\right) - \cos\left(\frac{\pi}{4} + \gamma\right)\right] \end{aligned} \right\} e^{j(\omega t + \alpha - \pi/4)} \tag{10.28}$$

The intensity of the light emerging from the analyzer is given by Eq. (9.21) as:

$$I \approx E_{ay} E_{ay}^{*} \tag{e}$$

Substituting Eq. (10.28) and its complex conjugate into Eq. (e) and combining terms through the use of suitable trigonometric identities yields:

$$I = K(1 - \cos 2\gamma \cos\Delta - \cos 2\alpha \sin 2\gamma \sin \Delta) \tag{10.29}$$

For a given angle of analyzer rotation γ, values of α and Δ required for maximum intensity or minimum intensity are obtained from:

$$\frac{\partial I}{\partial \alpha} = K(2 \sin 2\alpha \sin 2\gamma \sin \Delta) = 0 \tag{f}$$

$$\frac{\partial I}{\partial \Delta} = K(\cos 2\gamma \sin \Delta - \cos 2\alpha \sin 2\gamma \cos \Delta) = 0 \tag{g}$$

Values of α and Δ, satisfying Eqs. (f) and (g) simultaneously, which give minimum intensity are given by:

$$\alpha = \frac{n\pi}{2} \quad \text{and} \quad \Delta = 2\gamma \pm 2n\pi \qquad n = 0, 1, 2, 3, \ldots\ldots$$

This result for I = 0 indicates that one of the principal-stress directions must be parallel to the axis of the polarizer ($\alpha = 0, \pi/2, \ldots\ldots$). The fringe order is then given by:

$$N = \frac{n\pi}{2} = n \pm \frac{\gamma}{\pi} \tag{10.30}$$

Rotation of the analyzer through an angle γ (Tardy method of compensation) is widely used to determine fractional fringe orders at select points on a photoelastic model. A plane polariscope is employed first so that isoclinics can be used to establish the directions of the principal stresses at the point of interest, as illustrated in Fig. 10.16. The axis of the polarizer is then aligned with a principal-stress direction ($\alpha = 0$ or $\alpha = \pi/2$), and the other elements of the polariscope are oriented to produce a standard dark-field circular polariscope. The analyzer is then rotated until extinction occurs at the point of interest, as indicated by Eq. (10.30).

To illustrate the procedure, consider the hypothetical dark-field fringe pattern and points of interest shown in Fig. 10.16. At point P_1, which lies between fringes of orders 2 and 3, the value assigned to N is 2. As the analyzer is rotated through an angle γ, the second-order fringe will move toward point P_1 until extinction is obtained. The fringe order at P_1 is then given by $N = 2 + \gamma/\pi$. For point P_2 the value of N is also taken as 2, and the analyzer is rotated through an angle γ_1 until the second-order fringe produces extinction, giving a value for the fringe order of $N = 2 + \gamma_1/\pi$. In this instance, N could also be taken as 3, and the analyzer rotated in the opposite direction through an angle $-\gamma_2$ until the third-order fringe produced extinction at point P_2. In this instance, the fringe order would be given by $N = 3 - \gamma_2/\pi$, which should check the value of $N = 2 + \gamma_1/\pi$ obtained previously.

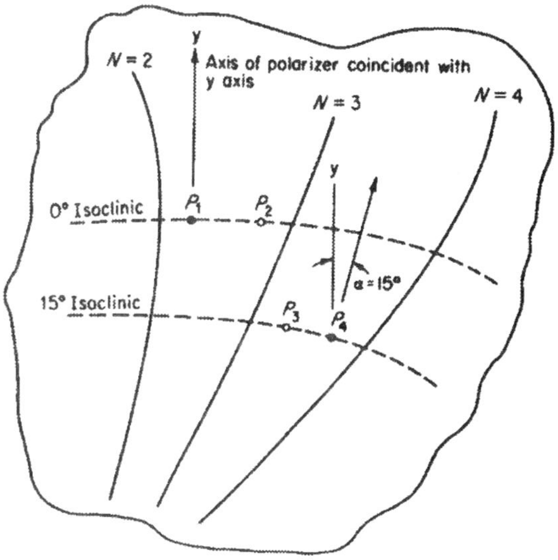

Figure 10.16 Locations of the points of interest relative to the isoclinic and isochromatic fringe patterns.

The Tardy method of compensation can be quickly and effectively employed to determine fractional fringe orders at arbitrary points in a model, provided isoclinic parameters are used to obtain the directions of the principal stresses. The accuracy of the method depends upon the quality of the quarter-wave plates employed in the polariscope; however, accuracies of ± 0.02 fringes usually can be achieved.

10.8 GREY FIELD PHOTOELASTICITY

Grey field photoelasticity was introduced by J. R. Lesnaik and Michael Zickel in 1997 [18, 19]. This method is similar to classical photoelasticity except for three important changes in the polariscope. First, the second quarter-wave plate has been eliminated. Second, the analyzer has been motorized to facilitate the recording of multiple images of the fringe patterns at different positions of the analyzer. Finally, the resulting images are recorded with a video camera and stored in the RAM memory of a personal computer.

The grey field polariscope is illustrated in Fig. 10.17. The light from a projector is passed through a polarizer and quarter-wave plate arranged to produce circularly polarized light. The circularly polarized light passes through the model before reaching the analyzer. At this point, the equations describing the light as it exits the model are identical with those presented in Section 10.6 [Eq. (d)]. This relation is rewritten below and identified as Eq. (10.31).

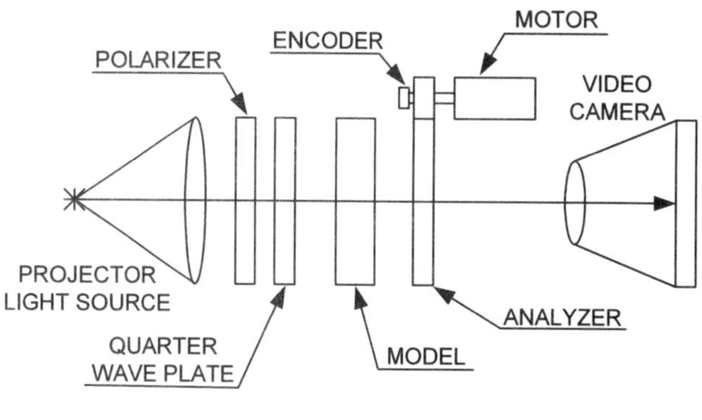

Figure 10.17 Optical arrangement in a grey-field polariscope.

$$E'_1 = \frac{\sqrt{2}}{2} k \cos(\omega t') \qquad E'_2 = \frac{\sqrt{2}}{2} k \sin(\omega t' - \Delta) \tag{10.31}$$

where $\omega t' = \omega t - \alpha - \pi/4$. This substitution does not affect the results and simplifies the trigonometry in the derivation shown below.

The orientation of the analyzer in the grey-field polariscope is a variable because it can be rotated to any number of different positions for any given load on the model. The angular position θ of the analyzer relative to the vertical position of the polarizer is shown in Fig. 10.18. Also shown in this drawing are the principal stresses σ_1 and σ_2. Note that the components of the light vector E'_1 and E'_2 are oriented along σ_1 and σ_2 directions, respectively.

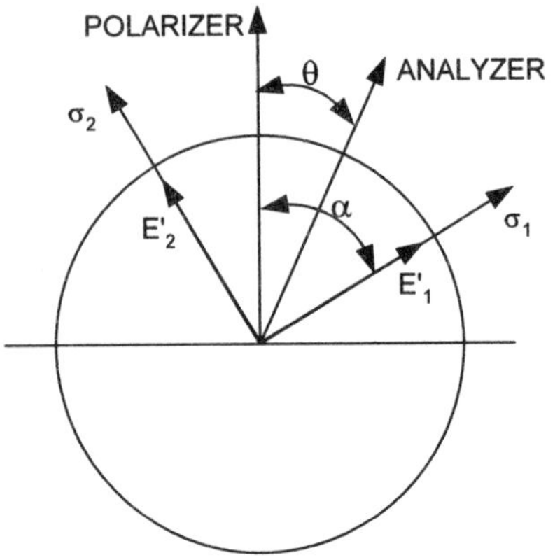

Figure 10.18 Orientation of the analyzer with respect to the principal stress directions and the polarizer.

Projecting the components E'_1 and E'_2 of the light vector onto the analyzer axis gives:

$$E_A = E'_1 \cos(\alpha - \theta) + E'_2 \sin(\alpha - \theta) \tag{10.32}$$

Substituting Eq. (10.31) into Eq. (10.32) and simplifying gives:

$$E_A = \frac{\sqrt{2}}{2} k \left\{ \cos\omega t' \left[\cos(\alpha-\theta) - \sin(\alpha-\theta)\sin\Delta \right] + \sin\omega t' \sin(\alpha-\theta)\cos\Delta \right\} \tag{10.33}$$

Square Eq. (10.33) to obtain the following relationship for the intensity I as:

$$I = \frac{k}{2} \left\{ \begin{aligned} &\cos^2\omega t' \left[\cos^2(\alpha-\theta) - 2\sin(\alpha-\theta)\cos(\alpha-\theta)\sin\Delta + \sin^2(\alpha-\theta)\sin^2\Delta \right] \\ &+ \sin\omega t'^2 \sin^2(\alpha-\theta)\cos^2\Delta + \sin 2\omega t' \sin(\alpha-\theta)\cos\Delta\left[\cos(\alpha-\theta) - \sin(\alpha-\theta)\sin\Delta \right] \end{aligned} \right\} \tag{10.34}$$

After considerable trigonometric manipulation, Eq. (10.34) can be rewritten as:

$$I = \frac{k}{4} \left\{ \begin{aligned} &1 - \cos 2\omega t' \left[\cos 2(\alpha-\theta)\cos^2\Delta + \sin^2\Delta - \sin 2(\alpha-\theta)\sin\Delta \right] - \sin 2(\alpha-\theta)\sin\Delta \\ &+ \sin 2\omega t' \sin(\alpha-\theta)\cos\Delta\left[\cos(\alpha-\theta) - \sin(\alpha-\theta)\sin\Delta \right] \end{aligned} \right\} \tag{10.35}$$

This equation appears to be complex, but recall that the frequency of light is of the order of 10^{15} Hz. Consequently, even a very short exposure time for recording the intensity I results in an integration of Eq. (10.35) with respect to time. Integrating eliminates the coefficients of the cos $2\omega t'$ and sin $2\omega t'$ terms. Hence:

$$I = C[1 - \sin(2\alpha - 2\theta)\sin\Delta] = C[1 + \sin(2\theta - 2\alpha)\sin\Delta] \tag{10.36}$$

where $C = k/4$ is a constant that is experimentally determined for the polariscope, the photoelastic model and the optical arrangement.

Examination of Eq. (10.36) shows that a uniform field where I = C is modulated by two sinusoidal functions. One is the retardation term Δ and the other contains the angles α and θ. Clearly this relation contains information pertaining to the difference in the principal stresses (or strains) and their directions. However, the form of the relation does not permit a direct interpretation of either the isoclinics or isochromatics as was the case for the plane and curricular polariscope arrangements. To obtain useful information in a grey field polariscope, it is necessary to record multiple images of the intensity with different analyzer positions.

Recall that an investigator can position the analyzer to arbitrarily vary θ. Consider the analyzer positioned at $\theta = 0, \pi/4, \pi/2$ and $\pi/6$. Substituting these four angles for θ into Eq. (10.36) yields:

$$I_0 = C - C \sin \Delta \sin 2\alpha \qquad \text{(a)}$$

$$I_{\pi/4} = C + C \sin \Delta \cos 2\alpha \qquad \text{(b)}$$

$$I_{\pi/2} = C + C \sin \Delta \sin 2\alpha \qquad \text{(c)}$$

$$I_{\pi/6} = C + \tfrac{1}{2} C \sin \Delta \left(\sqrt{3} \cos 2\alpha - \sin 2\alpha\right) \qquad \text{(d)}$$

From Eqs (a) and (c) the relation for C is given by:

$$C = \tfrac{1}{2} \left(I_0 + I_{\pi/2}\right) \qquad \text{(e)}$$

Substituting Eq. (e) into Eq. (b) yields:

$$\sin \Delta = (A - 1)/\cos 2\alpha \qquad (10.37)$$

where $A = 2I_{\pi/4}/(I_0 + I_{\pi/2})$ is an intensity ratio.

Substituting Eq (e) into Eq (d) and simplifying gives:

$$\tan 2\alpha = \sqrt{3} - 2(B - 1)/(A - 1) \qquad (10.38)$$

where $B = 2I_{\pi/6}/(I_0 + I_{\pi/2})$ is another intensity ratio.

Equations (10.37) and (10.38) give the difference in the principal stresses and their directions, respectively. In practice, the intensities at the four analyzer positions are measured at a large number of points on the specimen with a digital camera. The points on the specimen are identified with a pixel location and the intensities at each of these locations is stored in memory. A relatively simple computer program is utilized to perform the calculations indicated above. The method provides whole field data for Δ and α.

It is possible to manually rotate the analyzer of a conventional polariscope (with the second quarter wave plate removed) to the four different positions defined above. If the images are recorded with a digital camera, the relations derived above can be used to determine the principal stress differences and the principal directions. However, a grey field polariscope with a digital camera and software is commercially available. This commercial unit[2] is a complete system, which permits processing the data in real time. If tinted epoxies are used as birefringent coatings, this system is able to automatically measure the coating thickness and account for different thicknesses at different points on the model. The software handles the post processing and runs on Windows operating system.

---

[2] Stress Photonics, Inc., 3002 Progress Road, Madison, WI, 53716.

10.9 RED, GREEN AND BLUE (RGB) LIGHT PHOTOELASTICY

In recent years, several novel imaging processing techniques and automated polariscopes have been developed to reduce the time and the skill level necessary to complete a photoelastic analysis [20, 21]. Also, white light photoelastic methods of analysis involving spectral analysis of the resulting fringe pattern have been developed [22]. Many of these advances have been possible because of the availability of high-resolution CCD cameras, high-speed, low-cost computers and effective image analysis software.

One of the more recent developments is tricolor or red, green and (RGB) photoelasticity [23, 24]. This approach is based on the fact that the special positions of isochromatic fringes depend upon the wavelength of light; however, isoclinic fringes are independent of the wavelength of light. There are two possible approaches to design a plane polariscope adapted to three different wavelengths.

1. Employ light sources that provide three distinctly different wavelengths together with a monochromatic CCD camera.
2. Employ a white light source with a broad range of wavelengths together with a CCD camera capable of color imaging[3].

The tricolor plane polariscope developed by Yoneyama et al [24] is illustrated in Fig. 10.19. The halogen light source, when filtered, provides red light at a wavelength of 619 nm; the mercury light source, when filtered, provides two spectral lines at wavelengths of 436 (blue) and 546 nm (green), respectively. Beam splitters, neutral density filters and mirrors conditioned the three different light beams prior to their entry into a plane polariscope. The output from the polariscope is recorded with a color CCD camera equipped with three independent image sensors corresponding to the red, green and blue light. With this camera, color image data can be recorded as three separate monochromatic images during a single exposure.

Figure 10.19 Optical arrangement for a RGB polariscope with two light sources.

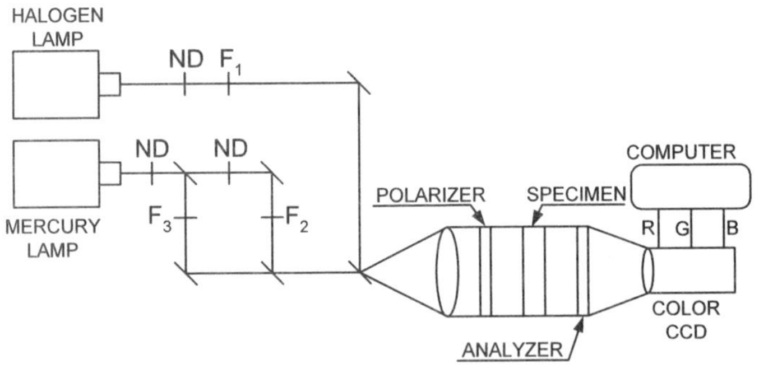

In the analysis of the data recorded from the plane polariscope, consider a modified form of Eq. (10.20) as given by:

$$I_i = K_i \sin^2 2\alpha \sin^2(N_i \pi) + Q_i \tag{10.39}$$

where K_i is a constant that depends on the intensity of light and the complete optical system.
Q_i is the background intensity—assumed to be independent of the wavelength of light[4].
The subscript i = R, G and B corresponding to the red, green and blue light.

Subtracting I_G from I_R and I_B from I_G gives:

---

[3] CCD cameras and their characteristics are described in more detail in Section 10.10.1.

[4] Neutral density filters are adjusted in each of the three light beams to insure that the background intensity Q_i is independent of wavelength.

$$I_R - I_G = [K_R \sin^2 (N_R\pi) - K_G \sin^2 (N_G\pi)] \sin 2\alpha \qquad \text{(a)}$$

$$I_G - I_B = [K_G \sin^2 (N_G\pi) - K_B \sin^2 (N_B\pi)] \sin 2\alpha \qquad \text{(b)}$$

Solve each of these relations for sin 2α to obtain:

$$\sin 2\alpha = \frac{I_R - I_G}{K_R \sin^2(N_R\pi) - K_G \sin^2(N_G\pi)} = \frac{I_G - I_B}{K_G \sin^2(N_G\pi) - K_B \sin^2(N_B\pi)} \qquad (10.40)$$

Expanding the equalities in Eq. (10.40) yields:

$$(I_G - I_B)\, K_R \sin^2 (N_R\pi) + (I_R - I_G)\, K_B \sin^2 (N_B\pi) + (I_B - I_R)\, K_G \sin^2 (N_G\pi) = 0 \qquad \text{(c)}$$

Rewriting Eq. (c) gives:

$$(I_G - I_B)\, K_R \sin^2 (CN_G\pi) + (I_R - I_G)\, K_B \sin^2 (DN_G\pi) + (I_B - I_R)\, K_G \sin^2 (N_G\pi) = 0 \qquad (10.41)$$

where $C = N_R/N_G$ and $D = N_B/N_G$. The constant C and D must be determined in calibration tests to account for the effects of dispersion.

Numerical methods are used to solve Eq (10.41) because this relation is non-linear in terms of the unknown N_G. The data for I_R, I_G and I_B are stored for each pixel location on the model; hence, the numerical routine to determine N_G is repeated for each pixel (point) to provide nearly[5] whole field results for the fringe order distribution. Finally, the difference in the principal stresses at each pixel location is determined from:

$$\sigma_1 - \sigma_2 = \frac{N_G f_{\sigma G}}{h} \qquad (10.42)$$

After establishing the value of N_G, the angle of the principal directions can be determined from any difference in the intensity. For example, the difference relations such as Eq (a) or Eq. (b) leads to:

$$\begin{aligned}
2\alpha &= \sin^{-1} \frac{I_R - I_G}{K_R \sin^2(CN_G\pi) - K_G \sin^2(N_G\pi)} \\
2\alpha &= \sin^{-1} \frac{I_G - I_B}{K_G \sin^2(N_G\pi) - K_B \sin^2(DN_G\pi)} \\
2\alpha &= \sin^{-1} \frac{I_B - I_R}{K_B \sin^2(DN_G\pi) - K_R \sin^2(CN_G\pi)}
\end{aligned} \qquad (10.43)$$

Because the system of equations for the principal angle α is overdetermined, it is possible to minimize errors by averaging the three values.

The red, green, blue plane polariscope equipped with three different light sources and a color sensitive CCD camera enables a researcher to automate the data collection process. The computer, with suitable software, permits the determination of whole field distributions of $\sigma_1 - \sigma_2$ and the principal angles α and $\alpha + \pi/2$.

---

[5] When $\sin^2 2\alpha = 0$ or when $\sigma_1 - \sigma_2 = 0$, Eq. (10.41) does not provide a solution for N_G. In these cases, interpolation from nearby data points enable accurate determination of the fringe order at these locations.

10.10 RECORDING PHOTOELASTIC FRINGE PATTERNS

In most photoelastic analyses, photographs are taken of the isochromatic and isoclinic fringe patterns to establish a permanent record of the test results. The photographs can be taken with a conventional camera using photographic film or with digital camera using a high pixel count CCD array to record the image. Of the two methods, the digital recording is preferred because the digital format enables the use of image analysis software to more effectively process the data. In this discussion, both methods for recording the image will be described.

10.10.1 Photoelastic Photography with a Digital Camera

A digital camera is similar to a conventional camera in that it utilizes a lens to focus an image on the image plane and a light proof housing that protects the image plane from stray light. The difference is that the film in a conventional camera has been replaced with an image sensor. This image sensor is fabricated from single crystal silicon using the same processing techniques employed in the production of microelectronic devices. The image sensors are based on Charge Coupled Devices (CCDs) arranged in a rectangular array on a silicon chip. A typical CCD implanted on the silicon chip is illustrated in Fig. 10.20. The dimensions of a CCD cell depends on the density of the array of cells (pixels) used in the image sensor. At the high end of the digital photography market a 16.8 mega pixel (4k × 4k) image sensor uses a 9-μm pixel in a 36 × 36 mm format. Over the next decade image sensors will transition to 36 mega pixels (6k × 6k) if the market demands large-format high-resolution sensors.

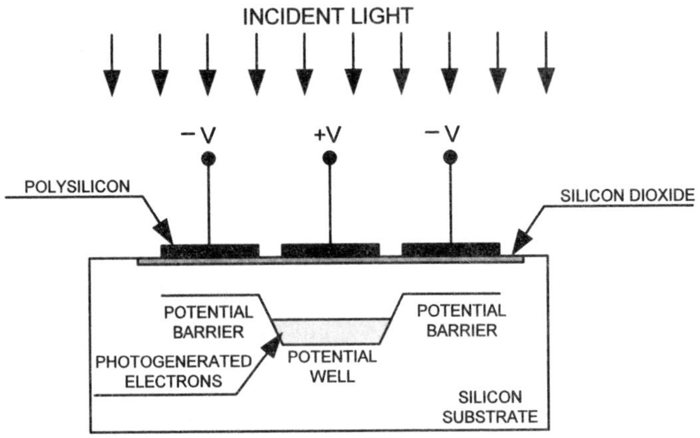

Figure 10.20 A CCD Metal Oxide Semiconductor (MOS) structure creating a pixel on an image sensor.

CCD imaging is performed in a four step process:

1. Exposure to light which converts photons into electronic charge at discrete sites called pixels.
2. Transfer of packets of charge within the silicon substrate.
3. Charge to voltage conversion and output amplification.
4. Analog to digital conversion.

An image is acquired when incident light (photons) fall on an array of pixels. The energy associated with each photon is absorbed in the silicon creating an electron. The number of electrons collected at each pixel depends on the light level, exposure time and wavelength of the light. When the electrons are generated, a charge develops at the center gate where the potential well exists. The center gate is surrounded by four adjacent gates which form a potential barrier that bound the potential well. Because adjacent gates are shared, two gates are required for each pixel (one with V+ and one with V−). The polysilicon gate electrodes are arranged in arrays so that they form long chains making up a column along one axis and a row along the other axis.

There are several methods of charge transfer and they all involve moving the charge from one pixel to the next either along a column of gates or a row of gates. These methods all use a scheme to vary the gate voltages sequentially to move the charge from a specified pixel along a column to the output sense nodes located at the end of each column.

The charge at the output nodes is converted to voltage by a charge to voltage amplifier such as a source follower. The voltage output from the source follower, an analog signal, is then converted to a digital word. This conversion also includes the identification of the pixel location in the array of pixels. Considering the large number of pixels involved, an extremely large amount of data is associated with the output from a single frame of a digital image.

There is a tradeoff between pixel area and pixel count. Increasing the number of pixels increases the resolution that can be achieved when recording an image. However, if pixel count is increased while decreasing pixel area responsivity, dynamic range and framing rate are degraded. Responsivity is a measure of the signal that each pixel can produce and it is directly proportional to the pixel area. This fact implies that more electrons (signal) can be collected for a fixed exposure time. It also implies that the signal to noise ratio is higher.

Pixel area also affects the dynamic range of the optical system. Larger pixels have higher charge capacity and produce higher signal levels. The increased charge capacity improves dynamic range because brighter objects will not saturate the pixel. Wide dynamic range is important because it permits areas of high and low brightness to be recorded in a single exposure without loss of clarity. Dynamic range is often characterized by the number of gray levels specified for the image sensor. The number of gray levels is a measure of the contrast that can be achieved in an image. An 8-bit sensor can display 256 gray levels and a 10-bit sensor can display 1024 gray levels.

It is possible to increase pixel count and pixel area simultaneously by increasing the size of the chip. A high quality 1400 × 1000 square pixel array with a pixel size of 9.3 μm requires a chip with a 16 mm sensor diagonal. Of course large chips are more expensive than smaller ones; hence, the cost of higher quality digital imaging systems depends not only on pixel count but also on pixel area and active chip area.

There are three basic types of CCD architecture—full-frame, frame-transfer and interline. The full-frame architectures incorporate a parallel CCD shift register, a serial CCD shift register and a signal sensing output amplifier as illustrated in Fig. 10.21. The image is projected by the lens onto the parallel array that acts as an image plane. The image is partitioned into discrete elements (pixels). The resulting rows of data are then shifted in a parallel fashion to the serial register that subsequently shifts the row of information to the output as a stream of data. The shifting continues row by row until all of the data is transferred off of the chip. The image is then reconstructed in a suitable computer and displayed on a monitor or printed out as hard copy.

The full frame systems employ a shutter and the light generates charge on all of the pixels in the array simultaneously. For a given pixel size, the full-frame CCDs usually will have higher charge capacity, better signal to noise ratio, and higher responsivity than interline or frame-transfer CCDs. The simplicity of the full-frame design yields CCD image sensors with the highest resolution and the highest density. Mechanical shutters are effective when acquiring static images with long exposure times when the light level is low.

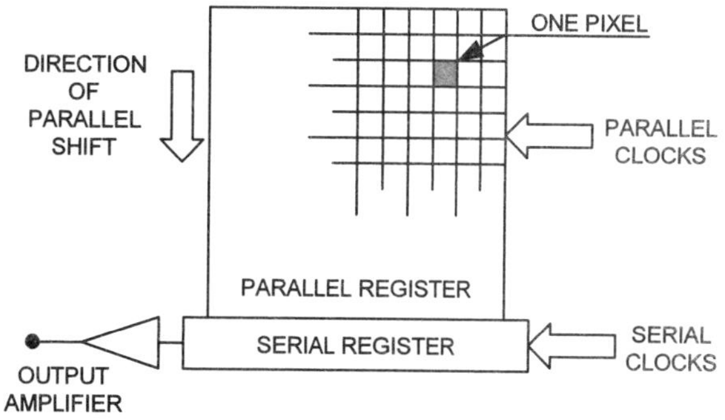

Figure 10.21 Architecture of a full-frame CCD imaging system.

Frame-transfer CCDs are similar to the full-frame architectures except that a separate parallel register called a storage array of equal size as the image array is added which is not light sensitive. The concept is to quickly shift the data from the image array to the storage array. The advantage of this design is that it enables higher framing rates. The disadvantage is that twice the silicon area is required to implement this architecture resulting in significantly higher cost.

Interline architectures separate photon detection and readout functions by forming isolated pixels between lines of shielded parallel readout CCD. After collecting the charge corresponding to an image, the signal collected on every pixel is transferred simultaneously to the light shielded parallel CCD. Transfer of this signal to the output register is then achieved in a manner similar to the frame-transfer CCD. The advantage is higher framing rate because of the simultaneous transfer of the charge from each pixel. The disadvantages are the higher cost of this architecture due to its complexity and its lower sensitivity due to the reduced pixel area. Additional information on CCD image sensors is described in detail in Reference [23-25].

Today there are many commercially available digital cameras and the technology for the image sensors is still improving at a rapid rate. The selection of the most suitable camera will depend on the application. For most optical measurements of stress or strain, framing rate is not the primary consideration. For this reason, it is suggested that framing rate be sacrificed for pixel size, dynamic range and resolution. A typical camera suitable for large-format, high-resolution imaging provides high-sensitivity 12-bit images with 4,000 × 4,000 pixel spatial resolution at up to 3 frames/s. Pixel size of 9 μm × 9 μm and aperture size of 36 mm × 24 mm are common. The cameras for scientific applications are programmable and data transfer is accomplished at rates up to 30 MHz/pixel.

When an image is acquired with a digital system, the data collected is transferred from the camera to a computer. To facilitate this transfer, the computer is usually equipped with a PCI image grabber (frame grabber card). A typical image grabber transfers the image either to the host personal computer (PC) memory for processing or to its graphics card for display. Generic image analysis software is available for processing the images captured during an experiment.

EXERCISES

10.1 Derive Eq. (10.10) beginning with Eq. (10.7).

10.2 A secondary plane is defined by an Ox'y'z' coordinate system where the z' axis is normal to the plane. If the Oxyz coordinate system is a principal system and the angles between the two coordinate systems are known, write the equations for the secondary principal stresses which lie in the Ox'y' plane.

10.3 If a particular point in a photoelastic model is examined in a polariscope with a mercury light source (λ = 548.1 nm) and a fringe order of 4.00 is established, what fringe order would be observed if a sodium light source (λ = 589.3 nm) were used in place of the mercury source?

10.4 The stress fringe value f_σ for a material was determined to be 35 kN/m when sodium light (λ = 589.3 nm) was used in its determination. What would the stress fringe value for the same material be if mercury light (λ = 548.1 nm) were used in place of the sodium light?

10.5 Derive the equations for light passing through a stressed model in a plane polariscope with the polarizer and analyzer in parallel positions. Under what conditions does extinction ($I = 0$) occur?

10.6 Derive the equations for light passing through a stressed model in a plane polariscope (polarizer and analyzer crossed). Use an exponential representation for the light wave.

10.7 Compare the frequency response of a polariscope with that of an oscilloscope.

10.8 Prepare sketches similar to Fig. 10.10 showing the four different arrangements (see Table 10.1) for the circular polariscope.

10.9 Derive the equations for light passing through a stressed model in a circular polariscope (arrangement C of Table 10.1). Use a trigonometric representation for the light wave.

10.10 Derive the equations for light passing through a stressed model in a circular polariscope (arrangement D of Table 10.1). Use an exponential representation for the light wave.

10.11 Determine the optical effects produced by light passing through a stressed model in a circular polariscope with imperfect quarter-wave plates. Use arrangements A and B of Table 10.1 for the polariscope and assume $\Delta = \pi/2 + \varepsilon$ for the imperfect quarter-wave plates.

10.12 Determine the optical effects produced by light passing through a stressed model in a circular polariscope with imperfect quarter-wave plates. Use arrangements C and D of Table 10.1 for the polariscope and assume $\Delta = \pi/2 + \varepsilon$ for the imperfect quarter-wave plates.

10.13 With Tardy compensation, the fractional fringe order can be determined at an arbitrary point in a stressed model by aligning a principal-stress direction at the point with the axis of the polarizer and rotating the analyzer for extinction. What effect would be produced by rotating the polarizer instead of the analyzer?

10.14 Investigate the error introduced in fractional fringe-order determinations by improper setting of the isoclinic angle ($\alpha = 0 + \varepsilon$ or $\pi/2 + \varepsilon$) in Tardy compensation. Numerically evaluate for $\varepsilon = \pm 10°$.

10.15 Investigate the error introduced in fractional fringe-order determinations by imperfect quarter-wave plates ($\Delta = \pi/2 + \varepsilon$) in Tardy compensation. Make a numerical evaluation for the case of quarter-wave plates for mercury light (λ = 548.1 nm) being used with:

(a) sodium light (λ = 589.3 nm)
(b) helium-neon laser light (λ = 632.8 nm)
(c) argon laser light (λ = 488.0 nm)

10.16 The following procedure has been proposed for establishing the fractional fringe order at a point in a stressed model. Establish the validity of the procedure.

1. Begin with a circular polariscope (arrangement A).
2. Remove the first quarter-wave plate.
3. Position the model so that the principal stress directions, at the point of interest, are oriented at ± 45° with respect to the axis of the polarizer.
4. Align the fast axis of the second quarter-wave plate with the axis of the polarizer.
5. Rotate the analyzer for extinction.

10.17 Establish a procedure for aligning a plane polariscope to obtain a dark field.

10.18 Establish a procedure for aligning a circular polariscope to give arrangement A. Then add to it to provide a methods to give arrangement B.

10.19 Beginning with Eq. (10.31), verify Eq. (10.36).

10.20 Beginning with Eq. (10.36), verify Eq. (10.37) and Eq. (10.38).

10.21 Assume that a grey polariscope is used in a photoelastic experiment. The four angles of the analyzer are set at 0, $\pi/6$, $\pi/4$ and $\pi/2$. If the intensity measurements are $I_0 = 13$, $I_{\pi/6} = 187$, $I_{\pi/4} = 50$ and $I_{\pi/2} = 187$, determine the angular retardation Δ and the direction of the principal stresses.

10.22 Assume that a grey polariscope is used in a photoelastic experiment. The four angles of the analyzer are set at 0, $\pi/6$, $\pi/4$ and $\pi/2$. If the intensity measurements are $I_0 = 8$, $I_{\pi/6} = 149$, $I_{\pi/4} = 165$ and $I_{\pi/2} = 192$, determine the angular retardation Δ and the direction of the principal stresses.

10.23 Beginning with Eq. (10.39), derive Eq. (10.41) and Eq. (10.43).

10.24 Write an engineering brief describing the role of the digital camera and the personal computer in the development of the grey and RGB polariscopes.

10.25 Perform an Internet search and locate a digital camera suitable for a laboratory equipped to perform photoelastic analysis in addition to several other optical methods of stress analysis. Write the specification for the camera and justify your selection of the camera's characteristics such as pixel count, pixel size, framing rate, frame transfer methods, etc. Show also the cost performance trade-offs.

BIBLIOGRAPHY

1. Jenkins, F. A., and H. E. White: Fundamentals of Optics, 4th Edition, McGraw-Hill Book Company, New York, 1976.
2. Coker, E. G., and L. N. G. Filon: A Treatise on Photoelasticity, Cambridge University Press, New York, 1931.
3. Maxwell, J. C.: On the Equilibrium of Elastic Solids, Transactions. Royal Society, Edinburgh, Vol. XX, Part 1, pp. 87-120, 1853.
4. Neumann, F. E.: Die Gesetze der Doppelbrechung des Lichts in comprimierten oder ungleichformig erwarmten unkrystallinischen Korpern, Abh. K. Acad. Wiss. Berlin, pt. II, pp. 1-254, 1841.
5. Favre, H.: Sur une nouvelle methode optique de determination des tensions interieures, Review Optics, Vol. 8, pp. 193-213, 241-261, 289-307, 1929.
6. Frocht, M. M.: Photoelasticity, John Wiley & Sons, Inc., New York, Vol. 1, 1941. Vol. 2, 1948.
7. Durelli, A. J., and W. F. Riley: Introduction to Photomechanics, Prentice-Hall, Englewood Cliffs, N.J., 1965.
8. Kuske, A., and G. Robertson: Photoelastic Stress Analysis, John Wiley & Sons, Inc., New York, 1974.
9. Mindlin, R. D.: A Review of the Photoelastic Method of Stress Analysis, Journal Applied Physics., Vol. 10, pp. 222-241, 273-294, 1939.
10. Vandaele-Dossche, M., and R. van Geen: La birefringence mecanique en lumiere ultra-violette et ses applications, Bull. Class Sci. Acad. R. Belgium, Vol. 50, No. 2, 1964.
11. Monch, E., and R. Lorek: A Study of the Accuracy and Limits of Application of Plane Photoplastic Experiments, Proceedings International Symposium. Photoelasticity, Pergamon Press, New York, 1963.
12. Mindlin, R. D.: Distortion of the Photoelastic Fringe Pattern in an Optically Unbalanced Polariscope, Journal Applied Mechanics, Vol. 4, pp. A170-172, 1937.
13. Mindlin, R. D.: Analysis of Doubly Refracting Materials with Circularly and Elliptically Polarized Light, Journal Optical Society America, Vol. 27, pp. 288-291, 1937.
14. Tardy, M. H. L.: Methode pratique d'examen de mesure de la birefringence des verres d'optique, Review Optics, Vol. 8, pp. 59-69, 1929.
15. Chakrabati, S. K., and K. E. Machin: Accuracy of Compensation Methods in Photoelastic Fringe-Order Measurements, Experimental Mechanics, Vol. 9, No. 9, pp. 429-431, 1969.
16. Lesniak, J. R., Bazile, D. J. and M. J. Zickel: "New Coating Techniques in Photoelasticity," Proceedings of the Society for Experimental Mechanics, June 1999.
17. Lesniak, J. R., Zickel, M. J., Welch, C. S. and D. F. Johnson: "An Innovative Polariscope for Photoelastic Stress Analysis," Proceedings of the Society for Experimental Mechanics, June 1997.
18. Patterson, E. A. and Wang, F. Z.: "Towards Full Field Automated Photoelastic Analysis of Complex Components," Strain, Vol 27, pp-49-56, 1991.
19. Morimoto, Y., Morimoto, Y., Jr, and Hayashi, T.: "Separation of Isochromatics and Isoclinics Using Fourier Transform," Experimental Techniques, Vol. 18, No. 5. pp. 13-17, 1994.
20. Redner, A. S.: "Photoelastic Measurements by Means of Computer Assisted Spectral Content Analysis," Experimental Mechanics, Vol. 25, pp. 148-153, 1985.
21. Ajovalasit, A., Barone, S. and G. Perucci: "Towards RGB Photoelasticity: Full-field Automated Photoelasticity in White Light," Experimental Mechanics, Vol. 35, pp 193-200, 1995.
22. Yoneyama, S., J. Gotoh and M. Takashi: "Tricolor Photoviscoelastic Technique and its Application to Moving Contact, Experimental Mechanics, Vol. 38, pp. 211-217, 1998.
23. Anon: Charge Coupled Device (CCD) Image Sensors, CCD Primer, MTD/PS-0218, Eastman Kodak Co., Rochester, NY, May 29, 2001.
24. Anon: "CCD Image Sensors and Analog-to-Digital Conversion," Texas Instruments, January 1993.
25. Benamati, B. L.: "In Search of the Ultimate Image Sensor," Photonics Spectra, September 2001.

CHAPTER 11

APPLIED PHOTOELASTICTY

11.1 INTRODUCTION

Applications of photoelasticity are usually divided into two broad classes depending on the geometry of the body or structure under analysis. If the body or structure is plane, in the sense that it can be represented by plane stress or plane strain, then two-dimensional methods of photoelasticity are employed. If the body is not plane, then the more complex and time consuming methods of three-dimensional photoelasticity are required. In some studies, a photoelastic coating is applied to a prototype and the requirement for a model is eliminated. The photoelastic response of the coating is measured in either a grey field or reflecting polariscope. These three different applications of photoelasticity will be described in this chapter.

11.2 TWO-DIMENSIONAL PHOTOELASTIC STRESS ANALYSIS

In conventional two-dimensional photoelastic analysis, a suitable model is fabricated, placed in a polariscope and loaded before the fringe pattern is examined and recorded. The next step in the analysis is the interpretation of the fringe patterns which, in reality, represent raw test data. The purpose of this section is to discuss the interpretation of the isoclinic and isochromatic fringe patterns, compensation techniques, separation methods and scaling of the stresses between the model and prototype in a typical stress analysis.

11.2.1 Isochromatic Fringe Patterns

The isochromatic fringe pattern obtained from a two-dimensional model gives lines along which the principal-stress difference $\sigma_1 - \sigma_2$ is equal to a constant. A typical example of a light-field isochromatic fringe pattern, which will be utilized to describe the analysis, is shown in Fig. 11.1. This photoelastic model represents a chain link subjected to tensile loads applied axially through roller pins. First, it is necessary to determine the fringe order at each point of interest in the model. In this example, the assignment of the fringe order is relatively simple because thirteen rather obvious 1/2- order fringes can quickly be identified. The two oval-like fringes located on the flanks of the teeth (labeled A) are 1/2-order fringes because the flank, with its geometry, cannot support high stresses. The four fringes located at points B on the pinholes can be identified if the model is viewed with white light because zero-order fringes appear black while higher-order fringes are colored. The irregularly shaped fringe labeled C near the center of the link is also a 1/2-order fringe. Explanation of the 1/2-order fringes at points D and E is covered in the exercises.

With these 1/2-order fringes established, it is a simple matter to determine the fringe order at any point in the model by progressively counting the fringes outward from the 1/2-order location. For instance, the order of the fringe at the flank fillets is 7 ½. When the fringe order at any point on the model has been established, it is possible to compute $\sigma_1 - \sigma_2$ from Eq. (10.13) as:

$$\sigma_1 - \sigma_2 = \frac{Nf_\sigma}{h} \qquad (10.13)$$

where $\sigma_1 - \sigma_2$ are the principal stresses in the plane of the model. The maximum shear stress is given by:

$$\tau_{max} = \frac{1}{2}(\sigma_1 - \sigma_2) = \frac{Nf_\sigma}{2h} \qquad (11.1)$$

provided $\sigma_1 - \sigma_2$ of opposite sign and $\sigma_3 = 0$; otherwise:

$$\tau_{max} = \begin{cases} \frac{1}{2}(\sigma_1 - \sigma_3) = \frac{1}{2}\sigma_1 & \text{if } \sigma_1 \text{ and } \sigma_2 \text{ are positive} \\ \frac{1}{2}(\sigma_3 - \sigma_2) = \frac{1}{2}\sigma_2 & \text{if } \sigma_1 \text{ and } \sigma_2 \text{ are negative} \end{cases} \qquad (11.2)$$

The difference between Eq. (11.1) and Eqs. (11.2) is presented graphically in Fig. 11.2, where Mohr stress circles for two cases are represented. When $\sigma_1 > 0$ and $\sigma_2 < \sigma_3 = 0$, the maximum shear stresses τ_{max} is one-half the value of $\sigma_1 - \sigma_2$ and can be determined directly from the isochromatic fringe pattern according to Eq. (11.1). However, when $\sigma_1 > \sigma_2 > \sigma_3 = 0$, the maximum shear stress does not lie in the plane of the model, and Eq. (11.1) gives τ_p (see Fig. 11.2) and not τ_{max}. To establish τ_{max} in this case, it is necessary to determine σ_1 individually. This is an important point because the maximum-shear theory of failure is often used in the design of machine components and high-performance parts for vehicles of all types.

Figure 11.1 Light-field isochromatic fringe pattern of a chain link subjected to axial tensile loads through the roller pins.

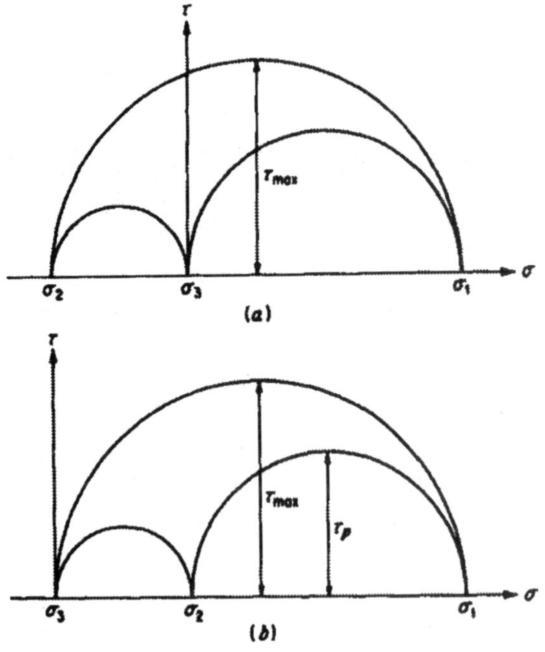

Figure 11.2 Mohr's circle for the state of stress at a point (a) $\sigma_1 > 0, \sigma_2 < \sigma_3 = 0$; (b) $\sigma_1 > \sigma_2 > \sigma_3 = 0$

On the free boundary of the model either σ_1 or σ_2 is equal to zero and the stress tangential to the boundary is given by:

$$\sigma_1 \text{ or } \sigma_2 = \frac{Nf_\sigma}{h} \tag{11.3}$$

The sign of σ_1 or σ_2 can usually be determined by inspection, particularly in the critical areas where the boundary stresses are a maximum. By referring to Fig. 11.1, it is apparent that the stresses in the flank fillets are tensile while the stresses along the back of the plate are compressive. On the free surface of the pinholes, the stresses on the horizontal diameter are tensile and the stresses on the vertical diameter are compressive (see Section 3.14). The stresses on that portion of the boundary of the hole where the pin is in contact cannot be determined by applying Eq. (11.3) because the boundary is not free and, in general, σ_1 or σ_2 will not approach zero. In this case, neither σ_1 nor σ_2 is known, a priori, and it is necessary to separate the stresses, i.e., individually determine the values of σ_1 and σ_2, at this region of contact on the pinhole. A numerical method to employ in separating the stresses is presented later in this section.

As another example of the interpretation of isochromatic fringe patterns, consider the photograph shown in Fig. 11.3. In this instance, a photoelastic model of a square conduit with a pressurized circular borehole was analyzed. A uniformly distributed load (the pressure) was applied to the circular hole of the model. The stresses along the outside edges of the model can be determined directly from Eq. (11.3) because these edges represent free boundaries of the model. Along the boundary of the circular hole, the surface is not free; however, in this case the stress acting normal to the boundary is known ($\sigma_2 = -p$, the distributed load). From Eq. (10.3) it is clear that:

$$\sigma_1 - \sigma_2 = \sigma_1 + p = \frac{Nf_\sigma}{h}$$

$$\sigma_1 = \frac{Nf_\sigma}{h} - p \tag{11.4}$$

Note that the pressure reading of p is considered as a positive quantity.

Figure 11.3 Dark-field isochromatic fringe pattern for a pressurized section of square tubing with a circular hole.

Some rules for interpreting isochromatic fringe patterns are:

1. The principal stress difference $\sigma_1 - \sigma_2$ can be determined at any point in the model by using Eq. (10.13).
2. If $\sigma_1 > 0$ and $\sigma_2 < 0$, $\sigma_1 - \sigma_2$ can be related to the maximum-shear stress through Eq. (11.1).
3. If $\sigma_1 > \sigma_2 > 0$ or if $0 > \sigma_1 > \sigma_2$, $\sigma_1 - \sigma_2$ cannot be related to the maximum-shear stress and it is necessary to determine σ_1 and σ_2 individually and relate τ_{max} to σ_1 or σ_2 by Eq. (11.2).
4. If the boundary can be considered free (that is, σ_1, or $\sigma_2 = 0$), the other principal stress can be determined directly from Eq. (11.3).
5. If the boundary is not free, but the applied normal pressure is known, the unknown tangential boundary stress can be determined by applying Eqs. (11.4)
6. If the boundary is not free and the applied load is not known, numerical methods must be applied to determine the boundary stresses.

11.2.2 Isoclinic Fringe Patterns

The isoclinic fringe pattern observed in the plane polariscope is used to determine the directions of the principal stresses at any point in the model. In practice, determining isoclinic parameters may be accomplished in one of two ways. First, a number of isoclinic patterns at different polarizer orientations can be obtained and combined to give a composite picture of the isoclinic parameters over the entire field of the model. Second, the isoclinic parameter can be determined at individual points of interest.

An example of a series of isoclinic fringe patterns is shown in Fig. 11.4, where a thick-walled ring has been loaded in diametral compression. The data presented in this series of photographs are combined to give the composite isoclinic pattern illustrated in Fig. 11.5. Several rules can be followed in establishing composite isoclinic patterns from the individual patterns. These rules are:

1. Isoclinics of all parameters must pass through isotropic or singular points.
2. An isoclinic of one parameter must coincide with any axis of symmetry which exists.
3. The parameter of an isoclinic intersecting a free boundary is determined by the slope of the boundary at the point of intersection.
4. Isoclinics of all parameters pass through points of concentrated load because they are singular points.

Inspection of Figs. 11.4 and 11.5 shows that isoclinics of all parameters pass through the isotropic points labeled A through J. At isotropic points, $\sigma_1 = \sigma_2$ and all directions are principal; therefore, isoclinics of all parameters pass through these points. Again, from Fig. 11.5, it is clear that the parameter of an isoclinic can be established from the slope of the boundary at its point of intersection. The reason for this behavior of isoclinics at free boundaries is based on the fact that the boundaries are isostatics or stress trajectories. Hence, the tangential stress at the boundary is principal, and as such, the isoclinic parameter must identify the slope of the boundary at the point of intersection. Also, axes of symmetry are principal directions; therefore, the isoclinics must identify them. The horizontal and vertical axes of the ring shown in Fig. 11.4 are axes of symmetry and are included in the 0° isoclinic family. Finally, points where a concentrated load is applied are singular, and the stress system in the local neighborhood of these points is principal in r and θ. Thus, the principal-stress directions will vary from 0 to 180° in this local region; therefore, isoclinics of all parameters will converge at the point of load application as illustrated in Fig. 11.5.

The isoclinics, lines along which the principal stresses have a constant inclination, give the principal-stress directions in a form that is often confusing. To improve interpretation, the principal-stress directions can be presented in the form of an isostatic or stress trajectory diagram, where the principal stresses are tangent or normal to the isostatic lines at each point. The isostatic diagram can be constructed directly from the composite isoclinic pattern by utilizing the procedure illustrated in Fig. 11.6.

Figure 11.4 Isoclinic fringe patterns for a circular ring subjected to a diametral compressive load.

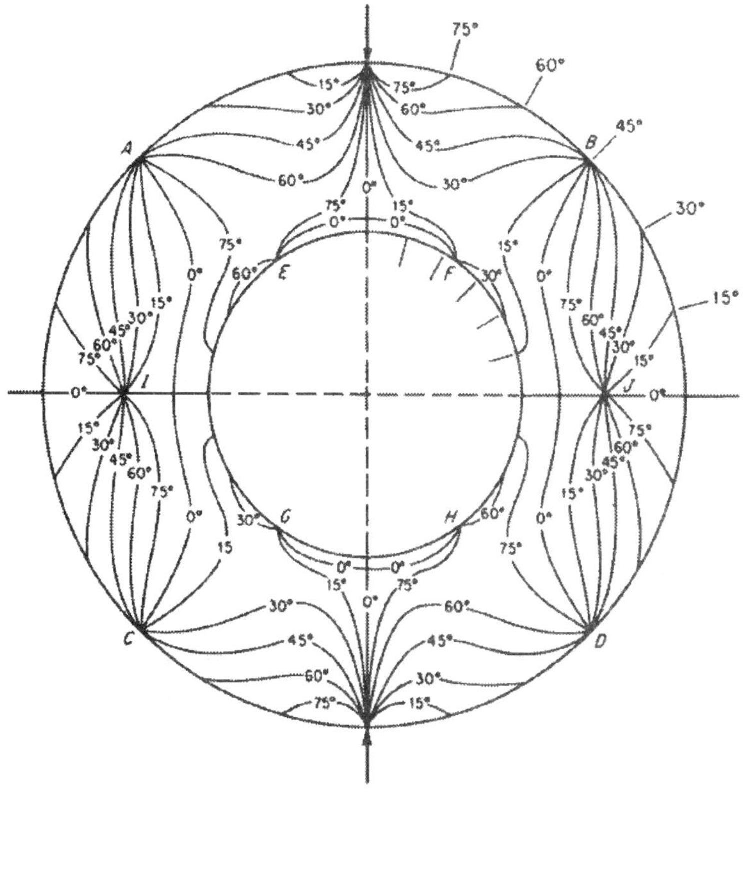

Figure 11.5 Composite isoclinic pattern for a circular ring subjected to a diametral compressive load.

In this construction technique, the stress trajectories are initiated on the 0° isoclinic at arbitrarily spaced points. Lines labeled (1) and oriented 0° from the normal are drawn through each of these arbitrary points until they intersect the 10° isoclinic line. The lines (1) are bisected, and a new set of lines (2) is drawn, inclined at 10° to the vertical to the next isoclinic parameter. Again these lines are bisected, and another set of construction lines (3) is drawn oriented at an angle of 20° to the vertical. The process is repeated until the entire field is covered. The stress trajectories are then drawn by using lines 1, 2, 3, etc. as guides. The stress trajectories are tangent to the construction lines at each isoclinic intersection, as illustrated in Fig. 11.6.

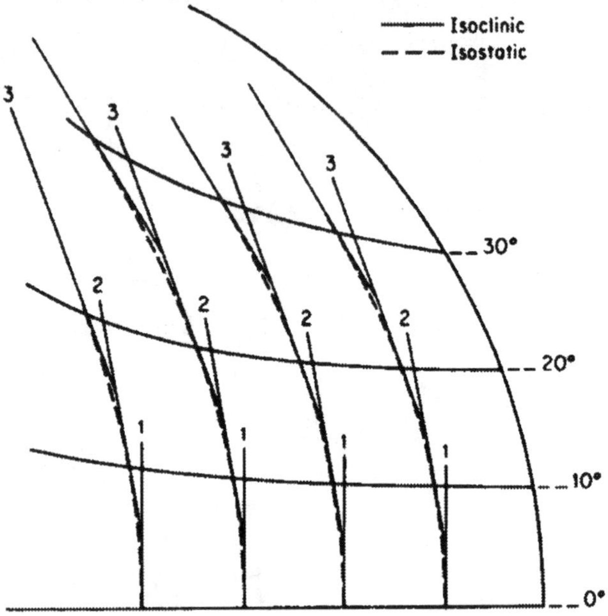

Figure 11.6 Illustration of a construction technique for converting isoclinics to isostatics.

The isoclinic parameters are also employed to determine the shear stresses on an arbitrary plane. First, recall that the isoclinic parameter gives the direction between the x axis of the coordinate system and the direction σ_1 or σ_2. Then refer to Eqs. (1.16) to show:

$$\tau_{xy} = -\frac{\sigma_1 - \sigma_2}{2}\sin 2\theta_2 = \frac{Nf_\sigma}{2h}\sin 2\theta_2 \tag{11.5a}$$

or

$$\tau_{xy} = -\frac{\sigma_1 - \sigma_2}{2}\sin 2\theta_1 = \frac{Nf_\sigma}{2h}\sin 2\theta_1 \tag{11.5b}$$

where θ_1 or θ_2 is the angle between the x axis and the direction of σ_1 or σ_2 as given by the isoclinic parameter. The combined isochromatic and isoclinic data represented by Eqs. (11.5) yield τ_{xy}.

11.2.3 Compensation Techniques

The isochromatic fringe order can be determined to the nearest 1/2 order by employing both the light- and dark-field fringe patterns. Further improvements on the accuracy of the fringe-order determination can be made by employing a digital camera to record the image. The intensity data associated with individual pixels (points) is used to establish fractional fringe orders. However, in some instances point-per-point compensation techniques are employed to establish the fringe order N.

The Tardy method of compensation is very commonly employed to determine the order of the fringe at any arbitrary point on the model. Actually, the Tardy method is often preferred over other compensation techniques because no auxiliary equipment is necessary and the analyzer of the polariscope serves as the compensating device. To employ the Tardy method, the axis of the polarizer is aligned with the principal direction of σ_1 at the point in question. Then, all other elements of the polariscope are rotated relative to the polarizer to give a standard dark-field polariscope. The analyzer is then rotated to produce extinction at the point in question. The interpretation of the angle of rotation of the analyzer in terms of fractional fringe order was discussed in Section 10.7.

11.2.4 Calibration Methods

In most photoelastic analyses, the stress distribution in a complex model is sought as a function of the load. To determine this stress distribution accurately requires the careful calibration of the material fringe value f_σ. The values of f_σ given in the technical literature are typical. The material fringe value of photoelastic materials varies with the supplier, the batch of resin, temperature and age. Consequently, it is necessary to calibrate each sheet of photoelastic material at the time of the experiment. The calibration method presented here is a simple and accurate way to determine f_σ.

In any calibration technique, one must select a body for which the theoretical stress distribution is well established. The model should also be easy to machine and simple to load. The calibration model is loaded in increments, and the fringe order and the loads are noted. From these data, the material fringe value f_σ can be determined.

The circular disk subjected to a diametral compressive load P is often employed as a calibration model because it is easy to machine and to load. The stress distribution along the horizontal diameter, where y = 0, is given by the following expressions:

$$\begin{aligned} \sigma_{xx} &= \sigma_1 = \frac{2P}{\pi hD}\left(\frac{D^2 - 4x^2}{D^2 + 4x^2}\right)^2 \\ \sigma_{yy} &= \sigma_2 = -\frac{2P}{\pi hD}\left(\frac{4D^4}{(D^2 + 4x^2)^2} - 1\right) \\ \tau_{xy} &= 0 \end{aligned} \tag{11.6}$$

where D and h are the diameter and thickness of the disk, respectively, and x is the distance along the horizontal diameter measured from center of disk.

The difference in the principal stresses is given by:

$$\sigma_1 - \sigma_2 = \frac{8P}{\pi hD}\left(\frac{D^4 - 4D^2x^2}{(D^2 + 4x^2)^2}\right) = \frac{Nf_\sigma}{h} \tag{11.7}$$

Solving Eq. (11.7) for f_σ gives:

$$f_\sigma = \frac{8P}{\pi DN}\left(\frac{D^4 - 4D^2x^2}{(D^2 + 4x^2)^2}\right) \tag{11.8}$$

Equation (11.8) can be employed to calibrate photoelastic materials if only a single load P is applied to the disk. In this case, the fringe order N is determined as a function of position x along the horizontal diameter. These values of N and x are then substituted into Eq. (11.8) to give several values of f_σ, which in turn are averaged to reduce errors in the reading of the fringe order N.

More often, however, the center point of the disk, that is, x = y = 0, is used for the calibration point, and several values of load are applied to the model. In this instance, Eq. (11.8) reduces to:

$$f_\sigma = \frac{8}{\pi D}\left(\frac{P}{N}\right) \tag{11.9}$$

Note that the value of f_σ is independent of the model thickness h. The ratio P/N in Eq. (11.9) is determined by plotting a line through several points of a graph of N versus P and then its slope P/N is measured.

11.2.5 Separation Methods

At interior regions of the model, individual values for the principal stresses σ_1 and σ_2 cannot be obtained directly from the isochromatic and isoclinic patterns without using supplementary data or employing numerical methods. The separation method emphasized here is an analytical approach based on the compatibility equation. In previous editions, the shear-difference method which is based on the integration of the equilibrium equations was described. The shear-difference method is tedious and error prone because it is based on differences ($\Delta\tau_{xy}$) drawn from experimental measurements of both isochromatic and isoclinic data at many points along a line. With modern computers, numerical methods, finite element codes, and hybrid techniques such as the analytic separation method are more suitable for determining unknowns at interior points in a model.

The Analytic Separation Method

The compatibility equation for plane stress or plane strain in terms of Cartesian stress components are given by:

$$\frac{\partial^2}{\partial x^2}\left(\sigma_{xx}+\sigma_{yy}\right)+\frac{\partial^2}{\partial y^2}\left(\sigma_{xx}+\sigma_{yy}\right)=0 \qquad (11.10)$$

With constant or zero body forces, equations of this form are known as **Laplace's equation**, and any function that satisfies this equation is said to be a **harmonic function**. In photoelasticity, interest in Laplace's equation arises from the fact that the value of this function can be uniquely determined at all interior points of a region if the boundary values are known. It was shown previously that the photoelastic isochromatics provide an accurate means for determining both the principal stress difference $\sigma_1 - \sigma_2$ at all interior points of a two-dimensional model and in many instances complete boundary-stress information. Knowledge of the principal-stress sum $\sigma_1 + \sigma_2$ throughout the interior, together with the principal-stress difference $\sigma_1 - \sigma_2$ provides an effective means for evaluating the individual principal stresses.

Solution of Laplace's equation by the method of separation of variables yields a sequence of harmonic functions which can be superimposed in a linear combination to give a series representation H that is an approximation of the first stress invariant I. A number of terms which are solutions to $\nabla^2 H = 0$ referred to Cartesian and polar coordinate systems are presented in Table 11.1.

Table 11.1
Solution of $\nabla^2 H = 0$ in Cartesian and polar coordinate systems

Coordinate system	Sequence of harmonic functions		
Cartesian	1	sinh kx sin ky	sinh ky sin kx
$\frac{\partial^2 H}{\partial x^2}+\frac{\partial^2 H}{\partial y^2}=0$	x	sinh kx cos ky	sinh ky cos kx
	y	cosh kx sin ky	cosh ky sin kx
	xy	cosh kx cos ky	cosh ky cos kx
Polar		$x = r\cos\theta$	$y = r\sin\theta$
$\frac{\partial^2 H}{\partial r^2}+\frac{1}{r}\frac{\partial H}{\partial r}+\frac{1}{r^2}\frac{\partial^2 H}{\partial \theta^2}=0$	1	$r^n \cos n\theta$	$r^n \sin n\theta$
	ln r	$r^{-n} \cos n\theta$	$r^{-n} \sin n\theta$

If the region of the model conforms to a regular coordinate system, determining the coefficient for each term in the series representation is considerably simplified. The sequence of harmonic functions reduces to a Fourier series on the boundaries of the region. The unknown coefficients in these cases are the Fourier coefficients obtained by integrating the prescribed boundary values.

If the region of the model does not conform to a particular coordinate system, Fourier analysis cannot be employed to determine the coefficients which satisfy the prescribed boundary conditions. Instead, the method of least squares is used to determine the coefficients. A finite number of harmonic functions is selected to form a truncated series solution for H. Coefficients for the terms in the series are then chosen such that the mean-square difference between the prescribed boundary values and the evaluation of the series along the boundary is minimized. If N harmonic functions F_1, F_2, ..., F_n are selected and the associated unknown coefficients are denoted as C_1, C_2, ..., C_n, the truncated series for H is given by:

$$H = \sum_{n=1}^{N} C_n F_n \tag{11.11}$$

If the function I(s) is used to represent the distribution of the first stress invariant along a boundary of total length L, the series approximation for the function H must be selected such that:

$$\int_0^L \left[I(s) - \sum_{n=1}^{N} C_n F_n \right]^2 ds \Rightarrow \text{minimum} \tag{11.12}$$

The N unknown coefficients of this series can be evaluated by using the method of least squares where the partial derivative of the integral with respect to each of the coefficients is set equal to zero to obtain:

$$\frac{\partial}{\partial C_k} \int_0^L \left[I(s) - \sum_{n=1}^{N} C_n F_n \right]^2 ds = 0 \qquad k = 1, 2, \ldots.., N$$

this can be reduced to:

$$\sum_{n=1}^{N} C_n \int_0^L F_n F_k ds = \int_0^L I(s) F_k ds \qquad k = 1, 2, \ldots.., N \tag{11.13}$$

Equation (11.13) yields N simultaneous equations in terms of the N unknown coefficients. Solution of this set of equations gives the coefficients which provide the best match of boundary values possible with the initial selection of N harmonic functions in the truncated series. By increasing the number of functions in the series for H, the fit can be made as accurate as the original photoelastic determination of I(s).

Other approaches for separating the individual values of σ_1 and σ_2 from the isochromatic data are given in References [7-9].

11.2.6 Scaling Model-to-Prototype Stresses

In the analysis of a photoelastic model fabricated from a polymeric material, the question of applicability of the results is often raised because the prototype is usually fabricated from metal. Obviously, the elastic constants of the photoelastic model are greatly different from those of the metallic prototype. However, the stress distribution obtained for a plane-stress or plane-strain problem by a photoelastic analysis is usually independent of the elastic constants, and the results for an elastic analysis are applicable to a prototype constructed from any isotropic material. This statement is established most readily by reference to the stress equation of compatibility for the plane-stress case. From Eqs. (3.15) and (3.17c) it is clear that:

$$\nabla^2\left(\sigma_{xx}+\sigma_{yy}\right)=-(\nu+1)\left(\frac{\partial F_x}{\partial x}+\frac{\partial F_y}{\partial y}\right) \tag{11.14}$$

This equation of stress compatibility is independent of the modulus of elasticity E and shows that the modulus of the model does not influence the stress distribution. The influence of Poisson's ratio depends on the nature of the body-force distribution. If $\partial F_x/\partial x + \partial F_y/\partial y = 0$, the stress distribution is independent of Poisson's ratio which implies that Poisson's ratio will not affect the results when:

1. $F_x = F_y = 0$ (the absence of body forces).
2. $F_x = C_1$, $F_y = C_2$ (the uniform body-force field, i.e., gravitational)
3. $F_x = C_1\,x$, $F_y = -\,C_1\,y$ (a linear body-force field in x and y).

There are two exceptions to this general law of similarity of stress distributions in two-dimensional bodies. First, if the two-dimensional photoelastic model is multiply connected, Eq. (11.14) does not apply. The multiply-connected body has a hole or series of holes, and the influence of Poisson's ratio will depend upon the loading on the boundary of the hole. If the resultant force acting on the boundary of the hole is zero, the stress distribution is again independent of Poisson's ratio. However, if the resultant force applied to the boundary of the hole is not zero, Poisson's ratio influences the distribution of the stresses. Fortunately, in those examples where the effect of Poisson's ratio has been evaluated, its influence on the maximum principal stress is usually small (less than about 7%).

The second exception to the laws of similitude is when the photoelastic model undergoes appreciable distortion under the action of the applied load. Local distortions are a source of error in analyzing notches, for example, because curvatures are modified and the stress-concentration factors are decreased. These model distortions can be minimized by selecting a model material with a high figure of merit and reducing the applied load to the lowest value consistent with adequate model response.

Because the photoelastic model may differ from the prototype in respect to scale, thickness and applied load, as well as the elastic constants, it is important to extend this treatment to include the scaling relationships. The literature abounds with scaling relationships employing dimensionless ratios and the Buckingham π theory. However, in most photoelastic applications, scaling the stresses from the model to the prototype is a relatively simple matter where the pertinent dimensionless ratios can be written directly. For instance, for a two-dimensional model with an applied load P, the dimensionless ratio for stress is $\sigma hL/P$ and for displacements $\delta Eh/P$. Thus the prototype stresses σ_p are written as:

$$\sigma_p=\sigma_m\frac{P_p}{P_m}\frac{h_m}{h_p}\frac{L_m}{L_p} \tag{11.15}$$

and the prototype displacements δ_p is written as:

$$\delta_p=\delta_m\frac{P_p}{P_m}\frac{E_m}{E_p}\frac{h_m}{h_p} \tag{11.16}$$

where h is the thickness and L is typical lateral dimension. The subscripts p and m refer to the prototype and the model, respectively.

It is clear that scaling between model and prototype is accomplished easily in most two-dimensional problems encountered by the experimentalist. The modulus of elasticity is never a consideration in determining the stress distribution unless the loading deforms the model and changes the load distribution, e.g., contact stresses. Also, Poisson's ratio need not be considered when the body is simply connected and the body-force field is either absent or uniform, i.e., gravity loading.

11.3 MATERIALS FOR TWO-DIMENSIONAL PHOTOELASTICITY

One of the most important factors in a photoelastic analysis is the selection of the proper material for the photoelastic model. Unfortunately, an ideal photoelastic material does not exist, and the investigator must select from the list of available polymers the one which most closely fits his or her needs. The quantity of photoelastic plastic used each year is not sufficient to entice a chemical company into the development and subsequent production of a polymer especially designed for photoelastic applications. As a consequence, the photoelastician must select a model material which is usually employed for some purpose other than photoelasticity. The following list gives properties which an ideal photoelastic material should exhibit. The material should:

1. Be transparent to the light employed in the polariscope.
2. Be sensitive to either stress or strain, as indicated by a low material fringe value in terms of either stress f_σ or strain f_ε.
3. Exhibit linear characteristics with respect to (a) stress-strain properties, (b) stress-fringe-order properties and (c) strain-fringe-order properties.
4. Exhibit mechanical and optical isotropy and homogeneity.
5. Not exhibit viscoelastic behavior.
6. Have a high modulus of elasticity and a high proportional limit.
7. Have sensitivities f_σ or f_ε that are essentially constant with small variations in temperature.
8. Be free of time-edge effects.
9. Be capable of being machined by conventional means.
10. Be free of residual stresses.
11. Be available at reasonable cost.

These criteria are discussed individually in the following subsections.

11.3.1 Transparency

In most applications, the materials selected for photoelastic models are transparent plastics. These plastics must be transparent to visible light, but they need not be crystal clear. This transparency requirement is not difficult to meet because most polymeric materials are colored or made opaque by the addition of fillers. The basic polymer, although not crystal clear, is usually transparent.

In certain special applications which involved a study of the residual stresses in normally opaque materials, e.g., germanium or silicon, infrared polariscopes are used. A few materials are transparent in either the ultraviolet region or the infrared region of the radiant-energy spectrum. Polariscopes can be constructed to operate in either of these regions if advantages can be gained by employing light with very short or very long wavelengths. However, for most conventional stress-analysis purposes, visible-light polariscopes are preferred.

11.3.2 Sensitivity

A sensitive photoelastic material is preferred because it increases the number of fringes which can be observed in the model. If the value of f_σ for a model material is low, a satisfactory fringe pattern can be achieved in the model with relatively low loads. This feature reduces the complexity and size of the loading fixture and limits the distortion of the model. In the case of birefringent coatings, which will be discussed later, a material with a low value of f_σ is essential to reduce problems due to the effects of coating thickness.

Photoelastic materials are available with values of f_σ which range from 0.035 kN/m to 350 kN/m. The situation regarding values of f_ε is not as satisfactory because materials with a sufficiently low value of f_ε are not yet available. Values of f_ε are usually in the range from 0.005 to 0.50 mm. A material with a lower sensitivity would enhance the applicability of the birefringent coating method of photoelasticity[1].

11.3.3 Linearity

Photoelastic models are normally employed to predict the stresses that occur in a metallic prototype. Because model-to-prototype scaling must be used to establish prototype stresses, the model material must exhibit linear stress-strain, optical-stress and optical-strain properties. Very few data are available in the technical literature on optical-strain relationships, but fortunately, the photoelastic method is usually employed to determine stress differences and the lack of data on strain behavior is not serious. The typical stress-strain and stress-fringe-order curves show that most polymeric photoelastic materials exhibit linear stress-strain and stress-fringe-order curves for the initial portion of the curve. However, at higher levels of stress, the material may exhibit nonlinear effects. For this reason, higher stress levels are to be avoided in photoelastic experiments associated with nonlinear polymeric materials.

11.3.4 Isotropy and Homogeneity

Many photoelastic materials are prepared from liquid polymers by casting between two glass plates which form a mold. When the photoelastic materials are prepared by a casting process, the molecular chains of the polymer are randomly oriented and the materials are essentially isotropic and homogeneous. However, some plastics are rolled, stretched or extruded during the production process. In these production processes the molecular chains are oriented in the direction of rolling, stretching or extrusion. Consequently, these materials will exhibit anisotropic properties (both mechanical and optical) and must be annealed to randomly orient the molecular chains to provide a material with isotropic properties.

11.3.5 Creep

Unfortunately, most polymers creep both mechanically and optically over the time required to conduct a photoelastic analysis. Because of the effects of mechanical and optical creep, polymers cannot be truly characterized as elastic materials but must be considered viscoelastic.

One of the first attempts to formulate a mathematical theory of photoviscoelasticity was made by Mindlin who considered a generalized viscoelastic model with both elastic elements and viscous elements. By assuming that the photoelastic effect resulted from only the deformation of the elastic elements, Mindlin showed that the relative retardation could be related to the stress and strain as:

$$\begin{aligned} \sigma_1 - \sigma_2 &= \frac{N}{h} f_\sigma(t) \\ \varepsilon_1 - \varepsilon_2 &= \frac{N}{h} f_\varepsilon(t) \end{aligned} \qquad (11.17)$$

where f_σ and f_ε are written as functions of time rather than as constants.

The results of Eqs. (11.17) are significant because they show that viscoelastic model materials can be employed to perform elastic stress analyses. Because of the viscoelastic nature of the model materials, the stress and strain material fringe values are functions of time. However, over the short time interval needed to record a fringe pattern, both f_σ and f_ε can be treated as constants. An experiment to measure the variation in material fringe value with time shows that with most polymers the stress fringe value f_σ

---

[1] Relatively new methods for measuring retardation and principal stress directions with either the RGB or grey field polariscopes enable accurate measurements from thin coatings using existing materials.

decreases rapidly with time immediately after loading but then tends to stabilize after about 30 to 60 min. In practice, the load is maintained on the model until the fringe pattern stabilizes. The pattern is then photographed, and the material fringe value associated with the time of the photograph is used for the analysis. It should be noted that photoelastic materials vary from batch to batch; hence, each sheet of material must be calibrated at the time of the photoelastic analysis to determine $f_\sigma(t)$. Also, in most photoelastic materials, the polymerization process continues, and f_σ changes with time on a scale of months. Hence, material stored for a year will, in general, exhibit a higher value of f_σ than newly processed material.

11.3.6 Modulus of Elasticity and Proportional Limit

The modulus of elasticity is important in the selection of a material for a photoelastic analysis because it controls the distortion of the model resulting from the applied loads. If the geometry of a boundary changes due to distortion, the photoelastic solution will be in error. For example, a strip with a very sharp notch will have a very high stress concentration at the root of the notch. If the model distorts under load, the sharpness of the notch is decreased and the experimentally determined stress concentration is reduced. Another case where model distortion influences the photoelastic results is the contact problem. In this instance, the stress distribution is a function of the bearing area between the two components and obviously, the modulus of elasticity is influential in controlling this quantity.

The factor which can be used to judge different photoelastic materials in regard to their ability to resist distortion is $1/f_\varepsilon$ or $E/f_\sigma(1 + \nu)$. The best photoelastic materials to resist distortion will exhibit low values for the material fringe value in terms of strain. Because Poisson's ratio for most polymers varies over the limited range between 0.36 and 0.42, the ratio $Q = E/f_\sigma$ is sometimes used to evaluate the merits of the materials. This factor Q is known as the **figure of merit**.

The proportional limit σ_{pl} of a photoelastic material is important in two respects. First, a material with a high proportional limit can be loaded to a higher level without endangering the integrity of the model (it should be noted here that some of the polymers employed in photoelasticity exhibit a brittle fracture and as such fail catastrophically when the ultimate load is reached). Second, a material with a high proportional limit can produce a higher-order fringe pattern which tends to improve the accuracy of the stress determinations. A sensitivity index for a model material can be defined as:

$$S = \frac{\sigma_{pl}}{f_\sigma} \tag{11.18}$$

Superior model materials exhibit high values for both the sensitivity index S and the figure of merit Q.

11.3.7 Temperature Sensitivity

If the material fringe value f_σ changes markedly with temperature, errors can be introduced in a photoelastic analysis by minor temperature variations during the experiment. Typical changes in the material fringe value f_σ with temperature are shown in Fig. 11.7. For most polymers, there is a linear region of the curve where f_σ decreases slightly with temperature. For a commonly used epoxy, the change in f_σ is only 0.022 kN/m-°C in this linear region. However, at temperatures in excess of 150 °F (65 °C), the value of f_σ begins to drop sharply with any increase in temperature, as shown in Fig. 11.7. For conventional two-dimensional photoelastic studies conducted at room temperature (75 °F or 24 °C), the slope of the curve in the linear region is the important characteristic. For most of the materials used in photoelastic experiments, the slope is modest at room temperature and variations in f_σ can be neglected if temperature fluctuations during the test period are limited to ±5 °F or ± 3 °C.

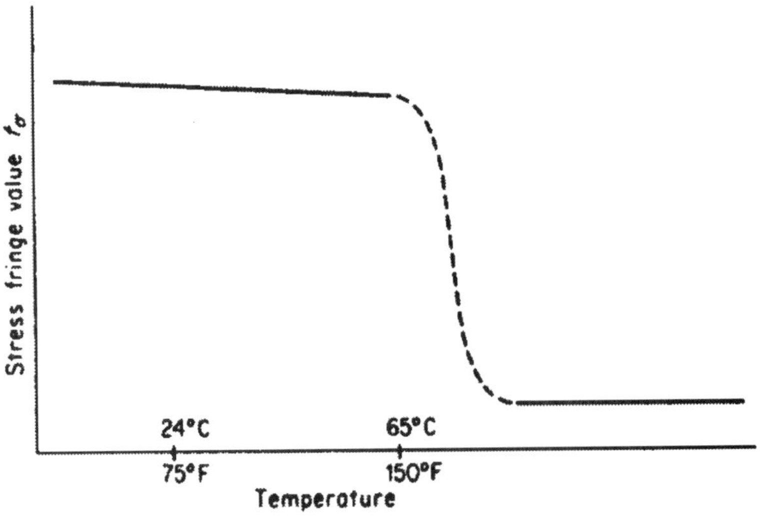

Figure 11.7 Typical curve showing the change in the stress fringe value with temperature.

11.3.8 Time-Edge Effect

When a photoelastic model is machined from a sheet of material and examined under no-load for a period of time, a time-edge stress is induced on the boundary of the model which produces a fringe or a series of fringes parallel to the boundary.

The influence of these time-edge stresses on a photoelastic analysis is quite important. The time-edge fringe pattern observed is due to the superposition of two states of stress, the first associated with the load and the second a result of the time-edge stresses. Because the time-edge stresses are large on the boundary, the errors introduced may be quite large in determining the extremely important boundary stresses.

It has been established that the time-edge stresses are caused by diffusion of water molecules from the air into the plastic or from the plastic into the air. For many plastics, the diffusion process is so slow at room temperature that it requires years to reach an equilibrium state. For this reason, a freshly machined edge of a model usually will be in a condition to accept water from the air (its central region has not been saturated), and time-edge stresses will begin to develop. The rate at which the time-edge stresses develop for a particular photoelastic polymer will depend upon the relative humidity of the air and the temperature.

The epoxy resins are somewhat different from most other polymers in that their diffusion rate is sufficiently high that a saturated condition can be established after 2 to 3 months. If a two-dimensional model is machined from a sheet of material that has been maintained at a constant humidity for several months so that it is in a state of equilibrium (concentration of water is uniform through the thickness of the sheet), and if the model is tested under these same humidity conditions, then time-edge stresses will not develop.

11.3.9 Machinability

Photoelastic materials must be machinable in order to form the complex models employed in photoelastic analyses. Ideally, it should be possible to turn, mill, rout and drill these plastics. Although machinability properties may appear to be a trivial requirement, it is often extremely difficult to machine a high-quality photoelastic model properly. The action of the cutting tool on the plastic often produces heat coupled with relatively high cutting forces. As a consequence, residual stresses due to machining can be introduced and locked in the boundary of the model, making it unsuitable for quantitative photoelastic analysis.

In machining photoelastic models, care must be taken to avoid high cutting forces and the generation of excessive amounts of heat. These requirements can best be accomplished by using sharp carbide-tipped tools, with air or oil cooling and light cuts coupled with a relatively high cutting speed. For two-dimensional applications, complex models may be routed from almost any thermosetting plastic. In this machining method a router motor (20,000 to 40,000/RPM) is used to drive a carbide rotary file. The photoelastic model is mounted to a metal template which describes the exact shape of the final model. The plastic is rough-cut with a jigsaw to within about 2 to 3 mm of the template boundary. The final machining operation is accomplished with the router, as illustrated in Fig. 11.8. The metal template is guided by an oversize-diameter pin which is coaxial with a rotary file. The rate of feed along the boundary of the model is carefully controlled by moving the model along the pin by hand. Successive cuts are taken by reducing the diameter of the stationary pin until it finally coincides with the cutter diameter. By using this technique, satisfactory two-dimensional models can be produced in less than one hour by a skilled technician.

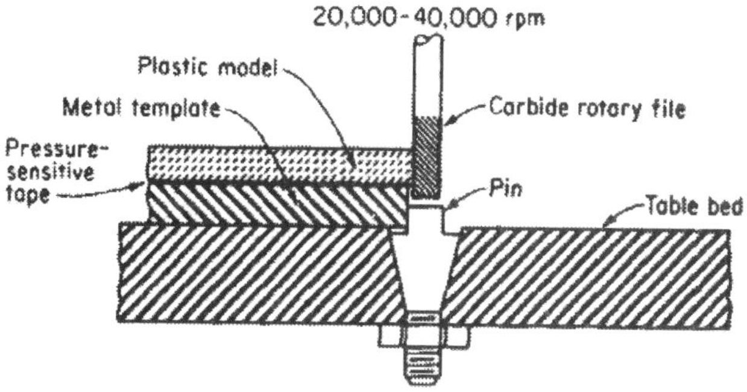

Figure 11.8 Method for machining a two-dimensional photoelastic model with a router.

11.3.10 Residual Stresses

Residual stresses are sometimes introduced into photoelastic plastics during casting and curing operations and almost always by rolling or extrusion processes. They can be observed simply by inserting the sheet of material into a polariscope and noting the order of the fringes in the sheet. The presence of residual stresses in photoelastic models is detrimental because they are superimposed on the true stress distribution produced by loading the model. Because it is difficult to subtract out the contribution due to the residual-stress distribution, their presence in the model material often introduces serious errors into a photoelastic analysis. In certain cases, it is possible to reduce the level of the residual stresses by thermally cycling the sheet above its softening point on a flat plate or in an oil bath.

11.3.11 Cost of Material

Normally the cost of the model material in a photoelastic analysis represents a very small percentage of the total cost of the investigation. For this reason, the cost of the materials should not be overemphasized, and the most suitable material should be selected on the basis of other parameters, regardless of the apparent difference in cost of the material on a pound or kilogram basis.

11.4 PROPERTIES OF COMMON PHOTOELASTIC MATERIALS

A brief examination of the photoelastic literature will show that almost all polymers exhibit temporary double refraction and that many different materials have been employed in photoelastic analyses. The list includes several types of glass, celluloid, gelatin, the glyptal resins, natural and synthetic rubber, fused silica, the phenolformaldehydes, polycarbonate, allyl diglycol (CR-39), and several compositions of epoxies and polyesters. Today, most elastic-stress analyses are conducted by employing one of the following materials:

1. Homalite 100
2. Polycarbonate
3. Epoxy resin
4. Urethane rubber

11.4.1 Homalite 100

Homalite 100 is a polyester which is cast between two plates of glass to form very large sheets. The surfaces of the commercially available sheets are of optical quality, and the material is free of residual stresses. Models can be machined by routing; however, because the material is extremely brittle, edge chipping can be a problem.

Homalite 100 does not exhibit appreciable creep; therefore, the material fringe value can be treated as a constant for loading times in excess of 10 to 15 min. Because moisture absorption is very slow in this material, time-edge effects do not become apparent for several days even under very humid test conditions. The material exhibits both a low figure of merit Q and a low sensitivity index S. High fringe orders cannot be achieved without fracturing the model.

11.4.2 Polycarbonate

Polycarbonate is an unusually tough and ductile polymer which yields and flows prior to fracture. It is known by the trade name Lexan® in the United States and as Makrolan® in Europe. Polycarbonate exhibits both a high figure of merit Q and a high sensitivity index S. It is relatively free of time-edge effects and exhibits very little creep at room temperature.

Polycarbonate is a thermoplastic and is produced in sheet form by an extrusion process. It is available in large sheets with reasonably good surface characteristics. Unfortunately, the extrusion process usually produces some residual birefringence in the sheets. Annealing for an extended period of time at or near the softening temperature is required to eliminate this residual birefringence. Polycarbonate is also difficult to machine. Any significant heat produced by the cutting tool will cause the material to soften and deform under the tool. Routing and side milling are possible only with a continuous flow of coolant at the tool-work-piece interface. Band sawing and hand filing are often required to produce satisfactory model boundaries. Because the material exhibits both yield and flow characteristics, it can also be employed for photoplasticity studies. The birefringence introduced in the plastic state is permanent and is locked into the material on a molecular scale. This behavior makes the material suitable for three-dimensional photoplasticity studies.

11.4.3 Epoxy Resin

Epoxy resins were first introduced in photoelastic applications in the mid-1950s, when they were employed predominantly as materials for three-dimensional photoelasticity. However, a brief review of their properties indicates that they are also quite suitable for a wide variety of two-dimensional applications. The commercial epoxy resins are condensation products of epichlorohydrin and a polyhydric phenol. The basic monomer can be polymerized by using acid anhydrides, polyamides or polyamines. In general, curing with the acid anhydrides requires higher temperatures than curing with the polyamides or polyamines.

A wide variety of epoxy materials can be cast into sheet form. The type of basic monomer, the curing agent, and the percentage of the curing agent relative to the basic monomer can be varied to give an almost infinite number of epoxy materials. The epoxies are usually characterized as brittle materials, but they are easier to machine than the polyesters and polycarbonate. Most of the epoxies exhibit better optical sensitivity than Homalite 100, but they are less sensitive than polycarbonate.

Although the material is susceptible to time-edge effects, the rate of diffusion of water into epoxy is sufficiently high to permit a saturation condition to be achieved in about 2 months. If the sheets are stored until saturated, the model can be cut from the conditioned sheet and little or no time-edge effect will be noted as long as the humidity is held constant. Finally, the material creeps at approximately the same rate as polycarbonate or Homalite 100.

11.4.4 Urethane Rubber

Urethane rubber is an unusual photoelastic material in that it exhibits a very low modulus of elasticity (three orders of magnitude lower than that of the other materials listed) and a very high sensitivity, $f_\sigma \approx$ 0.175 kN/m. The material can be cast between glass plates to produce an amber-colored sheet with optical-quality surfaces. Except for its very low figure of merit, the material ranks relatively well in comparison with the other materials listed. Its strain sensitivity is so low that time-edge effects are negligible; moreover, the material exhibits little mechanical or optical creep. The material can readily be machined on a high-speed router, but it must be frozen at liquid-nitrogen temperatures before its surfaces can be turned or milled.

The material is well-suited for demonstration models. Loads applied by hand are sufficient to produce well-defined fringe patterns, and the absence of time-edge effects permits the models to be stored for years. Also, the material is so sensitive to stress that it can be used to study body-force problems if suitable methods for analyzing fractional fringe orders are employed. Finally, urethane rubber can be used for models in dynamic photoelasticity, where its low modulus of elasticity has the effect of lowering the velocity of the stress wave to less than 90 m/s as compared with 2000 m/s in rigid polymers. The low-velocity stress waves in urethane-rubber models are easy to photograph with moderate-speed framing cameras (10,000 frames per second), which are common, while the high-speed stress waves in rigid polymers require high-speed cameras (200,000 frames per second or more) to produce satisfactory fringe patterns for analysis.

Table 11.2
Summary of the optical and mechanical properties of several photoelastic materials

Property	Homalite 100	Polycarbonate	Epoxy*	Urethane Rubber**
Time-edge-effects	Excellent	Excellent	Good	Excellent
Creep	Excellent	Excellent	Good	Excellent
Machinability	Good	Poor	Good	Poor
Modulus of Elasticity				
psi	560,000	360,000	475,000	450
MPa	3,860	2,480	3,275	3
Poisson's ratio, ν	0.35	0.38	0.36	0.46
Proportional limit, σ_{pl}				
psi	7,000	5,000	8,000	20
MPa	48.3	34.5	55.2	0.14
Stress fringe value♦ f_σ				
lb/in.	135	40	64	1
kN/m	23.6	7.0	11.2	0.18
Strain fringe value♦ f_ε				
in.	0.00033	0.00015	0.00018	0.00324
mm	0.0084	0.0038	0.0046	0.082
Figure of merit, Q				
1/in.	4,150	9,000	7,400	450
1/mm	163	354	292	17
Sensitivity index, S				
1/in.	52	125	125	20
1/mm	2.05	4.92	4.92	0.78

*ERL-2774 with 42 pph phthalic anhydride and 20 parts per hundred hexahydrophthalic anhydride.
**100 parts by weight Hysol 2085 with 24 parts by weight Hysol 3562.
♦ For green light (λ = 546.1 nm).

11.4.5 Model Material Summary

A summary of the mechanical and optical properties of the four photoelastic materials is presented in Table 11.2. It is clear by comparing the figure of merit Q and the sensitivity index S that polycarbonate and the epoxies exhibit superior properties. Unfortunately, the polycarbonate material is difficult to machine, and the epoxy resin materials require special precautions to minimize time-edge effects.

Homalite 100, with its low sensitivity, can be used in applications where high precision and low model distortion are not required. Homalite 100 has the advantage of being available in large sheets with optical-quality surfaces and exhibits low creep and little time-edge effect.

Urethane rubber is extremely useful in special-purpose applications such as demonstration models for instructional purposes. Because of its low material stress fringe value it is also useful for modeling where body forces due to gravity produce the loads. Finally, urethane rubber can be used to great advantage in dynamic photoelastic studies, where its low modulus of elasticity results in low-velocity stress waves which are relatively easy to photograph.

11.5 THREE DIMENSIONAL PHOTOELASTICITY - STRESS FREEZING

The applications of photoelasticity described in previous sections have been limited to two-dimensional techniques for determining stresses in plane models. However, many bodies exist which are three-dimensional in character and cannot be effectively approached by using two-dimensional techniques. For example, it is not practical to relate the integrated optical effects in a complicated three-dimensional model to the stresses in the model. It is possible, however, to construct and load a three-dimensional model and to analyze interior planes of the model by using frozen-stress methods. With the frozen-stress method, model deformations and the associated optical response are locked into a loaded three-dimensional model. Once the stress-freezing process is completed, the model can be sliced and analyzed in a polariscope to obtain interior-stress information.

The frozen-stress method for three-dimensional photoelasticity was initiated by Oppel in Germany. The foundation for the method is the process by which deformations are permanently locked in the model. The four different techniques for locking deformations in the loaded model include stress-freezing, creep, curing, and gamma-ray irradiation methods. In all these methods, the deformations are locked into the model on a molecular scale, permitting the models to be sliced without relieving the locked-in deformations. Because the stress-freezing process is by far the most effective and most popular technique for locking the deformations in the model, it will be the only technique described.

The stress-freezing method of locking in the model deformations is based on the diphase behavior of many polymeric materials when they are heated. The polymeric materials are composed of long-chain hydrocarbon molecules, as illustrated in Fig. 11.9. Some of the molecular chains are well bonded into a three-dimensional network of primary bonds. However, a large number of molecules are loosely bonded together into shorter secondary chains. When the polymer is at room temperature, both sets of molecular bonds, the primary and the secondary, act to resist deformation due to applied load. However, as the temperature of the polymer is increased, the secondary bonds break down and the primary bonds in effect carry the entire applied load. Because the secondary bonds constitute a very large portion of the polymer, the deflections which the primary bonds undergo are quite large yet elastic in character. If the temperature of the polymer is lowered to room temperature while the load is maintained, the secondary bonds will re-form between the highly elongated primary bonds and serve to lock them into their extended positions. When the load is removed, the primary bonds relax slightly, but a significant portion of their deformation is not recovered. The elastic deformation of the primary bonds is permanently locked into the body by the re-formed secondary bonds. Moreover, these deformations are locked-in on a molecular scale; thus the deformation and accompanying birefringence are maintained in small sections cut from the model.

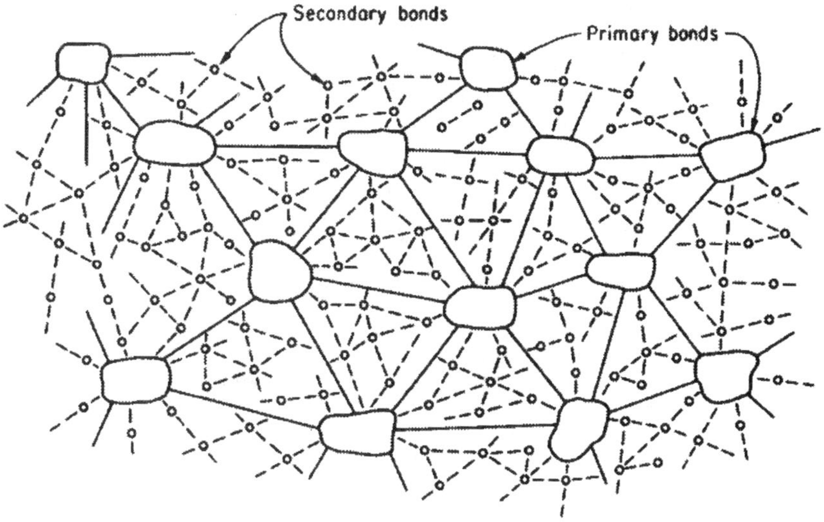

Figure 11.9 Primary and secondary molecular chains in a diphase polymer.

This diphase behavior of polymeric materials constitutes the basis of the stress-freezing process so often employed in three-dimensional photoelasticity. This process can be illustrated by considering the simple tensile specimen shown in Fig. 11.10. The specimen shown in Fig. 11.10a is first loaded at room temperature with an axially applied force P, and a displacement Δl_1 is produced, as shown in Fig. 11.10b. Next, the temperature is increased until the secondary bonds break down and the tensile specimen elongates by an additional amount Δl_2 (see Fig. 11.10c). The temperature is then reduced while the load is maintained until the secondary bonds reform. If thermal expansions and contractions are neglected, the tensile specimen does not change length during the cooling cycle. Finally, the load is removed and the specimen contracts by an amount Δl_1 while retaining a permanently locked-in deformation Δl_2, as illustrated in Fig. 11.10e.

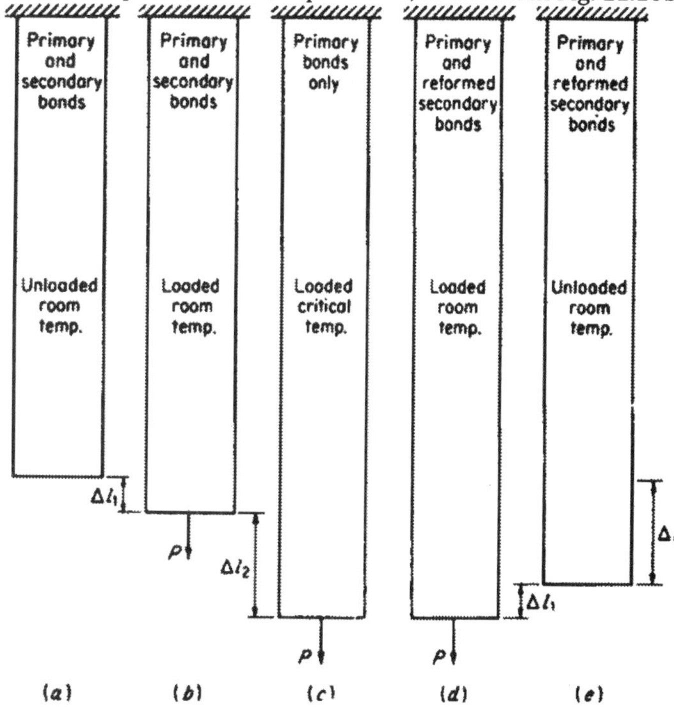

Figure 11.10 Stress freezing a simple tensile specimen.

After stress-freezing a photoelastic model, with its locked-in deformations and attendant fringe pattern, can be carefully cut or sliced without disturbing the character of either the deformation or the fringe pattern. This fact is due to the molecular nature of the locking process, where the deformed primary bonds are locked into place by the re-formed secondary bonds. The cutting process may relieve a molecular layer on each face of a slice cut from a model, but this relieved layer is so thin relative to the thickness of the slice that its effect cannot be observed. An example of a cut across a locked-in fringe pattern is shown in Fig. 11.11.

Figure 11.11 Illustration showing that careful cutting does not disturb the lock-in fringe pattern.

The temperature to which the polymer is heated to break down the secondary bonds is called the **critical temperature**. Actually, this terminology is somewhat unfortunate because the temperature required to break down the secondary bonds is not critical. Instead, degradation of the secondary bonds depends upon both the temperature and the time under load. If a load is applied to a tensile specimen and maintained at temperatures somewhat below the critical temperature, the deflection or fringe-order response will vary with time under load, as illustrated in Fig. 11.12. At temperatures greater than 95% of the critical temperature, the maximum fringe order or deflection is obtained in about 1 min. At the so-called critical temperature the response of the model is almost immediate, with the maximum fringe order or deflection attained in less than 0.1 min. For temperatures between 85 and 90% of the critical temperature, maximum response can be achieved, but the load must be maintained for several hours.

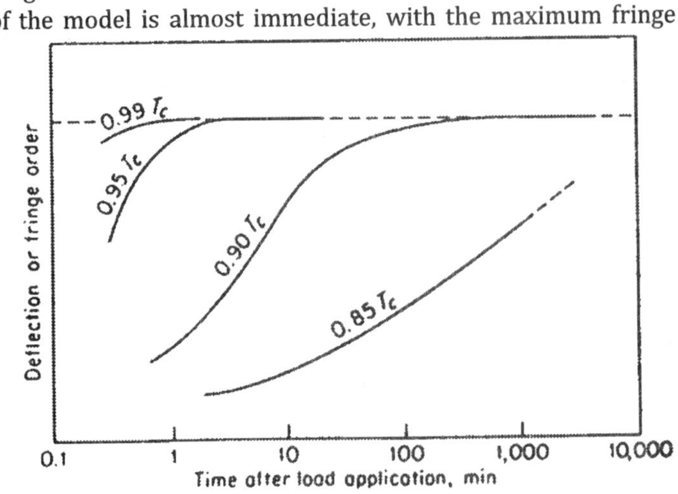

Figure 11.12 Deflection or fringe order as a function of time of load application for temperatures below the critical value.

Recently, Miyano systematically studied the viscoelastic response of an epoxy and its influence on the stress freezing process. His results indicated that a higher figure of merit (N/ε_{max}) was achieved if the stress freezing process was conducted at temperatures less than the glass transition temperature. For the particular epoxy studied, the maximum figure of merit was obtained at 120 °C which is 12 °C lower than the glass transition temperature of 132 °C. The time required for the process was 10^4 s, which implies that the model is held at temperature for about one week. The figure of merit drops rapidly if the stress freezing process is performed at temperatures exceeding the glass transition temperature, although large birefringence can be locked into the model at these elevated temperatures with the application of lower levels of stress.

The stress-freezing process is used today to lock in the fringe pattern in most three-dimensional photoelastic analyses. The method is extremely simple to apply, as temperature control to ± 5% of the critical temperature is usually sufficient. The model response is entirely elastic in character, and the model can be sliced into plane sections for analysis without disturbing the fringe pattern. The procedure for the stress-freezing process is described below.

1. Place the model in an oven with programmed temperature control.
2. Heat the model relatively rapidly until the critical temperature is attained.
3. Apply the required loads.
4. Soak the model at the critical temperature for 2 to 4 hours until a uniform temperature throughout the model is obtained.
5. Cool the model slowly to minimize temperature gradients.
6. Remove the load and slice the model.

11.6 MATERIALS FOR THREE-DIMENSIONAL PHOTOELASTICITY

Since three-dimensional photoelasticity became a reality late in the 1930s with the discovery of the stress-freezing process, a number of different model materials have been introduced and used. These materials include Catalin 61-893, Fosterite, Kriston, Castolite, and the epoxy resins. Of these five materials, only the epoxy-based resins will be discussed here because they have superior properties.

In the discussion of two-dimensional photoelastic materials, 11 requirements of an ideal material were listed. Two additional requirements should be added to this list when considering three-dimensional photoelastic materials:

1. The material should be castable in large sizes and to a final configuration.
2. The material should be cementable.

In three-dimensional analyses, the model often is rather large and intricate, and the need for large castings of the epoxy resin becomes very real. Also, in complex models it is very often desirable to cement parts together much as steel components are welded together to form a complex structure or to cast them to the final size in a precision mold. A complex multi-component model is shown in Fig. 11.13.

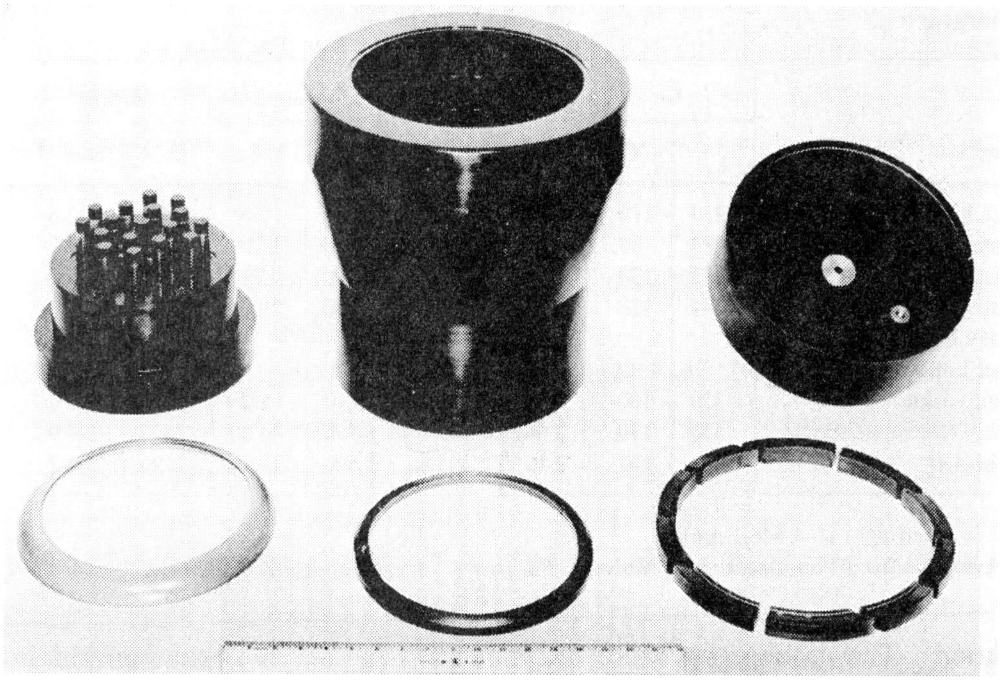

Figure 11.13 A complex model representing a reactor head. Courtesy of M. M. Leven.

The epoxies should be considered as a large family of resins because they cannot be characterized by a single molecular structure. Basically, they consist of condensation products of epichlorohydrin and a polyhydric phenol. The base monomer is commercially available from several sources in both the United States and Europe. The monomer is usually polymerized by employing either an amine or an acid anhydride curing agent. A very large number of these curing agents are commercially available. It is clear, then, that a very large number of polymerized epoxy resins can be produced by varying the type of monomer and the type of hardening agent as well as the percentage of each of these constituents.

Leven investigated the problem of compounding epoxy resins for the specific purpose of optimizing their three-dimensional photoelastic properties. From the results of his investigations, it appears that the choice of the basic monomer does not greatly influence the photoelastic properties of the polymerized resin. The basic monomer should be a liquid in which the hardening agent can easily be dissolved and then mixed. Also, the viscosity of the monomer should be low to permit both easy pouring and the release of bubbles produced by stirring. Finally, the monomer should be slow in reacting to the curing agents so that heat generated during curing can be minimized.

The selection of the hardening agent, on the other hand, is extremely important because it influences the photoelastic properties of the resin to an appreciable extent. Of the two general types of hardening agent, the anhydrides are superior to the amines. The amines are not suitable for large castings because of their extremely exothermic reactions. The heat generated in even modest-size castings of amine-cured epoxies is often sufficient to destroy the casting. Moreover, the locked-in fringe pattern in amine-cured epoxies tends to relax with time, and a decrease in the maximum fringe order of 10 to 20% within a year due to creep of the primary bonds may occur.

The anhydrides (phthalic or hexahydrophthalic) are recommended for the curing of large castings of epoxy resins because of their low exothermic reaction and the castings low susceptibility to time-edge stresses. Of the numerous anhydrides available for curing epoxy resins, Leven has recommended the following composition:

- Liquid epoxy, 100 parts by weight
- Phthalic anhydride, 42 parts by weight
- Hexahydrophthalic anhydride, 20 parts by weight

The pertinent three-dimensional photoelastic properties of this particular epoxy resin are:

- Critical temperature: 324 to 347 °F (162 to 175 °C)
- Effective modulus: 5300 to 6500 psi (37 to 45 MPa)
- Effective material fringe value: 2.48 to 2.84 lb/in (435 to 500 N/m)
- Figure of merit: 2100 to 2450 in^{-1} (83 to 96 mm^{-1})

These properties are compared with the photoelastic properties of other epoxies in Table 11.3.

Table 11.3
Properties of three-dimensional photoelastic materials at their critical temperature

Material	T_{cr}		$f_{effective}$*		$E_{effective}$		Q = E/f	
	°F	°C	lb/in.	N/m	psi	MPa	in.$^{-1}$	mm^{-1}
Epoxy (Cernosek)	285	141	2.00	350	2,900	20.0	1,450	57.1
Epoxy (Leven)	338	170	2.68	469	6,450	44.5	2,407	94.9
PLM-4B♦	248	120	2.20	385	2,500	17.2	1,136	44.7

*For green light (λ = 546.1 nm)
♦ Available from Vishay Measurements Group, Raleigh, NC.

The epoxies cured with acid anhydride agents require an extended time at elevated temperatures to polymerize. Also, the surfaces of the casting exhibit a rind effect due to a reaction between the epoxy and the atmosphere in the oven. The birefringence associated with this rind effect eliminates the possibility of producing "cast to size and shape" models. These two problems (long curing periods and the rind effect) were circumvented with the development of a new material by Cernosek. This epoxy, comprised of:

1. Liquid epoxy (Epon 828)
2. Phthalic anhydride (50% of epoxy weight)
3. Hardener CA-1 (0.67% of epoxy weight)

is a slight modification of the materials developed by Leven. However, the addition of hardener CA-1 (a proprietary blend of aromatic amines) accelerates the cure so that the casting cures in about 12 hrs. In addition to markedly shortening the curing period, the addition of amines eliminates the rind effect and permits casting of photoelastic models to exact "size and shape". The amine-modified-epoxy is less brittle and can be machined easily using standard tooling.

The procedure for mixing and then casting the amine-modified-epoxy involves the following steps. First, the constituents are heated, mixed and partially cured to initiate polymerization. After initial polymerization, the batch is mixed again to disperse the polymerization centers (called mottles) and then the entrapped air is removed by vacuum. Next, the mixture is poured into a precision mold and cured for about 8 hrs. The epoxy solidifies and can be easily stripped from the mold. Any machining that is required is performed at this time because the material is not brittle and relatively easy to mill or turn. Finally, the epoxy, now a completed model, is post cured for an additional 6 h to increase its modulus of elasticity and figure of merit. Properties of the Cernosek material listed in Table 11.3 show that it has a higher figure of merit than the commercially available PLM-4B but a significantly lower figure of merit than the epoxy developed by Leven.

The anhydride-cured epoxy resins exhibit a time-edge-stress behavior which is unique in comparison with the behavior of other plastics. As discussed previously, time-edge effect is due to the diffusion of water molecules from the air into the plastic. The rate at which the water molecules diffuse into the plastic depends upon the diffusion constant and the humidity of the air, i.e., the concentration of water in the air. For most plastics, the diffusion process is so slow that a state of equilibrium is not reached in a period of years; however, for the anhydride-cured epoxy resins, the rate of diffusion is sufficiently rapid to produce saturation in about 2 months. This ability of the anhydride-cured epoxy resins to saturate can be used to control the time-edge stresses in a slice taken from a three-dimensional model. The stress-freezing process drives off the water vapor stored in the model and when the model is sliced in preparation for the photoelastic examination, the water vapor begins to diffuse into the dry plastic. The concentration gradient produces time-edge stresses, which increase to a maximum after about 5 days and then decreases to zero in about 2 months as the slice becomes saturated, i.e., gradients of concentration of water goes to zero. Any changes in the humidity conditions upset the state of equilibrium and create a new set of time-edge stresses. The procedure employed in controlling the time-edge stresses in slices is to heat the slices to 130 °F (55 °C) for 1 to 2 days to drive off the absorbed water vapor. The slices must be examined immediately upon their removal from the oven and stored in a dedicator to avoid reabsorption of the water vapor.

11.7 SLICING THE MODEL AND INTERPRETING THE RESULTS

If a three-dimensional photoelastic model is observed in a polariscope, the resulting fringe pattern cannot, in general, be interpreted. The conditioned light passing through the thickness of the model integrates the stress difference ($\sigma_1' - \sigma_2'$) over the length of the path of the light so that little can be concluded regarding the state of stress at any interior point.

To circumvent this difficulty, the three-dimensional model is usually sliced to remove planes of interest which can then be examined individually to determine the stresses existing in that particular plane or slice. In studies of this type, it is assumed that the slice is sufficiently thin in relation to the size of the model to ensure that the stresses do not change in either magnitude or direction through the thickness of the slice.

The particular slicing plan employed in sectioning a three-dimensional photoelastic model will depend upon the geometry of the model and the information being sought in the analysis. There are, however, some general principles which can be followed in slicing the model as described below.

11.7.1 Surface Slices

The free surfaces of a three-dimensional model are principal because both the stress normal to the surface and the shearing stresses acting on the surface are zero. As an example of a surface slice, consider the flat head on a thick-walled pressure vessel, illustrated schematically in Fig. 11.14. In this example, a surface slice of thickness h is removed from the head and examined at normal incidence, in the polariscope. The fringe pattern observed can be interpreted to give:

$$\sigma_1 - \sigma_2 = \frac{N_z f_\sigma}{h} \quad \text{or} \quad \sigma_{\theta\theta} - \sigma_{rr} = \frac{N_z f_\sigma}{h} \tag{11.19}$$

Because of symmetry, $(\sigma_{\theta\theta} - \sigma_{rr}) = (\sigma_1 - \sigma_2)$. The application of Eq. (11.19) tacitly assumes that the value of $(\sigma_1 - \sigma_2)$ is constant through the thickness of the slice and that the directions of the principal stresses do not rotate along the z axis. To determine the accuracy of these assumptions or to correct for the errors introduced by changes in the stress distribution with slice thickness, the shaving method is often employed. The shaving method consists essentially in removing a thick slice, determining the fringe pattern associated with this slice, and then progressively decreasing the thickness of the slice and establishing the fringe pattern corresponding to each thickness. The results of the analysis are then plotted as a function of slice thickness and extrapolated to h = 0 to establish the surface stresses.

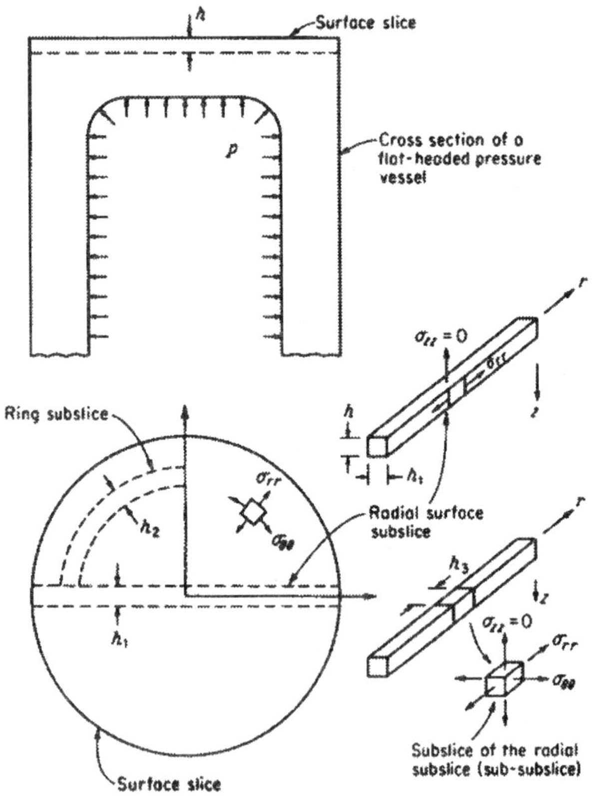

Figure 11.14 Surface slices on a flat-headed, thick-walled pressure vessel.

It is often advantageous to subslice the surface slice to obtain the individual values of σ_1 or σ_2. Two such subslices, a radial subslice, and a ring subslice, are illustrated in Fig. 11.14. The radial subslice is examined in the polariscope with the light passing through the subslice in the y or θ direction. The resulting fringe pattern gives:

$$\sigma_{rr} - \sigma_{zz} = \frac{N_\theta f_\sigma}{h_1}$$

Because $\sigma_{zz} = 0$ at z = 0,

$$\sigma_{rr} = \frac{N_\theta f_\sigma}{h_1} \tag{11.20}$$

Combining Eqs. (11.19) and (11.20) yields:

$$\sigma_{\theta\theta} = f_\sigma\left(\frac{N_z}{h} + \frac{N_\theta}{h_1}\right) \tag{11.21}$$

It should be noted in the analysis of the radial subslice that the influence of the $\sigma_{\theta\theta}$ stress, which is coincident with the direction of light, is not effective. Only the stresses which lie in the plane of the slice normal to the direction of light (that is, σ_{zz} or σ_{rr}) influence the fringe pattern. A value of $\sigma_{\theta\theta}$ will occur in this slice; however, this stress will not influence the nature of the fringe pattern.

An alternative procedure which can be employed in the sectioning of the model is the sub-subslice technique. This technique is illustrated in Fig. 11.14, which shows a small cube removed by sectioning the radial subslice. The cube has dimensions h in the z direction, h_1 in the θ direction, and h_3 in the r direction. By viewing the cube in the polariscope in all three possible directions, the following three relationships can be obtained:

$$\sigma_{\theta\theta} - \sigma_{rr} = \frac{N_z f_\sigma}{h}$$

$$\sigma_{\theta\theta} = \sigma_{\theta\theta} - \sigma_{rr} = \frac{N_r f_\sigma}{h_3} \tag{11.22}$$

$$\sigma_{rr} = \sigma_{rr} - \sigma_{zz} = \frac{N_\theta f_\sigma}{h_1}$$

where the subscript on N indicates the direction of the incident light.

This example shows that surface stresses can be determined by employing various slicing techniques with a three-dimensional model. Because more than one technique is often available, it is advisable to employ at least two different methods and cross-check the results obtained. In this manner, the results can be averaged and the errors minimized.

11.7.2 Principal Slices Other Than Surface Slices

Often, the three-dimensional model will contain planes of symmetry or other planes which are known to be principal. The flat-headed, thick-walled pressure vessel can also be used as an example to illustrate this topic. The meridional plane presented in Fig. 11.15 represents a plane of symmetry which is also known to be a principal plane. Moreover, in the cylindrical portion of the pressure vessel, the transverse or hoop planes are also known to be principal planes where the theoretical solution presented in Section 3.13 will apply. Thus, it is reasonable to section the three-dimensional model of the pressure vessel to obtain a meridional slice and several hoop or transverse slices along the axis of the cylinder. If these slices are examined in the polariscope at normal incidence, the resulting fringe patterns will provide the following data. For the meridional slice:

$$\sigma_1 - \sigma_2 = \frac{N f_\sigma}{h} \tag{10.13}$$

where $(\sigma_1 - \sigma_2) = (\sigma_a - \sigma_{rr})$ in the cylindrical portion of the model and $(\sigma_1 - \sigma_2) = (\sigma_{rr} - \sigma_{zz})$ in the head portion of the model. In the transition region between the cylinder and the head, the principal directions are not known because they differ from the axial or radial directions. Isoclinic results show the extent of the transition region and the directions of the principal stresses in this region.

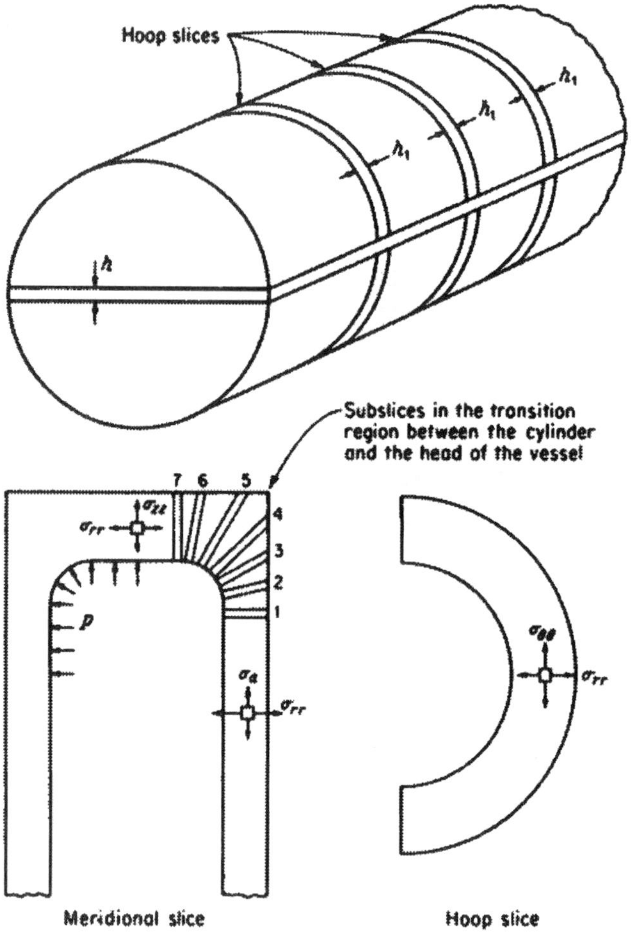

Figure 11.15 Slicing a thick-walled pressure vessel along the planes of principal stress.

On the external boundaries of the meridional slice shown in Fig. 11.15:

$$\sigma_a = \frac{Nf_\sigma}{h} \quad \text{on the vertical boundary}$$

$$\sigma_{rr} = \frac{Nf_\sigma}{h} \quad \text{on the horizontal boundary}$$

On the internal boundaries of the meridional slice:

$$\sigma_a = \frac{Nf_\sigma}{h} - p \quad \text{on the vertical boundary}$$

$$\sigma_{rr} = \frac{Nf_\sigma}{h} - p \quad \text{on the horizontal boundary}$$

At all interior points on the meridional plane, the value of the maximum shear stress τ_{max} is given by:

$$\tau_{max} = \frac{Nf_\sigma}{2h}$$

because one of the two principal stresses is always negative.

The hoop slices in the cylindrical portion of the pressure vessel provide fringe patterns which can be related to the circumferential and radial stresses. Thus:

$$\sigma_{\theta\theta} - \sigma_{rr} = \frac{Nf_\sigma}{h_1} \tag{10.13}$$

On the external boundary $\sigma_{rr} = 0$; therefore:

$$\sigma_{\theta\theta} = \frac{Nf_\sigma}{h_1}$$

On the internal boundary $\sigma_{rr} = -p$; therefore:

$$\sigma_{\theta\theta} = \frac{Nf_\sigma}{h_1} - p$$

and at interior points in the hoop slice, the maximum shear is given by:

$$\tau_{max} = \frac{\sigma_{\theta\theta} - \sigma_{rr}}{2} = \frac{Nf_\sigma}{2h_1}$$

It is clear that the meridional and hoop slices provide sufficient data to give the individual values of the principal stresses on the surfaces of the pressure vessel and the maximum shear stresses on these two planes. In the transitional region between the head and cylinder, it is frequently advisable to subslice the meridional slice (see Fig. 11.15) to determine the circumferential stress $\sigma_{\theta\theta}$ on the boundary. Although the subslices are not principal over their entire length, they are principal at the region near the interior boundary and provide results for $\sigma_{\theta\theta}$ on the interior boundary.

Separation of the principal stresses at interior points in this model is quite involved and requires using numerical methods. However, if the prototype material yields according to follows the maximum-shear-stress theory, the pressure required to fail the vessel in shear can be predicted without separating the stresses. The maximum shear stresses can be obtained directly from fringe patterns of the meridional and hoop slices.

11.7.3 The General Slice

In some instances it is necessary to remove and analyze a slice which does not coincide with a principal plane. For non-principal slices the stresses in the plane of the slice are secondary principal stresses σ_1' and σ_2', which do not coincide with the principal stresses σ_1, σ_2 or σ_3 as illustrated in Fig. 11.16.
The secondary principal stresses in the plane of the slice are given by Eqs. (1.12) as:

$$\sigma_1', \sigma_2' = \frac{\sigma_{xx} + \sigma_{yy}}{2} \sqrt{\left(\frac{\sigma_{xx} - \sigma_{yy}}{2}\right)^2 + \tau_{xy}^2} \tag{1.12}$$

Also, the angle which σ_1' makes with the x axis is given by:

$$\tan 2\theta = \frac{2\tau_{xy}}{\sigma_{xx} - \sigma_{yy}} \quad (1.14)$$

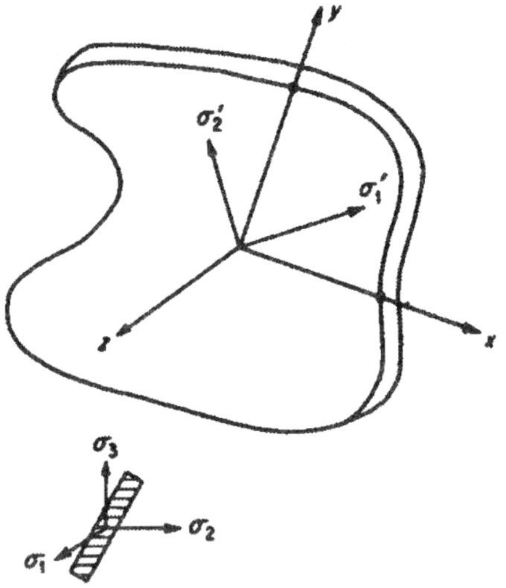

Figure 11.16 The general case where the secondary principal stresses σ_1' and σ_2' in the plane of the slice do not coincide with the principal stresses σ_1, σ_2 or σ_3.

If this slice is observed at normal incidence, with the light along the z axis, both the isoclinic and isochromatic fringe patterns will be due to stresses in the xy plane and will not be influenced by the z components of stress. The isochromatic fringe order is interpreted by employing:

$$\left(\frac{f_\sigma N_z}{h}\right)^2 = (\sigma_1' - \sigma_2')^2_{xy} = \left(\sigma_{xx} - \sigma_{yy}\right)^2 + 4\tau_{xy}^2 \quad (11.23)$$

The isoclinic parameters are related to the stresses in the xy plane by Eqs. (1.14). Thus:

$$\tan^2 2\theta_z = \frac{4\tau_{xy}^2}{\left(\sigma_{xx} - \sigma_{yy}\right)^2}$$

$$\cos^2 2\theta_z = \frac{\left(\sigma_{xx} - \sigma_{yy}\right)^2}{\left(\sigma_{xx} - \sigma_{yy}\right)^2 + 4\tau_{xy}^2} \quad (11.24)$$

$$\sin^2 2\theta_z = \frac{4\tau_{xy}^2}{\left(\sigma_{xx} - \sigma_{yy}\right)^2 + 4\tau_{xy}^2}$$

where θ_z is the angle between σ_1' and the x axis when light passes through the model along the z axis.

Determining the complete state of stress at an arbitrary interior point on a general slice is very involved and requires use of advanced numerical methods. The discussion presented in this Section represents only the interpretation of the fringe patterns obtained in a normal-incidence examination of a non-principal slice.

11.8 EFFECTIVE STRESSES

The data sought in a three-dimensional analysis depends to a great degree upon the specific problem being investigated. In certain instances, boundary stresses will be sufficient, and the methods presented in Section 11.7 can be applied. In other cases, a more complete solution is required, and an extensive analysis of a non-principal slice will be necessary to obtain the six Cartesian components of stress. However, the decision to determine the individual interior stresses should be given a great deal of consideration because application of numerical methods involving the integration of the stress equations of equilibrium is time-consuming and often prone to large errors.

One alternative to separating the individual stresses in three-dimensional problems is to determine the effective stress σ_e, which is defined as:

$$\sigma_e^2 = ½ [(\sigma_1 - \sigma_2)^2 + [(\sigma_2 - \sigma_3)^2 + [(\sigma_3 - \sigma_1)^2] \tag{11.25}$$

The effective stress σ_e is widely accepted as a criterion (von Mises) for plastic yielding or fatigue failure and is often much more significant than either shear or normal stresses.

The effective stress at a point can be determined from a cube of material, cut from the model at any random orientation, at that point, as illustrated in Fig. 11.17. The cube is observed at normal incidence on its three mutually orthogonal faces, and the fringe orders N_x, N_y, N_z and the isoclinic parameters θ_x, θ_y, and θ_z are recorded. Next, consider the following expansion of σ_e:

$$\sigma_e^2 = (\sigma_1^2 + \sigma_2^2 + \sigma_3^2) - (\sigma_1\sigma_2 + \sigma_2\sigma_3 + \sigma_1\sigma_3) \tag{a}$$

and note from Eqs. (1.9) that:

$$\sigma_1^2 + \sigma_2^2 + \sigma_3^2 = I_1^2 - 2I_2 \quad \text{and} \quad \sigma_1\sigma_2 + \sigma_2\sigma_3 + \sigma_1\sigma_3 = I_2 \tag{b}$$

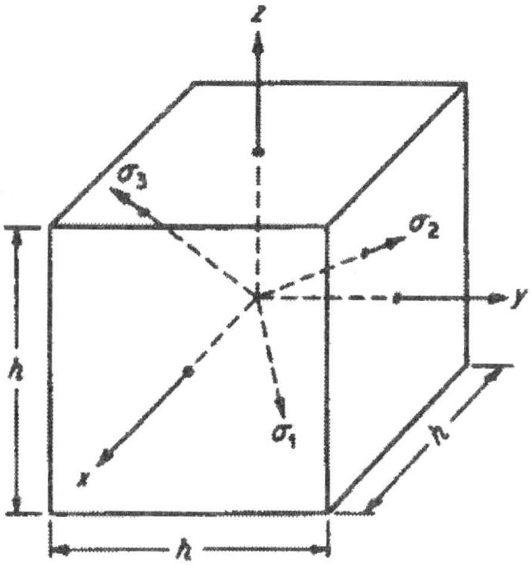

Figure 11.17 A cube removed from any interior point in a three-dimensional; photoelastic model. The x, y and z axes are at arbitrary angles relative to the principal coordinate system.

By combining Eqs. (a) and (b) and using Eqs. (1.8), it is clear that:

$$\sigma_e^2 = I_1^2 - 3I_2 = \sigma_{xx}^2 + \sigma_{yy}^2 + \sigma_{zz}^2 - \sigma_{xx}\sigma_{yy} - \sigma_{yy}\sigma_{zz} + \sigma_{zz}\sigma_{xx} + 3(\tau_{xy}^2 + \tau_{yz}^2 + \tau_{zx}^2) \tag{11.26}$$

It is evident from an inspection of Eqs. (11.23) and (11.26) that the effective stress may be expressed as:

$$\sigma_e^2 = \frac{1}{4}\left[\left(\frac{f_\sigma N_x}{h}\right)^2\left(2+\sin^2 2\theta_x\right) + \left(\frac{f_\sigma N_y}{h}\right)^2\left(2+\sin^2 2\theta_y\right) + \left(\frac{f_\sigma N_z}{h}\right)^2\left(2+\sin^2 2\theta_z\right)\right] \tag{11.27}$$

The value of σ_e obtained by using the cube technique with Eq. (11.27) permits the failure by fatigue or plastic yielding to be predicted for a wide variety of engineering materials employed in machine components or structures.

11.9 APPLICATION OF THE FROZEN-STRESS METHOD[2]

Perhaps one of the best examples of the use of three-dimensional photoelasticity in application to problems related to pressure-vessel design is the analysis of a reactor head. From the point of view of the engineer, there are two factors which make this type of problem difficult. First, the model must be made unusually large in order to scale all the important structural components. Second, the model is extremely complex. The size and complexity of the photoelastic model of the reactor head described in this analysis are shown in Fig. 11.18.

The model material used for the study was ERL-2774 with 50 parts by weight of phthalic anhydride. Standard procedures for stress-freezing with this material involve heating the model to a critical temperature of about 330 °F (165 °C). The bolts on the closure head were tightened against the compression springs to give a clamping force equal to 1.57 times the pressure thrust of the head when a pressure of 34.5 kPa was applied to the model. The model was slowly cooled at a rate of 0.7 °C/h to room temperature with the bolt and pressure loads acting on the model. The stresses and strains representing the elastic stress distribution due to these forces were permanently locked into the model upon completion of this stress-freezing cycle.

Figure 11.18 The assembled head of a pressure vessel ready for stress freezing.

After stress freezing, eight slices were removed from the closure head, as illustrated in Fig. 11.19. Four radial slices, two outer-surfaces slices, and two inner-surface slices were adequate to provide the necessary photoelastic data.

The isochromatic fringe patterns obtained from the radial slices, as shown in Figs. 11.20a and b, were interpreted by employing Eqs. (10.13) and (11.4) to determine the meridional stresses σ_m on the boundary. The circumferential stresses $\sigma_{\theta\theta}$ were obtained by using surface subslices cut from the radial slices, as shown in Fig. 11.19. These subslices give data for N_x and hence:

$$\frac{\sigma_m - \sigma_{\theta\theta}}{p} = \frac{N_z f_\sigma}{hp} \tag{11.28}$$

[2] The data and figures included in this section were provided through the courtesy of M. M. Leven.

Because σ_m is known from the examination of the radial slice, the data obtained from the radial subslice, when used with Eq. (11.28), will give $\sigma_{\theta\theta}$ directly.

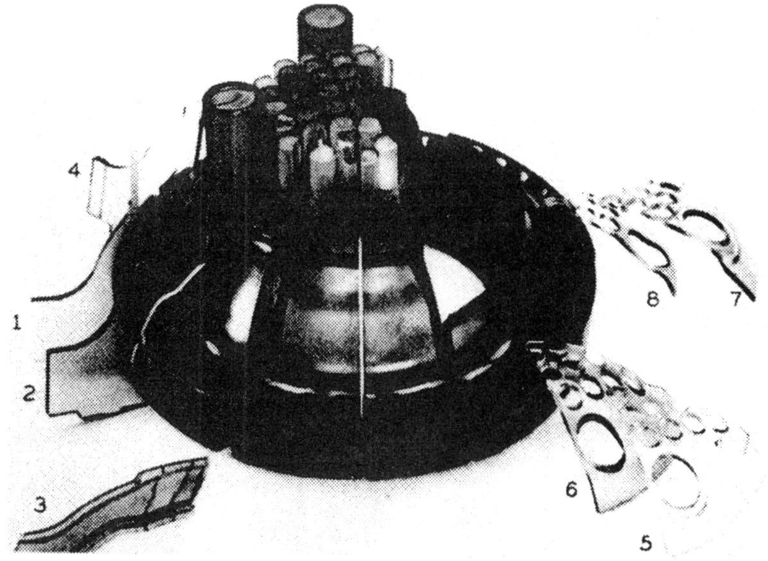

Figure 11.19 The pressure vessel head after stress freezing showing eight slices removed for analysis.

The distribution of the meridional and circumferential stresses in the transitional region of the closure head is shown in Fig. 11.21. The maximum tensile stress occurs on the outer surface at the knuckle between the head and the flange (see point A in Fig. 11.21). The maximum compressive stress occurs at point D on the interior surface of the model. The maximum tensile stress at the knuckle fillet A is equal to 10.2 times the pressure, or 175.8 MPa in the prototype based on a design pressure of 17.2 MPa.

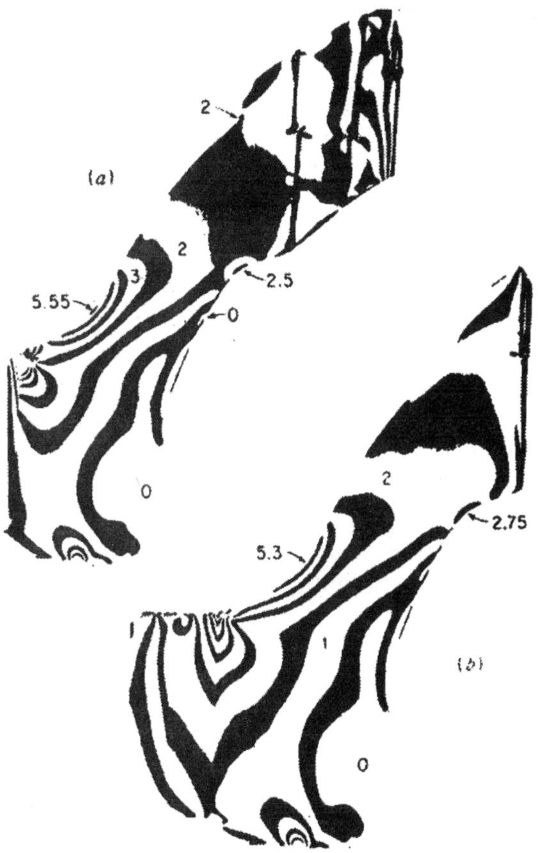

Figure 11.20 Isochromatic fringe patterns for two radial slices from the model.

A typical example of the fringe pattern obtained from one of the surface slices is shown in Fig. 11.22. A maximum fringe order of 9.8 occurred in this slice at refueling penetration V. This proved to be the point of maximum stress (19 times the pressure) in the entire reactor head.

This example problem shows the applicability of the photoelastic method in solving extremely complex design problems. The whole-field potential of the photoelastic method is quite advantageous in examining the region of the penetrations. It was possible to examine large symmetric portions of the head, select critical locations from among the many possible points of stress concentration, and precisely determine the magnitude of both the model stresses and the prototype stresses. In this example, photoelasticity was employed to verify the validity of a particular design. The method could also be employed to improve the design. The procedure would involve testing of a number of models after contour changes have been introduced to increase radii of curvature in regions of high stress.

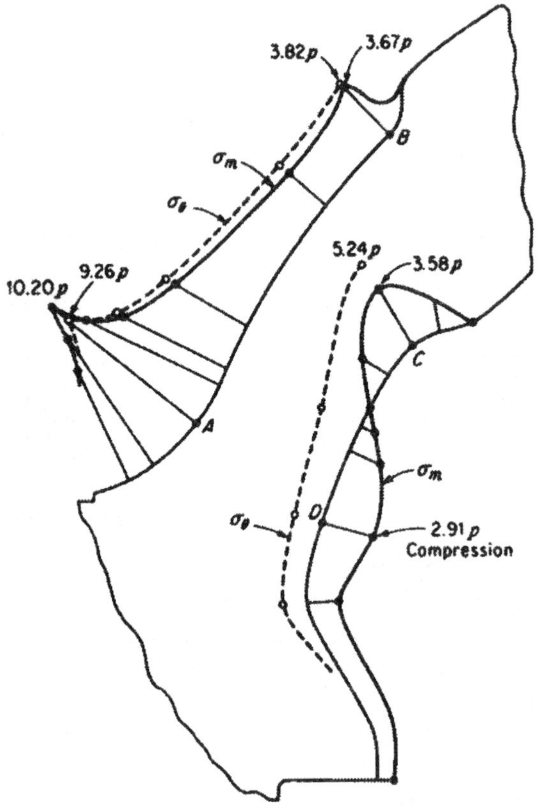

Figure 11.21 Meridional σ_m and circumferential $\sigma_{\theta\theta}$ stresses along the inner and outer surfaces of the pressure vessel head as determined from radial slice # 3 and subslices.

Figure 11.22 Isochromatic fringe patterns for an inner surface slice from the model.

11.10 BIREFRINGENT COATINGS

In the application of a coating, a thin layer of a reactive material is applied to the surface of the body that is to be analyzed. The thin coating is bonded to the surface and displacements at the coating-specimen interface are transmitted without amplification or attenuation. These displacements at the interface produce stresses and strains in the coating and the coating responds. The analyst observes the coating response and infers the stresses on the surface of the specimen based on the observed behavior of the coating.

The two most significant advantages of coatings are first, the capability of applying the coating directly to the prototype. As with strain gages, it is not necessary to model the specimen. The prototype is used under operational conditions. The second advantage is the **whole field response** of the coating. Strain gages respond over small regions of the field and give approximations to strain at a point. Coatings respond over the entire surface of the specimen and give field data rather than point data.

The method of birefringent coatings represents an extension of the procedures of photoelasticity to determine the surface strains in opaque two-and three-dimensional bodies. The coating is a thin sheet of birefringent material, usually a polymer, which is bonded to the surface of the prototype being analyzed. The coating is mirrored at the interface to provide a diffused reflecting surface for the light. When the prototype is loaded, the displacements on its surface are transmitted to the mirrored side of the coating to produce a strain field through the thickness of the coating. The distribution of the strain field over the surface of the prototype, in terms of principal-strain differences, is determined by employing a reflected-light polariscope to record the fringe orders, as illustrated in Fig. 11.23.

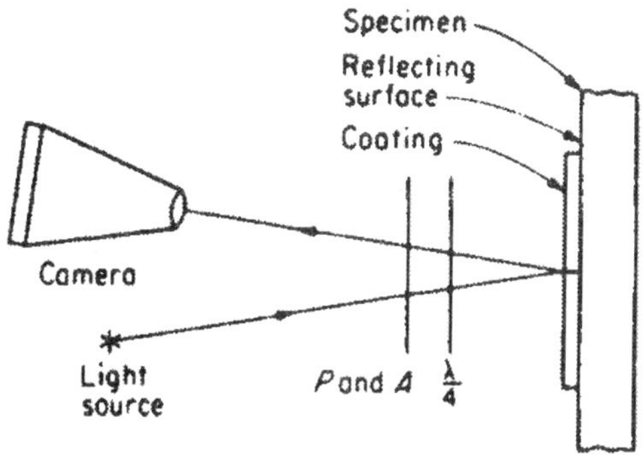

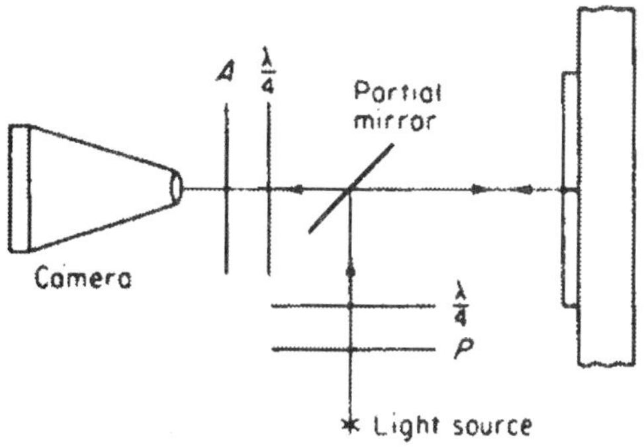

Figure 11.23 Reflection polariscopes commonly employed to record birefringent-coating data.

The birefringent-coating method has many advantages over other methods of experimental stress analysis. It provides full field data that enable the investigator to visualize the complete distribution of surface strains. The method is nondestructive, and because the coatings can be applied directly to the prototype, the need for models is eliminated. Through proper selection of coating materials, the method is applicable over a very wide range of strain. The method is also very useful in converting complex nonlinear stress-analysis problems in the prototype into relatively simple linear elastic problems in the coating. For instance, plastic and viscoelastic response of a prototype can be measured in terms of the elastic response of the birefringent coating. Similarly, the anisotropic response of composite materials can be examined in terms of an isotropic response of the coating.

The concept of birefringent coatings was first introduced by Mesnager in France in 1930, and soon thereafter by Oppel in Germany. These early efforts were not successful due to the lack of suitable adhesives, the low sensitivity of the available polymers and the reinforcing effects occurring when glass was employed as a coating. When the epoxies became available, in the early 1950s, the bonding and sensitivity problems were alleviated. Significant development of the method in the area of materials and technique were made by Fleury and Zandman in France in 1954 and later by D'Agostino, Drucker, Liu, and Mylonas in the United States and Kawata in Japan.

11.10.1 Stress and Strain Optic Relations for Birefringent Coatings

When the specimen is loaded, the surface displacements of the specimen at the specimen-coating interface are transmitted to the birefringent coating if the bond is adequate. As the coating responds to these transmitted displacements, stresses and birefringence are induced. Observing the coating in a reflection polariscope gives a fringe pattern which is related to the surface strains of the specimen. If it is assumed that the coating is sufficiently thin, the strains occurring on the surface of the specimen are transmitted to the coating without distortion. Within the framework of this assumption, it is clear that:

$$\sigma_3 = \sigma_{zz} = 0 \text{ in both coating and specimen}$$

and

$$\varepsilon_1^c\,(x, y) = \varepsilon_1^s\,(x, y) \qquad \varepsilon_2^c\,(x, y) = \varepsilon_2^s\,(x, y) \tag{11.29}$$

where the coordinates x and y lie in the plane of the coating and z is normal to the coating.

With these assumptions regarding the transmission of strains from the specimen to the coating it is easy to show that:

$$\begin{aligned} \sigma_{xx}^c &= \frac{E^c}{E^s\left(1-\nu^{c2}\right)}\left[\left(1-\nu^c\nu^s\right)\sigma_{xx}^s + \left(\nu^c-\nu^s\right)\sigma_{yy}^s\right] \\ \sigma_{yy}^c &= \frac{E^c}{E^s\left(1-\nu^{c2}\right)}\left[\left(1-\nu^c\nu^s\right)\sigma_{yy}^s + \left(\nu^c-\nu^s\right)\sigma_{xx}^s\right] \end{aligned} \tag{11.30}$$

If the Oxyz coordinate system is principal, it is evident from Eq. (11.30) that:

$$\sigma_1^c - \sigma_2^c = \frac{E^c\left(1+\nu^s\right)}{E^s\left(1+\nu^c\right)}\left(\sigma_1^s - \sigma_2^s\right) \tag{11.31}$$

Inspection of Eq. (11.31) shows that the difference in the principal stresses acting in the coating $\sigma_1^c - \sigma_2^c$ is linearly related to the difference in the principal stresses acting on the surface of the specimen $\sigma_1^s - \sigma_2^s$. The elastic constants E^c, E^s, ν^c and ν^s influence the magnitude of $\sigma_1^c - \sigma_2^c$. The photoelastic response of the coating is related to $\sigma_1^c - \sigma_2^c$ by employing Eq. (10.13) and noting that the optical path length is $2h^c$. Thus,

$$\sigma_1^c - \sigma_2^c = \frac{Nf_\sigma}{2h^c} = \frac{E^c\left(1+\nu^s\right)}{E^s\left(1+\nu^c\right)}\left(\sigma_1^s - \sigma_2^s\right) \tag{11.32a}$$

and the difference in the principal stresses for the specimen is given by:

$$\sigma_1^s - \sigma_2^s = \frac{E^s\left(1+\nu^c\right)}{E^c\left(1+\nu^s\right)}\frac{Nf_\sigma}{2h^c} \tag{11.32b}$$

It is clear that $\sigma_1^s - \sigma_2^s$ can be determined from the isochromatic fringe order in the birefringent coating provided E^c, E^s, ν^c, ν^s, f_σ, and h^c are known. In certain instances it may be preferable to work in terms of strain instead of stress. This transformation is quite simple because it has been assumed that $\varepsilon_1^c - \varepsilon_2^c = \varepsilon_1^s - \varepsilon_2^s$; hence:

$$\varepsilon_1^s - \varepsilon_2^s = \frac{Nf_\varepsilon}{2h^c} = \frac{(1+\nu^c)}{E^c}\frac{Nf_\sigma}{2h^c} \tag{11.33}$$

By using Eq. (11.33), the birefringent coating can be employed as a strain gage to give the difference in principal strains $\varepsilon_1^s - \varepsilon_2^s$.

The strain-optic relationship presented in Eq. (11.8) is often written in the form:

$$\varepsilon_1^s - \varepsilon_2^s = \varepsilon_1^c - \varepsilon_2^c = \frac{Nf_\varepsilon}{2h^c} = \frac{N}{2h^c}\frac{\lambda}{K} \tag{11.34}$$

where K is the strain coefficient for the coating and λ is the wavelength in μ in. of the light being used—that is, 21.5 μ in. or 546.1 nm for mercury green light. This alternative form of the strain-optic law is more general because the same strain coefficient K for a coating (provided by the manufacturer) can be employed with light sources having different wave-lengths. For a perfectly linear elastic photoelastic material the constants f_σ, f_ε, and K are related by the expressions:

$$\begin{aligned} f_\varepsilon &= \frac{\lambda}{K} = \frac{1+\nu^c}{E^c} f_\sigma \\ f_\sigma &= \frac{E^c}{1+\nu^c}\frac{\lambda}{K} = \frac{E^c}{1+\nu^c} f_\varepsilon \end{aligned} \tag{11.35}$$

The isochromatic and isoclinic data at a point in a coating is employed with Eq. (2.18) to establish the shearing strain γ_{xy}^s as:

$$\gamma_{xy}^s = \gamma_{xy}^c = (\varepsilon_1^c - \varepsilon_2^c)\sin 2\phi_1 \tag{11.36}$$

where ϕ_1 is the isoclinic parameter defining the angle between σ_1 and the x axis. From Eqs. (11.33) and (11.36), it is clear that:

$$\gamma_{xy}^s = \frac{Nf_\varepsilon}{2h^c}\sin 2\phi_1 \tag{11.37}$$

and from the stress-strain relationship expressed by Eq. (2.20), it is clear that:

$$\tau_{xy}^s = \frac{E^s}{2(1+\nu^s)}\frac{Nf_\varepsilon}{2h^c}\sin 2\phi_1 = \frac{E^s(1+\nu^c)}{E^c(1+\nu^s)}\frac{Nf_\sigma}{4h^c}\sin 2\phi_1 \tag{11.38}$$

The photoelastic data obtained from the isochromatic and isoclinic fringe patterns are not sufficient, in general, to determine the individual values of the principal stresses and strains. Auxiliary methods must be employed to determine σ_1, σ_2, and τ_{max}.

11.10.2 Coating Sensitivity

The response of a photoelastic coating to a stress field is controlled by a number of factors. The effects of these factors is evaluated by defining a stress sensitivity index $S_\sigma{}^s$ as:

$$S_\sigma^s = \frac{N}{\sigma_1^s - \sigma_2^s} \tag{11.39}$$

By substituting Eqs. (11.32b) and (11.35) into Eq. (11.39), the stress sensitivity index $S_\sigma{}^s$ is expressed in terms of coating properties and specimen properties as:

$$S_\sigma^s = \frac{2h^c}{f_\varepsilon}\frac{1+\nu^s}{E^s} = C_c C_s \tag{11.40}$$

where $C_c = 2h^c/f_\varepsilon$ is the coating coefficient of sensitivity.
$C_s = (1 + \nu^s)/E^s$ is the specimen coefficient of sensitivity.

The equation for the stress-sensitivity index indicates that optical response is increased, for a given coating and specimen material, only by increasing the coating thickness. Arbitrary increases in coating thickness are usually not possible, however, because of errors associated with thick coatings.

In applications of coatings to elastic-stress problems, the maximum response of the coating occurs when some point on the specimen yields. If the principal stresses in the specimen are of opposite sign, and if the material follows the Tresca yield criterion, the maximum stresses and strains at yielding are given by:

$$\left(\sigma_1^s - \sigma_2^s\right)_{max} = S_y \tag{11.41a}$$

and

$$\left(\varepsilon_1^s - \varepsilon_2^s\right)_{max} = C_s S_y \tag{11.41b}$$

where S_y is the yield strength of the specimen material. The maximum fringe order observed as the specimen begins to yield follows from Eqs. (11.39), (11.40), and (11.41a) as:

$$N_{max} = C_c C_s S_y = \frac{2h^c}{f_\varepsilon}\frac{1+\nu^s}{E^s} S_y \tag{11.42}$$

It is evident from Eq. (11.42) that N_{max} is a function of the three parameters S_y, C_s, and C_c. Typical values for N_{max} in a typical coating bonded to several engineering materials are listed in Table 11.4.

It is evident from Table 11.4 that the maximum optical response exhibited by a photoelastic coating depends strongly on the properties of the specimen material. Maximum fringe orders range over two orders of magnitude, with very low responses on concrete and glass and relatively high responses on high-strength steel, aluminum, and composite materials. The coating thickness and the methods employed to determine fringe orders in any elastic stress analysis will depend on the specific problem. Where the optical response is low, thick coatings, point-by point compensation methods, or special instruments such as the RGB or grey field polariscopes are required. With higher-strength materials, thin coatings can be employed, and the fringe orders can be photographed or observed directly on the model.

11.10.3 Coating Materials

One of the most important decisions in a photoelastic analysis is the selection of the proper material for the photoelastic model. Indeed, major advances in the application of two- and three-dimensional photoelasticity and birefringent coatings occurred only after the introduction of suitable materials. The physical properties which an ideal coating should exhibit include the following:

1. A high optical strain coefficient K to maximize coating response.
2. A low modulus of elasticity E^c to minimize reinforcing effects.
3. A high resistance to both optical and mechanical stress relaxation to insure stability of the measurement with time.
4. A linear strain-optical response to minimize data-reduction problems.
5. A good adhesive bond to ensure perfect strain transmission between coating and specimen.
6. A high proportional limit to increase the range of strain over which the coating can be utilized.
7. Sufficient pliability to permit use on curved surfaces of three-dimensional components.

Table 11.4
Yield strength, specimen coefficient of sensitivity, strain difference and maximum fringe order at yielding for a typical birefringent coating.

Material	S_y		C_s		$(\varepsilon_1^s - \varepsilon_2^s)_{max}$	N_{max} *
	ksi	MPa	10^{-8} /psi	10^{-11} m^2 /N	μm/m	Fringes
Steel						
HR 1020	35	240	4.3	6.2	1,500	1.86
CD 1020	45	310	4.3	6.2	1,940	2.40
HT 1040	80	550	4.3	6.2	3,450	4.26
HT 4140	130	900	4.3	6.2	5,600	6.92
HT 5210	180	1,240	4.3	6.2	7,700	9.58
Maraging (18 Ni)	250	1,720	4.9	7.1	12,200	15.20
Aluminum						
1100 H16	20	140	13.3	19.3	2,660	3.30
3004 H34	29	200	13.3	19.3	3,860	4.78
2024 T3	50	345	13.3	19.3	6,670	8.25
7075 T	73	500	13.3	19.3	9,700	12.00
Magnesium, AM 11	21	145	20.8	30.2	4,380	5.42
Cartridge Brass	63	435	8.7	12.6	5,480	6.80
Phosphor bronze	75	515	8.7	12.6	6,520	8.10
Beryllium copper	70	480	7.7	11.2	5,400	6.70
Glass	3	21	12.5	18.1	375	0.46
Concrete (compression)	4	28	42.0	60.9	1,680	2.10
Plastic, nylon 6-6	12	83	467	677	56,000	69.00
Glass-reinforced-plastic	120	830	31.8	46.1	38,200	47.40

In most instances, the selection of a coating material involves a compromise because no material exhibits all these characteristics. As an example, the first coating material employed by Mesnager was glass, which exhibited a relatively high sensitivity to strain. However, even a thin sheet of glass reinforces a component significantly and glass cannot be applied to curved surfaces. A number of polymers are available which can be employed as photoelastic coatings. Typical properties of these materials are listed in Table 11.5.

An examination of Table 11.5 shows that polycarbonate exhibits a superior combination of properties. Unfortunately, it is available only in sheet form and techniques have not yet been established to allow contouring of the sheets to three-dimensional surfaces with compound curvatures. The epoxies also exhibit good sensitivity and are preferred when contouring is necessary. The casting and contouring process with epoxy materials will be discussed in Section 15.10.4. For large strains, which are often encountered in plastic analyses, modified epoxies (a blend of rigid and flexible resins of a copolymer of epoxy and polysulfide) are employed as coatings because they exhibit a linear response over a large range of strain (30%). Polyurethane rubber is used in applications involving very large strains where sensitivity is not an issue but where the strains encountered range from 30 to 100%. Glass is most commonly used in optical transducers, where its excellent stability with time and the environment is an important characteristic.

Table 11.5
Properties of different photoelastic materials for coating applications.

Material	Modulus, E ksi	Modulus, E MPa	Sensitivity K	Sensitivity $1/f_\varepsilon$ fringes/in.*	Sensitivity $1/f_\varepsilon$ fringes/mm	Strain limit, %
Glass	10,000	69,000	0.14	6,500	256	0.10
Polycarbonate	320	2,210	0.16	7,300	287	10.0
PS-1	360	2,480	0.14	6,400	251	5.0
Epoxy with anhydride	475	3,280	0.12	5,400	213	2.0
Epoxy with amine	450	3,100	0.09	4,000	157	
PS-8	450	3,100	0.09	4,000	157	3.0
Homalite 100	560	3,860	0.04	1,950	78	1.5
Modified epoxy	2.8	19	0.02	1,000	39	15.0
PS-3	30	210	0.02	940	37	30.0
Polyurethane	0.5	3.5	0.008	380	15	
PS-4	0.5	4.0	0.009	430	17	40.0
PS-6	0.1	0.7	0.0006	28	1.1	100.0

With mercury green light (λ = 21.5 μin. = 546 nm). PS-1, PS-3, PS-4, PS-6 and PS-8 are commercially available from Vishay Measurements Group.

11.10.4 Bonding the Coating

A successful application of photoelastic coatings requires a perfect adhesive bond between the coating and the component. For this reason, careful attention must be given to component and coating surface preparation, the adhesive, and the bonding procedure. The surface of the component should be smooth and free of all foreign material. A satisfactory component surface can normally be obtained by sanding and cleaning with chemical agents and solvents. For plane surfaces, a pre-cured sheet of coating can be machined to size, cleaned, and then bonded to the component. During bonding, precise matching of the geometry of the coating to that of the test specimen is required. Alternatively, the sheet of coating can be bonded to the test specimen and after the adhesive cures, the specimen can be used as a template to machine the coating to the proper shape.

The epoxy adhesive normally used with birefringent coatings is filled with aluminum powder. The adhesive is spread evenly over the surface of the component, and the sheet of coating is applied. The coating should be positioned at one end and slowly rotated and pressed into position to work out the air bubbles. Pressure is not applied as the adhesive cures because residual stresses can be introduced in the coating by the clamping devices. The final layer of adhesive is between 0.003 and 0.010 in. (0.08 and 0.25 mm) thick, depending primarily on the flatness of the specimen surface.

The application of photoelastic coatings to curved surfaces is best achieved by using the contoured-sheet method illustrated in Fig. 11.24. A sheet of epoxy is cast to the desired thickness on a plate coated with Teflon or some other mold release agent. The polymerization process is made to proceed slowly by selecting proper amine curing agents. During the first stage of polymerization, the coating material slowly transforms from a liquid (stage A) to a rubbery solid (stage B). In stage B, the sheet is soft and pliable. Also, the strain-optic coefficient is very low; therefore, large deformations can be imposed on the sheet without introducing photoelastic response. At this stage, the sheet is stripped from the casting plate and contoured by hand to fit the curved surfaces of the test specimen.

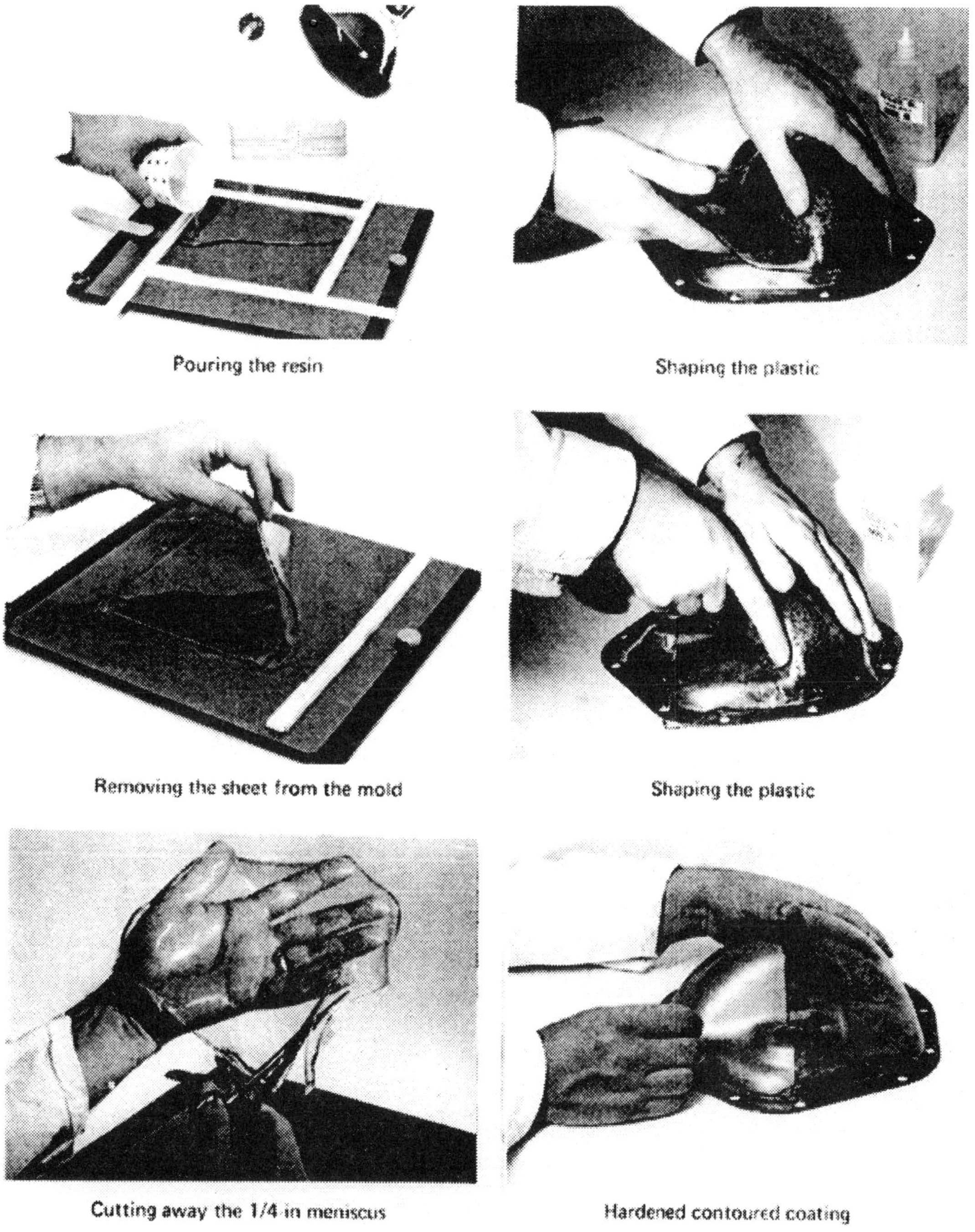

Figure 11.24 Sequence of operations in applying contour sheets.

During the next stage (C) of polymerization, as the coating becomes rigid and optically responsive, it should be maintained in contact with the surface of the test specimen. When the polymerization is completed, the rigid contoured shell of coating can be stripped from the specimen, cleaned and checked for thickness variations. The shell can then be trimmed to match adjoining sections of coating and cemented in place.

This contour-sheet method enables the coating to be applied to specimens of almost any shape. It permits the coating to be formed into stress-free shells of reasonably uniform thickness which conform closely to the surface to be coated. In actual operation, considerable experience is required to develop the skills associated with forming the shells used to coat a complex three-dimensional specimen.

11.10.5 Effects of Coating Thickness

When a photoelastic coating is bonded to a specimen, only in a few instances are the strains transmitted to the coating without some modification or distortion. More realistically, the coating is considered as a three-dimensional extension of the specimen which is loaded by means of shear and normal tractions at the interface. These tractions vary so that the displacements experienced by the coating and the specimen at the interface are identical (as dictated by perfect bonding). Thus, in the most general case:

1. The average strain in the coating does not equal the strain at the interface.
2. A strain gradient exists through the thickness of the coating.
3. The coating serves to reinforce the specimen.

It is evident that these thickness effects tend to vanish as the coating thickness approaches zero. However, coatings with finite thickness (usually 0.50 to 3.00 mm) are required to obtain a sufficiently high fringe count for accurate fringe-order determinations. As a result, the question naturally arises regarding the magnitude of the error associated with thickness effects in the application of the coating method.

The topic of thickness effects will be treated here by beginning with the simple model of the coating. The model of the coating will be made progressively more complex as additional factors influencing the behavior of the coating are introduced. Where possible, experimental verifications are used to justify assumptions and to minimize the complexity of the analysis.

Reinforcing Effects

When a birefringent coating is applied to a specimen and subjected to loads, the coating carries a portion of the load and consequently the strain on the surface of the specimen is reduced. It is possible in many cases to calculate the reinforcing effect due to the birefringent coating and to establish correction factors which are employed in a simple fashion to account for the reinforcement. In this section, the reinforcement due to the coating will be determined for plane-stress and flexural problems. For plane-stress, an element from a coated specimen is isolated as shown in Fig. 11.25.

A similar element from an uncoated specimen is also isolated, and the forces acting in the x direction on both elements can be equated to give:

$$h^s \, dy \, \sigma_{xx}^u = h^s \, dy \, \sigma_{xx}^s + h^c \, dy \, \sigma_{xx}^c$$

from which

$$\sigma_{xx}^u = \sigma_{xx}^s + \frac{h^c}{h^s} \sigma_{xx}^c \qquad (11.43a)$$

The corresponding expression for forces in the y direction is:

$$\sigma^{u}_{yy} = \sigma^{s}_{yy} + \frac{h^{c}}{h^{s}}\sigma^{c}_{yy} \qquad (11.43b)$$

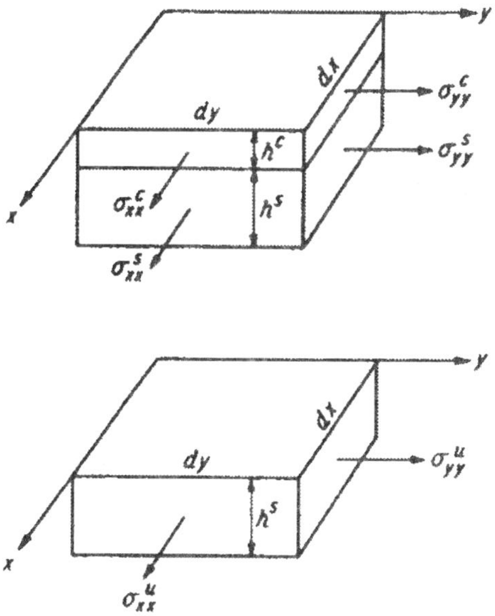

Figure 11.25 Comparison of stresses in coated and uncoated elements from a plane-stress specimen.

If it is again assumed that:

$\varepsilon_{xx}{}^{c} = \varepsilon_{xx}{}^{s}$ $\qquad$ $\varepsilon_{yy}{}^{c} = \varepsilon_{yy}{}^{s}$ $\qquad$ $\sigma_{zz}{}^{c} = \sigma_{zz}{}^{s} = 0$

both the coating and the specimen are in a state of plane stress, and from Eqs. (2.20) it is apparent that:

$$\frac{E^{s}}{1-\nu^{s2}}\left(\varepsilon^{u}_{xx} + \nu^{s}\varepsilon^{u}_{yy}\right) = \frac{E^{s}}{1-\nu^{s2}}\left(\varepsilon^{s}_{xx} + \nu^{s}\varepsilon^{s}_{yy}\right) + \frac{h^{c}}{h^{s}}\frac{E^{c}}{1-\nu^{c2}}\left(\varepsilon^{c}_{xx} + \nu^{c}\varepsilon^{c}_{yy}\right) \qquad (a)$$

$$\frac{E^{s}}{1-\nu^{s2}}\left(\varepsilon^{u}_{yy} + \nu^{s}\varepsilon^{u}_{xx}\right) = \frac{E^{s}}{1-\nu^{s2}}\left(\varepsilon^{s}_{yy} + \nu^{s}\varepsilon^{s}_{xx}\right) + \frac{h^{c}}{h^{s}}\frac{E^{c}}{1-\nu^{c2}}\left(\varepsilon^{c}_{yy} + \nu^{c}\varepsilon^{c}_{xx}\right) \qquad (b)$$

Subtracting Eq. (b) from Eq. (a) and simplifying yields:

$$\varepsilon^{u}_{xx} - \varepsilon^{u}_{yy} = \left(1 + \frac{h^{c}E^{c}(1+\nu^{s})}{h^{s}E^{s}(1+\nu^{c})}\right)\left(\varepsilon^{c}_{xx} - \varepsilon^{c}_{yy}\right) \qquad (11.44)$$

This equation may be rewritten as:

$$\varepsilon^{u}_{xx} - \varepsilon^{u}_{yy} = F_{CR}\left(\varepsilon^{c}_{xx} - \varepsilon^{c}_{yy}\right)$$

where

$$F_{CR} = \left(1 + \frac{h^{c}E^{c}(1+\nu^{s})}{h^{s}E^{s}(1+\nu^{c})}\right) \qquad (11.45)$$

The term F_{CR} represents a correction factor which must be applied to the strain difference ($\varepsilon_{xx}{}^{c} - \varepsilon_{yy}{}^{c}$) obtained from the birefringent coating to establish the true value of the principal strain difference in the uncoated specimen. The correction factor F_{CR} accounts for the reinforcement due to the presence of the birefringent coating.

A graph showing F_{CR} as a function of the thickness ratio h^c/h^s is presented for a number of different materials in Fig. 11.26. These results are based on values of E^c = 2.90 GPa and ν^c = 0.36, which are representative of rigid epoxy and polycarbonate coating materials. These results show that the correction factor is small for values of h^c/h^s less than one provided the specimen material is metallic. If, however, the specimen material has a low modulus like wood, concrete or plastic, the correction factor becomes appreciable.

Strain Variations through the Coating Thickness

A second example to illustrate the combined effect of strain variation through the thickness of the coating and the reinforcing effects is that of a plate subjected to a pure bending moment M. Consider an element from the central region of this plate, as indicated in Fig. 11.27. Because the strain distribution is linear and is transmitted through the specimen-coating interface, elementary plate theory gives the strain as a function of z in both the specimen and the coating as:

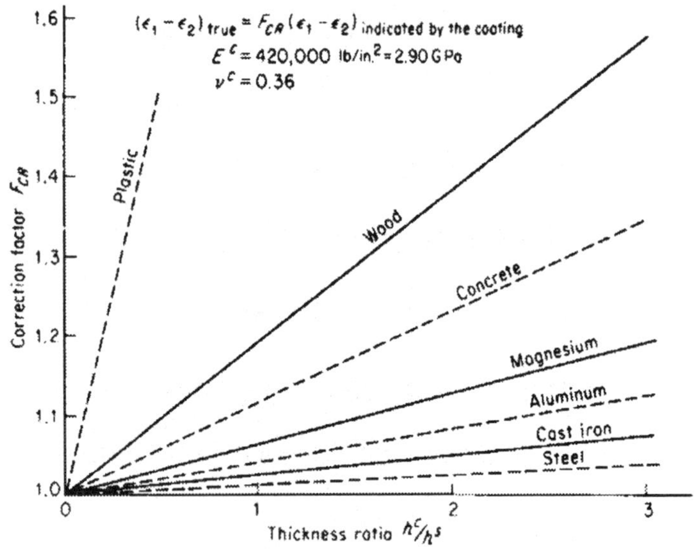

Figure 11.26 Correction factor F_{CR} for a number of different specimen materials.

$$\varepsilon^s_{xx} = \frac{z}{\rho} \qquad \text{for } (h^s - A) \leq z \leq A$$
$$\varepsilon^c_{xx} = \frac{z}{\rho} \qquad \text{for } A \leq z \leq (A + h^c) \qquad (11.46)$$
$$\varepsilon^s_{yy} = \varepsilon^c_{yy} = 0$$

where z is measured from the neutral axis of the composite plate, A is the distance from the neutral axis to the interface, and ρ is the radius of curvature.

Because σ_{zz} is assumed to vanish for all values of z, Eqs. (2.20) are employed with Eqs. (11.46) to express the stress σ_{xx} in terms of the strains.

$$\sigma^s_{xx} = \frac{E^s}{1-\nu^{s2}}\frac{z}{\rho} \qquad \text{for } (h^s - A) \leq z \leq A$$
$$\qquad (11.47)$$
$$\sigma^c_{xx} = \frac{E^c}{1-\nu^{c2}}\frac{z}{\rho} \qquad \text{for } A \leq z \leq (A + h^c)$$

The position of the neutral axis, defined by the symbol A, can be obtained by considering equilibrium of the plate in the x direction and is given by:

$$A = \frac{h^2\left(1 - BC^2\right)}{2(1 + BC)} \tag{11.48}$$

where

$$B = \frac{E^c\left(1 - \nu^{s2}\right)}{E^s\left(1 - \nu^{c2}\right)} \quad \text{and} \quad C = \frac{h^c}{h^s}$$

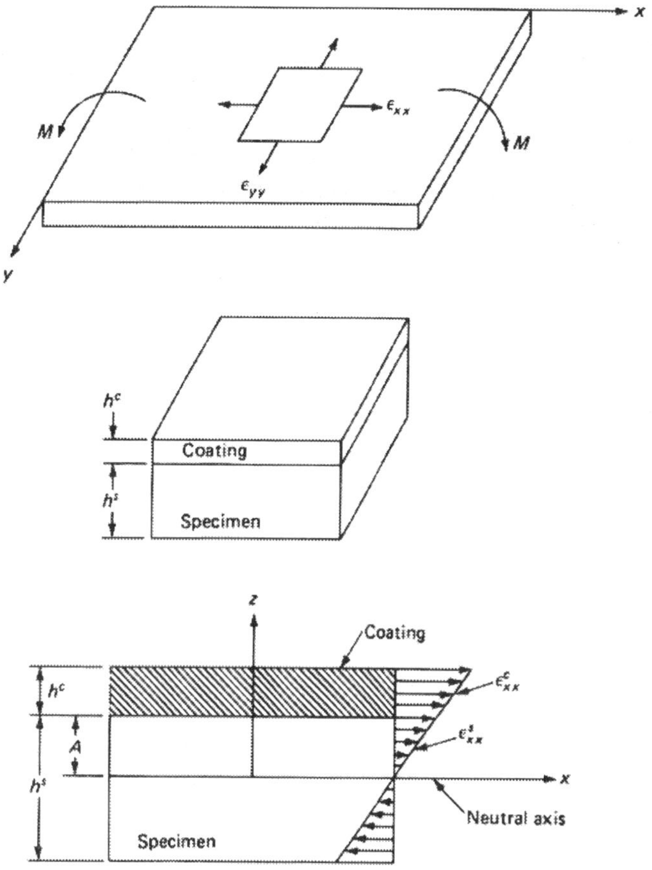

Figure 11.27 Element with a birefringent coating from the center of a wide plate in bending.

The radius of curvature ρ can be determined by considering equilibrium of the moments, to obtain:

$$\frac{1}{\rho} = \frac{12M}{H} \frac{1 - \nu^{s2}}{E^s h^{s3}} \tag{11.49}$$

where

$$H = 4\left(1 + BC^3\right) - \frac{3\left(1 - BC^2\right)^2}{1 + BC}$$

If the coating is examined in a reflection polariscope, the fringe pattern is proportional to the average of the strain difference $\varepsilon_{xx}^c - \varepsilon_{yy}^c$ through the coating thickness. The average strain can be determined from Eqs. (11.46), (11.48), and (11.49) as:

$$\left(\varepsilon_{xx}^{c}-\varepsilon_{yy}^{c}\right)_{ave}=\frac{12M}{H}\frac{1-\nu^{s2}}{E^{s}h^{s3}}\frac{1}{h^{c}}\int_{A}^{A+h^{c}}z\,dz$$

this yields:

$$\left(\varepsilon_{xx}^{c}-\varepsilon_{yy}^{c}\right)_{ave}=\frac{6M}{H}\frac{1-\nu^{s2}}{E^{s}h^{s2}}\frac{1+C}{1+BC} \qquad (11.50)$$

Because the true difference of strain on the surface of an uncoated plate is:

$$\left(\varepsilon_{xx}^{s}-\varepsilon_{yy}^{s}\right)_{true}=6M\frac{1-\nu^{s2}}{E^{s}h^{s2}} \qquad (11.51)$$

it is clear by comparison of Eqs (11.50) and (11.51) that the coating does not indicate the true difference in the surface strains. It is possible to correct the resulting error by introducing a bending correction factor defined by:

$$\left(\varepsilon_{xx}^{s}-\varepsilon_{yy}^{s}\right)_{true}=F_{CB}\left(\varepsilon_{xx}^{c}-\varepsilon_{yy}^{c}\right)_{ave}$$

where the bending correction factor F_{CB} is given by:

$$F_{CB}=\frac{H(1+BC)}{1+C}=\frac{1+BC}{1+C}\left[4\left(1+BC^{3}\right)-\frac{3\left(1-BC^{2}\right)^{2}}{1+BC}\right] \qquad (11.52)$$

The bending correction factor, which is a function of the dimensionless ratios B and C, is shown in Fig. 11.28.

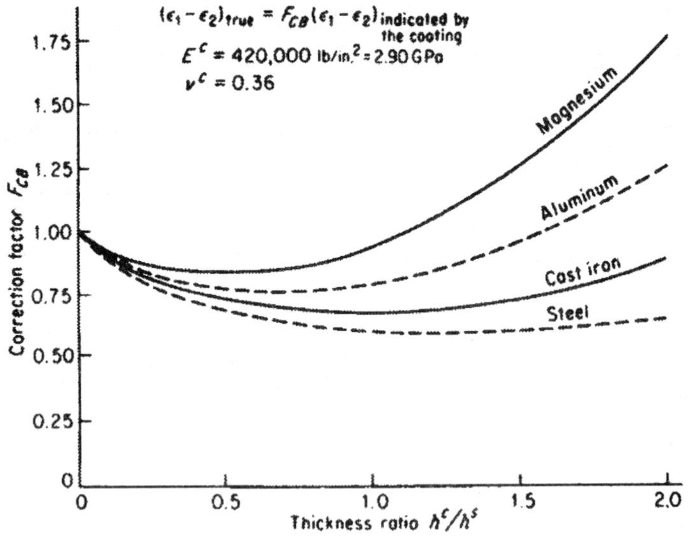

Figure 11.28 Correction factor F_{CB} for a photoelastic coating applied to a number of different specimen materials.

This correction factor accounts for the two thickness effects which occur in this example. The first is due to reinforcing which reduces the strain at the interface between the coating and the specimen in comparison with a plate without the coating. The second effect is due to the gradient of strain through the coating. The optical response of the coating is related to the average strain, which in this instance represents the strain at the midpoint in the coating. Because the average coating strain is higher than the interface strain, the effects of the strain gradient tends to offset the reinforcing effect of the coating. In applications of coatings on beams and plates, the coating thickness h^c is usually less than the specimen thickness h^s and the results of Fig. 11.28 indicate that the correction factor for $h^c/h^s < 1$ is appreciable and should not be neglected.

Poisson's Ratio Mismatch

In plane-stress problems, the errors arising from coating thickness effects, i.e., reinforcement, strain gradient, and curvature, are usually small. However, in almost all cases, a mismatch in Poisson's ratio occurs with ν^c usually greater than ν^s. This difference in Poisson's ratio produces a distortion of the displacement and strain fields through the thickness of the coating which is pronounced at the boundaries.

The significance of the Poisson's ratio mismatch effect can be examined by assuming that the strain ε_1 tangent to the boundary is transmitted without loss or amplification so that $\varepsilon_1{}^c = \varepsilon_1{}^s$. This assumption implies that the distortion of the strain field occurs in the transverse direction. At the interface, the transverse strain in the coating $\varepsilon_2{}^c$ is controlled by the specimen and that:

$$\varepsilon_2{}^c = \varepsilon_2{}^s = -\,\nu^s\,\varepsilon_1{}^s \qquad \text{(a)}$$

At the free surface of the coating, the transverse strain is controlled by the coating and it is clear that:

$$\varepsilon_2{}^c = -\,\nu^c\,\varepsilon_1{}^c = -\,\nu^c\,\varepsilon_1{}^s \qquad \text{(b)}$$

The average value of $\varepsilon_2{}^c$ through the thickness of the coating is bounded by these two limiting values; therefore, the fringe-order response of the coating at the boundary is bounded by:

$$\frac{(1+\nu^s)\varepsilon_1^s}{F_\varepsilon} < N < \frac{(1+\nu^c)\varepsilon_1^s}{F_\varepsilon} \qquad (11.53)$$

where $F_\varepsilon = f_\varepsilon / 2h^c$.

It is clear from this inequality that the magnitude of the distortion is controlled by the mismatch parameter $(1 + \nu^c)/(1 + \nu^s)$. For a constant value of $\nu^c = 0.36$ and variations in ν^s between 0 and 0.5, the mismatch parameter ranges from 1.36 to 0.90. Experiments with tensile specimens, illustrated in Fig. 11.29, indicate that the fringe orders on the boundary and in interior regions of the specimen are given by:

$$N = \frac{(1+\nu^c)\varepsilon_1^s}{F_\varepsilon} \quad \text{on the boundary}$$

$$N = \frac{(1+\nu^s)\varepsilon_1^s}{F_\varepsilon} \quad \text{in the interior} \qquad (11.54)$$

A transition zone exists near the boundary, where the relation between fringe order and strain in a tension specimen can be expressed as:

$$N = \left[1 + \nu^s + C_\nu(\nu^c - \nu^s)\right]\frac{\varepsilon_1^s}{F_\varepsilon} \qquad (11.55)$$

where C_ν is a correction factor accounting for the mismatch.

Experiments conducted with tension specimens with a large mismatch parameter (1.24) and with coatings ranging in thickness from 0.55 to 3.25 mm indicated that the width of the transition zone is about four times the thickness of the coating. The value of C_ν decreases from one at the boundary to zero at a position $4h^c$ from the boundary, as indicated in Fig. 11.30. For metallic specimens, $\nu^c - \nu^s$ is usually less than 0.06, which is relatively small in comparison with $1 + \nu^s \approx 1.3$ in Eq. (11.55). As a result, the effect of Poisson's ratio mismatch is often neglected in analyses of most metallic components.

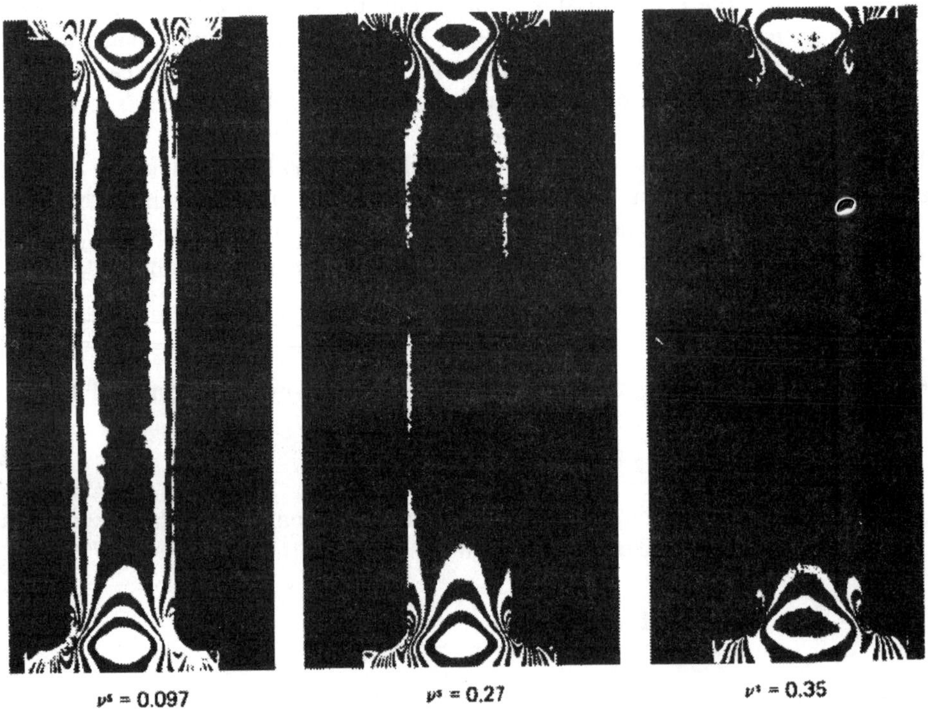

Figure 11.29 Effect of Poisson's ratio mismatch on the response of birefringent coatings, $\nu^c = 0.36$.

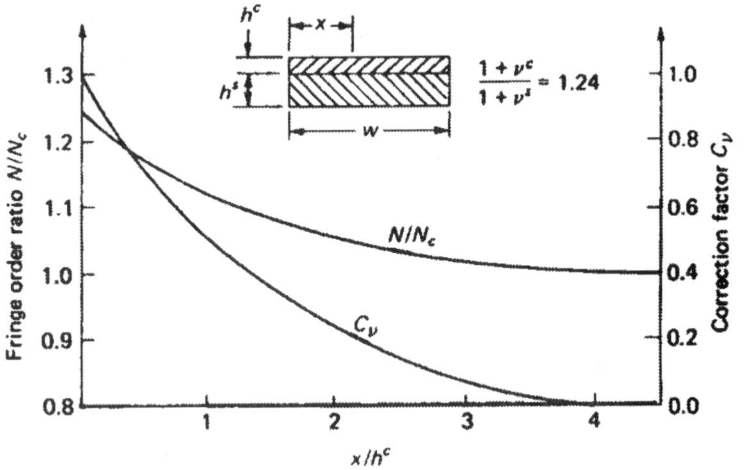

Figure 11.30 Correction factor C_ν as a function of x/h^c.

15.10.6 Fringe Order Determinations in Coatings

The methods employed to determine the order of fringes exhibited by the coating depends upon the response of the coating and the accuracy required in the analysis. If the response of the coating is large, (four or more fringes), monochromatic light can be used to obtain photographs of the light-and dark-field isochromatic fringe patterns. These fringes can usually be interpolated or extrapolated to ± 0.2 fringe, giving an accuracy of about 5% based on a maximum of four fringes. An example of a high-response fringe pattern which can be analyzed as indicated above is shown in Fig. 11.31.

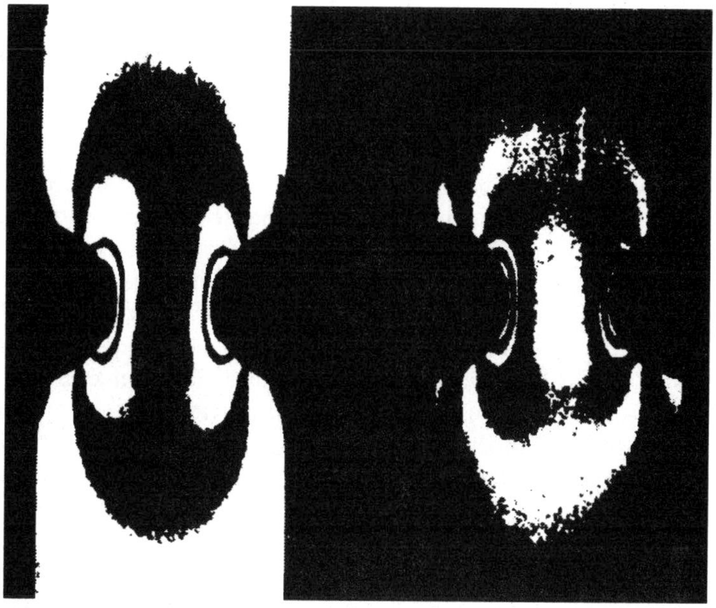

Figure 11.31 Fringe patterns from a thin coating on a glass-reinforced plastic specimen.

Table 11.6
Sequence of colors produced in a dark-field reflection polariscope with white light.

Color	Retardation, nm	Fringe order
Black	0	0
Gray	160	0.28
White	260	0.45
Yellow	350	0.60
Orange	460	0.79
Red	520	0.90
First, tint of passage*	577	1.00
Blue	620	1.06
Blue-green	700	1.20
Green-yellow	800	1.38
Orange	940	1.62
Red	1050	1.81
Second, tint of passage	1150	2.00
Green	1350	2.33
Green-yellow	1450	2.50
Pink	1550	2.67
Third, tint of passage	1730	3.00
Green	1800	3.10
Pink	2100	3.60
Fourth tint of passage	2300	4.00
Green	2400	4.13

*The tint of passage is a sharply divided zone between red and blue in the first order fringe, red and green in the second order fringe, and pink and green in the third-, fourth-, and fifth-order fringes. Beyond five fringes white light analysis is not practical because the colors become very pale and difficult to distinguish.

For fringe patterns exhibiting between two and four fringes, the use of colored patterns produced with white light is advantageous. The colored pattern is due to the attenuation and extinction of one or more colors from the white spectrum. The observed fringes represent the complementary color produced by the transmitted portion of the white-light spectrum. The sequence colored fringes produced by an increasing stress field is listed in Table 11.6. The exact shade of color will be a function of the energy distribution in the white-light spectrum and the recording characteristics of the film being used; however, the sequence listed in the table is adequate for direct visual observations.

It is evident from Table 11.6 that the use of white light substantially increases the number of fringes that can be identified. For example, in the interval $0 \leq N \leq 2$, 12 different color bands exist, which can be used to establish fractional fringe orders. Moreover, the polariscope can also be used with light-field settings to yield a second family of colored fringes to effectively double the amount of data available for estimating fractional fringe orders. Thus, the fringe order can be established to within ± 0.1 fringe, giving an accuracy of about ± 5% based on a maximum of 2 fringes.

For precise fringe-order determinations, where the maximum fringe order is less than 2 and accuracies of 5% or less are required in the analysis, it is necessary to use either compensation techniques or advanced methods using grey field or RGB polariscopes that were described in Chapter 10.

EXERCISES

11.1 Plot the fringe orders as a function of position across the horizontal centerline of the chain-link model shown in Fig. 11.1.

11.2 Determine the fringe orders associated with the tensile and compressive stress concentrations at the pinholes of the chain-link model shown in Fig. 11.1. Determine the stress concentration at these locations based on the maximum stress across the horizontal centerline of the chain-link.

11.3 For a state of plane stress with $\sigma_1 > 0 > \sigma_2$, plot a Mohr's circle and show the plane upon which the maximum shear stress acts.

11.4 For a state of plane stress with $\sigma_1 > \sigma_2 > 0$, plot a Mohr's circle and show the plane upon which the maximum shear stress acts.

11.5 Determine the distribution of stress $\sigma_{\theta\theta}$ on the boundary of the pinhole of the chain link shown in Fig. 11.1 if f_σ = 20 kN/m and h = 6 mm. Prepare a graph of this distribution which shows $\sigma_{\theta\theta}$ as a function of θ.

11.6 Plot the fringe orders as a function of position across the horizontal centerline of the pressurized square conduit shown in Fig. 11.3. Determine an approximate distribution for the stress σ_{yy} along this line if f_σ = 20 kN/m, h = 6 mm and p = 2.00 MPa.

11.7 Plot the fringe orders as a function of position across the diagonal of the pressurized square conduit shown in Fig. 11.3.

11.8 Plot the fringe orders as a function of position along the outer edge of the pressurized square conduit shown in Fig. 11.3.

11.9 Determine the stress distribution on the boundary of the circular hole of the pressurized square conduit shown in Fig. 11.3 if f_σ = 20 kN/m, h = 6 mm, and p = 2.00 MPa.

11.10 Construct the isostatics for one quadrant of a circular ring subjected to a diametral compressive load P by using the isoclinic data shown in Fig. 11.5.

11.11 Determine the maximum tensile and compressive stresses on the inner boundary of the thick-ring specimen shown in Fig. 10.14 if f_σ = 20 kN/m and h = 6 mm. The specimen is subjected to concentrated compressive loads at the ends of the vertical diameter.

11.12 Given a fringe order of 6, a model thickness of 8 mm, a material fringe value of 20 kN/m, and an isoclinic parameter of 30° defining the angle between the x axis and σ_1, determine the shear stress τ_{xy} and show the direction of the shear stress on the face of a small element.

11.13 Explain how a simple tension or compression specimen could be used as a compensator.

11.14 Describe how the Tardy method of compensation could be employed to determine the fractional fringe order at points P_3, and P_4 of Fig. 10.16.

11.15 If the circular disk shown in Fig. 11.11 has an outside diameter D = 100 mm and a thickness h = 10 mm, determine the material fringe value f_σ if a load P = 4.0 kN produced the fringe pattern.

11.16 A plane stress model with an axis of symmetry is placed in a circular polariscope. A fringe pattern recorded at normal incidence gives the fringe order distribution $N_0(y)$ along the axis of symmetry. The model is then rotated about the axis of symmetry by an angle θ (see Fig. E11.16) and the fringe order distribution $N_\theta(y)$ is recorded. If diffraction effects are eliminated by using an immersion tank, show that this oblique incidence technique gives:

$$\sigma_1 = \frac{f_\sigma \cos\theta}{h\sin^2\theta}(N_\theta - N_0\cos\theta)$$

$$\sigma_2 = \frac{f_\sigma}{h\sin^2\theta}(N_\theta\cos\theta - N_0)$$

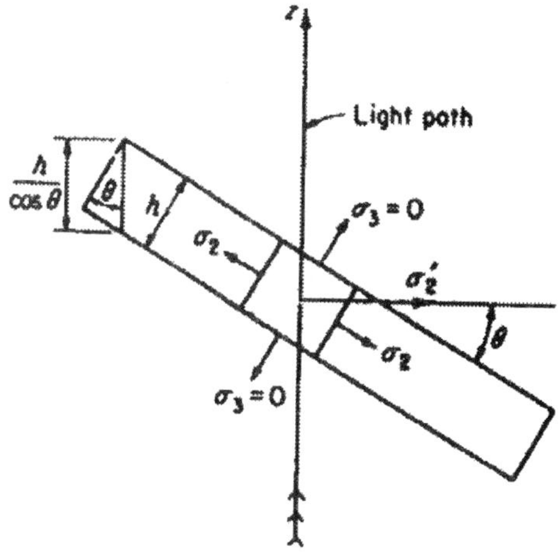

Figure E11.13

11.17 If the rotation angle θ in Exercise 11.16 is 15°, 30° and 45°, determine the corresponding relations for σ_1 and σ_2 for these three cases.

11.18 A load P = 1200 N is applied to a circular disk with a diameter D = 40 mm and a thickness h = 6 mm. The Tardy method of compensation is used to determine that the fringe order N at the center of the disk is 5.4. Determine the material fringe value f_σ.

11.19 Write a computer program for plane stress models based on the analytic separation method. The output of the program should give $\sigma_{xx} + \sigma_{yy}$, σ_1 and σ_2 over the entire field. The data input to the program must permit determination of $\sigma_{xx} + \sigma_{yy}$ on the boundary of the model.

11.20 Write an engineering brief to your manager (who is a business school major) explaining how a plastic photoelastic model 100 x 200 x 4 mm in size can be used to predict the stress distribution in a very large structural component fabricated from steel.

11.21 Write a test procedure describing techniques to be used to minimize error due to viscoelastic effects in a photoelastic model study of an elastic stress problem.

11.22 You are planning to conduct a photoelastic analysis of a series of deep beams subjected to bending produced by concentrated loads. From the materials listed in this chapter, select the most suitable model material and write an engineering brief justifying your selection.

11.23 You are planning to conduct a photoelastic analysis of a very large concrete structure where the loads are due to body forces. From the materials listed in this chapter, select the most suitable model material and write an engineering brief justifying your selection.

11.24 Determine the changes in length Δl_1 and Δl_2 for the tension strut shown in Fig. 11.10 if l_0 = 150 mm and the modulus of elasticity is 3500 MPa at room temperature and 35 MPa at elevated temperature. The stress applied to the strut was 350 kPa.

11.25 Prepare a slicing plan for a three-dimensional photoelastic model of a thick walled pressure vessel with a hemispherical head. The cylindrical region of the model has a length to diameter ratio of 3. The thickness to diameter ratio for both the head and the cylinder is 1:8. List the equations which hold for each slice and/or subslice.

11.26 Derive Eq. (11.27).

11.27 Write a computer program for determining the effective (von-Mises) stress from photoelastic data taken from a cube.

11.28 Write a test procedure which can be followed to determine the data necessary for program written for Exercise 11.27.

11.29 Verify Eq. (11.30).

11.30 Explain why a birefringent coating can be considered as a strain gage. Discuss the concept of the gage length associated with the coating.

11.31 A polycarbonate with K = 0.14 is used as a birefringent coating in a polariscope with a sodium light source. If the modulus of elasticity and Poisson's ratio of the polycarbonate are 2.48 GPa and 0.38, respectively, determine f_σ and f_ε for the coating.

11.32 Determine the specimen coefficient of sensitivity for the following materials:

(a) Mild steel AISI 1010 (b) High-strength steel AISI 4340
(c) Aluminum 24S (d) Aluminum 2S
(e) Hastelloy A (f) Inconel X
(g) Magnesium M1 (h) Red brass
(i) Titanium (j) Plexiglas

11.33 Verify Eq. (11.40).

11.34 For a coating with h^c = 1.50 mm and f_ε = 4.00 μm/fringe, determine the maximum fringe order N which could be developed in the materials listed in Exercise 11.32.

11.35 Specify a coating (material and thickness) to use for the analysis of the following steel (E^s = 207 GPa and ν^s = 0.30) structures:

(a) A curved beam where σ_{max} = 150 MPa.
(b) A spherical shell with radius R = 4.00 m and wall thickness t = 15 mm subjected to a pressure p of 1500 kPa.
(c) A cylindrical shell with radius R = 1500 mm and wall thickness t = 20 mm subjected to a pressure p of 5.00 MPa.

11.36 Determine the maximum fringe order developed in the coatings specified in Exercise 11.35.

11.37 Discuss the procedure to be used to install a coating on a panel with an elliptical hole.

11.38 Discuss the procedure to be used to install a coating at the intersection between two circular cylinders of the same diameter.

11.39 A coating is placed on an aluminum tension strip of known dimensions w and h. The tension strip is loaded with a known load P and the resulting fringe order N is measured with a reflection polariscope.

(a) Outline the procedure for determining the material fringe value f_ε for the coating from these data.
(b) If a sodium light source (λ = 589.3 nm) is used, determine the strain coefficient K for the coating.
(c) If a helium-neon laser light source (λ = 632.8 nm) is used, determine the strain coefficient K for the coating.

11.40 At an interior point on a steel specimen with a polycarbonate coating (h^c = 4.00 mm and f_ε = 5.10 μm/fringe), a value of N = 2.00 and ϕ_1 = 30° is measured. Describe the state of stress in the specimen at this point.

11.41 If the coating and measurements of Exercise 11.40 are associated with a point on the free boundary of a panel specimen with in-plane loads, describe the state of stress at the point.

11.42 Write an engineering brief describing the development of a contoured coating to match the geometry of a complex surface.

11.43 Verify Eq. (11.45).

11.44 Determine the correction factor F_{CR} needed to account for reinforcing effects of the coating for a plane-stress specimen fabricated from the materials listed in Exercise 11.32 if:

(a) $h^c/h^s = 0.1$ (b) $h^c/h^s = 0.2$ (c) $h^c/h^s = 0.5$
(d) $h^c/h^s = 1.0$ (e) $h^c/h^s = 1.5$ (f) $h^c/h^s = 2.0$

11.45 Verify Eq. (11.48).

11.46 Verify Eq. (11.49).

11.47 Verify Eq. (11.52).

11.48 For the materials listed in Exercise 11.32, determine the correction factor F_{CB} for wide plates in bending if:

(a) $h^c/h^s = 0.1$ (b) $h^c/h^s = 0.2$ (c) $h^c/h^s = 0.5$
(d) $h^c/h^s = 1.0$ (e) $h^c/h^s = 1.5$ (f) $h^c/h^s = 2.0$

Prepare a graph showing F_{CB} as a function of h^c/h^s for each material.

11.49 A birefringent coating (E^c = 2.50 GPa and ν^c = 0.36) is used on a plane-stress tensile specimen fabricated from glass-reinforced plastic (E^s = 27.5 GPa and ν^s = 0.20) to measure the stress concentration factor resulting from a centrally located hole. If N_{max} = 4.5 on the boundary of the hole and N_0 = 1.00 at an interior point well removed from the hole, determine the stress concentration factor due to the hole.

11.50 A polycarbonate coating (h^c = 2.5 mm) is used on an aluminum 2024T3 plane-stress specimen. The specimen is loaded (in plane) until certain regions have yielded. If $\varepsilon_2 < 0$ in these regions, determine the color of the fringe delineating the elastic plastic boundary. Assume that the observation is made with white light in a dark-field polariscope.

11.51 A specimen with a birefringent coating is subjected to a specified load P, and fringe orders $N_0(x,y)$ and isoclinics $\phi(x,y)$ are recorded over the field. Next, a slitting saw is used to cut a narrow channel through the thickness of the coating along a stress trajectory. A second fringe pattern is recorded to obtain the distribution of fringes N_1 along the edge of the saw cut. Derive a relation for the stresses σ_1^s and σ_2^s along the isostatic coinciding with the saw cut.

BIBLIOGRAPHY

1. Durelli, A. J. and W. F. Riley: Introduction to Photomechanics, Prentice-Hall, Englewood Cliffs, N.J., 1965.
2. Frocht, M. M.: Photoelasticity, John Wiley & Sons, Inc., New York, Vol. 1, 1941, Vol. 2, 1948.
3. Coker, E. G., and L. N. G. Filon: A Treatise on Photoelasticity, Cambridge University Press, London, 1931.
4. Kuske, A., and G. Robertson: Photoelastic Stress Analysis, John Wiley & Sons, Inc., New York, 1974.
5. Tardy, M. H. L.: "Methode pratique d'examen de mesure de la birefringence des verres d'optique," Review Optique., Vol. 8, pp. 59-69, 1929.
6. Dally, J. W., and E. R. Erisman: "An Analytic Separation Method for Photoelasticity," Experimental Mechanics., Vol. 6, No. 10, pp. 493-499, 1966.
7. Drucker, D. C.: "The Method of Oblique Incidence in Photoelasticity", Proceedings SESA, Vol. VIII, No. 1, pp. 51-66, 1950.
8. Doyle, J. F.: Modern Experimental Stress Analysis, John Wiley & Sons, New York, 2004.
9. Clutterbuck, M.: "The Dependence of Stress Distribution on Elastic Constants," British Journal Applied Physics, Vol. 9, pp. 323-329, 1959.
10. Young, D. F.: "Basic Principles and Concepts of Model Analysis," Experimental Mechanics, Vol. 11, No. 7, pp. 325-336, 1971.
11. Sanford, R. J.: "The Validity of Three-dimensional Photoelastic Analysis of Non-homogeneous Elastic Field Problems," British Journal Applied Physics, Vol. 17, pp 99-108, 1966.
12. Dundurs, J.: "Dependence of Stress on Poisson's Ratio in Plane Elasticity," International. Journal Solids Structures, Vol. 3, pp. 1013-1021, 1967.

13. Mindlin, R. D.: "A Mathematical Theory of Photoviscoelasticity," Journal Applied Physics, Vol. 20, pp. 206-216, 1949.
14. Leven, M. M: "Epoxy Resins for Photoelastic Use," Photoelasticity, Pergamon Press, New York, pp. 145-165, 1963.
15. Miyano, Y., T. Kunio and S. Sugimori: "A Study on Stress Freezing Method Based on Time and Temperature Dependent Photoviscoelastic Behaviors," IUTAM Symposium Advanced Optical Methods and Applications to Solid Mechanics, (ed.) A. Lagarde, Poitiers, France, Vol 1, pp. Mo-L5-p1. 1998.
16. Ito, K.: New Model material for Photoelasticity and Photoplasticity, Experimental Mechanics, Vol. 2, No. 12, pp 373-376, 1962.
17. Hetenyi, M.: "The Fundamentals of Three-dimensional Photoelasticity," Journal Applied Mechanics, Vol. 5, No. 4, pp. 149-155, 1938.
18. Cernosek, J.: "Three-dimensional Photoelasticity by Stress Freezing," Experimental Mechanics, Vol. 20, pp. 417-426, 1980.
19. Johnson, R. L.: "Model Making and Slicing for Three-dimensional Photoelasticity," Experimental Mechanics, Vol. 9, No. 3, pp. 23N-32N, 1969.
20. Nikola, W. E., and M. J. Greaves: "Construction of Complex Photoelastic Models Using Thin Molded-Epoxy Sheets," Experimental Mechanics, Vol. 10, No. 4, pp. 23N-30N, 1970.
21. Drucker, D., and R. D. Mindlin: "Stress Analysis by Three-dimensional Photoelasticity Methods," Journal Applied Physics, Vol. 11, pp. 724-732, 1940.
22. Mindlin, R. D., and L. E. Goodman: "The Optical Equations of Three-dimensional Photoelasticity," Journal Applied Physics, Vol. 20, pp. 89-95, 1949.
23. Drucker, D. C., and W. B. Woodward: "Interpretation of Photoelastic Transmission Patterns for Three-dimensional Models," Journal Applied Physics, Vol. 25, No. 4, pp. 510-512, 1954.
24. Leven, M. M.: "Quantitative Three-dimensional Photoelasticity," Proceedings SESA, Vol. XII, No. 2, pp. 157-171, 1955.
25. Brock, J. S.: "The Determination of Effective Stress and Maximum Shear Stress by Means of Small Cubes Taken from Photoelastic Models," Proceedings SESA, Vol. XVI, No. 1, pp. 1-8, 1958.
26. Frocht, M. M., and R. Guernsey, Jr.: "Studies in Three-dimensional Photoelasticity," Proceedings 1st U. S. National Congress Applied Mechanics, pp. 301-307, 1951.
27. Fleury, R., and F. Zandman: "Jauge d'efforts photoelastique," C. R. (Paris), vol, 238, p. 1559, 1954.
28. D'Agostino J., D. C. Drucker, C. K. Liu, and C. Mylonas: "Epoxy Adhesives and Casting Resins as Photo-elastic Plastics," Proceedings SESA, Vol. XII, No. 2, pp. 123-128, 1955.
29. Kawata, K.: "Analysis of Elastoplastic Behavior of Metals by Means of Photoelastic Coating Method", Journal Scientific Research Instruments, Tokyo, Vol. 52, pp. 17-40, 1958.
30. Ito, K.: "New Model Materials for Photoelasticity and Photoplasticity," Experimental Mechanics, Vol. 2, No. 12, pp. 373-376, 1962.
31. Melver, R. W.: "Structural Test Applications Utilizing Large Continuous Photoelastic Coatings," Experimental Mechanics, Vol. 5, No. 1, pp. 19A-25A, No. 2, pp. 19A-26A, 1965.
32. Zandman, F. S. Redner and J. W. Dally: Photoelastic Coatings, Iowa State University Press, Ames, 1977.
33. Zandman, F., S. S. Redner, and E. I. Riegner: "Reinforcing Effect of Birefringent Coatings," Experimental Mechanics, Vol. 2, No. 2, pp. 55-64, 1962.
34. Dally, J. W., and I. Alfirevich: "Application of Birefringent Coatings to Glass-fiber-reinforced Plastics," Experimental Mechanics, Vol. 9, No. 3, pp. 97-102, 1969.
35. Shukla, A., Rossmanith, H.P.: "Dynamic Photoelastic Investigation of Wave Propagation and Energy Transfer Across Contacts," with H.P. Rossmanith, Journal of Strain Analysis, Vol. 21, No. 4, pp. 213-218, 1986.
36. Shukla, A.: "Dynamic Photoelastic Studies of Wave Propagation in Granular Media," Journal of Optics and Lasers in Engineering, Vol. 14, pp. 165-184, 1991.
37. Durelli, A.J., Shukla, A.: "Identification of Isochromatic Fringes," with A.J. Durelli, Experimental Mechanics, Vol. 23, No. 1, pp. 111-119, 1983.

CHAPTER 12

INTERFEROMETRY AND HOLOGRAPHY

12.1 INTRODUCTION

The concept of constructive and destructive interference is described where an interference pattern is produced by two collinear waves (as in photoelasticity) or by two waves at oblique incidence (as in holography). The equation for the intensity of the light resulting from the superposition of these two waves is developed for both cases.

Newton's rings are described as an example of collinear interference. An optical system for measuring surface imperfections in flat plates using Newton's rings is presented. The generic, Michelson and Mach-Zehnder interferometers are described and a method to determine the sum of the principal stresses σ_1 and σ_2 based on the interferometric fringe pattern is introduced.

The study of two plane waves interfering at oblique incidence provides the basis for discussing holographic interferometry. This description is relatively brief and the more serious reader is referred to several books on the subject that are listed at the end of this chapter. The treatment, included here, begins with the method used to produce off-axis holograms. Methods for double exposing holograms and for viewing reconstructed virtual images from the holograms are presented. Relationships for the light intensity in the reconstructed holographic images are derived. These relations enable the researcher to uniquely determine the phase change due to specimen motion or deformation that occurs between exposures.

A more complete description of the relationship between phase change and specimen deformation is provided. Since the geometry involved in establishing the relations between phase change and deformation is complex, vector algebra is used in the derivations. Relations are developed for out-of-plane, in-plane displacements and for the general case that involves both, in- and out-of-plane displacements.

Finally, two special applications are introduced—real time and time averaged holographic interferometry. Real time holographic interferometry permits the investigator to observe the fringe pattern as it develops as the load is applied to the specimen or as its temperature changes. The time averaged holographic method enables the researcher to study vibratory motion, establish nodal points and to determine peak displacement amplitudes.

12.2 INTERFERENCE OF TWO PLANE COLLINEAR WAVES

To study interference of collinear waves, consider two plane waves of light propagating along the z axis as illustrated in Fig. 12.1. The light associated with both waves is coherent so that the two waves are in phase except for a small delay of wave #2 relative to wave #1 that is associated with z_0. Because both waves are produced by the same coherent source their wavelengths are identical.

From Eq. (13.8) it is possible to express the magnitude of the two light vectors as:

$$E_1 = a\cos\frac{2\pi}{\lambda}(z-ct)$$
$$E_2 = a\cos\frac{2\pi}{\lambda}(z-ct-z_0) \qquad (12.1)$$

Because the two waves are collinear, the two components E_1 and E_2 can be superimposed to obtain:

$$E' = a\left\{\cos\frac{2\pi}{\lambda}(z-ct) + \cos\frac{2\pi}{\lambda}(z-ct-z_0)\right\} \qquad (12.2)$$

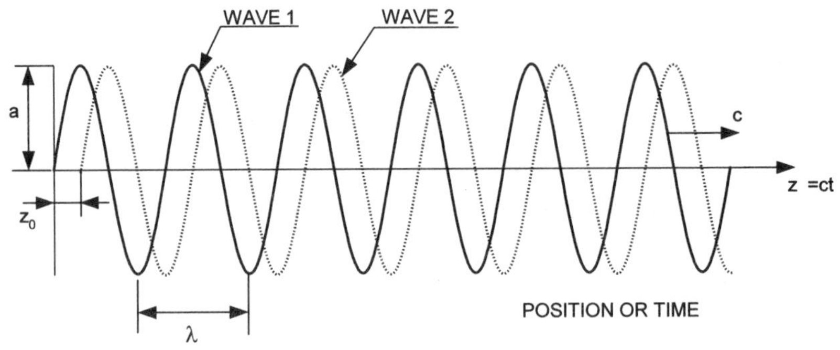

Figure 12.1 Two-plane collinear light waves propagating along the z axis.

Recall the trigometric identity:

$$\cos A + \cos B = 2\cos\left(\frac{A+B}{2}\right)\cos\left(\frac{A-B}{2}\right) \qquad (a)$$

Substitute Eq. (a) into Eq. (12.2) and simplifying gives:

$$E' = 2a\cos\frac{\pi z_0}{\lambda}\cos\frac{2\pi}{\lambda}(z-ct-z_0) \qquad (12.3)$$

Examination of Eq. (12.3) reveals that the amplitude of the superimposed waves is $2a\cos\frac{\pi z_0}{\lambda}$. Squaring the amplitude gives the intensity I as[1]:

$$I = 4a^2\cos^2\frac{\pi z_0}{\lambda} \qquad (12.4)$$

From Eq. (12.4) it is clear that the argument of the cosine function yields the maximum and the minimum values for the intensity I. The maximum intensity ($I = 4a^2$) occurs when the interference is constructive and the argument of the cosine function is given by:

$$z_0/\lambda = k \qquad (k = 0, 1, 2, ...) \qquad (12.5a)$$

The minimum intensity ($I = 0$) occurs when the interference is destructive and the argument of the cosine function is given by:

$$z_0/\lambda = (2k+1)/2 \qquad (k = 0, 1, 2, ...) \qquad (12.5b)$$

[1] This relation corresponds to Eq. (13.17) where the linear phase difference in the two waves $z_0 = \delta$.

12.2.1 Newton's Rings

The most commonly observed fringe pattern associated with the interference produced by two plane collinear waves are Newton's rings. Newton's rings are evident when two nearly flat plates of glass are brought into contact and observed in normal incidence. This interference phenomenon is illustrated with a diagram of the interfering light waves presented in Fig. 12.2.

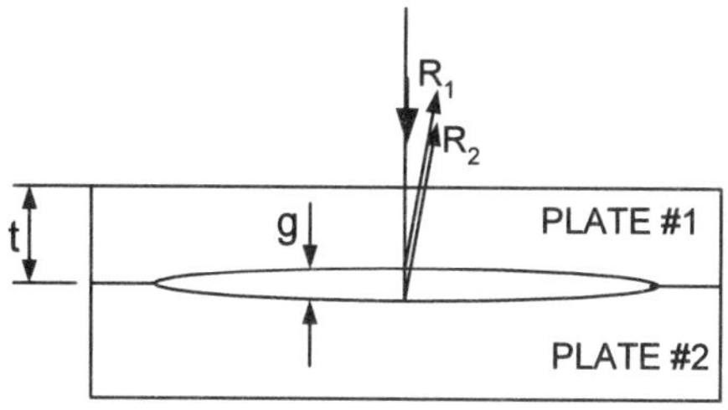

Figure 12.2 Light wave reflections from the contacting surfaces of two glass plates with a small gap.

Reference to Fig. 12.2 shows an incident light wave illuminating two glass plates at normal incidence. The partial reflection of the light from the top surface of plate #1 has been ignored because it dos not affect the results. However, the light waves reflected from the bottom surface of plate #1 and the top surface of plate # 2 are shown as R_1 and R_2, respectively. The axes of propagation of the two waves coincide; however, the two axes are inclined and separated in Fig. 12.2 to more clearly depict the reflecting surfaces. There is a space between the two reflecting surfaces that is filled with air. A gap g characterizes the distance between these reflecting surfaces.

The light waves reflecting from the two surfaces interfere to produce fringes that are known as Newton's rings. The theory associated with the formation of the rings is similar to that presented in developing Eq. (12.4) with one exception—there is a phase change associated with the reflection from the surface of plate #2. Because the light acts like a wave, it undergoes a phase change of π upon reflection, if the reflecting medium has a higher index of refraction than the medium in which the wave is propagating. Accordingly, reflection occurs at the lower surface of plate #1 without a phase change, but a phase change of π occurs upon reflection from plate #2.

Because of the phase change of π for the wave associated with R_2, it is easy to show that the intensity of the light produced by the interference of waves R_1 and R_2 is given by:

$$I = 4a^2 \sin^2 \frac{\pi z_0}{\lambda} \qquad (12.6)$$

From Eq. (12.6) it is clear that the argument of the sine function yields the maximum and the minimum values for the intensity I. The maximum intensity ($I = 4a^2$) occurs when the interference is constructive and the argument of the sine function is given by:

$$z_0 / \lambda = (2k + 1)/2 \qquad (k = 0, 1, 2, ...) \qquad (a)$$

Recognize that $z_0 = 2g$, because the light travels though the gap twice. Substitute this quantity into Eq. (a) to obtain:

$$g = \frac{2k+1}{4}\lambda \; (k = 0, 1, 2, ...) \qquad (12.7a)$$

The minimum intensity ($I = 0$) occurs when the interference is destructive and the argument of the sine function is given by:

$$z_0 / \lambda = k \qquad (k = 0, 1, 2, ...) \qquad (b)$$

The relation for the dimension of the gap which produces destructive interference (extinction) is given by:

$$g = \frac{k\lambda}{2} \qquad (k = 0, 1, 2, ...) \qquad (12.7b)$$

Clearly, the intensity of the light from the superimposed waves R_1 and R_2 provide a means or measuring small gaps between one transparent plate and another plate (of any material) with a reflecting surface.

Consider next the optical arrangement shown in Fig. 12.3, which enables an observer to measure surface irregularities using the interference patterns produced when two surfaces are in close proximity to one another.

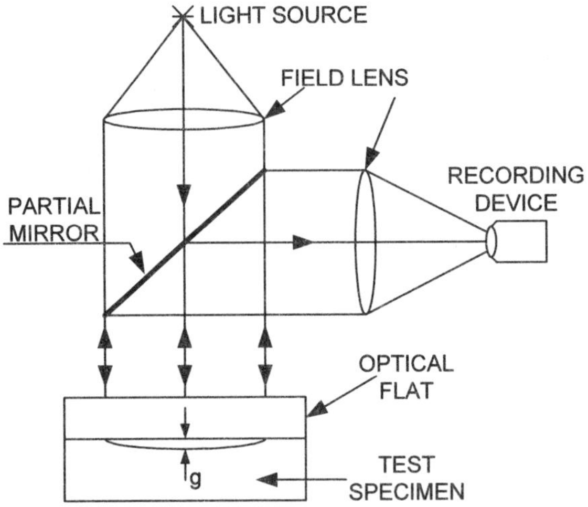

Figure 12.3 Optical arrangements for measuring surface imperfections using Newton's rings.

The light from the source is collected and collimated by a field lens and directed toward the test specimen at normal incidence. The incident light passes through a partial mirror and through an optical flat that serves as plate #1. The optical flat, as the name implies, is plane to within a fraction of a wavelength of light over its entire area. It serves as a reference plane from which surface deviations in the test specimen are measured. The incident light is reflected from the bottom surface of the optical flat and the top surface of the specimen. Both of these waves propagate to the partial mirror where they are reflected at an angle of 90°. This light is then focused by another field lens onto the lens of a camera where an image of the fringe pattern is recorded.

If the surface imperfection of the test specimen shown in Fig. 12.3 is represented by a spherical depression, the resulting fringe pattern would appear as illustrated in Fig. 12.4. The fringe order N corresponds to k with the zero order k = N = 0 corresponding to the outside ring where contact occurs between the optical flat and the test specimen. The maximum fringe order occurs in the center in this example with N = k = 3 where the gap $g = 3\lambda/2$.

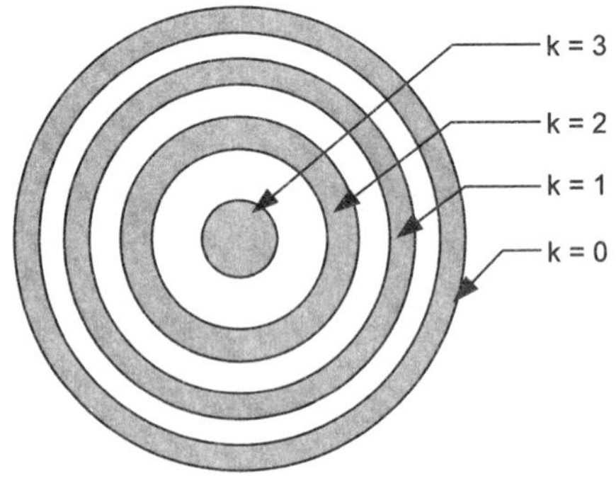

Figure 12.4 Appearance of Newton's rings for a test specimen with a spherical depression.

12.3 THE INTERFEROMETER

An interferometer is an optical instrument which is used to measure lengths or changes in length with great accuracy by means of interference fringes. The modulation of intensity of light by superposition of light waves was defined in Section 12.2 as an interference effect. The intensity of the wave resulting from the superposition of two waves of equal amplitude was shown in both Eqs. (13.7) and (12.4) to be a function of the phase difference between the waves. A fundamental requirement for the existence of well-defined interference fringes is that the light waves producing the fringes have a clearly defined phase difference which remains essentially constant during the recording interval. When light beams from two independent sources are superimposed, interference fringes are not observed since the phase difference between the beams varies in a random way (the beams are incoherent). Two beams from the same source, on the other hand, interfere, because the individual wavetrains in the two beams have the same phase initially (the beams are coherent). Any difference in phase, at the point of superposition, results solely from differences in optical paths, which occurs as the light beams propagate along different legs of an interferometer. In this treatment, optical-path length OPL is defined as:

$$OPL = \sum_{i=1}^{m} n_i L_i$$

where L_i is the mechanical-path length in a material having an index of refraction n_i.

12.3.1 The Generic Interferometer

Cloud [1] has introduced the concept of a generic interferometer, shown in Fig. 12.5, which illustrates the essential elements involved in the design of this optical instrument. The light for the instrument is provided by a coherent source—a helium-neon laser usually employed. The beam is immediately divided into two parts: one following the reference path and the other following the active path. Usually a neutral density filter (not shown in Fig. 12.5) is inserted into the reference path to adjust the intensity of the reference beam until the intensity of both beams is equal. The active beam passes through a specimen and undergoes retardation due to the optical characteristics of the specimen.

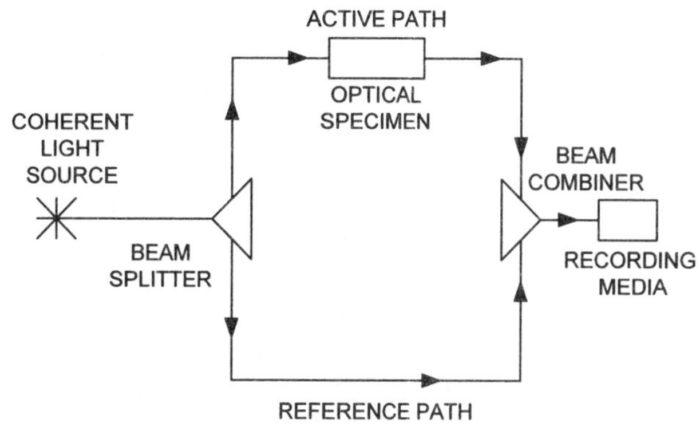

Figure 12.5 The concept of a generic interferometer (Cloud [1]).

The two beams arrive at an optical element that combines the two beams. In all probability, these two beams are out of phase because the optical path lengths between the active and reference paths may be different and the active beam is retarded by the test specimen. Because the two beams are out of phase, an interference pattern develops when the beams are combined (superimposed). This interference pattern, which is identical to the interference discussed in Section 12.2, develops when two collinear plane waves are superimposed. The interference pattern is monitored by some type of recording device such as a digital camera, film or a photodetector.

The interferometer provides a means to measure the difference in optical path length between the reference and active paths. In this sense, the interferometer is a type of micrometer capable of measuring in fractions of a wavelength of light. This capability is not fully realized because the recording methods (film, CCD image sensors, photodiodes, etc.) respond to the amplitude squared of the light and not to the phase angle produced by the differences between the active and reference optical path lengths. This difficulty is circumvented by forming an interference pattern that provides amplitude information, which is related to the phase difference between the two superimposed beams. Unfortunately, it is not possible to ascertain in a single measurement whether the two beams are out of phase by ϕ_1 or $k \pm \phi_1$ where k is any integer. To make accurate displacement measurements with an interferometer, it is necessary to conduct two experiments with small changes in one of the optical paths introduced in the second experiment. The interference patterns from each experiment are compared and from the differences it is possible to measure the small changes in displacement field introduced in the second experiment.

12.3.2 The Michelson Interferometer

The Michelson interferometer, invented in 1881 by A. A. Michelson, has been used by many physicists in important classical experiments that required extremely accurate measurements of length. A schematic diagram of this interferometer is presented in Fig. 12.6. A beam of light generated by a coherent source is collimated with a field lens then divided into two beams by a partial mirror positioned at A and inclined at an angle of 45° relative to the incoming beam. One beam is reflected vertically upward to the adjustable mirror at position B, and the second beam is transmitted through the partial mirror to the fixed mirror at position C. The reflected beam is reflected a second time from a movable mirror at position B, travels vertically downward, and is transmitted through the partial mirror. This beam is intercepted by a screen placed at position D.

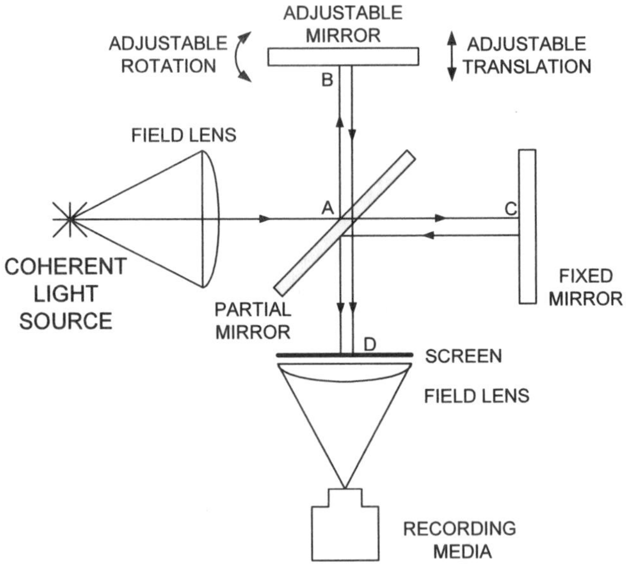

Figure 12.6 The Michelson interferometer.

The second beam, transmitted through the partial mirror at position A, is reflected from the mirror at position C and travels back to the partial mirror where it is reflected downward. This second beam is also intercepted by the screen at position D, where it is superimposed on the first beam producing a collinear interference pattern. The interference pattern is either observed by eye or recorded with a camera or some other optical detector.

The optical path length of the first and second beams is given by:

$$OPL_1 = AB + BA + AD$$
$$OPL_2 = AC + CA + AD \qquad \text{(a)}$$

The difference in the optical path length ΔOPL is given by:

$$\Delta OPL = OPL_1 - OPL_2 = 2(AB - AC) \qquad \text{(b)}$$

Reference to EQ. (12.5a) shows that the interference is constructive and a bright field is observed when $\Delta OPL = k\pi$ (k = 0, 1, 2, ...). Equation (12.5b) indicates that the interference is destructive and a dark field is observed when $\Delta OPL = (2k + 1)/2$ (k = 0, 1, 2, ...).

The Michelson interferometer is controlled by the adjustable mirror that can be translated or rotated (tilted). If this mirror is adjusted so that ΔOPL = 0, the intensity of the field will be uniform and bright. If the mirror is then tilted, a series of parallel fringes will be formed because the first wavetrain is intersecting the second wave train at oblique incidence.

If the interferometer is adjusted so that ΔOPL = 0 and a lens is inserted into the field, the resulting fringe pattern will indicate the thickness of the lens. For spherical lenses this fringe pattern should show concentric circles with fringe spacing dependent on the radius of curvature of the lens. Deviations from well established interference patterns show defects in the lenses.

12.3.3 The Mach-Zehnder Interferometer

The essential features of a Mach-Zehnder interferometer are illustrated in Fig. 12.7. The beam from a light source is divided into a reference beam and an active beam with a beam splitter (partial mirror). The beams are recombined after the active beam passes through the specimen by the second partial mirror which is adjusted to bring the optical axes of the two beams into concurrence. A lens is used to focus the recombined beams on a screen where an interference pattern is displayed.

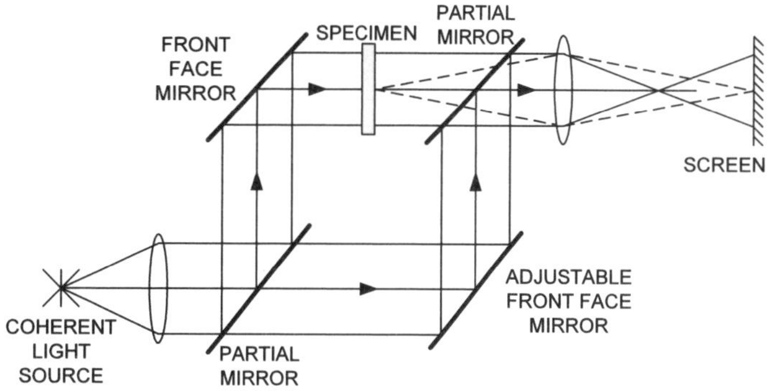

Figure 12.7 Light paths through a Mach-Zehnder interferometer.

If a laser with a coherent light output is used as a light source with the Mach-Zehnder interferometer, it is not necessary to employ a compensation specimen to adjust for the change in the optical path due to the insertion of the model. The coherent light from a laser permits the formation of very-high-order interference fringes.

A Mach-Zehnder interferometer is often employed to measure the change in thickness of a transparent specimen due to applied stresses. When a plane specimen is subjected to a plane state of stress, its thickness changes due to the strain ε_{zz}. Recall Eq. (2.16), note that σ_{zz} = 0 and write the expression for the strain ε_{zz} as:

$$\varepsilon_{zz} = (-\nu/E)(\sigma_{xx} + \sigma_{yy}) \tag{12.8}$$

From Eq. (12.8) it is evident that the change in thickness Δh of the plane specimen can be written as:

$$\Delta h = h\varepsilon_{zz} = (-h\nu/E)(\sigma_{xx} + \sigma_{yy}) \tag{12.9}$$

The changes in thickness Δh produces a change ΔOPL which in turn creates a whole field interference pattern with dark fringes occurring when:

$$\Delta h = \frac{2k+1}{2}\lambda \qquad k = 0, 1, 2, 3, ... \tag{a}$$

Substitute Eq (a) into Eq. (12.9), recall the first invariant of stress and write:

$$\sigma_{xx} + \sigma_{yy} = \sigma_1 + \sigma_2 = -\frac{E\lambda}{h\nu}\frac{2k+1}{2} \qquad k = 0, 1, 2, 3, ... \tag{12.10}$$

In practice, it is common to discover that the thickness of the specimen varies from point to point. Therefore, an initial no load fringe pattern is recorded to give $N_i(x, y)$. A load is applied to the specimen and a second fringe pattern is recorded to give $N_f(x, y)$. The isopachics ($\sigma_1 + \sigma_2$) or ($\sigma_{xx} + \sigma_{yy}$) are then determined from:

$$\sigma_{xx} + \sigma_{yy} = \sigma_1 + \sigma_2 = \frac{E\lambda}{h\nu}(N_f - N_i) \tag{12.11}$$

where N = 1/2, 3/2, 5/2, ...

12.4 INTERFERENCE OF TWO PLANE WAVES AT OBLIQUE INCIDENCE

Consider two plane polarized waves that are each propagating toward the z axis as shown in Fig. 12.8. Assume that the two waves have the same amplitude $a = a_1 = a_2$ and the same wavelength $\lambda = \lambda_1 = \lambda_2$.

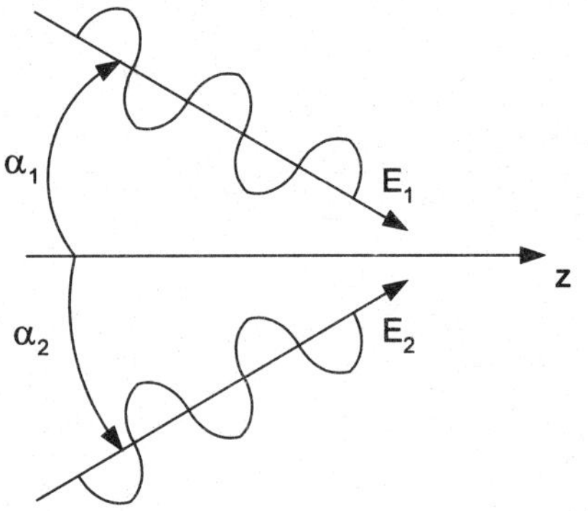

Figure 12.8 Two plane polarized waves interfering at oblique incidence.

The magnitude of the light vectors E_1 and E_2 can be expressed as:

$$E_1 = a \cos(\omega t - \phi_1)$$
$$E_2 = a \cos(\omega t - \phi_2) \qquad (12.12)$$

where the phase angle $\phi_1 = b \cos \alpha_1 + \phi_{01}$ and $\phi_2 = b \cos \alpha_2 + \phi_{02}$. Note that ϕ_{01} and ϕ_{02} are phase angles associated with the phase of the two waves at initiation.

Superimposing the two components yields:

$$E' = a\{\cos(\omega t - \phi_1) + \cos(\omega t - \phi_1)\} \qquad (a)$$

Using the trigometric identity introduced previously in Section 12.2, permits this relation to be written as:

$$E' = 2a\left[\cos\left(\frac{\phi_1 - \phi_2}{2}\right)\cos\left(\omega t - \frac{\phi_1 + \phi_2}{2}\right)\right] \qquad (12.13)$$

The amplitude term $2a\cos\left(\frac{\phi_1 - \phi_2}{2}\right)$ is squared to obtain the intensity of the interfering waves as:

$$I = 4a^2 \cos^2\left(\frac{\phi_1 - \phi_2}{2}\right) = 2a^2\left[1 + \cos(\phi_2 - \phi_1)\right] \qquad (12.14)$$

The resulting response is sinusoidal with respect to the argument $(\phi_2 - \phi_1)$. However, the contrast in the fringes produced by the interfering waves is reduced by the presence of the constant term (1) in Eq. (12.14). Note also that the angles of incidence α_1 and α_2 are not evident in this relation; however, they are involved in the argument $(\phi_2 - \phi_1)$.

A more effective technique for describing the interference pattern given in Eq. (12.14) is to graphically show the wave fronts as they interfere to produce a new intensity pattern on a plane normal to the z axis. This graphical representation of the interference process is illustrated in Fig. 12.9.

From geometric relations it can be shown that the perpendicular distance s between interference fringes is given by:

$$s = \frac{\lambda}{2\sin\left(\frac{\alpha_1+\alpha_2}{2}\right)} \qquad (12.15)$$

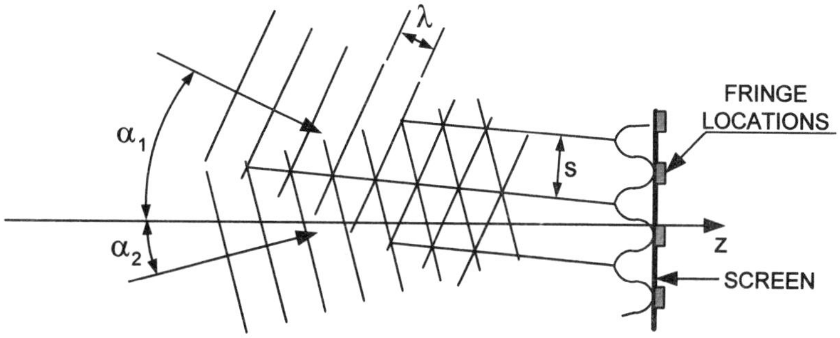

Figure 12.9 Geometric representation of oblique wave front interference to form interference fringes.

As an example, consider a series of fringes produced by a helium-neon light source with a wave length λ = 632.8 nm. If $\alpha_1 = \alpha_2 = 30°$ then the fringe spacing s is given by:

$$s = \frac{632.8}{2\sin(30°)} = 632.8 \text{ nm}$$

and the frequency of the fringes f_f is given by:

$$f_f = \frac{1}{s} = \frac{2\sin\left(\frac{\alpha_1+\alpha_2}{2}\right)}{\lambda} \qquad (12.16)$$

$$f_f = 1/s = 1/(632.8 \times 10^{-6}) = 1580 \text{ lines/mm}$$

It is clear from Eqs. (12.15) and (12.16) that the frequency of the line array formed by the interference fringes can be increased while the line spacing is decreased by increasing the angles α_1 and α_2 for the incident beams. The practical limit is established by the resolution of the recording device. If a high resolution photographic emulsion[2] is used to record the image, the limit is about 2000 lines/mm. If a digital camera is used to record the image, the resolution will depend on pixel size of the image sensor. With pixel frequencies of the order of 100 lines/mm, resolution of the high density interference fringes is not possible with the digital image sensors that are currently available.

12.5 HOLOGRAPHIC INTERFEROMETRY

In the discussion of interference patterns in Section 12.3, two well controlled (plane) waves were superimposed to produce constructive and destructive interference. Holographic interferometry is similar in that interference is produced by superimposing two waves; however, one wave (in the reference beam) is well controlled while the other wave (in the object beam) is produced by reflection from a diffuse surface. The combination of the two different types of beams does not cause difficulties if the characteristics of the object beam are taken into account. The optical arrangement for producing an off axis hologram is illustrated in Fig. 12.10.

[2] Both Kodak and Agfa-Gevaert market high resolution films and plates that will resolve 2,000 to 2,500 line pairs per millimeter.

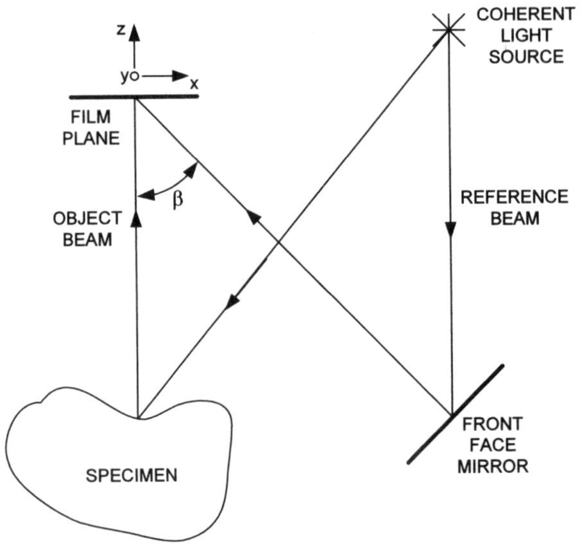

Figure 12.10 Optical arrangement for producing an off-axis hologram.

Before proceeding to derive the relations for the interference pattern produced on the film plane in Fig. 12.10, it is important to understand the behavior of light reflecting from a diffuse surface. Because the specimen is not polished and coated with a reflective material its surface is diffused. Consequently, each point on this diffused surface acts like a light source with spherical wave fronts propagating outward as illustrated in Fig. 12.11. Because there are many closely spaced light sources on the diffused surface, the waves superimpose to produce an irregular (non-planar) wave front that propagates to the film plane where it combines with the plane wavefront from the reference beam.

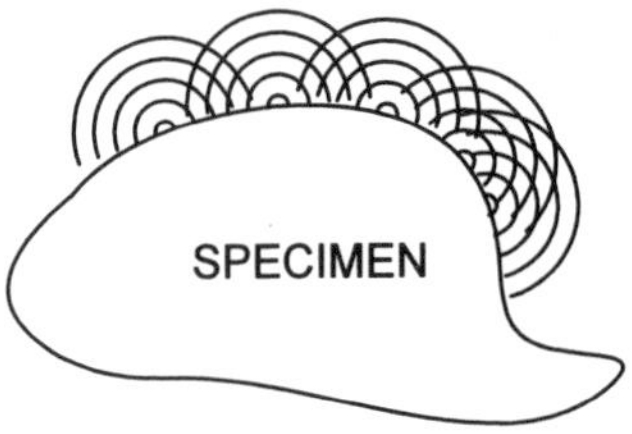

Figure 12.11 Light reflecting from a diffused surface exhibits many different point sources.

When the object beam meets the film plane at an angle of β relative to the reference beam, the variations in the angle β are negligible if the specimen is small and if it is located a large distance from the film plane.

With this assumption the equation for the light vector component in the object beam at the film plane is given by:

$$E_{01} = a_{01}(x,y)e^{j\phi(x,y)} \tag{12.17}$$

where $a_{01}(x, y)$ is the amplitude and $\phi(x, y)$ is the phase of the object beam. Both $a_{01}(x, y)$ and $\phi(x, y)$ are functions of the coordinates x and y on the film plane. Equation (12.17) is expressed in complex notation, and it is understood that the amplitude E of a light vector is the real part of the complex function.

The reference beam is controlled (a plane wavefront) and as such its amplitude a_R is constant (uniform) with respect to the film plane coordinates. Because the reference beam is inclined at an angle β with respect to the z axis, the phase of the wave is changed at the film plane by $(2\pi/\lambda)(x \sin \beta)$. Accordingly, the amplitude of the light vector E_{R1} associated with the reference beam is written as:

$$E_{R1} = a_{R1}e^{j[(2\pi/\lambda)(x\sin\beta)]} \tag{12.18}$$

The two waves superimpose at the film plane to give a combined amplitude E_1 as:

$$E_1 = E_{01} + E_{R1} = a_{01}(x, y)e^{j\phi(x,y)} + a_{R1}e^{j[(2\pi/\lambda)(x\sin\beta)]} \tag{12.19}$$

The intensity I_1 of the light at the film plane is obtained from Eq. (12.19) and Eq. (13.21) as:

$$I_1 = \left[E_{01} + E_{R1}\right]\left[E^*_{01} + E^*_{R1}\right] = E_{01}E^*_{01} + E_{01}E^*_{R1} + E_{R1}E^*_{01} + E_{R1}E^*_{R1} \tag{12.20}$$

where E^*_{01} and E^*_{R1} are the complex conjugates of the amplitudes of E_{01} and E_{R1}, respectively.

If film is placed in the film plane, properly exposed and then developed, the resulting hologram contains detailed information defining the specimen and the reference beam at each and every point on the film plane. An examination of the hologram is disappointing because the image of the specimen is not visible. Even the interference fringes cannot be observed with the naked eye. Instead, the film appears exposed without evidence of an image of either the specimen or the interference fringes.

The fringe pattern has been recorded (if the film used had sufficiently high resolution) and the fringes can be observed at high magnification (X500). The fringe density depends on the angle β as indicated in Eq. (12.16) and spatial frequencies of the fringes are commonly in the range of 1,000 to 2,000 fringes/mm.

The high spatial frequencies of the interference fringes place a severe constraint on the optical system. To properly record the fringes, every component in the optical system must be stable. A movement of a single component by a fraction of a wavelength of light during the exposure period will blur the fringe pattern. To achieve the required stability of the optical system shown in Fig. 12.10, the system is placed on a vibration isolation table. Air supports are placed in each leg of a typical vibration isolation table to mitigate floor induced vibrations. Shields are often placed around the optical system to prevent component movements or localized changes in temperature induced by air currents.

12.5.1 Double Exposed Holograms

There are many reasons for making a hologram; however, the most common purpose for a hologram in experimental mechanics is to determine surface displacements of the specimen when it is subjected to a defined system of loads. Measurement of surface displacements by holography requires two superimposed holograms—one taken before loading and the second after loading. Rather than taking two different holograms and dealing with a very severe registration problem, it is common practice to record both holograms on the same photographic plate. If the emulsion on the photographic plate is capable of about 2000 lines/mm, it is possible to record two or more holograms on the same plate without loss of the optical data stored in the combined images.

Prior to making the second exposure, the model is subjected to load. In applying the load, the specimen is fixed in the load frame so that the rigid body motion (both rotation and translation is prevented. The only motion permitted is due to the deformation of the specimen. With this constraint on loading the specimen, the relation for the superimposed waves for the second exposure is given by:

$$E_2 = E_{02} + E_{R2} \tag{12.21}$$

Clearly, the reference beam has not changed between exposures with regard to either amplitude or phase; hence: $E_{R1} = E_{R2} = E_R$ and $a_{R1} = a_{R2} = a_R$. For the objective beam, the amplitude has not changed to any appreciable degree with the very small displacements of the specimen's surface. For this reason, it is assumed that $a_{01}(x, y) = a_{02}(x, y) = a_0(x, y)$.

With these equalities, Eq. (12.21) can be expressed as:

$$E_2 = E_{02} + E_R \tag{12.22}$$

where

$$E_{02} = a_0(x, y)e^{j[\phi(x,y)+\Delta\phi(x,y)]} \tag{12.23}$$

and $\Delta\phi(x, y)$ is the phase change in the light wave due to the change in the optical path length of the object beam. The correspondence between $\Delta\phi(x, y)$ and the specimen deformations u, v or w will be developed in Section 12.5.3.

The intensity of the light on the film plane after the double exposure is given by:

$$I_{1+2} = E_1E_1^* + E_2E_2^* \tag{12.24}$$

where E_1^* and E_2^* are the complex conjugates of E_1 and E_2, respectively.

Write the complex conjugates as:

$$\begin{aligned} E_1^* &= a_0(x, y)e^{-j\phi} + a_R e^{-j\xi} \\ E_2^* &= a_0(x, y)e^{-j(\phi+\Delta\phi)} + a_R e^{-j\xi} \end{aligned} \tag{12.25}$$

where $\xi = (2\pi/\lambda)(x \sin\beta)$.

Substitute Eqs. (12.19), (12.21) and (12.25) into Eq. (12.24) and simplify to obtain:

$$I_{1+2} = 2\left(a_0^2(x, y) + a_R^2\right) + a_0(x, y)a_R\left[e^{j(\xi-\phi)} + e^{-j(\xi-\phi)} + e^{j(\xi-\phi-\Delta\phi)} + e^{-j(\xi-\phi-\Delta\phi)}\right] \tag{12.26}$$

Equation (12.26) contains the data describing the displacement of the specimen in the $\Delta\phi$ term; however, it is necessary to separate this data from the other terms in this relation. The separation is accomplished by using the double exposed hologram as a transmission filter in a reconstruction process that is presented in the next section.

12.5.2 Reconstruction from a Double Exposed Hologram

While Eq. (12.26) indicates that the double exposed hologram contains a wealth of information about the size and shape of the specimen as well as its displacements, the data is not in a form that is easily interpreted. To determine the $\Delta\phi$ term, which contains the displacement information, the double exposed hologram is used as a transmission filter in a reconstruction process. The optical arrangement used in reconstructing the image of the specimen from the hologram is presented in Fig. 12.12.

A coherent light source is required; however, it can be a different source than the one used to make the hologram. The optical elements are arranged so that the reference beam from the coherent source is inclined at the same angle β relative to a line normal to the hologram. In this arrangement, the double exposed hologram serves as a transmission filter through which the reference beam passes. Because of the high spatial frequency of the fringes on the hologram, it also serves as a diffraction grating. The reference beam is divided into many beams when it passes through the hologram (a diffraction grating). However, only the first order diffracted rays on either side of the undiffracted beam are examined to observe and record the image. The real image is located on the downstream side of the hologram in the real beam. A second (virtual) image is observed in the upstream direction of the virtual beam as indicated in Fig 12.12.

The virtual image of the specimen presented in Fig. 6.13 shows a well defined fringe pattern representing the displacement of the specimen.

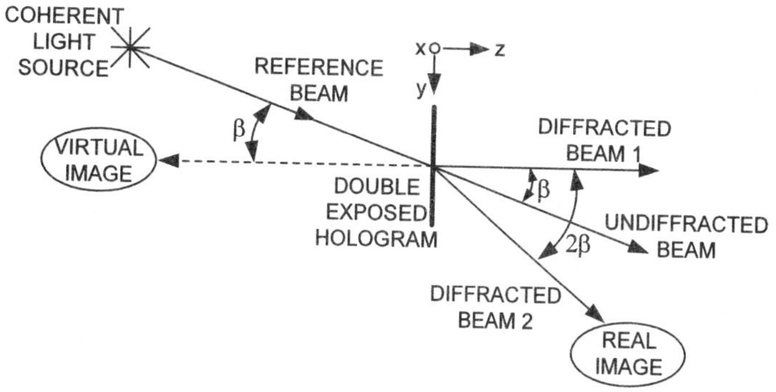

Figure 12.12 Optical arrangement for reconstructing an image from a double exposed hologram.

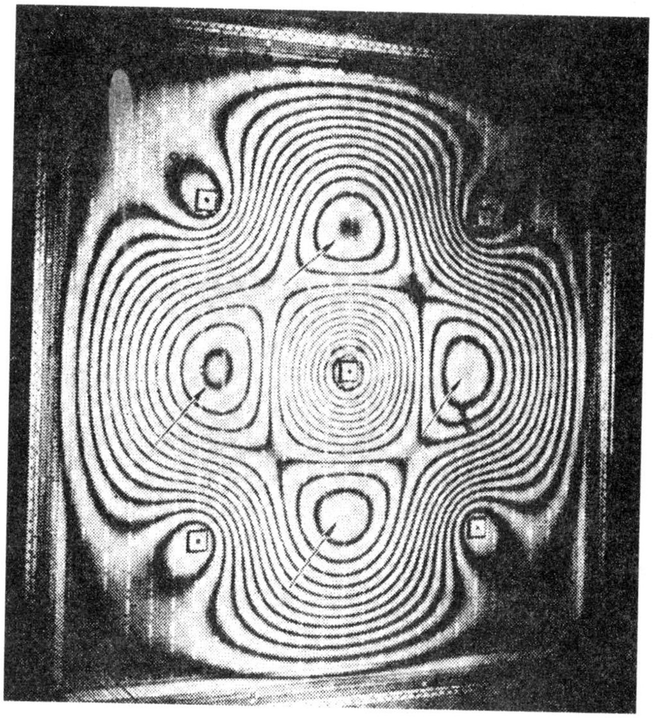

Figure 12.13 Holographic fringe pattern showing the deflection distribution for a strut-supported, flat-plate subjected to a pressure loading. Reprinted from Experimental Mechanics, Vol. 11, No. 5 with permission of the Society for Experimental Mechanics, Inc.

The amplitude of the light transmitted though the hologram is given by:

$$E_3 = I_{1+2}\, E_{R3} \qquad (12.27)$$

where $E_{R3} = a_{R3}\, e^{j\xi}$

The term a_{R3} is introduced because of the possibility that the amplitude of the laser used in the reconstruction process may be different from the one used to make the original hologram. The angle $\xi = \frac{2\pi}{\lambda} x \sin\beta$ is due to the inclination of the reference beam relative to the z direction.

Substituting this value for E_{R3} into Eq. (12.27) and multiplying gives:

$$E_3 = \left\{2\left(a_0^2(x,y) + a_R^2\right) + a_0(x,y)a_R\left[e^{j(\xi-\phi)} + e^{-j(\xi-\phi)} + e^{j(\xi-\phi-\Delta\phi)} + e^{-j(\xi-\phi-\Delta\phi)}\right]\right\}a_{R3}e^{j\xi} \qquad (a)$$

this is written as:

$$E_3 = 2\left(a_0^2(x,y) + a_R^2\right)a_{R3}e^{j\xi} + a_0(x,y)a_R a_{R3}\left[e^{j(2\xi-\phi)} + e^{j\phi} + e^{j(2\xi-\phi-\Delta\phi)} + e^{j(\phi+\Delta\phi)}\right] \quad (12.28)$$

Examination of Eq. (12.28) shows three types of terms that depend on the angle in the exponent of the exponential function. The first term E_{UN}, containing $e^{j\xi}$, describes the light amplitude in the undiffracted beam which is given by:

$$E_{UN} = 2\left(a_0^2(x,y) + a_R^2\right)a_{R3}e^{j\xi} \quad (12.29)$$

The terms in Eq. (12.28) with the exponent of the exponential term that are raised to the 2ξ power provide the real image with E_{Real} given by:

$$E_{Real} = a_0(x,y)a_R a_{R3}\left[e^{j(2\xi-\phi)} + e^{j(2\xi-\phi-\Delta\phi)}\right] \quad (12.30)$$

Finally terms in Eq. (12.28) where ξ vanishes in the exponent of the exponential term provide the virtual image with $E_{Virtual}$ given by:

$$E_{Virtual} = a_0(x,y)a_R a_{R3}\left[e^{j\phi} + e^{j(\phi+\Delta\phi)}\right] \quad (12.31)$$

The undiffracted beam does not contain data pertaining to the phase ϕ or the change in phase Δϕ; consequently, it is ignored. Both the virtual and the real images contain phase information as is evident from Eqs. (12.30) and (12.31). Because Eq. (12.31) is less complex that Eq. (12.30), it will be used to develop an explicit relation for the change in phase Δϕ.

The intensity of the virtual image is given by:

$$I_{Virtual} = E_{Virtual}\, E^*_{Virtual} \quad (12.32)$$

Substituting Eq. (12.31) into Eq. (12.32) and using the definition of the complex conjugate gives:

$$I_{Virtual} = \left[a_0(x,y)a_R a_{R3}\right]^2\left[e^{j\phi} + e^{j(\phi+\Delta\phi)}\right]\left[e^{-j\phi} + e^{-j(\phi+\Delta\phi)}\right] \quad (a)$$

Multiplication as indicated by Eq. (a) and subsequent simplification gives:

$$I_{Virtual} = 2\left[a_0(x,y)a_R a_{R3}\right]^2\left[1 + \cos\Delta\phi\right] \quad (12.33)$$

Examination of Eq. (12.33) shows that the intensity $I_{Virtual}$ goes to zero (extinction) when:

$$\text{Cos}\,\Delta\phi = -1 \qquad \text{or when } \Delta\phi = (2k+1)\pi \quad \text{where } k = 0, 1, 2, \ldots. \quad (12.34)$$

The reconstruction process provides a virtual image of the specimen with a superimposed fringe pattern related to the change in phase Δϕ as illustrated in Fig. 12.13. The phase change is due to a change in the optical path length of the object beam that occurred prior to making the second exposure of the hologram. Recall in the optical arrangement shown in Fig 12.10 that the path length of the reference beam was the same for both exposures. In the next section, the phase change Δϕ will be related to the deformation of the specimen which resulted in the change in the optical path length ΔOPL.

12.5.3 Specimen Deformations and Phase Change

Specimen movement clearly changes the optical path length (OPL) in the object beam. Reference to Fig. 12.10 shows that the OPL is the sum of two distances—the first from the light source to some point P on the specimen and the second is from that same point P on the specimen to the film plane. If the specimen moves either by deformation or rigid body motion, the OPL changes. Figure 12.14 is presented to better illustrate the different OPLs when point P on the specimen moves to point Q.

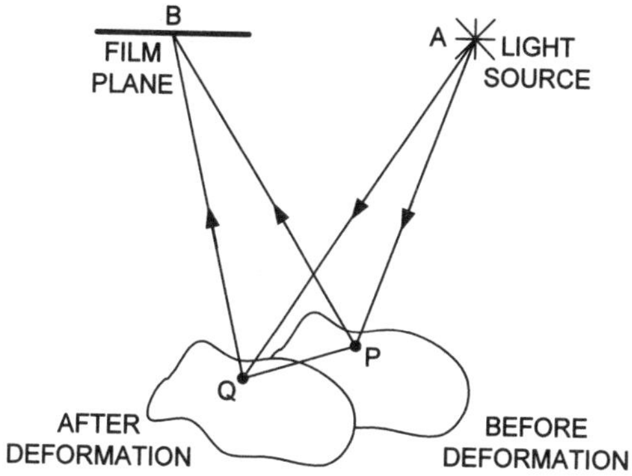

Figure 12.14 Diagram showing the change in the OPL in the object beam due to deformation PQ

The initial optical path length OPL_i is given by:

$$OPL_i = PA + PB = (\mathbf{PA} \bullet \mathbf{PA})^{1/2} + (\mathbf{PB} \bullet \mathbf{PB})^{1/2} \tag{12.35}$$

where the amplitudes of the vectors **PA** and **PB** are given by their dot products.

After deformation when point P moves to point Q the final OPL_f is given by:

$$OPL_f = QA + QB = (\mathbf{QA} \bullet \mathbf{QA})^{1/2} + (\mathbf{QB} \bullet \mathbf{QB})^{1/2} \tag{12.36}$$

The change in OPL is given by:

$$\Delta OPL = OPL_f - OPL_i \tag{a}$$

and the change in the phase is given by:

$$\Delta\phi = \frac{2\pi}{\lambda}(OPL_f - OPL_i) \tag{12.37}$$

While Eq. (12.37) is correct, it is not helpful because the displacement associated with the movement of point P to location Q is not shown explicitly. To circumvent this difficulty, the relations for **PA** and **PB** are written in a form to include the deformation term PQ. With this approach, the vectors **PA** and **PB** can be written as:

$$\mathbf{PA} = \mathbf{QA} + \mathbf{PQ}$$
$$\mathbf{PB} = \mathbf{QB} + \mathbf{PQ} \tag{12.38}$$

If it is assumed that the displacements are small, then $(PQ)^2$ is very small when compared to PQ and the higher order terms in PQ can be neglected without introducing serious error. Substituting Eqs. (12.38) into Eqs. (12.35) and (12.36) and after some manipulation with vector algebra, ΔOPL can be written as:

$$\Delta OPL = (\mathbf{PQ} \bullet \mathbf{u}_{PA} + \mathbf{PQ} \bullet \mathbf{u}_{PB}) \tag{12.39}$$

where $\mathbf{u}_{PA}$ and $\mathbf{u}_{PB}$ are unit vectors defining the directions of the vectors **PA** and **PB**.

Substituting Eq. (12.39) into Eq. (12.37) gives:

$$\Delta\phi = \frac{2\pi}{\lambda}(\mathbf{PQ} \bullet \mathbf{u}_{\mathbf{PA}} + \mathbf{PQ} \bullet \mathbf{u}_{\mathbf{PB}}) \tag{12.40}$$

Substituting Eq. (12.40) into Eq. (12.33) yields the relation for the intensity of the virtual image as:

$$I_{Virtual} = C\left[1 + \cos\frac{2\pi}{\lambda}(\mathbf{PQ} \bullet \mathbf{u}_{\mathbf{PA}} + \mathbf{PQ} \bullet \mathbf{u}_{\mathbf{PB}})\right] \tag{12.41}$$

where $C = 2[a_0(x, y)a_R a_{R3}]^2$

Out of Plane Displacement w

To show an example that involves less vector algebra, consider the optical arrangement shown in Fig. 12.15, which is used for out of plane displacement measurements. In this example, the displacement of the specimen is only along the z axis. The displacements u and v are zero. Moreover both PB and QP coincided and are normal to the film plane. With this optical arrangement, the deformation along the z axis PQ equals the displacement w.

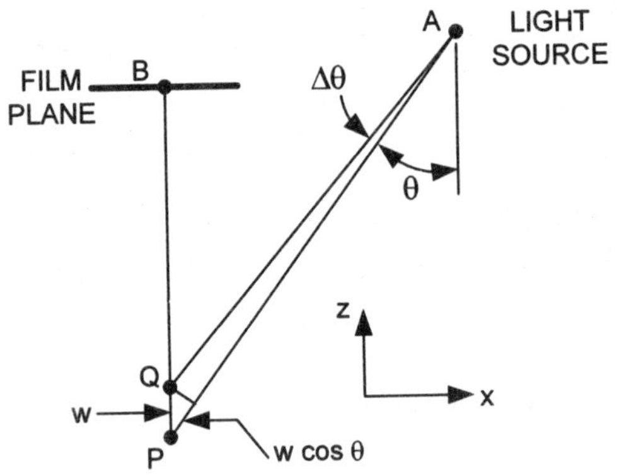

Figure 12.15 Changes in the object beam when the specimen undergoes a displacement w.

For this example, vector notation is not needed because it is clear that:

$$OPL_i = PA + PB = PA + QB + w$$

$$OPL_f = QA + QB \tag{a}$$

The change in the optical path length is given by:

$$\Delta OPL = OPL_i - OPL_f = PA + QB + w - QA - QB = PA + w - QA \tag{b}$$

If Δθ is small then:

$$PA = w \cos\theta + QA \tag{c}$$

Substitute Eq. (c) into Eq. (b) to obtain:

$$\Delta OPL = PA + w - QA = w \cos\theta + QA + w - QA = w\,(1 + \cos\theta) \tag{d}$$

Substitute Eq. (d) into Eq. (12.37) gives:

$$\Delta\phi = \frac{2\pi}{\lambda}[w(1 + \cos\theta)] \tag{e}$$

Then from Eq. (12.33)

$$I_{Virtual} = C\left\{1 + \cos\frac{2\pi}{\lambda}\left[w(1+\cos\theta)\right]\right\} \tag{12.42}$$

This result indicates that the fringe pattern is dependent upon the displacement w and angle of inclination θ. The angle θ is a fixed quantity because it is selected when arranging the optical elements used in making the hologram. Clearly as θ ⇒ 0, the number of fringes produced for a given displacement w increases.

Next consider the optical arrangement as shown in Fig. 12.15, but use Eq. (12.40) in the analysis. Recall Eq. (12.40):

$$\Delta\phi = \frac{2\pi}{\lambda}(\mathbf{PQ} \bullet \mathbf{u}_{PA} + \mathbf{PQ} \bullet \mathbf{u}_{PB}) \tag{12.40}$$

Note that:

$$PQ = w \qquad \mathbf{u}_{PQ} = \mathbf{k}$$
$$\mathbf{u}_{PA} = \mathbf{i}\sin\theta + \mathbf{k}\cos\theta \quad \text{and} \quad \mathbf{u}_{PB} = \mathbf{k} \tag{f}$$

Substituting Eq. (e) into Eq. (12.40) to obtain:

$$\Delta\phi = \frac{2\pi}{\lambda}PQ\left[\mathbf{k} \bullet (\mathbf{i}\sin\theta + \mathbf{k}\cos\theta) + \mathbf{k} \bullet \mathbf{k}\right] = \frac{2\pi}{\lambda}\left[w(1+\cos\theta)\right] \tag{g}$$

This result is the same as the relation derived previously in Eq. (e) using geometry and trigonometry rather that vector algebra.

In Plane Displacement u

Consider the point P on the specimen moves in the positive x direction and displacements v = w = 0. If the arrangement of the coherent source and the film plane, are as shown in Fig. 12.16, it is possible to write the unit vectors for **PA** and **PB** as:

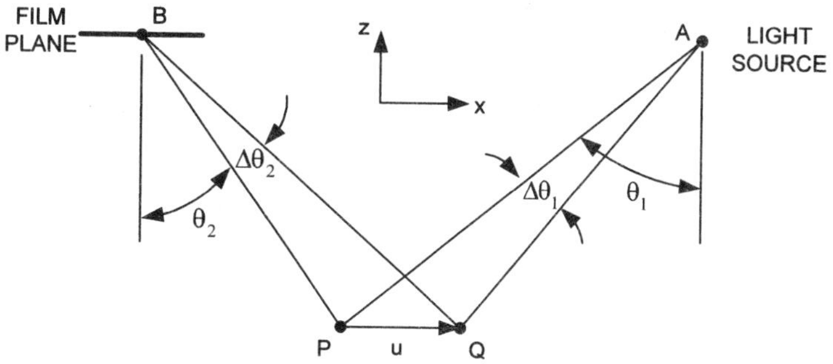

Figure 12.16 Changes in the object beam when the specimen undergoes a displacement u.

The expression for the unit vectors associated with Fig. 12.16 is given by:

$$PQ = u \qquad \mathbf{u}_{PQ} = \mathbf{i}$$
$$\mathbf{u}_{PA} = \mathbf{i}\sin\theta_1 + \mathbf{k}\cos\theta_1 \quad \text{and} \quad \mathbf{u}_{PB} = -\,\mathbf{i}\sin\theta_2 + \mathbf{k}\cos\theta_2 \tag{a}$$

Substituting Eq. (a) into Eq. (12.39) as:

$$\Delta\text{OPL} = \text{u}\mathbf{i} \bullet (\mathbf{i}\sin\theta + \mathbf{k}\cos\theta) - \text{u}\mathbf{i} \bullet (-\mathbf{i}\sin\theta + \mathbf{k}\cos\theta) \quad \text{(b)}$$

this reduces to:

$$\Delta\text{OPL} = \text{u}\,(\sin\theta_2 - \sin\theta_1) \quad \text{(c)}$$

The maximum change in the optical path length occurs when $\theta_1 = 0$. Then it is clear that

$$\Delta\phi = \frac{2\pi}{\lambda}\left[\text{u}\sin\theta_2\right] \quad (12.43)$$

when $\theta_1 = 0$.

The General Case

In the general case, the specimen's displacement as point P moves to Q involves all three displacement components—u, v and w. To interpret the holographic fringe patterns in this case express the displacement vector **PQ** as:

$$\mathbf{PQ} = \text{u}\,\mathbf{i} + \text{v}\,\mathbf{j} + \text{w}\,\mathbf{k} \quad \text{(a)}$$

where **i**, **j** and **k** are unit vectors along the axes of the coordinate system shown in Fig. 12.17.

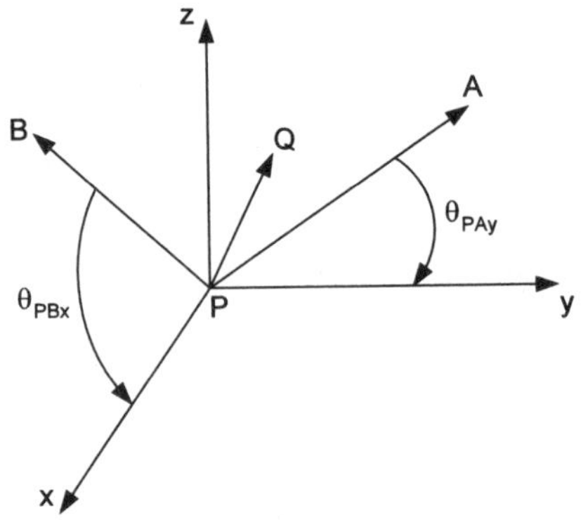

Figure 12.17 Cartesian coordinate system with its origin at point P.

The unit vectors $\mathbf{u_{PA}}$ and $\mathbf{u_{PB}}$ are defined by their direction cosines relative to this coordinate system. Hence:

$$\mathbf{u_{PA}} = \cos\theta_{PAx}\,\mathbf{i} + \cos\theta_{PAy}\,\mathbf{j} + \cos\theta_{PAz}\,\mathbf{k}$$

$$\mathbf{u_{PB}} = \cos\theta_{PBx}\,\mathbf{i} + \cos\theta_{PBy}\,\mathbf{j} + \cos\theta_{PBz}\,\mathbf{k} \quad \text{(b)}$$

Substituting Eq(a) and Eq. (b) into Eq. (12.40) and performing the vector algebra associated with the dot products yields:

$$\Delta\phi = \frac{2\pi}{\lambda}\left[(\cos\theta_{PAx} + \cos\theta_{PBx})\text{u} + (\cos\theta_{PAy} + \cos\theta_{PBy})\text{v} + (\cos\theta_{PAz} + \cos\theta_{PBz})\text{w}\right] \quad (12.44)$$

These results indicate that there are three unknowns u, v and w embedded in a single equation for the change in phase recorded on the hologram. The direction cosines are known as they depend upon the positions of the light source and the film plane relative to the specimen. To provide the additional data required to solve for the three unknown displacements, the usual practice is to record three holograms each with unique positions for the light source and the film plane. This approach can be employed when a zero order fringe exists on all three holograms (i.e. a point on the specimen where u = v = w = 0. This restriction is imposed by the requirement to count the fringe orders at each point on all three holograms to uniquely establish $\Delta\phi_1(x, y, z)$, $\Delta\phi_2(x, y, z)$ and $\Delta\phi_3(x, y, z)$.

12.5.4 Real Time and Time Averaged Methods in Holography

Real time holography is employed when it is advantageous to observe the interferometric fringe pattern that develops as load is slowly applied to the specimen. The procedure involves making a hologram of the specimen in its initial state using the optical arrangement presented in Fig. 12.10.

Next, the hologram (the photographic plate) with the image of the unloaded specimen plate is replaced precisely at its original position in the optical arrangement. This precise placement of the photographic plate is not a trivial task and registration of the plate with a few wavelengths of light is a critical step in the process. Precise placement is usually achieved by developing the photographic plate in a special plate holder that also serves as the development tank. In this manner, the hologram is developed without ever removing it from the plate holder.

With the holographic negative in its precise position, it is used as a transmission filter, as indicated in Fig. 12.18. The light from the reference and object beams pass through the hologram to reproduce an interference pattern similar to that obtained with the double exposure technique. Changes in the fringe pattern due to $\Delta\phi$ are recorded (or observed) as the load is slowly applied to the specimen.

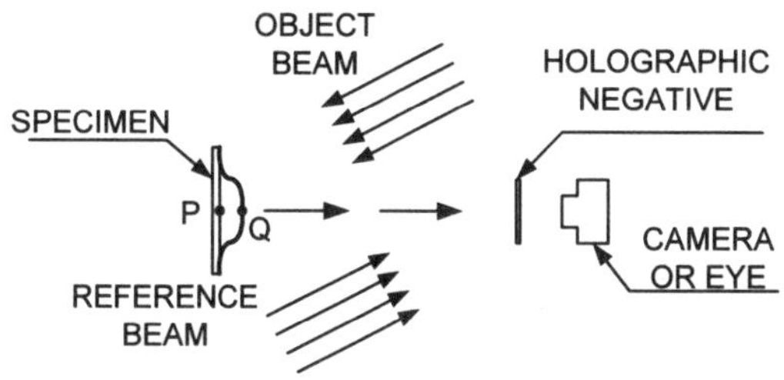

Figure 12.18 Optical arrangement for real time holographic measurement of displacement due to bending of flat plates.

Because the object beam from the specimen and the beam transmitted by the holographic negative interfere, a fringe pattern is produced. The analysis of this pattern is similar in many respects to the analysis of the double exposed hologram. The amplitude of the light vector associated with the image transmitted by the developed holographic negative can be expressed as:

$$E_{01} = -\, a_{01}\,(x,y)\; e^{j\phi(x,y)} \tag{12.45}$$

The negative sign in Eq. (12.45) is to account for the fact that the image on the photographic plate is a negative, which reverses the maxima and minima amplitudes.

The object wave from the specimen undergoes a phase change $\Delta\phi(x, y)$ due to the out-of-plane displacements. Accordingly, the amplitude of this light vector is given by:

$$E_{02} = a_{01}\,(x,y)\; e^{j[\phi(x,y) + \Delta\phi(x,y)]} \tag{12.46}$$

where it has been assumed that the amplitude of the two beams are equal, namely $a_{01}(x, y) = a_{02}(x, y) = a_0(x, y)$. It is important to balance the light intensity of the beams to provide equal amplitudes so as to enhance contrast in the fringe pattern developed.

Superimposing the two amplitudes in the interference process and substituting Eqs (12.45) and (12.46) into the result yields:

$$E = E_{01} + E_{02} = a_0\,(x, y)\; e^{j\phi(x,y)}\; [e^{j\Delta\phi(x,y)} - 1] \tag{12.47}$$

The intensity of the real time image I_{RT} is then determined from:

$$I_{RT} = E\,E^* \tag{a}$$

Substituting Eqs. (12.45) and (12.46) into Eq. (a) gives:

$$I_{RT} = a^2\left\{e^{j\phi}\left[e^{j\Delta\phi} - 1\right]e^{-j\phi}\left[e^{-j\Delta\phi} - 1\right]\right\} \tag{b}$$

where it has been assumed that the amplitude a_{01} (x, y) is constant over the field and equal to a.

Simplifying Eq. (b) gives:

$$I_{RT} = 2a^2\,(1 - \cos\Delta\phi) \tag{12.48}$$

Note that I = 0 (extinction) occurs when $\cos\Delta\phi = 1$; hence:

$$\Delta\phi = 2k\,\pi \qquad k = 0, 1, 2, \ldots \tag{12.49}$$

Time Averaged Holography

Holography is also useful in studying vibratory motion by using time averaged techniques. Consider time averaged holography with an example of a circular plate that is clamped about it perimeter. This plate is forced to vibrate in one of several different modes by applying excitation forces that do not disturb the optical arrangement for recording holograms. A single hologram of the vibrating plate is made; however, care is exercised in to insure that the exposure time for the photographic plate is long compared to the period of the vibratory motion. With such a long exposure time, it is clear that the interference pattern will move during the exposure interval and that the definition of the pattern will be compromised. However, during a vibration cycle, the body spends most of its time during a period at either the maximum or minimum displacement where the direction of motion is changing. For this reason, the single exposure records both the maximum and minimum displacements. Essentially the long exposure provides the equivalent of a double exposure with one image of the maximum amplitude of vibration and the other of its minimum amplitude. Of course there is an effect of the exposure time when the specimen is moving from the maximum to the minimum state, but this motion only reduces the contrast of the hologram.

With a single exposure, the amplitude of the light vector can be expressed as:

$$E_1 = a_0 e^{j\left[\phi(x,y)+2\left(\frac{2\pi}{\lambda}\right)w(x)\cos\omega t\right]} \tag{12.50}$$

where ω is the circular frequency of the vibratory motion.

a is the amplitude of the light vector taken as a constant over the field.

w is the out of plane displacement that is a function of only the coordinate x.

The formulation in Eq. (12.50) assumes that the object beam is perpendicular to the film plane and that the change in the optical path length is 2w as the plate moves from its maximum to its minimum states. The amplitude of the object wave is given by the time averaged value of E_1. Time averaging is obtained by integrating Eq. (12.50) to obtain:

$$E_1\big]_{Average} = \frac{a_0}{T}\int_0^T e^{j\left[\phi(x,y)+2\left(\frac{2\pi}{\lambda}\right)w(x)\cos\omega t\right]}dt \tag{12.51}$$

Integrating this relation gives:

$$E_1\Big]_{Average} = a_0 J_0^2\left[\left(\frac{4\pi}{\lambda}\right) w(x)\right] \tag{12.52}$$

where $J_0\left[\left(\frac{4\pi}{\lambda}\right) w(x)\right]$ is a zero order Bessel function with w(x) in its argument.

From Eq (12.52) it is clear that the intensity can be written as:

$$I_{Average} = a_0^2 J_0^2\left[\left(\frac{4\pi}{\lambda}\right) w(x)\right] \tag{12.53}$$

When $J_0\left[\left(\frac{4\pi}{\lambda}\right) w(x)\right] = 0$ extinction occurs and dark fringes are formed. These fringes are located where $\left[\left(\frac{4\pi}{\lambda}\right) w(x)\right] = 2.405, 5.520, 8.654, 11.792, \ldots$. Note also that the maximum intensity occurs when $J_0(0) = 1$ which corresponds to $w(x) = 0$. Hence, the nodal points of the vibrating specimen appear as bright spots on the reconstructed image of the hologram.

An example of the fringes representing the amplitudes of a rib stiffened plate vibrating at 725 and 986 Hz is presented in Fig. 12.19.

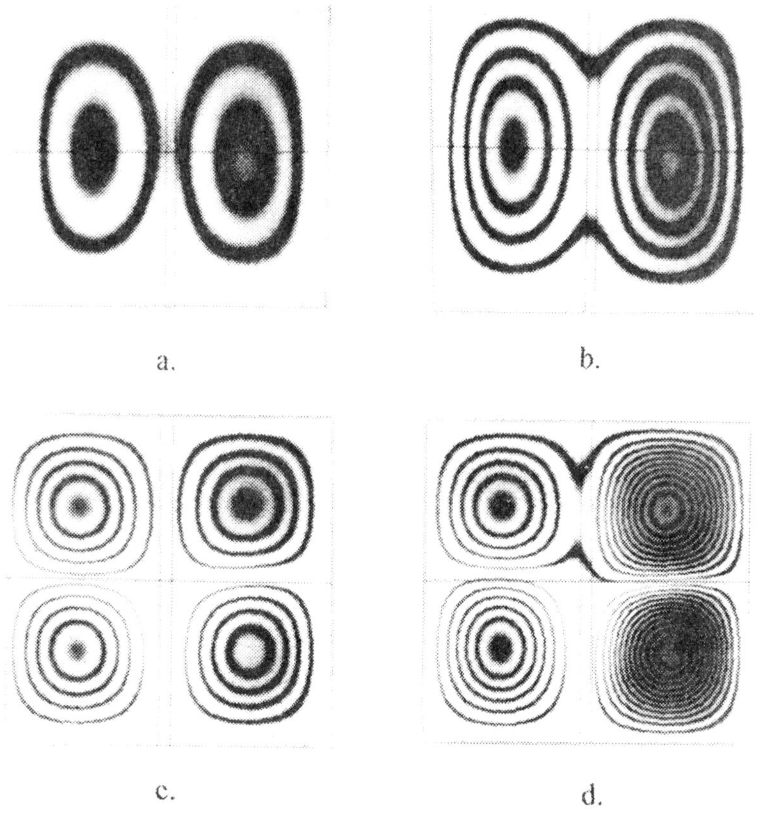

Figure 12.19 Time averaged holographic fringe pattern of a rib stiffened plate. (a) f = 725 Hz, (b) f = 725 Hz, but with increased amplitude. (c) f = 986 Hz, (d) f = 986 Hz, but with increased amplitude.
Reprinted from Experimental Mechanics, Vol. 11, No. 5 with permission of the Society for Experimental Mechanics, Inc.

EXERCISES

12.1 Verify Eqs. (12.3) and (12.4).

12.2 Considering the similarities in the derivations, why does Eq. (12.6) differ from Eq. (12.4)?

12.3 For the fringe pattern illustrated in Fig. 12.4, determine the maximum dimension of the gap if the wavelength of the light source was λ = 543 nm.

12.4 Prepare a sketch of a generic interferometer and describe the optical elements used in this arrangement.

12.5 Prepare a sketch of the Michelson interferometer and describe the optical elements used in this arrangement. Why is it necessary to incorporate an adjustable mirror in this arrangement?

12.6 Suppose the mirror in the Michelson interferometer is adjusted to provide a null field where $OPL_1 = OPL_2$. Then the adjustment is changed to tilt the mirror about its vertical diameter by 0.05°. If a helium neon laser is used as the light source, determine the number of fringes that will be observed if the field diameter is 150 mm.

12.7 Prepare a sketch of a Mach-Zehnder interferometer and describe the optical elements used in this arrangement. Compare the Michelson and Mach-Zehnder interferometers and cite advantages and disadvantages of each.

12.8 An interferometer is used to record the change in the thickness of a plane specimen when it is subjected to a plane state of stress. If $N_f = 12$ and $N_i = 3$ at a selected point in the specimen, determine the sum of the principal stresses at that point. The nominal thickness of the specimen is 0.6 mm. The specimen is fabricated from polycarbonate and the light source for the interferometer is provided by a helium-neon laser.

12.9 Beginning with Eq. (12.12) verify Eq. (12.14).

12.10 Using the geometric representation of oblique wave front interference presented in Fig. 12.9, derive Eq. (12.15) and Eq. (12.16).

12.11 Expand and simplify the equation for the light intensity given in Eq. (12.20).

12.12 Suppose you are to install a new holographic laboratory in a new ten story building. You are given your choice of rooms. Describe the characteristics of the room that you would select for the new laboratory.

12.13 Beginning with Eq. (12.22), verify Eq. (12.26).

12.14 Beginning with Eq. (12.26), derive Eq. (12.28).

12.15 Beginning with Eq. (12.31), derive Eq. (12.34)

12.16 Equation (12.43) was established using vector algebra. Derive the same relationship using only geometry and trigonometry.

12.17 Sketch a holographic arrangement that shows the changes in the object and reference beams when the specimen undergoes a displacement v. Derive a relation similar to Eq. (12.43) for $\Delta\phi$ in terms of the displacement v and the angles of inclination of the light beams.

12.18 Derive Eq. (12.44).

12.19 Why would a researcher prefer to employ real time holography instead of double exposure holography.

12.20 Prepare a sketch showing the optical arrangement for a real time holographic study of bending of flat plates.

12.21 Beginning with Eqs. (12.45) and (12.46), verify Eqs. (12.48) and (12.49).

12.22 Verify Eq. (12.51).

12.23 Use Eq. (12.53) to prepare a graph of $I_{Average}$ as a function of w(x) as it varies from 0 to 10λ.

BIBLIOGRAPHY

1. Cloud, G. L.: Optical Methods for Engineering Analysis, Cambridge University Press, New York, 1998.
2. Born M. and Wolf E.: Principles of Optics, 7th Edition, Cambridge University Press, New York, 1999.
3. Tolansky, S.: An Introduction to Interferometry, 2nd Edition, Halsted Press, New York, 1973.
4. Jenkins, F. A. and H. E. White: Fundamentals of Optics, 4th Edition, McGraw Hill, New York, 2001.
5. Vest, C. M.: Holographic Interferometry, John Wiley & Sons, New York, 1979.
6. Jones R. and C. Wykes: Holographic and Speckle Interferometry, 2nd Edition, Cambridge University Press, New York, 1989.
7. Stenson, K. A.: "Holographic Vibration Analysis," Holographic Nondestructive Testing, R. K. Erf, (ed.), Academic Press, New York, 1974.
8. Smith, H. M.: Principals of Holography, 2nd Edition, John Wiley & Sons, New York, 1975.
9. Waters, J. P.: "Holography" and "Interferometric Holography," Holographic Nondestructive Testing, R. K. Erf, (ed.), Academic Press, New York, 1974.
10. Leith, E. N. and J. Upatnieks: "Reconstructed Wavefronts and Communication Theory," Journal of the Optical Society, Vol. 52, pp. 1123-1130, 1962.
11. Stroke, G. W.: An Introduction to Coherent Optics and Holography, 2nd Edition, Academic Press, New York, 1969.
12. Ranson, W. F., M. A. Sutton, W. H. and Peters, "Holographic and Laser Speckle Interferometry," Handbook on Experimental Mechanics, 2nd Edition, Editor A. S. Kobayashi, Prentice-Hall, Englewood Cliffs, NJ, 1993.
13. Wilson, A. D., C. H. Lee, H. R. Lominac, and D. H. Strope: "Holographic and Analytic Study of a Semi-clamped Rectangular Plate Supported by Struts," Experimental Mechanics, Vol. 11, No. 5, pp. 229-234, 1971.
14. Hazell, C. R. and S. D. Liem: Vibration Analysis of Plates by Real-Time Stroboscopic Holography," Experimental Mechanics, Vol. 13, No. 8, pp. 339-334, 1973.
15. Sharpe, Jr., W. N. (Ed.): Springer Handbook of Experimental Solid Mechanics, Springer Science + Business Media LLC, New York, 2008.

CHAPTER 13

MOIRÉ AND MOIRÉ INTERFEROMETRY

13.1 INTRODUCTION

The word moiré is the French name for a fabric known as watered silk, which exhibits patterns of light and dark bands. This moiré effect occurs whenever two similar but not quite identical arrays of equally spaced lines or dots are arranged so that one array can be viewed through the other. Almost everyone has seen the effect in two parallel snow fences or when two layers of window screen are placed in contact.

The first practical application of the moiré effect may have been its use in judging the quality of line rulings used for diffraction gratings or halftone screens. In this application, the moiré fringes provide information about errors in spacing, parallelism and straightness of the lines in the ruling. All of these factors contribute to the quality of the ruling. Elimination of the moiré effect has always been a major problem associated with screen photography in the printing industry. In multicolor printing, for example, where several screened images must be superimposed, the direction of screening must be carefully controlled to minimize moiré effects.

Considerable insight into the moiré effect can be gained by studying the relationships which exist between the spacing and inclination of the moiré fringes in a pattern and the geometry of the two interfering line arrays which produced the pattern. This geometrical interpretation of the moiré effect was first published by Tollenaar [1] in 1945. Later Morse, Durelli, and Sciammarella [2] presented a complete analysis of the geometry of moiré fringes in strain analysis. A second method of analysis, in which moiré fringes are used to measure displacements, was presented by Weller and Shepard [3] in 1948. Dantu [4] followed the same approach in 1954 and introduced the interpretation of moiré fringes as components of displacements for plane elasticity problems. Sciammarella and Durelli [5] extended this approach into the region of large strains in a paper in 1961.

Moiré fringes have also been used by Theocaris [6, 7] to measure out-of-plane displacements and by Ligtenberg [8] to measure slopes and moment distributions in flat slabs. More recently, Post [9] has developed advanced methods using moiré interferometry which greatly improves the sensitivity of the method.

In the following sections of this chapter, both the geometrical and the displacement-field approaches to moiré fringe analysis will be outlined. The advantages and limitations of the approaches will be discussed. The more advanced methods of moiré interferometry and electronic moiré will be introduced at the end of the chapter.

13.2 MOIRÉ FRINGES PRODUCED BY MECHANICAL INTERFERENCE

The arrays used to produce moiré fringes may be a series of straight parallel lines, a series of radial lines emanating from a point, a series of concentric circles or a pattern of dots. In stress-analysis work, arrays consisting of straight parallel lines (ideally, opaque bars with transparent interspaces of equal width) are the most commonly used. Such arrays are frequently referred to as **grids**, **gratings**, or **grills**. In this book, the term **grid** has been used to denote a coarse array (10 lines per inch or less) of perpendicular lines that does not produce a moiré effect. Image analysis techniques are normally employed to measure changes in spacing between the intersecting points of a grid network before and after loading, and from these data, displacements and strains can be determined. In this book, the term **grating** will be used to denote a parallel-line array suitable for moiré work (50 to 1,000 lines per inch). When two perpendicular line arrays are used on a specimen, the term **cross-grating** is employed.

When two gratings are overlaid, moiré fringes are produced. The superposition of two gratings can be accomplished by either mechanical or optical means. In the following discussions, the two gratings will be referred to as the **model**, or **specimen, grating** and the **master**, or **reference, grating**. Often the model grating is applied by coating the specimen with a photographic emulsion and contact-printing through the master grating. In this way, the model and master gratings are essentially identical (matched) when the specimen is in the undeformed state. Model arrays can also be applied by bonding, etching, ruling, etc.

A typical moiré fringe pattern, obtained using transmitted light through the model and master gratings, is shown in Fig. 13.1. In this instance, both the master grating and the model grating before deformation had 40 lines per millimeter. The number of lines per unit length is frequently referred to as the **density of the grating**.

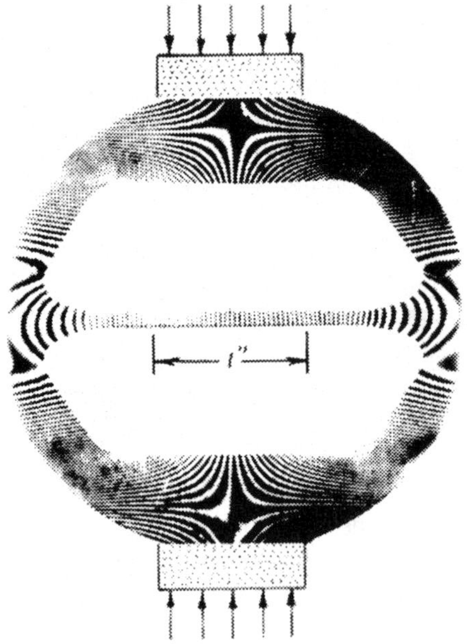

Figure 13.1 Moiré fringe patterns in a special tensile strength specimen. The primary direction of the reference grating was vertical. Courtesy of A. J. Durelli.

In the discussion which follows, the center-to-center distance between the master grating lines will be referred to as the **pitch of the grating** (reciprocal of the density) and will be designated by the symbol p. The center-to-center distance on the model grating in the deformed state will be designated by p′. The direction perpendicular to the lines of the master grating will be referred to as a **primary direction**. The direction parallel to the lines of the master grating will be referred to as a **secondary direction**.

The mechanism of formation of moiré fringes can be illustrated by considering the transmission of a beam of light through model and reference arrays, as shown in Fig. 13.2a. If the model and master gratings are identical, and if they are aligned such that the opaque bars of one grating coincide exactly with the opaque bars of the other grating, the light will be transmitted as a series of bands having a width equal to one-half the pitch of the gratings. However, due to the effects of diffraction and the resolution capabilities of the eye, this series of bands appears as a uniform grey field with an intensity equal to approximately one-half the intensity of the incident beam.

If the model is then subjected to a uniform deformation, like the one shown in the central tensile bar of Fig. 13.1, the model grating will exhibit a deformed pitch p′, as shown in Fig. 13.2b. The transmission of light through the two gratings will now occur as a series of bands of different width, with the width of the band depending on the overlap of an opaque bar with a transparent interspace. If the intensity of the emerging light is averaged over the pitch length of the master grating, to account for diffraction effects and the resolution capabilities of the eye, the intensity is observed to vary as a staircase function of position. The peaks of the function occur at positions where the transparent interspaces of the two gratings are aligned. A light band is perceived by the eye in these regions. When an opaque bar of one grating is aligned with the transparent interspace of the other grating, the light transmitted is minimum and a dark band known as a **moiré fringe** is formed.

Inspection of the opaque bars in Fig. 13.2b indicates that a moiré fringe is formed each time the model grating undergoes a deformation in the primary direction equal to the pitch p of the master grating. Specimen deformations in the secondary direction do not produce moiré fringes. In the case illustrated in Fig. 13.1, 32 fringes have formed in the 25-mm gage length indicated on the specimen. Thus, the change in length of the specimen in this 25-mm interval is given by:

$$\Delta l = np = 32(0.025) = 0.8 \text{ mm}$$

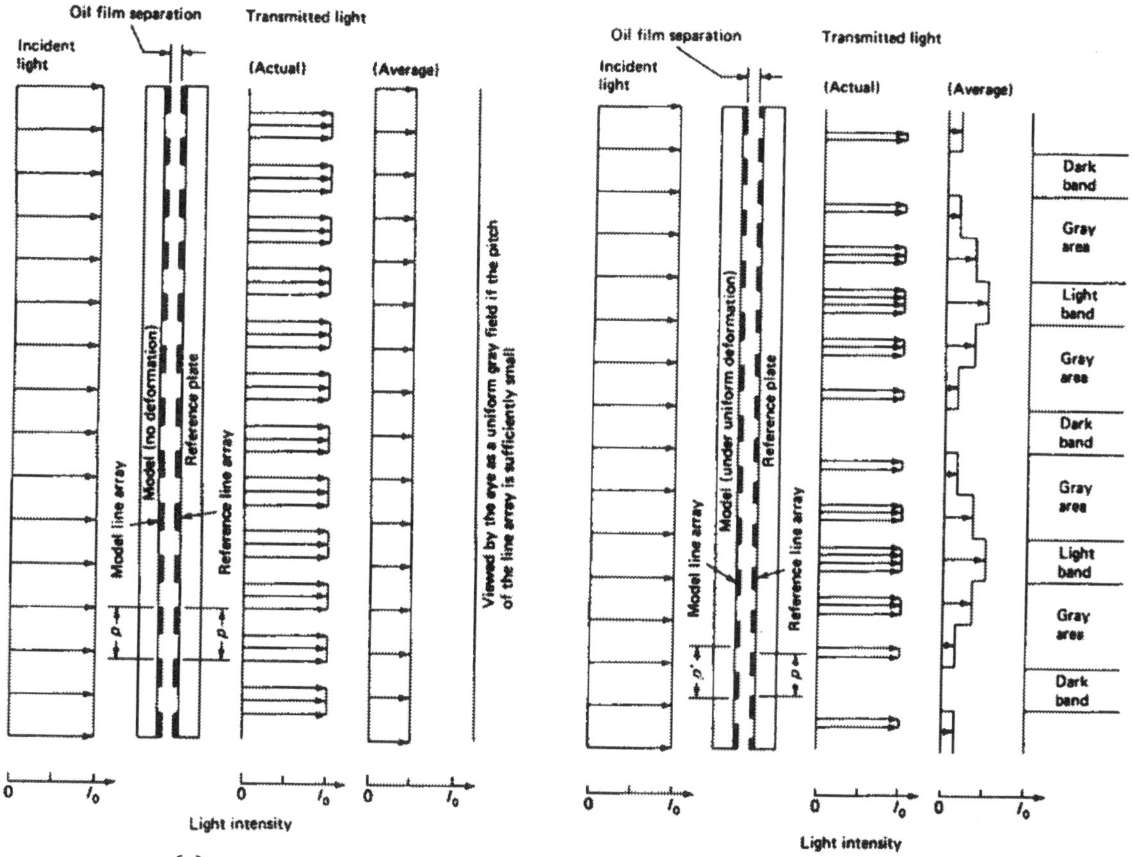

(a) (b)

Figure 13.2 (a) Light transmission through matched and aligned model and master gratings; (b) Formation of moiré fringes in a uniformly deformed specimen.

It should be noted in this illustration that the 25-mm gage length represents the final or deformed length of the specimen rather than the original length. Thus, the engineering strain in this interval could be expressed as:

$$\varepsilon = \frac{\Delta \boldsymbol{l}}{\boldsymbol{l}_0} = \frac{np}{\boldsymbol{l}_g - np} = \frac{0.8}{2.5 - 0.8} = 0.033$$

The results of the previous observations can be generalized to write expressions for either tensile or compressive strains over an arbitrary gage length for cases involving uniform elongation or contraction but without rotation as:

$$\varepsilon = \begin{cases} +\dfrac{\Delta l}{l_0} = +\dfrac{np}{l_g - np} & \text{for tensile strains} \quad (13.1) \\ -\dfrac{\Delta l}{l_0} = -\dfrac{np}{l_g + np} & \text{for compressive strains} \quad (13.2) \end{cases}$$

where p is the pitch of master grating and undeformed model grating.
n = number of moiré fringes in gage length and l_g is the gage length

On the horizontal bar of Fig. 13.1, the moiré fringes were formed by elongations or contractions of the specimen in a direction perpendicular to the lines of the master grating. Simple experiments with a pair of identical gratings indicates that moiré fringes can also be formed by pure rotations (without either elongations or contractions), as illustrated in Fig. 13.3. In this illustration, two gratings having a line density of 2.3 lines per millimeter have been rotated through an angle θ with respect to one another. Note that the moiré fringes have formed in a direction which bisects the obtuse angle between the lines of the two gratings. The relationship between angle of rotation θ and angle of inclination ϕ of the moiré fringes, both measured in the same direction and with respect to the lines of the master grating, can be expressed as:

$$\phi = \frac{\pi}{2} + \frac{\theta}{2} \tag{13.3}$$

or

$$\theta = 2\phi - \pi \tag{13.4}$$

Figure 13.3 Moiré Fringes formed by the rotation of one grating relative to the other.

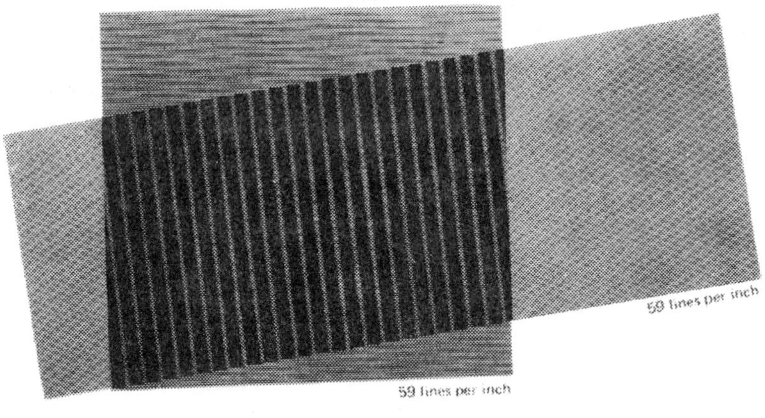

13.3 THE GEOMETRICAL APPROACH TO MOIRÉ-FRINGE ANALYSIS

In Section 13.2, it was shown that moiré fringes are produced either by rotation of the specimen grating with respect to the master grating or by changes in pitch of the specimen grating as a result of load induced deformations. At a general point in a stressed specimen, these two effects occur simultaneously and produce a fringe pattern similar to the one shown in Fig. 13.4. The information readily available from such a pattern is the angle of inclination ϕ of the fringes with respect to the lines of the master grating and the distance between fringes δ. The quantities to be determined at the point of interest are the angle of rotation θ of the specimen grating with respect to the lines of the master grating and the pitch p′ of the specimen grating in the deformed state. This method of analysis, which is very convenient when information is desired at only a few selected points, has become known as the geometrical approach. Because this approach gives rotations and strains that are average values between two fringes, such analyses should be limited to uniform strain fields or within very small regions of non-uniform fields.

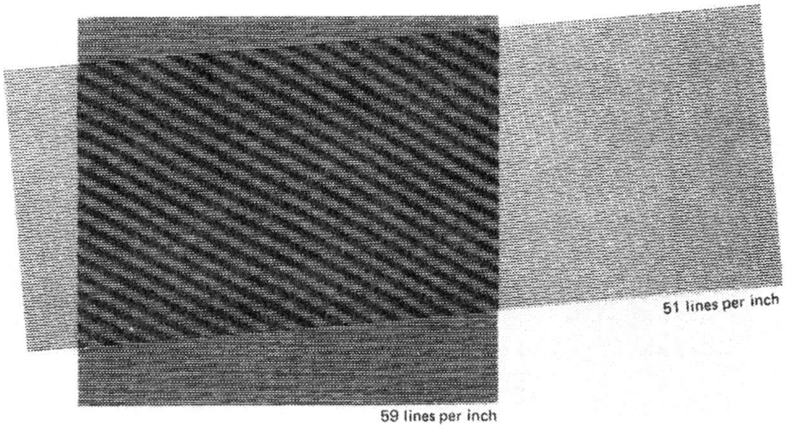

Figure 13.4 Moiré Fringes formed by a combination of rotation and difference in pitch.

Relationships between the master grating pitch p, the deformed specimen grating pitch p', the angle of rotation θ of the specimen grating with respect to the lines of the master grating, and the data available from a moiré fringe pattern can be obtained from a geometric analysis of the intersections of the lines of the two gratings as shown in Figs. 13.5a and b. In both of these figures, the opaque bars of the gratings are defined by their centerlines, and light rather than dark bands are referred to as fringes because they are easier to locate accurately.

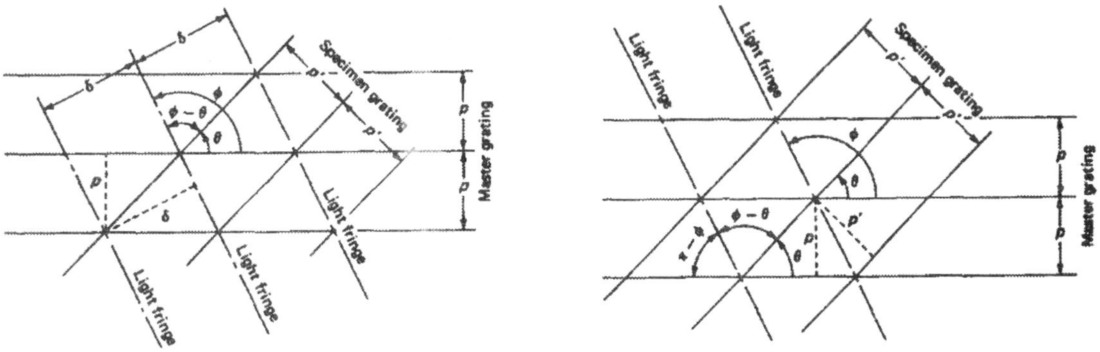

Figure 13.5 Geometry of moiré fringes:
(a) in terms of fringe spacing δ. (b) in terms of fringe inclination ϕ.

Consider first the distance δ between fringes as shown in Fig.13.5a where it is clear that:

$$\frac{p}{\sin\theta} = \frac{\delta}{\sin(\phi-\theta)} \tag{13.5}$$

Solving for the angle of rotation θ in terms of the pitch p of the master grating and the quantities δ and ϕ which can be obtained from the moiré fringe pattern gives:

$$\tan\theta = \frac{\sin\phi}{\delta/p + \cos\phi} \tag{13.6}$$

If gratings with a very fine pitch are used to measure small angles of rotation, then $\phi \approx \pi/2$ and Eq. (13.6) reduces to:

$$\theta \approx \tan\theta = p/\delta \tag{13.7}$$

In a similar manner, it is observed in Fig. 13.5b that:

$$\frac{p}{\sin(\pi-\phi)} = \frac{p'}{\sin(\phi-\theta)} \tag{13.8}$$

This reduces to:

$$p' = \frac{\delta\sin\theta}{\sin\phi} \tag{13.9}$$

The angle θ can be eliminated from Eq. (13.9) by using the trigonometric identity:

$$\sin\theta = \frac{\tan\theta}{\sqrt{1+\tan^2\theta}}$$

Then by using Eqs. (13.9) and (13.6), the deformed specimen pitch p' can be expressed in terms of the pitch p of the master grating and the quantities δ and ϕ as:

$$p' = \frac{\delta}{\sqrt{1+(\delta/p)^2+2(\delta/p)\cos\phi}} \tag{13.10}$$

In many instances, moiré fringe patterns will be evaluated in regions where rotations are small. In these cases, $\phi \approx \pi/2$, and applying the binomial expansion to Eq. (13.10) gives:

$$p' = \frac{p\delta}{p\pm\delta} \tag{13.11}$$

When the deformed specimen pitch p' has been determined, the component of normal strain in a direction perpendicular to the lines of the master grating can be written as:

$$\varepsilon = \frac{p'-p}{p} \tag{13.12}$$

13.4 THE DISPLACEMENT-FIELD APPROACH TO MOIRÉ-FRINGE ANALYSIS

Moiré fringe patterns can also be interpreted by relating them to a displacement field. In Section 13.2, it was shown that a moiré fringe is formed within a given gage length l_g in a uniformly deformed specimen each time the specimen grating within the gage length is extended (or shortened) in a direction perpendicular to the lines of the master grating by an amount equal to the pitch p of the master grating. Deformations in a direction parallel to the lines of the master grating do not affect the moiré fringe pattern.

The displacement-field concept can be extended to the general case of extension or contraction combined with rotation. In Fig. 13.6, the lines of a deformed specimen grating are shown superimposed on a master grating. One of the lines of each of the gratings is shown dotted for easy reference, and it will be assumed that these lines coincided in the undeformed state. Both gratings are assumed to have had an initial pitch p. Interference between the master grating and the deformed specimen grating produces the light moiré fringes. In this pattern it should be noted that the intersection of the two dotted lines has been used to locate the zero-order fringe. This point on the specimen did not displace in the vertical direction as the specimen deformed. Similar intersections of specimen and master grating lines which initially coincided are also located on the zero-order fringe. The intersection of the dotted line on the specimen with the line above the dotted line on the master grating lies on the fringe of order 1. This point on the specimen has moved a distance p in the vertical direction from its original position. Similarly, a point lying on the fringe of order 4 has moved a distance 4p in the vertical direction from its original position. Thus, a moiré fringe is a locus of points exhibiting the same component of displacement in a direction perpendicular to the lines of the master grating.

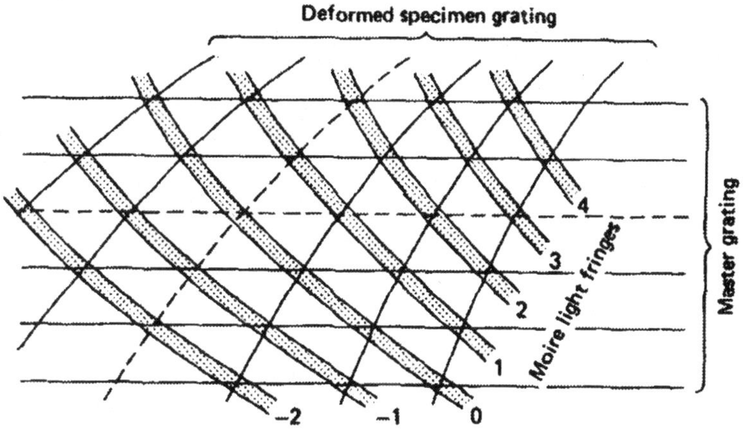

Figure 13.6 Moiré fringes at an arbitrary point in a stressed specimen.

For purposes of analysis, the moiré fringe pattern can be visualized as a displacement surface where the height of a point on the surface above a plane of reference represents the displacement of the point in a direction perpendicular to the lines of the master grating. A simple illustration of this concept is shown in Fig. 13.7. Similar patterns and displacement surfaces can be obtained for any other orientation of the specimen and master gratings.

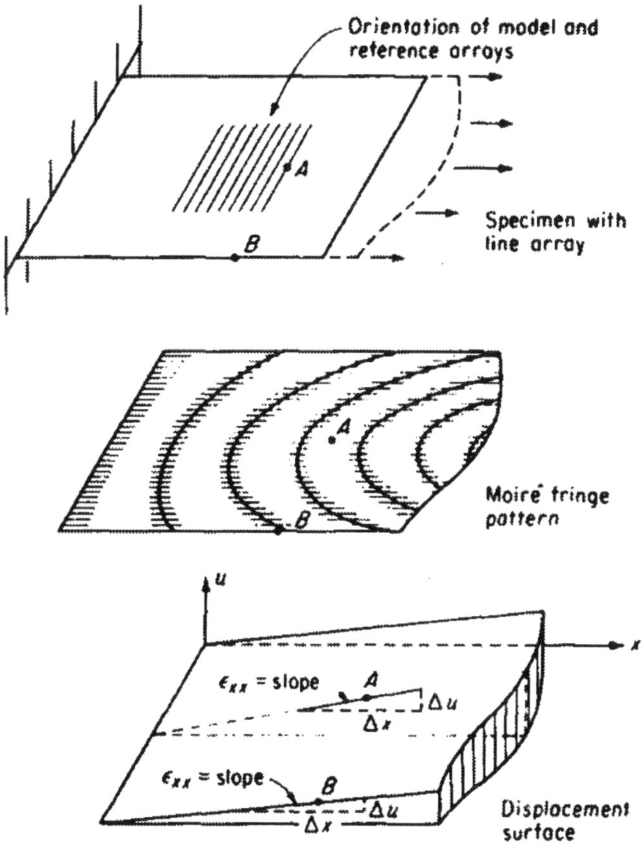

Figure 13.7 Moiré fringe pattern and associated displacement surface for a non uniform strain field.

When u and v displacement surfaces have been established by using line arrays perpendicular to the x and y axes of a specimen, respectively, the Cartesian components of strain can be computed from the derivatives of the displacements (slopes of the displacement surfaces). Two models are normally required for these determinations unless an axis of symmetry exists in the specimen so that mutually perpendicular arrays can be placed on the two halves of the specimen. For the case of large strains, the relationships between displacements and strains are given by Eqs. (2.2) and (2.3) as:

$$\varepsilon_{xx} = \sqrt{1 + 2\frac{\partial u}{\partial x} + \left(\frac{\partial u}{\partial x}\right)^2 + \left(\frac{\partial v}{\partial x}\right)^2 + \left(\frac{\partial w}{\partial x}\right)^2} - 1$$

$$\varepsilon_{yy} = \sqrt{1 + 2\frac{\partial v}{\partial y} + \left(\frac{\partial v}{\partial y}\right)^2 + \left(\frac{\partial w}{\partial y}\right)^2 + \left(\frac{\partial x}{\partial y}\right)^2} - 1$$

$$\gamma_{xy} = \arcsin \frac{\dfrac{\partial u}{\partial y} + \dfrac{\partial v}{\partial x} + \dfrac{\partial u}{\partial x}\dfrac{\partial u}{\partial y} + \dfrac{\partial v}{\partial x}\dfrac{\partial v}{\partial y} + \dfrac{\partial w}{\partial x}\dfrac{\partial w}{\partial y}}{(1+\varepsilon_{xx})(1+\varepsilon_{yy})} \qquad (13.13)$$

where u, v, and w are the displacement components in the x, y, and z directions, respectively. When products and powers of derivatives are small enough to be neglected, Eqs. (13.13) reduce to:

$$\varepsilon_{xx} = \frac{\partial u}{\partial x} \qquad \varepsilon_{yy} = \frac{\partial v}{\partial y} \qquad \gamma_{xy} = \frac{\partial v}{\partial x} + \frac{\partial u}{\partial y} \qquad (13.14)$$

The displacement gradients $\partial u/\partial x$ and $\partial v/\partial y$ are obtained from the slopes of the two displacement surfaces in a direction perpendicular to the lines of the master gratings. The displacement gradients $\partial u/\partial y$ and $\partial v/\partial x$ are obtained from the slopes of the displacement surfaces in a direction parallel to the lines of the master gratings. The displacement gradients $\partial w/\partial x$ and $\partial w/\partial y$ are not considered in moiré analysis of in-plane deformation fields.

In application of the moiré method two moiré fringe patterns are usually obtained with specimen and master gratings oriented perpendicular to the x and y axes. A schematic illustration of one of these fringe patterns is shown in Fig. 13.8. Lines along the x and y axes, say AB and CD, are drawn, and displacements u and v along each of these lines are plotted by noting that:

$$u, v = np \qquad (13.15)$$

where n is the order of the moiré fringe at the point and p is the pitch of the master grating. Tangents drawn to these curves give $\partial u/\partial x$ and $\partial u/\partial y$, as shown in Fig. 13.8. Similar moiré fringe patterns with an x oriented model array are used to determine $\partial v/\partial x$ and $\partial v/\partial y$. Equations (13.13) or (13.14) can then be used to determine the strains ε_{xx}, ε_{yy} and γ_{xy}. If lines similar to AB and CD are drawn at all critical areas of the specimen, a complete description of the strain over the entire field of the model can be achieved.

The displacement-field approach to moiré fringe analysis, as previously outlined, is based on an accurate determination of the displacement derivatives $\partial u/\partial x$, $\partial u/\partial y$, $\partial v/\partial x$ and $\partial v/\partial y$. In practice, when the u and v moiré fringe patterns are obtained from separate models, from the two halves of a symmetric model, or with separate u and v master gratings on a crossed model grating (an array of orthogonal lines), the direct derivatives $\partial u/\partial x$ and $\partial v/\partial y$ can usually be obtained with acceptable accuracy but the cross derivatives $\partial u/\partial y$ and $\partial v/\partial x$ cannot. The error in the cross derivatives is due to slight errors in alignment of either the specimen or master gratings with the x or y axes. Misalignment produces a fringe pattern due to rotation (see Fig. 13.3) in addition to the load-induced pattern.

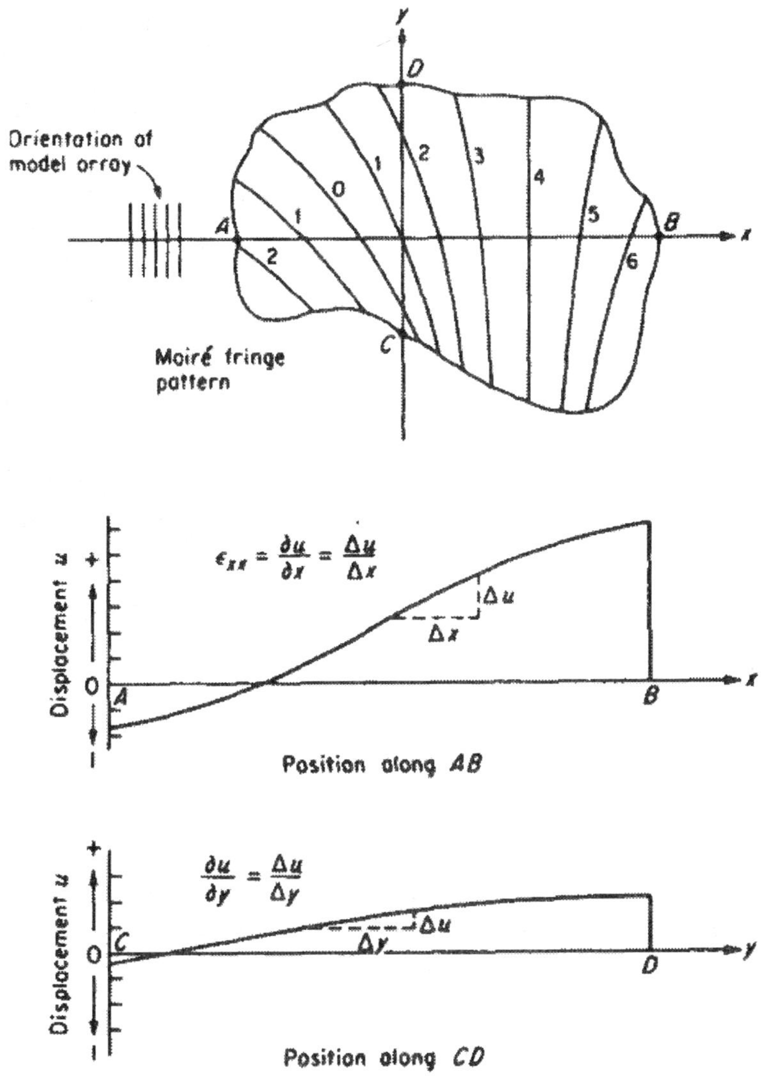

Figure 13.8 Displacement-position graphs used to determine ∂u/∂x and ∂u/∂y.

One method proposed to eliminate shear-strain error makes use of crossed gratings on both the specimen and master to obtain simultaneous displays of the u and v displacement fields. Because any rotational misalignment is then equal for the two fields, its contribution to the cross derivatives is equal in magnitude but opposite in sign and thus cancels in the shear-strain determination.

In 1948, Weller and Shepard [3] recognized the possibility of displaying two moiré fringe patterns simultaneously with crossed gratings but recommended against its use because of the interweaving between the two families of fringes. Post [11] finally resolved this difficulty by using crossed gratings with slightly different pitches on the specimen and master to produce moiré fringe patterns with two distinct families as illustrated in Fig. 13.9.

The problem of shear-strain errors can also be solved by using the strain-rosette concept commonly employed with electrical-resistance strain gages. With this method, model gratings are employed perpendicular to the x, n (usually 45° with respect to the x axis) and the y axes. Once the three normal strains ε_{xx}, ε_n and ε_{yy} are determined at a point, the complete state of strain at the point can be calculated by means of the strain rosette equations of Chapter 8.

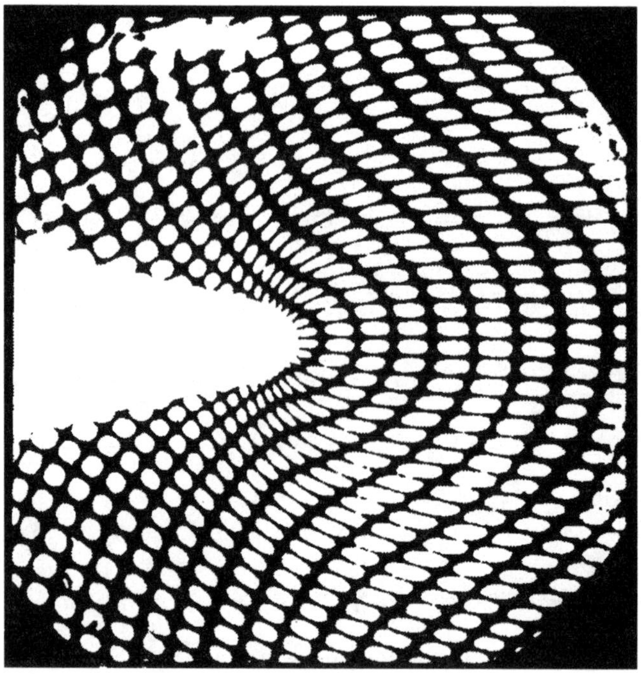

Figure 13.9 Moiré fringe pattern with crossed gratings of different pitch on the master and the specimen. Courtesy of Daniel Post.

13.5 OUT-OF-PLANE DISPLACEMENT MEASUREMENTS

The moiré fringe methods discussed in the previous sections of this chapter have been concerned with the determination of in-plane displacements u and v, rigid-body rotations θ_x and the strains ε_{xx}, ε_{yy} and γ_{xy}. In select plane-stress problems and in a wide variety of problems involving laterally loaded plates, out-of-plane displacements w become important considerations. A moiré method for determining out-of-plane displacements w has been developed by Theocaris [7, 8] and applied to a number of these problems. The essential features of the method are illustrated in Fig. 13.10.

For out-of-plane displacement measurements, a master grating is placed in front of the specimen, and a collimated beam of light is directed at oblique incidence through the master grating and onto the surface of the specimen. The shadow of the master grating on the surface of the specimen serves as the specimen grating. When the specimen is viewed at normal incidence, moiré fringes form as a result of interference between the lines of the master and the shadows. Use of a matte surface on the specimen ensures distinct shadows and improves the quality of the moiré fringe patterns.

From the geometry illustrated in Fig. 13.10 it can be seen that the difference in distance between the master grating and the specimen surface at two adjacent fringe locations can be expressed as:

$$\Delta w = d_2 - d_1 = \frac{p}{\tan\alpha}$$

where p is the pitch of the master grating and α is the angle of incidence of the collimated light beam.

In practice, the master grating is located a small distance away from the specimen to accommodate any surface displacements toward the master grating and to serve as a datum plane for the measurement of load-induced, out-of-plane displacements. Any distribution of moiré fringes appearing with the master grating in this initial position will represent irregularities in the surface of the specimen. The presence of any irregularity must be accounted for in the final determination of the out-of-plane displacements.

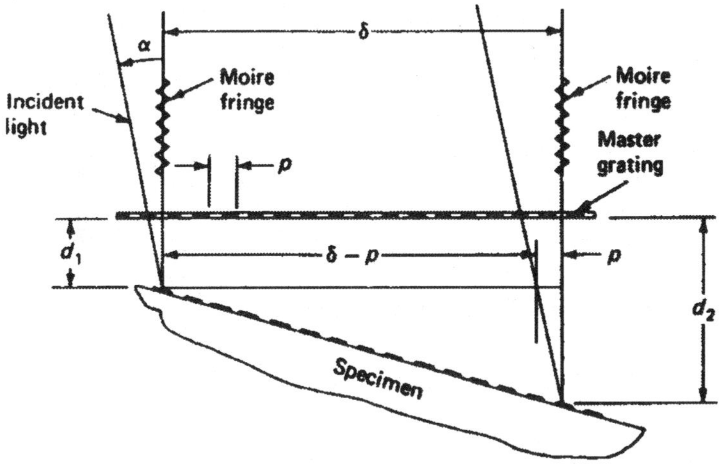

Figure 13.10 Moiré method for measuring out-of-plane displacements.

If a point of zero out-of-plane displacement is known to exist at some location on the specimen, e.g., from theoretical considerations, the master grating can be positioned to center a moiré fringe, identified as the zero-order fringe, over this point. At all other fringe locations, the out-of-plane displacement w can then be expressed as:

$$w = \frac{np}{\tan\alpha} \tag{13.16}$$

where n is the order of the moiré fringe at the point of measurement.

If no point exists in the specimen where w = 0, it will be necessary to measure the displacement at some convenient point by another experimental method. Then, the displacements at all other points on the surface can be referred to this reference point.

13.6 OUT-OF-PLANE SLOPE MEASUREMENTS

The determination of stress distributions and deflections in laterally loaded plates is a difficult but important engineering problem. From the theory of elasticity, it is known that the stresses at a point in the plate due to bending moments can be expressed in terms of the local curvatures of the plate as:

$$\begin{aligned}\sigma_x &= \frac{Ez}{1-\nu^2}\left(\frac{1}{\rho_x} + \nu\frac{1}{\rho_y}\right)\\ \sigma_y &= \frac{Ez}{1-\nu^2}\left(\frac{1}{\rho_y} + \nu\frac{1}{\rho_x}\right)\end{aligned} \tag{13.17}$$

The deflections are related to the curvatures by the approximate expressions:

$$\frac{1}{\rho_x} = -\frac{\partial^2 w}{\partial x^2} \qquad \frac{1}{\rho_y} = -\frac{\partial^2 w}{\partial y^2} \tag{13.18}$$

In theory, the out-of-plane displacement-measuring technique discussed in the previous section can provide the required curvatures for a solution to the stress problem. However, in practice, double differentiations cannot be performed with sufficient accuracy to provide suitable values for the curvatures. To overcome this experimental difficulty, Ligtenberg [9] has developed a moiré method for measuring the partial slopes $\partial w/\partial x$ and $\partial w/\partial y$. A single differentiation of these slopes then provides reasonably accurate values of the second derivatives needed to determine the curvatures, $1/\rho$.

The essential features of the Ligtenberg method are illustrated in Fig. 13.17. The equipment consists of a fixture for holding and loading the plate, a large cylindrical surface with a coarse line grating, and a camera for recording the moiré fringe patterns. The cylinder segment with the coarse grating is fabricated from a transparent sheet of plastic and the incident light passes through this shell. The surface of the plate is polished or coated to make it reflecting because the camera views the image of the grating on the surface of the plate. Because the image does not depend on the angle of incidence of the light, collimated light is not required for this method. The moiré fringe pattern is formed by superimposing grating images before and after loading. Double-exposure photography is usually used to superimpose the two images.

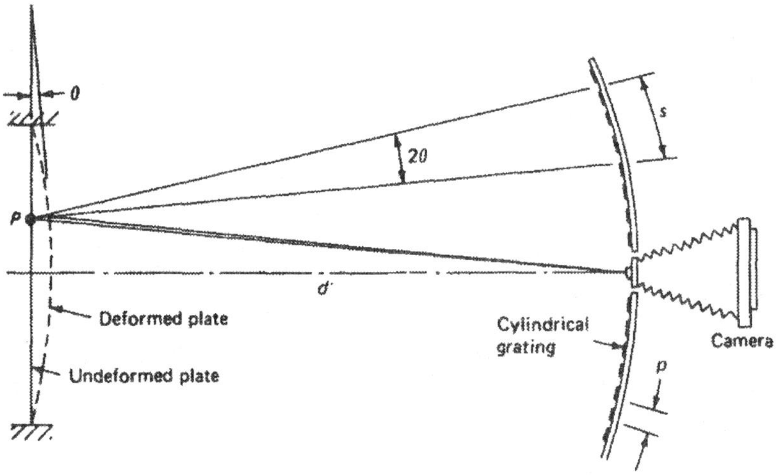

Figure 13.11 Moiré method for measuring out-of-plane slopes.

From the geometry of Fig. 13.11, it is evident that the location on the grating being viewed by the camera, as a result of reflections from a typical point P on the surface of the specimen, shifts as the plate deflects under load. The moiré fringe pattern formed by the superposition of the two images provides a measure of this shift. Observe in Fig. 13.11 that the shift can be expressed in terms of the local slope of the plate as:

$$s = 2\theta d \tag{13.19}$$

where s is the magnitude of shift, θ is the local slope of plate at point P' and d is the distance between plate and grating.

A moiré fringe will form upon superposition of the two images if the shift s is equal to the pitch p of the grating. Thus, the order of the moiré fringe can be expressed as:

$$n = (2\theta d)/p \quad \text{or} \quad \theta = np/2d \tag{13.20}$$

The separation distance d should be large to minimize the effects of out-of-plane displacements w on the shift distance s. Ligtenberg has also shown that the cylindrical grating should have a radius of the order of 3.5d. The angle θ given by Eq. (13.20) is the partial slope $\partial w/\partial x$ or $\partial w/\partial y$ depending on the orientation of the grating relative to the specimen. Two moiré patterns give the two slopes that are needed to solve the plate problem completely. The second pattern is obtained by rotating the grating 90° after the first pattern is recorded.

13.7 MULTIPLICATION OF MOIRÉ FRINGES

Application of geometric moiré methods to the study of deformations and strains in the elastic range of material response is usually limited by the lack of sensitivity of the method with respect to other methods, e.g., electrical resistance strain gages. Both the geometrical and displacement-field approaches to moiré fringe analysis, which were discussed in Sections 13.3 and 13.4, require accurate determinations of either the fringe spacings or the fringe gradients at the point of interest in the specimen. In a typical moiré fringe pattern obtained with line gratings having a density of 40 lines/mm or less, only a few fringes are normally present in studies of elastic strain fields. Therefore, the spacings or gradients cannot be established with the required accuracy. It is not practical to increase the line densities of the gratings, beyond 40 lines/mm, when mechanical methods are used to form the moiré fringe pattern. For this reason, several methods have been introduced to improve the sensitivity of the moiré method. Moiré fringe-multiplication methods, which enhance the sensitivity of mechanically formed images, will be discussed in this section. Moiré interferometry, a method which uses very high density gratings, will be described in Section 13.8.

The methods of moiré fringe formation discussed previously are called geometrical or mechanical moiré because the formation of the moiré pattern is based on a simple geometric-optics (or ray-optics) treatment of the passage of light through superimposed gratings. However, methods of fringe multiplication utilize the diffraction effects associated with light passing through the narrow slits formed between the grating lines.

The diffraction of light from a narrow slit was considered in Section 13.6. When light consisting of a series of plane wavefronts is used to illuminate a grating, each of the transparent spaces in the grating acts as a slit. In accordance with Huygens' principle, the transparent space in the grating emits light in the form of a series of secondary cylindrical wavelets. A number of wavefronts are possible and are referred to as the specific diffraction orders. For example, the zero-order beam is formed by wavelets which have identical phase. This wavefront is collinear to the incident plane wavefront and propagates along the axis of the optical system as shown in Fig. 13.12. The first-order beams, which are deflected to both sides of the zero-order beam, are formed by adjoining wavelets which are out of phase by one wavelength. Similarly, the second-order beams are formed by adjoining wavelets which are out of phase by two wavelengths, and so on. The diffraction of a plane wavefront by a grating which contains many slits is illustrated in Fig. 13.12. The rays emerging from each of the slits of the grating are associated with the different diffraction orders. The image formed by collecting all the diffraction orders from the grating with a lens is known as the diffraction spectrum of the grating. A photograph of an actual diffraction spectrum of a 12 line/mm grating is shown in Fig. 13.13. The dots of the diffraction spectrum are images of the point source and the intensity of the light associated with a dot decreases with its diffraction order.

Because the wavelets forming the first-order beam are out of phase by one wavelength, the angle θ_1 defined in Fig. 13.12 can be determined from the expression:

$$\sin \theta_1 = \theta_1 = \lambda/p$$

Note that $\sin \theta_1 = \theta_1$ because the angles are small. Similarly for the other diffraction orders, it is possible to write:

$$\sin \theta_n = \theta_n = (n\lambda)/p \tag{13.21}$$

The distance d between any two dots is then given simply as:

$$d = f\,\theta_1 = (f\lambda)/p \tag{13.22}$$

where f is the focal length of the second lens.

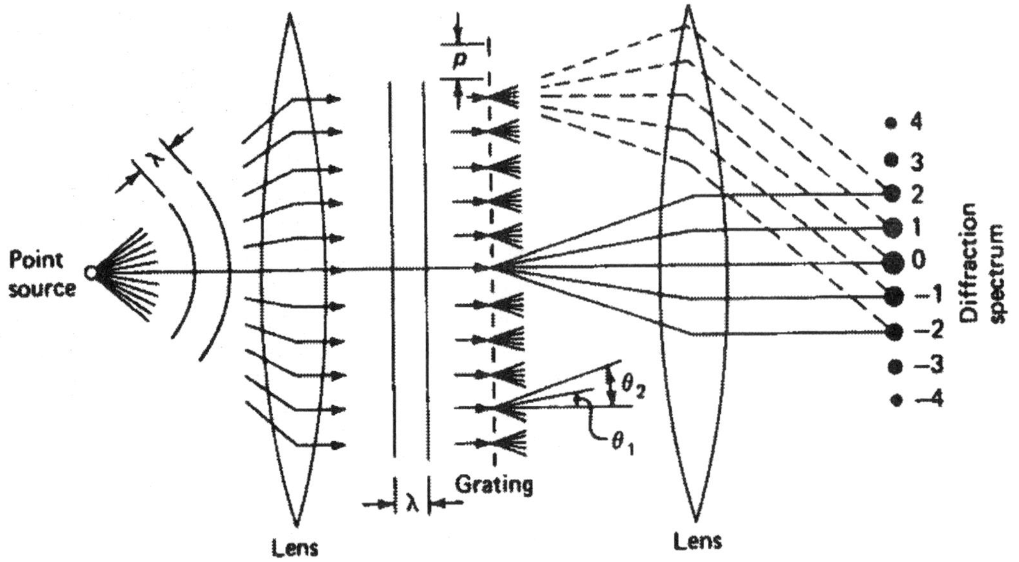

Figure 13.12 Mechanism for formation of the diffraction spectrum of a grating.

Figure 13.13 Diffraction spectrum from a 12 line/mm grating. Courtesy of Fu-pen Chiang.

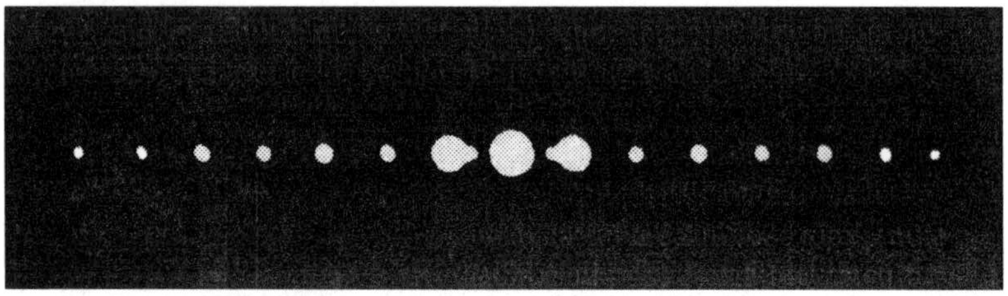

One of the methods used to produce moiré fringe multiplication is based on the diffraction phenomenon illustrated in Fig. 13.12. When both a model grating and a master grating are inserted in series in the collimated light beam between the two lenses, the plane wavefronts generated by the first grating are diffracted again by the second grating to produce a diffraction spectrum associated with the superimposed pair. If the complete diffraction spectrum is collected with a camera lens (an ideal situation not realized in practice), the image recorded by the camera will be an exact reproduction of the grating pair and their moiré pattern. However, if all the diffraction orders are not collected by the camera lens, the image recorded on the film plane will be modified. The theory associated with these modifications is beyond the scope of this book, but images recorded with certain individual diffraction orders can be shown to provide moiré fringe multiplication.

An optical system used for moiré fringe multiplication is illustrated in Fig. 13.14. Note the presence of the light stop, or aperture, which is used to isolate the particular diffraction order entering the camera lens. A selection of photographs recorded using different diffraction orders is shown in Fig. 13.15. The first photograph shows the image recorded if a large number of diffraction orders enter the camera lens. The moiré fringes and the lines of the gratings are visible. This pattern is similar to the ones previously shown, which were obtained with two coarse gratings in contact using diffused light (mechanical moiré). The second photograph shows the image recorded with the zero diffraction order. The moiré fringes in this image have been broadened and the grating lines have been eliminated. The third photograph shows the image produced by the +1 diffraction order. In this instance, the moiré fringes are sharpened and the lines of the gratings remain extinguished. It can be shown that at least two diffraction orders must enter the camera before a crude grating-like image is produced. In the fourth, fifth, and sixth photographs, the images produced by the + 2, + 3, and + 5 diffraction orders, respectively, are shown. In each of these photographs, moiré fringe multiplication (2 x, 3 x, and 5 x, respectively) is indicated. Higher levels of fringe multiplication can be achieved with the higher diffraction orders, but the intensity of light associated with these orders is very small.

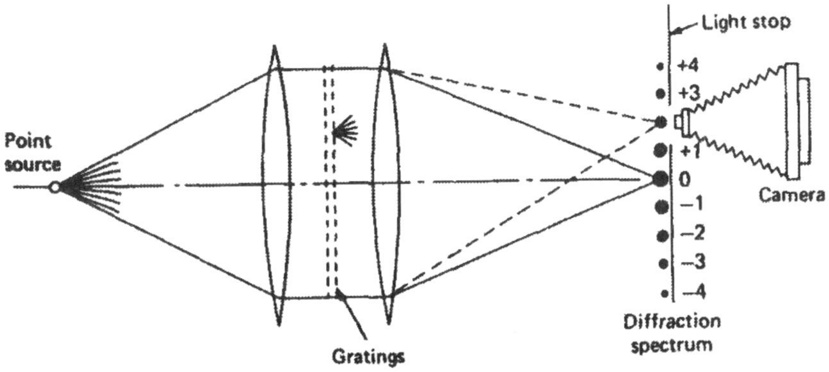

Figure 13.14 Optical system employed for moiré fringe multiplication.

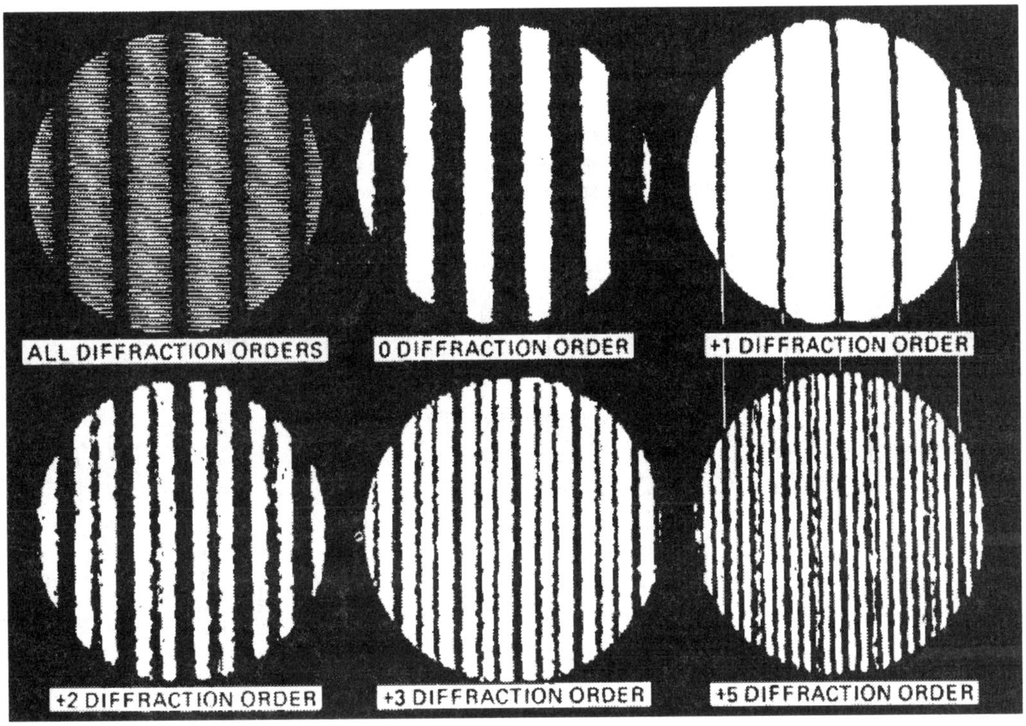

Figure 13.15 Moiré fringe patterns produced by different diffraction orders. Courtesy of A. J. Durelli.

In utilizing the diffraction method for forming the moiré images and increasing the sensitivity, light rays from several different diffraction orders are combined as indicated in Fig 13.16. In this figure, three different order groups (of many possible groups) are shown focused at the first spot. In a similar manner, multiple beams from several different diffraction orders are combined at each of the focus spots. The apparent fringe multiplication depends on the diffraction order of focus spots as indicated in Fig. 13.15. Recently, Graham has studied the effect of imperfections of the diffraction gratings and showed that if both the specimen and reference gratings had a 50% transmission ratio (i.e. the line width and space width are equal) then the added fringes are correctly positioned between the original (non-multiplied) fringes. However, if the transmission ratio of the gratings is different from 50%, the fidelity of the multiplication is impaired and error is introduced even in the location of the first-order fringes. Intensity-position curves for grating pairs with 40, 45, and 55% transmission are shown in Fig. 13.17.

Note first that the periodic sinusoidal form of these curves has been seriously distorted. More importantly, the positions of minimum intensity, which locates the fringes, are not at the theoretical locations. The effect of imperfections (even uniform deviations in the transmission ratio) is to produce serious errors in location of the zero order and the higher order moiré fringes. One observes the correct number of fringes but they are not at the correct positions. Indeed, a close examination of the third and fifth diffraction order fringes in Fig. 13.15 shows lack of uniformity in fringe width due to the distortion in the sinusoidal intensity-position relation. Also, the fringe spacing of the third and fifth diffraction orders does not represent a uniform displacement field as it should.

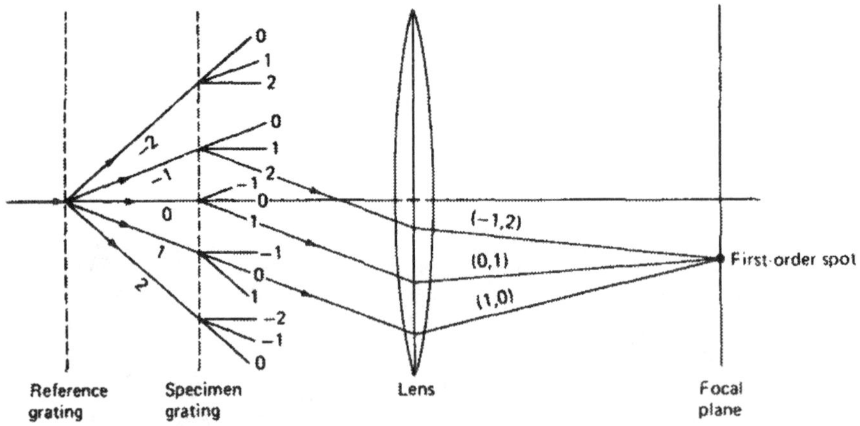

Figure 13.16 Optical rays showing the combination of diffraction orders in the first order spot on the focal plane.

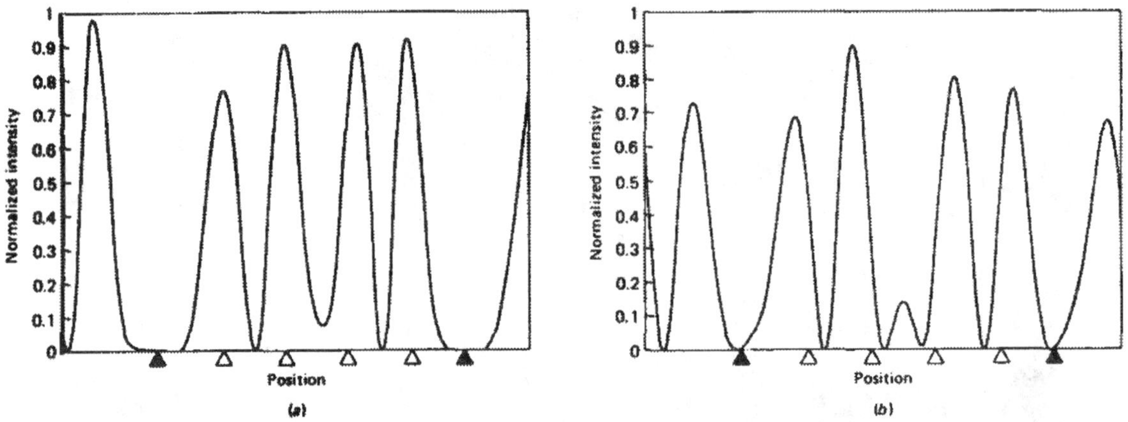

Figure 13.17 The influence of transmission ratio on the moiré fringe location for two gratings with fringe multiplication of 5. (a) transmission ratio of 40%; (b) transmission ratio of 45%; Δ indicates theoretical locations of fringes due to multiplication; Δ indicates location of first order fringes.

13.8 EXPERIMENTAL PROCEDURE AND TECHNIQUE

Master gratings for moiré work with (12, 20, 40, and 80 lines/ mm) are readily available from several commercial suppliers. Indeed, with the availability of high quality lithographic equipment used extensively in the electronics industry, custom master gratings can be procured at relatively low cost. Because of the fragility of the master grating (glass) and the risk involved in its use, high-quality duplicates are usually made on film for routine stress analysis work. If proper photographic procedures are followed, little difficulty is encountered in producing good-quality duplicates on high-resolution film.

Printing or etching the grating on the model, on the other hand, is usually more difficult. Work on the development of techniques to apply gratings to the various model materials is constantly in progress, and some interesting procedures have been developed for specific applications. Two factors are extremely important in the production of a satisfactory model array: (1) the photographic emulsion must adhere well to the model material and have a high resolution; (2) the exposure must be carefully controlled to produce an array with proper line width and spacing from the master array.

Because materials, techniques and procedures associated with moiré work are constantly being improved, no attempt will be made to outline specific procedures. For the latest materials, procedures, and techniques, the interested reader should consult the current technical literature. In particular, the handbook on Experimental Mechanics edited by A. S. Kobayashi [26] should be consulted for more detail on many of the topics discussed in this chapter.

13.9 MOIRÉ INTERFEROMETRY

13.9.1 Two-Beam Interference

In Section 13.3.2, the superposition of two waves with the same frequency ω propagating along the z axis was described. It was shown that the two waves combine and that the intensity of light, when the amplitudes $a_1 = a_2 = a$, is given by:

$$I = 4a^2 \cos^2 \frac{\pi\delta}{\lambda} \tag{13.17}$$

The ratio δ/λ, the relative phase, indicates the position of one wave with respect to the other. When the ratio equals an integer value n,

$$\frac{\delta}{\lambda} = n \qquad n = 0, 1, 2 \ldots \tag{13.23a}$$

then $I = 4a^2$ which is the maximum intensity. In this case, the two waves reinforce one another and the interference effect is constructive. However, if the ratio δ/λ is:

$$\frac{\delta}{\lambda} = \frac{n+1}{2} \qquad n = 0, 1, 2 \ldots \tag{13.23b}$$

then $I = 0$ which is the minimum intensity. In this instance, the waves acted to cancel one another and the interference effect is destructive. Because both of the waves are parallel to the z axis, the interference is the same at every point in space (x, y, z) where both beams of light exist.

A second example of two-beam interference involves two nonparallel beams (either diverging or intersecting) as illustrated in Fig. 13.18. The two beams are produced by a single coherent light source and emerge from the mirror arrangement, used to orient the beams, with the same phase ($\delta = 0$). At some point z_0 down stream from the mirror arrangement, the two waves interfere with a phase difference δ that is given by the equation:

$$\delta = 2y \sin\theta \tag{13.24}$$

When δ/λ is an integer value, constructive interference occurs and a light fringe is formed. On the other hand, when δ/λ is a half-order value, destructive interference occurs and a dark fringe is formed. The resulting interference fringe pattern is an array of alternating light and dark fringes as shown in Fig. 13.19. The distance p between fringes is determined from Eq. (13.24) by noting that $y = p$ when $\delta = \lambda$ so that:

$$p = \frac{\lambda}{2\sin\theta} \tag{13.25}$$

The fringe gradient is representative of a spatial frequency f given by:

$$f = \frac{1}{p} = \frac{2\sin\theta}{\lambda} \tag{13.26}$$

where f is expressed in terms of lines/mm.

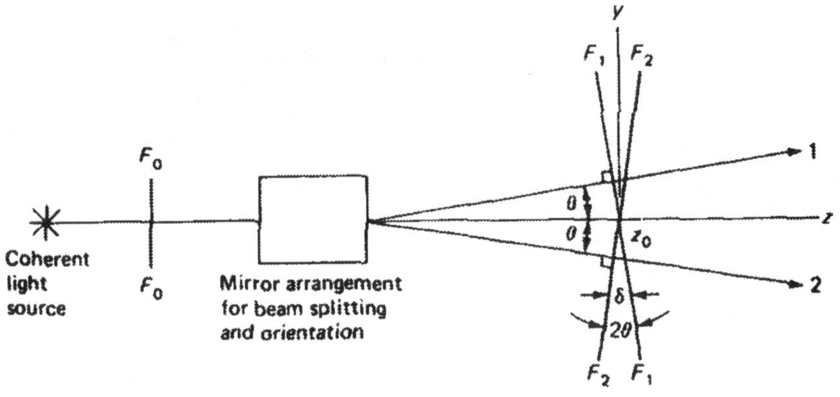

Figure 13.18 Interference of two wave fronts F_1 and F_2 associated with diverging waves.

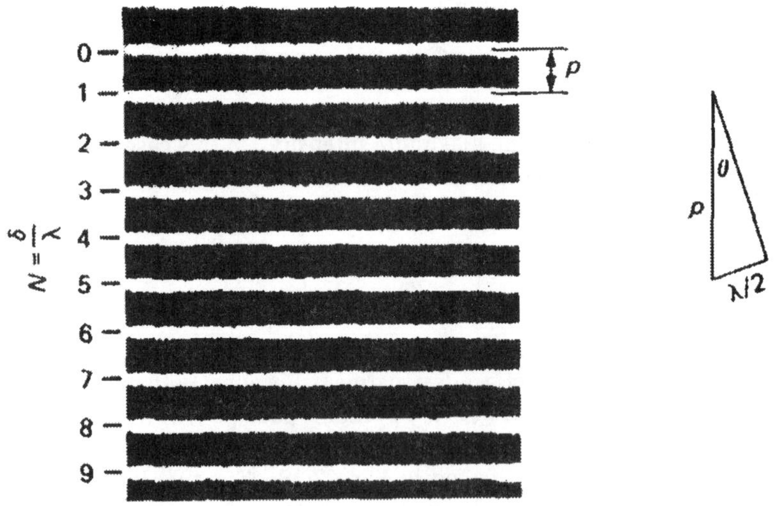

Figure 13.19 Interference fringe pattern produced by intersecting wavefronts F_1 and F_2. Courtesy of Daniel Post.

The two intersecting beams are used to produce high-frequency gratings which are employed as the specimen gratings in moiré interferometry. The optical arrangement recommended by Post is shown in Fig. 13.20. Because the frequency of the interference pattern is very high (1,000 – 3,000 lines/mm), production of the gratings is not a routine operation. An excellent vibration isolation table is required to maintain stability of the optical elements during the exposure interval. Also, forming the image of the grating on the film plane requires a very-high-quality process lens. Finally, glass plates with a very-high-resolution emulsion must be used to photograph the high density interference fringes formed by the intersecting beams of coherent light.

The theoretical limit of the frequency f_t of the interference pattern, produced with the two intersecting beams, is given when $\theta = \pi/2$ by Eq. (13.26) as:

$$f_t = \frac{2}{\lambda} \qquad (13.27)$$

Thus, the theoretical frequency f_t depends on the wave length of the light source employed. For a helium-neon laser with λ = 632.8 nm, f_t = 3,160 lines/mm and for an argon laser with λ = 488.0 nm, f_t = 4,098 lines/mm. It is not possible to achieve the theoretical frequency because of difficulties encountered in forming the interference pattern when $\theta \Rightarrow \pi/2$ and the two beams interfere at grazing incidence. Post has achieved 97.6% of the theoretical limit (4,000 lines/mm) by using θ = 77.4° with λ = 488.0 nm.

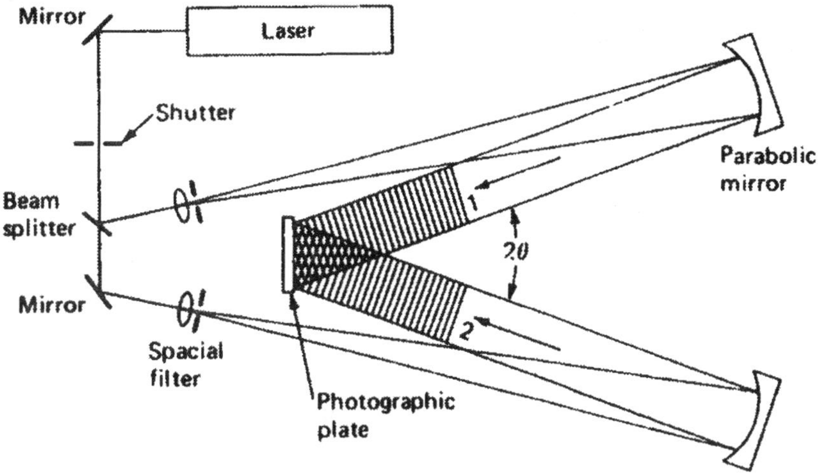

Figure 13.20 Optical arrangement used to produce high-frequency gratings.

13.9.2 Specimen Gratings

Moiré interferometry requires a specimen grating which is similar to the grating used in mechanical moiré. The primary difference between these two gratings is in the frequency or pitch of the line arrays employed. With moiré interferometry, the frequency of the specimen grating is usually in the range from 1,000 to 2,000 lines/mm; whereas, with mechanical moiré, the frequency rarely exceeds 80 lines/mm.

The placement of these very-high-frequency gratings on the surface of a specimen is accomplished by using a replication technique. The grating is first produced on a glass photographic plate by using the optical arrangement described in Fig. 13.20. In developing the photographic emulsion, the alternating bands of exposed and unexposed silver halide produce a wave pattern which is due to non-uniform shrinkage of the gelatin-based emulsion (see Fig. 13.21). The waved surface of the photographic plate is coated with a very thin coating of either gold or aluminum using a high-vacuum deposition process.

Figure 13.21 Gelatin-based photographic emulsions produce a surface wave pattern representing the interference fringe pattern.

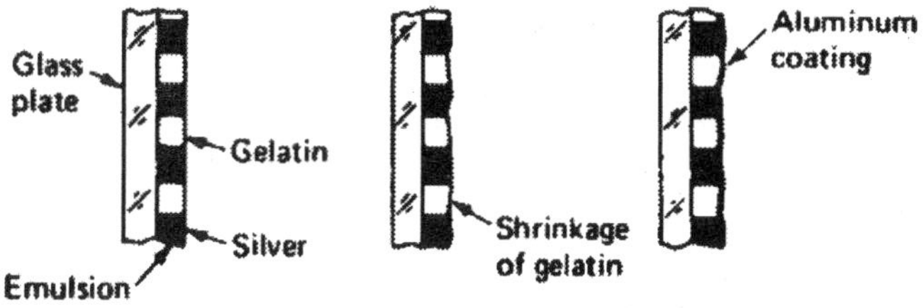

The metal coated photographic plate serves as a mold in the replication process used to transfer the grating to the specimen surface. This process, illustrated in Fig. 13.22, involves pressing a small quantity of an adhesive (epoxy) between the mold and the specimen. After the adhesive has cured, the mold is stripped from the specimen surface. The gelatin-metal interface is the weakest link in the mold-specimen bonding chain, and separation occurs at this interface. The result of the replication process is a thin film (about 0.025 mm thick) with a reflective high-frequency, phase-type diffraction grating as the exposed surface bonded to the surface of the specimen.

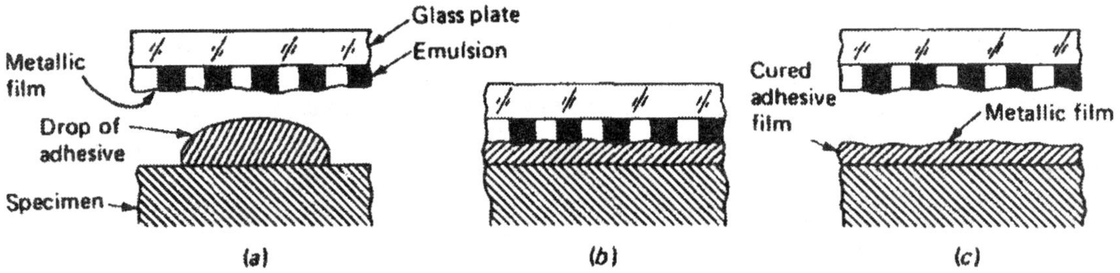

Figure 13.22 The replication process used in transferring the grating with a reflective metallic film to the specimen.

13.9.3 Moiré Interference

Moiré interferometry is identical to mechanical moiré in the sense that a moiré image is formed by light passing through two gratings—one on the specimen and the other, a reference grating, usually located adjacent to the specimen. The difference in moiré interferometry, in addition to the frequency of the gratings discussed previously, is with the reference grating. In mechanical moiré, the reference grating is real consisting of an array of lines on a sheet of film placed in contact with the specimen grating as illustrated in Fig. 13.2a. With moiré interferometry, the reference grating is imaginary consisting of a virtual image of an interference pattern produced with mirrors.

An optical arrangement that is used to develop the virtual reference grating is shown in Fig. 13.23. The frequency of the reference grating f_r, follows directly from the derivation of Eq. (13.26) which gives:

$$f_r = 2\sin\frac{\alpha}{\lambda} \tag{13.28}$$

where α is the angle of the incident light.

Figure 13.23 Optical arrangement for producing a virtual reference array.

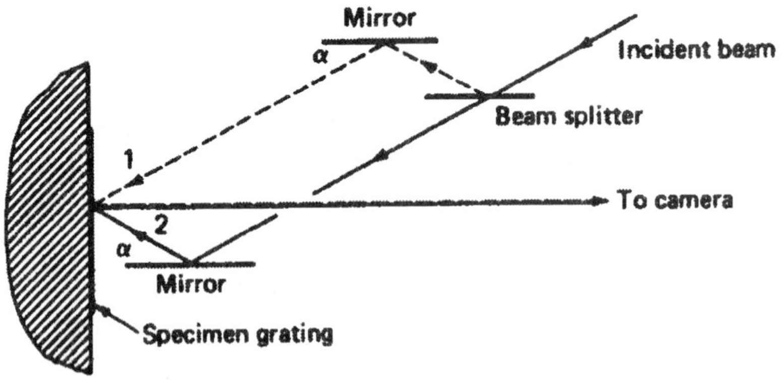

A more detailed description of a moiré interferometer is illustrated in Fig. 13.24. The two beams of the interferometer are incident to the surface of the specimen at symmetrical angles $\pm\ \alpha$. The beams are reflected from the diffraction grating of frequency f_s which has been replicated on the surface of the specimen. Upon reflection, the light is diffracted and several diffraction orders are produced as indicated in Fig. 13.24. It is important to consider certain combinations of these diffraction orders because emerging light rays collinear with the z axis are combined to produce interference which results in a moiré fringe pattern.

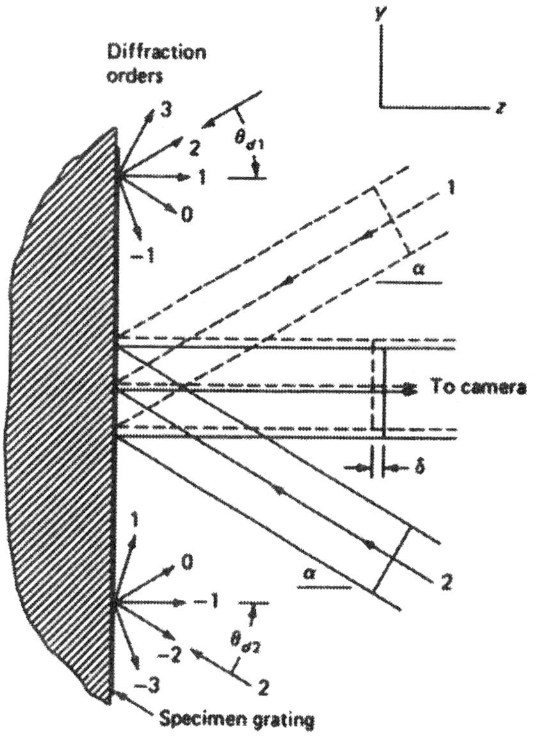

Figure 13.24 Diffraction of the light beams as they interact with the diffraction grating on the specimen.

Reference to Eq. 13.38a shows that the angle θ_d between the incident ray and the reflected ray associated with the first diffraction order is given by:

$$\sin \theta_d = \lambda / b \tag{13.29a}$$

In this case, the width b of the slit in the diffraction grating is equivalent to the pitch p in the specimen grating. Recall that $f_s = 1/p$ and rewrite Eq. (13.29a) to give:

$$\sin \theta_d = f_s \lambda \tag{13.29b}$$

Furthermore, if:

$$f_s = f_r / 2 \tag{13.30a}$$

then it is clear from Eqs. (13.28) and (13.29b) that:

$$\theta_{d1} = \theta_{d2} = \alpha \tag{13.30b}$$

Equation (13.30b) shows that the first diffraction order from incident beam (1) propagates along the z axis. Similarly, the negative first diffraction order from incident beam (2) also propagates along the z axis. These two collinear beams produce a field of uniform intensity—the null field.

Next, consider that the specimen is subjected to a strain ε_{yy}. This strain deforms the diffraction grating on the specimen and the new frequency f_s' becomes:

$$f_s' = \frac{f_s}{1 + \varepsilon_{yy}} \tag{13.31}$$

With the new frequency of the specimen grating, the angles θ_{d1} and θ_{d2} are changed by a very small amount. From Eq. (13.29b), it is evident that the new diffraction angles are given by:

$$\sin\theta_{d1} = \sin\theta_{d2} = \frac{f_s\lambda}{1+\varepsilon_{yy}} \tag{13.32}$$

Upon emergence these two beams do not propagate along the z axis. The beam associated with the first diffraction order of beam (1) emerges at an angle $\Delta\theta_{d1}$ relative to the z axis which is given by:

$$\Delta\theta_{d1} = -\frac{f_r\lambda\varepsilon_{yy}}{2} \tag{13.33}$$

The beam associated with the first diffraction order of beam (2) emerges at an angle $\Delta\theta_{d2}$ relative to the z axis which is given by:

$$\Delta\theta_{d2} = \frac{f_r\lambda\varepsilon_{yy}}{2} \tag{13.34}$$

This angle $\Delta\theta$ of the emerging beam is the same as that shown in Fig. 13.21. Indeed, the two beams emerge as divergent or intersecting beams (depending on the sign of ε_{yy}) and produce an interference pattern like the one illustrated in Fig. 13.22. The fringe gradient $\partial N/\partial y$ is the same as the frequency f_m of this interference (moiré) pattern that is given by Eq. 13.26 as:

$$\frac{\partial N}{\partial y} = f_m = \frac{2\sin\Delta\theta}{\lambda} \tag{13.35}$$

Because $\Delta\theta$ is small $\sin\Delta\theta = \Delta\theta$, and Eq. (13.35) can be combined with Eq. (13.34) to give:

$$\varepsilon_{yy} = \frac{1}{f_r}\frac{\partial N}{\partial y} = \frac{f_m}{f_r} \tag{13.36}$$

This result shows that the strain can be determined from the frequency of the moiré pattern relative to the frequency of the reference pattern. Equation (13.36) can also be written in terms of pitch as:

$$\varepsilon_{yy} = p_r/p_m \tag{13.37}$$

where p_m is the distance, measured on the moiré fringe pattern, between two adjacent fringes with orders N and N ± 1.

In practice, the angle $\Delta\theta$ is very small. For example, with λ = 488 nm, f_r = 2,000 lines/mm, and ε_{yy} = 0.002, Eq. (13.34), gives $\Delta\theta$ = 0.0559°. The corresponding moiré fringe pattern frequency or fringe gradient is 4 lines/mm. This example shows the validity of the small angle assumption and clearly indicates the enhancement of the sensitivity of the moiré method by using very-high-density diffraction gratings. Indeed, moiré fringe patterns developed using interferometry methods often have fringe densities so high that they are difficult to reproduce using conventional lithographic techniques.

When the specimen is subjected to a general state of deformation, where $\mathbf{u} = \mathbf{u}(x, y)$, the frequency of the specimen grating changes over the entire field. The previous analysis is valid for any given point P(x, y) in the field if the gradient of the fringes $\partial N/\partial y$ can be determined at the point. Generally, both the u and v displacement fields are required to provide sufficient data for determining the complete strain field $\varepsilon_{xx}(x, y)$, $\varepsilon_{yy}(x, y)$ and $\gamma_{xy}(x, y)$ as described in Section 13.4.

Post [27] has used a four-beam interferometer with a crossed diffraction grating (see Fig. 13.25) to produce moiré fringe patterns for both the u and v displacement fields. The light in the incident beam is partitioned into four regions as indicated in this illustration. Light from regions A_1 and A_2 is reflected from mirrors A_1 and A_2 onto the specimen with angles of incidence of $\pm\alpha$. These two beams combine to form a virtual reference grating with its lines parallel to the x axis. This reference grating interacts with the lines parallel to the x axis in the specimen grating to form the moiré pattern representative of the v displacement field. The light from regions B_1 and B_2 of the incident beam reflect from mirrors B_1 and B_2 to form a reference grating with lines parallel to the y axis. This reference grating interacts with lines parallel to the y axis in the specimen grating to form the moiré pattern which gives the u displacement field.

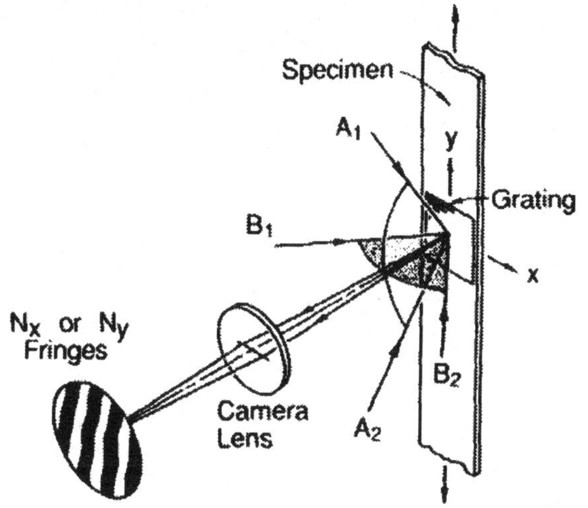

Figure 13.25 Schematic diagram of a four-beam moiré interferometer used to produce u and v moiré fringe patterns. Beams A_1 and A_2 create a virtual reference grating which interacts with the horizontal lines of the specimen grating to create the N_y fringe pattern. Beams B_1 and B_2 interact with the vertical lines to create the N_x fringe pattern. Courtesy of Daniel Post.

The two moiré fringe patterns are obtained separately with two different exposures. On the first exposure, the light over regions B_1 and B_2 is blocked and on the second exposure the light over regions A_1 and A_2 is blocked. On each exposure, the camera is adjusted to focus the image of the moiré pattern on the film plane. An example of fringe patterns for u and v displacement fields is shown in Fig. 13.26. The specimen, a beam with a bonded stiffener, was subjected to pure bending. Both the beam and stiffener were fabricated from a graphite-epoxy composite. The two elements were bonded together with a thick layer of an epoxy adhesive. The fringe patterns were obtained for the critical corner region where the load is transferred from the beam to the stiffener. The high density of the fringe patterns permits an accurate determination of both the displacement and the strain fields in this critical region.

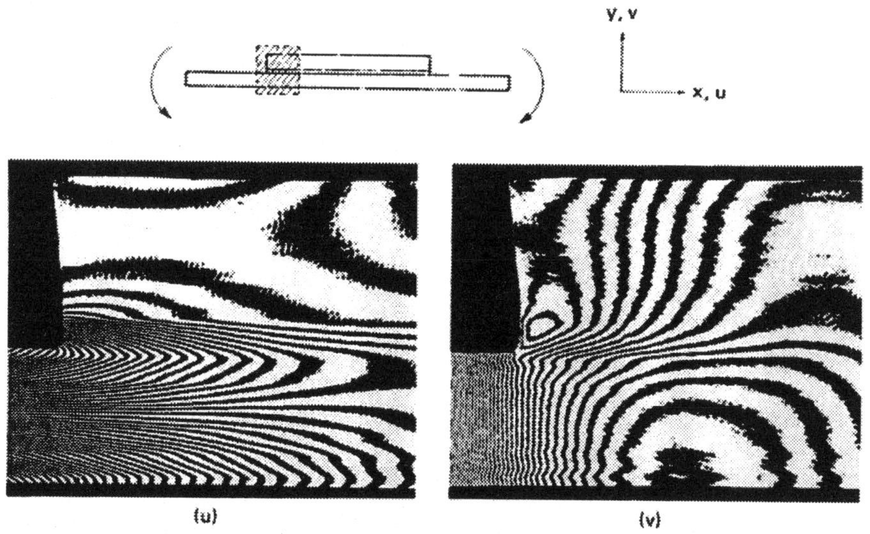

Figure 13.26 Moiré fringe patterns N_x and N_y showing the in-plane u and v displacement fields in the critical corner region of a bonded joint between two quasi-isotropic, graphite-epoxy members subjected to bending f = 2,400 lines/mm. Courtesy of D. Post.

13.10 ADVANCED APPLICATIONS OF MOIRÉ INTERFEROMETRY

In the past decade significant progress has been made in applying moiré interferometry to develop a better understanding of many problems of current interest in mechanics and materials. Page count considerations limit the coverage in this section to only a few of the many contributions by researchers here and abroad.

13.10.1 Microscopic Moiré

In recent years, there has been an increased emphasis in the design of very small components and electronic devices. Indeed, many engineers are actively engaged in one aspect or another of nanotechnology where length scales are measured in nanometers. As the size of components, structures or sub-structural elements decreases, the difficulties associated with measuring forces, stresses, strains and displacements increases significantly.

A viable approach to mechanical studies of nano-scale components is to employ microscopic moiré, where a high-powered microscope is used to view the specimen. Of course, viewing the specimen through a microscope markedly reduces the field of view. Hence, the displacements within this small field are small even for reasonably large strain fields. To circumvent this problem with moiré interferometry, Post et al [28] has developed two approaches to enhance the sensitivity of this method of measurement—an immersion moiré interferometer and digital fringe multiplication. Both of these techniques will be describe in the sub-sections presented below.

An Immersion Interferometer

One of the five different immersion interferometers designed by Post et al [28] is illustrated in Fig. 13.27. The device is relatively simple in that it consists of a prism cut from optical glass. The vertical side of the prism is coated with either aluminum or silver to produce a mirror. The opposite side is cut at an angle to produce a surface normal to the incoming light beam from a short-wavelength laser. One part of the incoming light is reflected from the mirror and superimposed on the other part of the incoming beam. The reflected and the incident beams of light interfere and produce a virtual grating on the base side of the prism.

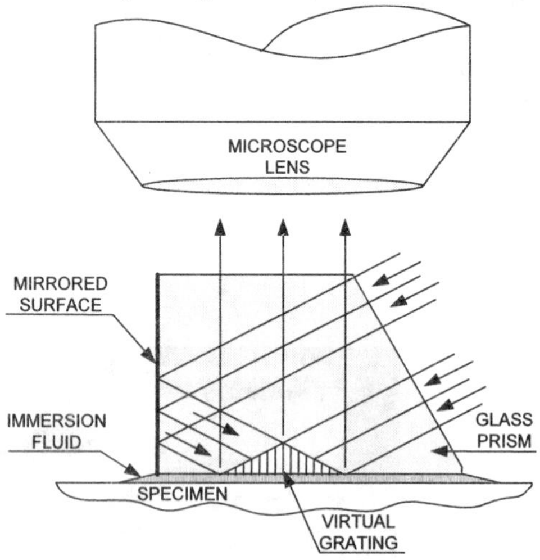

Figure 13.27 Schematic illustration of a moiré immersion interferometer.

The prism is placed over the specimen grating that is formed using the procedure described previously in Section 13.9.2. The prism is coupled to the surface of the specimen with an immersion fluid having an index of refraction that matches that of the glass in the prism. The light beam diffracted from the surface of the specimen exits the prism in the vertical direction and is captured by the lens of a microscope. The resulting fringe pattern is recorded by a digital camera positioned at the eye piece of the microscope.

The moiré immersion interferometer is more sensitive than a moiré interferometer operating in air. Recall that the measure of sensitivity of a moiré interferometer is defined by the frequency of the moiré fringes f_m. For an ordinary moiré interferometer the frequency f_m is given by Eq. (13.35) as:

$$f_m = \frac{2\sin\theta}{\lambda} \qquad (13.35)$$

However, if the light beams propagate in a refractive medium such as glass, the wavelength of the light changes to:

$$\lambda_g = \lambda/n \qquad (13.38)$$

where λ_g is the wavelength of the light propagating in glass and n is the index of refraction of glass.

Substituting Eq. (13.38) into Eq. (13.35) yields:

$$f_m = \frac{2n\sin\theta}{\lambda} \qquad (13.39)$$

Equation (13.39) indicates that the sensitivity of the moiré interferometry has been increased by a factor of n. If the immersion interferometer is fabricated from optical glass with n = 1.52, the sensitivity of the measurement is increased by 52%.

Increasing Sensitivity by Fringe Multiplication

If the number of fringes from the moiré immersion interferometer is not sufficient to perform an accurate analysis, it is possible to increase the response by shifting the reference (virtual) grating with respect to the specimen grating. Shifting the reference grating is accomplished by physically moving the immersion interferometer (the prism) through a distance s. Note that s is a fraction of the pitch p_m of the moiré pattern. Accordingly, the translation s of the prism is given by:

$$s = p_m/k \qquad (13.40)$$

where k is the multiplication factor usually taken as an even number such as 2, 4 or 6.

Shifting the immersion interferometer produces a phase change of $\Delta\phi$ in the output. This phase change is due to an increase in the optical path length of the light beam reflected from the mirrored surface of the prism. The light rays involved in the increase in the optical path length ΔOPL as the interferometer is shifted by a distance s are shown in Fig. 13.28.

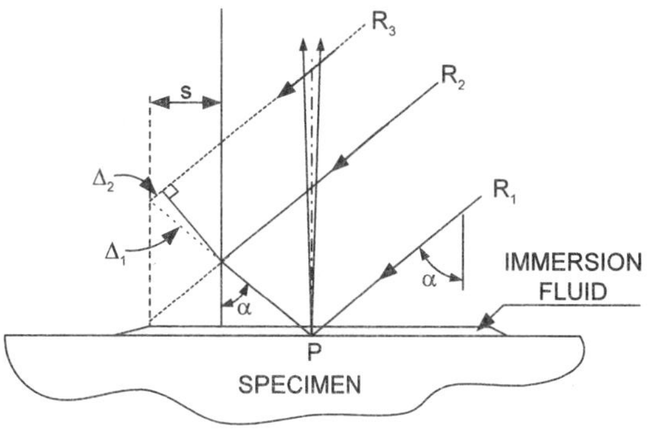

Figure 13.28 Illustration showing the effect of shifting the interferometer by a distance s on the optical path length.

Reference to Fig. 13.28 shows that the ray R_1 is not affected by the movement of the interferometer as the light in this ray will always propagate the same distance to reach point P on the specimen. However, the shift through a distance s moves the mirror to the position of the dotted vertical line. Hence, ray R_2 upon reflection will not fall on point P. Instead, ray R_3 reflects from the new position of the mirror and falls upon point P. The mechanical path length of ray R_3 is longer than that of ray R_2 by $(\Delta_1 + \Delta_2)$ and the change in the optical path length is given by:

$$\Delta OPL = n(\Delta_1 + \Delta_2) \quad (13.41)$$

The phase difference due to the shifting of the interferometer is given by substituting Eq. (13.41) into Eq. (16.37) to give:

$$\Delta\phi = \frac{2\pi}{\lambda}\Delta OPL = \frac{2\pi}{\lambda}n(\Delta_1 + \Delta_2) \quad (13.42)$$

The quantities Δ_1 and Δ_2 are shown together with the reflection angle β and the incidence angle α in the expanded drawing of the reflected ray presented in Fig. 13.29. From the geometry evident in this drawing, it is clear that:

$$s = \Delta_1 \cos\beta = \Delta_2 \sin\alpha \quad (a)$$

$$\Delta_2 = \Delta_1 \cos 2\beta \quad (b)$$

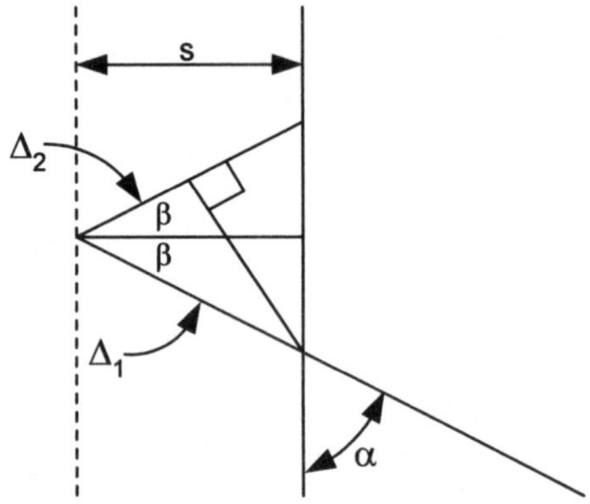

Figure 13.29 Expanded drawing of the reflection of ray R_3 after the mirror it has been shifted by a distance s.

From Eqs. (a) and (b), it is evident that:

$$\Delta OPL = \Delta_1 + \Delta_2 = 2s \sin\alpha \quad (13.43)$$

Substituting Eq. (13.43) into Eq. (13.42) yields:

$$\Delta\phi = 2\pi s\, f_m \quad (13.44)$$

Note that $p_m = sk$ or $f_m = 1/p_m = 1/sk$, then Eq. (13.44) can be rewritten as:

$$\Delta\phi = 2\pi/k \quad (13.45)$$

Clearly, the change in phase $\Delta\phi$ depends on the multiplying factor k with smaller changes in $\Delta\phi$ resulting from higher multiplication factors.

The shift s of the interferometer is very small and the translation must be accomplished with precision. Post has employed a piezoelectric actuator mounted to the stage of the microscope to move the interferometer in a direction perpendicular to the lines on the specimen grating. In cases where a cross grating was employed on the specimen, Post translated the interferometer at an angle of 45° to the grating directions. The extension of the piezoelectric actuator is controlled by adjusting the voltage applied across its electrodes. After each incremental shift, a digital photograph is recorded so as to produce k images that provide intensities $I_0(x, y)$, $I_1(x, y)$, ... $I_{k-1}(x, y)$. Each image contains the intensity of the light over the entire field of view, usually measured on 256 bit gray scale. These images, stored in memory on a personal computer, are used to produce maps of sharpened and multiplied fringes. The data processing to accomplish the fringe sharpening and multiplication is described in the next subsection.

Data Processing of Fringe Shifted Moiré Fringe Patterns

Suppose for example that k = 6, implying that intensity distributions are recorded with the digital camera for six different positions of the immersion interferometer to obtain intensity images I_0, I_1, I_2 ... I_5 . It is not necessary to record the image I_6 because I_0 and I_6 are identical. Note that the interferometer has been moved by the piezoelectric actuator by the incremental distance s between each exposure.

The intensity of the light in the interference pattern produced by the incident and the reflected beams in the interferometer is given by[1]:

$$I_0 = I_i + I_r + 2\sqrt{I_i I_r}\cos\phi \tag{13.46}$$

where I_0 is the intensity pattern recorded prior to translating the interferometer. I_i and I_r are the intensities associated with the incident and reflected beams, respectively. Both I_i and I_r are functions of the position (x, y) because the intensities of the two beams are not usually equal and not uniform over the field of view. Of course, the phase angle ϕ varies over the field and is given by:

$$\phi = 2\pi N \tag{13.47}$$

where N is the fringe order.

After translating the interferometer by a distance s, the intensity can be expressed as:

$$I_j = I_i + I_r + 2\sqrt{I_i I_r}\cos\left(\phi + \frac{2j\pi}{k}\right) \tag{13.48}$$

where j = 0, 1, 2 ... (k – 1).

When j = 0, the results of Eq. (13.48) are identical with those of Eq. (13.46). However, when j = k/2, Eq. (13.48) becomes:

$$I_{k/2=3} = I_i + I_r - 2\sqrt{I_i I_r}\cos\phi \tag{13.49}$$

because $\Delta\phi = \pi$ which is the equivalent of a half of a fringe order. It is evident that the phase change $\Delta\phi = \pi$ when I_1 is compared to I_4, when I_2 is compared to I_5 and when I_3 is compared to $I_6 = I_0$.

Next, use Eq. (13.48) and subtract these intensity pairs with $\Delta\phi = \pi$ to obtain:

$$I_\Delta = I_j - I_{j+k/2} = 4\sqrt{I_i I_r}\cos\left(\phi + \frac{2j\pi}{k}\right) \tag{13.50}$$

The subtraction term obviously improves the contrast of the fringe pattern because the term $(I_i + I_r)$ vanishes. Clearly extinction $I_\Delta = 0$ occurs when:

$$\phi + \frac{2j\pi}{k} = \left(\frac{2n+1}{2}\right)\pi \qquad \text{where } n = 0, 1, 2 \ldots \tag{a}$$

Solving Eq. (a) for ϕ yields:

$$\phi = 2\pi\left(\frac{2n+1}{4} - \frac{j}{k}\right) \tag{b}$$

[1] Equation (13.46) is different than Eq. (13.17) because the intensities I_i and I_r are not equal. The derivations using Eq. (13.46) are more realistic because of the difficulty of matching the intensities of the two beams in a moiré interferometer over its entire field of view.

Substituting Eq. (b) into Eq. (13.47) gives:

$$N = \frac{\phi}{2\pi} = \frac{2n+1}{4} - \frac{j}{k} \tag{13.51}$$

Examination of Eq. (13.51) shows a multitude of extinction conditions. For example, when j = 0, the subtraction of $I_0 - I_3$ yields quarter order fringes, which are counted with half order increments as 1/4, 3/4, 5/4, ... as n varies from 0, 1, 2, ... When j = 1, the subtraction of $I_1 - I_4$ yields 1/12 order fringes, which are counted with half order increments as 1/12, 7/12, 13/4, ... as n varies from 0, 1, 2, ...

Additional data processing is employed to enhance the quality of the fringe patterns. All of the values of I_Δ that result in negative numbers are converted to absolute (positive) numbers. This process inverts the negative peaks of the sinusoidal function of Eq. (13.50) into positive peaks as illustrated in Fig. 13.30. Next the intensity at each pixel location is tested against a threshold value I_{th}. When the absolute value of $I_\Delta \leq I_{th}$, the intensity is set to zero; however, when $I_\Delta > I_{th}$, the intensity is set to one. When the resulting intensity distribution is plotted as a function of position, a series notches occurs to identify the zero values that locate the fringes.

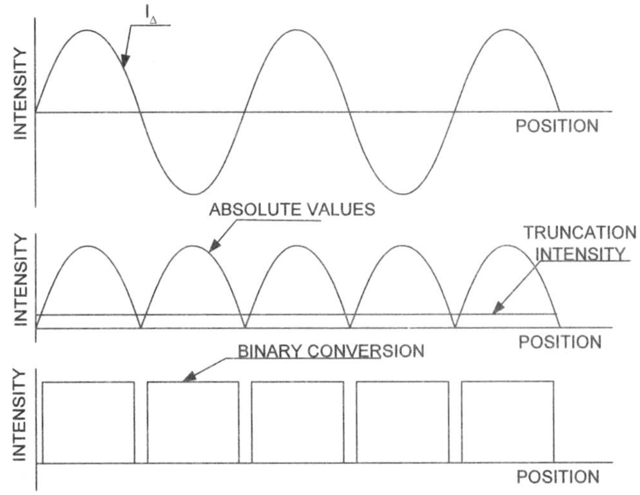

Figure 13.30 Illustration of the steps taken in processing phase shifted moiré data.

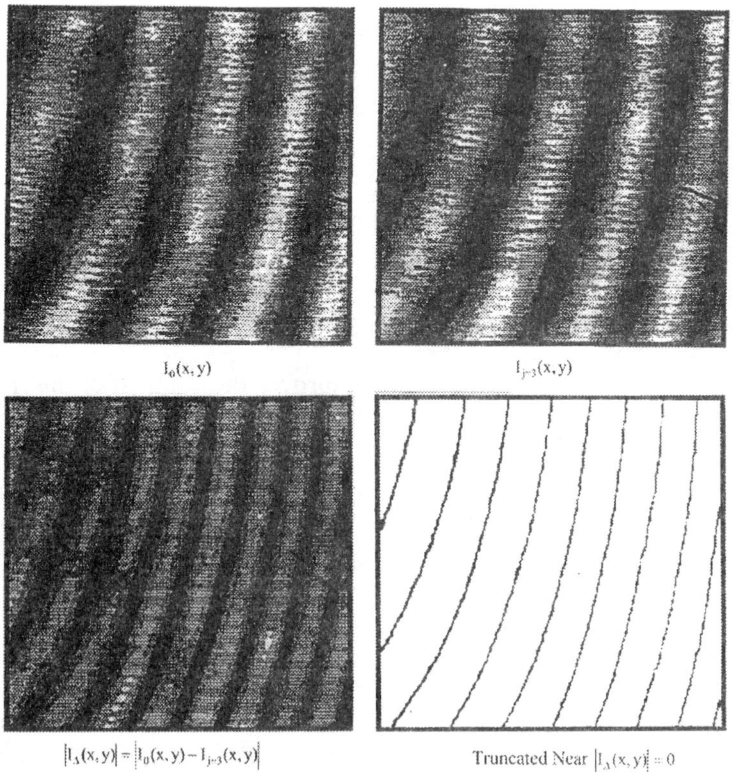

Figure 13.31 Fringe multiplication by a factor of 2 and fringe sharpening by digital data processing.

This process is repeated for each of the intensity pairs where $\Delta\phi = \pi$. The pixel locations where the conditioned value of $I_\Delta = 0$, are stored in memory. These locations are then plotted to produce a graph of lines representing fringes of fractional orders. The number of lines on the composite map is equal to the k times the original number of moiré fringes. An example of the fringe multiplication and fringe sharpening by this technique is presented in Fig. 13.31.

13.10.2 Displacement Field Determinations in Electronic Packaging

Electronic packaging involves many different aspects of designing and producing commercial products containing many electronic and electronic-mechanical components. In addition to carefully packing hundreds if not thousand of electronic components into a stylish and functional cabinet, packaging involves the protection of these components from the environment. Heat management is probably the most important of the environmental considerations because the operating temperature of electronic components markedly affects their reliability and the performance of the system. Electronic components often fail due either to excessively high temperature or due to thermal cycling. Thermal cycling between low and high temperatures, when the electronic systems are turned-on and powered, produce thermal stresses due to the temperature differentials ΔT. In many cases, the thermal stresses are difficult to measure because of the small size of the features on the electronic components. In addition, the components have complex geometries and are fabricated from different materials, which further complicate the measurements. However, moiré interferometry has been employed to measure thermally induced deformations in a number of different electronic components, because it can be adapted to measure strains on components with very small features. An interesting study of thermally induced strains in solder-ball joints is described in the next subsection.

Measuring Thermal Strains in Solder Ball Connections

Small solder balls with diameters less than 1.0 mm are sometimes used to attach[2] chip carriers to printed circuit boards (PCB). The difficulty arises due to the materials used in this application. The chip carriers are fabricated from ceramic so that their temperature coefficient of expansion closely matches that of the silicon chips which it houses. The PCB is fabricated from a glass-epoxy composite that has a much higher coefficient of expansion than the ceramic used to fabricate the chip carrier. An illustration of solder balls used to connect a ceramic chip carrier to a PCB is presented in Fig. 13.32

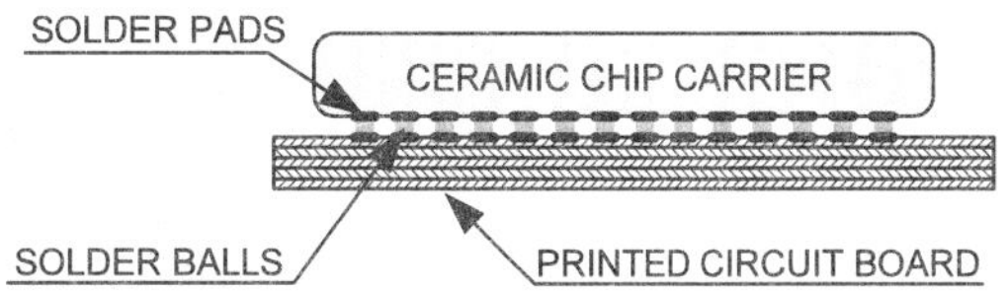

Figure 13.32 Connecting a ceramic chip carrier to a PCB with solder balls.

When this assembly undergoes an increase in temperature the PCB board expands more than the ceramic chip carrier because of the mismatch[3] in their coefficients of expansion. The solder balls are subjected to a shearing action by the differential thermal expansion of the two materials. To study the thermal strains induced in the solder balls by this action, Guo et al [29, 30] used a diamond saw to section an electronic assembly to obtain a specimen similar to that shown in Fig. 13.32. The face of this section was ground and polished to produce a smooth, flat surface. The ceramic chip carrier was 25 mm square and 2.8 mm thick. It was attached to a 1.6 mm thick PCB with 0.89 mm solder balls placed on 1.27 mm centers. After the soldering operation (a reflow process), the added solder on the pads increased the height of the solder joints to 0.97 mm.

---

[2] The attachment serves to mechanically fasten the chip carrier to the PCB and to electrically connect the solder pads on the chip carrier to the solder pads on the PCB.

[3] The temperature coefficients of expansion were 6.5×10^{-6}/°C and 19×10^{-6}/°C for the ceramic chip carrier and glass epoxy PCB, respectively.

The specimen grating is formed at elevated temperature using a technique similar to the one described in Section 13.9.2. An epoxy mold on an ultra low expansion glass substrate is made of a cross grating using a replication technique. This mold is coated with two very thin layers of aluminum which serves as a mold release later in a second replication process. This mold is placed in an oven and its temperature is adjusted to match the elevated temperature (82 °C) that the ceramic chip carrier encounters in service.

A drop of high-temperature-curing, low viscosity epoxy was placed on the surface of the mold and a piece of lens tissue was used to spread the epoxy over its surface. The lens tissue also served to remove the excess epoxy leaving a very thin uniform coating. The specimen was pressed into the epoxy coated mold and aligned so that one array of lines in the grating coincided with the plane of the PCB, and the other set was normal to this plane. After the epoxy had cured at elevated temperature, the specimen was stripped from the mold. One of the two layers of aluminum used as a mold release, remained on the specimen grating.

The specimen was then removed from the oven and cooled to room temperature (22 °C). It was maintained at room temperature for an hour before the moiré measurements were made to permit the solder joints to creep and undergo time dependent plastic deformation. The moiré analysis of the thermally induced deformations was made at room temperature (ΔT = 60 °C).

A four beam moiré interferometer was used to record fringe patterns for both the u and v displacement fields. The frequency of the virtual reference grating was adjusted to match frequency of the grating transferred to the specimen thereby eliminating the null field. Provisions were also incorporated into the specimen holder so the specimen could be rotated to precisely align the specimen and reference gratings.

The fringe patterns recorded for the u and v displacement fields are presented in Fig. 13.33.

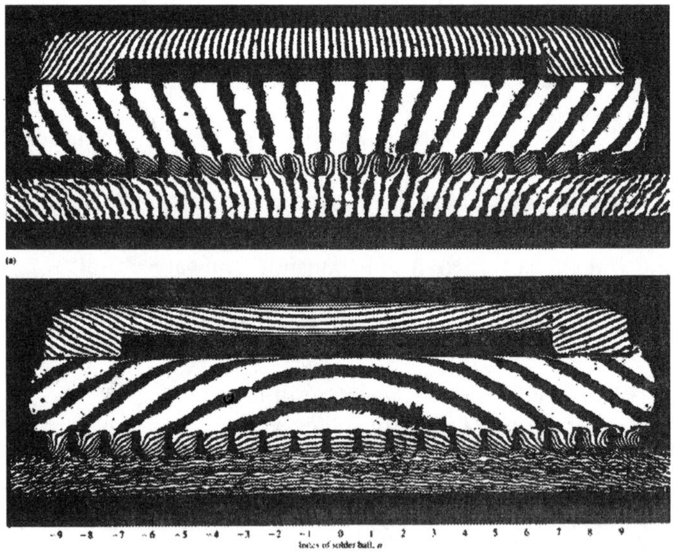

Figure 13.33 Fringe patterns for the u and v displacement fields induced by ΔT = 60 °C. Courtesy of Y. Guo.

Examination of the fringe patterns presented in Fig. 13.33 shows that the fringe density in the ceramic chip carrier is low indicating very small deformations. This result is to be expected because the ceramic material has a low thermal coefficient of expansion and a very high modulus of elasticity. In the PCB, the fringe density shows that the v displacement is much larger than the u displacement. This difference is a result of the anisotropic properties of the PCB. The glass cloth is layered to provide reinforcement in the x and z directions, but not the y (perpendicular) direction. The temperature coefficient of expansion of the PCB is also anisotropic. The PCB board also experienced bending as is evident from the inclination of the fringes in the u field. The fringe density in the solder joints indicates that the joints located near the center of the assembly exhibit very little deformation. However, the solder balls located near the end of the ceramic chip carrier undergo appreciable deformation and are the ones that fail first in service.

A detailed analysis of these fringe patterns gave the displacement in the x and y directions as indicated in Fig. 13.34 and Fig. 13.35. These results also showed that an individual solder joints exhibit stress concentrations at the interfaces where the solder ball connects to the solder pads on both the ceramic chip carrier and the PCB.

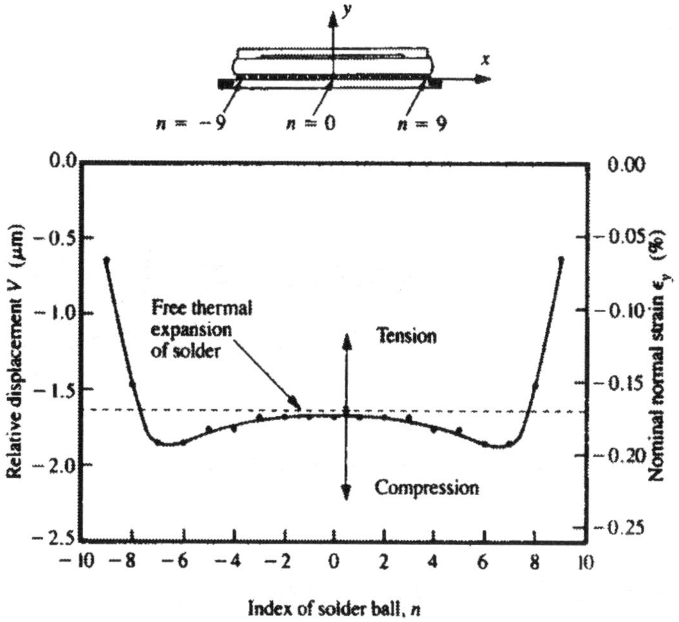

Figure 13.34 Distribution of displacement between the PCB and the ceramic chip carrier, and the nominal normal strain ε_{yy} in each solder joint. Courtesy of Y. Guo.

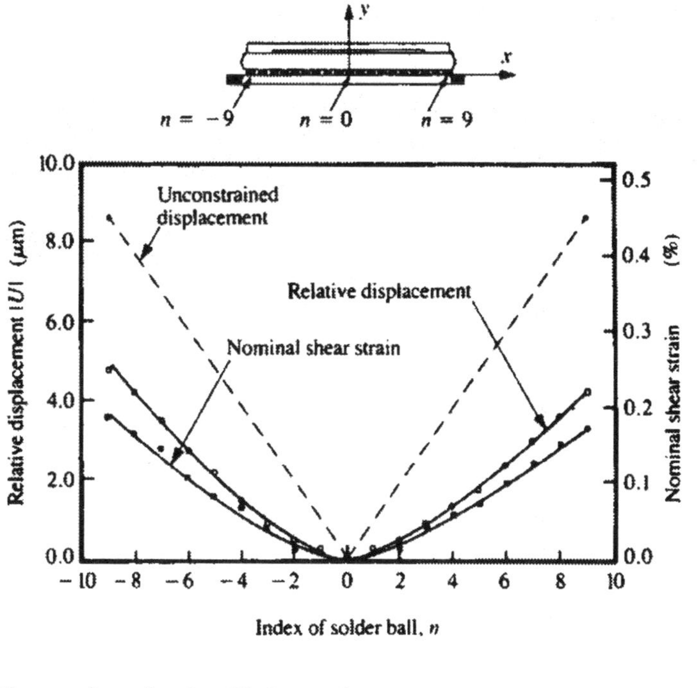

Figure 13.35 Relative displacement between the PCB and the ceramic chip carrier, and the nominal shearing strain in each solder joint. Courtesy of Y. Guo.

Measuring the Coefficient of Expansion

In the electronics industry, it is important to know the coefficient of expansion of all of the materials used in populating a printed circuit board (PCB) with microelectronic devices. When the power is turned on or off, an electronic system undergoes a temperature change where ΔT often exceeds 50° C. The temperature changes induce thermal stresses in the solder joints used to attach the components to the PCB, and thermal cycling often introduces fatigue cracks. Of course, the reliability of the electronic system is impaired over its service life as these fatigue cracks grow with each thermal cycle.

Analysis of the reliability of an electronic system depends on the knowledge of the thermal coefficient of expansion α as well as the change in temperature ΔT during a typical cycle of usage. Unfortunately measuring α of a PCB is difficult because the circuit bards are not homogeneous. Instead they are multilayered structures of glass cloth, epoxy, and thin-in-plane copper sheeting. They also contain large number of plated through holes drilled on close centers. These holes are plated with copper and are often filled with solder. The result is a heterogeneous structure with a coefficient of expansion that varies from point to point. A typical example of a multilayer PCB with plated through holes is shown in Fig. 13.36.

Y. Guo [31] has employed moiré interferometry to measure the coefficient of expansion at several different locations on the PCB. The procedure employed was similar to that described in the previous section. A cross line specimen grating with approximately 1,200 lines/mm was installed over a thick sample of multilayer PCB that was 50 mm square. The installation of the specimen grating was performed at a temperature of 72 °C and the measurement of α at different locations on the board was performed at a temperature of 22 °C. The coefficient of expansion is determined from the moiré fringe pattern by:

$$\alpha = \frac{1}{f_r \Delta T}\frac{dN}{dx} \tag{13.52}$$

where dN/dx is the slope of the of the fringe pattern associated with the u displacement filed. A similar expression can be written for the fringes associated with the v displacement field.

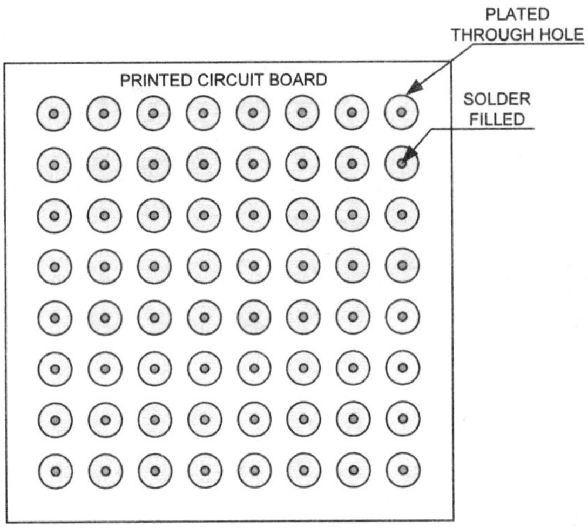

Figure 13.36 Typical construction of a multilayer printed circuit board (PCB) with plated through holes on close centers.

The measurements, made over a gage length of 5 mm, showed that α varied from 18.5 × 10^{-6} near the edge of the PCB, to a maximum of 21.0 × 10^{-6} at the center of the array of plated through holes. No attempt was made to measure the temperature coefficient of expansion over a plated through hole, although this measurement would be possible by using an immersion interferometer with fringe shifting to enhance the response by a factor of 6 to 10. Fringe shifting would permit the gage length to be reduced without sacrificing accuracy thereby permitting better measurements at local inhomogeneities.

13.10.3 Studies of Free-edge Effects in Laminated Composites

Post [32] has employed moiré interferometry to study the free-edge effects of composites due to uniaxial compressive loading. The study was performed with coupons, 13 × 13 × 25 mm, in size cut from 200 mm diameter, graphite epoxy cylinders fabricated with quasi-isotropic and cross ply fiber orientations. One side of each coupon, with exposed fiber ends was ground and polished to produce a smooth and flat surface. A specimen grating with a line frequency of 1,200 lines/mm was placed on this surface using the replica technique described previously. The coupons were subjected to a uniaxial compressive load of 22.2 kN. The moiré fringe patterns due to the compressive stresses were recorded with a four beam interferometer. The results for the quasi-isotropic composite coupon are presented in Figs. 13.37 and Fig. 13.38.

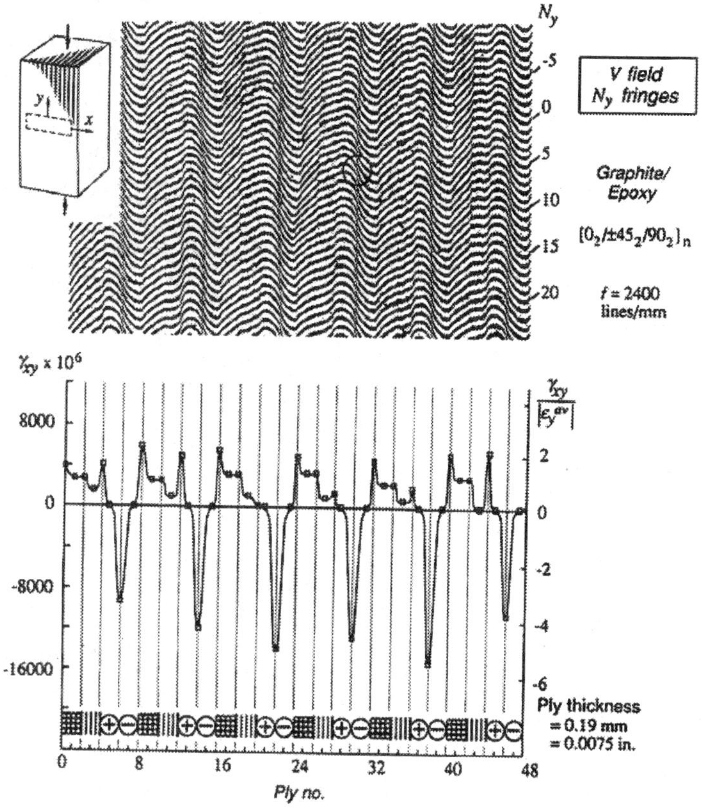

Figure 13.37 Moiré fringe pattern for the v displacement field in a quasi-isotropic specimen. Also shown is the shear strain distribution γ_{xy} as a function of ply number across the thickness of the coupon.

The fringe pattern shown in Fig 13.37 corresponds to the v displacement field. From Eq. (13.14) it is evident that:

$$\varepsilon_{yy} = \frac{\partial v}{\partial y} = \frac{1}{f_r}\frac{\Delta N_y}{\Delta y} \tag{13.53}$$

Inspection of the fringe pattern in Fig. 13.37 shows that the fringes have nearly a constant pitch implying that the fringe gradient $\Delta N_y/\Delta y$ and the strain ε_{yy} are essentially constant over the field. A value of ε_{yy} = 2,700 $\mu\varepsilon$, determined from this fringe gradient, is shown as a dashed line in Fig. 13.38.

The moiré fringes in the v displacement field exhibits significant variations in the x direction. These variations are due to the shear strains induced by the variation in the elastic properties (E and ν) from one ply to the next. This shearing strain can be determined from Eq. 13.14 as:

$$\gamma_{xy} = \frac{\partial u}{\partial y} + \frac{\partial v}{\partial x} = \frac{1}{f_r}\left[\frac{\Delta N_x}{\Delta y} + \frac{\Delta N_y}{\Delta x}\right] \tag{13.54}$$

The term $\Delta N_y/\Delta x$ is measured from the v field fringe pattern shown in Fig. 13.37 and is given by:

$$\frac{\Delta N_y}{\Delta x} = \frac{1}{\tan\beta}\frac{\Delta N_y}{\Delta y} \tag{13.55}$$

where β is the angle of the tangent to the fringe with the x axis.

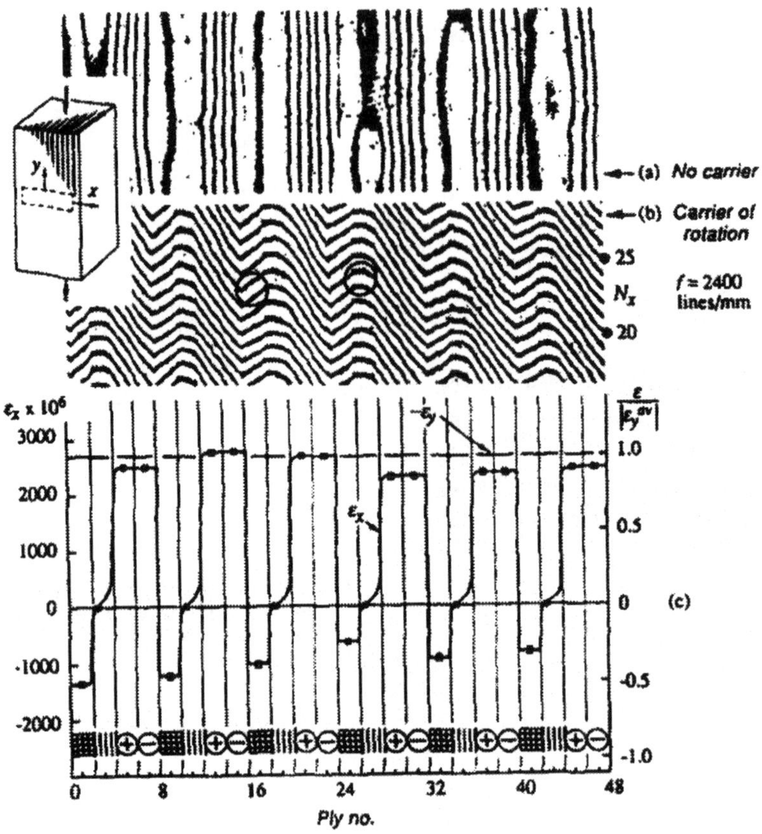

Figure 13.38 Moiré fringe pattern for the u displacement field in a quasi-isotropic specimen. Also shown is the distribution of ε_{xx} as a function of ply number across the thickness of the coupon.

The term $\Delta N_x/\Delta y$ is measured from the u field fringe pattern shown in Fig. 13.38. In this instance, $\Delta N_x/\Delta y \approx 0$ and Eq. (13.54) reduces to:

$$\gamma_{xy} = \frac{1}{\tan\beta \, f_r} \frac{\Delta N_y}{\Delta y} \qquad (13.56)$$

The distribution of the shearing strain determined from Eq. (13.56) is presented in Fig 13.37. The shearing strains exhibit significant peaks at the interface between the + 45° and – 45° plies with a magnitude of about 12,000 με. These large strains contribute to the delamination process which is a common mode of failure in composite materials.

The strain ε_{xx}, due to the u displacements, was determined from Eq. (13.14) as:

$$\varepsilon_{xx} = \frac{\partial u}{\partial x} = \frac{1}{f_r} \frac{\Delta N_x}{\Delta x} \qquad (13.57)$$

Inspection of the fringe pattern for N_x, presented in Fig. 13.38, shows relatively few fringes over some of the plies; consequently, the fringe gradient $\Delta N_x/\Delta x$ cannot be measured accurately. To circumvent this difficulty, a system of carrier fringes was superimposed on the original fringe pattern. The carrier fringes were generated by rotating the specimen about its x axis. This rigid body rotation generated a large number of uniformly spaced carrier fringes each parallel to the x-axis. The carrier fringes did not affect the fringe gradient $\Delta N_x/\Delta x$, because they were parallel to the x axis. The fringe pattern due to the u field and the rigid body rotation is also shown in Fig. 13.38. Due to the presence of the carrier fringes, the fringe gradient $\Delta N_x/\Delta x$ over the entire section can now be measured without difficulty.

The distribution of ε_{xx} from ply to ply is presented in Fig. 13.38. This transverse strain, along with the shearing strains, produces delamination of the plies in a composite material. Note that the transverse strain ε_{xx} is almost as large as the strain ε_{yy} which is in the direction of the uniaxial loading.

It is evident from this example that moiré interferometry can be used to study strain distribution in composite materials were abrupt changes in the elastic properties of the constituents cause concentrations of strain in very local neighborhoods.

13.11 e BEAM MOIRÉ

Since its introduction by Weller and Shepard [1] in 1948, the moiré method of displacement and strain analysis has been improved many times by introducing new techniques and new technologies. Since 1980, most of the advances involved developments in moiré interferometry where diffraction gratings are produced by interference of two plane beams from a coherent light source. Moiré interferometry represented a major advance because the frequency of the specimen grating was increased by a factor of about 30 to about 2,400 lines/mm. This increase improved the sensitivity of the method and extended the applicability of the moiré method to a wider range of problems.

Because additional increases in the frequency of the specimen gratings are limited by the wavelength of light, it is necessary to employ electronic methods to achieve line gratings with frequencies of 10,000 line/mm or higher. Our colleagues in high resolution lithography have already solved this problem by writing very fine lines using an electron beam. Kishimoto et al [33] were the first to introduce e-beam moiré and to demonstrate its application to study micro-deformation. The moiré fringe patterns were produced by interference of the lines in the specimen grating with the scanning lines in a conventional scanning electron microscope (SEM). This method of moiré fringe formation is similar to video-scanning moiré introduced by Morimoto et al [34] in 1984.

This section describes a study of e-beam writing methods to produce very high-frequency line and dot gratings (10,000 lines/mm) which are suitable for moiré applications. Examples of line and dot gratings are presented to show the quality of the arrays which can be produced with this method. The ability to write patterns on both homogenous and heterogeneous materials such as glass-fiber-reinforced plastics is demonstrated. The application of e-beam moiré at high magnification (1,900X), which is required in nano-mechanics, is emphasized. Finally, a moiré pattern from a high frequency line array (10,000 lines/mm), which represents the displacement field at interfaces in a GFRP specimen, is described.

13.11.1 e-Beam Generation and Control

A typical scanning electron microscope (SEM), shown in Fig. 13.39, is used to generate and control an electron beam. The electrons are generated by thermionic emission at a heated tungsten filament, which has a V-shaped tip about 200 μm in diameter. The filament is maintained at a high negative voltage during operation. The electrons emitted are accelerated to the anode (ground) with acceleration voltages that are adjustable from about 1 to 50 kV.

Conventional electromagnetic lenses are positioned in the SEM column to focus electron beam by the interaction of the magnetic field of the electronic lens on the moving electrons. The condenser and objective lens are used to reduce the beam diameter by a factor of 1,000 or more, to the final spot diameter of 5 to 20 nm when it reaches the specimen.

The objective lens is used to focus the electron beam at different specimen working distances, which usually range from about 10 to 40 mm. The spot size is minimized by reducing the working distance to improve resolution. However, the depth of field is increased by increasing the working distance, which produces a smaller divergence angle of the beam. The depth of field is also dependent on the beam limiting aperture used in the SEM as indicated in Table 13.1.

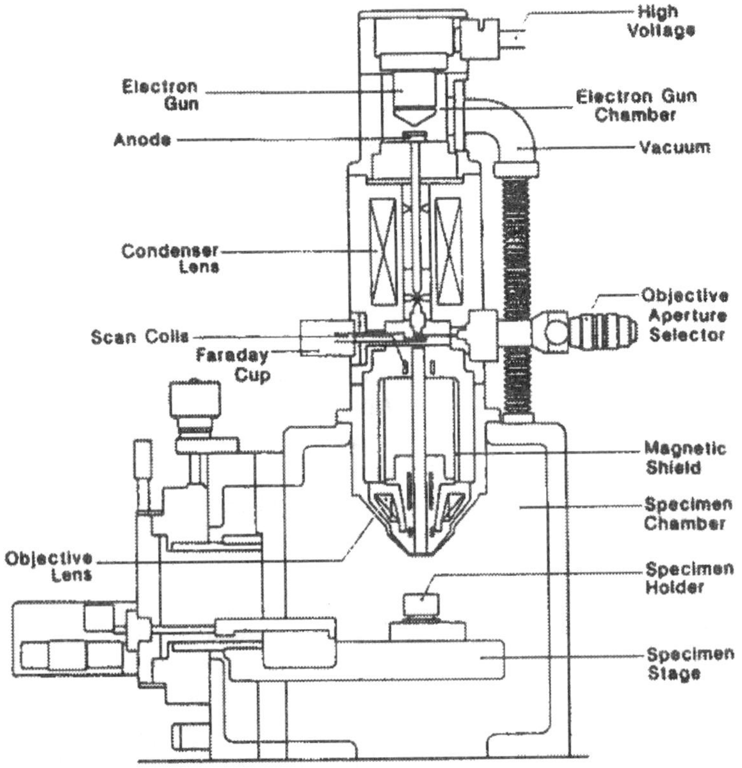

Fig. 13.39 Schematic diagram of elements in the vacuum column of an SEM

Table 13.1
Typical depth of field with a working distance of 10 mm

Magnification	Aperture Diameter (μm)			
	50	100	200	500
Depth of Field (μm)				
10^1	8000	4000	2000	800
10^2	800	400	200	80
10^3	80	40	20	8
10^4	8	4	2	0.8

The diameter of the e-beam is of critical importance in writing either line or dot patterns to produce a high resolution moiré grating. The variables which affect the beam diameter include the filament material, the accelerating voltage and the probe current. The probe current is the electrical current transmitted as the electron beam from the filament to the specimen. To minimize beam diameter, low probe currents (10 to 50 pA) are employed with higher accelerating voltages (20-30 kV). The filament material was tungsten, although filaments from lanthanum hexaboride produce a tighter beam. The beam diameter ranges from about 10 to 20 nm as the probe current increases from 10 to 50 pA with an accelerating voltage of 30 kV. The beam diameter increases as the accelerating voltage is decreased.

It is important to note the difference between the filament current and the probe current. The filament current, about 200 to 240 μA, heats the tungsten filament and produces the electrons at the gun. The probe current is the incident current produced by the electron beam striking the specimen. The probe current is adjusted with the condenser lenses (for a given aperture). Increasing the magnification of a condenser lens increases the divergence angle, and a larger portion of the electron beam is blocked by the beam-limiting aperture, which reduces the current in the beam.

A beam of electrons with a diameter in the range of 10 to 20 nm can be produced in a typical SEM. Moreover, by controlling the scanning coils, this beam can be directed over the surface of the specimen to trace any specified pattern on an x-y plane. This beam can be employed to write high density line and dot gratings for application in nano-mechanics.

13.11.2 e-Beam PMMA Interactions

The interaction of an electron beam with solids has been employed since 1968 to write intricate patterns required to fabricate very dense microelectronic devices. Certain plastics, such as polymethylmethacrylate (PMMA), undergo a chemical change during electron bombardment. The electrons sensitize the exposed material and the coating is then etched in a suitable solvent. The etching rate is controlled by the electron dose, and the etch boundaries are determined by contours of electron energy deposition. The electrons involved in sensitizing the PMMA come from the incident electron beam, the backscattered electrons and the secondary electrons. The shape of the etch pit in monolithic PMMA resembles a pear because of inelastic and elastic scattering of the electrons. The depth and width of the pear-shaped cavity depends on the electron dose, number of electrons per unit volume, and the developing time in the solvent.

Writing in a thin uniform layer of PMMA on a substrate is more difficult than writing in monolithic PMMA, because backscattered electrons and secondary electrons from the substrate also contribute to the electron dose. The incident beam is the most important source of the electron dose. Indeed, if the incident beam were the only source of electrons, it would be possible to write line and dot arrays with a pitch of 20 to 40 nm, because incident beam diameters in the range of 10 to 20 nm can be achieved. The effect of elastic and inelastic scattering of the electrons as the incident beam enters the PMMA is to effectively increase the diameter of the beam. However, this effect is minimized by reducing the thickness of the layer of PMMA to 100 nm or less.

13.11.3 e-Beam Writing Parameters

Experiments were performed to determine the parameters to employ in writing line and dot gratings[4] and to establish the minimum pitch which could be achieved. Aluminum, brass and GFRP were used as the substrate materials. The surfaces of the specimens were ground smooth and flat and then polished with 1-μm diamond particles. The substrates were cleaned in acetone in an ultrasonic tank.

The specimens were coated with PMMA within a few hours of polishing. The PMMA resist used for coating is a two-percent solution of PMMA, 950,000 relative molecular weight with chlorobenzene as the solvent. The PMMA resist was spun on the specimens and then the specimens were then baked to produce a layer of PMMA resist approximately 100-nm thick. The PMMA is applied in a class-100 clean room.

The specimens were mounted on a specimen stage in a digitally controlled SEM that was equipped with a Faraday cup and a beam blanker. The SEM was operated with an accelerating voltage of 20 kV and a 50-μm aperture. The specimen surface was 23 mm from the objective lens. A Faraday cup was inserted into the electron beam, and the beam current was adjusted to a specified value in the range from 10 to 40 pA. The e-beam in the SEM is computer controlled, and patterns are written in the PMMA resist in accordance with programmed instructions. When a line is written, the electron beam is moved from point to point, and the line is produced by many closely spaced points. To write a line with a series of points, the center-to-center distance is about 1/4 the final line width. To properly expose the PMMA, the line dose and the area dose are adjusted to accommodate the center-to-center distance and the pitch. The line dose D_L is given by:

$$DL = I_B t_e/cc \tag{13.58a}$$

and the area dose DA is given by:

---

[4] Cross gratings are produced by writing a rectangular array of dots (holes in the coating) spaced uniformly in the x and y directions.

$$DA = (I_B t_e)/(cc)(p) \quad (13.58b)$$

where I_B is the beam current; t_e is the exposure time per point; cc is the center-to-center distance and p is the pitch of the line grating. The units commonly used for DL and DA are nC/cm and μC/cm². The quantities I_B, t_e, cc and the pitch p when specified give the line and area doses required for exposing the PMMA resist.

After the patterns are written, the specimen is developed in a solution consisting of three parts (by volume) of isopropyl alcohol to one part of methyl isobutyl ketone. The specimen is immediately rinsed in isopropyl alcohol followed by a second rinse in deionized water. Next, the specimen is blown dry with a clean gas. The development of the resist is an etching process, and precise control of all aspects of the process is critical. Finally, the specimen was coated with a very thin layer (10 to 20 nm) of a gold-palladium alloy by plasma sputtering. This metallic coating is necessary to provide a conductive surface that prevents a surface charge from developing when viewing the specimen in the SEM.

Experiments were conducted to write lines and square arrays of dots with pitches of 400, 200, 100 and 75 nm. Line and area dosage was varied for each pitch to establish writing parameters. A typical line grating with a pitch of 200 nm, examined at a magnification of 100,000X, is presented in Figs. 13.40. In this illustration, the dark regions represent trenches where the PMMA resist has been etched away, and the light areas show ridges where the resist is intact. The line and area dosages were 1.25 nC/cm and 62.5 μC/cm², which resulted in a trench width of 85 nm and a ridge width of 115 nm. The waviness of the edges between the trenches and the ridges is due to slight oscillations of the e beam due to 60 cycle electrical noise that affected the stability of the lenses in the SEM.

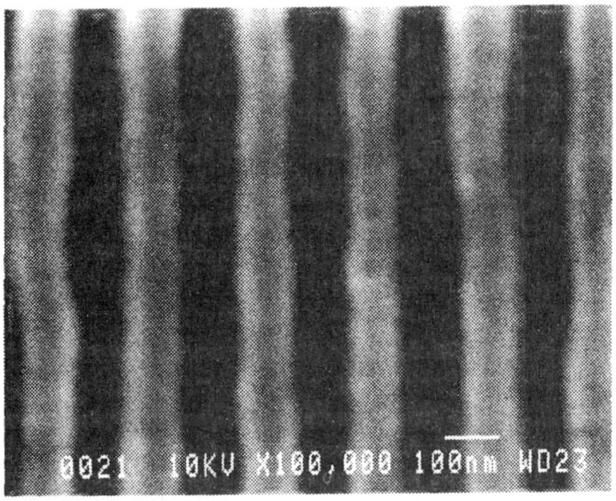

Figure 13.40 Demonstration of quality of a line array with a 200-nm pitch.

The line array shown in Fig. 13.41 is at a pitch of 100 nm. In this experiment, the exposure was set near its lower limit for a grating with a 100-nm pitch. Because the trenches are only 25 nm wide, this result indicates the possibility of producing line gratings with frequencies of 20,000 lines/mm.

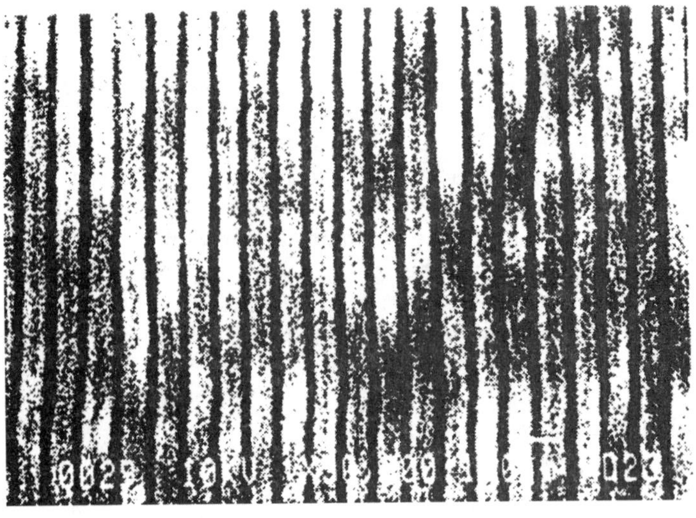

Figure 13.41 Trenches cut in the PMMA are narrow because of lower exposures.

The limit on the pitch of the line arrays that was achieved in this series of experiments was 75 nm. An inspection of the SEM images of gratings with pitches less than 100 nm recorded at 120,000X shows transverse electron beam oscillations with an amplitude of ± 10 nm as it moves along from point to point in writing a line.

Sufficient data was obtained to establish the electron beam line dosage required to write line arrays of various pitches as indicated in Fig. 13.42. For line pitches of 400 nm, the exposure band is very wide—ranging from 2.5 to 4.0 nC/cm. Although the ratio of the trench-to-ridge width varies across this band, the line arrays are suitable for moiré applications. However, as the pitch decreases, the exposure band decreases markedly. For gratings with pitches of 100 nm, the line dosage must be controlled within a narrow band ranging from 0.55 to 0.75 nC/cm.

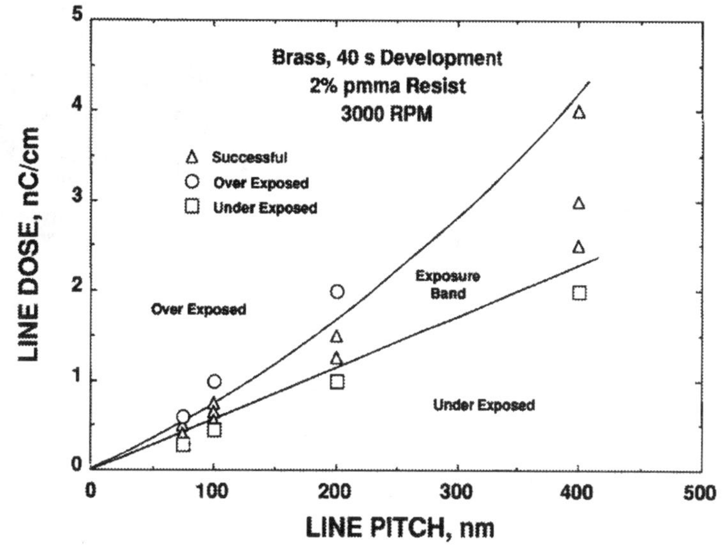

Fig. 13.42 Line dosage as a function of line pitch

13.11.4 Moiré Fringe Formation

The line or dot patterns can be interrogated either by line counting that involves a fast-Fourier transform (FFT) of a digital image or by producing scanning moiré fringes. In this presentation, moiré fringe patterns produced by the scanning lines in a SEM are considered. The scanning lines have a pitch p_r, which depends on the magnification and the scan rate selected, and is given by:

$$p_r = s/(mM) \qquad (13.59)$$

where s is a characteristic length (about 90 mm) dependent on the SEM, m is the number of scan lines in the image and M is the magnification.

The reference pitch can be varied by adjusting either m or M; however, the variation is not continuous since the choices of m are very limited because the magnification is varied from 10 to 300,000X in discrete steps.

Moiré fringe patterns are observed when:

$$p_s = kp_r\,(1 + \alpha) \qquad (13.60)$$

where p_s is the pitch of the lines on the specimen; k is a multiplication factor and α is a mismatch factor usually varying from 0 to 0.1

Examples of typical moiré fringe patterns are shown in Fig. 13.43. This fringe pattern has formed over a 50 x 50 μm square line grating on 100-nm centers. The irregularities in this pattern are due to surface imperfections that were frequently encountered on brass surfaces. These pits usually vary in size from 1 to 5 μm.

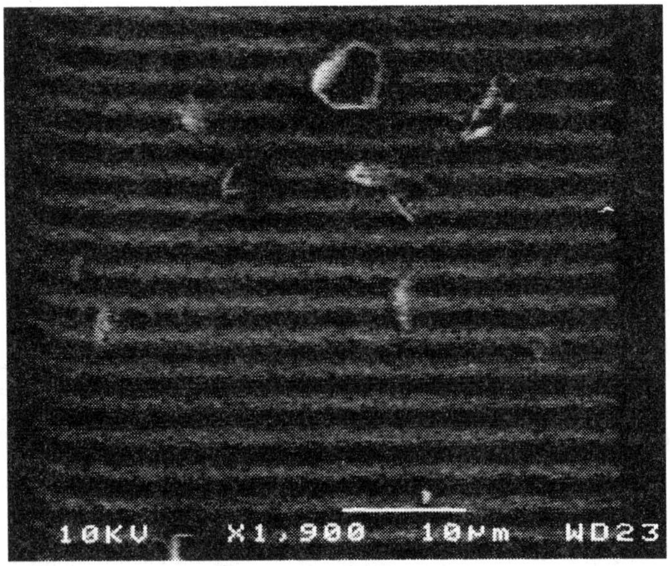

Figure 13.43 Moiré fringe formation in a SEM from a 100-nm line grating on brass.

13.11.5 Discussion

Electron beam moiré avoids the limits imposed on optical methods of moiré by the wavelength of light. Both line and dot patterns can be written on metallic and composite specimens with spatial frequencies as high as 13,300 lines/mm. Moiré fringe patterns can be formed using the scanning lines of an SEM as the reference gratings at magnifications as high as 1,900X with 100-nm pitch line patterns.

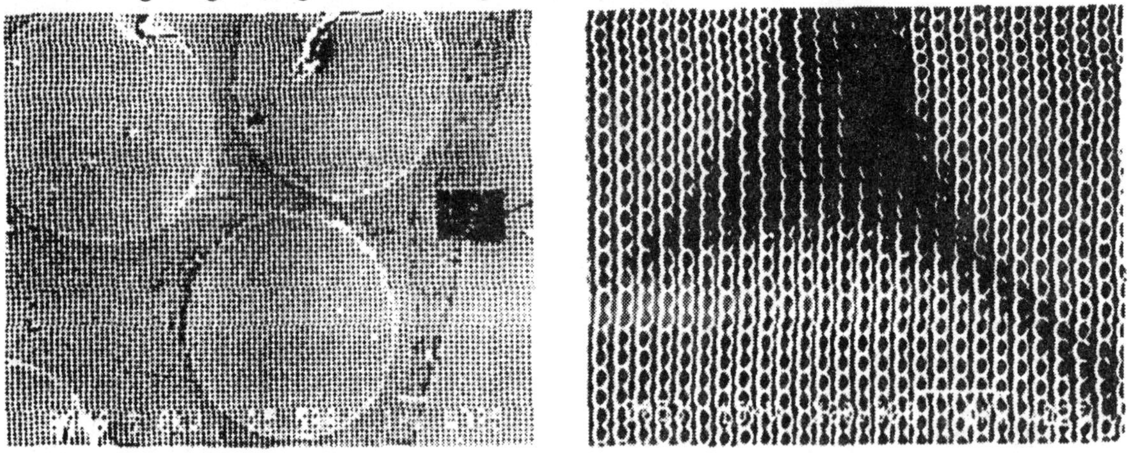

Figure 13.44 Cross gratings on 200 nm centers on a cross ply GFRP specimen with 8 μm fibers viewed at 5,500X and 20,000X.

The potential applications of e-beam moiré in micromechanics are evident from the line frequencies which can be achieved and the magnifications involved in forming the image of the moiré fringes. To illustrate the nano-mechanics potential of the method, a dot pattern, on 200-nm centers, is shown in Fig. 13.44. In Fig. 13.44a at 5500X, the dot pattern covers one complete fiber end, parts of three others, and part of a longitudinal fiber. About 40 rows of dots are written over the end of a fiber 8 μm in diameter. The triangular matrix region between three fibers is shown in Fig. 13.44b where the dot pattern is represented at a magnification of 20,000X.

A moiré fringe pattern obtained from a 100-nm pitch grating array positioned at an interface between a longitudinal ply and a cross ply is illustrated in Fig. 13.45. The specimen in this case was fabricated from a cross-ply glass-fiber-reinforced epoxy composite. The specimen was loaded in tension within the vacuum chamber of the SEM using a specially adapted tensile loading system.

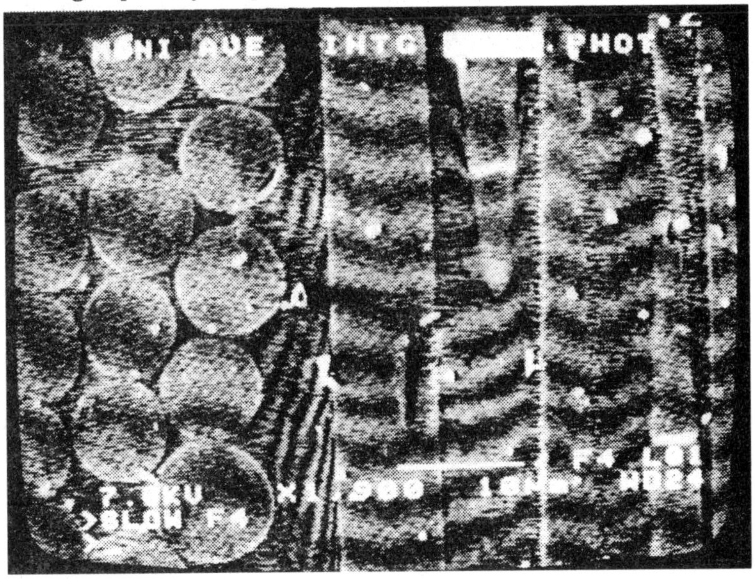

Figure 13.45 Moiré fringe pattern at an interface between longitudinal and cross plies (100-nm line grating).

The fringe pattern in Fig. 13.45 is due to both mismatch and imposed strain. A significant fringe distortion at the interface between the cross ply and the longitudinal ply is evident. The inclination of the fringes between the plies is due to relatively high shearing strains (γ_{xy} = 9.3%) that appeared suddenly as the load on the tension specimen was increased. At lower loads the fringes across the interface did not show either the gradient or the inclinations that are observed in Fig. 13.45. The cause of the sudden change in the fringe pattern at the interface between the cross and longitudinal plies occurred due to cross-ply cracking. The location of the cross ply cracks relative to the area of observation where the moiré fringe pattern was recorded is shown in Fig. 13.46. This image of the specimen, taken at a magnification of 220X, shows a two cross ply cracks—one above the area of observation and the other below it. Also evident in this figure are a large number of cracks in the longitudinal fibers that did not appear to play a major role in the subsequent failure of the specimen.

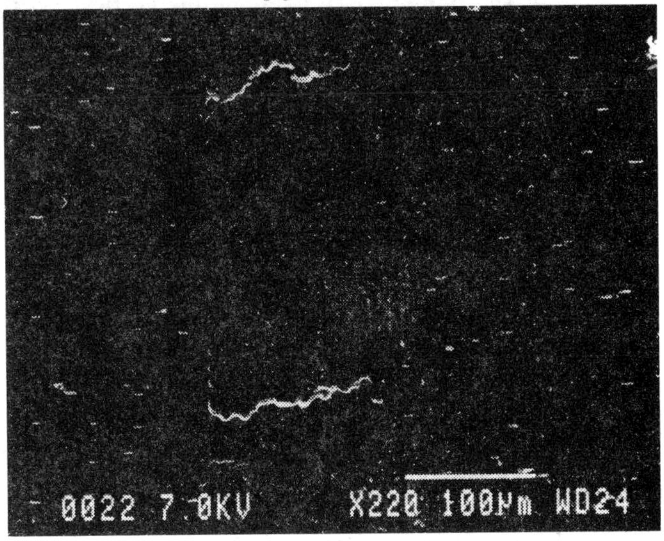

Figure 13.46 SEM image of the region about the observation area.

These studies with GFRP show another significant advantage of e-beam moiré. The e-beam penetrates the grating and it is possible to observe the structure of the material beneath the grating at high magnification. In this example, the individual fibers in both the longitudinal and cross plies could be identified. The resin rich area between plies could be examined in detail. Reducing the magnification factor enabled the field of view to be expanded and the reason for the sudden increase in shearing strain at the interface could be established.

EXERCISES

13.1 Determine the change in length ΔD of the horizontal diameter of the theta specimen (θ-shaped ring) shown in Fig. 13.1. The pitch of the vertical grid is 0.025 mm.

13.2 Equations (13.1) and (13.2) give the relations for engineering strain where l_0 is the original length. Write the corresponding relations for true strain.

13.3 Beginning with Eq. (13.5) verify Eqs. (13.7) and (13.9).

13.4 Determine the angle of rotation θ from the moiré fringe pattern shown in Fig. 13.3.

13.5 Verify Eqs. (13.10) and (13.11).

13.6 In many moiré fringe patterns, the perpendicular distance δ between fringes is difficult to measure accurately while the distance δ_p between fringes in a direction perpendicular to the master grating lines is easy to measure. For this case, develop an expression for the deformed-specimen grating pitch p' in terms of the distance δ_p, the angle of inclination ϕ of the fringes, and the master grating pitch p.

13.7 In many moiré fringe patterns, the perpendicular distance δ between fringes is difficult to measure accurately while the distance δ_s between fringes in a direction parallel to the master grating lines is easy to measure. For this case, develop an expression for the deformed-specimen grating pitch p' in terms of the distance δ_s, the angle of inclination ϕ of the fringes, and the master grating pitch p.

13.8 Use the moiré fringe pattern shown in Fig. E13.8 to prepare a plot of the displacement v as a function of position along the vertical axis of symmetry of the disk. Note that v = 0 at the center of the disk. The grating lines are horizontal and have a density of 12 lines per millimeter. The original diameter of the disk was 100 mm. From the displacement-position curve, determine the distribution of strain ε_{yy} along the vertical axis of symmetry of the disk.

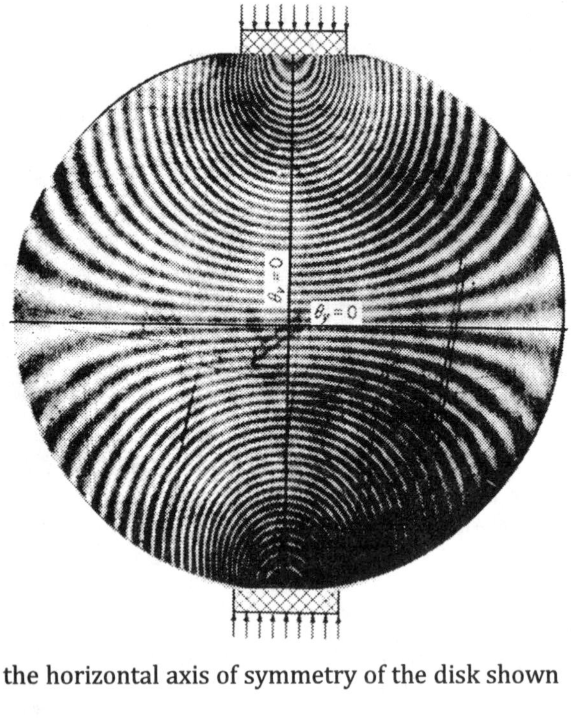

Figure E13.8

13.9 Determine the distribution of strain ε_{yy} along the horizontal axis of symmetry of the disk shown in Fig. E13.8.

13.10 Using the geometry defined in Fig. 13.10 derive the expression for the out-of-plane displacement w given by Eq. (13.16).

13.11 A circular plate of radius R_0 and thickness h with a transverse load P applied at the center undergoes a displacement:

$$w = \frac{P}{16\pi D}\left[\frac{3+\nu}{1+\nu}\left(R_0^2 - r^2\right) + 2r^2 \log\frac{r}{R_0}\right]$$

where $D = Eh^3/12(1-\nu^2)$ is the flexural rigidity of the plate.

Sketch the moiré fringe pattern observed if the optical arrangement of Fig. 13.10 is used in the displacement analysis.

13.12 Repeat Exercise. 13.11 by assuming that the Ligtenberg method, illustrated in Fig. 13.11, is used to produce the moiré fringe pattern.

13.13 When the fringe multiplication method illustrated in Fig. 13.14 is used, what requirement is placed on both moiré gratings? Discuss the errors involved if this requirement is not satisfied.

13.14 Design a moiré grating for use in analyzing one-dimensional axisymmetric problems where r is the only variable of significance. Write the specifications for this grating.

13.15 Derive the equation for the spatial frequency f of the interference pattern produced by two diverging beams of coherent light as shown in Fig. 13.18.

13.16 Prepare a graph showing the frequency f of the interference pattern of Exercise 13.14 as a function of angle θ with the wave length λ light as a display parameter. Use helium-neon and argon laser light.

13.17 Describe the replication process used to produce a diffraction grating on a model that is to be analyzed using moiré interferometry.

13.18 What is a virtual reference grating?

13.19 A diffraction grating with a frequency f of 2000 lines/mm is placed on an aluminum (E = 71 GPa and ν = 0.33) specimen with the lines oriented in the x direction. If the specimen is subjected to a uniaxial stress σ_{xx} = 50 MPa, determine the new frequency f′ of the grating.

13.20 Beginning with Eq. (13.28), derive Eq. (13.37).

13.21 Write an engineering brief describing the operating features of Post's four-beam interferometer shown in Fig. 13.25.

13.22 Use the moiré interference patterns shown in Fig. 13.26 to determine the strain distribution at select locations in the adhesive joints and adherents. Note f = 2400 lines/mm.

13.23 Discuss the concept of gage length when displacement data from moiré interferometry is used for the measurement of strain. In the discussion, consider strain fields with magnitudes of 1, 10, 100, and 1,000 $\mu\varepsilon$. Compare the moiré "gage length" with the gage length used in electrical resistance strain gages.

13.24 Design an immersion interferometer for use in a microscope where the distance between the stage and the lens of the microscope is only 20 mm.

13.25 Repeat Exercise 13.24 if the working distance is reduced to 15 mm.

13.26 Using the diagram of the light rays in Fig. 13.28, show that $\Delta OPL = 2s \sin \alpha$.

13.27 Describe the system employed to shift the immersion interferometer a closely controlled incremental amount.

13.28 Beginning with Eq. (13.46), derive Eq. (13.50).

13.29 If an immersion interferometer is used to multiply fringes by a factor k = 6, identify all of the fringe orders that are obtained after the subtraction process defined in Eq. (13.50) is used in conditioning the data.

13.30 Write an engineering brief, for a computer programmer, describing the process to condition the digital data taken during a phase shifting operation. You are expecting the programmer to prepare a program for you to automatically process the data.

13.31 Describe why it is often difficult to make meaningful measurement of strains occurring on electronic assemblies.

13.32 If the frequency of the reference grating used to produce the fringe pattern in Fig. 13.32 is 2,440 lines/mm, determine the strain ε_{xx} in the lid on the ceramic chip carrier and the ceramic chip carrier. Use the ceramic chip carrier's dimension of 25 mm to scale the illustration.

13.33 Repeat Exercise 13.33 except determine the strain ε_{yy} in these two components.

13.34 If the frequency of the reference grating used to produce the fringe pattern in Fig. 13.32 is 2,440 lines/mm, determine the shearing strain γ_{xy} in the center solder ball and in the solder balls at the ends. Use the ceramic chip carrier's dimension of 25 mm to scale the illustration.

13.35 Describe the technique used to measure the temperature coefficient of expansion for heterogeneous material using moiré interferometry. Would phase shifting with an immersion interferometer be beneficial in reducing the gage length required for accurate measurements? Explain in detail how you would employ phase shifting to improve the measurement method.

13.36 Beginning with Eq. (13.54), derive Eq. (13.56).

13.37 Describe the components in an SEM and its operation to produce highly magnified images of objects exposed to the scanning electron beam.

13.38 Describe the process used to write line gratings and cross gratings. Is the thickness of the layer of electron resist important? Why?

13.39 Use the results presented in Fig. 13.41, determine the line dose required for a line grating with a pitch of 50, 75, 100, 200 and 500 nm. Also select the beam current, exposure time and center-to-center distance for each line grating.

13.40 What is a critical limitation of a line grating that can be written with an electron beam with a high frequency? What is the size of the observation area shown in Fig. 13.45? How does this compare to the cross sectional area of one of your hairs?

13.41 What is a significant advantage of the electron beam moiré method when studying materials with heterogeneous characteristics?

BIBLIOGRAPHY

1. Weller, R. and Shepard, B.M.: "Displacement Measurement by Mechanical Interferometry," Proceedings SESA, 6 (1), 35-38 (1948).
2. Morse, S., A. J. Durelli, and C. A. Sciammarella: "Geometry of Moiré Fringes in Strain Analysis," Journal Engineering Mechanics Division, ASCE, Vol. 86, No. EM 4, pp. 105-126, 1960.
3. Weller, R., and B. M. Shepard: "Displacement Measurement by Mechanical Interferometry," Proceedings SESA, Vol. VI, No. 1, pp. 35-38, 1948.
4. Dantu, P.: "Recherches diverses d'extensometrie et de determination des contraintes," Anal. Contraintes, Mem. GAMAC, Tome II, No. 2, pp. 3-14, 1954.
5. Sciammarella, C. A., and A. J. Durelli: "Moiré Fringes as a Means of Analyzing Strains," Journal Engineering Mechanics Division, ASCE, Vol. 87, No. EM1, pp. 55-74, 1961.
6. Theocaris, P. S.: "Isopachic Patterns by the Moiré Method," Experimental Mechanics, Vol. 4, No. 6, pp. 153-159, 1964.
7. Theocaris, P. S.: "Moiré Patterns of Isopachics," Journal Scientific Instruments, Vol. 41, pp. 133-138, 1964.
8. Ligtenberg, F. K.: "The Moiré Method: A New Experimental Method for the Determination of Moments in Small Slab Models," Proceedings SESA, Vol. XII, No. 2, pp. 83-98, 1954.
9. Post, D.: "Moiré Interferometry in White Light, Applied Optics," Vol. 18, No. 24, pp. 4163-4167, 1979.
10. Parks, V. J., and A. J. Durelli: "Various Forms of the Strain Displacement Relations Applied to Experimental Strain Analysis," Experimental Mechanics, Vol. 4, No. 2, pp. 37-47, 1964.
11. Post, D.: "The Moiré Grid-Analyzer Method for Strain Analysis, Experimental Mechanics," Vol. 5, No. 11, pp. 368-377, 1965.
12. Dantu, P.: "Extension of the Moiré Method to Thermal Problems," Experimental Mechanics, Vol. 4, No. 3, pp. 64-69, 1964.
13. Post, D.: "Sharpening and Multiplication of Moiré Fringes," Experimental Mechanics, Vol. 7, No. 4, pp. 154-159, 1967.
14. Holister, G. S.: "Moiré Method of Surface Strain Measurement," Engineer,1967.

15. Post, D.: "Analysis of Moiré Fringe Multiplication Phenomena," Applied Optics, Vol 6, No. 11, pp. 1039-1942, 1967.
16. Post, D.: "New Optical Methods of Moiré Fringe Multiplication," Experimental Mechanics, Vol. 8, No. 2, pp. 63-68, 1968.
17. Sciammarella, C. A.: "Moiré-Fringe Multiplication by Means of Filtering and a Wave-front Reconstruction Process," Experimental Mechanics, Vol. 9, No. 4, pp. 179-185, 1969.
18. Chiang, F.: "Techniques of Optical Spatial Filtering Applied to the Processing of Moiré-fringe Patterns, Experimental Mechanics," Vol. 9, No. 11, pp. 523-526, 1969.
19. Durelli, A. J., and V. J. Parks: "Moiré Analysis of Strain,' Chapter 16, Prentice-Hall, Inc., Englewood Cliffs, N.J., 1970.
20. Graham, S. M.: "Stress Intensity Factors for Bodies Containing Initial Stress," PhD. Dissertation, Mechanical Engineering Department, University of Maryland, 1988.
21. Holister, G. S., and A. R. Luxmoore: "The Production of High-Density Moiré Grids," Experimental Mechanics, Vol. 8, No. 5, pp. 210-216, 1968.
22. Chiang, F.: "Discussion of the Production of High-Density Moiré Grids," Experimental Mechanics, Vol. 9, No. 6, pp. 286-288, 1969.
23. Zandman, F.: "The Transfer-Grid Method: A Practical Moiré Stress-Analysis Tool," Experimental Mechanics, Vol. 7, No. 7, pp. 19A-22A, 1967.
24. Durelli, A. J., and V. J. Parks: "Moiré Analysis of Strain," Chapters 14 and 15, Prentice-Hall, Inc., Englewood Cliffs, N.J., 1970.
25. Theocaris, P. S.: "Moiré Fringes in Strain Analysis," Chapter 11, Pergamon Press, New York, 1969.
26. Handbook on Experimental Mechanics, 2nd Edition, A. S. Kobayashi, (ed.) Prentice-Hall, Englewood Cliffs, N.J., 1993.
27. Post, D.: "Moiré Interferometry," Handbook on Experimental Mechanics, 2nd Edition, A. S. Kobayashi, (ed.) Chapter 7, Prentice-Hall, Englewood Cliffs, N.J., 1993.
28. Post, D., B. Han and P. Ifju: High Sensitivity Moiré, Springer-Verlag, New York, 1994.
29. Guo, Y., C. K. Lim, W. T. Chen and C. G. Woychic: "Solder Ball Connection (SBC) Assemblies under Thermal Loading: I. Deformation Measurement via Moiré and Its Interpretation," IBM Journal of Research and Development, Vol. 37, No. 5, 1993.
30. Guo, Y., D. Post and R. Czarnek: "The Magic of Carrier Patterns in Moiré Interferometry, Experimental Mechanics, Vol. 29, No. 2, pp 169-173, 1989.
31. Guo, Y., W. T. Chen and C. K. Lim: "Experimental Determination of Thermal Strains in Semiconductor Packaging Using Moiré Interferometry," Proceedings 1992 Joint ASME/JSME Conference on Electronic Packaging, ASME, New York, pp. 779-784, 1992.
32. Post, D., J. Morton, Y. Wang and F. L. Dai: "Interlaminar Compression of a Thick Composite," Proceeding American Society for Composites, 4th Annual Meeting, October 1989.
33. Kishimoto, S., M. Egashira, N. Shinya, and R. A. Carolan: "Local Micro-Deformation Analysis by Means of Micro grid and Electron Beam Moiré Fringe Method," Proceedings 6th International Conference on Mechanical Behavior of Materials, M. Jono and T. lnoue, (eds.), Vol. 4, pp. 661-666, 1991.
34. Morimoto, Y. and T. Hayashi: "Deformation Measurement during Powder Compaction by a Scanning Moiré Method," Experimental Mechanics, Vol. 24, pp. 112-116, 1984.
35. Walker, C. A., J. McKelvie, and A. McDonach: "Experimental Study of Inelastic Strain Patterns in a Model of a Tube-Plate Ligament Using an Interferometric Moiré Technique," Experimental Mechanics, Vol. 23, No. 1, pp. 21-29, 1983.
36. Weissman, E. M., and D. Post: "Moiré Interferometry near the Theoretical Limit," Applied Optics, Vol. 21, No. 9, pp.1621-1623, 1982.
37. Post, D.: "High Sensitivity Displacement Measurements by Moiré Interferometry," Proceedings 7th International Conference Experimental Stress Analysis, Haifa, Israel, pp. 397-408, 1982.
38. Smith, C. W., D. Post, and G. Nicoletto: "Experimental Stress-Intensity Factors in Three-Dimensional Cracked-Body Problems," Experimental Mechanics, Vol. 23, No. 4, pp. 378-381, 1983.

39. Baschore, M. L., and D. Post: "High-Frequency, High-Reflectance Transferable Moiré Gratings," Experimental Techniques, Vol. 8, No. 5, pp. 29-31, 1984.
40. Read, D. T and J. W. Dally: "Electron Beam Moiré Study of Fracture of a Glass Fiber Reinforced Plastic Composite," Journal of Applied Mechanics, Vol. 61, pp. 402-409, 1994.
41. Dally J. W. and D. T. Read: "Electron Beam Moiré" Experimental Mechanics, Vol. 33, pp. 270-277, 1993.
42. Post, D., "Moiré Interferometry: Advances and Applications," Experimental Mechanics, 31, 276-280 (1991).
43. Hailer, L, Hatzakis, M. and Srinivasan, R.: "High Resolution Positive Resists for Electron-beam Exposure," IBM Journal Research and Development, Vol. 12, p. 251, 1968.
44. Sharpe, Jr., W. N. (Ed.): Springer Handbook of Experimental Solid Mechanics, Springer Science + Business Media LLC, New York, 2008.

CHAPTER 14

SPECKLE METHODS

14.1 INTRODUCTION

The advent of the laser as a light source for photographers proved to be a significant disappointment. Photographs of objects illuminated with coherent laser light were poorly defined because of a field of bright and dark spots distributed in a random pattern over the image. These spots, known as laser speckle, limited the clarity and resolution that could be achieved and effectively eliminated the use of laser light sources for classical photography.

The speckle phenomenon is the basis for two optical methods for measuring specimen displacement—speckle photography and speckle interferometry. A speckle is a spot with a unique size, shape and intensity. It is formed by the interference of two or more rays of coherent light reflecting from a diffuse surface of a specimen. The unique properties of a specific speckle are due to the local microscopic surface imperfections on the specimen. If the specimen undergoes a displacement, the speckles move. This speckle movement provides a non-contact method for measuring small displacements.

Both speckle methods—speckle photography and speckle interferometry—are very similar to geometric moiré. Speckle interferometry is randomly generated moiré resulting from the subtraction of two intensity distributions that are created by interference within each speckle. Two beams of coherent light are used to create superimposed speckle patterns. Double exposure of these speckle patterns produces a poorly defined fringe pattern related to either the in-plane or out-of plane displacement fields depending on the optical arrangement of the speckle interferometer. Fringe patterns are produced even though the "gratings" are random because the changes in the displacements distributions are orderly (smooth).

Speckle photography and speckle interferometry will be described in the following sections. Also discussed is the use of digital cameras for recording speckle patterns. This approach enables enhancements of the fringe patterns by signal processing the digital images.

14.2 FORMING AND SIZING SPECKLES

Speckle patterns are produced by interference of a multitude of rays of coherent light that reflect from the surface of an object. Recall from Section 12.5 that surfaces on an object, not polished or coated with a reflective material, are diffuse. Consequently, each point on the object's diffuse surface acts as a light source with rays propagating radially outward in all directions. This scattering phenomenon is illustrated in Fig. 14.1.

Rays from these many wavefronts cross and interfere in space and on the image plane. The interference can be constructive, producing a bright speckle, or destructive, creating a dark one. The number of speckles produced is huge, because of the large number of surface irregularities present on the typical object and because the speckles form in space. The intensity (brightness) of each speckle varies from bright, through all shades of gray to dark depending on the optical path length of all of the rays that intersect at a specific point on the object plane. This type of random interference is known as **objective speckle**.

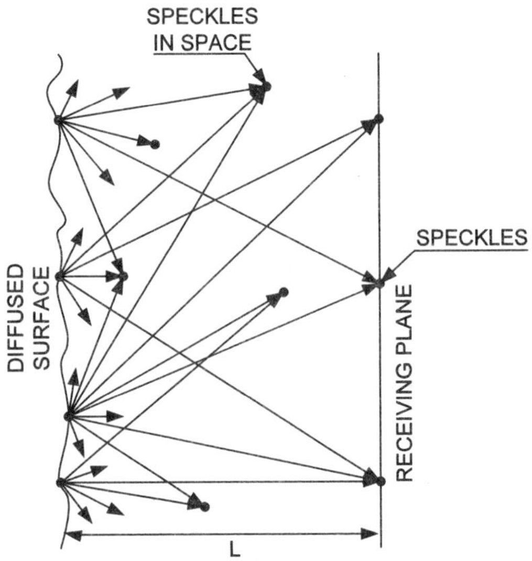

Figure 14.1 Interfering rays from a diffuse surface that produce speckles in space and on a receiving plane.

Because the process depends on the random distribution of surface imperfections, the size of the speckles formed varies and it is necessary to characterize their size statistically. The average size of an objective speckle S_0 is given by:

$$S_0 = 1.22\,\lambda\,(L/D) \tag{14.1}$$

where L is the distance from the object's surface to the receiving plane and D is the diameter of the illuminated area on the surface of the object.

Objective speckle patterns are recorded on high resolution photographic plates placed at the receiving plane. Objective speckle patterns cannot be observed directly by eye because the lens in the eye influences the speckle size. Speckle patterns recorded with a lens system are known as **subjective speckle**. Insertion of a lens into the system changes the process in three ways. First, the lens usually focuses on a smaller area on the object surface that characterized by the diameter D. Second, the lens enables the image to be magnified. Finally, the resolution of the lens limits the ability of the optical system to capture all of the rays that formed the objective speckle pattern. The average size of subjective speckles depends on the characteristics of the lens and its position relative to the diffuse surface and the image plane. The average size of subjective speckles S_s on the image plane is given by:

$$[S_s]_{\text{Image Plane}} = 1.22\,\lambda\,(1 + M)\,(f/a) \tag{14.2a}$$

where M is the magnification ratio (taken as a positive number in this relation), f is the focal length of the lens and a is the diameter of its aperture.

Comparison of Eqs. (14.1) and (14.2), even with a magnification ratio of M = 1, shows that the average subjective speckle size will be larger than the average objective speckle size if a < D/2. Most lenses used in optical arrangements have apertures that are very small compared to the incident beam diameter D[1]. The speckle size on the object is given by:

[1] The exception is in the study of astronomy where lenses with extremely large apertures are common. Laser speckle is not a problem in these studies as the illumination comes from the stars and it is only partially coherent. However, it is interesting that speckle effects from this partially coherent light have been used to study the existence and separation of binary stars.

$$[S_s]_{object} = 1.22\,\lambda\,(1 + M)\,(f/aM) \tag{14.2b}$$

In most optical arrangements, the magnification ratio is considerably less than 1; hence the size of the speckle on the object is much smaller than the corresponding speckle on the image plane.

14.3 INTENSITY DISTRIBUTIONS

Speckle patterns are formed by the interference produced by intersecting light rays from coherent sources. The intensity of each speckle is dependent on the number of rays intersecting at a speckle location and their relative optical path lengths. Clearly, this process is statistical; hence, it is necessary to characterize the intensity I of an individual speckle with a probability function given by:

$$p(I) = e^{-I/I_0} \tag{14.3}$$

where I_0 is the average intensity of the speckles in the pattern.

Examination of the term e^{-I/I_0} shows that p(I) varies from p = 1.0 for I = 0, to p = 0.0183 for $I = 4I_0$. These results indicate the presence of many more dark speckles than bright ones when the speckles are produced by a single object beam.

In some applications, the speckle field is produced by the superposition of two beams such as object and reference beams. The addition of the reference beam changes the distribution of the speckle intensity in the patterns. The probability function for the speckles produced by the combined coherent beams is given by:

$$p(I) = 2e^{-(1+2I/I_0)}J_0\left[2\left(\frac{2I}{I_0}\right)^{1/2}\right] \tag{14.4}$$

where J_0 is the zero order Bessel function.

While this mathematical expression differs from Eq. (14.3), its results are similar. There is a high probability (p = 0.7358) of zero intensity (dark) speckles and a much lower probability for bright speckles.

In other experiments, speckles are produced from coherent light reflected from two different diffused surfaces. In this case, two separate speckle patterns are superimposed and the resulting probability function is written as:

$$p(I) = \frac{4I}{I_0}e^{-2I/I_0} \tag{14.5}$$

In this case, the probability of a dark speckle is zero, and the probability of a very bright speckle is very low. The function p(I) = 0.7357 is a maximum where $I = I_0/2$; hence, the field will have a large number of darker gray speckles.

14.4 SPECKLE DECORRELATION

Measurement methods involving speckle patterns require the patterns to be recorded for two different positions of the specimen. Measurements of specimen displacements using speckle interferometry require that these two speckle patterns remain correlated. If one of the speckle patterns moves too far (due to excessive specimen displacement), the resulting speckle pattern is not comparable to the original one and the pattern is classified as decorrelated. When correlation is lost, the optical effects, described later, which enable speckle patterns to be used in metrology, are no longer valid.

Decorrelation occurs due either to phase change across a speckle or due to insufficient overlap of one speckle relative to the other. Decorrelation due to insufficient overlap is illustrated in Fig. 14.2. The maximum displacement of one speckle relative to the other is a fraction of the speckle diameter.

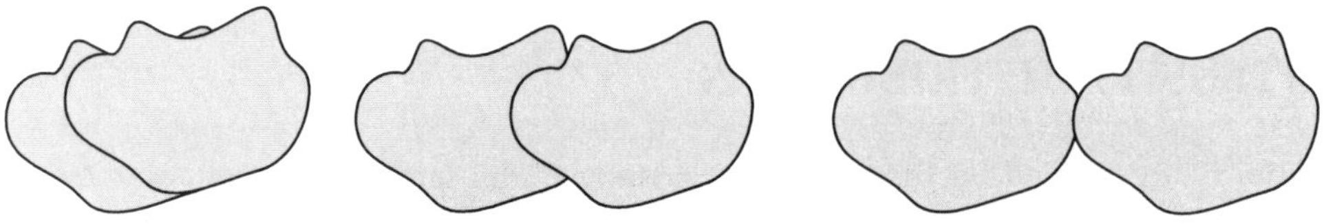

Figure 14.2 Illustration of speckle movement allowable for speckle correlation.

Decorrelation also occurs due to phase change across a speckle. Out of plane rotations of a specimen, which can be large relative to in-plane displacements, may change the optical path length between the rays forming the speckle. If a phase change of about 2π occurs across the diameter of a speckle, correlation is lost.

While correlation limits the magnitude of specimen displacements when interferometric methods of measurement are used, displacement measurement using speckle photography do not require speckle correlation. Indeed, speckle photography requires sufficient speckle movement so that the two speckle patterns form a diffraction grating of randomly distributed speckle pairs.

14.4.1 Decorrelation of Speckle Patterns Due to Specimen Movement

Changes in the speckle pattern, due to several different cases of specimen movement, are considered in this section to show the magnitude of displacements that can be tolerated prior to the onset of decorrelation. Specimen displacement, either translation or rotation, may cause decorrelation because of excessive speckle movement (on the specimen) or a phase change of 2π across a speckle. To determine the allowable specimen movement before the onset of decorrelation, consider the optical arrangement illustrated in Fig. 14.3.

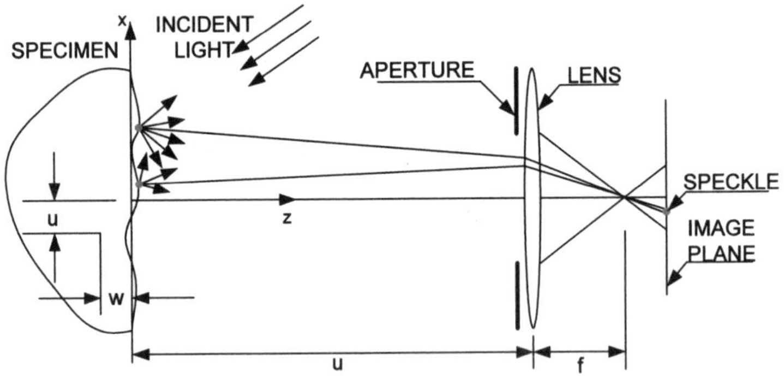

Figure 14.3 Specimen movement of w and u affects the speckle pattern on the image plane.

By determining the difference in optical path length before and after the out-of-plane displacement w and using Eq. 14.2b, it can be shown that a phase change of 2π occurs across as speckle (on the object) when:

$$w_{2\pi} = \frac{Mau^2}{1.22(1+M)fx} \qquad (14.6)$$

where u is the distance from the specimen to the lens; x is the distance from the optical axis of the lens to the center of the speckle on the specimen and M, a and f were defined previously with reference to Eq. (14.2a).

A uniform displacement w in the z direction will clearly cause the speckle on the specimen to move. When the speckle moves a distance equal to the original speckle size decorrelation occurs. This out-of-plane displacement is given by:

$$w_S = 1.22(1+M)\frac{f\,u\lambda}{a\,xM} \tag{14.7}$$

To show a typical magnitude of the out-of-plane displacement w that will produce decorrelation, consider an optical arrangement with u = 300 mm, x = 25.4 mm, M = 1/10, λ = 632.8 nm and a numerical aperture of f/a = 8. Substituting these values into Eqs. (14.7) and (14.8) yields:

$w_{2\pi} = 40.26$ mm and $w_S = 0.8024$ mm

This example shows that excessive movement of the speckle (memory loss) limits the allowable displacement. However, relatively large out-of-plane displacements can be tolerated before the onset of decorrelation.

Next, consider a uniform in-plane displacement u in the x direction as indicated in Fig. 14.3. An analysis of the change in the optical path length produced by this displacement shows that phase decorrelation occurs when:

$$u_{2\pi} = \frac{Mau}{1.22(1+M)f} \tag{14.8}$$

Decorrelation due to memory loss occurs when:

$$u_S = 1.22(1+M)\frac{f\lambda}{Ma} \tag{14.9}$$

Using the same optical parameters that were used in the previous example gives:

$u_{2\pi} = 2.794$ mm and $u_S = 67.94\ \mu m$

Again, speckle decorrelation occurs due to excessive speckle movement which produces memory loss. In this case, the in-plane displacement that causes decorrelation is relatively small (about the diameter of a human hair).

Decorrelation also occurs due to rigid body rotations of the specimen. Consider first the out of plane rotation θ_x about the x-axis as shown in Fig. 14.3. The speckles will decorrelate when:

$$\theta_x = \frac{\lambda}{S_s|_{Object}} = \frac{Ma}{1.22(1+M)f} \tag{14.10}$$

For the same optical parameters that were used in the previous example, it is evident that decorrelation occurs when the out-of-plane rotation angle $\theta_x = 0.5337°$. Clearly, very small rigid body rotations result in a loss of correlation.

Finally, consider an in-plane rigid body rotation θ_z about the optical axis. A point P(x, y) a distance r from the optical (z) axis moves a distance $s = r\theta_z$. Decorrelation occurs when $S_s|_{Object} = r\theta_z$ which can be written as:

$$\theta_z = \frac{S_s|_{Object}}{r} = \frac{1.22(1+M)\lambda f}{aMr} \tag{14.11}$$

For the same optical parameters that were used in the previous example and with r = 25.4 mm, it is evident that decorrelation occurs when the in-plane rotation angle $\theta_z = 0.1532°$. This result provides for a small in-plane rotation prior to the loss of correlation.

This example illustrates that decorrelation usually is caused by excessive speckle movement (memory loss). Also, small in-plane displacements produce decorrelation while relatively large out-of-plane displacements do not.

14.5 SPECKLE PHOTOGRAPHY

Speckle photography is an optical method used to measure in-plane displacements—u and v. The method is based on the fact that a specific speckle on the image plane is uniquely defined by the local surface imperfections within a small region on the specimen that can be resolved by the lens. The size of the speckle depends on the geometry of the optical system and the lens parameters. In the previous section, it was shown that speckle movement occurred when the specimen was subjected to either in-plane or out-of-plane displacements. In this sense the motion of a specific speckle, which can be recorded, provides a means to measure a corresponding displacement at some point P on the specimen.

As the name—speckle photography—implies, a photograph of the speckle pattern is taken using an optical arrangement similar to the one shown in Fig. 14.4. Coherent light is provided by a laser, which is conditioned with a spatial filter. The size of the speckle that controls the sensitivity of the measurement is established by selecting the numerical aperture f/a of the lens and the magnification ratio as indicated by Eq. (14.2)

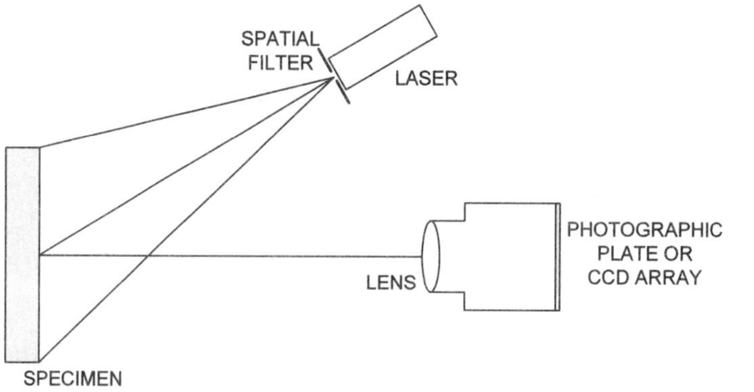

Figure 14.4 A simple optical arrangement to record speckle patterns from a specimen.

In an experiment, a load exposure is made with a small load to establish the initial speckle pattern. Additional load is then applied to the specimen producing a displacement u, v or a combination of both u and v. Next, a second exposure is recorded on the same photographic plate. The result is a photographic plate with two superimposed speckle patterns, which are identical, except one pattern is shifted relative to the other by some displacement. The negative on the photographic plate will exhibit a density D given by:

$$D(x, y) = D_0 + D_1 [I(x, y) + I(x + u, y + v)] \qquad (14.12)$$

where D_0 is the background density and D_1 is a function of the characteristics of the photographic emulsion.

A typical speckle pattern formed by double exposure techniques, shown in Fig. 14.5, does not appear to provide information pertaining to the displacement field. The speckles do occur as matched pairs, but it is not practical[2] to measure the displacement u and v from the positions of any given pair of speckles. A better approach is to use the specklegram as a transmission filter and to examine the diffraction pattern that is produced.

[2] The speckles are poorly defined and randomly shaped. Finding the centers of two of these speckles, necessary to measure displacement, is an invitation to introduce large errors in the results.

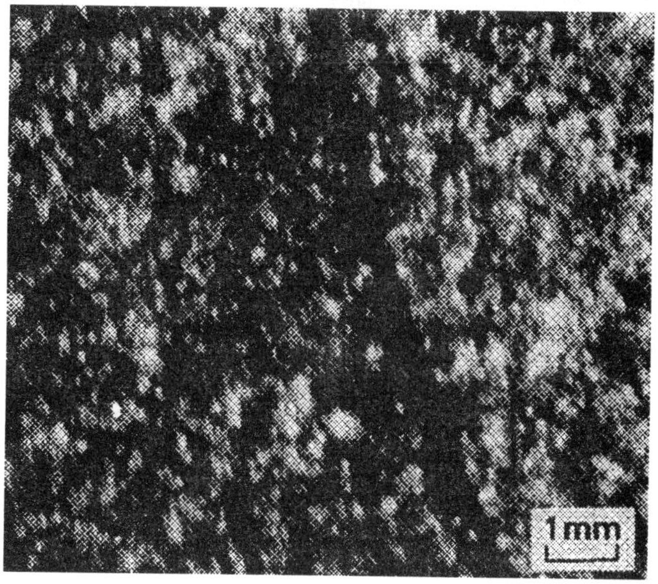

Figure 14.5 A typical double exposure speckle pattern. Courtesy of Gary Cloud.

14.5.1 Determining Displacements from a Specklegram

The two speckle patterns on the double exposed photographic plate have the same spatial distribution I (x, y) except that the second pattern has been shifted due to the displacement components u and v to give I(x + u, y + v). As a consequence, the two patterns produce a specklegram that acts like a diffraction grating. The pitch p of the diffraction grating is related to the displacements u and v.

The specklegram (diffraction grating) can be interrogated by transmitting a narrow beam of coherent light though it, as shown in Fig 14.6. The light is diffracted and a fringe pattern with characteristics similar to the one shown in Fig. 14.7 appears on a screen placed a distance L from the specklegram. The spacing of the speckle pairs represents the pitch p on the specklegram. The pitch d of the diffraction pattern is related to the pitch p on the specklegram by recalling Eq. (10.38b).

$$p = \lambda(L/d) \tag{14.13}$$

In this case, the pitch p represents the total displacement due to u and v on the specimen; hence:

$$p = (u^2 + v^2)^{1/2}. \tag{14.14}$$

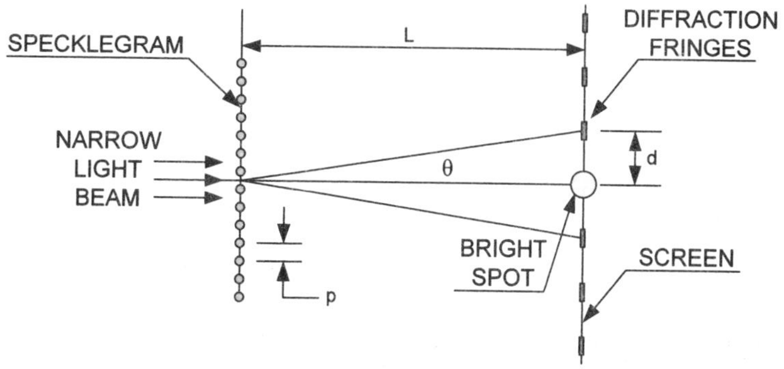

Figure 14.6 Geometric parameters involved in the development of a the diffraction pattern.

The components of the displacement u and v at the point of interrogation can be determined from the diffraction fringe pattern. Reference to Fig 14.7 shows a field of uniformly spaced fringes with a pitch d. The fringes are inclined at some angle relative to the x, y coordinates with the angle of inclination dependent upon the relative magnitudes of u and v. Usually, the displacement components u and v are determined by measuring the fringe spacing in the x and y directions using an average value over a reasonable number of fringes.

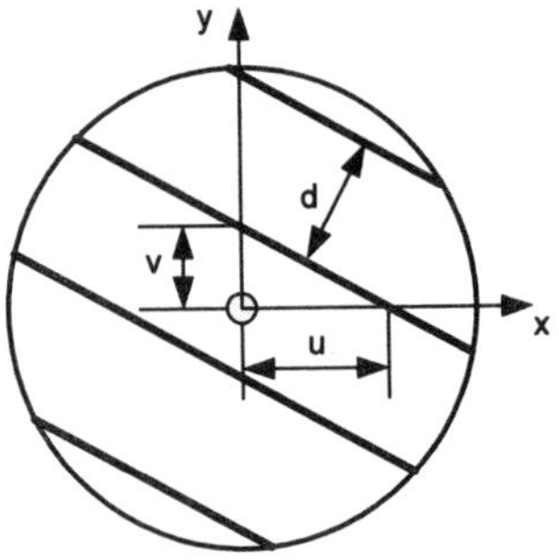

Figure 14.7 Fringe pattern produced on the diffraction plane.

14.5.2 Whole Field Measurements of Displacements

The technique described in Section 14.5.1 pertained to measuring the displacements at a particular point on the specklegram. This measurement must then be related to the point on the specimen that produced the corresponding speckle region. This is a tedious process and a much better approach is to employ a whole field method. Clearly, the specklegram incorporates whole field data if the entire area of the specimen was illuminated as illustrated in Fig. 14.4. Each small region of the specklegram maps to a corresponding region on the specimen. If the displacement of the specimen varies from point to point, the spacing of the matched speckle pairs will vary from point to point on the image plane (or photographic plate).

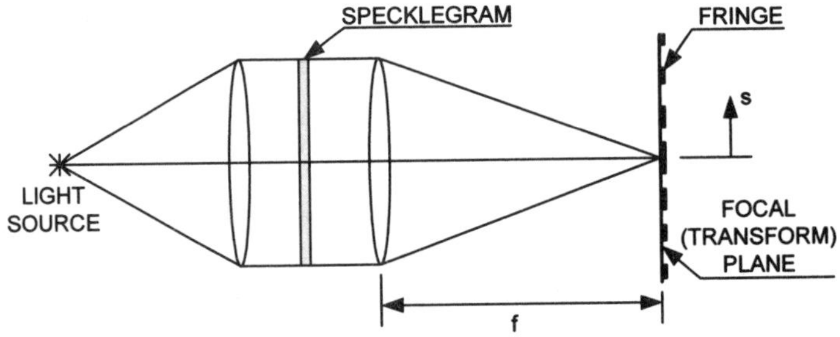

Figure 14.8 Optical arrangement for producing a diffraction pattern from a specklegram.

To acquire data pertaining to the whole displacement field, the specklegram is again used as a transmission filter; however, the entire specklegram is illuminated, as shown in Fig 14.8. A diffraction pattern is projected onto a screen that is located at the focus point of the second field lens. The pitch d of the fringes[3] in the diffraction pattern was previously derived in Section 13.7 as:

$$d = f\theta_1 = (f\lambda)/p \tag{13.22}$$

where f is the focal length of the second field lens.

Note that the pitch d is a function of x and y on the transform plane because the pitch p on the specklegram varies from point to point. Indeed, the fringe pattern shown in Fig. 14.7 represents a special case. It corresponds to a spatially filtered double exposed speckle photograph recorded on a specimen subjected to a uniform displacement. In the more general case with displacements that vary over the field of the specimen, the fringes are not well defined. Instead, a "diffraction halo" consisting of a large number of bright diffraction bands which cross one another is observed.

A fringe will be observed at a distance s measured from the optical axis on the focal (transform) plane when:

$$s = Nd = (Nf\lambda)/p \tag{14.15a}$$

---

[3] In Section 13.7, the diffraction pattern for a moiré line grating, with a constant pitch over its total area, consisted of a row of dots; however the specklegram generates fringes, not dots, because it is comprised of many matched speckle pairs with a broad continuous spectrum of pitches with different orientations over its entire area.

or

$$p = (Nf\lambda)/s \qquad (14.15b)$$

where N is the order of the fringe (an integer).

Equation (14.15b) shows that the displacement could possibly be determined[4] over the whole field by measuring s from the diffraction fringe pattern and determining the corresponding pitch p. However, this gives the total displacement at any point on the specimen as indicated in Eq. (14.14), and it is often necessary to determine individual displacements u and v.

The final step in the process is the separation of the displacement components u and v. An inverse transform is employed to determine the displacement components in the x and y directions. Mathematically inverse transforms are difficult; however, they are easy to implement in an optical arrangement. If a small hole (representing a spatial filter) is made in the screen representing the focal plane, and only the light that is passing though this hole is used to record the image of the specklegram, then the result is an inverse transform.

Now suppose that the small hole is placed in the screen a distance y_1 measured from the optical axis along the y (vertical) axis. The only regions on the specklegram that generate light rays that pass through this hole are those regions with a vertical (v) displacement where $p = (Nf\lambda)/y_1$. With the spatial filter hole placed at a position y_1 along the y axis the displacement v is given by:

$$v = p = (Nf\lambda)/y_1 \qquad (14.16)$$

The fringe pattern can be observed with one's eye through the spatial filter hole on the focal plane. When creating and recording the fringe pattern with a camera, a spatial filter is placed in front of the camera lens, the camera is positioned at the appropriate location on focal plane (the screen is removed) and the camera is focused through the second field lens onto the specklegram. When the camera, with its attached spatial filter, is positioned at $x = 0$ and $y = y_1$ on the focal plane, the fringe pattern represents lines along which the displacement v is a constant. When the camera, with its attached spatial filter, is positioned at $x = x_1$ and $y = 0$ on the focal plane, the fringe pattern represents lines along which the displacement u is a constant. The sensitivity of the measurement is controlled by x_1 or y_1. As the offset from the optical axis is increased, the sensitivity is increased as is evident from Eq. (14.16).

An example of whole field displacement fringes recorded in this manner is presented in Fig. 14.9.

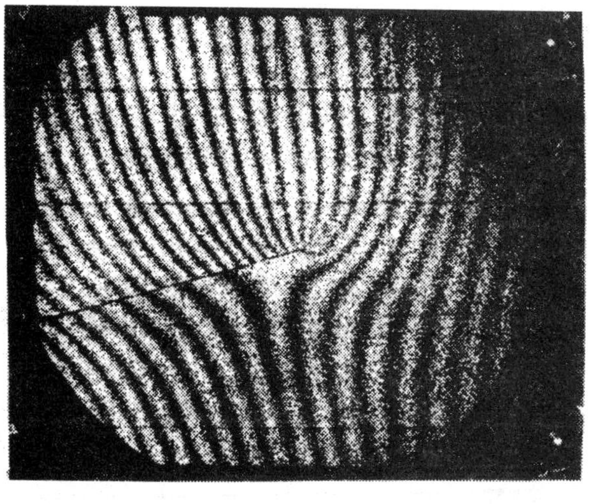

Figure 14.9 Whole field displacement fringes recorded using spatial filtering on the focal plane. Courtesy of Fu-pen Chiang.

[4] The determination would depend on the clarity of the fringe pattern. For uniform displacement fields or for fields where the variation is minimal, the definition should be adequate to make the necessary measurements.

14.5.3 Calibration

To convert from fringe order to pitch as prescribed by Eq. (14.16) requires the knowledge of a sensitivity constant for the filtering arrangement shown in Fig. 14.7. It is evident from Eq. (14.16) that the sensitivity of the measurement of the displacement (pitch of the fringes) is controlled by the ratio of $f\lambda/y_1$. It is possible to establish this sensitivity factor by measuring f and y_1, but it is usually more accurate to calibrate the optical system used for filtering the specklegram by determining the calibration factor as a lumped constant.

Cloud [1] has described a very effective procedure for calibrating the optical filtering arrangement. He begins with a calibration disk that can be rotated about its center to produce small but well-controlled in-plane displacements. The displacement field over the disk varies from zero at its center to a maximum at it periphery. Moreover, the gradient of the displacement field is constant and dependent upon the amount of rotation. The calibration disk is used as a specimen and several double exposed specklegrams are made of the disk with different amounts of rotation. These specklegrams are the calibration standards that are used in establishing $f\lambda/y_1$ for the optical system used in the filtering operations described in Section 14.5.2.

The calibration process is simple. First, fringe patterns from the specklegram for an unknown specimen are recorded using the optical filtering process. Next, the calibration specklegram is placed in the optical arrangement at exactly the same location. Then the calibration fringe pattern is recorded using precisely the same settings for the camera and the location of the spatial filter. Because the pitch (gradient) of the fringes is known on the calibration disk, a sensitivity factor in terms of μm/fringe can be established directly for the filtering system. This sensitivity factor is then used in place of $f\lambda/y_1$ to determine the unknown displacements in the specimen.

14.5.4 Conclusions

Speckle photography is very similar to geometric moiré where the specklegram replaces a pair of cross gratings. The double exposure of the speckle patterns to produce a specklegram is equivalent to the use of a specimen grating and a reference grating. Because the speckles are not nearly as well defined as a typical cross grating, the definition of the speckle-generated fringe patterns is poor when compared to those generated in a moiré experiment. Nevertheless, accurate displacement measurements can be made with this method by using the optical filtering technique arrangement described in section 14.5.2.

Decorrelation is not an issue with speckle photography. In fact, the speckles must move a sufficient distance to produce the separation between speckle pairs that is necessary to generate a low quality diffraction grating.

The sensitivity of the method is controlled by the speckle size in recording the specklegram. The highest sensitivity is achieved with the smallest speckles. Reference to Eq. (14.2) indicates that the speckle size is reduced as the numerical aperture (f/a) is reduced (large aperture and small focal length lens). If film resolution is limiting the smallness of the speckle size, sensitivity can also be gained by increasing the magnification ratio to produce larger speckles at the image plane.

Optical filtering described in Section 14.5.2 enables the generation of fringe patterns from the specklegrams that provide whole field displacement results. The optical filtering has the advantage that individual displacement components can be determined from separate fringe patterns. Also, the sensitivity of the measurement can be changed, within limits, by selecting the position of the spatial filter hole on the focal plane prior to recording the fringe patterns.

A significant advantage of the method is that it is not necessary to place a grating on the specimen to obtain a moiré-like fringe pattern.

14.6 SPECKLE INTERFEROMETRY

Displacement measurements made with speckle photography employed a single beam of coherent light and a double exposure of the speckle image. Speckle movement, due to specimen displacements imposed between exposures, produced a low quality diffraction grating containing information regarding the displacement field. Speckle interferometry differs in several ways. First, two beams of coherent light are employed—the object beam and the reference beam—using optical arrangements similar to those used to produce a hologram. Second, two different speckle patterns, one from the object beam and the other from the reference beam, are superimposed on the image plane. This superposition process enables phase information in the object beam to be retained in the image. Finally, the superpositioning of the speckles from the two beams improves the sensitivity of the measurement of out-of-plane displacements.

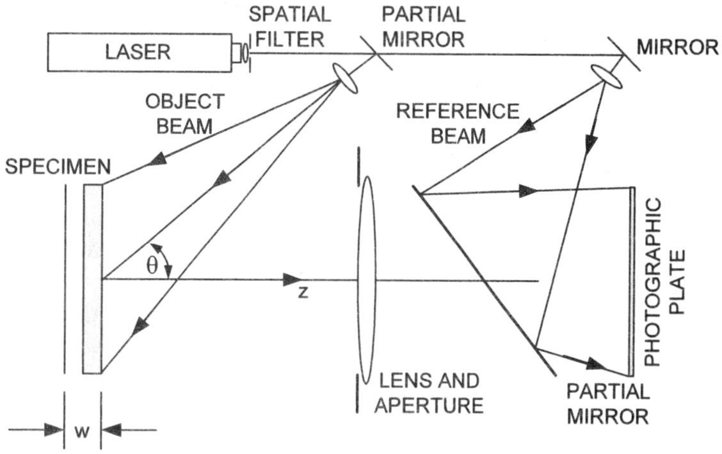

Figure 14.10 A speckle interferometer for out-of-plane measurements.

14.6.1 Measuring Out-of-Plane Displacements

An optical arrangement often employed to measure out-of-plane displacements is presented in Fig. 14.10. In this arrangement, the object beam reflects from the surface's diffuse surface and is focused on the image plane by a long focal length lens. The reference beam is reflected from a partial mirror so that it impinges on the image plane at normal incidence. When the specimen undergoes a displacement w in the z direction, the optical path length changes for the light rays in the object beam, which result in an angular phase change given by:

$$\Delta\phi = \frac{2\pi}{\lambda}\left[w(1+\cos\theta)\right] \tag{14.17}$$

This equation was derived previously in Section 12.5.3 in the discussion of measuring out-of-plane displacements using holography.

Obtaining, processing and interpreting data in speckle interferometry is similar to holography because fringes formed from speckle patterns can be observed in real-time, time-averaged or double exposure. For this discussion, consider an experiment performed using real-time observation of fringe patterns due to an out-of-plane displacement w. To begin, a speckle pattern of the specimen under a small initial load is recorded. The photographic plate is developed and placed back in its original location maintaining perfect registration. Since the developed photographic plate is a negative, registration can be checked by insuring that all of the bright speckles in the image plane line-up with all of the dark speckles on the negative. The negative, placed on the image plane, will block all of the light from both beams if it is in perfect registration.

After the adequacy of the registration has been verified, load is applied to the specimen, out-of-plane displacements occur and an angular phase change takes place at each point on the specimen as indicated by Eq. (14.17). The real time image viewed through the initial load negative[5] will show regions of extinction when the angular phase change is some multiple of 2π.

$$\Delta\phi = 2\pi k \tag{14.18}$$

where k is an integer k = 0, 1, 2 .....

Substituting Eq. (14.18) into Eq. (14.17) gives an expression of the out-of-plane displacement w as:

$$w = \frac{N\lambda}{1+\cos\theta} \tag{14.19}$$

where N = k is the fringe order.

Examination of Eq. (14.19) shows that the sensitivity of the measurement is increased as the angle of inclination θ of the illuminating beam goes to zero.

In developing Eq. (14.19) it has been assumed that the viewing angle for the object beam was zero. However, the object beam reflected from the specimen's diffuse surface is focused by a lens onto the image plane. As a consequence, the viewing angle is not zero over the entire field of the image and Eq. (14.19) does not account for this fact. To minimize the relatively small errors resulting from the lens-induced viewing angles, long focal length lenses are recommended for the imaging optics.

14.6.2 Measuring In-Plane Displacements

The optical arrangement for measuring in-plane displacements is presented in Fig. 14.11. Again two beams are employed in developing the speckle pattern; however, in this instance both beams illuminate the specimen. One incident beam comes from above and the other from below. Both of the angles of inclination equal θ. The reflections of both beams from the diffuse surface of the specimen are focused on the image plane by a long focal length lens.

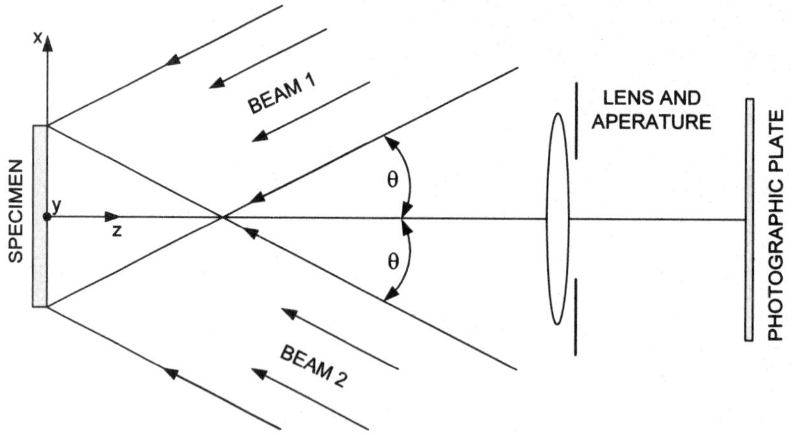

Figure 14.11 Optical arrangement of measuring in-plane-displacements.

The intensity of light corresponding to the initial state of the specimen at any speckle located at point P on the image plane can be written as:

$$I(x, y) = a_1^2(x, y) + a_2^2(x, y) + a_1(x, y)\, a_2(x, y) \cos\xi \tag{14.20}$$

[5] This process causes the subtraction of the original intensity distribution of the speckle pattern (recorded on the negative) from the new intensity distribution—speckle by speckle. The fringes are not formed by interferometric effects due to the optical path length changes as is the case in holography where two plane wavefronts interfere.

where $a_1(x, y)$ and $a_2(x, y)$ are amplitudes of the beams number 1 and 2, respectively and ξ is the initial phase difference between the two waves.

Next consider that the specimen is loaded and that the resulting displacement field **d** can be written as:

$$\mathbf{d} = u\,\mathbf{i} + v\,\mathbf{j} + w\,\mathbf{k} \tag{14.21}$$

where **i**, **j**, and **k** are unit vectors and u, v and w are the Cartesian displacement components.

The change in the phase angle $\Delta\phi_1$ of beam 1 can be determined from an expression similar to Eq. (12.40).

$$\Delta\phi_1 = \frac{2\pi}{\lambda}(\mathbf{k} - \mathbf{u}_i)\bullet\mathbf{d} \tag{14.22}$$

where $\mathbf{u}_i$ is a unit vector defining the direction of the incident light beam and **k** is the unit vector defining the direction of the light from the specimen to the image plane, which coincides with the z axis. For beam 1, $\mathbf{u}_1$ is written as:

$$\mathbf{u}_1 = -\sin\theta\,\mathbf{i} - \cos\theta\,\mathbf{k} \tag{14.23}$$

Substituting Eq. (14.23) into Eq. (14.22) and taking the dot product gives:

$$\Delta\phi_1 = \frac{2\pi}{\lambda}\left[u\sin\theta + w(1+\cos\theta)\right] \tag{14.24}$$

For beam 2, $\mathbf{u}_2$ is written as:

$$\mathbf{u}_2 = \sin\theta\,\mathbf{i} - \cos\theta\,\mathbf{k} \tag{14.25}$$

Substituting Eq. (14.25) into Eq. (14.22) and taking the dot product gives:

$$\Delta\phi_2 = \frac{2\pi}{\lambda}\left[-u\sin\theta + w(1+\cos\theta)\right] \tag{14.26}$$

The change in the phase angle $\Delta\phi$ due to the specimen displacement is given by:

$$\Delta\phi = \Delta\phi_1 - \Delta\phi_2 = \frac{4\pi}{\lambda}u\sin\theta \tag{14.27}$$

Extinction occurs when $\Delta\phi = 2\pi k$ where k is an integer 0, 1, 2, ... Then, from Eq. (14.27), it is evident that the expression for the displacement component u may be written as:

$$u = \frac{N_u\lambda}{2\sin\theta} \tag{14.28}$$

where $N_u = k = 0, 1, 2, ...$ is the fringe order.

With the incident beam oriented from above and below the specimen with equal inclination angles θ, the fringe pattern is produced by the superposition of the speckles from the two beams. The fringe pattern is proportional to the in-plane displacement component u. The sensitivity of the measurement can be varied by adjusting the inclination angle θ, with increasing sensitivity (N_u is increased) with larger values of θ.

Changing the inclination of the incident beams from above and below to impinge on the specimen from the left and the right produces a second fringe pattern, which corresponds to the displacement v. Following the same procedure for the derivation of the change in the phase angle $\Delta\phi$, due to the displacement component v is given by:

$$v = \frac{N_v \lambda}{2\sin\theta} \tag{14.29}$$

It is clear that by adjusting the planes containing the two object beams that illuminate the specimen that the in-plane components of displacement can be measured from the real-time fringe patterns.

14.6.3 Discussion

The methodology of speckle interferometry is similar to holography because fringe patterns due to either in-plane or out-of-plane displacements can be observed using real-time, time-averaged or double exposure techniques. The requirements placed on the optical system are similar—a high quality vibration isolation table is required together with rigid fixtures for holding all of the optical components. High resolution photographic emulsions are recommended to improve the contrast between the light and dark speckles. The resolution requirements for the photographic emulsion depend on the speckle size, but with larger speckles (20 to 50 μm) very high resolution films are not necessary.

The fringes observed using speckle interferometry are poorly defined—they lack contrast and the intensity of the fringe varies from point to point. The loss of definition of the fringes is due to the effect of a large number of speckles with different intensities that produce the fringes. Recall, in the derivation of the conditions for the formation of a fringe (extinction), that the first two terms in Eq. 14.20 were ignored. These terms produce a speckle intensity that is not extinguished as the phase angle changes. The superposition of the two speckle patterns, which produces the fringes, gives rise to many shades of gray, whereas other optical methods such as photoelasticity, holography and moiré produce high contrast fringe patterns with clearly distinguished regions of black and white.

While several techniques have been described to improve fringe contrast in these speckle fringe patterns, a better approach is to record the patterns with a digital camera and to enhance the contrast of the images with well known image processing techniques. The use of digital methods in processing data from a speckle interferometer is described in the next section.

14.7 DIGITAL ACQUISITION AND PROCESSING OF SPECKLE IMAGES

As the pixel size and costs of CCD arrays continue to decrease, the advantages of replacing the conventional camera and film with a digital camera has become apparent. The key issue in the decision to employ either digital or conventional methods is the resolution requirements. Resolution with a conventional camera depends on the quality of the lens and the capability of the film. With a digital camera, the size of the cell in the CCD camera determines its resolution. Holographic methods are the most demanding of resolution because the pitch of the gratings that must be recorded is of the order of 0.5 to 1.0 μm. Even with the high pixel count digital cameras available today, the dimensions of the individual cells on a CCD array are about 10 μm by 10 μm. Clearly, the pixel size in the CCD arrays is too large by a factor of about 10 to resolve the line arrays required for holographic measurements.

Speckle size can be varied, as indicated by Eq. 14.2, and the average size of a speckle can be adjusted to equal or exceed the size of a cell on a CCD array. Consequently, digital cameras can resolve the speckle patterns at the image plane of a speckle interferometer. The reader is referred to Section 14.10; where a discussion of both conventional and digital photography is presented.

14.7.1 Incorporating a Digital Camera in a Speckle Interferometer

The optical arrangement for a speckle interferometer equipped with a digital camera is similar to the arrangement described in Fig. 14.10. Additions to this system to adapt it to a digital recording system include a piezoelectric actuator and a computer and monitor as shown in Fig 14.12. Both the reference and object beams are usually passed through spatial filters to remove optical noise and to create a well-conditioned spherical wave front. The piezoelectric actuator moves a prism by incremental amounts to adjust the path length of the reference beam. This feature enables the operator to adjust the phase angle ϕ by controlled amounts. The two speckle patterns due to the object and reference beams are superimposed on the CCD array in the image plane of the camera. The optical image is converted into a voltage in the digital camera. This voltage is converted to a digital code by an analog-to-digital converter. The digital code is downloaded to a frame grabbing circuit in the computer and then stored in memory and displayed on the monitor.

14.7.2 Fringe Contrast Enhancement by Signal Manipulation

With digital recording, the intensity of the speckle pattern on the image plane is measured at a huge number of points[6]. Digital codes with 8, 10 or 12 bit resolution provide the intensity (gray level) at each pixel location of the CCD array. This intensity data are stored in memory and is available for subsequent signal processing. Two signal processing approaches are used to improve the quality of the speckle fringe patterns so that they are more useful—signal subtraction and addition. Signal subtraction, the more important of the two techniques, will be described in the following subsection.

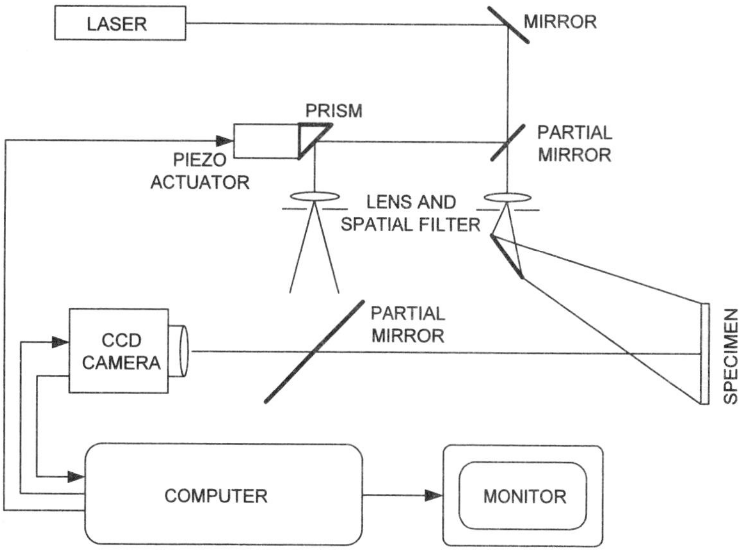

Figure 14.12 A speckle interferometer equipped with a digital camera and a computer.

Signal Subtraction

Consider the intensity of a speckle pattern produced by the object and reference beams focused on the image plane. The intensity $I_i(x, y)$ of the superimposed speckle patterns prior to loading the specimen can be expressed as:

$$I_i(x,y) = I_0(x,y) + I_R(x,y) + 2\sqrt{I_0(x,y)I_R(x,y)}\cos\phi(x,y) \tag{14.30a}$$

---

[6] A large-format, high-resolution digital camera is capable of excellent spatial resolution with $3072 \times 2048 = 6.29 \times 10^6$ pixels.

where $I_0(x, y)$ and $I_R(x, y)$are gray scale values of the intensity for each speckle in the pattern and $\phi(x, y)$ is the phase angle between the two beams.

After the load is applied, the specimen undergoes a displacement and the intensity $I_d(x, y)$ of the speckle patterns change to:

$$I_d(x,y) = I_0(x,y) + I_R(x,y) + 2\sqrt{I_0(x,y)I_R(x,y)}\,\cos[\phi(x,y) + \Delta\phi(x,y)] \qquad (14.30b)$$

where $\Delta\phi(x, y)$ is the phase angle difference due to the displacement field.

Subtracting Eq. (14.30a) from Eq. (14.30b) yields:

$$\Delta I = 2\sqrt{I_0 I_R}\,\left[\cos(\phi + \Delta\phi) - \cos\phi\right] \qquad (a)$$

Recall the trigonometric identity:

$$\cos A - \cos B = -2\sin\frac{1}{2}(A + B)\sin\frac{1}{2}(A - B) \qquad (b)$$

Using Eq. (b) to modify the form of Eq. (a) gives:

$$\Delta I = -4\sqrt{I_0 I_R}\,\sin\left(\phi + \frac{\Delta\phi}{2}\right)\sin\frac{\Delta\phi}{2} \qquad (14.31)$$

Inspection of Eq. (14.31) indicates that negative values for ΔI probably will occur for many values of x and y. This result causes difficulties because negative values cannot be displayed on the computer monitor. To avoid this problem, the absolute values of ΔI are taken and Eq. (14.31) is rewritten as:

$$\Delta I = \left|4\sqrt{I_0 I_R}\,\sin\left(\phi + \frac{\Delta\phi}{2}\right)\sin\frac{\Delta\phi}{2}\right| \qquad (14.32)$$

Expanding the trigometric function in Eq. (14.32) yields:

$$\Delta I = \left|4\sqrt{I_0 I_R}\left[\sin\phi\sin\frac{\Delta\phi}{2}\cos\frac{\Delta\phi}{2} - \cos\phi\sin^2\frac{\Delta\phi}{2}\right]\right| \qquad (14.33)$$

Dark fringes are formed when ΔI goes to zero or when sin $(\Delta\phi/2) = 0$. Hence:

$$\frac{\Delta\phi}{2} = k\pi \quad \text{where } k = 1, 2. 3 \ldots. \qquad (14.34a)$$

Bright fringes are formed when ΔI is a maximum, which occurs when sin $(\Delta\phi/2) = 1$. Hence:

$$\frac{\Delta\phi}{2} = \frac{(2k+1)\pi}{2} \quad \text{where } k = 0, 1, 2, \qquad (14.34b)$$

It is interesting to note that:

$$\Delta I_{max} = 4\sqrt{I_0 I_R}\,\cos\phi \qquad (14.35)$$

Contrast is affected by the value of ϕ because the initial phase difference appears as the argument of the cosine function in Eq. (14.35). In some cases, contrast can be improved by adjusting the value of ϕ in the initial speckle pattern by minor changes to the optical path length in the reference beam with the piezo actuator so that $\phi \Rightarrow 0$ and the $\cos \phi \Rightarrow 1$.

14.7.3 Speckle Methods Applied to Vibration Studies

Speckle interferometry is also useful in studying the vibration of plates. In this application the procedure is similar to that used in holography where time-averaged exposures of the plate motion are made. Recall from Section 12.5.4 that with the time-averaged approach, the exposure time for recording the pattern (speckles or line gratings) is long compared to the period of vibration of the plate. This approach is successful because during a vibration cycle the plate is at its extreme positions, where the direction of motion is changing, most of the time.

Measurements of the vibratory amplitude of a plate experiencing harmonic motion where $w(t) = a \sin \omega t$ is made with dual beam speckle interferometer similar to the one shown in Fig. 14.12. This interferometer is arranged to provide a measurement of the out-of-plane amplitude of the vibration. The intensity of light on the image plane of the CCD camera is given by Eq. (14.30b) as:

$$I(t) = I_0 + I_R + 2\sqrt{I_0 I_R}\cos[\phi + \Delta\phi] \qquad (14.30b)$$

where I(t) is the intensity at some time t during the exposure interval.

$\Delta\phi = \dfrac{4\pi}{\lambda} w(t)$ and $w(t) = a \sin \omega t$ where a is the amplitude of the vibratory motion.

Eq. (14.30b) is integrated and averaged to account for the effect of the long exposure time. Hence, the average intensity $\bar{I}$ is given by:

$$\bar{I} = \frac{1}{t_e}\int_0^{t_e}\left[I_0 + I_R + 2\sqrt{I_0 I_R}\cos[\phi + \Delta\phi]\right]dt \qquad (14.36)$$

where t_e is the exposure time.

Performing the integration gives:

$$\bar{I} = I_0 + I_R + 2\sqrt{I_0 I_R}\, J_0^2\left(\frac{4\pi\, w(t)}{\lambda}\right)\cos\phi \qquad (14.37)$$

where J_0 is the zero order Bessel function.

Comparing Eq. (14.37) with Eq. (12.52) shows that the speckle fringe pattern will be similar to the holographic fringe pattern except for the presence of the $I_0 + I_R$ terms and the $\cos \phi$ multiplier. These terms combine to reduce the contrast in the speckle fringe pattern.

When J = 1, $(4\pi a/\lambda) = 0$ and the amplitude a = 0. Accordingly, the speckle pattern will exhibit a maximum intensity given by:

$$\bar{I}_{max} = I_0 + I_R + 2\sqrt{I_0 I_R}\cos\phi \qquad (14.38)$$

because $J_0 < 1$ for all other values of the amplitude.

The minimum intensities occur when $J_0 = 0$:

$$\bar{I}_{min} = I_0 + I_R \quad (14.39)$$

this corresponds to $(4\pi a/\lambda) = 2.405, 5.520, 8.654, 11.792$, etc.

It is clear that Eq. (14.37) proves that a speckle fringe pattern will form showing the vibration amplitude over the illuminated area of the plate. However, the reader is cautioned that the correspondence between the fringe order N and the amplitude of the vibration is **not** linear.

An example of a fringe pattern recorded using time averaged speckle interferometry is presented in Fig 14.13.

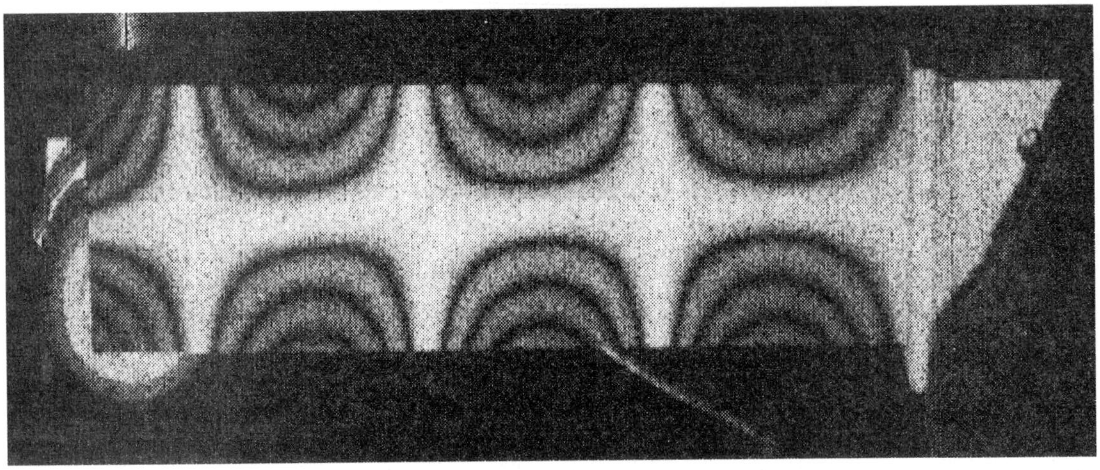

Figure 14.13 A fringe pattern representing the amplitude of a vibrating plate recorded with time-averaged speckle interferometry. Courtesy of Karl Stetson and HoloMetrology, LLC.
The image depicts the displacements due to the ninth mode of a flat titanium plate clamped at its right-hand side.

14.7.4 Discussion

Speckle size can be controlled within limits so that a compromise can be achieved between sensitivity and the ability to resolve speckles with a high-pixel-count digital camera. With speckle size of 10 to 20 μm on the image plane, CCD cameras with light sensing cells about 10 μm by 10 μm or smaller are adequate for use in real time or time-average modes of operation.

There are several advantages gained by converting from a conventional camera and film to a digital camera. First, the process is much faster as it is not necessary to develop film prior to viewing the results. With a digital system the fringe patterns or the speckle patterns can be viewed almost immediately on the computer monitor. This eliminates the dark room and the necessity for keeping film, photographic paper, developer and other chemicals in inventory. Second and more important, the fringe patterns can be enhanced with signal processing. While only signal subtraction was considered, several other image processing techniques can be employed to sharpen the fringes and to automatically extract data along specified lines and/or boundaries. Finally, exposure times are usually shorter with the CCD cameras and the isolation requirements for the optical table and fixtures are less stringent.

The disadvantage is that software is required for processing the image, which adds to the cost of the system and in time to learn to use the software effectively. However, this is a small inconvenience when measured in terms of the advantage of doing away with film, developers and the dark room.

14.8 PHASE STEPPING

Phase stepping was previously introduced in Section 13.10 where the optical response of a moiré fringe pattern was enhanced by shifting the reference (virtual) grating with respect to the specimen grating by a controlled amount. A similar phase stepping technique is employed in electronic speckle pattern interferometry to provide additional data that is extremely useful in determining the phase angle ϕ between the object and reference beams.

Phase stepping with speckle interferometry is accomplished by changing the optical path length of the reference beam by a controlled incremental amount. This approach is feasible only with electronic speckle pattern interferometry because of the ability of the operator to rapidly acquire a large number of phase stepped intensity distributions of the speckle pattern and to process them on a computer.

A typical optical arrangement for implementing phase stepping is illustrated in Fig. 14.14. The piezo-actuator is used to move the front surface mirror, which produces a change in the optical path length of the reference beam by very small controlled increments. The change in the optical path length results in a change in the phase angle ϕ between the reference and object beams. The intensity distribution of the speckle pattern is recorded after each phase step with a digital camera equipped with a suitable CCD array. Care must be exercised in recording the intensity of each speckle so that variations produced by phase stepping are measured with sufficient accuracy. The fidelity of the digital record depends on the pixels used to sample the intensity of the light for the individual speckles. If the speckle size is large relative to the pixel size of the CCD array, the sample is adequate and the fidelity of the record of the intensity distribution is ensured.

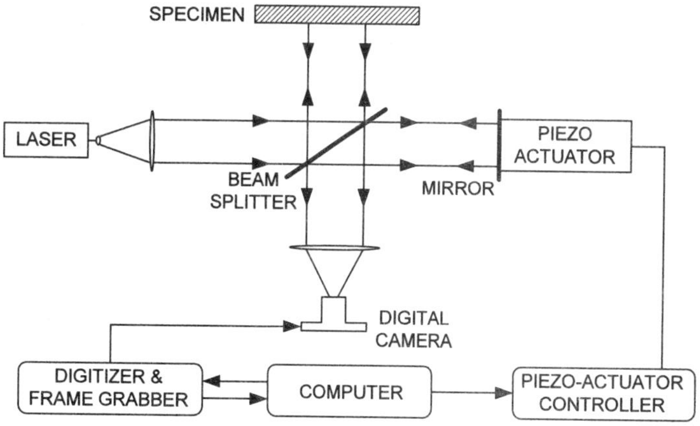

Figure 14.14 A speckle interferometer equipped with a piezo-actuated mirror to phase step the reference beam.

The experimental approach is to make several digital records of the intensity distribution with no load on the specimen. After each record, the piezo actuator is employed to adjust the length of the reference beam so that each record corresponds to a slightly different phase angle. The specimen is loaded and the process is repeated. The data from the digital camera is downloaded on a computer and analyzed to give the displacement of the surface of the specimen over the area illuminated by the object beam.

14.8.1 Data Analysis Methods

There are several mathematical techniques for analyzing the digital data acquired when phase stepping is used to provide additional intensity distributions. All of the techniques are based on the fundamental intensity equation for the two beam speckle interferometer, which is given by:

$$I(x,y) = I_0(x,y) + I_R(x,y) + 2\sqrt{I_0(x,y)I_R(x,y)}\cos\phi(x,y) \qquad (14.30a)$$

There are three unknowns in Eq. (14.30a)—$I_0(x,y)$, $I_R(x,y)$ and the phase angle between the reference and object beams ϕ (x,y); consequently, a minimum of three measurements of the distribution of the light intensity I(x,y) are necessary to solve for the three unknowns. Of course, only the phase angle ϕ is related to the unknown displacement distributions.

Two methods of analysis will be described in the following subsections. The first method is used to analyze three intensity patterns, I_1, I_2 and I_3 with three equal and known phase steps. The second method is used to analyze four intensity patterns. In this case, the phase is stepped by a constant value α between consecutive intensity records.

The Three Step Method

Equation (14.30a) is the basis for all of the mathematical approaches to measure the phase angle ϕ. It is clear that the unknowns I_0, I_R and ϕ are all functions of the coordinates x and y. However, to reduce the length of the equations that will be derived in subsequent paragraphs, the function indicator (x,y) is omitted and Eq. (14.30a) is written as:

$$I_i = I_0 + I_R + 2\sqrt{I_0 I_R}\cos\phi \tag{14.40}$$

where I_1 is the i th intensity distribution recorded with a phase step of α_i.

Consider the three intensity patterns recorded with $\alpha_1 = \pi/4$, $\alpha_2 = 3\pi/4$ and $\alpha_3 = 5\pi/4$. The intensity distributions for these three records are given by:

$$\begin{aligned} I_1 &= I_0 + I_R + 2\sqrt{I_0 I_R}\cos(\phi + \pi/4) \\ I_2 &= I_0 + I_R + 2\sqrt{I_0 I_R}\cos(\phi + 3\pi/4) \\ I_3 &= I_0 + I_R + 2\sqrt{I_0 I_R}\cos(\phi + 5\pi/4) \end{aligned} \tag{14.41}$$

Subtract I_2 from I_3 to obtain:

$$I_3 - I_2 = A\left[\cos(\phi + 5\pi/4) - \cos(\phi + \pi/4)\right] = \sqrt{2}A\sin\phi \tag{a}$$

where $A = 2\sqrt{I_0 I_R}$

Subtract I_2 from I_1 to obtain:

$$I_1 - I_2 = A\left[\cos(\phi + \pi/4) - \cos(\phi + 3\pi/4)\right] = \sqrt{2}A\cos\phi \tag{b}$$

Dividing Eq. (a) by Eq. (b) yields:

$$\phi = \tan^{-1}\frac{I_3 - I_2}{I_1 - I_2} \tag{14.42}$$

This relation provides a means for determining the phase between the object and the reference beams. However, it is necessary to adjust the phase angle ϕ because to account for the signs of $(I_3 - I_2)$ and $(I_1 - I_2)$ that determines the quadrant for the tangent function in Eq. (14.42). The adjustments to the phase angle are summarized in Table 14.1.

Table 14.1
Adjusting the Phase Angle

$I_3 - I_2$	$I_3 - I_2$	Adjusted Phase Angle ϕ	Range of ϕ
Positive	Positive	ϕ	0 to $\pi/2$
Positive	Negative	$\pi - \phi$	$\pi/2$ to π
Negative	Positive	$\pi + \phi$	π to $3\pi/2$
Negative	Negative	$2\pi - \phi$	$3\pi/2$ to 2π
Zero	Any Value	π	π
Positive	Zero	$\pi/2$	$\pi/2$
Negative	Zero	$3\pi/2$	$3\pi/2$

The adjusted phase angle will be used later to determine the surface displacements due to a specified load applied to the specimen.

The Four Step Method

This method is due to Carré [14] and differs from the three step method in that the phase steps may be any value. However, the steps must be equal to a constant value α and consecutive. Four intensity distributions are recorded with the intensities given by:

$$\begin{aligned} I_1 &= I_0 + I_R + 2A\cos(\phi - 3\alpha/2) \\ I_2 &= I_0 + I_R + 2A\cos(\phi - \alpha/2) \\ I_3 &= I_0 + I_R + 2A\cos(\phi + \alpha/2) \\ I_4 &= I_0 + I_R + 2A\cos(\phi + 3\alpha/2) \end{aligned} \tag{14.43}$$

Subtracting I_3 from I_2 yields:

$$I_2 - I_3 = A[\cos(\phi - \alpha/2) - \cos(\phi + \alpha/2)] = \sqrt{2}A\sin\phi \sin\alpha/2 \tag{a}$$

Subtracting I_4 from I_1 yields:

$$I_1 - I_4 = A[\cos(\phi - 3\alpha/2) - \cos(\phi + 3\alpha/2)] = \sqrt{2}A\sin\phi \sin 3\alpha/2 \tag{b}$$

Form the quantity $[3(I_2 - I_3) - (I_1 - I_4)]$ from Eqs. (a) and (b) to obtain:

$$[3(I_2 - I_3) - (I_1 - I_4)] = 2A[3\sin\phi \sin\alpha/2 - \sin\phi \sin 3\alpha/2] \tag{c}$$

This reduces to:

$$[3(I_2 - I_3) - (I_1 - I_4)] = 8A\sin\phi \sin^3\alpha/2 \tag{d}$$

Form the quantity $[(I_2 - I_3) + (I_1 - I_4)]$ from Eqs. (a) and (b) to obtain:

$$[(I_2 - I_3) + (I_1 - I_4)] = 2A[\sin\phi \sin\alpha/2 + \sin\phi \sin 3\alpha/2] \quad \text{(e)}$$

This reduces to:

$$[(I_2 - I_3) + (I_1 - I_4)] = 8A\sin\phi \sin\frac{\alpha}{2}\cos^2\frac{\alpha}{2} \quad \text{(f)}$$

Form the quantity $\{[3(I_2 - I_3) - (I_1 - I_4)][(I_2 - I_3) + (I_1 - I_4)]\}^{1/2}$ from Eqs. (d) and (f) and derive:

$$\{[3(I_2 - I_3) - (I_1 - I_4)][(I_2 - I_3) + (I_1 - I_4)]\}^{1/2} = 8A\sin\phi \sin^2\frac{\alpha}{2}\cos\frac{\alpha}{2} \quad (14.44)$$

Add I_2 to I_3 to obtain:

$$I_2 + I_3 = 2(I_0 + I_R) + A[\cos(\phi - \alpha/2) + \cos(\phi + \alpha/2)] \quad \text{(g)}$$

This reduces to:

$$I_2 + I_3 = 2(I_0 + I_R) + 2A\cos\phi\cos\frac{\alpha}{2} \quad \text{(h)}$$

Add I_1 to I_4 to obtain:

$$I_1 + I_4 = 2(I_0 + I_R) + A[\cos(\phi - 3\alpha/2) + \cos(\phi + 3\alpha/2)] \quad \text{(i)}$$

This reduces to:

$$I_1 + I_4 = 2(I_0 + I_R) + 2A\cos\phi\cos\frac{3\alpha}{2} \quad \text{(j)}$$

Form the quantity $[(I_2 + I_3) - (I_1 + I_4)]$ from Eqs. (h) and (i) and derive:

$$[(I_2 + I_3) - (I_1 + I_4)] = 8A\cos\phi \cos\frac{\alpha}{2}\sin^2\frac{\alpha}{2} \quad (14.45)$$

Divide Eq. (14.44) by Eq. (14.45) and simplify to obtain:

$$\phi = \tan^{-1}\frac{\sqrt{[3(I_2 - I_3) - (I_1 - I_4)][(I_2 - I_3) + (I_1 - I_4)]}}{(I_2 + I_3) - (I_1 + I_4)} \quad (14.46)$$

14.8.2 Calculating the Displacement

The adjusted phase angles are determined from records of the intensity distributions of the speckle patterns using the methods described in the previous subsections. The change in the phase angle due to the loading on the specimen is then given by:

$$\Delta\phi(x,y) = \phi_2(x,y) - \phi_1(x,y) \quad (14.47)$$

where ϕ_1 and ϕ_2 are the adjusted phase angles before and after loading, respectively.

The out-of-plane displacement w is related to $\Delta\phi$ by:

$$w = \frac{\Delta\phi(x,y)\lambda}{4\pi} \tag{14.48}$$

The factor 4 in the denominator of Eq. (14.48) assumes the optical arrangement shown in Fig. 14.14, where the direction of the incoming light and the light used in recording the pattern are both collinear with the outer normal of the specimen surface.

EXERCISES

14.1 Describe the mechanism for the formation of speckles on an object plane. Why do speckles form in space? What is the general shape of the speckles that form in space?

14.2 A photographic plate is positioned a distance L = 1.4 m from a specimen with a flat diffuse surface. A helium-neon laser projects a beam of light that illuminates an area on the plate with a diameter of 140 mm. Determine the size of the objective speckles formed on the photographic plate.

14.3 For the conditions in Exercise 14.2, determine the subjective speckle size. The lens used to photograph the speckle pattern has an aperture number f/a = 16. The lens is positioned to give a magnification ratio of M = 1.

14.4 Prepare a drawing showing the position of the lens relative to the specimen and the image plane for the speckle pattern if M = 1.0, 1.5 and 2.0.

14.5 Prepare a graph that characterizes the probability p(I) of the intensity of a speckle due to a single beam of coherent light reflecting from a diffuse surface. Reference this probability to the average intensity I_0 of the speckles in the pattern.

14.6 Prepare a graph that characterizes the probability p(I) of the intensity of a speckle due to a single beam of coherent light reflecting from a diffuse surface combined with a reference beam of coherent light. Reference this probability to the average intensity I_0 of the speckles in the pattern.

14.7 Prepare a drawing showing a pair of speckles that are separated but in correlation and another drawing of the speckle pair after decorrelation.

14.8 Determine the uniform out-of-plane displacement w that will cause decorrelation due to excessive speckle movement. The optical arrangement utilizes an argon laser, a lens with a numerical aperture of 1/16, a magnification ratio of 1/5, and a distance from the lens to the specimen of 400 mm. Consider a point on the specimen positioned a distance of 75 mm from the optical axis.

14.9 Determine the uniform in-plane displacement u that will cause decorrelation due to excessive speckle movement. The optical arrangement utilizes an argon laser, a lens with a numerical aperture of 1/16, a magnification ratio of 1/5, and a distance from the lens to the specimen of 400 mm.

14.10 Determine the out-of-plane rigid body rotation θ_x that will cause decorrelation due to excessive speckle movement. The optical arrangement utilizes an argon laser, a lens with a numerical aperture of 1/16, a magnification ratio of 1/5, and a distance from the lens to the specimen of 400 mm.

14.11 Determine the in-plane rigid body rotation θ_z that will cause decorrelation due to excessive speckle movement. The optical arrangement utilizes an argon laser, a lens with a numerical aperture of 1/16, a magnification ratio of 1/5, and a distance from the lens to the specimen of 400 mm.

14.12 Write an engineering brief describing the procedure for determining the displacement field of a plane specimen subjected to in plane loading using speckle photography.

14.13 What is a specklegram?

14.14 Describe the procedure for determining individual displacement components u and v from a specklegram.

14.15 Design the optical arrangement for making the measurements of u and v from a specklegram. Specify all of the optical components required and locate them on an optical table.

14.16 Design a calibration specimen that is to be used to prepare a set of double exposed specklegrams, which will be used as standards. These standards establish $f\lambda/y_1$ for the optical filtering procedure which gives the displacement components u and v.

14.17 Prepare a drawing of the optical arrangement of a typical speckle interferometer used for out-of-plane displacement measurements.

14.18 Design a speckle interferometer to measure both in plane and out-of plane displacements.

14.19 Using purchasing information available on the Internet, prepare a parts list with model numbers and prices for all of the components specified in the design of the speckle interferometer in Exercise 14.18.

14.20 Beginning with Eq. (14.21), verify Eq. (14.28).

14.21 A speckle interferometer arranged for out-of-plane measurements is employed to produce a fringe speckle fringe pattern where the fringe order N is proportional to the displacement component w. Determine this proportionality constant if a helium-neo laser is used as the coherent light source and the angle of inclination of the object beam in the interferometer is 45°.

14.22 A speckle interferometer, arranged for in-plane measurements, is employed to produce a speckle fringe pattern where the fringe order N is proportional to the displacement component u. Determine this proportionality constant if a helium-neo laser is used as the coherent light source and the angle of inclination of the two beams in the interferometer is 30°.

14.23 Describe the differences between speckle photography and speckle interferometry.

14.24 Describe the characteristics of a CCD camera and list the parameters that should be considered before purchasing a camera suitable for a wide range of speckle measurements. Describe the reasons for including each item on the listing.

14.25 The speckle interferometer with the digital additions, shown in Fig. 14.12, is arranged for out-of-plane measurements of the displacement fields. Modify this drawing so that the interferometer is arranged for in-plane measurements.

14.26 Begin with Eq. (14.30a), derive Eq. (14.32). Explain why signal subtraction improves the contrast of the speckle fringe pattern.

14.27 Discuss the effect of the cos ϕ term in Eq. (14.35). Is this a parameter that can be changed to improve the contrast? If so, how?

14.28 Given a choice of signal subtraction or signal additions to improve the contrast of a speckle fringe pattern, which technique would you select? Why?

14.29 Describe other image processing techniques that could be used with speckle methods to improve the quality of the optical images.

14.30 Equation 14.37, shown below, gives the intensity of the speckles in a time-averaged exposure.

$$\bar{I} = I_0 + I_R + 2\sqrt{I_0 I_R}\, J_0^2\left(\frac{4\pi\, w(t)}{\lambda}\right)\cos\phi$$

This equation is non-linear in the out-of-plane displacement being measured. As a consequence, a simple proportionality constant cannot be derived to establish the relation between the fringe order N and the amplitude of the vibratory motion. Prepare a table showing the constant that relates N to the amplitude of the vibratory motion for N = 0, 1, 2, 3 and 4. Recall that w(t) = a sin ωt and assume that a helium-neon laser was used as a light source.

14.31 Prepare an engineering brief for your instructor citing the advantages and disadvantages of digital photography applied to moiré, holographic and speckle measurements of displacements. Also, include photoelastic measurements of stresses in your discussion.

14.32 Beginning with Eq. (14.40), verify Eq. (14.42) for the three step method.

14.33 Beginning with Eq. (14.40), verify Eq. (14.46) for the four step method.

BIBLIOGRAPHY

1. Cloud, G. L.: Optical Measurements of Engineering Analysis, Cambridge University Press, Chapters 18-21, 1998.
2. Ennos, A.: "Speckle Interferometry," Laser Speckle and Related Phenomena, Topics in Applied Physics, Chapter 6, Vol. 9, Ed. J. C. Dainty, Springer-Verlag, Berlin, 1975.
3. Goodman, J. W.: "Some Fundamental properties of Speckle," Journal Optical Society of America, Vol. 66, pp. 1145-1150, 1976.
4. Cloud, G. A.: "Practical Speckle Interferometry for Measuring In-Plane Deformations," Applied Optics, Vol. 14, pp. 878-884, 1975.
5. Stetson, K. A.: "A review of Speckle Photography and Interferometry, Optical Engineering, Vol. 14, pp. 482-489, 1975.
6. Jones, R. and C. Wykes: Holographic and Speckle Interferometry, 2nd Edition, Cambridge University Press, Cambridge, 1989.
7. Vest, C. M.: Holographic Interferometry, John Wiley & Sons, New York, 1979.
8. Parks, V. J.: "The Range of Speckle Metrology," Experimental Mechanics, Vol. 20, pp. 181-191, 1980.
9. Leendertz, J. A.: "Interferometric Displacement Measurements on Scattering Surfaces Utilizing Speckle Effect," Journal of Physics E: Scientific Instruments, Vol. 3, p. 214, 1970.
10. Stetson, K. A.: New Design for Laser Image Speckle Interferometer," Optics and Laser Technology, Vol. 2, pp. 179-181, 1970.
11. Lu, B, H. Abendroth and H. Eggers: "Time Averaged Subtraction Method in Electron Speckle Pattern Interferometry," Optics Communication, Vol. 70, pp. 177-180, 1989.
12. Slettemoen, G. A.: "Optical Signal Processing in Electronic speckle Pattern Interferometry," Optic Communications, Vol. 23, pp. 213-216, 1977.
13. Hogmoen, K. and O. J. Lokberg: "Detection and Measurement of Small Vibrations Using Electronic Speckle Pattern Interferometry, Applied Optics, Vol. 16, pp. 1869-1875, 1977.
14. Carré, P.: "Installation et Utilisation du Comparateur Photoelectric et Interferentiel du Bureau International des Poids et Measures," Metrologia, Vol. 2, No. 1: pp. 13-23, 1966.

CHAPTER 15

DIGITAL IMAGE CORRELATION

15.1 INTRODUCTION

Several experimental methods including holographic interferometry, moiré and moiré interferometry and speckle methods for measuring in-plane and out-of-plane displacements have been described in previous chapters. Digital image correlation represents a different approach that is less demanding optically—ordinary incoherent light is sufficient, a vibration isolation table is not required, and the optical components such as beam splitters, prisms, spatial filters, piezoelectric actuators, etc. are eliminated. However, digital image correlation does require at least one high-resolution digital camera and for three-dimensional displacement measurements, two cameras are required. These cameras must be equipped with distortion free lenses to avoid lens induced errors. Also, a suitable computer with adequate memory and a frame grabbing circuit card to digitize the output from the camera is required.

In concept the method is simple. A digital camera[1] is employed to photograph the surface of a plane (two-dimensional) specimen as indicated in Fig. 15.1. The image is downloaded from the camera to a frame grabbing circuit card where the analog signals from the CCD array are digitized. These data are then stored in memory for subsequent processing.

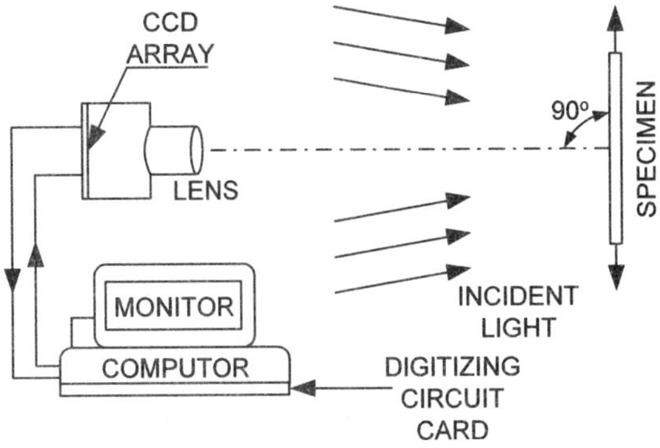

Figure 15.1 Typical arrangement of equipment used for displacement measurements with digital image correlation.

The surface of the specimen is covered with a target pattern, which is photographed before and after the specimen is deformed. The digital image of the specimen will contain intensity measurements at each pixel location on the CCD array of the specimen's surface before and after deformation. If target features are sufficient to identify a number of unique points and their precise locations in each of the images, the displacement field can be established. The correlation method has been developed and refined by a number of different researchers to locate unique points and to establish their locations in the x-y plane to within about 0.02 pixels.

[1] Usually a digital camera with a CCD array is used to capture the image. CCD arrays with small photosensitive cells and high pixel count have significant advantages.

15.2 LOCATING A TARGET FEATURE ON A PIXEL ARRAY

A digital camera with a CCD array records the intensity of light falling on a pixel. The array in a high quality digital camera is rectangular with a thousand or more pixels per line and a thousand or more lines per image. An example of a square[2] pixel array with 16 pixels in the x direction and 16 pixels in the y direction is shown in Fig. 15.2. A square target that is 3 × 3 pixels in size is shown at two different locations on this pixel array.

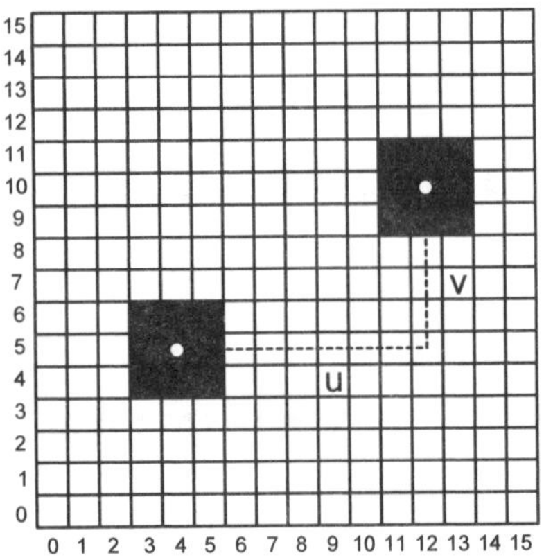

Figure 15.2 A 16 × 16 square pixel array with a 3 × 3 square target.

The location of the square target in the lower left represents the position of the target before deformation. Its center, shown with a small white dot, is located at x = 4 pixels and y = 5 pixels. After deformation, the target has moved in the pixel plane and its center is located at x = 12 pixels and y = 10 pixels. The displacement components are clearly u = 12 − 4 = 8 pixels and v = 10 − 5 = 5 pixels. This simple example shows that an image of a target on the pixel plane before and after deformation contains sufficient information to determine the in-plane displacement components u and v.

This example was simple because the target remained centered on the pixels grid and did not change shape during the deformation process. A more realistic example is considered next where the target does not remain centered on the pixel grid and the target moves during the deformation process. The target, shown in Fig. 15.3, is a plus sign formed by two 2 × 8 blocks of pixels crossed with respect to each other. The investigator observes a large black plus sign on a 10 × 10 field of pixels. The signal from the CCD array after it has been digitized gives a reading of the light intensity for each pixel. In Fig. 15.3, the intensity readings are shown as 0 for dark pixels and 100 for light pixels[3].

100	100	100	100	100	100	100	100	100	100
100	100	100	100	0	0	100	100	100	100
100	100	100	100	0	0	100	100	100	100
100	100	100	100	0	0	100	100	100	100
100	0	0	0	0	0	0	0	0	100
100	0	0	0	0	0	0	0	0	100
100	100	100	100	0	0	100	100	100	100
100	100	100	100	0	0	100	100	100	100
100	100	100	100	0	0	100	100	100	100
100	100	100	100	100	100	100	100	100	100

Figure 15.3 Left: Plus sign target centered on a 10 × 10 pixel array.
Right: Intensity readings at each pixel location that identify the plus sign.

---

[2] A square pixel is one with equal dimensions in the x and y directions. Most pixels are not square but have some aspect ratio.

[3] The actual reading will depend on the number of bits of resolution for the digital system. Eight bits gives 256 gray levels, 10 bits gives 1024 gray levels, etc.

Suppose that the specimen is deformed and the plus sign moves one pixel downward and one pixel to the left relative to the original 10 × 10 pixel array. The image of the plus sign after this movement is shown in Fig. 15.4. The new set of light intensity readings for each pixel which corresponds to the new location of the plus sign is also presented in Fig. 15.4.

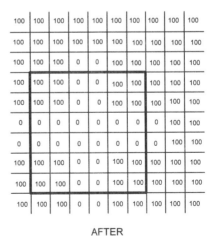

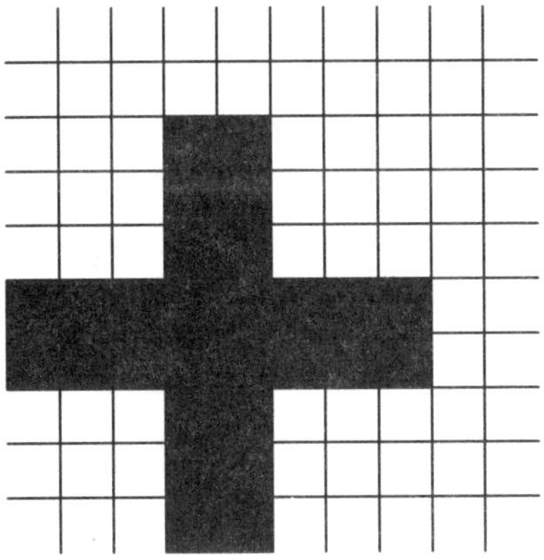

100	100	100	100	100	100	100	100	100	100
100	100	100	100	100	100	100	100	100	100
100	100	100	0	0	100	100	100	100	100
100	100	100	0	0	100	100	100	100	100
100	100	100	0	0	100	100	100	100	100
0	0	0	0	0	0	0	0	100	100
0	0	0	0	0	0	0	0	100	100
100	100	100	0	0	100	100	100	100	100
100	100	100	0	0	100	100	100	100	100
100	100	100	0	0	100	100	100	100	100

Figure 15.4 Left: Plus sign target after moving one pixel downward and one pixel to the left. Right: Intensity readings at each pixel location that identify the plus sign.

The problem is to locate the new location of the plus sign from the intensity data on the pixel array. This is usually accomplished with an intensity pattern matching method called **correlation**. This is usually started by defining a subset of pixels that surround the key features of the target in the initial intensity pattern as illustrated in Fig. 15.5. The subset used in this example is a 6 × 6 block of pixels that surrounds the center of the plus sign. The image recorded after the plus sign moved is then searched to find a corresponding 6 × 6 block of pixels with the same intensity pattern. After the 6 × 6 block has been located in the second image, the center of the target can be ascertained and the displacement components u and v measured in terms of pixels. There is systematic method for determining the location of the target as it moves due either to strain induced deformation or rigid body movement. However, before introducing the correlation method, it is important to discuss the requirements for the target used in these studies.

100	100	100	100	100	100	100	100	100	100
100	100	100	100	0	0	100	100	100	100
100	100	100	100	0	0	100	100	100	100
100	100	100	100	0	0	100	100	100	100
100	0	0	0	0	0	0	0	0	100
100	0	0	0	0	0	0	0	0	100
100	100	100	100	0	0	100	100	100	100
100	100	100	100	0	0	100	100	100	100
100	100	100	100	0	0	100	100	100	100
100	100	100	100	100	100	100	100	100	100

INITIAL

100	100	100	100	100	100	100	100	100	100
100	100	100	100	100	100	100	100	100	100
100	100	100	0	0	100	100	100	100	100
100	100	100	0	0	100	100	100	100	100
100	100	100	0	0	100	100	100	100	100
0	0	0	0	0	0	0	0	100	100
0	0	0	0	0	0	0	0	100	100
100	100	100	0	0	100	100	100	100	100
100	100	100	0	0	100	100	100	100	100
100	100	100	0	0	100	100	100	100	100

AFTER

Figure 15.5 A 6 × 6 block of pixels which surrounds the central region of the plus sign defines an intensity pattern on both images. Matching the two patterns establishes the target displacement.

15.3 TARGET FEATURE REQUIREMENTS

In the previous examples, a square and a plus sign were used as targets. Both of these shapes are excellent targets if the displacement is to be determined at a single point on the specimen. However, if whole field displacement data is required, these target shapes are not suitable. Suppose a set of 1000 squares were printed onto the surface of the specimen and digital images were taken before and after deforming the specimen. The experiment would be a failure because it would be impossible to distinguish one square from the other 999 squares. To identify specific points on the surface of the specimen, the target at that point must have a unique shape. Otherwise the image matching techniques, upon which the correlation method is based, cannot distinguish one target from another. Information (variations in intensities) useful in establishing the location of the target on both images is not available.

A speckle pattern[4], which covers the region of the surface of the specimen that is of interest, provides a whole field of unique targets. Each speckle has a unique shape and intensity and serves as an ideal target. A typical speckle pattern is illustrated in Fig. 15.6. There are several techniques for creating speckle patterns. The technique depends on the size of the speckles needed to conduct the experiment. If a small area (15 × 15 mm) on the specimen is to be examined, the image will be magnified and small speckles will be required. In this case, white paint is sprayed onto the surface of the specimen, and before the paint is dried carbon particles (copy machine toner) are sprinkled on the surface. These particles clump together in a random way to provide a unique set of targets.

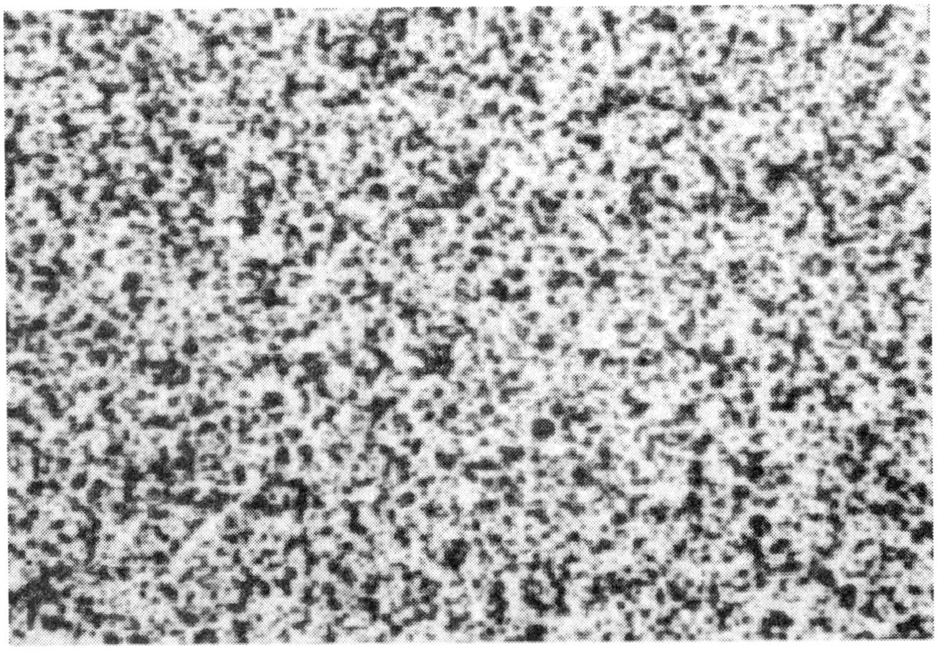

Figure 15.6 A speckle pattern serves to provide unique targets for image analysis.
Courtesy of Correlated Solutions, Inc.

If a moderate size area (50 × 50 mm) on the specimen is to be examined, the image will be about the same size as the CCD pixel array moderate size speckles will be required. In this case, white paint is sprayed onto the surface of the specimen and permitted to dry. Then a light pass of black paint is sprayed onto the surface to produce a suitable speckle pattern.

---

[4] While speckles formed by the interference of coherent light may be used for targets, easier methods for producing speckle patterns are employed. The usual procedure is to create a speckle pattern that can be viewed in ordinary white light by using paint spray or splatter.

If a large area (150 × 150 mm) on the specimen is to be examined, the image will be smaller that the object (magnification ratio less than one) and larger speckles will be required. In this case, white paint is sprayed onto the surface of the specimen and permitted to dry. The nozzle of the sprayer for the black paint is modified by increasing (doubling) its diameter. The spray from this modified nozzle produces a pattern of larger speckles more consistent with the requirements.

If a very large area (450 × 450 mm) on the specimen is to be examined, the image will be much smaller that the object (magnification ratio much less than one) and very larger speckle will be required. In this case, white paint is sprayed onto the surface of the specimen and permitted to dry. The black speckles are applied with a paint brush. A small quantity of paint is transferred from the brush to the specimen with a quick whipping motion. This motion generates a splatter of paint that forms relatively large speckles. Another approach is to print a speckle pattern on a sheet of plastic and then adhesively bonded to the surface of the specimen.

The size of the speckle is also important. The speckle size, measured in terms of the pixel dimensions on the CCD array, is dependent on the size of the pixel subset used in the sampling process. An illustration of different size speckles relative to a 3 × 3 pixel subset is presented in Fig. 15.7. The target (speckle) in Fig. 15.7a is too small as it is contained in one pixel. Because the gray value is determined by the average intensity over the area of a pixel, the deformation will not produce a change in the intensity of the pixel if the target remains within its boundaries. The size of the speckle in Fig. 15.7b is an improvement as it partially covers six of the nine pixels in the 3 × 3 pixel subset. The size of the speckle in Fig. 15.7c is even better as it totally covers one pixel and partially covers the remaining eight pixels in the subset. The size of the speckle in Fig. 15.7d is too large as a portion of the speckle falls outside the boundaries of the 3 × 3 pixel subset.

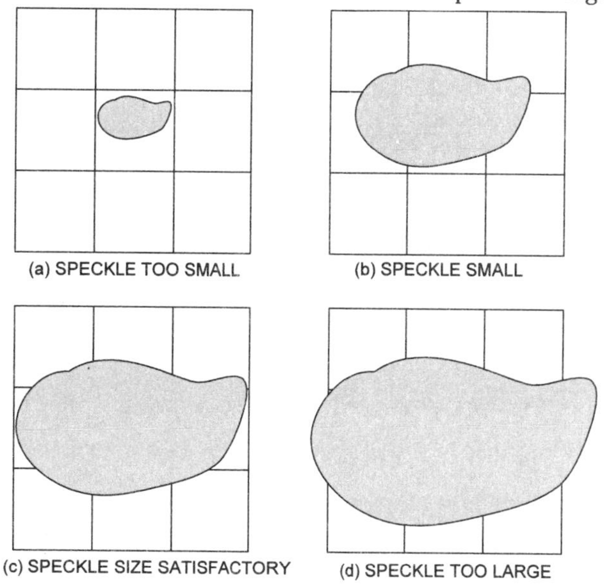

Figure 15.7 Illustration of speckle size relative to a 3 × 3 pixel subset.

The general rule is that the speckle or target size should cover most of the area of a 3 × 3 pixel array. Because both the speckle and pixel size are known prior to initiating an experiment, the magnification used to record the images can be adjusted to properly size the speckle on the camera's CCD array.

A final point that is important to consider in establishing the concept of speckle pattern matching—the algorithm used is not based on matching the pattern about a single speckle in its two states. Instead, the algorithm is based on matching the intensity field of many speckles (targets) over the field of view. The reason for this strategy in structuring the algorithm is evident by inspecting the patterns in Fig. 15.8 and Fig. 15.9. In Fig. 15.8, a single speckle is shown before and after deformation states on a 10 × 10 pixel array. Intensity data from most of the 100 pixels is not helpful as only 9 pixels show intensities less than the maximum value. In Fig. 15.9, three speckles are shown on the same 10 × 10 pixel array. In this case, the intensity data is much richer because intensity measurements differ from the maximum value on 27 of the 100 pixels shown. The additional data improves ones ability to determine the precise location of each of the three speckles to within a small fraction of a pixel. The correlation method employed to locate the positions of the speckles is presented in the next section.

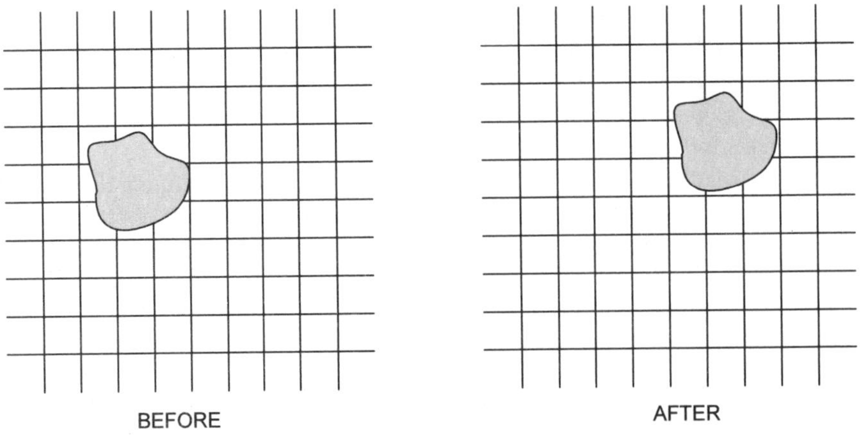

Figure 15.8 A single speckle before and after deformation referenced to a 10 × 10 pixel array.

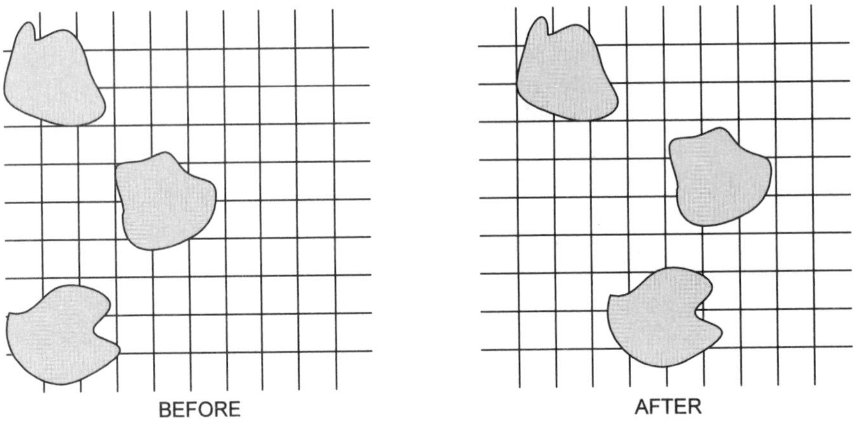

Figure 15.9 A speckle pattern before and after deformation referenced to a 16× 16 pixel array.

15.4 THE CORRELATION ALGORITHM

Digital image correlation involves comparisons of intensity data from two pixel fields. The first is the reference field representing the initial (undeformed) image. The second is a pixel field showing the deformed image of a specimen. A grid of nodes is established on the reference image and the displacement over the whole field is mapped with respect to these nodes. The calculations involved in the correlation process are performed at the node level.

15.4.1 Mapping Displacements

Fro a two dimensional displacement field, a point P(x, y) in the reference field is mapped into a corresponding point $P^*(\tilde{x}, \tilde{y})$ in the deformed field. Mapping the coordinates between the two fields is accomplished by using:

$$\begin{aligned} \tilde{x} &= x + u(x, y) \\ \tilde{y} &= y + v(x, y) \end{aligned} \qquad (15.1)$$

where u and v are the displacement components.

If it is assumed that the displacement components u and v can be approximated by a second order Taylor series expansion about some point $P(x_0, y_0)$, then Eq. (15.1) can be expressed as:

$$\tilde{x} = x_0 + u_0 + \frac{\partial u}{\partial x}\Delta x + \frac{\partial u}{\partial y}\Delta y + \frac{1}{2}\frac{\partial^2 u}{\partial x^2}\Delta x^2 + \frac{1}{2}\frac{\partial^2 u}{\partial y^2}\Delta y^2 + \frac{\partial^2 u}{\partial x \partial y}\Delta x \Delta y$$
$$\tilde{y} = y_0 + v_0 + \frac{\partial v}{\partial x}\Delta x + \frac{\partial v}{\partial y}\Delta y + \frac{1}{2}\frac{\partial^2 v}{\partial x^2}\Delta x^2 + \frac{1}{2}\frac{\partial^2 v}{\partial y^2}\Delta y^2 + \frac{\partial^2 v}{\partial x \partial y}\Delta x \Delta y \qquad (15.2)$$

where $\Delta x = x - x_0$ and $\Delta y = y - y_0$.

Examination of Eq. (15.2) shows that 12 mapping parameter were introduced in the Taylor series expansion. These include the displacement components u_0 and v_0 at point (x_0, y_0), the first order displacement gradients (slopes) $\frac{\partial u}{\partial x}, \frac{\partial v}{\partial x}, \frac{\partial u}{\partial y}$ and $\frac{\partial v}{\partial y}$ and the second order displacement gradients $\frac{\partial^2 u}{\partial x^2}, \frac{\partial^2 v}{\partial x^2}, \frac{\partial^2 u}{\partial y^2}, \frac{\partial^2 v}{\partial y^2}, \frac{\partial^2 u}{\partial x \partial y}$ and $\frac{\partial^2 v}{\partial x \partial y}$.

15.4.2 Smoothing the Intensity Data

The digital image of both the reference and deformed specimen consist of thousands of pixels with many different gray-scale values. Usually these values change abruptly from one pixel to the next as illustrated in Fig. 15.10.

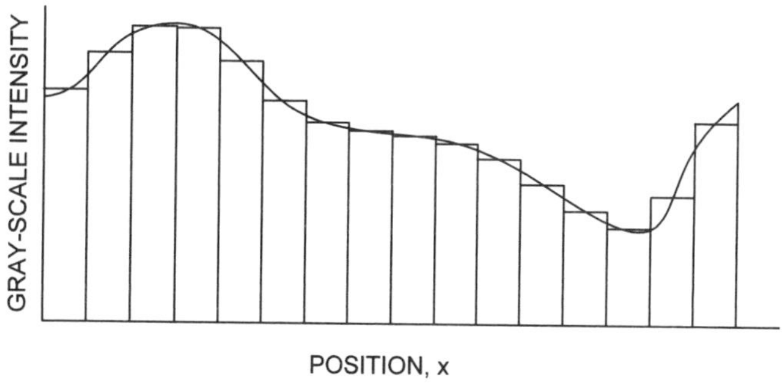

Figure 15.10 Fitting a smoothing function to discrete gray-scale values in the digital image.

These abrupt changes in the gray scale values from one pixel to the next produce significant mathematical difficulties when attempting to determine the mapping parameters in Eq. (15.2). To circumvent these difficulties, the data is smoothed over the entire field in both images. Several smoothing functions have been employed by different researchers including bi-linear interpolation, bi-cubic interpolation and bi-cubic spline interpolation. In this treatment, the bi-cubic spline interpolation is described.

Bi-cubic spline interpolation is a piecewise interpolation process where a set of fitting coefficients are determined for each interpolation region (a small subset of pixels). The gray-scale value of the intensity at any point in the interpolated region of the reference image is calculated from:

$$g(x, y) = \sum_{m=0}^{3} \sum_{n=0}^{3} a_{mn} x^m y^n \qquad (15.3)$$

where a_{mn} are the fitting coefficients for the bi-cubic spline.

For the deformed specimen, the bi-cubic spline interpolation function for its gray-scale image is written as:

$$h(\tilde{x}, \tilde{y}, P) = \sum_{m=0}^{3} \sum_{n=0}^{3} b_{mn} \tilde{x}^m \tilde{y}^n + \alpha \tag{15.4}$$

where b_{mn} are the fitting coefficients for the bi-cubic spline, α is an additional mapping parameter[5] and P is a vector with the 13 mapping parameters as its components.

15.4.3 The Correlation Coefficient

The 13 mapping parameters are determined by using a least squares **correlation coefficient**, which is defined as:

$$C = \frac{\sum_{S_p \in S} \{g(S_p) - h(S_p, P)\}^2}{\sum_{S_p \in S} g^2(S_p)} \tag{15.5a}$$

where S represents all of the points in the subset surrounding a single point S_p in that subset.

An alternate form of the equation for the correlation coefficient is given by:

$$C = 1 - \frac{\sum_{S_p \in S} g(S_p) h(S_p, P)}{\left[\sum_{S_p \in S} g^2(S_p) \sum_{S_p \in S} h^2(S_p, P)\right]^{1/2}} \tag{15.5b}$$

When the mapping parameters u, v, their first and second order derivatives and the brightness term α are exactly correct, the correlation coefficient $C \Rightarrow 0$. The best estimate of these mapping parameters is established by minimizing the value of C.

The Newton-Raphson method is used to solve for the 13 mapping parameters. To apply this method, an initial estimate P_0 is made for the 13 mapping parameters. The next step is to write C(P) as a truncated Taylor series around P_0 as:

$$C(P) = C(P_0) + \nabla C[P_0]^T [P - P_0] + ½ [P - P_0]^T \nabla\nabla C(P_0)[P - P_0] \tag{15.6}$$

Taking the gradient of Eq. (15.6) $\nabla C(P_0)$ and setting the result equal to zero is a mathematical operation for finding a solution for P_0 that minimizes C. Performing this operation on Eq. (15.6) yields the Newton Raphson equation as:

$$[\nabla\nabla C(P_0)(P - P_0)] = - [\nabla C(P_0)] \tag{15.7}$$

The gradient of C is given by:

$$\nabla C = \left(\frac{\partial C}{\partial P_i}\right)_{i=1,2,\ldots 13} = -\frac{2}{\sum_{S_p \in S} g^2(S_p)} \left\{\sum_{S_p \in S} g(S_p) - h(S_p, P) \frac{\partial h(S_p, P)}{\partial P_i}\right\}_{i=1,2,\ldots 13} = 0 \tag{15.8}$$

---

[5] This parameter, first introduced by Vendroux and Knauss, is used to adjust the brightness of the deformed image to more closely match that of the reference image. This term α accounts for differences in the intensity of the two images that may occur between the two different exposures.

The second order gradient of the correlation coefficient C(P), known as the Hessian matrix is given by:

$$\nabla\nabla C = \left(\frac{\partial^2 C}{\partial P_i \partial P_j}\right)_{i=1,13;\, j=1,13} = \begin{Bmatrix} -\dfrac{2}{\sum\limits_{S_p \in S} g^2(S_p)} \sum\limits_{S_p \in S} g(S_p) - h(S_p, P) \dfrac{\partial^2 h(S_p, P)}{\partial P_i \partial P_j} \\ + \dfrac{2}{\sum\limits_{S_p \in S} g^2(S_p)} \sum\limits_{S_p \in S} \dfrac{\partial h(S_p, P)}{\partial P_i} \dfrac{\partial h(S_p, P)}{\partial P_j}) \end{Bmatrix}_{i=1,13;j=1,13} \tag{15.9}$$

The initial estimate for the mapping parameters is based on the assumption that $g(x, y) \approx h(x, y, P)$, which leads to:

$$\sum_{S_p \in S} g(S_p) - h(S_p, P) \frac{\partial^2 h(S_p, P)}{\partial P_i \partial P_j} \approx 0 \tag{15.10}$$

By substituting Eq. (15.10) into Eq. (15.9), the Hessian matrix reduces to:

$$\nabla\nabla C = \left(\frac{\partial^2 C}{\partial P_i \partial P_j}\right)_{i=1,13;\, j=1,13} = \left\{ \frac{2}{\sum\limits_{S_p \in S} g^2(S_p)} \sum_{S_p \in S} \frac{\partial h(S_p, P)}{\partial P_i} \frac{\partial h(S_p, P)}{\partial P_j}) \right\}_{i=1,13;j=1,13} \tag{15.11}$$

The partial derivatives of the function $h(S_p, P)$ are functions of the displacement mapping and the bi-cubic interpretation of the intensity information in the deformed image. These partial derivatives for each of the 13 mapping parameters are determined by using the chain rule to write:

$$\frac{\partial h(S_p, P)}{\partial P_i} = \frac{\partial h(\tilde{x}, \tilde{y}, P)}{\partial \tilde{x}} \frac{\partial \tilde{x}(S_p)}{\partial P_i} + \frac{\partial h(\tilde{x}, \tilde{y}, P)}{\partial \tilde{y}} \frac{\partial \tilde{y}(S_p)}{\partial P_i} + \frac{\partial h(\tilde{x}, \tilde{y}, P)}{\partial P_i} \tag{15.12}$$

The relations used in determining the gradient $\dfrac{\partial h(S_p, P)}{\partial P_i}$ for each point S_p are given by:

$$\begin{gathered} \frac{\partial h}{\partial u} = \frac{\partial h}{\partial \tilde{x}} \quad \frac{\partial h}{\partial v} = \frac{\partial h}{\partial \tilde{y}} \quad \frac{\partial h}{\partial u_x} = \frac{\partial h}{\partial \tilde{x}} \Delta x \quad \frac{\partial h}{\partial v_x} = \frac{\partial h}{\partial \tilde{y}} \Delta x \\ \frac{\partial h}{\partial u_y} = \frac{\partial h}{\partial \tilde{x}} \Delta y \quad \frac{\partial h}{\partial v_y} = \frac{\partial h}{\partial \tilde{y}} \Delta y \quad \frac{\partial h}{\partial u_{xx}} = \frac{1}{2} \frac{\partial h}{\partial \tilde{x}} \Delta x^2 \quad \frac{\partial h}{\partial v_{xx}} = \frac{1}{2} \frac{\partial h}{\partial \tilde{y}} \Delta y^2 \\ \frac{\partial h}{\partial u_{yy}} = \frac{1}{2} \frac{\partial h}{\partial \tilde{x}} \Delta y^2 \quad \frac{\partial h}{\partial v_{yy}} = \frac{1}{2} \frac{\partial h}{\partial \tilde{y}} \Delta y^2 \quad \frac{\partial h}{\partial u_{xy}} = \frac{\partial h}{\partial \tilde{x}} \Delta x \Delta y \quad \frac{\partial h}{\partial v_{xy}} = \frac{\partial h}{\partial \tilde{y}} \Delta x \Delta y \quad \frac{\partial h}{\partial \alpha} = 1 \end{gathered} \tag{15.13}$$

where

$$u_x = \frac{\partial u}{\partial x}, v_x = \frac{\partial v}{\partial x}, u_{xx} = \frac{\partial^2 u}{\partial x^2}, v_{xx} = \frac{\partial^2 v}{\partial x^2}, u_{yy} = \frac{\partial^2 u}{\partial y^2}, v_{yy} = \frac{\partial^2 v}{\partial y^2}, u_{xy} = \frac{\partial^2 u}{\partial x \partial y}, v_{xy} = \frac{\partial^2 v}{\partial x \partial y}.$$

The terms $\frac{\partial h}{\partial \tilde{x}}$ and $\frac{\partial h}{\partial \tilde{y}}$ in Eq. (15.13) are related to the fitting coefficients used in the bi-cubic spline interpolation of the intensity data in the deformed image. Taking the gradients of Eq. (15.4) yields:

$$\begin{aligned}
\frac{\partial h}{\partial \tilde{x}} &= b_{10} + b_{11}\tilde{y} + b_{12}\tilde{y}^2 + b_{13}\tilde{y}^3 + 2b_{20}\tilde{x} + 2b_{21}\tilde{x}\tilde{y} + 2b_{22}\tilde{x}\tilde{y}^2 + 2b_{23}\tilde{x}\tilde{y}^3 \\
&\quad + 3b_{30}\tilde{x}^2 + 3b_{31}\tilde{x}^2\tilde{y} + 3b_{32}\tilde{x}^2\tilde{y}^2 + 3b_{33}\tilde{x}^2\tilde{y}^3 \\
\frac{\partial h}{\partial \tilde{y}} &= b_{01} + 2b_{02}\tilde{y} + 3b_{03}\tilde{y}^2 + b_{11}\tilde{x} + 2b_{12}\tilde{x}\tilde{y} + 3b_{13}\tilde{x}\tilde{y}^2 \\
&\quad + b_{21}\tilde{x}^2 + 2b_{22}\tilde{x}^2\tilde{y} + 3b_{23}\tilde{x}^2\tilde{y}^2 + b_{31}\tilde{x}^3 + 2b_{32}\tilde{x}^3\tilde{y} + b_{33}\tilde{x}^3\tilde{y}^2
\end{aligned} \tag{15.14}$$

Lu and Cary [7] have tested this method[6] using both first and second order deformations. In both tests, they used 8-bit gray scale bit map images. A subset consisting of a 41 × 41 square array of pixels centered about the nodal point under consideration was selected. For the first order deformation test, the second derivatives of the displacements vanished. The results for the displacements were resolved within ± 0.005 pixels and the displacement gradients were determined within ± 0.0002. The second order deformation test was more demanding, but the higher order theory with 13 mapping parameters enable Lu and Cary to predict second order displacement gradients to within ± 0.0002 per pixel. The strains were measured to within 0.13% of their true value.

Clearly, the digital correlation method can provide an accurate method of measuring displacements and strains using the mathematical approach described above. The mathematics may appear to be challenging for some readers, however, commercially available software[7] is available that performs the matrix manipulations required in the Newton-Raphson minimization process.

15.5 CALIBRATION

Calibration of the camera-specimen arrangement is required to provide numerical factors necessary in the analysis of the data and for the conversion of pixel count to units of displacement. For two-dimensional studies, four calibration factors must be determined.

1. The aspect ratio λ for the camera-digitizing circuit board combination.
2. The magnification M used in the experiment.
3. The location of the center of the lens relative to the center of the specimen.
4. The distortion coefficient for the lens.

15.5.1 Determining the Aspect Ratio λ

The aspect ratio is related to the vertical and horizontal dimensions of the light sensitive cells in the CCD array. It also depends on the design of the digitizing circuit board; however, λ is a constant for a specific combination of camera and digitizing circuit board. Hence, it is only necessary to measure this parameter one time.

---

[6] The images used in the test were "theoretical" and did not contain errors typically introduced when the images are acquired in a real experiment. Nevertheless, the tests did establish the power of the algorithms to establish the displacements and their first and second order gradients with great accuracy.

[7] Software is available from Correlated Solutions, Inc. West Columbia, SC and from LaVision Inc. Ypsilanti, MI.

Of the several techniques, which may be used to measure λ, two will be described here. The first technique involves the preparation of a target plate with a notch as shown in Fig. 15.11. Also shown in Fig. 15.11 is a group of pixels illustrating the size, shape and spacing of the light sensitive cells in the CCD array. The size of the target is such that it will cover (block the light) of several pixels as it is translated.

The target plate is mounted on a precision x-y stage[8] so that it can be translated by controlled amounts in the x and y directions. Images are recorded after a series of translations of Δx in the x direction and a series of translation of Δy in the y direction. Comparisons of the gray level intensities recorded in the pixels covered by translations of Δx and Δy, provide the data needed to determine the aspect ratio λ.

The second method for determining λ utilizes a calibration grid placed on the plane of the specimen. A contact measuring reticule, with a 10 × 10 mm grid divided into 100 squares, serves as a suitable grid. Accuracy of the spacing of the intersecting grid lines on a quality reticule is typically ± 4 μm with line thickness of about 30 μm. Many images of the grid are acquired at relatively high magnification so that the grid area fills several cells on the CCD sensor array. By using image analysis software, the locations of the grid points are established from image threshold intensities [9] and least square curve fitting. A large number (about 20) of these image points are selected, and a 5 × 5 pixel subset that surrounds each of these image points is analyzed. The value of the ratio of Δy/Δx of alternate diagonal vertices in the 5 × 5 subsets is measured. An example of the dimensions Δx and Δy are shown in Fig. 15.12. The value of the aspect ratio λ is then determined by averaging as indicated by:

$$\lambda = \frac{1}{k}\sum_{1}^{k}\frac{\Delta y}{\Delta x} \qquad (15.15)$$

where k is the number of image points considered in determining the average.

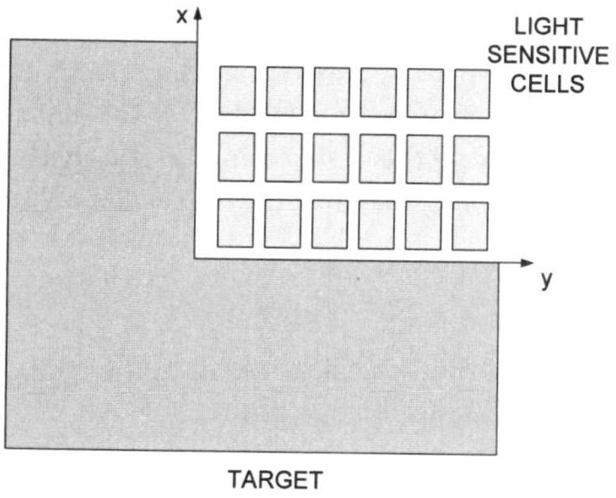

Figure 15.11 Notch target on a translating plate.

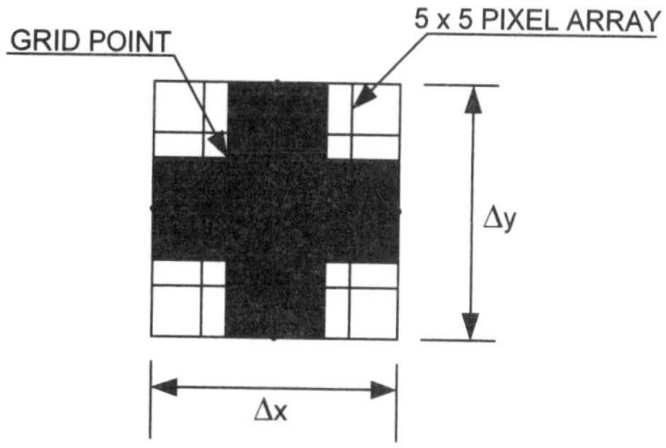

Figure 15.12 Dimensions of the 5 × 5 pixel array surrounding the grid point.

15.5.2 Determining the Magnification Factor

The magnification factor M relates a dimension on the specimen to a corresponding dimension on the image plane of the camera. Again, two methods of determining M are employed. One method uses a target plate with a high contrast speckle pattern that is translated by known increments in the x and y directions. Images of the speckle pattern are taken after each translation. The images are analyzed to determine the distance moved by a specific point before and after motion. This motion measured in

[8] Precision translation stages are capable of incremental translations Δx or Δy with a least count accuracy of ± 1.0 μm.

pixels is compared to the known translations and a magnification factor giving the number of pixels per mm of translation is determined.

A second approach is to attach a standard scale to the surface of the specimen. An image of the scale is recorded and an edge detecting routine [9] is employed to locate the pixel positions of two marks on the scale which have a known spacing. The grid on the measuring reticule described in Section 15.5.1 serves well as a standard scale to measure the magnification factor in both the x and y directions.

15.5.3 Locating the Lens Center

To reduce the effect of any lens distortion, the center of the camera lens should coincide with the center of the area of interest on the specimen. Centering the lens is easily accomplished by placing a black single point target on a white background at the center of the area of interest on the specimen. An image of the single point target is acquired and the intensities of the pixels near the center of the CCD array are monitored. The center pixels in the array should show the lowest intensities. If the centers do not coincide, either the camera or the specimen should be translated until the two centers coincide.

15.5.4 Lens Distortion

To minimize the effect of lens distortion, high-quality lenses with relatively long focal lengths are employed with the CCD cameras. Lenses with focal length ranging from 200 to 300 mm are commonly employed to minimize the effect of distortion. If sufficient light is available, the lens is stopped down so that only its central region is used in focusing the image on the CCD sensor array. Limiting the use of the lens to its central region greatly reduces the distortion effects.

15.5.5 Errors Due to out-of-Plane Displacement

As shown in Fig. 15.1, the specimen is mounted so that it is perpendicular to the optical axis and its plane is parallel to the CCD sensor array on the image plane of the camera. It is assumed that the specimen does not undergo any displacement or rigid body motions in the z direction as it is loaded. If there is a change in the distance of the camera to the specimen due to motion in the z direction, error in the measurement of pixel locations on the x-y plane is produced. This error E, measured in pixels, is given by:

$$E = r(w/L) \tag{15.16}$$

where w is the out of plane displacement; r is the radial distance from the optical axis to a point on the image plane measured in pixels and L is the distance from the specimen to the image plane.

15.6 DISCUSSION OF DIGITAL IMAGE CORRELATION—2-D

Digital image correlation is a promising non-contacting method for measuring whole field displacements. The correlation method described above, with the bi-cubic spline interpolation to smooth the gray scale levels and a correlation coefficient that accommodates 13 mapping parameters, provides an effective means to convert the digital images to whole field displacement maps. The mathematics involved for a typical experimentalist is challenging; however, when a suitable computer code is available the mapping process becomes automatic.

The accuracy of the method in the determination of displacements is often quoted at ± 0.02 pixels for each displacement component; however, use of the bi-cubic spline interpolation together with a mapping function that includes the second derivatives of the displacement field results in improved resolution. Lu and Cary [7], in their studies with theoretical images that included both first and second order derivatives of the displacements, achieved much better accuracies (± 0.005 pixels for translations)[9]. Also, with the expanded mapping function, the first derivatives of the displacements are obtained with significantly improved accuracies. Thus, strains can be measured with confidence when the second derivatives of the displacement field are included in the mapping function.

The variation of the intensity and distribution of the light on the surface of the specimen between exposures can cause problems in achieving good correlation. The variation of the intensity (not the distribution) is accommodated with the addition of the α term in the bi-cubic spline interpolation function shown in Eq. (15.4). However, care should be exercised to maintain a uniform field of light on the surface of the specimen that remains constant during the time required to complete the loading of the specimen and to acquire all of the digital images.

The method has many advantages, which include:

1. Preparation of the specimen is relatively simple as the speckle patterns used as unique targets are made by spray painting its surface.
2. While speckles are used as targets, decorrelation is not an issue as the speckle pattern is not generated by coherent light.
3. Ordinary white light is used to illuminate the surface of the specimen.
4. Specimen size is not an issue as the method can be applied to both large and small specimens. Specimen size is accommodated by changing the magnification used in recording the image.
5. The method is not overly sensitive to vibration and a vibration isolation table is not necessary. However, it is essential that both the specimen and the camera be stable during the time required to record the image.
6. Large strains or significant rigid body movement does not cause difficulties if the specimen does not move out of the field of view of the camera.

15.7 DIGITAL IMAGE CORRELATION—3-D

The measurement of the three displacement components on a specimen is much more difficult than two dimensional measurements. Stereo imaging techniques, which require two digital cameras, are employed. The typical camera arrangement is illustrated in Fig. 15.13. This drawing shows the pan angles θ_{p1} and θ_{p2} of the cameras, their spacing d_{12} and the distance s from the cameras to the specimen. Also defined in Fig. 15.13 are three coordinate systems—the one centered at the specimen O-XYZ is the reference system. The other two are coordinate systems are centered on the two camera lenses at C_1 and C_2.

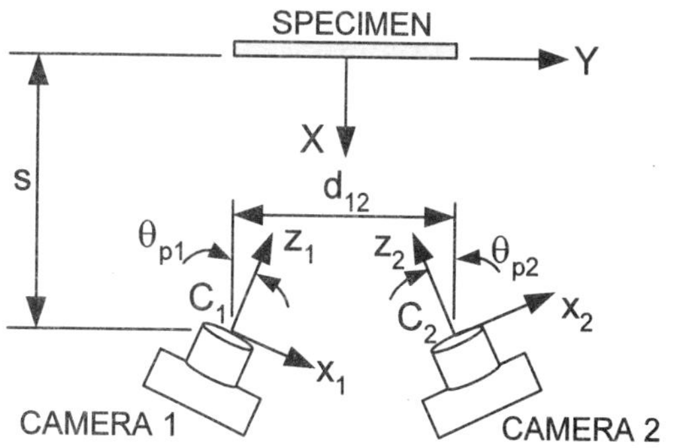

Figure 15.13 Stereo camera arrangement for 3-dimensional displacement measurements.

The procedure is to record images with both cameras of the initial state of the specimen and then to load the specimen and to acquire images with both cameras after the specimen has deformed. Before beginning the digital image correlation process for the 3-dimensional data recorded as gray scale images, it is necessary to develop the stereo imaging equation.

[9] The overall accuracy of the digital correlation method is a function of several parameters including the order of the interpolation scheme, the lens distortion, uniformity of the light distribution, out-of-plane displacements, quality of the speckle pattern, etc. Accuracies of 0.005 pixels are rarely achieved in actual experiments.

This development is due to P. F. Luo, Y. J. Chao, M. A. Sutton and W. H. Peters, which is described in References [10, 11].

15.8 STEREO IMAGING RELATIONS

The position of the camera relative to the specimen in the two-dimensional measurements was simple. The plane of the specimen was parallel to the image plane in the camera and the camera was centered on the area of interest on the specimen. However, for three-dimensional measurements, the simplicity of the camera-specimen arrangement vanishes and is replaced with a more complex arrangement with camera spacing, pan angles, camera stand-off and seven coordinate systems. A drawing showing the coordinate systems, specimen and camera images is presented in Fig. 15.14.

Figure 15.14 Drawing showing the coordinate systems used in analyzing stereo images.

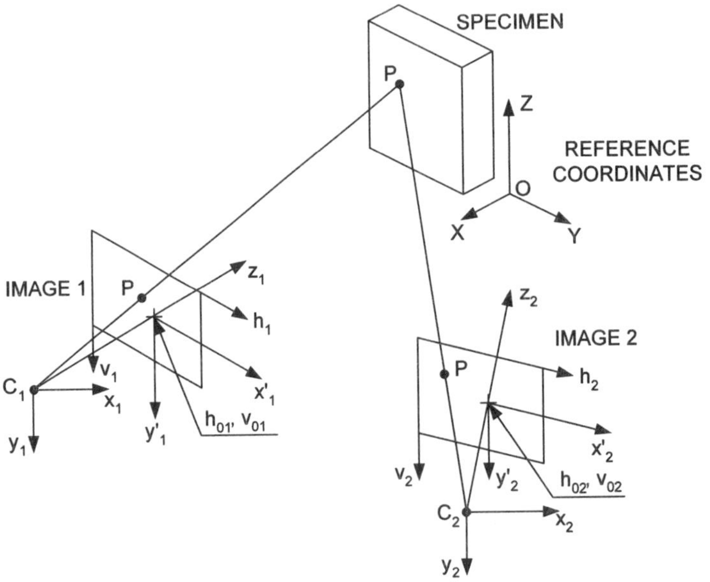

The drawings in Figs. 15.13 and 15.14 differ to some degree because the latter contains more detail. Definitions are provided below to clarify the various coordinates shown in Fig. 15.14.

The reference coordinates O-XYZ define all points in space. These coordinates may be positioned anywhere in space although it is common practice to place their origin on the specimen.

A coordinate system with its origin at C_i is defined for each camera, where C_i is the lens center of the camera. For camera #1, the coordinates are x_1, y_1 and z_1. Each point in the image plane of camera #1 has coordinates relative to C_1 of x_1, y_1 and f, where f is the focal point of the system. An identical coordinate system is established for camera #2.

Another coordinate system is placed on the image plane with coordinates x'_i, y'_i and 0. This system is identical to the camera system of coordinates except it is shifted in the positive z direction by the distance f.

A second coordinate system on the image plane with coordinates (h_i, v_i) is introduced with its origin at the upper left hand corner of the pixel array. This is the standard coordinate system for computer-image systems. The origin of the first image plane coordinate system for camera #1 relative to the computer image coordinate system is h_{01} and v_{01} as shown in Fig. 15.14.

Pixel coordinates for each camera are converted into physical dimension on the specimen by determining two magnification factors M_x and M_y for each camera.

15.8.1 The Rotation Tensor

Before interpreting the image from a camera that is rotated and translated relative to a line perpendicular to the specimen surface, it is necessary to develop equations that show the position of the point P on the image plane as defined in Fig. 15.14. To develop these relations, consider the coordinates and the position vectors for a single camera as shown in Fig. 15.15.

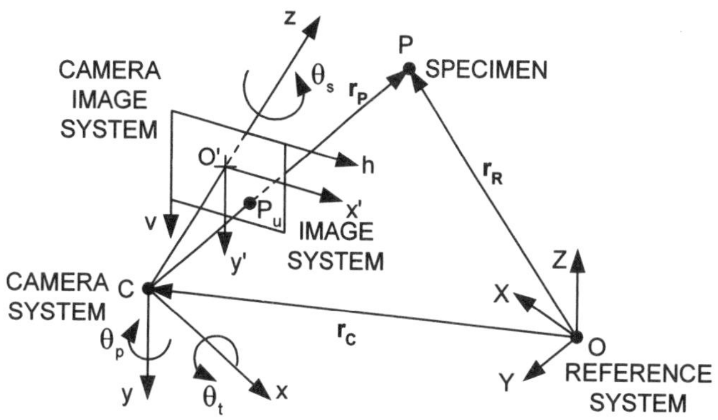

Figure 15.15 Position vectors defining an undistorted point P_u on the image plane.

The coordinate systems defined in Fig. 15.15 are the same as those shown in Fig. 15.14. However, in Fig. 9.14 only one camera is shown and the position vectors $\mathbf{r_R}$, $\mathbf{r_C}$ and $\mathbf{r_P}$ are defined. Also shown are the three angles of rotation of the camera—θ_p the pan angle or the angle of rotation about the y axis, θ_t the tilt angle or angle of rotation about the rotated x axis and θ_s the swing angle or the angle of rotation about the twice rotated z axis.

The position vectors for the point P on the specimen and the lens center C for the camera in terms of the reference coordinates are given by:

$$\mathbf{r_R} = \begin{Bmatrix} X_P \\ Y_P \\ Z_P \end{Bmatrix} \quad \text{and} \quad \mathbf{r_C} = \begin{Bmatrix} X_C \\ Y_C \\ Z_C \end{Bmatrix} \tag{15.17}$$

The position vector $\mathbf{r_P}$ is given in terms of camera coordinates as:

$$\mathbf{r_P} = \begin{Bmatrix} x \\ y \\ z \end{Bmatrix} \tag{15.18}$$

It is clear from Fig. 15.15 and Eq. (15.17) that $\mathbf{r_R} - \mathbf{r_C}$ is a position vector (in terms of reference coordinates) with its origin moved to the lens center C. To accommodate the shifting of the origin from O to C and the rotation of the coordinate system, a rotation matrix [R] is used. With this matrix, the position vectors may be written as:

$$\{\mathbf{r}\} = [R]\begin{Bmatrix} X \\ Y \\ Z \end{Bmatrix} \tag{15.19}$$

where X, Y and Z are the coordinates of a point in the reference system.

The four different rotation matrices are defined as:

$$[R_p] = \begin{bmatrix} \cos\theta_p & 0 & -\sin\theta_p \\ 0 & 1 & 0 \\ \sin\theta_p & 0 & \cos\theta_p \end{bmatrix}$$

$$[R_t] = \begin{bmatrix} 1 & 0 & 0 \\ 0 & \cos\theta_t & \sin\theta_t \\ 0 & -\sin\theta_t & \cos\theta_t \end{bmatrix}$$

$$[R_s] = 0\begin{bmatrix} \cos\theta_s & \sin\theta_s & 0 \\ -\sin\theta_s & \cos\theta_p & 0 \\ 0 & 0 & 1 \end{bmatrix} \qquad (15.20)$$

$$[R_r] = [R_t][R_s]$$

The matrix $[R_r]$ is employed to align the reference system with the image system of coordinates. For the coordinate directions shown in Figs. (15.14) and (15.15), it is clear that θ_t and θ_s are both equal to – 90°; hence, it can be shown that $[R_r] = 1$.

Because the relationship between the camera system and the reference system of coordinates is a rotation and a translation for any object point X_P, Y_P, and Z_P, it is possible to write:

$$\begin{Bmatrix} x \\ y \\ z \end{Bmatrix} = [R]\left(\begin{Bmatrix} X_P \\ Y_P \\ Z_P \end{Bmatrix} - \begin{Bmatrix} X_C \\ Y_C \\ Z_C \end{Bmatrix}\right) \qquad (15.21)$$

Equation (15.21) gives the coordinates of the point P in the camera's coordinate system in terms of the location of the point P on the specimen (X_P, Y_P, Z_P) and the location of the camera (X_C, Y_C, Z_C) relative to the reference system coordinates.

The final form of the relation between the camera coordinates and the reference coordinates is written as:

$$\begin{Bmatrix} x \\ y \\ z \end{Bmatrix}_{3\times1} = \begin{bmatrix} [R]_{3\times3} & [R]_{3\times3}\begin{Bmatrix} -X_C \\ -Y_C \\ -Z_C \end{Bmatrix} \end{bmatrix}_{3\times4} \begin{Bmatrix} X_P \\ Y_P \\ Z_P \\ 1 \end{Bmatrix}_{4\times1} \qquad (15.22)$$

The usual practice in the analysis of a three dimensional vision field is to use a pinhole camera model that assumes that all points are focused on the image plane through a pinhole centered at C. Then the point P on the specimen with coordinates X_P, Y_P, Z_P is imaged at position x'_u, y'_u, 0 on the image coordinate system as shown in Fig. 15.15. The transformation from the camera system to the image system of coordinates is given by:

$$\begin{Bmatrix} x'_u \\ y'_u \\ 1 \end{Bmatrix} = \frac{1}{z}\begin{bmatrix} f & 0 & 0 \\ 0 & f & 0 \\ 0 & 0 & f \end{bmatrix}\begin{Bmatrix} x \\ y \\ z \end{Bmatrix} \qquad (15.23)$$

In the three dimension applications, it is common practice to account for radial distortion due to imperfect lenses. The new positions x'_d and y'_d, which are corrected for image distortion, are given by:

$$x'_d = \frac{2x'_u}{1+\left[1-4\kappa\left(x'^2_u+y'^2_u\right)\right]^{1/2}} \quad \text{and} \quad y'_d = \frac{2y'_u}{1+\left[1-4\kappa\left(x'^2_u+y'^2_u\right)\right]^{1/2}} \tag{15.24}$$

where κ is the radial distortion coefficient for the camera lens.

Solving Eq. (15.24) for x'_u and y'_u yields:

$$x'_u = \frac{x'_d}{1+\kappa\left(x'^2_u+y'^2_u\right)} \quad \text{and} \quad y'_u = \frac{y'_d}{1+\kappa\left(x'^2_u+y'^2_u\right)} \tag{15.25}$$

Finally, the position of the point P can be written in terms of computer image coordinates as:

$$\begin{aligned} h &= M_x\, x'_d + h_0 \\ v &= M_y\, y'_d + v_0 \end{aligned} \tag{15.26}$$

The expression between the computer image coordinates and the reference coordinates includes:

1. Magnification factors M_x and M_y
2. The focal length of the lens f
3. The radial distortion coefficient κ
4. The computer image coordinates h_0 and v_0
5. The position of the camera lens X_C, Y_C, Z_C
6. The camera angles (θ_p, θ_t and θ_s).

The magnification factors M_x and M_y, the focal length of the lens f, the radial distortion coefficient κ, the computer image coordinates h_0 and v_0, are called intrinsic parameters because they depend only on the camera and its lens.

The position of the camera lens (for each of the two cameras) X_C, Y_C, Z_C and the camera angles (θ_p, θ_t and θ_s) are called extrinsic factors because they are dependent on the positioning of the camera and its orientation which are determined by the operator.

The preceding development was for a single camera although two cameras were illustrated in Fig. 15.14. The second camera is necessary because mapping a three-dimensional scene onto a two-dimensional image plane does not provide a unique solution. The image point (h, v) corresponds to a set of collinear three-dimensional points that lie on the projecting ray from the lens center through point (h, v). The missing depth information is obtained by using a second camera and stereoscopic imaging methods.

15.9 CALIBRATING THE STEREO IMAGING SYSTEM

The two cameras used in a stereo imaging system are illustrated in Fig. 15.13. The equations describing the mapping are the same for both cameras except that subscripts 1 and 2 are used with the camera and image coordinate systems to associate them with either camera #1 or #2. The stereo vision system enables the investigator to determine the coordinates (X, Y, Z) of a point P from its image points (h_1, v_1) and (h_2, v_2).

To calibrate the stereo imaging system, a plane specimen with a speckle pattern is mounted to a precision translation stage that enables incremental displacements u, v and w to be applied to the specimen. The procedure listed below is then followed to calibrate the system.

1. The speckle pattern is placed in the field of view of both cameras.
2. The center point of a 20 × 20 pixel region of the speckle pattern that is close to the center of the image for camera #1 is selected as the "calibration" point.
3. The location of the center point of the 20 × 20 pixel region[10] is established as the origin O of the reference system of coordinates.
4. Acquire an image of the speckle pattern on the specimen with each camera.
5. Translate the specimen by a known amount in each direction and record images in both cameras after each translation.
6. Repeat step number 5 N times to obtain statistically significant data that can be used in a digital correlation process.

To apply the two-dimensional image correlation process, first define (H_{1m}, V_{1m}) for camera #1 as the origin for the computer image coordinate system of the center point of the 20 × 20 pixel region. With this notation, the subscript m identifies the various positions of the subset after the m th translation of the specimen. The point (H_{2m}, V_{2m}) is defined as the origin for the computer image coordinate system of the center point of the same 20 × 20 pixel region in camera #2. Both of these positions correspond to the same point P(X_m, Y_m, Z_m) on the specimen with an accuracy of ± 1 μm, if a precision translation stage is used in the calibration process.

Next, choose the 20 × 20 subset centered at (H_{1m}, V_{1m}) and define it as S_1. Correlate it with the 20 × 20 pixel region for camera # 2 to determine its center point relative to camera #1. From this correlation, establish the distance (translation) between cameras and then compute the location of (H_{2m}, V_{2m}) on the computer coordinate system in camera #2.

After translating the specimen by a known amount, correlate S_1 with the 20 × 20 pixel region from the 2nd image of camera #1 to obtain the translation. Then compute the new location (H_{12}, V_{12}) of S_1. Next, correlate S_1 with the 20 × 20 pixel region from the 2nd image of camera #2 to determine its center point translation relative to camera #1. Using this translation, the location (H_{22}, V_{22})of S_1 in the computer coordinate system of camera #2 is calculated.

Repeat the process described in the paragraph above moving the specimen and correlating S_1 from image #1 with image #N from camera #1 to obtain (H_{1N}, V_{1N}). Also, continue to correlate S_1 with image #N from camera #2 to determine (H_{2N}, V_{2N}).

Experience has shown that the image coordinates of the subset S_1 is measured with an accuracy of about ± 0.05 pixels in each direction. When the set of N measurements of the position of point P on the specimen is established, then the directions of the axes in the reference coordinate system are determined.

The results of the N calibration measurements are employed with a non-linear least-squares method to determine the camera parameters for each camera. The error function to be minimized for each camera is given by:

$$E(k) = \sum_{m=1}^{N} \left[(H_{km} - h_m)^2 + (V_{km} - v_m)^2 \right] \qquad (15.27)$$

where k is the camera number and h_m and v_m are functions of the unknown camera parameters and the known positions of the calibration point P(X_m, Y_m, Z_m).

---

[10] The size of the 20 × 20 pixel region was selected to minimize the effects of image distortion, reduce the correlation time and to obtain a pattern that was statistically different from its neighbors.

This procedure provides a solution for the 12 camera parameters; however, some of the unknowns are coupled such as fM_x and fM_y. Hence, it is not possible to determine individual values of M_x, M_y and f.

15.10 DISPLACEMENT MEASUREMENTS WITH A STEREO IMAGING SYSTEM

After the calibration process for a two camera arrangement has been completed, it is possible to measure displacements components u, v and w of the specimen. The procedure is outlined below:

1. Select an arbitrary location (H_1, V_1) of subset S_1 in camera #1 as the first point of interest.
2. Using a digital image correlation program, determine the location (H_2, V_2) of the center point of subset S_1 in the second camera.
3. By using Eqs. (15.20) to Eq. (15.26) with the camera parameters and (H_1, V_1), it is possible to determine the point of intersection in the Y-Z plane of a vector from the origin C_1 that passes through the point (H_1, V_1).
4. Another point of intersection in the Y-Z plane is determined in a similar manner using the vector that passes through the origin C_2 and the point (H_2, V_2).
5. Steps 3 and 4 generate two vectors in space. Each vector initiates at the origin C_i and passes through an intersection point on the Y-Z plane. Unfortunately, due to small measurement errors, these two vectors do not intersect the Y-Z plane at precisely the same point.
6. The position of point P on the specimen is defined as the position in space where there is a minimum separation between these two vectors.
7. Steps numbered 1 through 6 are repeated to establish the position of the same point P before and after specimen deformation. The displacement of this point P is measured by subtracting its initial position from the position measured for the deformed specimen.

15.11 CONCLUSIONS

Whole field displacement measurements may be made that include the in-plane displacement components u and v as well as the out-of-plane displacement component w. The optical arrangement includes a second camera to provide stereo imaging capability. The digital images acquired by the two cameras are processed using the two-dimensional correlation method described in Section 15.4. However, relatively complex transformation equations are needed to locate the position of a point P on the specimen relative to the reference coordinate system. Deformations are obtained by establishing the location of point P before and after deformation and subtracting one set of reference coordinates from the other. Investigators who initially developed the system [10, 11] report accuracies of the calibration process of about 0.08 pixels with a standard deviation of about 0.06 pixels.

While the mathematics associated with stereo imaging equations is challenging, the calibration process is tedious and the number of unknown camera parameters (12) is large, the method can be applied with relative ease because of the availability of suitable commercial software. This software enables calibration to be performed automatically from the images collected during the calibration process. It also includes a program that performs the digital correlation, which locates a large number of points on the image planes of the two cameras. The next program included in the software performs the coordinate transformations that are needed to determine the location of points before and after deformation in terms of the reference coordinates. Finally, the results of the analysis, which includes displacement contours, and three-dimension graphs of displacement components and strains, can be displayed or printed out as graphs to include in engineering reports.

EXERCISES

15.1 Your manager asks you to procure the equipment to conduct a series of experiments to measure in-plane displacements on specimens that range in size from 50 × 50 mm to 250 × 250 mm. Using the information available on the Internet regarding the components shown in Fig. 15.1, prepare a list of the equipment that you would purchase. This list should include the specifications for the important features of each item of equipment. Also, include a discussion that justifies your selection of each of the components.

15.2 Prepare an engineering brief that describes the features required for the targets placed on the specimen's surface if the digital correlation method is used to measure displacement fields.

15.3 What is the most suitable size of a target (speckle) to be used with the digital correlation method to measure in-plane displacements? Discuss why the target size is important.

15.4 Using Eq. (15.2) as an example, write a first order Taylor series expansion about some point P(x, y) for the displacement components u and v relative to the mapping coordinates $\tilde{x}$ and $\tilde{y}$. How many mapping parameters are introduced in this expansion?

15.5 For the first order Taylor series expansion written as a solution to Exercise 15.4, describe the order of the displacement fields that can be mapped without introducing error.

15.6 Using Eq. (15.2) as an example, write a third order Taylor series expansion about some point P(x, y) for the displacement components u and v relative to the mapping coordinates $\tilde{x}$ and $\tilde{y}$. How many mapping parameters are introduced in this expansion?

15.7 For the third order Taylor series expansion written as a solution to Exercise 15.6, describe the order of the displacement fields that can be mapped without introducing error.

15.8 Explain why it is necessary to smooth the gray scale values in the digital image from a CCD camera.

15.9 Verify Eq. (15.14) by taking the derivatives of Eq. (15.4).

15.10 Discuss the merits of a commercially available software program that can be used to perform the mathematics associated with the two-dimensional digital image correlation method. Use the Internet to find the companies offering this product. Also, ascertain the order of the Taylor series expansion used in their algorithms and the smoothing function for converting the discrete data in the digital image to smooth data.

15.11 Design an experiment to determine the aspect ratio λ for the digital camera selected in Exercise 15.1.

15.12 Design an experiment to determine the magnification factor for a camera equipped with a 150 mm focal length lens. The specimen 75 mm × 75 mm in size is located 600 mm in front of the lens. Lens tubes may be employed to focus the image of the specimen on the CCD array.

15.13 Design a procedure to bring the center point of the area of interest on the specimen into coincidence with the center pixel in the image of the specimen on the CCD array.

15.14 During an experiment, to determine the in-plane displacement components, the specimen undergoes a small rigid body rotation of 1.4°. Determine the maximum error if the specimen is 100 mm wide, 200 mm long and is positioned 800 mm from the camera lens.

15.15 Prepare a PowerPoint presentation to be presented to your manger that supports a decision to develop an experimental capability to determine in-plane displacements using digital image correlation.

15.16 Prepare a PowerPoint presentation to be presented to your manger that opposes a decision to develop an experimental capability to determine in-plane displacements using digital image correlation.

15.17 What is the fundamental reason for the difficulty in making out-of-plane displacement measurements using digital image correlation?

15.18 Prepare a sketch of the experimental arrangement for three-dimensional digital image correlation measurements of both in-plane and out-of-plane displacement measurements.

15.19 Evaluate the rotational matrix $[R_p]$ if θ_p = 10°, 20°, 30° and 45°.

15.20 Show that $[R_r] = 1$ if $\theta_t = \theta_s = -90°$.
15.21 Describe the purpose of Eqs. (15.17) to (15.26).
15.22 Describe the procedure for calibrating a stereo imaging system.
15.23 Describe the procedure for making displacement measurements with a stereo imaging system.
15.24 Compare two- and three-dimensional digital image correlation methods for measuring displacements. Cite the advantage and disadvantage of each method. Compare the digital image correlation methods with other experimental methods for measuring both in-plane and out-of-plane measurements.

BIBLIOGRAPHY

1. Peters, W. H. and W. F. Ranson, "Digital Imaging Techniques in Experimental Stress Analysis," Optical Engineering, Vol. 21, pp. 427-432, 1982.
2. Sutton, M. A., M. Cheng, W. H. Peters, Y. J. Chao, and S. R. McNeil, "Applications of an Optimized Digital Image Correlation Method to Planar Deformation Analysis," Image Vision Computing, Vol. 4, pp. 143-150, 1986.
3. Bruck, H. A., S. R. McNeil, M. A. Sutton, and W. H. Peters, "Digital Image Correlation Using Newton Raphson Method of Partial Differential Correction," Experimental Mechanics, Vol. 29, pp. 261-267, 1989.
4. Vendroux, G. and W. G. Knauss, "Submicron Deformation Field Measurements: Part 2, Improved Digital Correlation," Experimental Mechanics, Vol. 38, pp. 86-91, 1998.
5. Spath, H., Two Dimensional Spline Interpolation Algorithms, A. K. Peters, Wellesley, MA, 1995.
6. Gerald, C. F. and P. O. Wheatley, Applied Numerical Analysis, Addison-Wesley, Reading, MA, 1994.
7. Lu, H. and P. D. Cary, "Deformation Measurements by Digital Image Correlation: Implementation of a Second-order Displacement Gradient, Experimental Mechanics, Vol. 40, pp. 393-400, 2000.
8. Chu, T. C., W. F. Ranson, M. A. Sutton, and W. H. Peters, "Applications of Digital Image Correlation Techniques to Experimental Mechanics, Experimental Mechanics, Vol. 25, pp. 232-244, 1985.
9. Morimoto, Y., "Digital Image Processing," Chapter 21, Handbook on Experimental Mechanics, 2nd Revised Edition, Edited by A. S. Kobayashi, VCH Publishers, New York, pp. 969-1027, 1993.
10. Lou, P. F., Y. J. Chao, M. A. Sutton and W. H. Peters, "Accurate Measurements of Three-Dimensional Deformations in Deformable and Rigid Bodies Using Computer Vision," Experimental Mechanics, Vol. 33, pp. 123-132, 1993.
11. Lou, P. F., Y. J. Chao and M. A. Sutton, "Application of Stereo Vision to Three-Dimensional Deformation Analysis in Fracture Experiments," Optical Engineering, Vol. 33, pp. 991-990, 1994.
12. Sobel, I. E., "ON Calibrating Computer Controlled Cameras for Perceiving 3D Scenes," Artificial Intelligence, Vol. 5, pp. 185-198, 1974.
13. Kahn-Jetter, Z. L. and T. C. Chu, "Three-dimensional Displacement Measurement Using Digital Image Correlation and Photogrammic Analysis," Experimental Mechanics, Vol. 30, pp. 10-16, 1990.
14. Tsai, R. Y., Á Versatile Camera Calibration Technique for High Accuracy 3D Machine Vision Metrology Using Off-the-shelf TV Cameras and Lenses," IEEE Journal of Robotics and Automation, Vol. 4, pp. 323-344, 1987.
15. Anon, Manual of Photogrammetry, 4th Edition, American Society of Photogrammetry, Falls Church, VA, 1980.
16. Duda, R. O. and P. E. Hart, "Perspective Transformations," Chapter 10, Pattern Classification on Scene Analysis, John Wiley & Sons, New York, pp. 379-404, 1973.
17. Chao, Y.J. and M. A. Sutton, "Accurate Measurement of Two and Three-Dimensional Surface Deformations for Fracture Specimens by Computer Vision," Chapter 3, Experimental Techniques

in Fracture, Vol. 3, J. Epstein, (ed.), Society of Experimental Mechanics, VCH Publishers, New York, pp. 59-93, 1993.

18. Sutton, M. A., S. R. McNeill, J. D. Helm and Y. J. Chao, "Advances in Two-Dimensional and Three-Dimensional Computer Vision," Photomechanics, Editor, P. K. Rastogi, Springer Verlag, pp. 323-372, 1999.
19. Sharpe, Jr., W. N. (Ed.): Springer Handbook of Experimental Solid Mechanics, Springer Science + Business Media LLC, New York, 2008.
20. Sutton, M.A., J. Orteu and H.W. Schreier, Image Correlation for Shape, Motion and Deformation Measurements, Springer Science + Business Media, New York, 2009.

CHAPTER 16

OPTICAL METHODS FOR DETERMINING FRACTURE PARAMETERS

16.1 INTRODUCTION

When a structure or machine component contains a flaw, such as a crack, stresses in the local neighborhood of the crack tip are singular. Because these local stresses exceed both the yield strength and the ultimate tensile strength of the material for very small loads, the usual approach for predicting failure loads based on von Mises or Tresca failure theories is not feasible. Instead, one determines a fracture parameter, such as a stress intensity factor K_I, and compares this parameter to the toughness of the material K_c as described previously in Section 4.1. This method of structural analysis uses fracture mechanics to predict if the crack will be stable under an applied load or if it will become unstable by initiating and then propagating to cause an abrupt failure.

In applying fracture mechanics in structural analysis it is necessary to determine the pertinent fracture parameter, such as K_I, as a function of the applied load. In some simple bodies, K_I can be determined by using handbook [1] relationships. For more complex structures, it is necessary to use either finite-element methods or experimental procedures to determine the fracture parameter-load relationship. The use of strain gages in determining K_I and K_{II} was covered in Sections 8.6 to 8.8. In this chapter, optical methods for determining the fracture parameters will be described. In this treatment, emphasis is placed on the photoelastic method because the isochromatic fringe pattern in the local region near the crack tip (see Fig. 16.1) provides a rich field of data that enables an accurate determination of the fracture parameter to be made.

Figure 16.1 Isochromatic fringes at the crack tip in an SEN specimen.

16.2 REVIEW OF IRWIN'S METHOD TO DETERMINE K_I

Post and Wells [2, 3] were the first researchers to show the application of photoelasticity to fracture mechanics. However, Irwin [4], in a discussion of reference [3], developed a relation for the opening-mode, stress-intensity factor K_I in terms of the geometric characteristics of the fringe loops near the tip of a crack as illustrated in Fig. 16.2. Irwin began by modifying the very-near-field relations, Eq. (4.36),

$$\begin{aligned}\sigma_{xx} &= \frac{K_I}{\sqrt{2\pi r}}\cos\frac{\theta}{2}\left(1-\sin\frac{\theta}{2}\sin\frac{3\theta}{2}\right)-\sigma_{0x}\\ \sigma_{yy} &= \frac{K_I}{\sqrt{2\pi r}}\cos\frac{\theta}{2}\left(1+\sin\frac{\theta}{2}\sin\frac{3\theta}{2}\right)\\ \tau_{xy} &= \frac{K_I}{\sqrt{2\pi r}}\cos\frac{\theta}{2}\sin\frac{\theta}{2}\cos\frac{3\theta}{2}\end{aligned} \tag{16.1}$$

where the stress σ_{0x} was subtracted from the expression for σ_{xx} to provide another degree of freedom in bringing the theoretical fringe loops into correspondence with the experimentally observed loops.

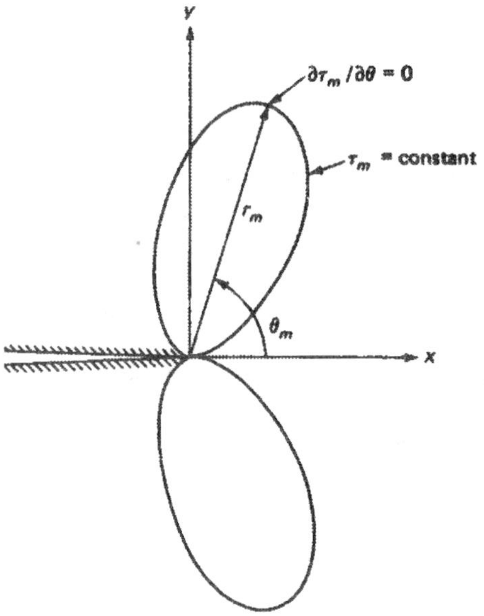

Figure 16.2 Characteristic geometry of an isochromatic fringe loop near the crack tip. Opening mode loading.

The maximum shear stress τ_m is expressed in terms of the Cartesian stress components as:

$$(2\tau_m)^2 = (\sigma_{yy}-\sigma_{xx})^2 + (2\tau_{xy})^2 \tag{16.2}$$

From Eqs. (16.1) and (16.2), it is apparent that:

$$(2\tau_m)^2 = \frac{K_I^2}{2\pi r}\sin^2\theta + \frac{2\sigma_{0x}K_I}{\sqrt{2\pi r}}\sin\theta\sin\frac{3\theta}{2} + \sigma_{0x}^2 \tag{16.3}$$

Next, Irwin observed the geometry of the fringe loops and noted that:

$$\frac{\partial\tau_m}{\partial\theta} = 0 \tag{16.4}$$

at the extreme position on the fringe loop where $r = r_m$ and $\theta = \theta_m$. Differentiating Eq. (16.3) with respect to θ and using Eq. (16.4) gives:

$$\sigma_{0x} = \frac{-K_I}{\sqrt{2\pi r}}\left[\frac{\sin\theta_m \cos\theta_m}{\cos\theta_m \sin(3\theta_m/2) + \frac{3}{2}\sin\theta_m \cos(3\theta_m/2)}\right] \tag{16.5}$$

The two unknown parameters K_I and σ_{0x} are determined from the complete solution of Eqs. (16.3) and (16.5) as:

$$\sigma_{0x} = \frac{-2\tau_m \cos\theta_m}{\cos(3\theta_m/2)\left[\cos^2\theta_m + \frac{9}{4}\sin^2\theta_m\right]^{1/2}} \tag{16.6}$$

and

$$K_I = \frac{2\tau_m\sqrt{2\pi r}}{\sin\theta_m}\left[1+\left(\frac{2}{3\tan\theta_m}\right)^2\right]^{-1/2}\left[1+\frac{2\tan(3\theta_m/2)}{3\tan\theta_m}\right] \tag{16.7}$$

The term τ_m in Eq. (16.7) is determined from the isochromatic data since the maximum shear stress is given by:

$$\tau_m = \frac{Nf_\sigma}{2h} \tag{16.8}$$

Etheridge and Dally [5] showed that Irwin's two-parameter (K_I and σ_{0x}) method predicted K_I to within 5% of the solution for the central crack problem provided $73° < \theta_m < 139°$ and the data used in Eqs. (16.6) and (16.7) was exact. Outside this range for θ_m, the errors increase rapidly and the Irwin approach is not applicable because Eqs. (16.1) do not represent the stress field.

Irwin's approach is sometimes called the **apogee method** and requires only a single data point where N, r_m and θ_m are prescribed for a given load. Unfortunately, it is difficult to measure r_m and θ_m with precision and small errors in these dimensions can lead to large errors in determining K_I.

16.3 MODIFICATIONS OF IRWIN'S TWO-PARAMETER METHOD

There have been several modifications of Irwin's original approach to improve accuracy by using data from more than one fringe loop. Bradley and Kobayashi [6] let $\sigma_{0x} = \delta K_I/\sqrt{\pi a}$, which permitted Eq. (16.3) to be simplified to:

$$2\tau_m = K_I\, g(\theta, r, a) \tag{16.9}$$

where

$$g(\theta, r, a) = \left[\sin^2\theta + 2\delta\left(\frac{2r}{a}\right)^{1/2}\sin\theta\sin\left(\frac{3\theta}{2}\right) + \frac{2r\delta^2}{a}\right]^{1/2}$$

and 2a is the length of the crack as shown in Fig. 4.1.

Data is then taken from two fringe loops (r_1, θ_1) and (r_2, θ_2) with the fringe order $N_2 > N_1$. Substituting this data into Eq. (16.9) and solving for K_I gives:

$$K_I = \frac{f_\sigma (2\pi r_1 r_2)^{1/2} (N_1 - N_2)}{h\left(g_2\sqrt{r_1} + g_1\sqrt{r_2}\right)} \tag{16.10}$$

The relation for g(θ, r, a) is often simplified by letting δ = 1; however, this simplification has never been justified.

A different modification of the Irwin method was developed by Schroedl and Smith [7] by restricting the data to a line defined by θ = 90°. With this restriction Eq. (16.3) reduces to:

$$(2\tau_m)^2 = \frac{K_I^2}{2\pi r} + \frac{\sqrt{2}K_I\sigma_{0x}}{\sqrt{2\pi r}} + \sigma_{0x}^2 \tag{16.11}$$

Solving Eq. (16.11) for K_I and retaining only the positive root from the quadratic formula gives:

$$K_I = \sqrt{\pi r}\left[\left(8\tau_m^2 - \sigma_{0x}^2\right)^{1/2} - \sigma_{0x}\right] \tag{16.12}$$

Smith simplified Eq. (16.12) by neglecting σ_{0x}^2 relative to $8\,\tau_m^2$ to obtain:

$$K_I = \sqrt{\pi r}\left[\sqrt{2}(2\tau_m) - \sigma_{0x}\right] \tag{16.13}$$

By adopting the Bradley-Kobayashi differencing technique, Smith uncouples the K_I and σ_{0x} relation. Using τ_m from the ith and jth fringe loops gives:

$$K_I = \sqrt{\pi r_i}\,\frac{(2\tau_m)_i - (2\tau_m)_j}{1 - \left(r_i/r_j\right)^{1/2}} \tag{16.14}$$

In application, K_I is determined from Eq. (16.14) for all possible permutations of pairs of fringe loops to give several different values of K_I. Then the mean $\overline{K}_I$ and standard deviation S_K are determined. The results are conditioned by eliminating all the values of K_I outside ± S_K and recomputing $\overline{K}_I$ from the remaining K_I values.

All three of the two-parameter methods are applicable for determining K_I in the range 73° < θ_m < 139° provided r_m/a < 0.03. If no measurement errors are made in r_m or θ_m, the two parameter methods will predict K_I with an accuracy of ± 5 percent.

When two or more fringe loops occur at the crack tip and if the radii can be measured with better than 2% accuracy, then the Bradley-Kobayashi shear-stress differencing method provides the most accurate results. When the measurement errors exceed 2%, the differencing technique magnifies these errors and Irwin's method produces more accurate estimates of K_I. However, the Irwin method is slightly more sensitive to errors in the measurement of θ_m than the Bradley-Kobayashi method.

The errors in θ_m determinations can be eliminated in both of the differencing methods since any fixed value of θ can be employed. However, at least part of this advantage is offset by the error introduced in making the second r measurement.

If only one fringe loop is available for analysis, Irwin's method must be used because it is the only two parameter method which uses data from a single fringe loop.

16.4 HIGHER PARAMETER METHODS FOR DETERMINING K_I

The use of the two-parameter methods requires data to be taken close to the crack tip with $r_m/a < 0.03$ so that Eq. (16.1) is a valid representation of the stress field. There are two difficulties associated with restricting $r_m/a < 0.03$. First, for the region $r/h < 0.5$ the stress state is three dimensional and the plane-stress assumptions used in deriving Eqs. (16.1) and (16.3) are not valid. To avoid errors due to the three-dimensional state of stress at the crack tip, it is important that $0.03a > r_m > h/2$. Clearly, the requirement on model thickness ($h < 0.06a$), for a valid region restricts where data can be taken. This concept of a valid region for the two-parameter methods of analysis is illustrated in Fig. 16.3.

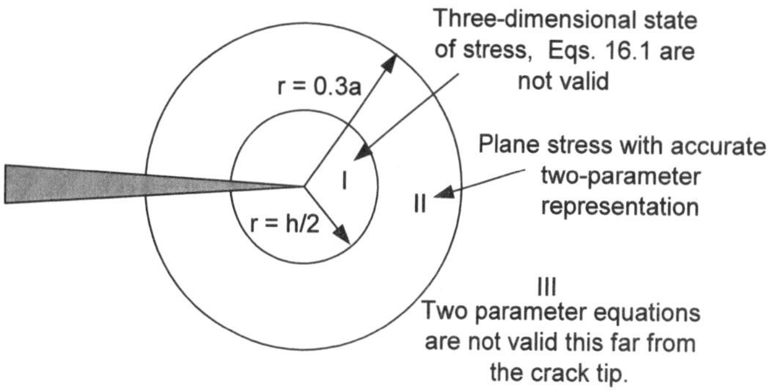

Figure 16.3 Concept of three regions near a crack tip. Only data taken from region II can be analyzed.

The second difficulty is associated with the measurement of the position coordinates r_i and θ_i locating a specific data point. Both the width of the fringe and the poor definition of the origin lead to errors in measuring r_i and θ_i. For fringes very near the crack tip where r_i is small, the relative errors $\Delta r_i/r_i$ and $\Delta\theta_i/\theta_i$ can be large. Examination of Fig. 16.1 illustrates the width of the fringe loops and the uncertainties in locating the origin at the tip of the crack.

To circumvent these difficulties, a higher-order representation of the stress field is utilized and Eqs. (16.1) are replaced with:

$$\begin{aligned}\sigma_{xx} &= A_0 r^{-1/2}\cos\frac{\theta}{2}\left(1-\sin\frac{\theta}{2}\sin\frac{3\theta}{2}\right)+2B_0+A_1 r^{1/2}\cos\frac{\theta}{2}\left(1+\sin^2\frac{\theta}{2}\right)\\ &\quad+2B_1 r\cos\theta+A_2 r^{3/2}\left(\cos\frac{3\theta}{2}-\frac{3}{2}\sin\theta\sin\frac{\theta}{2}\right)+2B_2 r^2(\sin^2\theta+\cos 2\theta)\\[2ex] \sigma_{yy} &= A_0 r^{-1/2}\cos\frac{\theta}{2}\left(1+\sin\frac{\theta}{2}\sin\frac{3\theta}{2}\right)+A_1 r^{1/2}\cos\frac{\theta}{2}\left(1-\sin^2\frac{\theta}{2}\right)\\ &\quad+A_2 r^{3/2}\left(\cos\frac{3\theta}{2}+\frac{3}{2}\sin\theta\sin\frac{\theta}{2}\right)+2B_2 r^2\sin^2\theta\\[2ex] \tau_{xy} &= A_0 r^{-1/2}\cos\frac{\theta}{2}\sin\frac{\theta}{2}\cos\frac{3\theta}{2}-A_1 r^{1/2}\sin\frac{\theta}{2}\cos^2\frac{\theta}{2}\\ &\quad-2B_1 r\sin\theta-3A_2 r^{3/2}\sin\frac{\theta}{2}\cos^2\frac{\theta}{2}-2B_2 r^2\sin 2\theta\end{aligned} \tag{4.41}$$

These equations provide six terms to represent the stress field, and with the additional terms, it is possible to extend significantly the boundary between regions II and III of Fig. 16.3. This representation permits larger values of r to be employed in selecting data from the valid field (region II) near the crack tip.

The coefficients A_0, A_1, A_2, B_0, B_1, and B_2 are unknown and must be determined. Recall from Eq. (4.28) that:

$$K_I = \sqrt{2\pi}\, A_0 \tag{16.15a}$$

and

$$\sigma_{0x} = -\, 2B_0 \tag{16.15b}$$

The other coefficients are not used in any fracture mechanics based analysis. They are included only to improve the accuracy of the determination of K_I. Substituting Eq. (4.48) into Eqs. (16.2) and (16.8) leads to:

$$\left(\frac{Nf_\sigma}{2h}\right)^2 = \left(\frac{\sigma_{xx} - \sigma_{yy}}{2}\right)^2 + \tau_{xy}^2 = D^2 + T^2 \tag{16.16}$$

where

$$D = \frac{\sigma_{xx} - \sigma_{yy}}{2} = \sum_{n=0}^{2}\left(n - \frac{1}{2}\right)A_n r^{n-1/2} \sin\theta \sin\left(n - \frac{3}{2}\right)\theta + \sum_{m=0}^{2} B_m r^m \left[m \sin\theta \sin(m\theta) + \cos(m\theta)\right] \tag{16.17}$$

$$T = \tau_{xy} = -\sum_{n=0}^{2}\left(n - \frac{1}{2}\right)A_n r^{n-1/2} \sin\theta \cos\left(n - \frac{3}{2}\right)\theta - \sum_{m=0}^{2} B_m r^m \left[m \sin\theta \cos(m\theta) + \sin(m\theta)\right] \tag{16.18}$$

When Eqs. (16.17) and (16.18) are substituted into Eq. (16.16), a higher-order equivalent to Eq. (16.3) is obtained, which is to be solved for the unknown coefficients. Unfortunately, Eq. (16.16) is nonlinear in the unknown coefficients and the matrix methods associated with linear algebra cannot be applied. In the next section, an overdeterministic method capable of solving nonlinear equations will be developed.

The use of overdeterministic methods is important in this application because these methods provide a means of statistically averaging the results from many data points. This averaging process improves the accuracy in K_I by accommodating for random error in the measurement of r_i and θ_i.

16.5 AN OVERDETERMINISTIC METHOD TO SOLVE NONLINEAR EQUATIONS

The unknown coefficients A_n and B_m in Eq. (16.16) appear as nonlinear terms (i.e. A_0^2, A_0B_0, etc.) and this nonlinearity complicates the approach used to solve this relation. Moreover, many data points from region II are used so that the results for K_I are statistically averaged. The use of many data points (more than the number of unknowns) leads to a set of overdetermined relations, each of the form of Eq. (16.16). One then seeks a least-squares type of solution to these relations. The solution is obtained by defining a function g based on Eq. (16.16) so that:

$$g_k = D_k^2 + T_k^2 - \left(\frac{N_k f_\sigma}{2h}\right)^2 = 0 \tag{16.19}$$

where the subscript k indicates the value of g evaluated at (r_k, θ_k) with a fringe order N_k in region II.

Because D_k and T_k are both dependent on A_n and B_m, the correct values for these constants will give g_k = 0 for all values of k. One initially makes an estimate of the coefficients and computes g_k only to find $g_k \neq 0$. The initial estimates of A_n and B_m are in error and must be corrected. The correction process involves an iterative equation, based on a Taylor-series expansion of g_k, which is given by:

$$g_k^{i+1} = g_k^i + \frac{\partial g_k^i}{\partial A_0}\Delta A_0 + \frac{\partial g_k^i}{\partial A_1}\Delta A_1 + \cdots + \frac{\partial g_k^i}{\partial B_0}\Delta B_0 + \frac{\partial g_k^i}{\partial B_1}\Delta B_1 + \cdots \tag{16.20}$$

where the superscript i shows the iteration step and ΔA_0, ΔA_1, ... ΔB_0, ΔB_1, ...are corrections to the previous estimates of A_0, A_1 .... B_0, B_1, ..., respectively.

Since Eq. (16.19) indicates $g_k = 0$, it is clear that the correction terms in Eq. (16.20) can be written as a system of linear equations of the form:

$$-g_k^i = \frac{\partial g_k^i}{\partial A_0}\Delta A_0 + \frac{\partial g_k^i}{\partial A_1}\Delta A_1 + \cdots + \frac{\partial g_k^i}{\partial B_0}\Delta B_0 + \frac{\partial g_k^i}{\partial B_1}\Delta B_1 + \cdots \tag{16.21}$$

In matrix notation, Eq. (16.21) becomes:

$$[\mathbf{g}] = [\mathbf{c}][\Delta] \tag{16.22}$$

where:

$$[\mathbf{g}] = \begin{bmatrix} -g_1 \\ -g_2 \\ \cdots\cdots \\ \cdots\cdots \\ \cdots\cdots \\ -g_L \end{bmatrix} \quad [\Delta] = \begin{bmatrix} \Delta A_0 \\ \cdots\cdots \\ \cdots\cdots \\ \Delta A_N \\ \Delta B_0 \\ \cdots\cdots \\ \cdots\cdots \\ \Delta B_M \end{bmatrix} \quad [\mathbf{c}] = \begin{bmatrix} \frac{\partial g_1}{\partial A_0} \cdots \frac{\partial g_1}{\partial A_N} & \frac{\partial g_1}{\partial B_0} \cdots \frac{\partial g_1}{\partial B_M} \\ \cdots \quad \cdots & \cdots \quad \cdots \\ \cdots \quad \cdots & \cdots \quad \cdots \\ \frac{\partial g_L}{\partial A_0} \cdots \frac{\partial g_L}{\partial A_N} & \frac{\partial g_L}{\partial B_0} \cdots \frac{\partial g_L}{\partial B_M} \end{bmatrix} \tag{16.23}$$

L is the total number of data points used, and N and M are the upper limits of the truncated series. Usually it is sufficient to take N = M = 2 to give a six-term-series representation with L = 5(N + M + 2) data points.

Matrix **[c]** involves the evaluation of L(M + N) partial derivatives; however, the task is made easy by noting that the functional form of each column is identical. Only the coordinates (r_k, θ_k) used to evaluate the row elements are changed. Moreover, the partial derivatives are obtained from Eq. (16.19) as:

$$\frac{\partial g_k}{\partial A_n} = 2D_k \frac{\partial D_k}{\partial A_n} + 2T_k \frac{\partial T_k}{\partial A_n} \tag{16.24}$$

Since the functions D_k and T_k are linear in A_n and B_m, the partial derivatives in Eq. (16.24) are obtained from Eqs. (16.17) and (16.18) by inspection. The implementation of this approach is a straightforward algebraic exercise which yields relations for each element in the **[c]** and **[g]** matrices.

The overdetermined system, given in Eq. (16.22), is solved in a least-squares sense. In matrix notation, the least-squares minimization process is accomplished by multiplying, from the left, both sides of Eq. (16.22) by the transpose of matrix **[c]** to give:

$$[\mathbf{c}]^T [\mathbf{g}] = [\mathbf{c}]^T [\mathbf{c}] [\Delta] \tag{16.25}$$

Next define:

$$[\mathbf{a}] = [\mathbf{c}]^T [\mathbf{c}] \tag{16.26}$$

and then solve Eq. (16.25) for **[Δ]** to obtain:

$$[\Delta] = [a]^{-1} [c]^T [g] \tag{16.27}$$

The solution of Eq. (16.27) yields corrections to be applied to the previous estimates of A_n and B_m. As indicated in Eq. (16.20), iteration is used with the corrections repeated until a sufficiently accurate set of coefficients is obtained. The steps in the iterative process include:

1. From the fringe pattern, in region II, select L data points which characterize the distinguishing features of the pattern. Record (r_k, θ_k, N_k) for each of the L points.
2. Assume initial values for the unknown constants A_0, A_1, ... A_N, B_0, B_1, ... B_M. The algorithm is not sensitive to the accuracy of the initial estimates.
3. Calculate the elements of **[c]** and **[g]**.
4. Solve Eq. (16.27) and correct the estimates of the unknown constants by using the expressions:

$$\begin{aligned} A_0^{i+1} &= A_0^i + \Delta A_0 \\ &\cdots\cdots \\ A_N^{i+1} &= A_N^i + \Delta A_N \\ B_0^{i+1} &= B_0^i + \Delta B_0 \\ &\cdots\cdots \\ B_M^{i+1} &= B_M^i + \Delta B_M \end{aligned} \tag{16.28}$$

5. Repeat steps 3 and 4 until the corrections ΔA_0, ... , ΔB_M become sufficiently small. Convergence is rapid and good results are usually obtained in less than 10 iterations.
6. Determine K_I from A_0 by using Eq. (16.15a).

An example of a fringe pattern which has been analyzed by using this method is shown in Fig. 16.4. The isochromatic pattern was obtained from a beam in three-point bending. The crack length to beam depth ratio a/w was 0.6. The fringe pattern in the region near the crack tip was enlarged as shown in Fig. 16.5a. The data points used in the analysis (about 80 points) is shown in Fig. 16.5b. Note the priority given in selecting points close (but not too close) to the crack tip. The small circle about the crack tip defines region I, which is excluded from the data field.

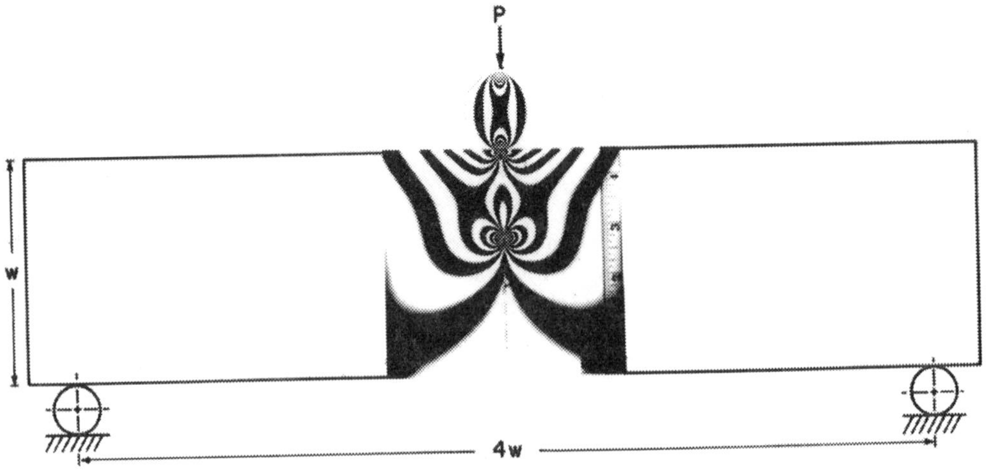

Figure 16.4 Fringe pattern for a cracked beam subjected to three point bending. Crack length to beam depth ratio = a/w = 0.6. Courtesy of R. J. Sanford.

The analysis employed a six-term-series representation with N = M = 2, and the six unknown coefficients were determined using the overdeterministic method. These coefficients were then substituted into Eqs. (16.16), (16.17) and (16.18) and the fringe orders were determined over region II. A plotting program was written and these computer-generated fringes were reconstructed as shown in Fig. 16.5c. A comparison of the experimental pattern with the computer-reconstructed pattern shows the adequacy of the data analysis method in representing the data. A good match between the two fringe patterns gives confidence in the accuracy of the K_I determination.

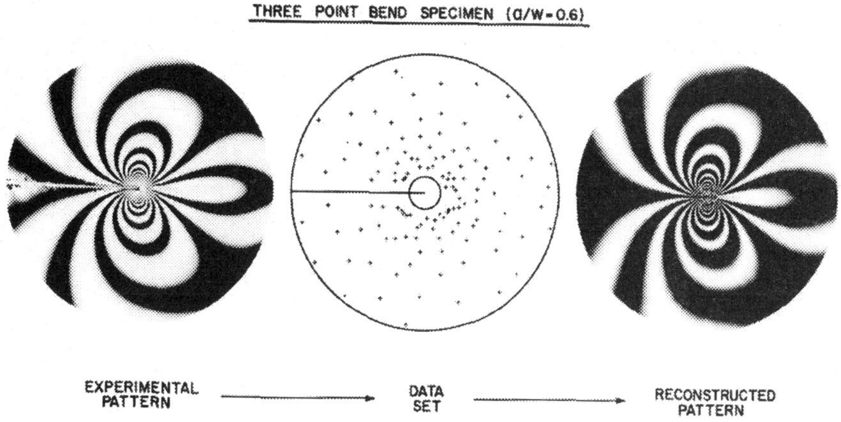

Figure 16.5 Development of the analysis in the overdeterministic method for determining K_I. Courtesy of R. Chona.

16.6 DETERMINING K_I FROM ISOPACHIC FRINGE PATTERNS

A typical isopachic fringe pattern in the region of a crack tip is shown in Fig. 16.6. This pattern was recorded using double-exposure interferometry with an optical arrangement that responds to out-of-plane displacements w. Since the model was a sheet of Plexiglas loaded in plane stress, the out-of-plane displacements are related to the change in thickness Δh of the model. Recall Eq. (10.42) and note that the sum of the in-plane stresses $(\sigma_{xx} + \sigma_{yy})$ is given by:

$$\sigma_{xx} + \sigma_{yy} = -\frac{E}{\nu}\varepsilon_{zz} = -\frac{E}{\nu}\frac{\Delta h}{h} \qquad (16.29)$$

In the holographic interferometer, the fringe order N is proportional to Δh and the corresponding stress-optic law is given by:

$$\sigma_{xx} + \sigma_{yy} = \frac{Nf_p}{h} \qquad (16.30)$$

where f_p is an optical constant. The fringes in the pattern shown in Fig. 16.6 are called isopachics, which are contour lines representing a constant value of $(\sigma_{xx} + \sigma_{yy})$.

The isopachic fringe pattern gives full-field data (in the region near the crack tip) which can be analyzed to determine the opening-mode stress intensity factor. The approach is similar to that previously described for isochromatic fringe patterns. Begin with Eqs. (4.46) and develop the relation for $(\sigma_{xx} + \sigma_{yy})$ and then use Eq. (16.30) to obtain:

$$\frac{\sigma_{xx}+\sigma_{yy}}{2}=\sum_{n=0}^{N}A_n r^{n-1/2}\cos\left(n-\frac{1}{2}\right)\theta+\sum_{m=0}^{M}B_m r^m\cos m\theta=\frac{Nf_p}{2h} \tag{16.31}$$

Figure 16.6 An isopachic fringe pattern representing ($\sigma_{xx}+\sigma_{yy}$) in the region near the crack tip.

Inspection of Eq. (16.31) shows that the unknown coefficients A_n, B_m appear as linear terms. This is an important feature of the analysis of isopachic patterns since it simplifies the mathematical methods used in an overdetermined analysis.

Dudderar and Gorman [10] described a simple approach for extracting K_I from Eq. (16.31). In this simplification of Eq. (16.31), the series was truncated at N = M = 0 and data was selected only along the line $\theta = 0$. With these two restrictions, Eq. (16.31) reduces to:

$$A_0 r^{-1/2}+B_0=\frac{Nf_p}{2h} \tag{16.32}$$

Following Irwin's approach, both A_0 and B_0 are treated as unknowns in this two-parameter representation. The unknown constant B_0 is eliminated by using fringe orders N_1 and N_2 located at positions r_1 and r_2 along the x axis. By writing Eq. (16.32) twice (once for each data point), subtracting one from the other, and using Eq. (4.35) gives:

$$K_I=\left(\frac{\pi}{2}\right)^{1/2}\frac{(r_1 r_2)^{1/2}}{r_2^{1/2}-r_1^{1/2}}\left[\frac{(N_1-N_2)f_p}{h}\right] \tag{16.33}$$

Again, this relatively simple approach is effective if the data is taken from a region where the two-parameter representation of ($\sigma_{xx}+\sigma_{yy}$) is valid and if very accurate measurements of r_1 and r_2 are made.

Accuracy in the determination of K_I from isopachic patterns can be improved by using the overdeterministic method which uses whole-field data and permits error to be minimized in a least-squares sense. In application of the overdeterministic method, data (N_k, r_k, θ_k) are collected from region II (see Fig. 16.3) and then each data set is substituted into Eq. (16.31). The result is an overdetermined system of equations that are solved by using a least-squares approach. In matrix notation, the system of Eqs. (16.31) is written as:

$$[N] = [c]\,[U] \tag{16.34}$$

where **[N]** is a column matrix of the fringe orders; **[c]** is a matrix of the coefficients of the unknown constants and **[U]** is a column matrix of the unknown constants A_n and B_m.

To square the matrix, multiply through Eq. (16.34), from the left, with the transpose of **[c]** to obtain:

$$[c]^T [N] = [c]^T [c] [U] = [a] [U] \tag{16.35}$$

where

$$[a] = [c]^T [c]$$

Solving Eq. (16.35) for **[U]** by using ordinary methods of matrix inversion gives:

$$[U] = [a]^{-1} [c]^T [N] \tag{16.36}$$

Solution of Eq. (16.36) is easily accomplished with a short computer program which can be executed on a personal computer. Of the 6 to 8 unknown coefficients usually employed in an analysis, only A_0 is retained. It is converted to K_I by using the relation $K_I = \sqrt{2\pi}\,A_0$.

16.7 DETERMINING K_I FROM MOIRÉ FRINGE PATTERNS

A typical pair of moiré fringe patterns representing the in-plane displacements u and v in the region near a crack tip is shown in Fig. 16.7. Recall that the moiré fringe orders N are related to the displacement field by:

$$u, v = N\,p \tag{13.15}$$

where p is the pitch of the master grating. The moiré fringe pattern provides the data necessary to determine K_I. The relation between the displacements and the unknown constants A_n and B_m is obtained by integrating Eqs. (4.42) to obtain:

For the u field:

$$\begin{aligned} N_u E p = {} & \sum_{n=0}^{N} A_n \frac{r^{n+1/2}}{n+\frac{1}{2}}\left[(1-\nu)\cos\left(n+\frac{1}{2}\right)\theta - (1+\nu)\left(n+\frac{1}{2}\right)\sin\theta\sin\left(n-\frac{1}{2}\right)\theta\right] \\ & + \sum_{m=0}^{M} B_m \frac{r^{m+1}}{m+1}\left[2\cos(m+1)\theta - (1+\nu)(m+1)\sin\theta\sin m\theta\right] \end{aligned} \tag{16.37}$$

For the v field:

$$\begin{aligned} N_v E p = {} & \sum_{n=0}^{N} A_n \frac{r^{n+1/2}}{n+\frac{1}{2}}\left[2\sin\left(n+\frac{1}{2}\right)\theta - (1+\nu)\left(n+\frac{1}{2}\right)\sin\theta\cos\left(n-\frac{1}{2}\right)\theta\right] \\ & + \sum_{m=0}^{M} B_m \frac{r^{m+1}}{m+1}\left[(1-\nu)\sin(m+1)\theta - (1+\nu)(m+1)\sin\theta\cos m\theta\right] \end{aligned} \tag{16.38}$$

These relations assume a symmetrical loading of the cracked body and a fixed origin at the crack tip. In producing moiré fringe patterns in the laboratory, it is very difficult to produce a symmetrical fringe pattern because of rigid-body motions. These motions, which usually involve both a rotation and translation of the origin, produce parasitic fringes that are not due to the strain. This contribution of the rigid-body motion to the fringe field may be written as:

$$T = P\,R\cos\theta + Q\,r\sin\theta + R \qquad (16.39)$$

where P, Q and R are unknown constants that describe the parasitic fringe field.

In application of the overdeterministic method, either Eq. (16.37) or Eq. (16.38) can be employed. Usually the fringe pattern corresponding to the v displacement field is selected because this pattern exhibits a higher-order fringe pattern. Regardless of the choice of patterns, the analysis requires that the term T of Eq. (16.39) be added to the right hand side of either Eq. (16.37) or (16.38) to account for the effect of rigid-body motions. The superposition of the term T adds three additional unknowns P, Q, and R to the unknown constants A_n, B_m; however, the overdeterministic method can accommodate these additional unknowns.

In solving the modified form of say Eq. (16.38) we note that the unknowns A_n, B_m, P, Q and R are all linear. This fact implies that the matrix manipulations described in Eqs. (16.34) to (16.36) can be applied to determine the unknowns in the column matrix **[U]**.

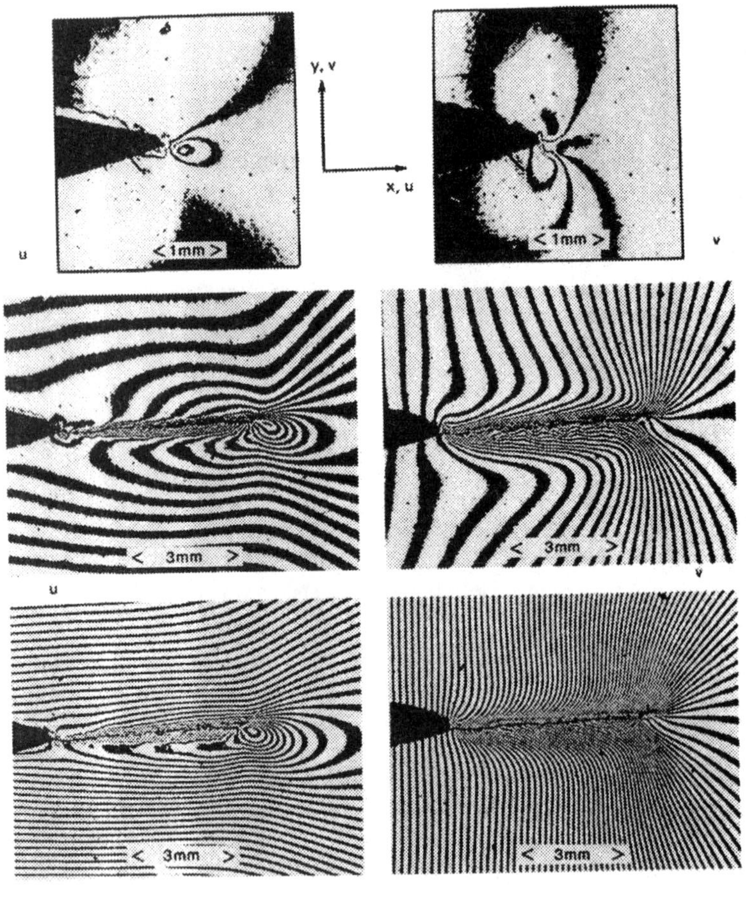

Figure 16.7 Moiré fringe patterns N_u and N_v representing the in-plane u and v displacement fields around a fatigue crack in 7075-T6 aluminum. Top and center patterns show residual deformations; lower patterns show the live load plus residual deformations. Courtesy of D. Post.

16.8 DETERMINING K FROM THE METHOD OF CAUSTICS

The method of shadow caustics was developed in the 1970s and extensively used by Manogg, Theocaris, and Kalthoff [16, 17, 18]. The principle of the method, illustrated in Fig. 16.8, is simple in concept. The incident light, which can be ordinary or coherent, is collimated to produce a system of parallel rays. These rays encounter a transparent specimen with its plane oriented so that the light is at normal incidence. When the specimen is subjected to in-plane stresses, a change in thickness $\Delta h(x,y)$ occurs, which is given by the expression:

$$\Delta h = -\frac{h\nu}{E}\left(\sigma_{xx} + \sigma_{yy}\right) \tag{16.40}$$

Figure 16.8 Optical system used to form the shadow spot.

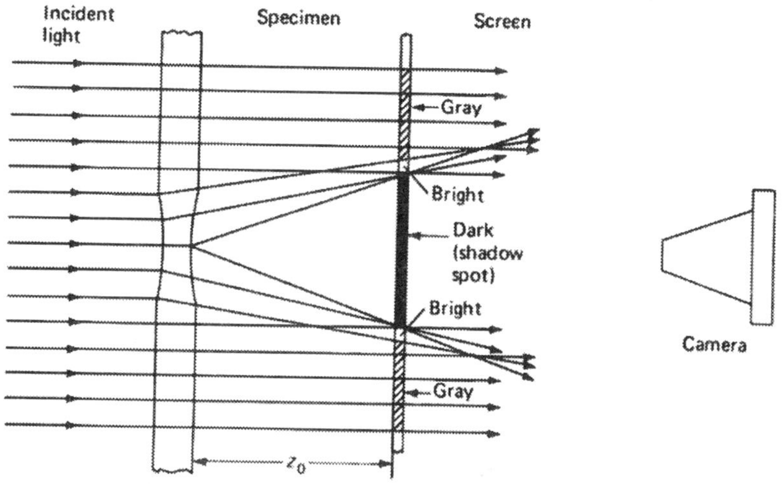

Because $\sigma_{xx} + \sigma_{yy}$ varies with x and y, the front and back surfaces of the specimen deform as shown in Fig. 16.8. The deformed surfaces cause the light rays to deflect (like a lens), and upon exiting from the specimen, the rays are not parallel. If a screen is placed at some location z_o down stream from the specimen, the light rays produce an interesting optical pattern of gray, dark, and bright regions. The gray regions correspond to locations where the gradient of the sum of the in-plane stresses ($\sigma_{xx} + \sigma_{yy}$) is small and the parallel rays pass through the model without significant deflection. The dark areas give the shadow spot after which the method is named. These dark regions are the result of the deflection of the light rays from this local area on the screen. The light regions are due to the added intensity of the normal rays plus those deflected rays which impinge on this area of the screen. This pattern of gray, bright, and dark regions as it appears on the screen is photographed to provide an image that in effect describes the gradient of the sum of the in-plane stresses.

The optical system presented in Fig. 16.8 is the transmission arrangement used with transparent specimens. If the specimen is opaque, a similar arrangement can be employed to develop shadow spots if one surface of the specimen is mirrored. With the specimen acting as a mirror, the reflected light produces a pattern of bright, dark, and gray regions. However, the reflected pattern is observed as a virtual image which is formed on the opposite side of the specimen from that shown in Fig. 16.8.

16.8.1 Mapping

The mapping of the specimen plane π_s on the image plane π_i for a transparent model, which forms a real image, is shown in Fig. 16.9. The light ray deflection is due to surface curvature which produces a divergent lens effect in the local region near the root of a notch. The point P_s is placed in this local region near the geometric discontinuity. The deflected light ray propagates down the optical bench a distance z_o and impinges on a screen to form an image at point P_i. The location of point P_i on the image plane π_i is given by the vector $\mathbf{r_i}$, which is written as:

$$\mathbf{r_i} = \mathbf{r_s} + \mathbf{g} \tag{16.41}$$

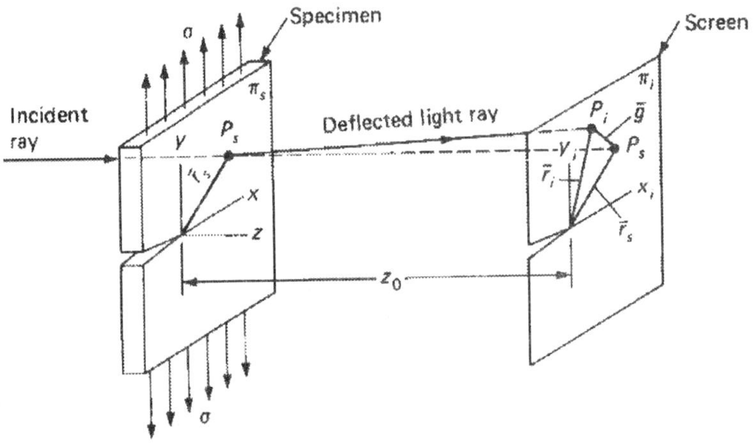

Figure 16.9 Coordinate system for developing the mapping equation.

The movement of the point P_s on the specimen plane to P_i on the image plane is described by the vector **g**, which is related to the screen position and the local slope of the specimen surface as:

$$\mathbf{g} = -z_o \nabla(\Delta s) \tag{16.42}$$

where $\nabla = \mathbf{i}(\partial/\partial x) + \mathbf{j}(\partial/\partial y) + \mathbf{k}(\partial/\partial z)$ is a vector operation and Δs is the change in the optical path length due to stress induced changes in the thickness of the model and the index of refraction n. For a specimen with a uniform thickness h, the change in optical path length Δs is:

$$\Delta s = (n - 1)\Delta h + h\,\Delta n \tag{16.43}$$

where the change in thickness Δh is given by Eq. (16.40) for plane stress conditions ($\sigma_{zz} = 0$).

The changes in the index of refraction are related to the principal stresses by the Maxwell relations described in detail in Section 10.3. By using the Maxwell equations, Δn is related to the stresses by:

$$\Delta n_1 = c_1 \sigma_1 + c_2 \sigma_2$$
$$\Delta n_2 = c_1 \sigma_2 + c_2 \sigma_1 \tag{16.44}$$

where c_1 and c_2 are material-dependent optical constants.

Substituting Eqs. (16.40) and (16.44) into Eq. (16.43) yields:

$$\Delta s_1 = h(a\sigma_1 + b\sigma_2)$$

$$\Delta s_2 = h(a\sigma_2 + b\sigma_1) \qquad (16.45a)$$

where

$$a = c_1 - \frac{(n-1)\nu}{E}$$

$$b = c_2 - \frac{(n-1)\nu}{E} \qquad (16.45b)$$

It is possible to recast Eq. (16.45a) in terms of the sum and difference of the principal stresses as:

$$\Delta s_1 = C_1 h[(\sigma_1 + \sigma_2) + C_2 (\sigma_1 - \sigma_2)] \qquad (16.46a)$$

$$\Delta s_2 = C_1 h[(\sigma_1 + \sigma_2) - C_2 (\sigma_1 - \sigma_2)] \qquad (16.46b)$$

where

$$C_1 = \frac{c_1 + c_2}{2} - \frac{(n-1)\nu}{E} \qquad (16.47a)$$

$$C_2 = \frac{c_1 - c_2}{c_1 + c_2 - [2(n-1)\nu/E]} \qquad (16.47b)$$

The constants c_1, c_2, n, C_1, and C_2 for a number of transparent polymers are presented in Table 16.1

Table 16.1
Constants for Transparent Materials (after Kalthoff) Reference 18

Material	Elastic constants			General optical constants		Shadow optical constants	
	E (MN/m²)	ν	n	c_1 (μm² /N)	c_2 (μm² /N)	C_1 (μm² /N)	C_2
Araldite B	3,660	0.392	1.592	– 5.6	– 62.0	– 97.0	– 0.288
CR-39	2,580	0.443	1.504	– 16.0	–52.0	– 120.0	– 0.148
Plate glass	73,900	0.231	1.517	+ 0.32	–2.5	– 2.7	– 0.519
Homalite 100	4,820	0.310	1.561	– 44.4	–67.2	– 92.0	– 0.121
PMMA	3,240	0.350	1.491	– 5.0	–57.0	– 108.0	0

The fact that Eq. (16.46) shows two different relations for Δs indicates that a double caustic is formed in the image plane. The formation of the second caustic, due to Δs_2, is the result of the optical anisotropic nature of the model material with $\Delta n_1 \neq n_2$. When the model material is optically isotropic, $\Delta n_1 = n_2$ and it is clear from Eqs. (16.44) and Eqs. (16.47) that:

$$c_1 = c_2 = c$$

$$C_1 = c - \frac{(n-1)\nu}{E} \qquad (16.48)$$

$$C_2 = 0$$

For optically isotropic materials, it is possible to substitute Eqs. (16.48) into Eqs. (16.46) to obtain:

$$\Delta s = \Delta s_1 = \Delta s_2 = C_1 h(\sigma_1 + \sigma_2) \qquad (16.49)$$

These results show that a single caustic is formed since $\Delta s_1 = \Delta s_2$. An example of the single and double caustic patterns formed by the surface deformation of a specimen containing a crack is shown in Fig. 16.10.

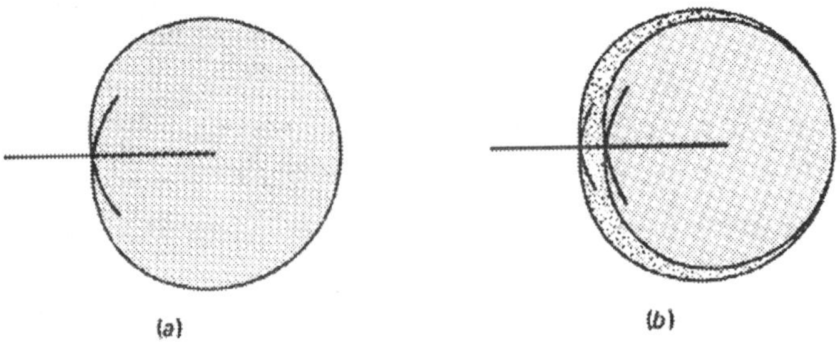

Figure 16.10 Single and double caustics.
(a) Optically isotropic material.
(b) Optically anisotropic material (birefringent)

16.8.2 Caustics Due to Stress Singularities

The formation of an optical pattern in a caustic arrangement is dependent on the gradient of the stresses (see Eq. 16.42). Higher gradients produce larger deflections of the light rays and images with distinguishing characteristics. This sensitivity of the method of caustics to stress gradients permits an effective approach to stress analysis of plane bodies with stress singularities. The application described here pertains to a plane body containing a crack with opening mode loading; however, the method can be applied to many other problems where singular stresses occur.

Consider a plane specimen with a crack of length a loaded in the opening mode as shown in Fig. 4.1. The point P_s is taken in the very near field where the stresses are represented by Eqs. (4.36). Substituting Eqs. (4.36) into the stress equations of transformation Eq. (1.11) and using Eq. (4.35) to relate K_I with A_0 gives:

$$\sigma_{rr} = \frac{K_I}{4\sqrt{2\pi r}}\left[5\cos\frac{\theta}{2} - \cos\frac{3\theta}{2}\right]$$
$$\sigma_{\theta\theta} = \frac{K_I}{4\sqrt{2\pi r}}\left[3\cos\frac{\theta}{2} + \cos\frac{3\theta}{2}\right] \qquad (16.50)$$
$$\tau_{r\theta} = \frac{K_I}{4\sqrt{2\pi r}}\left[\sin\frac{\theta}{2} + \sin\frac{3\theta}{2}\right]$$

The mapping of point P_s on the image plane is dependent on the specimen material. Consider an optically isotropic material which produces a single caustic. Substituting Eqs. (16.50) into Eqs. (16.41) and (16.42) leads to:

$$x_i = r\cos\theta + \frac{K_I}{\sqrt{2\pi}} z_o C_1 h r^{-3/2} \cos\frac{3\theta}{2}$$
$$(16.51)$$
$$y_i = r\sin\theta + \frac{K_I}{\sqrt{2\pi}} z_o C_1 h r^{-3/2} \sin\frac{3\theta}{2}$$

If the point P_s is moved over the very near field with many values of (r, θ), then a complete family of light rays are generated. These rays form a shadow "cone" down stream from the specimen plane. The surface of this "cone" is called the caustic surface and the intersection of the caustic surface with the image plane produces the caustic curve. The caustic pattern for this case, presented in Fig. 16.11, shows the caustic curve as the line between the dark shadow zone and the bright band. This transition occurs when:

$$\frac{\partial x_i}{\partial r}\frac{\partial y_i}{\partial \theta} - \frac{\partial x_i}{\partial \theta}\frac{\partial y_i}{\partial r} = 0 \qquad (16.52)$$

From Eqs. (16.51) and Eq. (16.52) we obtain a relation for r which locates the loci of points P_s on the specimen that correspond to the caustic curve on the image plane. This relation

$$r = \left[\frac{3K_I}{2\sqrt{2\pi}} z_o C_1 h \right]^{2/5} \equiv r_o \tag{16.53}$$

shows that the initial curve is a circle with a radius r_o. Note that the location of the initial circle on the model can be varied by changing the position of the screen (z_o) or by changing the load (K_I).

Figure 16.11 A caustic pattern produced by a crack in an optically isotropic material loaded in mode I. Courtesy of J. F. Kalthoff.

For a fixed K_I and z_o, we locate the initial circle according to Eq. (16.53). Substituting this value for r into Eq. (16.52) gives the description of the caustic curve as an image of the initial curve. Thus:

$$\begin{aligned} x' &= r_o\left[\cos\theta + \frac{2}{3}\,\mathrm{sgn}(z_o C_1)\cos\frac{3\theta}{2}\right] \\ y' &= r_o\left[\sin\theta + \frac{2}{3}\,\mathrm{sgn}(z_o C_1)\sin\frac{3\theta}{2}\right] \end{aligned} \tag{16.54}$$

where

$$\mathrm{sgn}(z_o C_1) = \begin{Bmatrix} 1 & \text{if } (z_o C_1) > 0 \\ 0 & \text{if } (z_o C_1) = 0 \\ -1 & \text{if } (z_o C_1) < 0 \end{Bmatrix}$$

Equations (16.54) generate a caustic curve which is classified as a generalized epicycloid. The maximum diameter D of the caustic and the radius r_o of the initial circle are shown in Fig. 16.12. The diameter D is related to r_o by

$$D = 3.17\, r_o \tag{16.55}$$

Finally, substituting Eq. (16.55) into Eq. (16.53) and solving for K_I gives:

$$K_I = 0.0934 \frac{D^{5/2}}{z_o C_1 h} \tag{16.56}$$

It is clear from Eq. (16.56) that the opening mode stress intensity factor K_I can be determined simply by measuring the major diameter D of the caustic since z_o, C_1, and h are known parameters in a typical experiment. Care should be taken to insure that $r_o \geq h/2$ so that the plane stress assumption made in developing all of the relations shown here is valid. If $r_o < h/2$, then the radius of the initial curve is so close to the crack tip that measurement is made in a region where the stress state is three dimensional. If $r_o < h/2$ the use of Eq. (16.56) will give results for which are much lower than the true values.

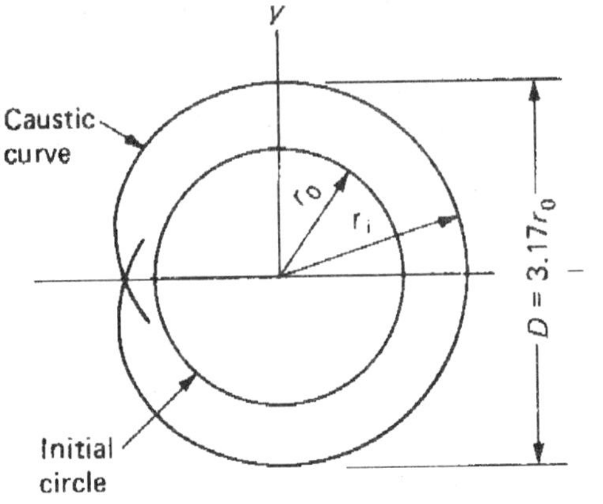

Figure 16.12 The initial circle and the caustic curve (mode I).

When the model material is optically anisotropic, two caustics are formed as shown in Fig. 16.12. Both caustics, the inner and outer, provide diameters D_i and D_o. In this case, two estimates of K_I are possible by using a relation similar to Eq. (16.56) which is written as:

$$K_I = \frac{2\sqrt{2\pi}\, D_{o/i}^{5/2}}{3 f_{o/i}^{5/2}\, z_o C_1 h} \tag{16.57}$$

where $f_{o/i}$ is a linear function of the constant C_2 shown in Fig. 16.13. Note that

$$D_{o/i} = f_{o/i}\, r_o \tag{16.58}$$

The restriction $r_o < h/2$ is also required for Eq. (16.57).

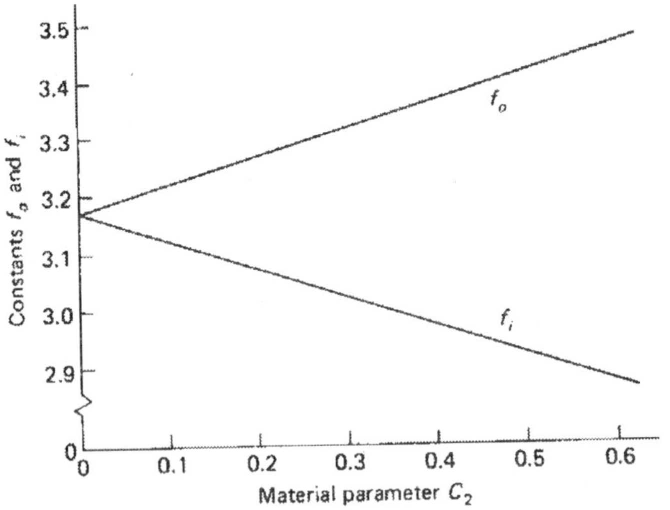

Figure 16.13 Numerical constants f_o and f_i used in Eq. (16.58)

The advantage of the caustic method in experimental studies of fracture parameters such as the stress intensity factors K_I or K_{II} is its simplicity. The optical bench has relatively few and inexpensive components. The analysis to determine K_I involves knowledge of the experimental variables z_o, C_1, and h and the measurement of the major diameter D of the caustic.

A complete coverage of the method of caustics is beyond the scope of this text. We have attempted a logical extension by including several exercises which require the development of the mapping equations for the use of caustics in reflection. We have also included exercises dealing with the shearing mode in fracture mechanics. For a more complete treatment of this interesting experimental method, the reader is referred to reference [18].

16.9 DETERMINING K FROM COHERENT GRADIENT SENSING METHOD

Coherent Gradient Sensing (CGS) is a relatively new technique developed about twenty years ago by Tippur and Rosakis [19,20,21]. In this technique, the gradients of the out of plane displacements are measured. These gradients are then used in the determination of the fracture mechanics parameters. The technique employs basic principles of moiré deflectometry [22]. The CGS method involves a simple optical set-up that is relatively insensitive to vibrations and rigid body motions, and produces high contrast optical fringe patterns. The method has been used in recent years to study quasi-static and dynamic fracture problems in a variety of materials, including, polymers, metals, bimaterials and functionally graded materials.

16.9.1 Experimental Arrangement

The schematic of the experimental arrangement is shown in Fig. 16.14. This reflection mode arrangement can be easily modified if a transparent object is to be studied by placing the collimated laser beam behind the transparent object. In the reflection mode, the object waves are reflected from the specularly reflective and optically flat object surface. These waves are then transmitted through a pair of parallel high density Ronchi gratings separated by a distance Δ. The principal directions of the rulings on the grating are coincident with either the x or y co-ordinate axes to obtain $(\partial w/\partial x)$ or $(\partial w/\partial y)$, respectively, where w denotes the out of plane surface displacement. The diffracted wave fronts are collected by a positive lens and the resulting diffraction spectrum is registered on its focal plane as shown Fig. 16.14. An aperture is then used to block all but the required diffraction orders. The interference patterns are finally recorded using a high resolution digital camera for analysis. It should be emphasized that the camera system comprising of the lens and the photo sensor array or film needs to be focused on the object surface during initial optical alignment.

The gratings used in CGS are typically chrome-on-glass line-pair depositions commonly referred to as Ronchi gratings. The choice of Ronchi grating plates is based on commercial availability or ease of fabrication but the method should function equally well with other grating profiles such as sinusoidal rulings. Anti-reflection coatings on the gratings are essential for minimizing optical noise in the form of 'ghost fringes' which would otherwise occur due to multiple reflections in the cavity between the gratings. Other parameters such as grating pitch, grating separation distance and focal length of the imaging lens are chosen based on the need for:

1. Producing sufficient spatial shearing of the object wave front without unduly compromising the accuracy of derivative representation of the out-of-plane displacements.
2. resolving the diffraction spots sufficiently on the back focal plane of the imaging lens for easy filtering.

The use of micro-positioning devices to achieve in-plane and out-of-plane parallelism between the two gratings limit the smallest grating separating distance to about 15 mm as reported in most studies. The shearing interferometer needs to be aligned to ensure planarity of the incident laser beam used for interrogating the object surface and parallelism of the Ronchi gratings G_1 and G_2 as well as the grating lines. For details on alignment see Tippur (21).

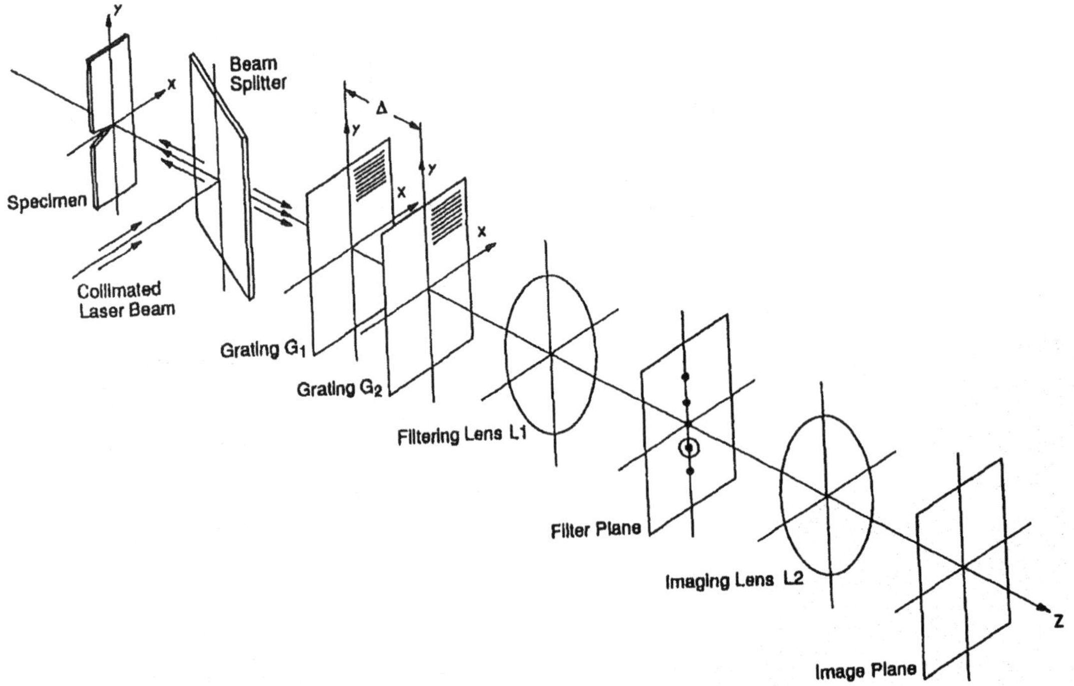

Figure 16.14 Diagram of the experimental arrangement for the CGS system [19].

16.9.2 Working Principle

The working principle of the CGS system is illustrated in Fig. 16.15. Consider an object wave, making an angle ϕ with the optical axis in the y-z plane, transmitted through a pair of Ronchi gratings G_1 and G_2 of pitch p with principal grating direction, say, along the y – axis (Fig. 16.15). The separation distance between the two gratings along the optical axis (the z-axis) is Δ. The diffracted light emerging from the first grating consists of a zero and several odd diffraction orders, each denoted by the corresponding complex amplitude distribution E. For simplicity, only consider diffraction orders E_i (i = 0, ±1) after the first grating G_1. These waves are propagating in discretely different directions according to the diffraction equation

$$\theta = \sin^{-1}(\lambda/p) \approx (\lambda/p) \text{ for small angles} \tag{16.59}$$

Each of these diffracted wave fronts diffract once more at the second grating G_2 plane. The corresponding wave fronts, propagating in several discrete directions, are denoted by $E_{(i,j)}$ (i = 0, ±1, j = 0, ±1) where the two subscripts correspond to diffraction order at the first and the second grating, respectively. The wave fronts $E_{\pm1,0}$ and $E_{0,\pm1}$ contribute to the ±1 diffraction spot on the focal plane, and diffracted wave fronts $E_{0,0}$, $E_{+1,-1}$ and $E_{-1,+1}$ contribute to the 0th order. By letting ±1 diffraction order to pass through the filtering aperture, interference fringes resulting from the corresponding complex amplitudes can be evaluated.

Let l_0 and $l_{\pm1}$ denote the optical path lengths of E_0 and $E_{\pm1}$, respectively, between the two gratings. Then, complex amplitudes E_0 and $E_{\pm1}$ can be represented as

$$E_0 = A_0 \exp(ikl_0), \qquad E_{\pm1} = A_{\pm1}\exp(ikl_{\pm1}) \tag{16.60}$$

where A's denote amplitudes, k the wave number, $(= 2\pi/\lambda)$ and $i = \sqrt{-1}$.

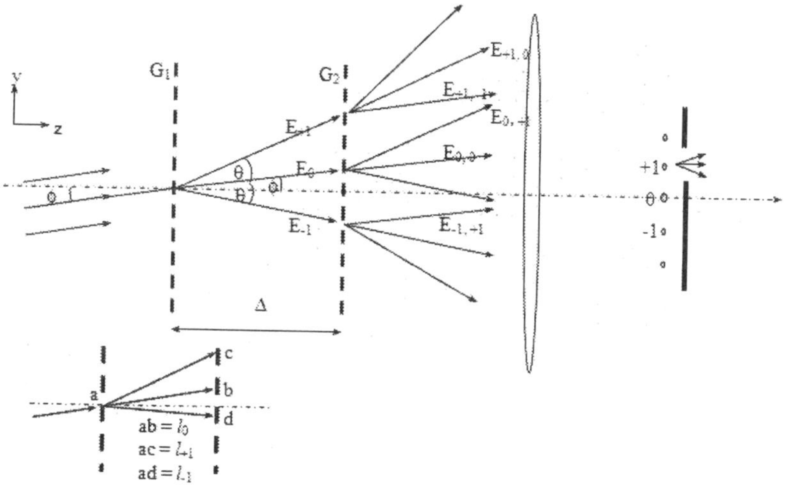

Figure 16.15 Schematic representation of working principle for CGS system [21].

Noting that no additional path difference occurs beyond the grating G_2, the intensity distribution on the image plane for a 1:1 magnification is proportional to,

$$I_{\pm 1} = \left(E_{0,\pm 1} + E_{\pm 1,0}\right)\left(E_{0,\pm 1} + E_{\pm 1,0}\right)^* = A_0^2 + A_{\pm 1}^2 + 2A_0A_{\pm 1}\cos k\left(l_0 - l_{\pm 1}\right) \tag{16.61}$$

when the ±1 diffraction spot is permitted to pass through the aperture at the spectrum plane. In Eq. (16.61)—( )* denotes the complex conjugate. Based on geometrical considerations, the optical path lengths l_0 and $l_{\pm 1}$ can be expressed as,

$$l_0 = \frac{\Delta}{\cos\phi} = \frac{\Delta}{\left(1 - \frac{\phi^2}{2!} + \frac{\phi^4}{4!} - \ldots\right)}, \qquad l_{\pm 1} = \frac{\Delta}{\cos(\theta \pm \phi)} = \frac{\Delta}{\left(1 - \frac{(\theta \pm \phi)^2}{2!} + \frac{(\theta \pm \phi)^4}{4!} - \ldots\right)} \tag{16.62}$$

by expanding Cos ϕ in the neighborhood of ϕ = 0. For small angles and by neglecting higher order terms, Eq. (16.62) becomes,

$$l_0 = \frac{\Delta}{\left(1 - \frac{\phi^2}{2!}\right)} = \Delta\left(1 + \frac{\phi^2}{2!}\right), \qquad l_{\pm 1} = \frac{\Delta}{\left(1 - \frac{(\theta \pm \phi)^2}{2!}\right)} = \Delta\left(1 + \frac{(\theta \pm \phi)^2}{2!}\right) \tag{16.63}$$

By calculating the path difference $(l_0 - l_{\pm 1})$ from Eq. (16.61), the expression for intensity distribution Eq. (16.61) on the image plane when ±1st diffraction order is allowed to pass through the filtering aperture is given by:

$$I_{\pm 1} \approx A_0^2 + A_{\pm 1}^2 + 2A_0A_{\pm 1}\cos\left[k\Delta\left(\frac{t^2}{2} \pm \theta\,\phi\right)\right] \tag{16.64}$$

The constructive interference occurs when the argument of the cosine term in Eq. (16.64) is $2N_d\pi$ where, N_d = 0, ±1, ±2, ... is the fringe order. After substituting for the wave number we obtain,

$$\frac{\Delta}{\lambda}\theta\left(\frac{\theta}{2} \pm \phi\right) = N_d \tag{16.65}$$

When the specimen is in the undeformed ($\phi = 0$) state, the fringe order of the uniform bright fringe of the undeformed object can be denoted by

$$N_u = \frac{\Delta}{p}\left(\frac{\theta}{2}\right) = \text{constant} \tag{16.66}$$

with $N_u = 0, \pm1, \pm2, \ldots$, and where $\theta = \frac{\lambda}{p}$ utilized.

By incorporating the expression for N_u, the magnitude of angular deflection of light at a point on the surface can be written as,

$$\phi = (N_d - N_u)\frac{p}{\Delta} = N\frac{p}{\Delta}, \qquad N = (N_d - N_u) = 0,\ \pm1,\ \pm2, \ldots \tag{16.67}$$

Evidently, the sensitivity of the interferometer for measuring angular deflections of light rays depends on the ratio of the grating pitch p and the grating separation distance Δ. This offers experimental flexibility for controlling measurement sensitivity by suitably choosing p and Δ. Clearly this is an added experimental advantage over other methods used for measuring fracture parameters. Also, it should be noted that the interference fringes obtained by filtering all but +1 or −1 diffraction orders can be interpreted as forward and backward differences of the optical signal.

The angular deflections of light rays can be related to surface deformations of the object using a first order approximation. Let the propagation vector of the object wave front be

$$\mathbf{d} = \frac{\partial(\delta S)}{\partial x}\mathbf{e_x} + \frac{\partial(\delta S)}{\partial y}\mathbf{e_y} + \mathbf{e_z} \tag{16.68}$$

where δS denotes optical path difference and $\mathbf{e_i}$ is a unit normal in the *i*-th direction. For light rays in the

y-z plane, $\frac{\partial(\delta S)}{\partial y} \approx \phi$ when the small angles approximation is invoked.

When a transparent object is studied using transmission-mode CGS, the optical path S through the specimen can be expressed as

$$S = (n-1)h \tag{16.69}$$

where n is the refractive index of the material and h is the thickness in the unstressed state.

Upon deformation the optical path length becomes

$$S + \delta S \tag{16.70}$$

where $\delta S = C_\sigma h(\sigma_x + \sigma_y)$ and $C_\sigma = \left(-\frac{\nu}{E} + C_0\right)(n-1)$ is the elasto-optic coefficient and C_0 is the stress-

optic constant of the material.

It should be noted that C_σ accounts for stress induced refractive index changes as well as thickness changes due to Poisson effect. Thus, angular deflections of light rays in transmission mode CGS relate to the mechanical fields as,

$$\phi \approx \frac{\partial(\delta S)}{\partial y} = C_\sigma h\frac{\partial(\sigma_x + \sigma_y)}{\partial y} = N\frac{p}{\Delta}, \qquad N = 0,\ \pm1,\ \pm2, \ldots. \tag{16.71}$$

When an optically opaque object is studied using reflection-mode CGS, the path difference is δS = 2w where w is out-of-plane displacement along the z-direction. Then, angular deflections can be related to the mechanical fields (surface slopes) as,

$$\phi \approx \frac{\partial(\delta S)}{\partial y} = 2\left(\frac{\partial w}{\partial y}\right) = N\frac{p}{\Delta}, \qquad N = 0,\ \pm 1,\ \pm 2,\ .... \tag{16.72}$$

where w is the out-of-plane deflection in the z-direction.

16.9.3 Application

Tippur et al. were the first to show the application of CGS to fracture mechanics problem. They utilized a three point bend geometry shown in Fig. 4.8. The specimen was fabricated from 4340 steel. The surface of the specimen was lapped and polished to provide an optically flat surface. The specimen was statically loaded to produce mode I loading. The CGS contours obtained during the experiment and representing constant values of $(\partial u_3/\partial x_1)$ and $(\partial u_3/\partial x_2)$ for a given load are shown in Fig. 16.19. By using Eqs. (4.31) and (16.65), one can show that the mode I stress intensity factor is given by

$$K_I = \frac{2E\sqrt{2\pi}\, r^{3/2}}{\nu h \cos\left(\frac{3\theta}{2}\right)}\left[\frac{Np}{2\Delta}\right] \tag{16.73}$$

where E is the elastic modulus, ν is the Poisson's ratio, N is the fringe order, h is the specimen thickness, $r = \sqrt{x^2 + y^2}$ and $\theta = \tan^{-1}\frac{y}{x}$

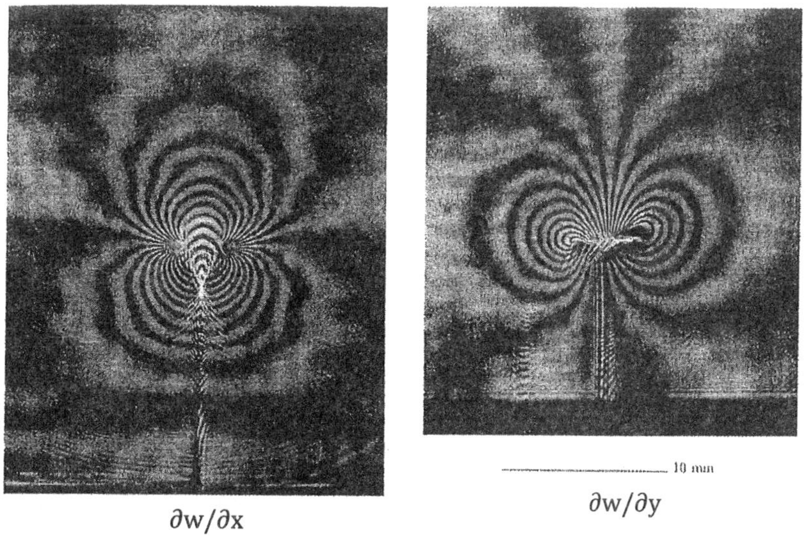

Figure 16.19 Contours of (∂w/∂x) and (∂w/∂y) in the three point bend specimen (*courtesy H.V. Tippur*).

For further details of this technique refer to [19, 20, 21].

16.10 METHODS TO DETERMINE K_I IN THREE-DIMENSIONAL BODIES

Methods for determining the opening-mode stress-intensity factor for two-dimensional bodies under plane-stress conditions with through thickness cracks have been described in the previous sections. However, in many engineering problems, the components are three dimensional and the cracks only extend part of the way into the body. This part-through crack in a three-dimensional body is a very difficult problem to solve using either analytical or numerical methods. The most effective experimental method used to determine K_I involves the stress-freezing process of photoelasticity. A three-dimensional model of the body containing a crack is loaded and the temperature is cycled to lock-in the deformations on a molecular scale. The model is then sliced so that the crack front is essentially perpendicular to the plane of the slice. If the crack front is curved, several slices are removed at various locations along the front of the crack as shown in Fig. 16.17.

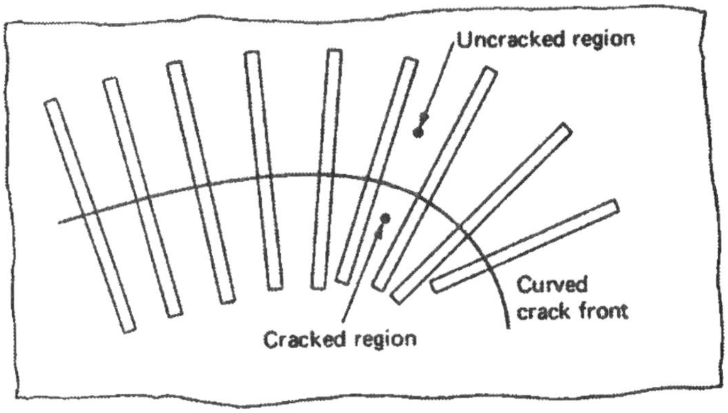

Figure 16.17 Slices removed from a three-dimensional photoelastic model oriented so that the plane of the slice is normal to the crack front.

The slices containing the crack-front segments are examined in a standard circular polariscope and fringe patterns are recorded as illustrated in Fig. 16.18. Smith and his associates use an extrapolation technique that is based on Eq. (16.13). In the extrapolation technique, an apparent stress intensity factor K^*_{AP} is defined as:

$$K^*_{AP} = \sqrt{2\pi r}(2\tau_m) = \sqrt{2\pi r}\frac{Nf_\sigma}{h} \qquad (16.74)$$

where the effect of σ_{0x} in Eq. (16.13) has been dropped in the definition of K_I. Data along the $\theta = 90°$ line is taken to give (N_i, r_i) at many different locations. Using Eq. (16.40) and experimental information regarding the loading σ on the model and the crack length a, Smith determines $K^*_{AP}/\sigma(\pi a)^{1/2}$ for each (N_i, r_i) and then plots these values as a function of $(r/a)^{1/2}$ as illustrated in Fig. 16.19.

Figure 16.18 Isochromatic fringe patterns for a slice removed from a three-dimensional model with a crack. Courtesy of C. W. Smith.

Experience has shown that the data in Fig. 16.19 falls into three regions. For small values of r, the points falls under the extrapolation line. This is probably due to the effects of crack-tip blunting which occurs during the stress-freezing process. For intermediate values of r (the solid points in Fig. 16.19), $K^*_{AP} / \sigma(\pi a)^{1/2}$ increases linearly with $(r/a)^{1/2}$. These points define the linear region of the response and provide the basis for the extrapolation procedure. Finally, for large values of r, the data points fall above the extrapolation line. This deviation is probably due to the inadequacies of the two-parameter method used in the formulation of Eq. (16.13).

A second approach is to treat the slice as a principal plane and use the overdeterministic method described in Sections 16.4 and 16.5 to determine K_I. Hyde and Warrior [15] have evaluated the accuracies of several different methods of determining K_I from slices taken from frozen stress photoelastic models. They concluded that the multiple-point overdeterministic method was simple to apply and provided very accurate results.

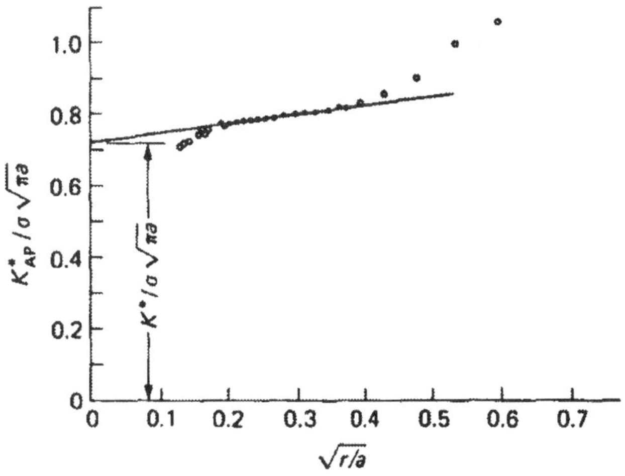

Figure 16.19 Extrapolation of $K^*_{AP} / \sigma(\pi a)^{1/2}$ to determine $K_I / \sigma(\pi a)^{1/2}$ from the intercept.

16.11MIXED-MODE STRESS INTENSITY-FACTORS K_I AND K_{II}

In many instances, the loading on a body with a crack is not symmetric and the crack is subjected to either shear loading or a combination of shear- and opening-mode loading. In situations where only the shearing mode exists, the behavior of the crack is assessed using the stress intensity factor K_{II}. However, if the body is subjected to mixed-mode loading with both opening and shearing components of the load, then both K_I and K_{II} must be determined to assess the stability of the crack.

Several experimental methods can be employed to determine K_I, K_{II}, and other stress parameters such as σ_{0x}. The photoelastic method, which yields isochromatic fringe patterns, is an excellent approach since the shape of the fringes in the pattern is markedly affected by these three quantities. Computer reconstructed fringe patterns, presented in Fig. 16.20, show the effects of varying K_I, K_{II}, and σ_{0x} on the shape of the classical fringe loops. In Fig. 16.20a, the loading is pure mode I and the fringe loops are symmetric about both the x and y axes. The fringe loops close at the origin which is located at the tip of the crack. The fringe pattern for pure mode II loading, shown in Fig. 16.20b, is symmetric again relative to both the x and y axes. Note that the fringe loops appear to be continuous across the crack line and do not close on the crack tip.

As indicated in Fig. 16.20c, the application of K_{II} destroys symmetry relative to the x axis. The application of the shear loading has rotated the two sets of fringe loops by the same angle and the pattern is symmetric about a rotated coordinate system 0x'y'. The final example, presented in Fig. 16.20d, shows the effect of the uniform field stress σ_{0x} on a mixed mode fringe pattern. All aspects of symmetry are destroyed by σ_{0x}. Moreover, the orders of the fringes above and below the crack line are markedly different.

Figure 16.20 Computer reconstructed fringe patterns showing the effects of K_I, K_{II}, and σ_{0x} on the characteristic shape of the isochromatic fringe patterns.

When the isochromatic fringe pattern is not symmetric relative to the crack line (i.e. the x axis), the stress state is due to a combination of both opening- and shear-mode loading. Two approaches are described to treat the mixed-mode case. The first approach relates the stress field to three unknowns; namely, K_I, K_{II} and σ_{0x}. The second approach describes the stress field in terms of many unknowns A_n, B_m, C_n, D_m, where K_I and K_{II} are defined in terms of A_0 and C_0.

Consider the first approach, which is valid only in the very-near field, as described in Fig. 4.12 and superimpose Eqs. (4.36) and (4.43) to write:

$$
\begin{aligned}
\sigma_{xx} &= (2\pi r)^{-1/2}\left[K_I \cos\frac{\theta}{2}\left(1-\sin\frac{\theta}{2}\sin\frac{3\theta}{2}\right) - K_{II}\sin\frac{\theta}{2}\left(2+\cos\frac{\theta}{2}\cos\frac{3\theta}{2}\right)\right] - \sigma_{0x} \\
\sigma_{yy} &= (2\pi r)^{-1/2}\left[K_I \cos\frac{\theta}{2}\left(1+\sin\frac{\theta}{2}\sin\frac{3\theta}{2}\right) + K_{II}\sin\frac{\theta}{2}\cos\frac{\theta}{2}\cos\frac{3\theta}{2}\right] \qquad (16.75) \\
\tau_{xy} &= (2\pi r)^{-1/2}\left[K_I \sin\frac{\theta}{2}\cos\frac{\theta}{2}\cos\frac{3\theta}{2} + K_{II}\cos\frac{\theta}{2}\left(1-\sin\frac{\theta}{2}\sin\frac{3\theta}{2}\right)\right]
\end{aligned}
$$

Substituting Eqs. (16.75) and (16.8) into Eq. (16.2) gives the stress-optic relation for the limited-parameter, mixed-mode condition as:

$$\left(\frac{Nf_\sigma}{h}\right)^2 = (2\pi r)^{-1}\left[(K_I \sin\theta + 2K_{II}\cos\theta)^2 + (K_{II}\sin\theta)^2\right]$$
$$+2\sigma_{0x}(2\pi r)^{-1/2}\sin\frac{\theta}{2}\left[K_I\sin\theta(1+2\cos\theta) + K_{II}\left(1+2\cos^2\theta+\cos\theta\right)\right] + \sigma_{0x}^2 \qquad (16.76)$$

This relation is nonlinear in terms of the three unknowns K_I, K_{II} and σ_{0x} and the solution is more difficult because of the nonlinear terms. A method of solution of Eq. (16.76) involves selecting data along one or two lines which constrains θ and simplifies Eq. (16.42). For example, consider the case where $K_{II} < 0$ and $\sigma_{0x} \neq 0$. The isochromatic fringes do not form closed loops but instead they intersect the upper crack boundary as illustrated in Fig. 16.12. Along this selected line (θ = π) Eq. 16.76 reduces to:

$$\left(\frac{Nf_\sigma}{h}\right)^2 = (2\pi r)^{-1}\left(4K_{II}^2\right) + (2\pi r)^{-1/2}\left(4K_{II}\sigma_{0x}\right) + \sigma_{0x}^2 \qquad (a)$$

Note that Eq. (a) is independent of K_I. Rewriting Eq. (a) gives:

$$\frac{Nf_\sigma}{h} = \pm\left(\frac{2K_{II}}{\sqrt{2\pi r}} + \sigma_{0x}\right) \qquad (16.77)$$

The choice of signs in Eq. (16.77) is based on the fact that the isochromatic fringes intersect the θ = π line when $K_{II} < 0$ and intersect the θ = – π line when $K_{II} > 0$. Consider two different fringes, $N_1 > N_2$, both of which intercept the upper edge of the crack (θ = π) at positions r_1, r_2, respectively, as shown in Fig. 16.21. Taking the negative sign option in Eq. (16.77) and eliminating σ_{0x} by substitution leads to:

$$K_{II} = \frac{f_\sigma}{h}\sqrt{\frac{\pi}{2}}\left(\frac{\sqrt{r_1 r_2}}{\sqrt{r_1} - \sqrt{r_2}}\right)(N_1 - N_2) \qquad (16.78)$$

Note that $K_{II} < 0$, as required, since $N_1 < N_2$ and $r_2 < r_1$. Next, σ_{0x} is obtained from Eqs. (16.77) and (16.78) as:

$$\sigma_{0x} = -\left(\frac{N_1 f_\sigma}{h}\right) - \frac{2K_{II}}{\sqrt{2\pi r_1}} \qquad (16.79)$$

To determine K_I reevaluate Eq. (16.76) using θ = π/2, N_3 and r_3 to obtain:

$$\left(\frac{N_3 f_\sigma}{h}\right)^2 = \frac{1}{2\pi r_3}\left(K_I^2 + K_{II}^2\right) + \frac{\sigma_{0x}}{\sqrt{\pi r_3}}\left(K_I + K_{II}\right) + \sigma_{0x}^2 \qquad (16.80)$$

Equation (16.80) is quadratic in terms of K_I and thus:

$$K_I = \frac{-b + \sqrt{b^2 - 4ac}}{2a} \qquad (16.81)$$

where

$$a = \frac{1}{(2\pi r_3)} \qquad b = \frac{\sigma_{0x}}{\sqrt{\pi r_3}} \qquad c = \frac{K_{II}^2}{2\pi r_3} + \frac{K_{II}\sigma_{0x}}{\sqrt{\pi r_3}} + \sigma_{0x}^2 - \left(\frac{N_3 f_\sigma}{h}\right)^2 \tag{16.82}$$

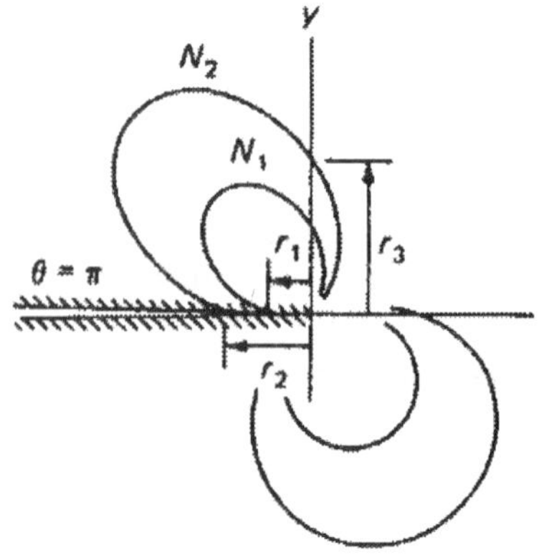

Figure 16.21 Isochromatic fringes intersect the upper crack line ($\theta = \pi$) when $K_{II} < 0$.

The terms N_3 and r_3 are defined in Fig. 16.21. The plus sign is selected in front of the radical in Eq. (16.81) because $K_I > 0$.

This example illustrates the selected-line method where the data was constrained to two lines $\theta = \pi$ and $\theta = \pi/2$ to simplify Eq. (16.76). Other selections of lines are possible as other choices of θ exist which simplify Eq. (16.76). However, care should be exercised in selecting θ so as to define lines which intersect one or more fringes as nearly normal as possible. Near-normal intersections improve the accuracy in determining the radial position r of the intersection point.

The second approach to solve for K_I and K_{II} in a mixed-mode loading situation is to use the near-field equations which describe the stresses in region 2 of Fig. 4.12. Superposition of Eqs. (4.48) for the opening mode and Eqs. (4.52) for the shearing mode gives series equations for σ_{xx}, σ_{yy} and τ_{xy}. The series equations contain unknown coefficients A_n, B_m, C_n, and D_m. The number of unknowns depends upon the complexity of the mixed-mode pattern. In most analyzes, three terms in each series for a total of eleven terms (recall D_0 does not occur) is more than sufficient to accurately represent the stresses in the near field.

The relations for the superimposed stresses are given by:

$$\sigma_{xx} = (\sigma_{xx})_I + (\sigma_{xx})_{II} \tag{16.83}$$

Similar relations for σ_{yy} and τ_{xy} are substituted into Eq. (16.16) to obtain expressions for D and T. Thus,

$$\begin{aligned} D = \frac{\sigma_{yy} - \sigma_{xx}}{2} &= \sum_{n=0}^{2}\left(n-\frac{1}{2}\right)A_n r^{n-1/2}\sin\theta\sin\left(n-\frac{3}{2}\right)\theta + \sum_{m=0}^{2} B_m r^m\left[m\sin\theta\sin(m\theta)+\cos(m\theta)\right] \\ &-\sum_{n=0}^{N} C_n r^{(n-1/2)}\left[\left(n-\frac{1}{2}\right)\sin\theta\cos\left(n-\frac{3}{2}\right)\theta + \sin\left(n-\frac{1}{2}\right)\theta\right] - \sum_{m=0}^{M} D_m r^m \sin\theta\cos(m-1)\theta \end{aligned} \tag{16.84}$$

and

$$\begin{aligned} T = \tau_{xy} &= -\sum_{n=0}^{2}\left(n-\frac{1}{2}\right)A_n r^{n-1/2}\sin\theta\cos\left(n-\frac{3}{2}\right)\theta - \sum_{m=0}^{2} B_m r^m\left[m\sin\theta\cos m\theta + \sin m\theta\right] \\ &+\sum_{n=0}^{N} C_n r^{(n-1/2)}\left[\cos\left(n-\frac{1}{2}\right)\theta - \left(n-\frac{1}{2}\right)\sin\theta\sin\left(n-\frac{3}{2}\right)\theta\right] + \sum_{m=0}^{M} D_m r^m\left[m\sin m\theta\sin(m-1)\theta\right] \end{aligned} \tag{16.85}$$

Substituting Eqs. (16.84) and (16.85) into Eq. (16.16) yields a very long and tedious stress-optic relation. This relation is solved for the 11 unknown coefficients by using the overdeterministic methods described in Section 16.5. The values of A_0 and C_0 are related to stress intensity factors K_I and K_{II} by:

$$K_I = \sqrt{2\pi}\, A_0 \tag{4.86}$$

$$K_{II} = \sqrt{2\pi}\, C_0 \tag{16.87}$$

16.12 BIREFRINGENT COATINGS IN FRACTURE MECHANICS

The use of birefringent coatings in fracture mechanics has been limited for a variety of different reasons. In the classical two-dimensional elasto-static problem there is little motivation to use coatings because the problem can be modeled accurately and either transmission photoelasticity or finite-element methods provide more direct solutions. In the two-dimensional elasto-dynamic problem, numerical methods are difficult to implement and modeling is not possible. Experiments to measure initiation toughness K_{Ic} and propagation toughness K_{ID} are possible using either strain gages [28] or birefringent coatings [27]. Of the two experimental approaches the strain gage methods are preferred over the birefringent coating method because of equipment requirements.

It appears that there are two technical areas where birefringent coatings have significant advantage over other experimental approaches. The first area is in the development of optical gages (which sense say K_I) that can be used to monitor flawed structures over extended periods of time. The second is with static problems where elastic modeling is not possible but where surface measurements are sufficient to give the data necessary to characterize the fracture process. This is an important category of problems (ductile fracture) where the yield zone near the crack tip has enlarged and the normal procedure of characterizing fracture with a stress intensity factor K_I is not valid. Instead, other fracture parameters such as the J-integral are required to describe the plastic field at the crack tip.

16.12.1 Birefringent Coatings for K_I Determinations

When a birefringent coating is applied to cover the near-field region, the coating should be cut as shown in Fig. 16.22. This cut prevents the coating from bridging the crack and avoids the response due to the crack opening displacements. The isochromatic fringe loops which form in the coating are related to K_I. Note from Eq. (11.32b) and Eq. (11.35) that:

$$\sigma_1^s - \sigma_2^s = \frac{E^s(1+\nu^c)}{E^c(1+\nu^s)}\frac{Nf_\sigma}{2h^c} = \frac{E^s}{1+\nu^s}\frac{Nf_\varepsilon}{2h^c} \tag{16.88}$$

Since $2\tau^s_m = \sigma^s_1 - \sigma^s_2$, it is clear that:

$$\tau_m = \frac{E^s}{1+\nu^s}\frac{Nf_\varepsilon}{4h^c} \tag{16.89}$$

and that Eq. (16.16) holds if it is modified to read:

$$\left(\frac{E^s}{1+\nu^s}\frac{Nf_\varepsilon}{4h^c}\right)^2 = D^2 + T^2 \tag{16.90}$$

where D and T are given by Eqs. (16.17) and (16.18), respectively.

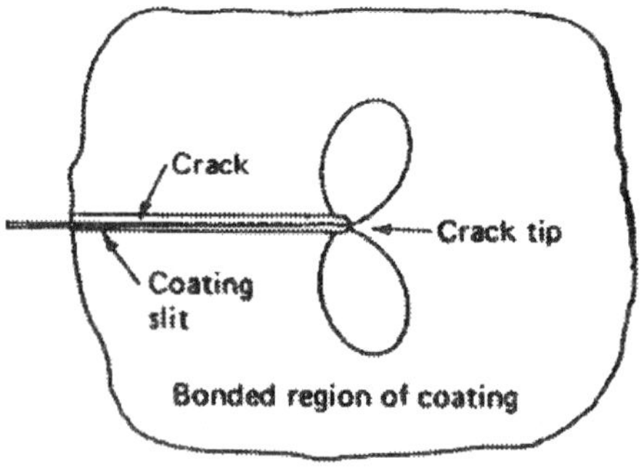

Figure 16.22 Placement of a slit birefringent coating to avoid bridging the crack.

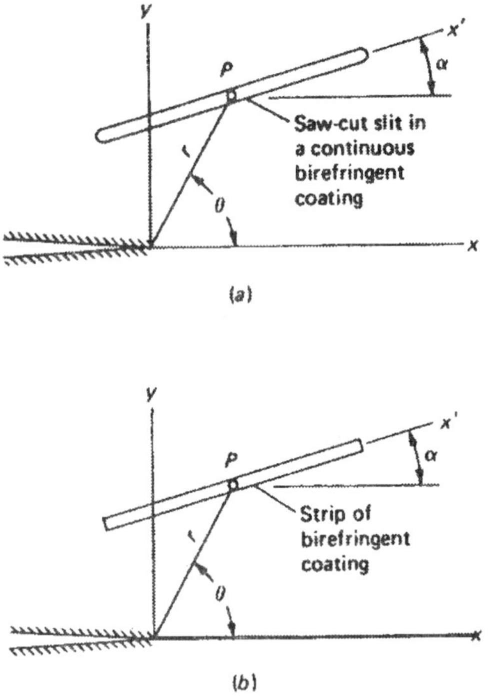

Figure 16.23 Discontinuous coatings in the near field region.
(a) Coating with a straight slit oriented at an angle α.
(b) A birefringent strip oriented at an angle α.

The stress intensity factor K_I is contained in D and T as described previously in Sections 16.4 and 16.5. Because the unknowns in Eq. (16.90) appear as nonlinear terms, this approach with birefringent coatings is not recommended. A much easier approach is possible if the continuous coating is slit or if a coating strip is employed as shown in Fig. 16.23. We will describe the advantage of using either slit or strip coatings in the Section 16.12.2.

16.12.2 Birefringent Strip or Slit Coatings

At the edge of a slit, the stress in the coating tangent to the edge is uniaxial and in the direction of the slit. Also, the coating stress normal to the slit is zero. The same condition holds for the birefringent strip providing the strip is thin compared to its thickness h^c. This fact implies that:

$$\varepsilon^c_{x'x'} = \varepsilon^s_{x'x'} \tag{16.91}$$

and $\sigma^c_{y'y'} = 0$ even when $\sigma^s_{y'y'} \neq 0$. Clearly, Eq. (16.91) shows that the slit and/or strip acts like a row of strain gages spaced on infinitely close centers.

From Eq. (4.42) and Eq. (2.6a), it is clear that:

$$2\mu\varepsilon_{x'x'} = A_0 r^{-1/2}\left[k\cos\frac{\theta}{2} - \frac{1}{2}\sin\theta\sin\frac{3\theta}{2}\cos 2\alpha + \frac{1}{2}\sin\theta\cos\frac{3\theta}{2}\sin 2\alpha\right]$$
$$+ B_0(k + \cos 2\alpha) + A_1 r^{1/2}\cos\frac{\theta}{2}\left[k + \sin^2\frac{\theta}{2}\cos 2\alpha - \frac{1}{2}\sin\theta\sin 2\alpha\right] \tag{16.92}$$

where α is defined in Fig. 16.23, $\mu = E/2(1 + \nu)$ is the shear modulus, and

$$k = (1 - \nu)/(1 + \nu) \tag{16.93}$$

Equation (16.57) is a three-term series representing $\varepsilon_{x'x'}$ that is linear in the unknowns. By using slits and/or strips and restricting the data taken to these lines, it is not necessary to measure $\sigma_1 - \sigma_2$ or $\varepsilon_1 - \varepsilon_2$ which gives rise to the nonlinear terms A_0^2, B_0^2, A_0B_0, etc.

Next, examine Eq. (16.92) and note that the coefficient of the B_0 term $(k + \cos 2\alpha)$ will vanish if:

$$\cos 2\alpha = -k = -(1 - \nu^s)/(1 + \nu^s) \tag{16.94}$$

Also, the coefficient of the A_1 term in Eq. (16.92) vanishes if:

$$\tan(\theta/2) = -\cot 2\alpha \tag{16.95}$$

These restrictions on α and θ depend entirely on Poisson's ratio of the specimen material as indicated in Table 16.2.

Table 16.2
Angles α and θ as a function of Poisson's Ratio ν

ν^s	θ, Degrees	α, Degrees
0.250	73.74	63.43
0.300	65.16	61.43
0.333	60.00	60.00
0.400	50.76	57.69
0.500	38.97	54.74

If the slit and/or strip is positioned with α according to Eq. (16.94) and the fringe order measured at point P which gives the angle θ used in Eq. (16.95), then at this point (and only this point):

$$2\mu\varepsilon_{x'x'} = A_0 r^{-1/2}\left[k\cos\frac{\theta}{2} + \frac{k}{2}\sin\theta\sin\frac{3\theta}{2} + \frac{1}{2}\sin\theta\cos\frac{3\theta}{2}\sin 2\alpha\right] \tag{16.96}$$

In this case, Eq. (10.17) for a birefringent coating reduces to:

$$\varepsilon_{x'x'} = \frac{Nf_\varepsilon}{2h} \tag{16.97}$$

Substituting Eqs. (16.97) and (16.15a) into Eq. (16.96) yields:

$$K_I = \frac{\sqrt{2\pi}\mu\left(Nf_\varepsilon/h^c\right)r^{1/2}}{k\cos\frac{\theta}{2} + \frac{k}{2}\sin\theta\sin\frac{3\theta}{2} + \frac{1}{2}\sin\theta\cos\frac{3\theta}{2}\sin 2\alpha} \tag{16.98}$$

where N is measured at point P(r,θ) with restrictions on θ as defined in Table 16.2. Inspection of Eq. (16.98) shows that the opening-mode stress-intensity factor K_I is a linear function of the fringe order N. Application of Eq. (16.98) to a direct reading K_I gage is discussed in Section 16.12.3.

16.12.3 A Direct Reading K_I Gage.

Consider a slit and/or strip positioned on an aluminum plate ($\nu = 1/3$) with the orientation $\alpha = 60°$. Note also from Table 16.1 that the angle θ, which defines the measurement point P, is also 60°. This fact that $\alpha = \theta$ implies that any point on a strip or on the edge of a slit can be used to measure K_I. With an aluminum specimen ($\mu = 26.2$ GPa, $\nu = 1/3$, $k = 1/2$, $\theta = \alpha = 60°$) and a polycarbonate coating ($f_\varepsilon = 40 \times 10^{-4}$ mm/fringe), Eq. (16.98) reduces to:

$$K_I = 2.355\frac{Nr^{1/2}}{h} \quad (16.99)$$

where K_I is given in units of MPa $\sqrt{m}$

The results shown in Eq. (16.99) suggest the use of a scale to measure K_I directly on the specimen without using instrumentation (other than a bonded circular polarizer and a light source). Consider a coating with h = 1 mm and observe the first two fringe orders N = 1 and N = 2. Equation (16.99) reduces to:

$$\begin{aligned} K_I &= 2.355\, r^{1/2} && \text{for } N = 1 \\ K_I &= 4.710\, r^{1/2} && \text{for } N = 2 \end{aligned} \quad (16.100)$$

where r is measured in mm. These two equations permit the two scales for the strip/slit to be determined, as shown in Fig. 16.24.

The range of the K_I gage shown in Fig. 16.24 varies from about 30 to 140 MPa$\sqrt{m}$. Of course, strips or slits covering different ranges can be fabricated by changing the coating thickness. More sensitive gages measuring lower values of K_I employ thicker strips and less sensitive gages employ thinner strips. It is important to select the thickness so that the fringe orders used to scale K_I are located at:

$$\frac{h^s}{2} < r < \frac{a}{2} \quad (16.101)$$

where a is the crack length.

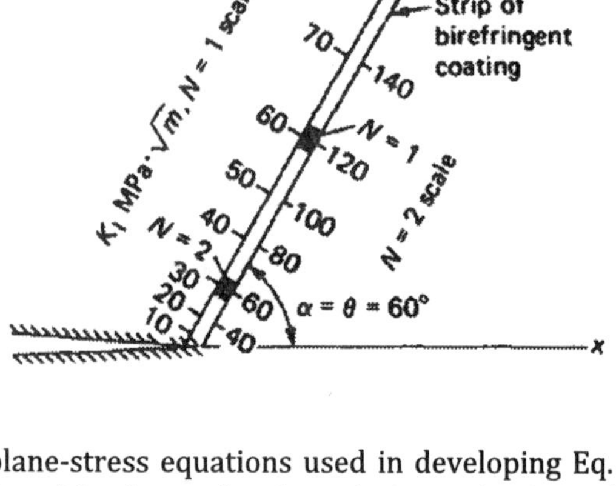

Figure 16.24 Scaling a birefringent strip for a direct reading K_I gage. ($h^c = 1$ mm, $\alpha = 60°$, $\nu = 1/3$, $f_\varepsilon = 40 \times 10^{-4}$ mm/fringe.)

If r is larger than half the specimen thickness, then the plane-stress equations used in developing Eq. (16.92) are valid. If r is less than one fifth the crack length and far from other boundaries and points of load application, then Eq. (16.92), with its three- term series representation, gives a close approximation to the infinite series solution.

16.12.4 Determining J in Power-Law Hardening Materials

Hutchinson [30] and Rice and Rosengren [31] considered materials with power-law hardening where the uniaxial stress-strain relation is defined by:

$$\frac{\varepsilon}{\varepsilon_0} = \alpha\left(\frac{\sigma}{\sigma_0}\right)^n \tag{16.102}$$

where σ_0 is the yield stress; ε_0 is the strain at yield given by (σ_0/E);α is a material constant; n is the strain hardening coefficient.

These authors developed the HRR theory, applicable for ductile materials, which gives the stress and strain distributions in the J-dominated region near the crack tip, as illustrated in Fig. 16.25. These relations for stress and strain are:

$$\sigma_{ij} = \sigma_0\left(\frac{J}{\alpha\sigma_0\varepsilon_0 I_n r}\right)^{1/(n+1)} S_{ij}(n,\theta) \tag{16.103a}$$

$$\varepsilon_{ij} = \varepsilon_0\left(\frac{J}{\alpha\sigma_0\varepsilon_0 I_n r}\right)^{n/(n+1)} E_{ij}(n,\theta) \tag{16.103b}$$

where I_n is a dimensionless constant which varies with n, ranging from 5 to 2.57 as n varies from 1 to ∞. S_{ij} and E_{ij} are functions of n and θ. These functions I_n, S_{ij}, and E_{ij} are tabulated by Shih in reference [32].

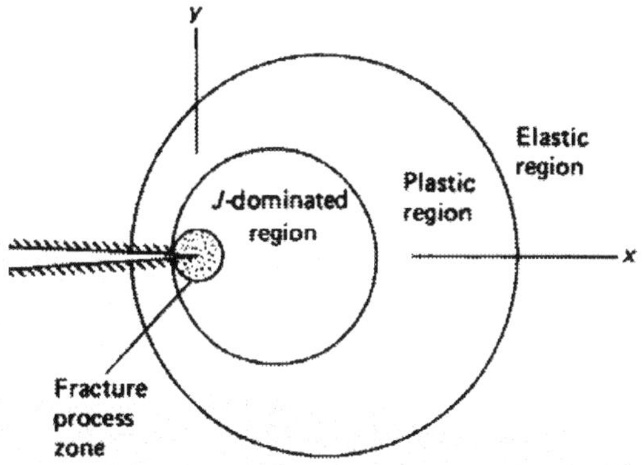

Figure 16.25 Schematic illustration of the J-dominated region embedded in the plastic region.

Next, examine the strain as expressed by Eq. (16.103) in the radial and tangential directions along the x axis. First, consider the distribution of strain along the positive x-axis where it is evident that $\varepsilon_{xx} = \varepsilon_{rr}$, $\theta = 0$ and $E_{rr}(n,\theta) = E_{rr}(n,0)$. Substituting these equalities into Eq. (16.103) gives:

$$\varepsilon_{xx} = \alpha\varepsilon_0\left(\frac{1}{\alpha\sigma_0\varepsilon_0 I_n}\right)^{n/(n+1)}\left(\frac{J}{x}\right)^{n/(n+1)} E_{rr}(n,0) \tag{16.104}$$

For a given material, α, ε_0, σ_0 and n are fixed and we may simplify the expressions for the strains by defining a material constant Q as:

$$Q = \alpha\varepsilon_0\left(\frac{1}{\alpha\sigma_0\varepsilon_0 I_n}\right)^{n/(n+1)} \tag{16.105}$$

Substituting Eq. (16.105) into Eq. (16.104) gives:

$$\varepsilon_{xx} = Q\, E_{rr}(n,0)\left(\frac{J}{x}\right)^{n/(n+1)} \tag{16.106}$$

This expression is valid along the positive x-axis. The strain ε_{yy} along the same x-axis is:

$$\varepsilon_{yy} = \varepsilon_{\theta\theta} = Q\, E_{\theta\theta}(n,0)\left(\frac{J}{x}\right)^{n/(n+1)} \tag{16.107}$$

Note that the x axis is a principal axis and that $\varepsilon_1 = \varepsilon_{yy}$ and $\varepsilon_2 = \varepsilon_{xx}$ for the opening-mode loading used in the HRR theory.

Consider next a patch of birefringent coating positioned in front of the crack as shown in Fig. 16.26. The coating will respond to the strains developed in the plastic and J-dominated regions and give the fringe order N as a function of position x. To interpret this data in terms of J (see Section 4.9.3 for a description of the J integral), substitute Eqs. (16.106) and (16.107) into Eq. (11.19) to obtain:

$$\frac{J}{x} = \left\{\frac{N f_\varepsilon}{2h^c Q\left[E_{\theta\theta}(n,0) - E_{rr}(n,0)\right]}\right\}^{(n+1)/n} \tag{16.108}$$

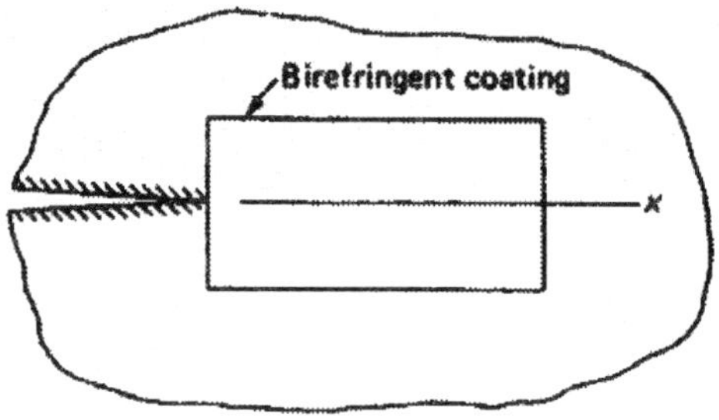

Figure 16.26 Patch of birefringent coating located in front of a crack covering both the J-dominated region and the plastic region.

If N is recorded at a number of locations along the x axis, J can be determined at a number of points in the plastic zone in front of the crack. However, only those values of the fringe order N measured in the J-dominated region are valid. It is possible to extend this analysis to determine the bounds of the J-dominated region by taking the logarithm of both sides of Eq. (16.108) to obtain:

$$\log_{10}\left(\frac{J}{x}\right) = \frac{(n+1)}{n}\log_{10}\left\{\frac{N f_\varepsilon}{2h^c Q\left[E_{\theta\theta}(n,0) - E_{rr}(n,0)\right]}\right\} \tag{16.109}$$

This relation is the basis for the graph shown in Fig. (16.27) where three regions are defined. Region I corresponds to low values of N which occur at large x and relatively low values of J/x. These measurements, close to the origin in Fig. 16.27, are taken at locations beyond the J-dominated region. Region III corresponds to high order fringes, with small x and large J/x. Locations of these measurements are too close to the crack tip for the HRR theory to be valid. Region II is between these extremes, and the results shown on the log-log graph are linear with a slope S = (n + 1)/n. The range of the linear portion of the curve of Fig. 16.27 defines the bounds of the J-dominated region along the x axis.

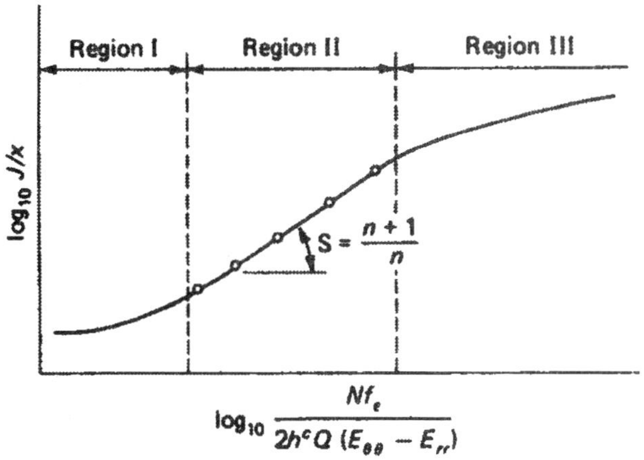

Figure 16.27 A log-log graph of J/x shows a linear relation in region II, which is J dominated where the HRR theory is valid.

16.12.5 Discussion

In the analysis of specimens containing cracks, the form of the solution for the stresses and strains in the local neighborhood of the crack tip is known in terms of some theory. In this coverage, the generalized Westergaard theory was used to represent the stress and strain distributions for the elastic or small-scale yielding examples. The HRR theory was used to give the plastic strain relations near the crack tip where the loading produced larger plastic zones. Because the theory describes the form of the solution, the experiment is only necessary to establish the validity of the theory and to determine the scaling constants used to fit the theory to the observed behavior as either elastic or plastic (i.e. K_I or J).

There are many experimental methods, which may be employed in checking the validity of the theory and in determining the scaling constants. The birefringent coatings, strips/slits and patches described here, have the advantage of simplicity since no instrumentation is necessary and the analysis is straight forward in all but the overdeterministic approaches. A second advantage is the availability of nearly continuous data over the region of interest. This is particularity important in ductile fracture where the plastic zone size is large and the available theories for the strain distribution are valid only over relatively small regions within the plastic zone.

EXERCISES

16.1 Verify Eq. (16.3).

16.2 Verify Eqs. (16.6) and (16.7) by using Eq. (16.3) together with the fact that $\partial\tau_m/\partial\theta = 0$.

16.3 From the fringe pattern shown in Fig. 16.1, determine K_I, if $f_\sigma = 7.0$ kN/m and h = 6.35 mm. Use the method developed by Irwin, which uses a single data point.

16.4 From the fringe pattern shown in Fig. 16.1, determine K_I if $f_\sigma = 7.0$ kN/m and h = 6.35 mm. Use a two-point deterministic approach and determine K_I and σ_{0x} without using $\partial\tau_m/\partial\theta = 0$.

16.5 From the fringe pattern shown in Fig. 16.1, determine K_I if $f_\sigma = 7.0$ kN/m and h = 6.35 mm. Use the two-fringe differencing method of Bradley and Kobayashi.

16.6 From the fringe pattern shown in Fig. 16.1, determine K_I if $f_\sigma = 7.0$ kN/m and h = 6.35 mm. Use C.W. Smith's approach which uses data along the line defined by $\theta = 90°$.

16.7 Write an engineering brief describing the difficulties encountered in measuring N, r and θ from Fig. 16.1 as required in Exercises 16.3, 16.4, 16.5, and 16.6.

16.8 Develop a technique for measuring K_I based on data taken along the line $\theta = 60°$.

16.9 Expand the relations for D and T in Eqs. (16.17) and (16.18) for θ = 0, 30, 45, 60 and 90°.

16.10 Write a computer program for determining K_I by using the overdeterministic method described in Section 16.5.

16.11 Use the computer program written for Exercise 16.10 together with 40 data points taken from Fig. 16.1 to determine K_I. Use M = N = 2.

16.12 Write a computer program which takes, as input, the results from an overdeterministic analysis for A_0, A_1, A_2, B_0, B_1, and B_2 and gives, as output, a reconstructed pattern similar to that shown in Fig. 16.5.

16.13 Use the program from Exercise 16.12 and the results from Exercise 16.11 to reconstruct a fringe pattern. Compare this reconstructed pattern with the fringe pattern of Fig. 16.1 which is the experimentally determined input data.

16.14 Verify Eqs. (16.32) and (16.33).

16.15 From the fringe pattern shown in Fig. 16.6, determine the quantity $(K_I h/f_p)$ by using the method of Dudderar and Gorman.

16.16 Consider a six-parameter expansion of Eq. (16.31) and write a definition of each matrix **[N]**, **[c]** and **[U]** in Eq. (16.34).

16.17 From the fringe pattern shown in Fig. 16.6, determine the quantity $(K_I h/f_p)$ by using the overdeterministic method outlined in Section 16.6.

16.18 Develop a simple method for determining K_I from moiré data which uses only two data points; one from the u field and the other from the v field.

16.19 Repeat Exercise 16.18, but select both data points from the u field.

16.20 Repeat Exercise 16.18, but select both data points from the v field.

16.21 Verify Eqs. (16.46) and then let the model material be optically isotropic and verify Eq. (16.49).

16.22 Begin with Eqs. (4.36) verify Eq. (16.48) and Eq. (16.49).

16.23 A caustic due to a stress singularity at a crack tip is formed on a screen located 1 m from a 4 mm thick model fabricated from PMMA (Plexiglas). The diameter D_o of the shadow spot is 10 mm. Determine K_I and then the radius r_o of the initial curve.

16.24 If the screen is moved to a new position in Exercise 10.38 so that z_o = 800 mm. Determine the diameter D_o of the shadow spot. Also, find the new radius r_o of the initial curve.

16.25 For mode II loading of a crack tip, show that the mapping relation (similar to Eq. 16.51) is given by:

$$x' = r\cos\theta - \frac{K_I}{\sqrt{2\pi}} z_o C_1 h r^{-3/2} \sin\frac{3\theta}{2}$$

$$y' = r\sin\theta + \frac{K_{II}}{\sqrt{2\pi}} z_o C_1 h r^{-3/2} \cos\frac{3\theta}{2}$$

16.26 Show that the crack tip caustic curve for pure mode II loading is of the form shown in Fig. E16.26.

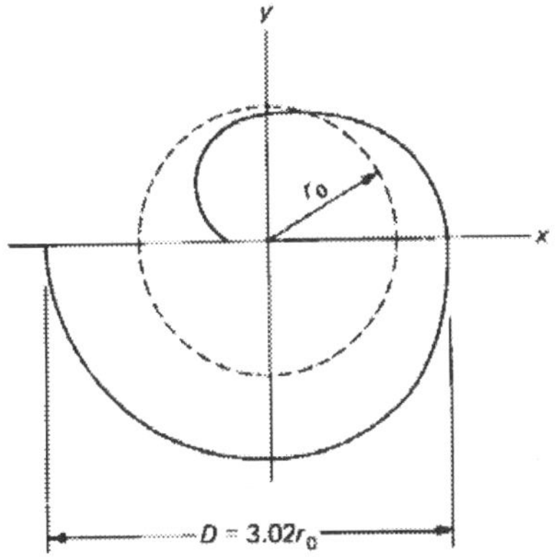

Figure E16.26

16.27 Show for pure mode II loading that

$$r_o = \left[\frac{3}{2}\frac{K_{II}}{\sqrt{2\pi}} z_o C_1 h\right]^{2/5}$$

16.28 Show for pure mode II loading that

$$K_{II} = \frac{2\sqrt{2\pi}}{3(3.02)^{5/2} z_o C_1 h} D^{5/2}$$

Hint, note that D = 3.02 r_o as defined in Fig. E16.26.

16.29 Sketch the shape of a mixed-mode caustic.
16.30 Starting with Eq. (16.65) derive Eq. (16.66).
16.31 You have a slice taken from a three-dimensional frozen stress model which yields a fringe pattern similar to that shown in Fig. 16.18 when the slice is examined in a circular polariscope. Can you use the extrapolation technique to determine $K^*_{AP}/\sigma\sqrt{\pi a}$? How will you generate the additional data points, which are necessary for the success of the extrapolation method?
16.32 Write a computer program that will generate a photoelastic fringe pattern for a crack in a large plane body subjected to mode II loading.
16.33 Write a computer program that will generate an isopachic fringe pattern for a crack in a large plane body subjected to mode II loading.
16.34 Write a computer program that will generate a u type moiré pattern for a crack in a large plane body subjected to mode II loading.
16.35 Write a computer program that will generate a v type moiré pattern for a crack in a large plane body subjected to mode II loading.
16.36 Verify Eqs. (16.41) and (16.42).
16.37 Equation (16.43) is valid only if the fringe order N is determined along the line $\theta = \pi$. Derive similar relations for the cases where the fringe order N is determined along the lines $\theta = \pm \pi/2$. Show a procedure for determining K_I, K_{II}, and σ_{0x} from these relations.
16.38 Verify Eqs. (16.47) and (16.48).
16.39 Verify Eqs. (16.53), (16.54), and (16.56).
16.40 Verify Eq. (16.57).
16.41 Verify Eq. (16.63).
16.42 Prepare a scale for N = 1 and N = 2 corresponding to a direct reading K_I gage exactly like the one described in Fig. 16.15 if h = 1.5 mm.
16.43 Repeat Exercise 16.32 but let h = 0.5 mm.

16.44 If you place a continuous birefringent coating on a through cracked panel as shown in Fig. 16.13 and observe a mode I isochromatic fringe loop of N = 2 with r_{max} = 5 mm and θ_{max} = 80°, estimate $(K_I h/f_\sigma)$.

16.45 Plot several stress-strain curves using Eq. (16.67) to describe the functional relationship between σ and ε for steel. Let the strain hardening coefficient vary between 1 and 30. Take σ_0 = 50,000 psi, ε_0 = 1666 $\mu\varepsilon$ and α = 1. Vary n in the process.

16.46 Verify Eq. (16.73).

16.47 Write an engineering brief describing an experiment using birefringent coatings to determine the validity of the HRR theory for a new material which has not been classified as ductile or brittle.

BIBLIOGRAPHY

1. Tada, H., Paris, P. C., Irwin, G.R.: Stress Analysis of Cracks Handbook, 3rd Edition, ASM International, Ohio, 2000.
2. Post, D.: "Photoelastic Stress Analysis for an Edge Crack in a Tensile Field," Proceedings SESA, Vol. 12, No. 1, pp. 99-116, 1954.
3. Wells, A. and D. Post: "The Dynamic Stress Distribution Surrounding a Running Crack - A Photoelastic Analysis," Proceedings SESA, Vol. 16, No. 1, pp. 69-92, 1958.
4. Irwin, G. R.: Discussion of Reference 3, Proceedings SESA, Vol. 16, No. 1, pp 93-96, 1958.
5. Etheridge, J. M. and J. W. Dally: "A Critical Review of Methods for Determining Stress Intensity Factors from Isochromatic Fringes," Experimental Mechanics, Vol. 17, pp 248-254, 1977.
6. Bradley, W. B. and A. S. Kobayashi: "An Investigation of Propagating Cracks by Dynamic Photoelasticity," Experimental Mechanics, Vol. 10, pp 106-113, 1970.
7. Schroedl, M. A. and C. W. Smith: "Local Stress Near Deep Surface Flaws Under Cylindrical Bonding Fields," Progress in Flaw Growth and Fracture Toughness Testing, ASTM STP 536, pp 45-63, 1973.
8. Sanford, R. J. and J. W. Dally: "A General Method for Determining Mixed Mode Stress Intensity Factors from Isochromatic Fringe Patterns," Engineering Fracture Mechanics, Vol. 11, pp 621-633, 1979.
9. Sanford, R. J.: "Determining Fracture Parameters with Full-field Optical Methods," Experimental Mechanics, Vol. 29, pp 241-247, 1989.
10. Dudderar, T. D. and H. J. Gorman: "The Determination of Mode I Factors by Holographic Interferometry," Experimental Mechanics, Vol. 13, No. 4, pp 145-149, 1973.
11. Dudderar, T. D. and R. O'Regan: "Measurement of the Strain Field Near a Crack Tip in Polymethylmethacrylate by Holographic Interferometry," Experimental Mechanics, Vol. 11, No. 2, pp. 49-56, 1971.
12. Barker, D. B., R. J. Sanford, and R. Chona: "Determining K_I and Related Stress-Field Parameters from Displacement Fields," Experimental Mechanics, Vol. 25, No. 12, pp. 399-407, 1985.
13. Smith, C. W.: "Use of 3-D Photoelasticity in Fracture Mechanics," Experimental Mechanics, Vol. 13, pp. 539-544, 1973.
14. Smith, C. W. and A. S. Kobayashi: "Experimental Fracture Mechanics," 2nd Edition, Handbook on Experimental Mechanics, Ed. A. S. Kobayashi, Prentice Hall, Englewood Cliffs, New Jersey, pp. 905-968, 1993.
15. Hyde, T. H. and N. A. Warrior: "A Critical Assessment of Photoelastic Methods of Determining Stress Intensity Factors," Applied Solid Mechanics - 2, A. S. Tooth and J. Spence, (eds.), Elsevier Applied Science, New York, pp. 23-40, 1987.
16. Manogg, P., Schattenoptische Messung der spezifischen Bruchenergie hrend des Bruchvorgangs bei Plexiglas, Proc. Int. Conf. Phys. Non-Crystalline Solids, Delft, The Netherlands, pp. 481-490, 1964.
17. Theocaris, P. S. and N. Joakimides, Some Properties of Generalized Epicycloids Applied to Fracture Mechanics, J. Appl. Mech., vol. 22, pp. 876-890, 1971.
18. Kalthoff, J. F., Shadow Optical Method of Caustics, Handbook on Experimental Mechanics, Ed. A. A. Kobayashi, Chapter 9, pp. 430-500, Prentice-Hall, Inc. 1987.
19. Tippur, H. V., S. Krishnaswamy and A. J. Rosakis, "A Coherent Gradient Sensor for crack tip deformation Measurements: Analysis and Experimental Results," International Journal of Fracture, Vol. 48, pp. 193-204, 1991.
20. Tippur, H. V., S. Krishnaswamy and A. J. Rosakis, "Optical Mapping of Crack Tip Deformations Using the Method of Transmission and Reflection Coherent Gradient Sensing: A Study of Crack Tip K-Dominance," International Journal of Fracture, Vol. 52, pp. 91-117, 1991.
21. Tippur, H. V., "Optical Methods for Dynamic Fracture Mechanics," Dynamic Fracture Mechanics, Editor Arun Shukla, World Scientific Publishing Company, ISBN 981-256-840-9, 2006.
22. Kafri, O., "Noncoherent Method for Mapping Phase Objects" Optics Letters, Vol. 5, pp. 555-557, 1980.

23. Dally, J. W. and R. J. Sanford: "Classification of Stress-Intensity Factors from Isochromatic-fringe Patterns," Experimental Mechanics, Vol. 18, No. 12, pp. 441-448, 1978.
24. Smith, D. G. and C. W. Smith: "Photoelastic Determination of Mixed Mode Stress Intensity Factors," Engineering Fracture Mechanics, Vol. 4, No. 2, pp. 357-366, 1972.
25. Gdoutos, E. E. and P. G. Theocaris: "A Photoelastic Determination of Mixed Mode Stress Intensity Factors," Experimental Mechanics, Vol. 18, No. 3, pp. 87-97, 1978.
26. Der, V. K., D. B. Barker, and D. C. Holloway: Mechanics Research Communications, Vol. 5, No. 6, pp. 313-318, 1978.
27. Kobayashi, T. and J. W. Dally: "Dynamic Photoelastic Determination of the $\dot{a}$ -K Relation for 4340 Alloy Steel," Crack Methodology and Application, G. T. Hahn and M. F. Kanninen (eds.), ASTM STP 711, pp 189-210, 1980.
28. Dally, J. W. and R. J. Sanford: "On Measuring the Instantaneous Stress Intensity Factor for Propagating Cracks," Proceedings 7th International Conference on Fracture, ICF7 Houston, TX. March 20-24, 1989.
29. Dally, J. W., J. R. Berger, and Y-C. Ham: "On the Use of Birefringent Coatings in Fracture Mechanics," Proceedings SEM Spring Conference, pp. 513-520, 1989.
30. Hutchinson, J. W.: "Singular Behavior at the End of a Tensile Crack in a Hardening Material," Journal of the Mechanics and Physics of Solids, Vol. 16, pp. 13-31, 1968.
31. Rice, J. R. and G. F. Rosengren: "Plane Strain Deformation Near a Crack Tip in a Power Law Hardening Material," Journal of the Mechanics and Physics of Solids, Vol. 16, pp. 1-12, 1968.
32. Shih, C. F.: "Tables of Hutchinson-Rice-Rosengren Singular Field Quantities," Brown University Report MRL E-147, Material Research Laboratory, Brown University, June 1983.
33. Sharpe, Jr., W. N. (Ed.): Springer Handbook of Experimental Solid Mechanics, Springer Science + Business Media LLC, New York, 2008.
34. Nigam, H. and A. Shukla: "A Note on the Stress Intensity Factor and Crack Velocity Relationship for Homalite 100," Engineering Fracture Mechanics, Vol. 25, No. 1, pp. 91-102, 1986.
35. Shukla, A. and R. Chona: "The Stress Field Surrounding a Rapidly Propagating Curving Crack: An Experimental Study," ASTM STP 945, pp. 86-99, 1987.
36. Nigam, H. and A. Shukla: "Comparison of the Techniques of Transmitted Caustics and Photoelasticity as Applied to Fracture," Experimental Mechanics, Vol.28, No.2, pp. 123-135, 1988.
37. Shukla, A., R.K. Agarwal and H. Nigam: "Dynamic Fracture Studies of 7075-T6 Aluminum and 4340 Steel Using Strain Gages and Photoelastic Coatings," Engineering Fracture Mechanics, Vol.31, No.3, pp.501-515, 1988.

PART IV

MEASUREMENTS UNDER HIGH STRAIN RATES AND SMALL LENGTH SCALES

CHAPTER 17

DYNAMIC MEASUREMENTS

17.1 INTRODUCTION

Materials and structures are often subjected to dynamic loading. Examples of dynamic loading include the force applied on a package when it drops to the ground, a car body being impacted during an accident, armor penetration by a bullet; a building collapsing due to an explosion, etc. During the loading process, the mechanical quantities such as force, displacement, stress, strain, etc. vary rapidly with time. To understand this dynamic response of materials and structures, the real time variation of these mechanical quantities must be measured. The equipment and techniques necessary to implement real time dynamic measurements are described in this chapter.

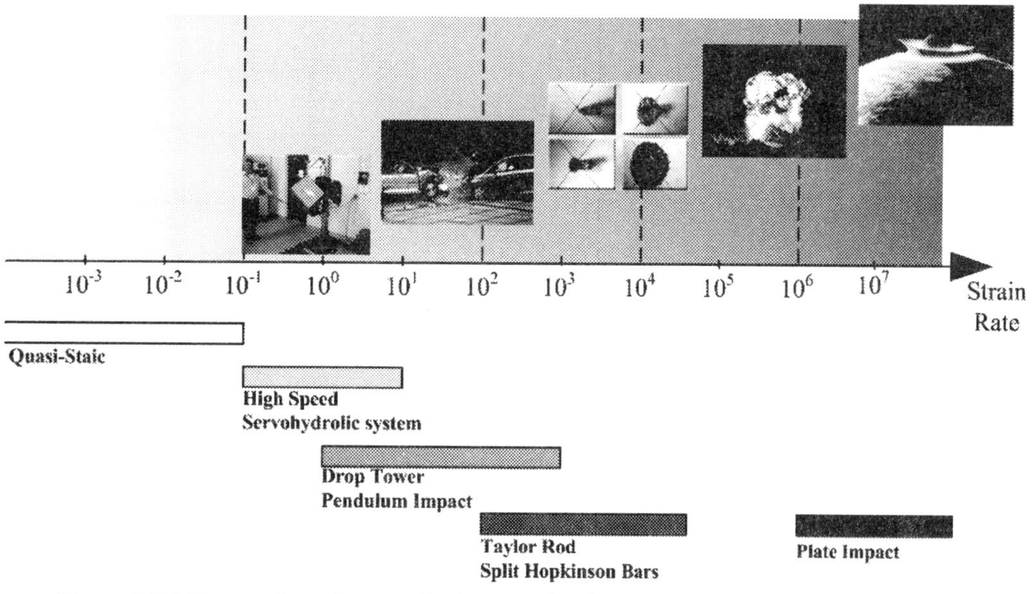

Figure 17.1 Range of strain rates in dynamic loadings and measurement techniques

In order to perform dynamic measurements and understand dynamic response, it is helpful to understand the concept of strain rate and also develop a sense of the range of strain rate values encountered in dynamic loadings. Strain rate is defined as the change in strain with respect to time. The strain rate defines how quickly the material deforms under dynamic loading. The range of strain rate values for typical events and experimentation techniques to measure material response are shown in Fig. 17.1. For quasi-static loading, the strain rates are less than 10^{-1} ε/s. For crash events, the strain rates have a range from 10^{-1} to 10^{2} ε/s; for ballistic impacts, the strain rates range from 10^{2} to 10^{4} ε/s; for blast events, the strain rates range from 10^{4} to 10^{6} ε/s. During an asteroid impact on the earth or a supernova explosion, the strain rates can reach values greater than 10^{7} ε/s.

In this chapter, classical dynamic measurement techniques are described based on the strain rate range. These techniques include high speed servo-hydraulic systems ($10^{-1} \sim 10^{2}$ ε/s), drop towers ($10^{0} \sim 10^{3}$ ε/s), the Taylor rod ($10^{2} \sim 10^{4}$ ε/s), pendulum impact ($10^{0} \sim 10^{3}$ ε/s), split Hopkinson bars ($10^{2} \sim 10^{4}$ ε/s), and plate impact methods ($10^{6} \sim 10^{8}$ ε/s). Some non-contact optical methods used in dynamic measurements are also introduced in the final section.

17.2 HIGH-SPEED SERVO-HYDRAULIC SYSTEM

Since World War II, servo-hydraulic systems have been widely used to apply mechanical loads on materials and specimens of interest. Today, most of the quasi-static material test machines are based on servo-hydraulic systems due to the advantages they offer. These advantages include high reliability, high accuracy, high stiffness, short positioning time, etc. Many companies such as MTS, Instron and others manufacture these quasi-static servo-hydraulic systems. If appropriately designed, servo-hydraulic systems can be extended to conduct controlled, high-speed material tests. Dynamic loading using a servo-hydraulic system is limited to inter-mediate strain rates ($10^{-1} \sim 10^{2}$ ε/s) due to the inertia of the hydraulic fluid and the actuator piston.

It is helpful to understand the fundamental aspects of servo-hydraulic systems. In simple terms, the servo-hydraulic system is an auto-controlled, auto-adjusted and liquid-driven system. The word "servo" means "feedback control". The control system of the machine constantly tracks feedback information from the transducer, adjusts the driver element in the machine and verifies that the machine is performing properly. The word "hydraulic" means that the mechanism is operated by the pressure transmitted when a liquid is forced into an actuator.

Figure 17.2 Sketch of a typical servo-hydraulic system

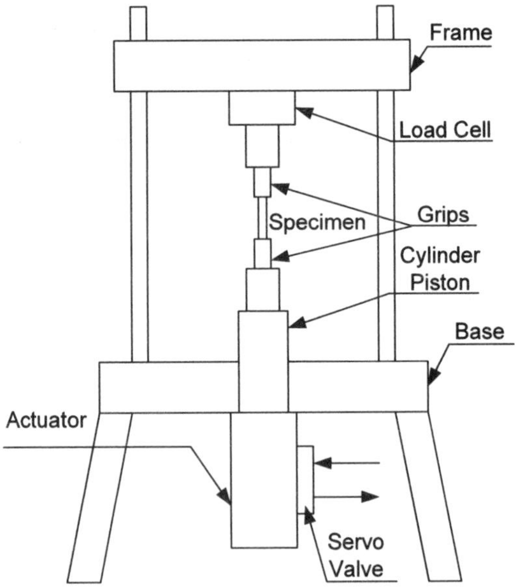

A sketch of a typical servo-hydraulic system for both low-speed and high-speed tests is presented in Fig. 17.2. A photograph of a high-speed servo-hydraulic system is shown in Fig. 17.3. The specimen is clamped between the upper and lower grips, where the upper grip is fixed and the lower grip is controlled by an actuator, which is driven by a servo-hydraulic power source. A block diagram depicting the system for controlling the application of the load or displacement is presented in Fig. 17.4. The control signal is usually generated using system software and a personal computer. The control signal is sent to a controller where the signal current is amplified prior to its transmission to the servo control valve. The servo control valve controls the direction of flow as well as the flow rate of hydraulic fluid moving into the actuator in order to achieve a motion with a constant velocity of up to 30 m/s. The test specimen is mounted in a loading frame. Sensors that are incorporated with a system usually include a load cell and a displacement transducer. One of these sensors provides a feedback signal that is transmitted to the controller. The feedback signal is compared to the input signal and the difference is amplified and transmitted to the servo-valve where it moves the spool in the valve, which in turn adjusts the flow rate of fluid entering the actuator. This closed loop control system enables the application of high-rate loads or displacements to materials and test specimens of interest.

(a) (b)

Figure 17.3 A high-speed servo-hydraulic system (From Axel Products, Inc)
(a) Complete arrangement (b) Specimen, grips and loading head

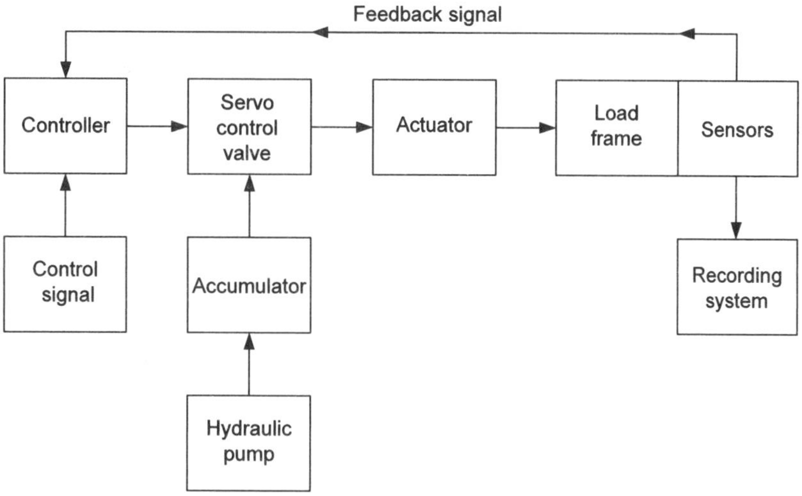

Figure 17.4 Diagram of a servo-controlled closed-loop loading system.

17.2.1 Dynamic and Inertia Effects

During the loading process, only two parts are moving: the lower loading head and the piston in the actuator. In high-speed loading, the inertia causes two problems. First, the inertia force of the loading head on the actuator, due to the acceleration, reduces the accuracy of the force measurement. Second, it is a progressive process to accelerate the piston, which has a large mass. In this process, it takes time to reach the required velocity and to stabilize the piston at this value. Therefore, the strain rate may not be constant when the specimen is loaded.

There are two ways to minimize the effects of the inertia force. The first method is passive. Because the inertia is proportional to the mass, reducing the head mass will increase the accuracy of the force measurement. The head mass can be reduced by either using metals with high strength to weight ratios, such as Titanium, or by using more compact designs of the head. The second method is active. A built-in accelerometer is used to measure the real time acceleration of the loading head[1]. Next, through an automatic calculation process, the force data, obtained from the force transducer, can be compensated for these inertia effects.

To reduce the time required to reach and stabilize the velocity, a slack adapter is placed between the loading head and the piston[2]. The actual unit and a simple schematic of the mechanism are shown in Fig. 17.5. The track rod is able to slide freely within the cylinder, which is connected to the piston directly. When the experiment begins, the piston is first accelerated to the required velocity. During this time, the track rod slides in the cylinder but no load is applied on the specimen. When the track rod reaches the end of the cylinder, the motion of the piston, which is moving with constant velocity, is transmitted to the specimen.

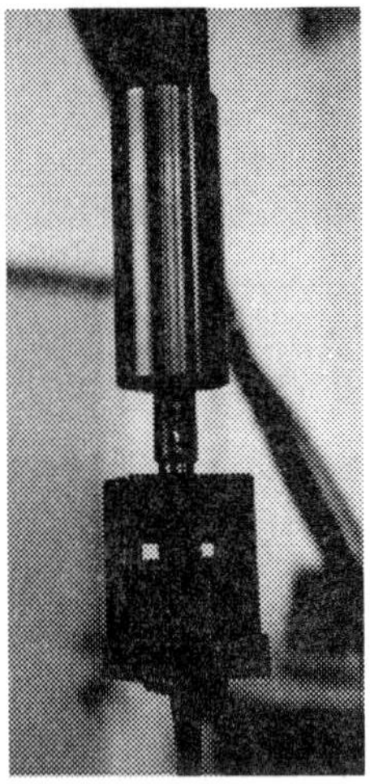

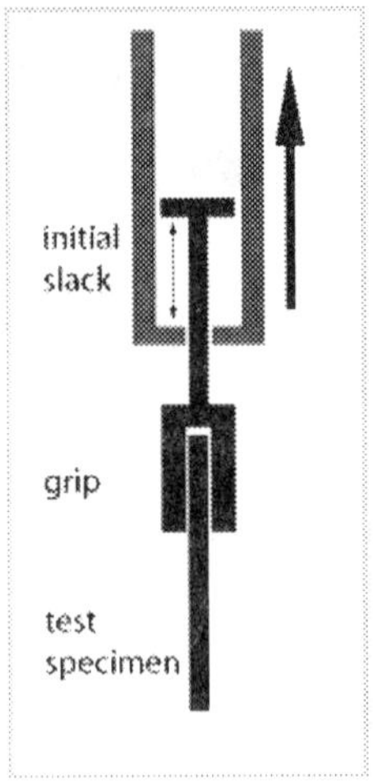

Figure 17.5 A slack adapter (From Axel Products, Inc)

17.2.2 Displacement and True Strain Measurement

The displacement of the actuator piston is easy to measure with a displacement transducer, which is mounted on the piston. When using a slack adapter, this displacement can only be used to control the loading rate. For measuring the true strain in the specimen during a high-speed loading process, a common mechanical extensometer cannot be used because its response frequency is too low and the high acceleration to which it is exposed causes it to fail. Therefore, non-contacting optical measuring methods are preferred for high-speed tests.

---

[1] The Instron company employs this approach.

[2] Axel Products and the MTS companies use this approach.

There are several ways to implement true strain measurement. One method is based on tracking the position of points of light, as shown in Fig. 17.6a. The ends of two optical fibers are mounted on the specimen. Laser light is injected into the optical fibers from their other ends. Two optical detectors monitor the positions of the emitted light. Through the positions or relative positions of the two light beams, the deformation of the specimen is measured. Though optical fibers are light, this attached measurement method is invasive and may affect the material behavior. Thus, a non-contact strain measuring system, shown in Fig. 17.6b, is used to measure the high rate strain in the specimen. This method does not require attachments on the specimen. Instead, two reference points or strips are made on the specimen surface. Then the deformation of the specimen can be determined by tracking the position of these reference points or strips. Interferometric techniques have also been used to measure strains on the specimen by directing laser beams on diamond indentations on the specimen and subsequently combining the reflected light beams to create displacement patterns. Recently, high-speed Digital Image Correlation (DIC) system has also been used to record the dynamic strain. This method will be discussed later in Section 17.9.

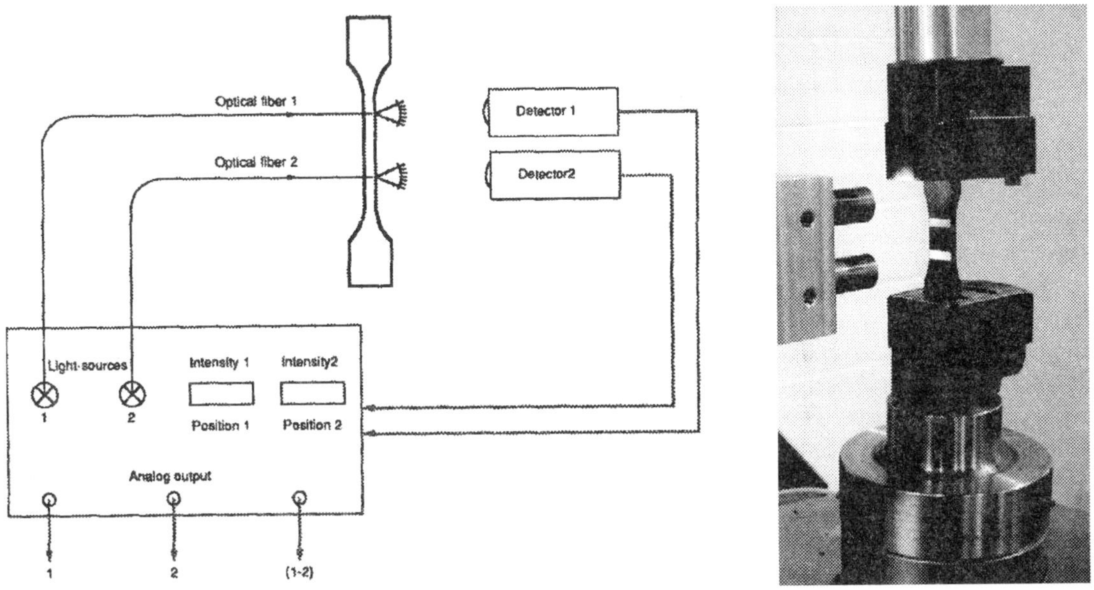

(a) (b)

Figure 17.6 Optical high-speed true strain measuring methods (a) point light tracking [4]
(b) non-contact strain measuring system (From Axel Products, Inc)

17.2.3 Force Measurement

A force measuring transducer is used to measure the force transmitted through the specimen. Piezoelectric force sensors are generally chosen as a force transducer. Though these sensors drift with time and are usually not applicable for quasi-static tests, their high frequency response makes them a good choice for dynamic force measurement.

17.2.4 Example of Experimental Data

The results from high-speed tensile tests on cold rolled steel at different strain rates are shown in Fig. 17.7. The smooth, relatively noise free, curves demonstrate that high-speed servo-hydraulic systems provide a means to reliably test materials at an inter-mediate strain rate range.

Figure 17.7 True tensile stress-strain curves of cold rolled steel at different intermediate strain rates [5]

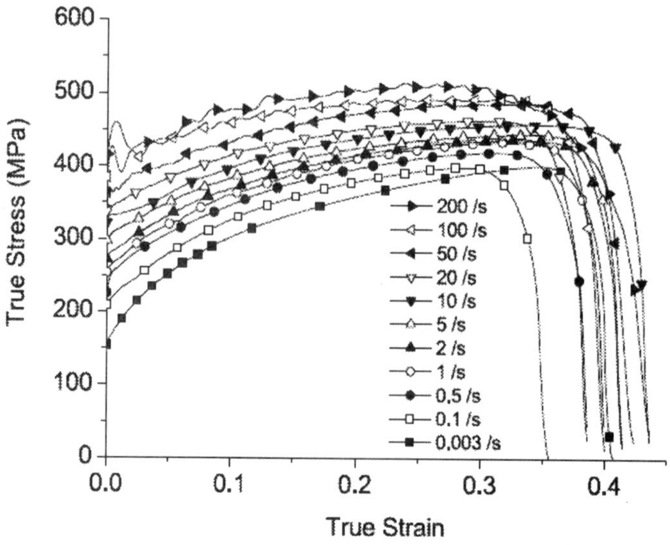

17.3 DROP WEIGHT IMPACT TEST

The drop weight impact test is one of the most widely used dynamic test method to simulate very low-velocity impact loading. As its name implies, this method involves dropping a hammer or weight, which strikes the specimen in order to produce dynamic loads used in the investigation of the dynamic behavior of materials and structures. The advantages of this experimental technique include easy implementation, control and good repeatability. Consequently, this method is widely applied in the transportation, aerospace, packaging and defense industries. Depending on the impact velocity, the strain rates produced induced by a drop weight impact tower vary from 10^0 to 10^3 ε/s.

17.3.1 System Configuration and Experimental Procedure

Generally, a drop weight impact test apparatus is comprised of two systems, a mechanical system and a data acquisition system. The mechanical system includes the drop tower frame, the hammer lifting and releasing equipment as well as the mechanism which supports and holds the specimen. The second part consists of the data acquisition and transmission system. A sketch of a typical drop weight impact test system and photograph of a commercial test system are shown in Fig. 17.8.

When beginning a test, the hammer is lifted to a preset height. After initializing the data acquisition system, the hammer is released, falls along a guided path until it strikes the specimen, which is held in the support fixture. The data acquisition element is triggered when the hammer contacts the specimen and transmits the force-time signal to the computer system.

17.3.2 Input Dynamic Loading

The kinetic energy of the drop weight is the sole contributor to the dynamic loading of the specimen in this test system. Recall the inertia kinetic energy E_k for a moving rigid body is:

$$E_k = \frac{1}{2} m v^2 \qquad (17.1)$$

where m is the mass of the rigid body, v is the velocity of the rigid body.

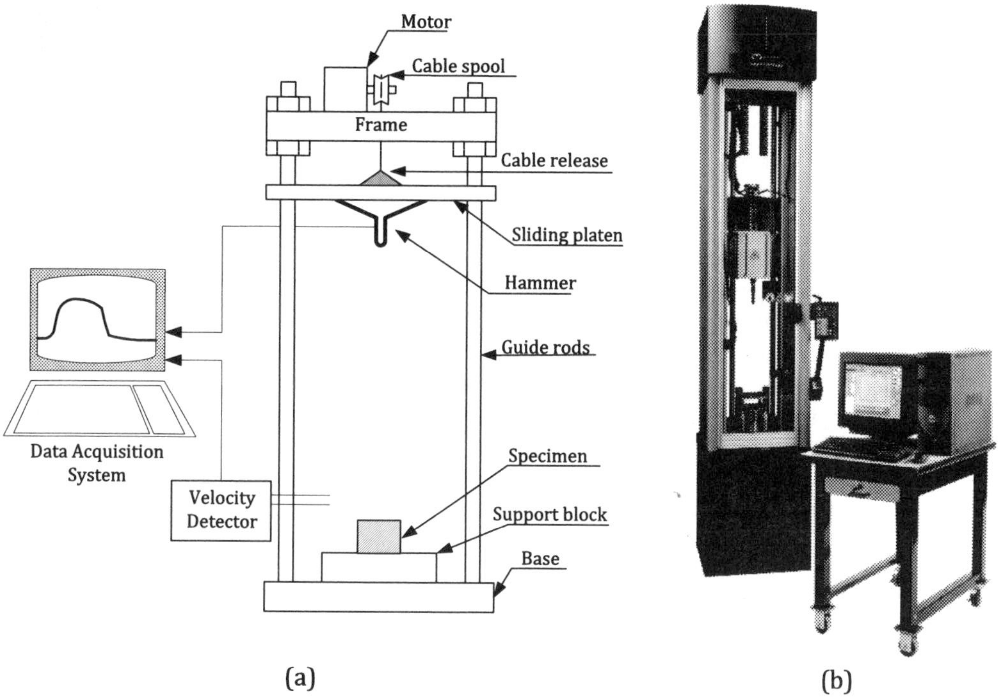

Figure 17.8 Drop weight impact test system (a) Sketch (b) Instron Dynatup 9250 system

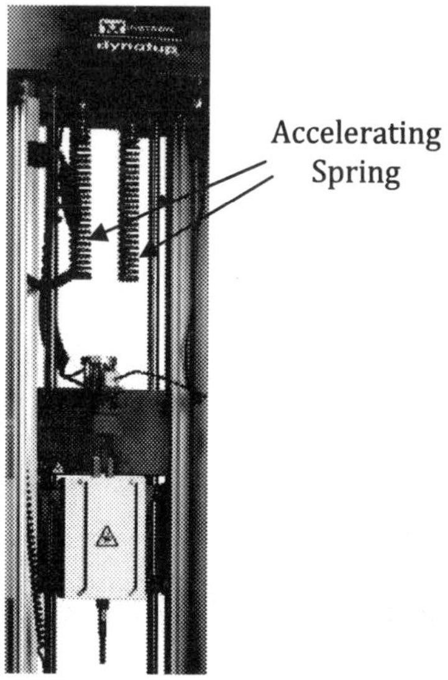

Figure 17.9 Accelerating spring system[3] .

To adjust the kinetic energy, the mass and the velocity must be varied. In the drop weight impact test system, the mass can be adjusted by adding weights to the hammer. The velocity can be adjusted by changing the drop height. Due to the limitation of the height of the system, very high velocities cannot be reached. For example, to achieve an impact velocity of 20 m/s, a frame with a height of more than 20 m is required. It is not always feasible to build a frame of this height. In this case, a spring system can be installed between the frame and the hammer to accelerate the hammer to a higher velocity. Some

[3] The Instron Dynatup 9250 system for accelerating the falling weight.

commercial systems, have implemented this system, which can accelerate the hammer to 20 m/s from a 1.25 m drop height. The accelerating spring system used in the Instron Dynatup 9250 system is shown in Fig. 17.9.

Reference to Eq. (17.1) shows that the same input energy can be achieved both with a large mass impacting at a low velocity as with a small mass impacting with a high velocity. However, they must be considered to represent different states, because they have different impact parameters despite having the same input energy.

17.3.3 Experimental Data Acquisition

The experimental data is acquired by the transducer system located in the hammer. Two important parameters, force and displacement, must be measured in order to obtain useful information in a drop weight impact test. If a force-time trace can be recorded, the impact energy and hammer velocity can be also determined. This experimental data can be used to identify the low-velocity impact resistance of the plate and to calculate the dynamic fracture toughness, and tear strength of the specimen material.

Two different transducer systems used to obtain the force and displacement as a function of time after impact. The first system employs an accelerometer embedded in the hammer to record the acceleration-time data and a velocity transducer to measure the initial impact velocity v_0. With the mass of the hammer known, the force F(t) can be obtained by Newton's second law:

$$F(t) = ma(t) \tag{17.2}$$

where F is the force, m is the mass, a is the acceleration and t is the time.

Using acceleration data, the displacement can be obtained by integrating the acceleration twice with respect to time.

$$s(t) = v_0 t + \iint a(t)dt \tag{17.3}$$

where s is the displacement, a is the acceleration and t is the time.

The impact energy and hammer velocity with respect to time can also be obtained by using following equations:

$$E(t) = \int F(t)s(t)dt \tag{17.4}$$

$$v(t) = v_0 + \int a(t)dt \tag{17.5}$$

The second system uses a dynamic force transducer to determine the initial impact velocity v_0 as well as the force history in the hammer directly. Through Newton's second law, Eq. (17.2), the acceleration history can be calculated. Subsequently, the history of the displacement, velocity and impact energy can also be obtained.

17.3.4 Example of Experimental Data

A typical displacement-time, force-time and energy-time curves for an E-glass polyester face sheet with a CoreCell A800 foam core sandwich structure under impact loadings is shown in Fig. 17.10.

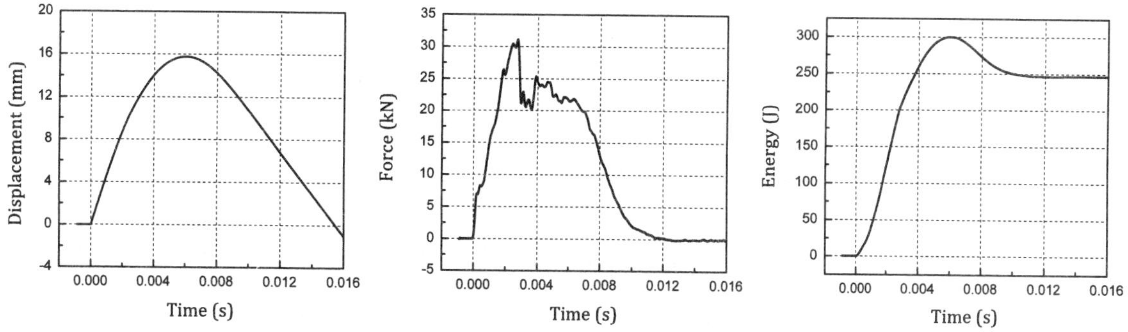

(a) Displacement-Time curve (b) Force-Time curve (c) Energy-Time curve

Figure 17.10 Drop tower experimental data of E-glass poly ester face sheet with a CoreCell A800 foam core sandwich structure. Hammer mass equal 26.16 kg and an impact velocity of 4.7 m/s.

17.4 THE DOUBLE PENDULUM

The double pendulum, first proposed by Volterra [68] and illustrated in Fig. 17.11, is comprised of two long bars suspended with wires. The specimen, mounted between the two bars, is impacted by the swinging bar, which is travelling with a velocity v. The surfaces of the ends of the two bars are lubricated to reduce the friction between the bar surfaces and the specimen. The double pendulum is employed to determine stress-strain curves of relatively low modulus (soft) materials such as rubber or plastics. The strain rate can be varied by changing the impact velocity v, which is achieved by increasing or decreasing the height h.

Figure 17.11 Illustration of the experimental arrangement for the double pendulum.

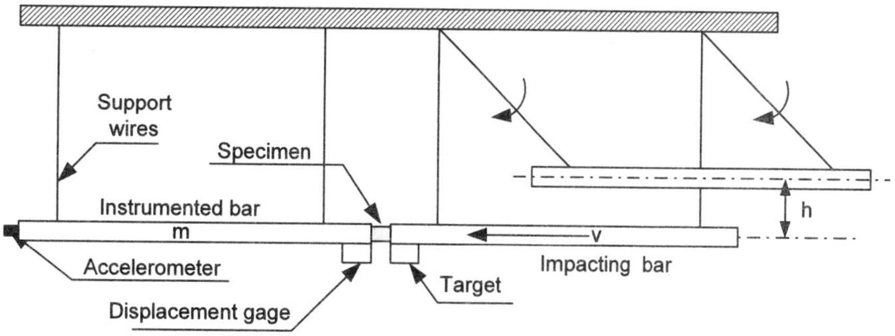

The instrumented bar on the left is equipped with an accelerometer usually placed at its free end for easy mounting. A displacement gage and a target are arranged to measure the spacing between the two bars as a function of time during the impact event. An eddy current gage is often used for this measurement.

Upon impact, the accelerometer senses the acceleration $a(t)$ of the instrument bar and the displacement gage senses the change in length of the specimen $\delta(t)$. For specimens having low moduli of elasticity, the impact event is sufficiently long to eliminate the effect of stress wave propagation in the pendulum bars and the specimen. It is assumed that the specimen is subjected to two compressive forces that are equal in magnitude and opposite in direction. The force is determined from the measured acceleration of the instrumented bar using $F(t) = ma(t)$, where m is the mass of the instrumented bar and the attached transducers.

It is assumed that the specimen is subjected to axial compressive forces that are equal at both ends. With this assumption, the stress can be written as:

$$\sigma(t) = F(t)/A \tag{17.6}$$

where A is the cross sectional area of the specimen.

Substituting Eq. (17.2) into Eq. (17.6) gives:

$$\sigma(t) = m\, a(t)/A \tag{17.7}$$

The strain is determined from:

$$\varepsilon(t) = \delta(t)/L \tag{17.8}$$

where L is the initial length of the specimen.

The elastic modulus E(t) is determined as a function of time during the dynamic event from Eqs. (17.7) and (17.8) as:

$$E(t) = \frac{\sigma(t)}{\varepsilon(t)} = \frac{mLa(t)}{A\delta(t)} \tag{17.9}$$

A typical stress strain curve of a low modulus material is presented in Fig. 17.12.

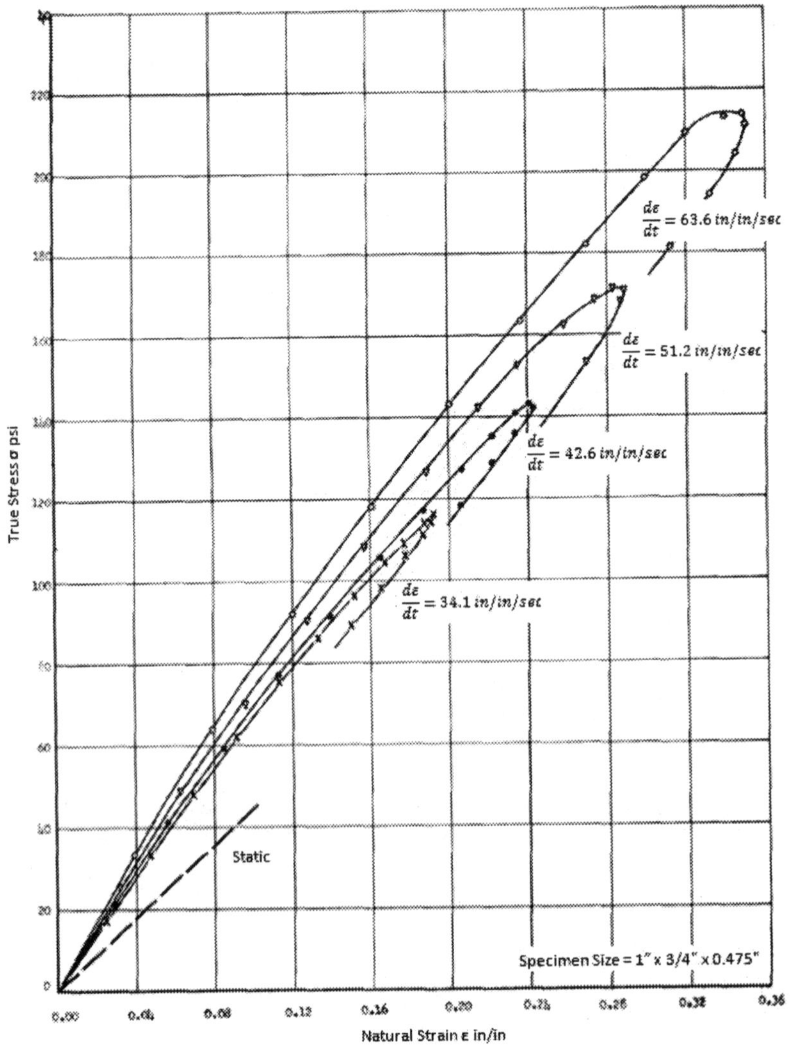

Figure 17.12 A stress strain diagram for a urethane rubber [78].

The measurements taken using a double pendulum provide the stress strain curve for both the loading and unloading portions of the dynamic event. The velocity of impact, which controls the initial strain rate, can be increased by increasing the height h of the impacting pendulum. Increasing the impact velocity also increases the maximum force and stress applied to the specimen. A typical experimental series to characterize a low-modulus material includes tests conducted at several strain rates and several different levels of applied stress.

17.5 PENDULUM IMPACT TEST

The pendulum impact test, such as the Charpy V-notch test, is a standard method endorsed by the American Society for Testing and Materials (ASTM) to achieve high strain rates. In this test, a pendulum swings to impact a specimen which allows the user to study materials under high rate loading. Similar to the drop weight impact test, the inertial kinetic energy, which is transformed from the potential energy of the pendulum, produces the dynamic loading of the specimen. The pendulum impact test is widely used in industry because it is easy to prepare and conduct, and results can be obtained quickly and inexpensively. Depending on impact energy, the strain rates induced by pendulum impact tests vary from 10^0 to 10^3 ε/s.

17.5.1 System Configuration

A pendulum impact test system consists of a base frame with a pendulum and a specimen support fixture. The pendulum swings through an axis located on the base frame towards a specimen that is supported in the fixture. There are various pendulum and support fixture designs available to ensure proper boundary and loading conditions depending on the application. A sketch of a pendulum impact test system and a commercial 50 Joule system are shown in Fig. 17.13.

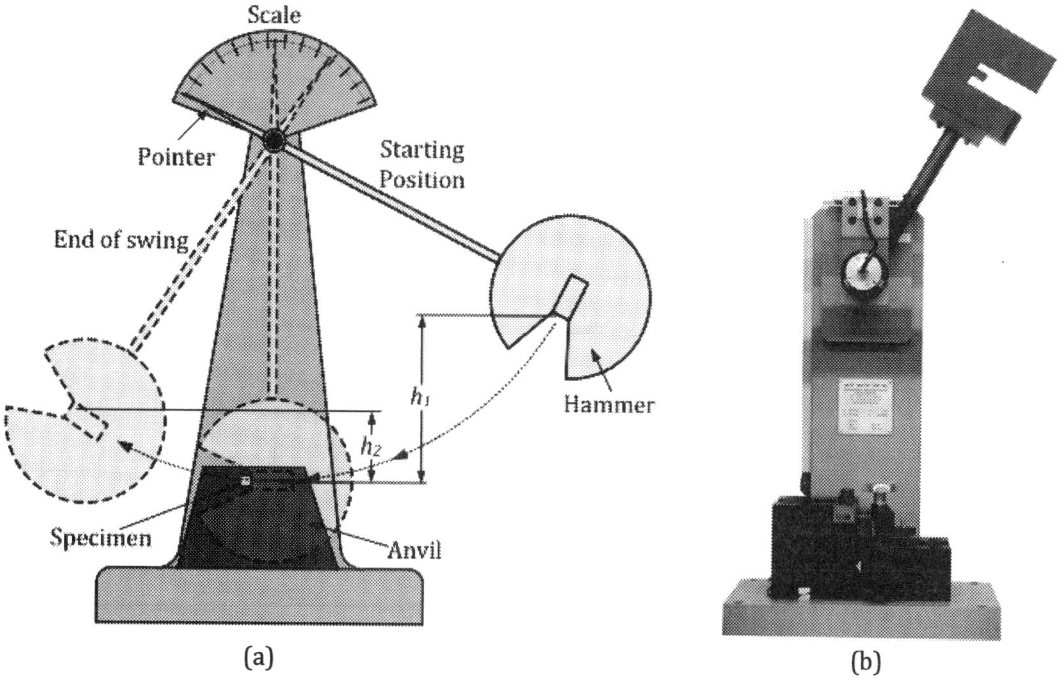

Figure 17.13 Pendulum impact test system (a) Sketch (b) Walter+Bai ag PH 50 Joule system

17.5.2 Data Acquisition

The initial height (h_1 in Fig. 17.13a) of the pendulum and the maximum height that the pendulum reaches after it impacts the specimen (h_2 in Fig. 17.13a) are the primary data recorded in tests. If friction is neglected, the impinging kinetic energy of the pendulum is equal to the potential energy of the pendulum at the position when it is released. The kinetic energy remaining in the pendulum is equal to the potential energy of the pendulum at its highest position after the impact. These kinetic energies are:

$$E_{k\_initial} = mgh_1 \tag{17.10}$$

$$E_{k\_remaining} = mgh_2 \tag{17.11}$$

where, m is the mass of the pendulum, g is the local acceleration of gravity.

The energy loss, defined as the difference of the initial and the remaining kinetic energies, is equal to the work of deformation and fracture, which is given by:

$$E_k = mg(h_1 - h_2) \tag{17.12}$$

This value of E_k is widely used to determine the impact strength of the specimen.

17.5.3 Charpy V-notch Impact Test

The Charpy V-notch test was developed in 1905 by the French scientist Georges Charpy. During World War II, there was a major problem with ship hulls fracturing for unknown reasons. This test was pivotal in understanding the brittle fracture associated with these failures. Today, the Charpy V-notch test is used in many industries for testing building and construction materials used in pressure vessels and bridges. Because the specimen size, shape, and notch configuration can affect the test results, a standard test method has been described in ASTM E23 standard.

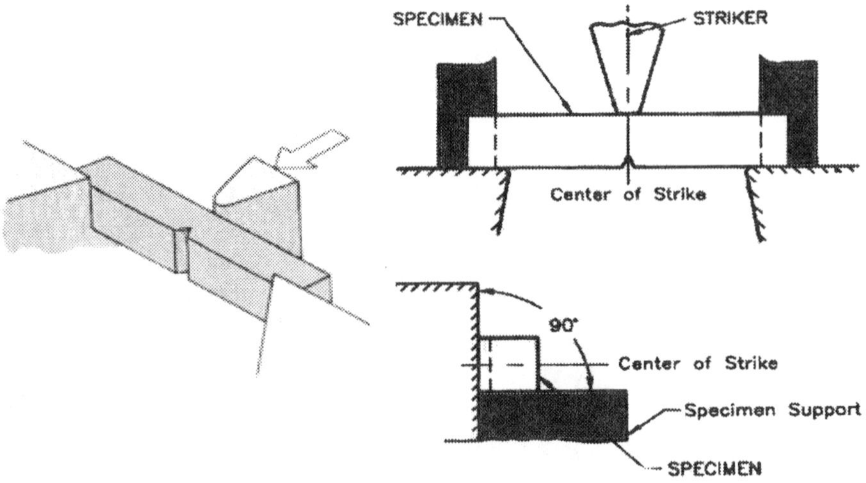

Figure 17.14 Supporting and loading setup for Charpy v-notch test [18].

A typical arrangement of the supporting and loading procedure for a Charpy V-notch test is illustrated in Fig. 17.14. The specimen, in a shape of a bar of square cross section with a V notch, is placed in the supporting fixture with the V notch facing the support boundaries. The pendulum then strikes the back of the specimen causing it to fracture at the notch.

The energy loss E_k calculated by Eq. (17.12) from the experimental data is called impact energy (also known as notch toughness). It can be used to calculate the impact resistance $\mathfrak{R}$ as:

$$\mathfrak{R} = E_k / A \qquad (17.13)$$

where, A is the cross sectional area of the specimen where the notch is located. These parameters represent the resistance of the materials to dynamic fracture.

The Charpy notch test is easy to implement at low and high temperatures. Thus, it is often used to obtain the impact resistance of materials at different temperatures. The impact energy as a function of temperature for steel is shown in Fig. 17.15. These curves are used to determine the temperature at which the transition from brittle to ductile fracture occurs for a given material.

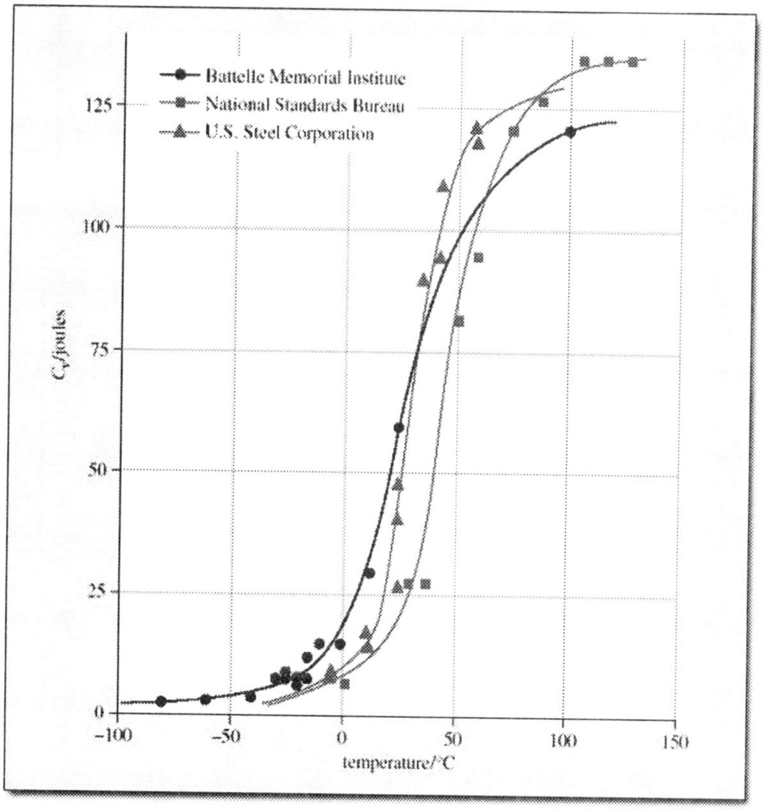

Figure 17.15 Charpy v-notch impact data of steel at different temperatures [25].

17.5.4 Izod Impact Test

The Izod impact test is named after the English engineer Edwin Gilbert Izod (1876-1946), who introduced the experimental arrangement in 1903. It differs from the Charpy impact test in that the sample is mounted as a cantilevered beam as opposed to three point bending. The typical method of supporting and loading the notched specimen in an Izod test is shown in Fig. 17.16.

The specimen, in a shape of a bar of square cross section with a V notch, is held as cantilevered beam with the V notch positioned at precisely the clamped end of the bar and facing the striker. The pendulum strikes causing the specimen to fracture at its notch. The impact strength, defined by E_k/L, is calculated from the experimental data, where E_k is the impact energy and L is the net section length ahead of the notch. The Izod impact strength for thermoplastic polyesters PBT PlANAC at different temperatures is shown in Fig. 17.17.

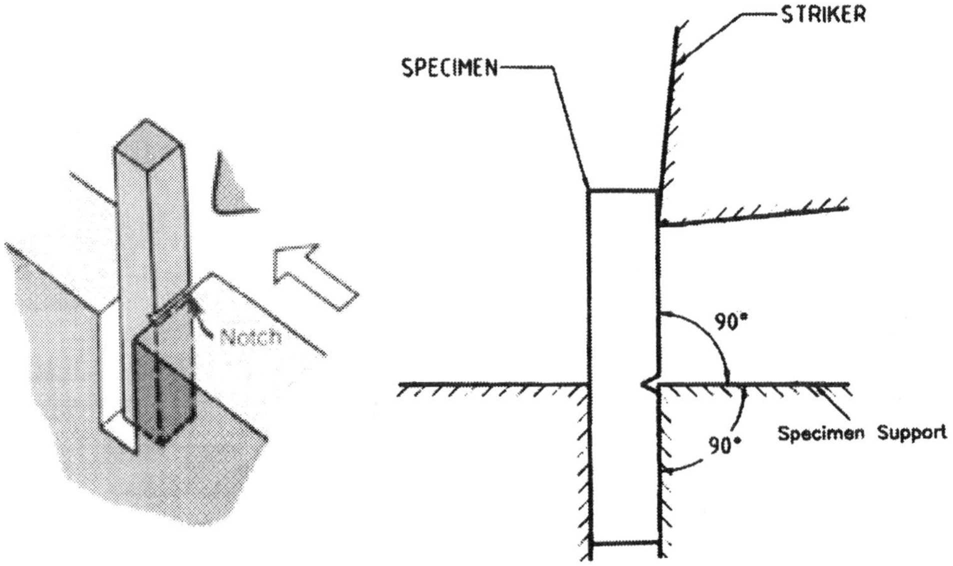

Figure 17.16 Supporting and loading setup for an Izod test [18].

Figure 17.17 Izod impact strength of thermoplastic polyesters PBT PIANAC at different temperatures [23].

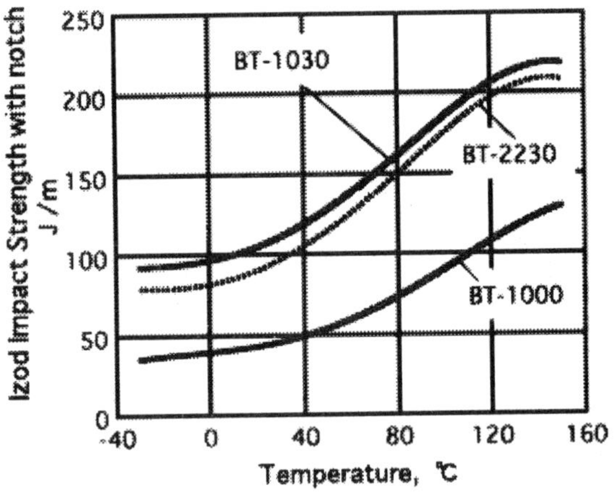

17.6 TAYLOR IMPACT TEST

The Taylor impact test was developed by G.I. Taylor during the 1930s. The simple experimental arrangement and data analysis procedure makes this a favorite technique of laboratories that investigate the behavior of materials at high deformation rates. This technique involves firing a long cylinder of the material against a massive, rigid anvil. The kinetic energy of the specimen is transformed into dynamic plastic deformation in the material. The plastic deformation of the material extends in a stepwise manner from the specimen's contact end (between the specimen and the anvil). Therefore, a long cylindrical specimen will show two regions: undeformed and plastically deformed regions. Measurement of the length of the original and deformed specimen gives relatively accurate results for the dynamic yield stress of the material within approximately ± 10% error. Depending on the impact velocity, the strain rates induced by Taylor impact tests vary from 10^2 to 10^4 ε/s. In this section, the experimental arrangement, experimental phenomenon and basic method of data analysis are described.

17.6.1 Experimental Arrangement and Procedure

The experimental arrangement used in a Taylor impact test is very simple as indicated by the sketch shown in Fig. 17.18. The experimental arrangement consists of mechanical and measurement systems. The mechanical test system includes a gas gun with a launch tube and a massive solid anvil. The gas gun fires a long cylindrical specimen. The massive solid anvil is fixed, implementing a rigid wall condition. The impact velocity is measured with a laser beam and laser detector system. A series of laser beams and laser detectors are placed in the path of the specimen. When the specimen cuts the laser beams, the output signals from the laser detectors show a sharp drop or rise. The average velocity of the specimen is calculated from these recorded drop or rise times. High speed photography systems have also been utilized in Taylor impact tests to acquire the real time deformation of the specimen for a more thorough understanding of the failure mechanisms.

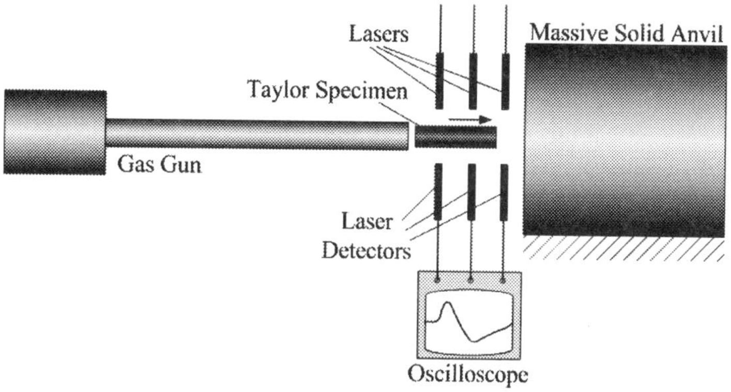

Figure 17.18 A sketch of typical Taylor impact test

When performing a test, the original dimensions of the cylindrical specimen are measured. Then, the specimen is loaded into the gas gun and fired towards the anvil. The instrument measuring the velocity is triggered when the specimen passes through the first laser beam. After the experiment, the deformed specimen is recovered and the postmortem deformation length is measured.

The aspect ratio (length/diameter) of the long cylindrical specimen and the impact velocity are two parameters which can be modified in the Taylor impact test. Note that a standard aspect ratio of the specimen has not been established. A specimen with a small aspect ratio may not exhibit an un-deformed part, which is needed to perform the data analysis. Therefore, when choosing the aspect ratio it is important to ensure there will be an undeformed portion of the cylindrical test specimen after the impact test. The impact velocity v is controlled by using different gas pressures in the gas gun.

17.6.2 Theory[4]

The theory involved in the derivation of the relations for the determination of the flow stress from the geometry of a deformed cylindrical rod is due to House et al [66] and Jones et al [67]. Consider an element of length dx from the front of the rod at some time t greater than zero but less than T, as shown with the darkened square in Fig. 17.19. The properties of the element change from rigid to plastic during the time increment dt. The element also undergoes discontinuous changes in area from A to A_o and in velocity from u to u_p, where u_p is the particle velocity of the material in the plastic element. The rigid-plastic interface is moving away from the anvil with a velocity v which is given by v = dh/dt. The force at the rigid-plastic interface is denoted as F. During the time interval dt, the change in the linear momentum of the segment of undeformed rod equals the net impulse. Hence we may write:

---

[4] The symbols used in the development of the relations are the same as those used by Sir G. I. Taylor in his original paper.

$$\rho A_o xu - \rho A_o(x+dx)(u+du) + \rho A_o dx(u_p + du_p) = (F+dF)dt \qquad (17.14)$$

where ρ is the mass density of the material that is assumed to be constant during the deformation.

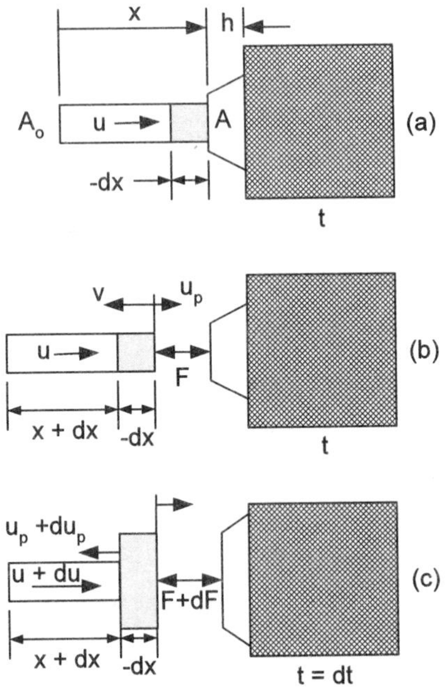

Figure 17.19 Diagrams showing the force and velocities at the rigid-plastic interface at two different times during the impact event.

Dividing by $\rho A_o dt$ reduces Eq. (17.14) to:

$$\frac{1}{dt}\left[xu - (x+dx)(u+du) + dx(u_p + du_p)\right] = \frac{(F+dF)}{\rho A_o} \qquad (17.15)$$

and eliminating higher order terms gives:

$$-x\dot{u} - \dot{x}u + \dot{x}u_p = \frac{F}{\rho A_o} \qquad (17.16)$$

where the dots represent the first derivative of the quantity with respect to time.

It is clear that the force F at the interface is given by:

$$F = S\,A \qquad (17.17)$$

where S is the flow stress

Substituting Eq. (17.17) into Eq. (17.16) yields:

$$-\left(\frac{S}{\rho}\right)\left(\frac{A}{A_o}\right) = \dot{x}(u - u_p) + \dot{u}x \qquad (17.18)$$

Next, consider the rod element undergoing transformation from the rigid to plastic state. If we consider the geometry of this element at time t and (t + dt), shown in Fig. 17.20, it is clear that:

$$-dx = (u + v)dt \tag{17.19}$$

This can be rewritten as:

$$\frac{dx}{dt} = \dot{x} = -(u+v) \tag{17.20}$$

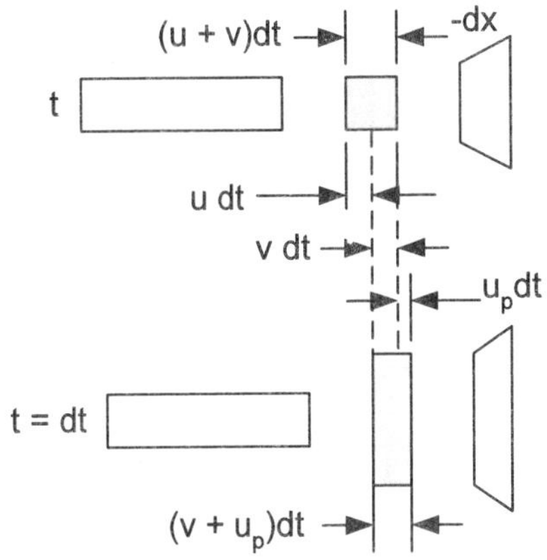

Figure 17.20 Geometry of the deformed element at two different times in terms of the velocities at the rigid-plastic interface.

The length of the plastic element L_p after a time increment dt is given by:

$$L_p = (v + u_p)dt \tag{17.21}$$

Conservation of mass leads to:

$$\rho A_o(u+v)dt = \rho A(v+u_p)dt \tag{17.22}$$

This reduces to:

$$\frac{A}{A_o} = \frac{(u+v)}{(v+u_p)} \tag{17.23}$$

Substituting Eqs. (17.20) and (17.23) into Eq. (17.18) yields:

$$-(u+v)(u_p - u) - x\dot{u} = \frac{S}{\rho}\left(\frac{u+v}{v+u_p}\right) \tag{17.24}$$

It is assumed that the particle velocity $u_p = 0$ in the plastic region and that the velocity of the rigid-plastic interface v = C is a constant. With these two assumptions, Eq. (17.24) reduces to:

$$x\dot{u} = u(u+C) - \frac{S}{\rho}\left(\frac{u+C}{C}\right) \tag{17.25}$$

Recall that $\dot{x} = -(u+C)$ and with the chain rule, time can be eliminated from Eq. (17.25) to give:

$$x\frac{du}{dx} = -u + \frac{S}{\rho C} \tag{17.26}$$

Separating the variables x and u and integrating gives:

$$\int_L^x \frac{dx}{x} = -\int_U^u \frac{du}{u + S/(\rho C)} \tag{17.27}$$

Completing the integration of Eq. (17.27) yields:

$$\ln\left(\frac{x}{L}\right) = \ln\left[\frac{S/(\rho C) - U}{S/(\rho C) - u}\right] \tag{17.28}$$

At the end of the impact event t = T, u = 0 and x = X. Substituting these quantities into Eq. (17.28) yields:

$$\frac{X}{L} = \frac{S/\rho C - U}{S/\rho C} = 1 - \frac{U\rho C}{S} \tag{17.29}$$

Rearranging Eq. (17.29) yields:

$$\frac{S}{\rho} = \frac{UC}{1 - X/L} \tag{17.30}$$

The velocity of the rigid-plastic interface v = C, is determined by:

$$C = \frac{L_1 - X}{T} \tag{17.31}$$

Substituting Eq. (17.31) into Eq. (17.30) gives:

$$\frac{S}{\rho} = \frac{U(L_1 - X)L}{T(L - X)} \tag{17.32}$$

This relation enables one to determine the flow stress from measurements of the Taylor rod before and after plastic deformation. However, it requires the impact time T to be measured. To eliminate the requirement for the measurement of the impact time, note that the impact time is related to the geometry of the Taylor rod by:

$$L - L_1 = \int_0^T u(t)dt \tag{17.33}$$

In his analysis in 1948, Taylor assumed that the velocity of the rigid portion of the rod decreased as a linear function of time with;

$$u(t) = U(1 - t/T) \tag{17.34}$$

Substituting Eq. (17.34) into Eq. (17.33), integrating and rearranging terms leads to an approximation of the time of the impact event as:

$$T = \frac{2(L - L_1)}{U} \tag{17.35}$$

Substituting Eq. (17.35) into Eq. (17.32) gives:

$$\frac{S}{\rho U^2} = \frac{1}{2}\left[\frac{L(L_1 - X)}{(L - L_1)(L - X)}\right] \tag{17.36}$$

If the impact velocity U is known, Eq. (17.36) permits the determination of the flow stress S using dimensions of the Taylor rod before and after deformation. An extensive series of experiments was conducted by House et al with low strength steel, two copper alloys and two aluminum alloys. These investigators found that the Eq. (17.36) underestimated the flow stress for all the materials tested. House has modified Eq. (17.36) to provide a closer estimate of the flow stress as:

$$\frac{S}{\rho U^2} = k\left[\frac{L(L_1 - X)}{(L - L_1)(L - X)}\right] \qquad (17.37)$$

where k is a fraction that varies from 0.58 to 0.70. House determined that for k = 0.63 the use of Eq. (17.37) predicted the flow stress to be within ±10% in all cases and within ±5% in 91% of the cases.

With modern instrumentation, it is possible to accurately measure the impact velocity, the time of the dynamic event and the dimensions of the rod before and after impact. Thus, the use of either Eq. (17.37) or Eq. (17.32) should provide close estimates of the flow stress of ductile materials that behave as rigid-plastic materials.

17.6.3 Example of Experimental Data

Real-time deformation images of a polycarbonate cylinder recorded with high speed photography during a Taylor impact test are shown in Fig. 17.21. The impact velocity was 187 m/s. In the images, the specimen travels from the left to the right and the anvil is located at the right end of the images. It is evident that the plastic deformation is a continuous process throughout the impact event. A postmortem image of the tested polycarbonate cylinder is shown in Fig. 17.22. The specimen shows an undeformed region and a plastic region with a complex profile.

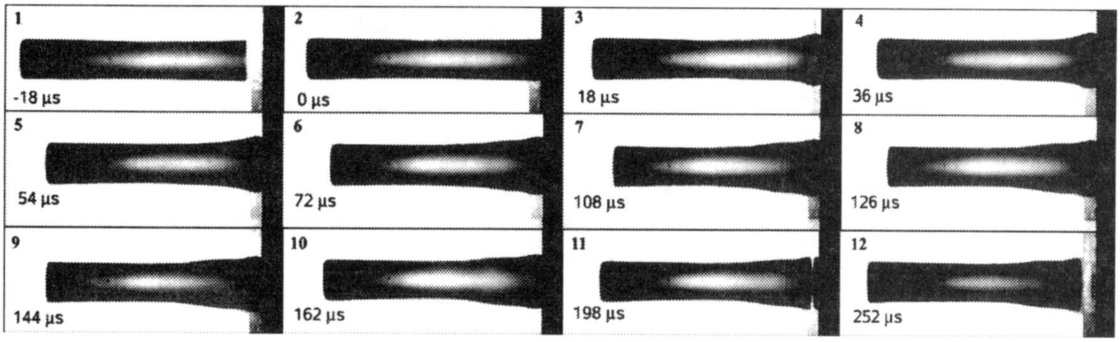

Figure 17.21 High-speed silhouette photographs during a Taylor impact test [30]. Specimen diameter D = 12.7 mm and L = 76.2 mm.

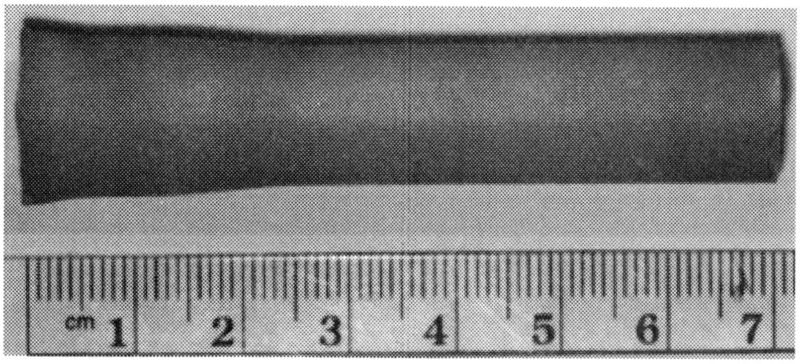

Figure 17.22 Postmortem image of the polycarbonate cylinder after a Taylor impact test [30].

17.7 SPLIT HOPKINSON (KOLSKY) BAR TECHNIQUE

The split Hopkinson bar test, also called the Kolsky bar test, is a classical experimental technique in the dynamic measurement domain. It was first proposed by B. Hopkinson in 1914 to measure a stress pulse propagation in a metal bar. Then, in 1949, H. Kolsky extended Hopkinson's technique by using two long bars and placing the specimen between them. The great advantage of Kolsky's arrangement is that it succeeds in decoupling the inertial effect and the high strain rate effect, which are generally coupled when materials are subjected to dynamic loading. This makes it possible to determine the mechanical properties of materials at high strain rates. After 60 years, this technique is widely used and well developed not only for compression loading but also for tension, torsion or combinations of all of these loading. The strain rates induced in a SHPB test vary from 10^2 to 10^4 ε/s.

17.7.1 Split Hokinson Pressure Bar (SHPB)

Experimental Arrangement and Procedure

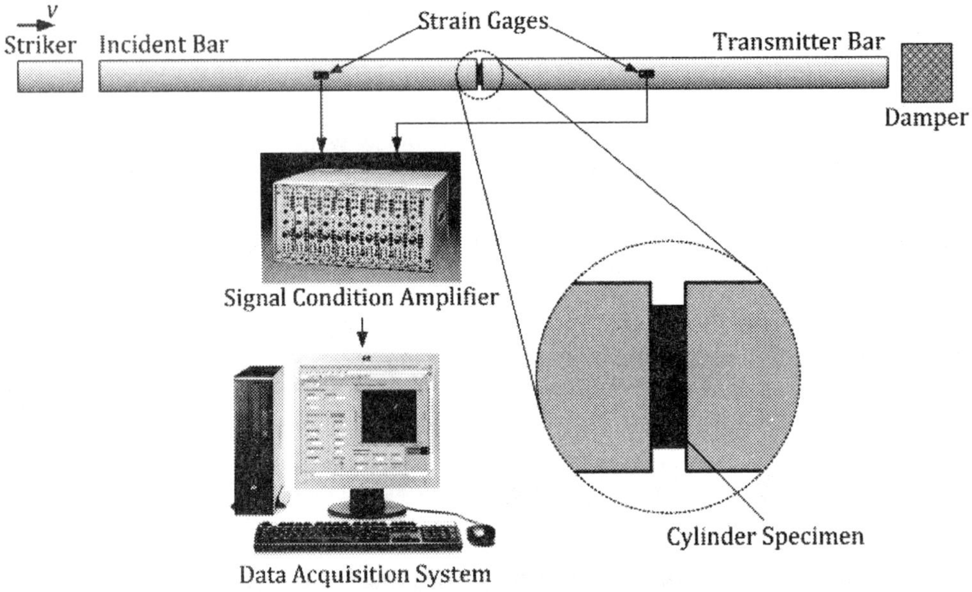

Figure 17.23 A sketch of typical split Hopkinson pressure bars

A sketch of a typical SHPB test setup is shown in Fig. 17.23. The experimental arrangement involves mechanical and measurement systems. The mechanical component includes a gas gun and two long bars—incident and transmitter bars. The two bars are aligned along a single axis and are designed so that the stresses imposed during impact remain within the elastic limit throughout the test. Depending on the application, the bar materials can be high-strength steel, aluminum or some type of polymer. A thin specimen (usually cylindrical) is placed between the two bars for testing. The gas gun, also aligned axially with the bars, launches a striker bar that impacts the incident bar.

The measurement system is comprised of the strain gages, a dynamic signal conditioning amplifier and a signal acquisition system. The strain gages are attached to both the incident and the transmitter bar. The signals from the strain gages are transferred to the dynamic signal conditioning amplifier and finally captured by an oscilloscope or a PC system.

When conducting a test, a well lubricated specimen is placed between the incident and transmitter bars. The striker is launched by the gas gun and strikes the incident bar, which in turn generates a stress pulse. The stress level of the pulse is related to the impact velocity and the length of the pulse depends on the length of the striker. The stress pulse propagates through the incident bar and loads the specimen. An incident pulse, a reflected pulse and a transmitted pulse are generated and then recorded by the data acquisition system.

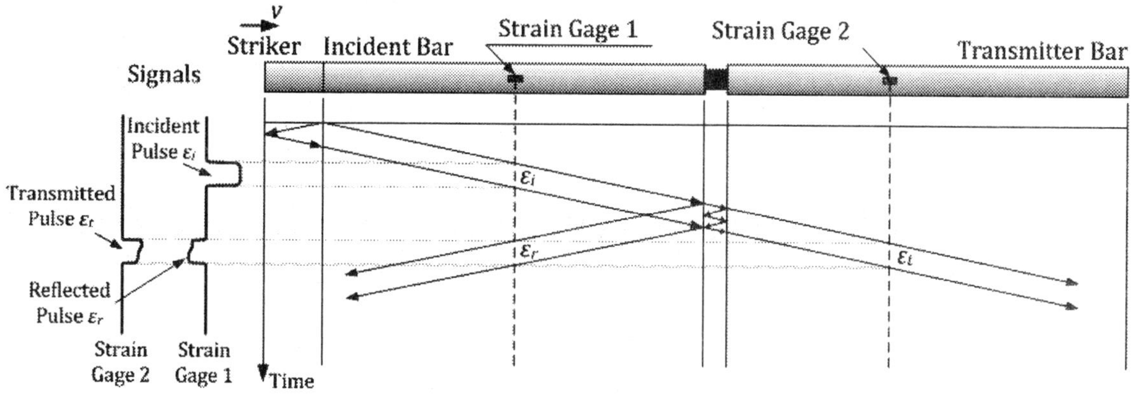

Figure 17.24 Stress pulse generation and propagation process

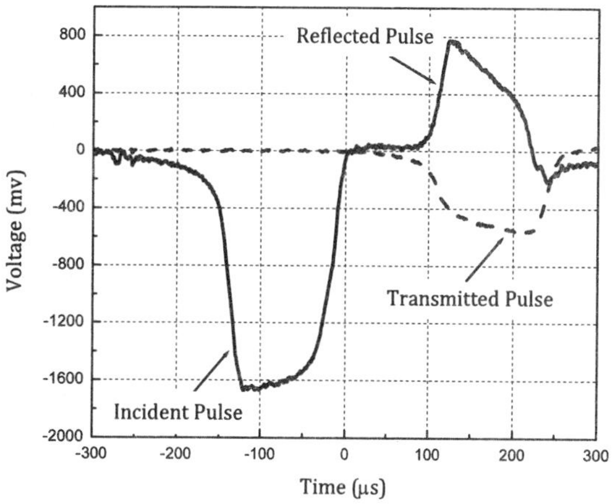

Figure 17.25 Typical pulse profiles

The generation of the strain pulse and its propagation along a pressure bar are presented in Fig. 17.24. When the striker bar impacts the incident bar, two compression stress waves are generated in the incident and striker bars. The stress wave in the incident bar propagates along the bar and generates a negative strain (compression) when it reaches the strain gage on the incident bar. When this stress wave reaches the specimen, a reflected tension stress wave and a transmitted compression wave are generated, which propagate in opposite directions. These stress waves give a positive strain (tension) on the strain gage located on the incident bar and a negative strain (compression) on the strain gage located on the transmitter bar. The stress wave in the striker reflects from its free end and unloads the striker bar. The loading and unloading stress waves generate the pulse profiles, as shown on the left side of Fig. 17.24 and in Fig. 17.25.

17.7.1.2 Conditions

SHPB technique is valid when the following two conditions are satisfied.

1. The elastic waves in the bars must be one-dimensional longitudinal waves.
2. The specimen must deform uniformly.

To achieve the first condition, the length of the stress pulse must be much larger than the bar diameter: usually it is at least 10 times the size of the bar diameter. For the second condition, there are two considerations: longitudinal and radial deformation. To achieve uniform longitudinal deformation, the stress pulse must be long enough to create several (at least 3~10) reverberations within the length of the specimen. To achieve uniform radial deformation, the contact interfaces between the specimen and bars must be well lubricated. For adequate coupling of longitudinal and radial deformation, the length l_s and the diameter d_s of the specimen are governed by:

$$\frac{l_s}{d_s} - \sqrt{\frac{3}{4}\nu_s} \qquad \text{(a)}$$

where ν_s is the Poisson ratio of the specimen material.

Experimental Data Analysis

Consider a longitudinal elastic stress wave in a homogeneous bar with density ρ and elastic modulus E_b as shown in Fig. 17.26. The classical D'Alembert displacement solution for a wave equation can be written as,

$$u = f(x - ct) + g(x + ct) \qquad (17.38)$$

where c is the longitudinal wave velocity in a bar, $c = \sqrt{E_b/\rho}$, and f and g are arbitrary functions determined by initial conditions.

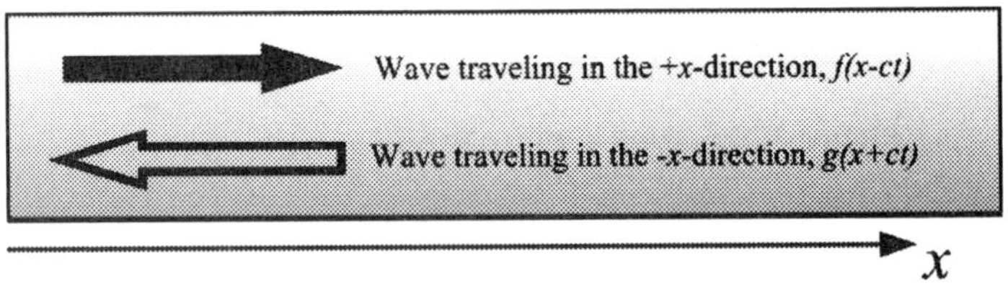

Figure 17.26 Wave solutions in a homogenous bar

The strain, stress and particle velocity can be derived from the displacement as,

$$\varepsilon = \frac{\partial u}{\partial x} = \frac{\partial f}{\partial x} + \frac{\partial g}{\partial x} = f' + g' \qquad (17.39)$$

$$\sigma = E\varepsilon = E(f' + g') \qquad (17.40)$$

$$\dot{u} = \frac{\partial u}{\partial t} = \frac{\partial f}{\partial t} + \frac{\partial g}{\partial t} = -cf' + cg' = c(-f' + g') \qquad (17.41)$$

Let the subscript i denote the parameters related to the incident pulse, r denote the parameters related to the reflected pulse and t denote the parameters related to the transmitted pulse. The displacements at the interfaces between the specimen and two bars are shown in Fig. 17.27.

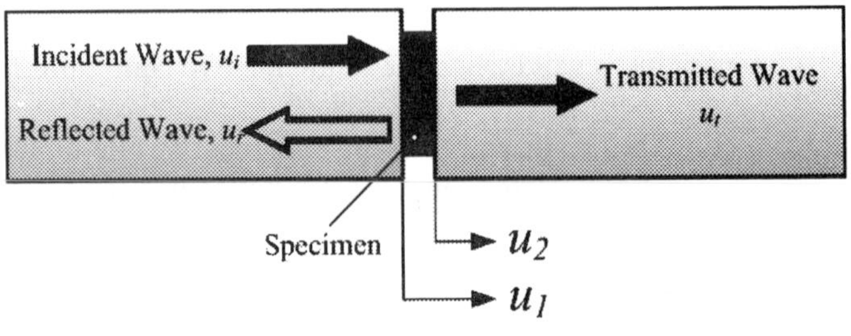

Figure 17.27 Motion analysis at the interfaces of SHPB

From Eq. (17.38) to Eq. (17.41), the displacement, strains, stresses and particle velocities of two interfaces can be described as,

$$u_1 = f_1(x - ct) + g_1(x + ct) = u_i + u_r \qquad (17.42)$$

$$u_2 = f_2(x - ct) = u_t \qquad (17.43)$$

$$\varepsilon_1 = f_1' + g_1' = u_i' + u_r' = \varepsilon_i + \varepsilon_r \qquad (17.44)$$

$$\varepsilon_2 = f_2' = u_t' = \varepsilon_t \qquad (17.45)$$

$$\sigma_1 = E_b(\varepsilon_i + \varepsilon_r) \qquad (17.46)$$

$$\sigma_2 = E_b\,\varepsilon_2 = E_b\,\varepsilon_t \qquad (17.47)$$

$$\dot{u}_1 = c\left(-f_1' + g_1'\right) = c(-\varepsilon_i + \varepsilon_r) \qquad (17.48)$$

$$\dot{u}_2 = c\left(-f_1'\right) = -c\varepsilon_t \qquad (17.49)$$

The load on the specimen is equal to the average of the two loads at the two interfaces. Then the nominal stress σ_s in the specimen can be derived as,

$$\sigma_s = \frac{\sigma_1 A_b + \sigma_2 A_b}{2A_s} = \frac{E_b A_b}{2A_s}(\varepsilon_i + \varepsilon_r + \varepsilon_t) \qquad (17.50)$$

where, A_b is the cross sectional area of the bar and A_s is the cross sectional area of the specimen.

The nominal strain rate in the specimen is,

$$\dot{e}_s = \frac{\dot{u}_2 - \dot{u}_1}{l_s} = \frac{c}{l_s}(-\varepsilon_t + \varepsilon_i - \varepsilon_r) \qquad (17.51)$$

where, l_s is the length of the specimen.

The nominal strain in the specimen is then written as,

$$e_s = \int_0^t \dot{\varepsilon}_s dt = \frac{c}{l_s}\int_0^t(-\varepsilon_t + \varepsilon_i - \varepsilon_r)dt \qquad (17.52)$$

If the specimen deforms uniformly then,

$$\varepsilon_i + \varepsilon_r = \varepsilon_t \qquad (17.53)$$

Therefore, Eqs. (17.50) ~ (17.52) can be modified as,

$$\sigma_s = \frac{E_b A_b}{A_s}\varepsilon_t = \frac{E_b A_b}{A_s}(\varepsilon_i + \varepsilon_r) \qquad (17.54)$$

$$\dot{e}_s = -\frac{2c}{l_s}\varepsilon_r = \frac{2c}{l_s}(+\varepsilon_i - \varepsilon_t) \tag{17.55}$$

$$e_s = -\frac{2c}{l_s}\int_0^t \varepsilon_r dt = \frac{2c}{l_s}\int_0^t (\varepsilon_i - \varepsilon_t)dt \tag{17.56}$$

The dynamic nominal stress-strain relationship of the materials is obtained from Eqs. (17.54) and (17.56) at the strain rate calculated by Eq. (17.55). The true stress, true strain rate and true strain in the specimen are then given by,

$$\sigma_s = S_s(1 - e_s) \tag{17.57}$$

$$\dot{\varepsilon}_s = \frac{\dot{e}_s}{1 - e_s}\sigma_s \tag{17.58}$$

$$\varepsilon_s = -\ln[1 - e_s] \tag{17.59}$$

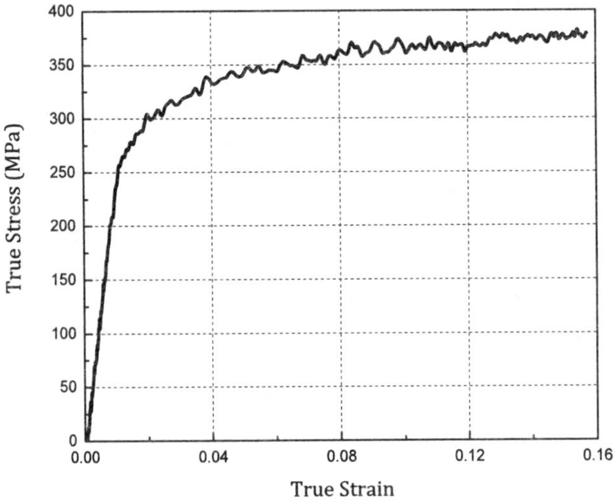

Figure 17.28 The stress-strain curve for 6061-T6 aluminum at room temperature and strain rate of 1,700 s^{-1}

Example of Experimental Data

A stress-strain curve for 6061-T6 aluminum at room temperature and strain rate of 1,700 s^{-1} is presented in Fig. 17.28. Though the dynamic stress-strain curve shows elastic and plastic regions of the materials, the initial elastic region is not valid due to the delay in achieving the equilibrium deformation of the specimen.

17.7.2 Split Hopkinson Tension Bar

The split Hopkinson tension bar uses the same data analysis procedure as that of the split Hopkinson pressure bar, which has been described previously. Therefore, data analysis techniques are not discussed in this section. However, the experimental arrangement and an example of experimental data obtained from a split Hopkinson tension bar experiment are presented.

Experimental Arrangement

The main function of a split Hopkinson tension bar test is to generate a tension stress wave using a striker bar that initially produces a compression wave. There are three common experimental arrangements used today for this purpose.

In the first arrangement, the tension stress wave is generated by the reflection of a compression stress wave at the free end of a bar. The experimental arrangement is very similar to that shown in Fig. 17.23. The only difference between the two is the method for mounting the specimen, which is shown in Fig. 17.29. Both ends of the specimen are threaded to allow the specimen to be screwed into the bars. This technique allows the bars to apply a tension load on the specimen. A similar technique, used in testing softer materials, employs adhesive to replace the threaded ends of the specimen.

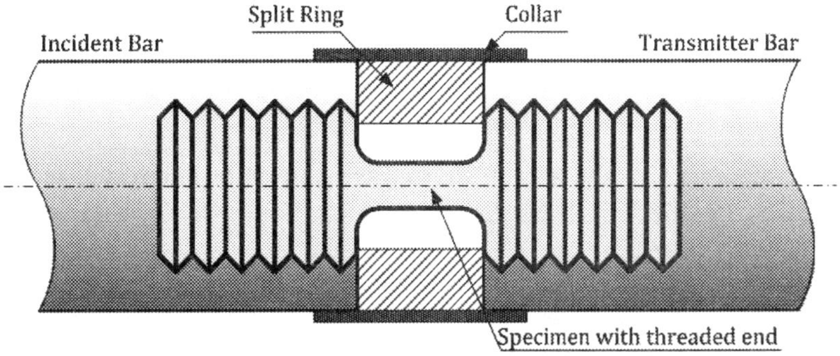

Figure 17.29 Installing a threaded specimen in the slit Hopkins tension bar,

The generation of a tension wave and the stress wave reflection process with a specimen held by threads is illustrated in Fig. 17.30. After a striker impacts the incident bar, a compression stress wave is generated and propagates through the incident bar. Note that there is a split ring between the incident and transmitter bars. This split ring is specifically designed to allow the compression stress wave to propagate through it without reflection and to maintain the specimen within the elastic region. When the compression wave reaches the split ring, it continues to propagate into the transmitter bar. At the free end of the transmitter bar, the compression stress wave is reflected as a tension stress wave. This tension wave is used as the incident pulse. When this tension wave reaches the specimen, the split ring is separated from the bar because it cannot transmit a tension pulse. The total tension stress pulse is applied on the specimen and generates a reflected and transmitted pulse within the bars. These pulses are used to analyze the high strain rate tension behavior of the materials using Eqs. (17.54) ~ (17.59).

In the second experimental arrangement, the incident tension stress pulse is applied directly in the specimen, as shown in Fig. 17.31. A flange plate is attached to the end of the incident bar. A tubular striker is launched at the flange plate, which generates a compression wave that is reflected from the free surface of the flange plate as a tension stress wave.

The third experimental arrangement is shown in Fig. 17.32. The incident bar is first clamped (fixed) at one position. An axial actuator pulls on the left side of the incident bar storing tensile strain within the left side of the incident bar. When the clamp is suddenly released, the stored tensile strain energy generates a tensile stress pulse that applies a tensile pulse on the specimen. For experiments conducted with these two arrangements, the data acquired is analyzed using the equations developed in Section 17.7.1.

Example of Experimental Data

Typical pulse signals and high strain rate stress-strain curves for Epon 828/ T-403 composites obtained with the second experimental arrangement for split Hopkinson tension bar are shown in Fig. 17.33.

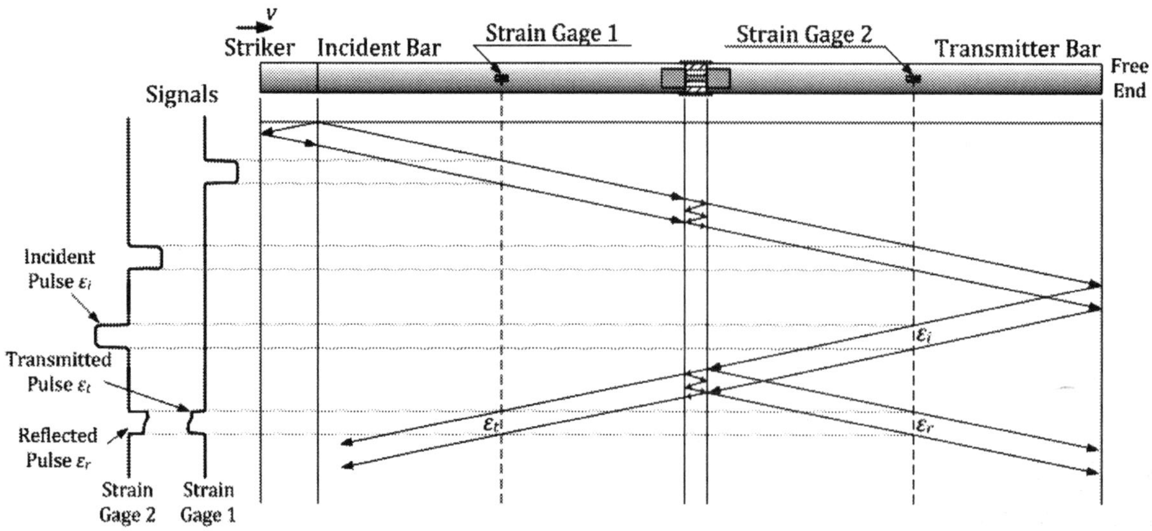

Figure 17.30 The tension stress wave generation and test process with the threaded specimen

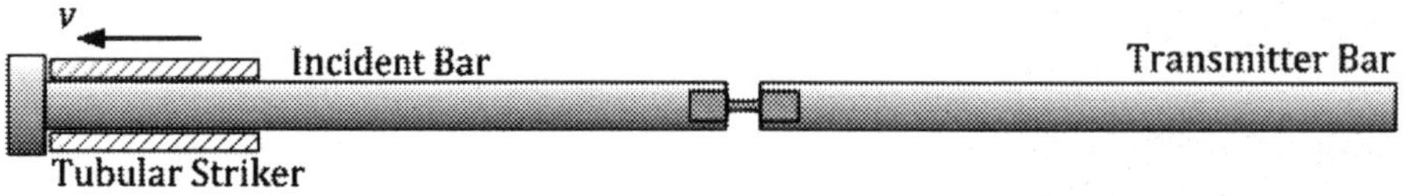

Figure 17.31 Sketch of second experiment arrangement for the split Hopkinson tension bar.

Figure 17.32 Sketch of third experiment arrangement for split Hopkinson tension bar.

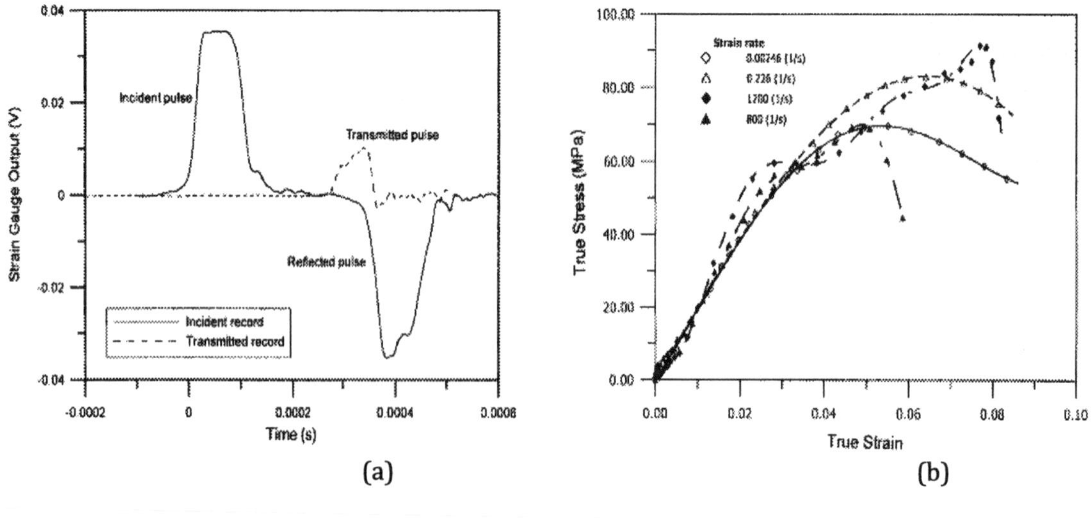

Figure 17.33 Experimental data with the second experimental arrangement.
(a) Tensile stress wave pulses; (b) Tensile stress-strain curves for Epon 828/ T-403 composites [74]

17.7.3 Split Hopkinson Torsion Bar

Experimental Setup

A typical experimental arrangement of the split Hopkinson torsion bar is presented in Fig. 17.34. This arrangement is similar the arrangement presented in Fig. 17.32. The incident bar is first clamped (fixed) at one position. A torque is applied to one side of the incident bar to store a large torsional strain within the section of the incident bar to the left side of the clamp. After suddenly releasing the clamp, the stored torsional strain energy generates a torsional stress pulse, applying a torsional load on the specimen.

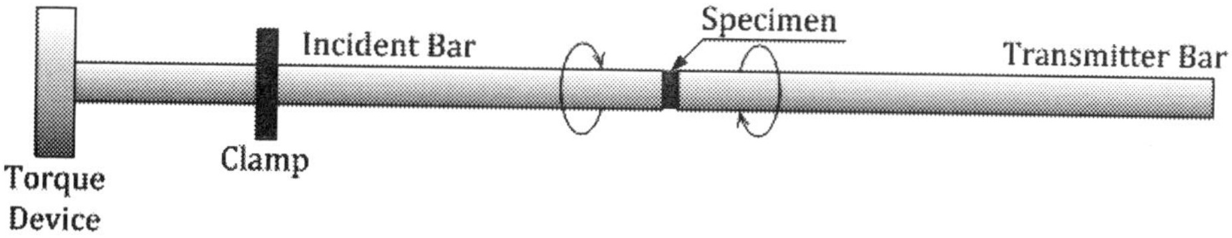

Figure 17.34 Experimental setup of the split Hopkinson torsion bar test

Figure 17.35 Thin walled tubular specimen

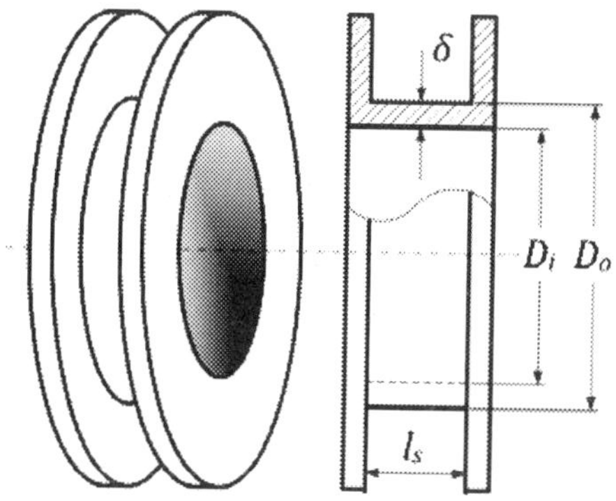

A thin walled tubular specimen is adhesively bonded between the incident and the transmitter bar. A typical sketch of the thin walled tubular specimen is shown in Fig. 17.35. The wall thickness δ, gage length l_s, inner diameter D_i and outer diameter D_o all affect the strain rate during experiments. Usually, the mean diameter of the specimen, $D_s = (D_i + D_O)/2$, is used in data analysis. The detrimental effects of inertia and friction are not present in torsional testing of thin-walled tubular specimens. Furthermore, geometric dispersion is not an issue when a torsional pulse travels along an elastic bar.

Experimental Data Analysis

Consider a torsional elastic stress wave in a homogeneous bar with density ρ, diameter D_b and shear modulus G_b. Following the analysis procedure already described in Section 17.7.1, the longitudinal displacement u is replaced by the angular displacement θ. The longitudinal strain ε is replaced by shear strain γ. The longitudinal stress σ is replaced by shear stress τ. The longitudinal force F is replaced by torque T. Then the relationships (Eqs. (17.38) ~ (17.41)) among the parameters in a bar are changed to:

$$\theta = f(x - c_t t) + g(x + c_t) \tag{17.60}$$

$$\gamma = \frac{D_b}{2}\frac{\partial \theta}{\partial x} = \frac{D_b}{2}(f' + g') \tag{17.61}$$

$$\tau = G_b\gamma = \frac{G_b D_b}{2}(f' + g') \tag{17.62}$$

$$\dot{u} = \frac{\partial u}{\partial t} = c_t(f' + g') \tag{17.63}$$

where c_t is the longitudinal wave velocity in a bar, $c_t = \sqrt{G_b/\rho}$, and f and g are arbitrary functions determined by initial conditions.

The relation between torque T and shear strain γ in the bar is given,

$$T = \frac{G_b J_b}{D_b/2}\gamma = G_b J_b(f' + g') \tag{17.64}$$

where J_b is the polar moment of inertia of the cross section of a bar, $J_b = \pi G_b^4/32$.

By replacing the corresponding parameters in Eq. (17.42) to (17.49), the shear stress, shear strain rate and shear strain in the specimen can be derived as:

$$\tau_s = \frac{D_s/2}{J_s}\left(\frac{T_1 + T_2}{2}\right) = \frac{D_s G_b J_b}{4J_s}(\gamma_i + \gamma_r + \gamma_t) = \frac{D_b^3 G_b}{16D_s^2\delta}(\gamma_i + \gamma_r + \gamma_t) \tag{17.65}$$

$$\dot{\gamma}_s = -\frac{D_s c_t}{D_b I_s}(\gamma_i - \gamma_r - \gamma_t) \tag{17.66}$$

$$\gamma_s = -\frac{D_s c_t}{D_b I_s}\int_0^t (\gamma_i - \gamma_r - \gamma_t)dt \tag{17.67}$$

Similarly, based on the equilibrium condition in the specimen,

$$g_i + g_r = g_t \tag{17.68}$$

Therefore, Eqs. (17.65) ~ (17.67) can be modified as,

$$\tau_s = \frac{D_b^3 G_b}{8D_s^2\delta}\gamma_t = \frac{D_b^3 G_b}{8D_s^2\delta}(\gamma_i + \gamma_r) \tag{17.69}$$

$$\dot{\gamma}_s = \frac{2D_s c_t}{D_b I_s}\gamma_r = -\frac{2D_s c_t}{D_b I_s}(\gamma_i - \gamma_t) \tag{17.70}$$

$$\gamma_s = \frac{2D_s c_t}{D_b I_s}\int_0^t \gamma_r dt = -\frac{2D_s c_t}{D_b I_s}\int_0^t (\gamma_i - \gamma_t)dt \tag{17.71}$$

The dynamic nominal shear stress-shear strain relationship of the materials is obtained from Eqs. (17.69) and (17.71) at the strain rate calculated with Eq. (17.70).

Example of Experimental Data

Typical torsion pulse signals and high strain rate shear stress-shear strain curves of tungsten heavy alloys from a split Hopkinson torsion bar test are shown in Fig. 17.36.

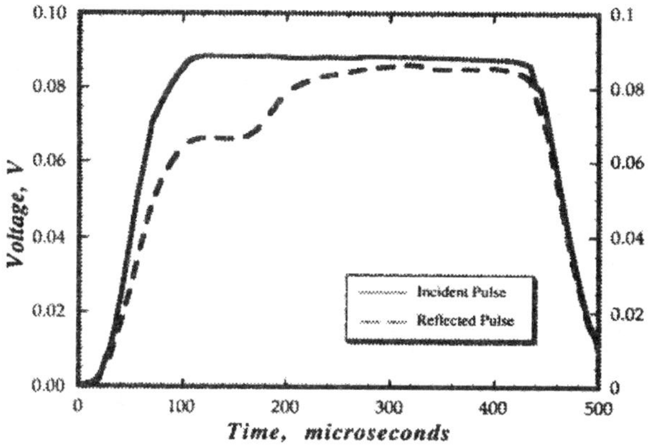

(a)

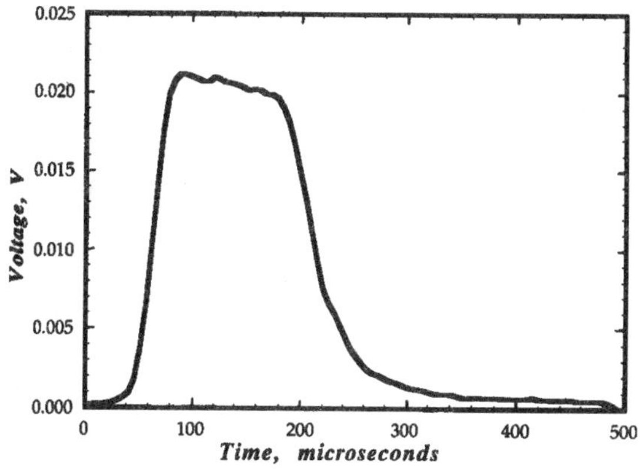

(b)

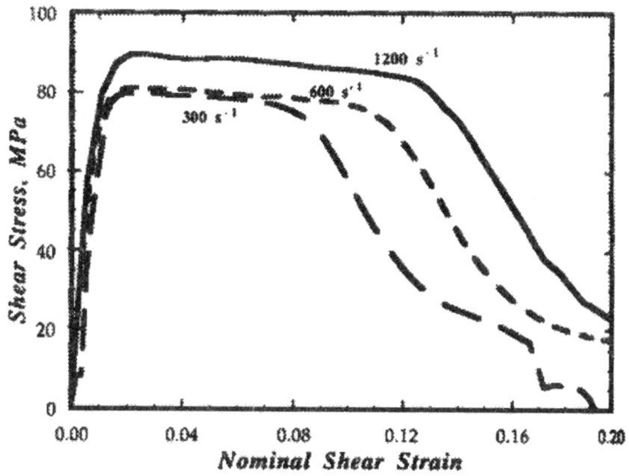

(c)

Figure 17.36 Torsion stress wave pulses and stress-strain curves of tungsten heavy alloys from a high strain rate torsion test (a) Incident and reflected pulses (b) transmitted pulse (c) shear stress-strain curve at different strain rates [75].

17.8 PLATE IMPACT TEST

In many dynamic events, materials deform and fail under an extremely high three-dimensional stress state. The plate impact test was developed to study the dynamic behavior of materials under such conditions. As the name implies, a planar flyer plate is launched by a gas gun or an explosive and impacts a stationary planar target plate with a velocity up to a few kilometers per second. This high velocity impact induces extremely high rate deformation in the materials. Due to the inertia of the material, the deformations, which are perpendicular to the impact direction, are constrained. Therefore, the materials only experience deformation in the impact direction. In other words, the materials are subjected to a one-dimensional strain state and a three-dimensional stress state. Shock waves are produced in both the impactor and target materials during impact. Normal impact of the flyer plate results in only compression shock waves. An oblique plate impact results in compression-shear shock waves. The stresses, particle velocities, shock wave velocities, etc., are measured during the experiments for subsequent data analysis. The strain rates induced by a plate impact test vary from 10^6 to 10^8 ε/s, which are the highest strain rates that can be achieved in the laboratory.

17.8.1 Experimental Setup and Procedure

A sketch of the experimental arrangement for normal plate impact is shown in Fig. 17.37. The apparatus consists of two systems, a test system and a measurement system. The test system includes a gas gun, a flyer plate, a holder (sabot), and a target plate. The flyer plate and the target plate are highly polished (lapped flat) and aligned. The flyer plate is always held by a sabot to maintain its alignment. For an oblique plate impact, the sabot has a key that slides through a keyway in the gun barrel preventing the flyer plate from rotating.

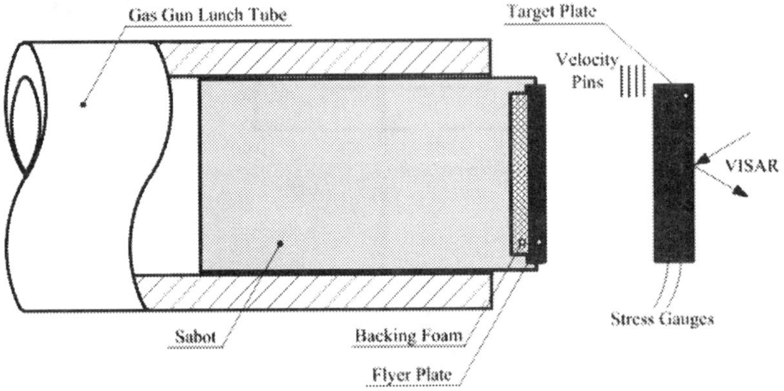

Figure 17.37 A sketch of normal plate impact test

The measurement system includes stress gages, particle velocity sensors and a signal acquisition system. Manganin stress gages are the most popular stress gages used in plate impact tests as their output is linear to pressure and stable with time. In addition, these stress gages have high sensitivity to pressure while having low sensitivity to temperature, composition, and manufacturing techniques. Recently, piezoelectric stress gages, including quartz and PVDF, have also been used in plate impact tests. One or more of these gages are placed within the target plate, as shown in Fig. 17.37. Several velocity pins are appropriately positioned in the path of the flyer in order to measure the initial impact velocity. An interferometer, for example VISAR, is utilized to measure the free surface velocity of the target plate.

When conducting an experiment, the flyer plate, held by a sabot, is launched through a gas gun barrel towards the target plate. When the flyer plate reaches the first velocity pin, the data acquisition system is triggered. Measurements of the stress and the back face velocity of the target plate are recorded as a function of time during the impact event. The shock wave speed can be calculated using several points in the stress history taken from different gages and distances between them.

17.8.2 Shock Physics in Solid Materials and Data Analysis

The dynamic response of a solid material under intensive impulse loading, such as shock wave loading, can be described by the Rankine-Hugoniot relation, or the Rankine-Hugoniot curve (Duvall, 1971). It can be expressed as the relationship between the inner pressure or stress p and the particle velocity u of a solid. The slope of a Rankine-Hugoniot curve, shown in Fig. 17.38, indicates the material's wave impedance with steeper the slopes corresponding to higher wave impedance of a material.

The p-u status when a flyer plate impacts a target plate is presented in Fig. 17.39. Assume that the flyer has an initial impact velocity u_0. The initial state of the target material is p = 0 and u = 0 (denoted by ① in Fig. 17.39b); the initial state of the flyer material is p = 0 and u = u_0 (denoted by ② in Fig. 17.39b). After the impact, two shock waves are produced and propagate in opposite directions, with one wave traveling into the flyer plate and the other into the target plate, as shown in Fig. 17.39a. The contact region of the flyer and target has the same stress p_1 and particle velocity u_1, as denoted by ③ in

Fig. 17.39b. State ③ must lie on the point of intersection of the right and left going Rankine-Hugoniot curve of the target and the flyer material, respectively.

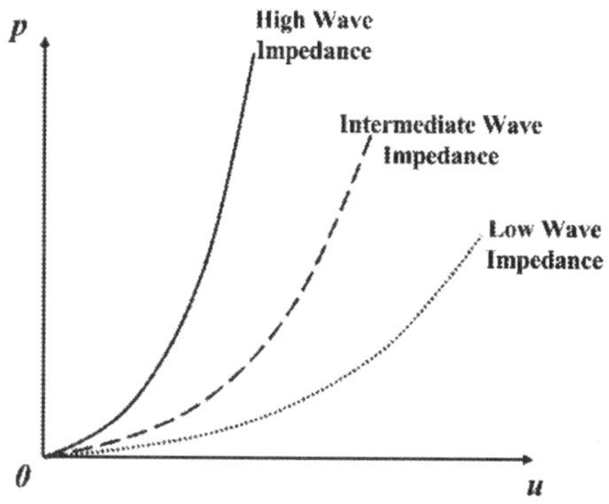

Figure 17.38 Rankine-Hugoniot curves for solid materials

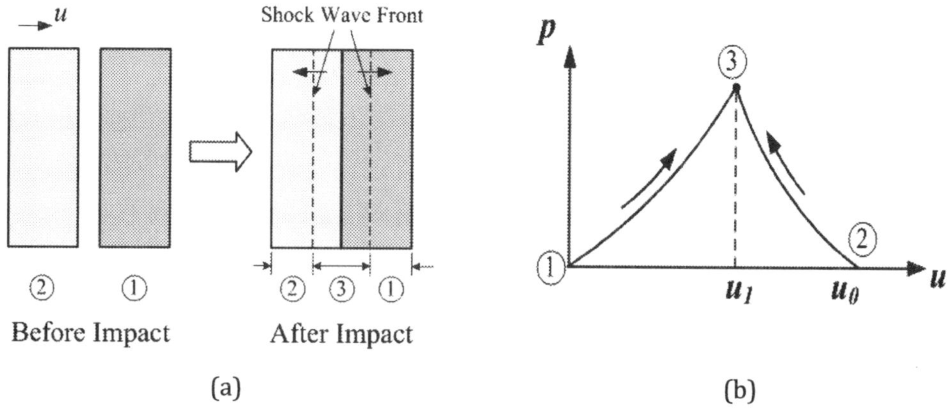

Figure 17.39 Impact process and *p-u* curves

A plate impact is called a symmetric impact shock wave test if the flyer plate material is the same as the target material. It can be used to determine the Rankine-Hugoniot curve of the materials. Based on the analysis depicted in Fig. 17.39, the states for a symmetric impact shock wave test are as shown in Fig. 17.40. Because the materials of the flyer and the target plates are same, their Rankine-Hugoniot curves are identical and the final particle velocity is $u_0/2$. Then, point ③ on the Rankine-Hugoniot curve of this material can be established by measuring the initial impact velocity of the flyer plate and the final (maximum) pressure in the target plate. Further, if we know the Rankine-Hugoniot curve of the flyer plate material, we can also determine the Rankine-Hugoniot curve of the target material based on the analysis depicted in Fig. 17.39.

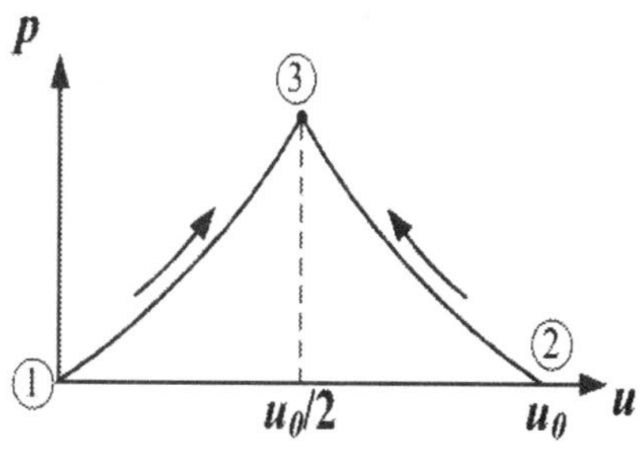

Figure 17.40 p-u curves in the symmetric impact

17.8.3 Example of Experimental Data

The experimental data acquired for the Rankine-Hugoniot relation for Ti-6Al-4V alloy are shown in Fig. 17.41. A least square fit curve is used to explore the Rankine-Hugoniot relation over a wide stress range.

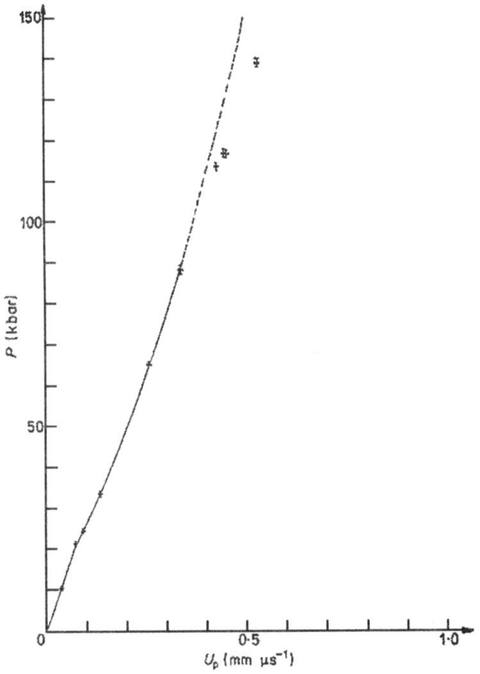

Figure 17.41 The Rankine-Hugoniot relations for Ti-6Al-4V alloy [76]

17.9 OPTICAL METHODS FOR DYNAMIC MEASUREMENT

Optical methods are generally non-intrusive and can give full field information of a specimen. Thus, optical methods are preferred over other techniques to measure full field stresses, strains or displacements. Generally, these methods are coupled with high speed photography to visualize the required information. In this section, dynamic photoelasticity and digital image correlation (DIC) methods for dynamic measurement are introduced.

17.9.1 Photoelasticity

The optical method of photoelasticity has already been discussed in detail in Chapter 11. When this method is used in conjunction with high speed photography it provides valuable details of the dynamic phenomenon being studied. Typical experimental arrangements for transmission and reflection dynamic photoelasticity experiments are shown in Figs. 17.42a and 17.42b. Two examples of the results obtained using dynamic photoelasticity are shown in Figs. 17.43 and 17.44.

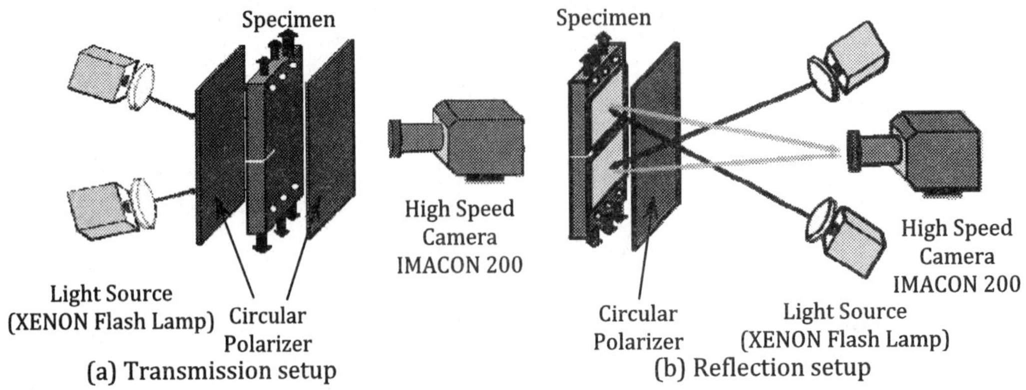

Figure 17.42 Sketch of dynamic photoelastic experimental arrangements

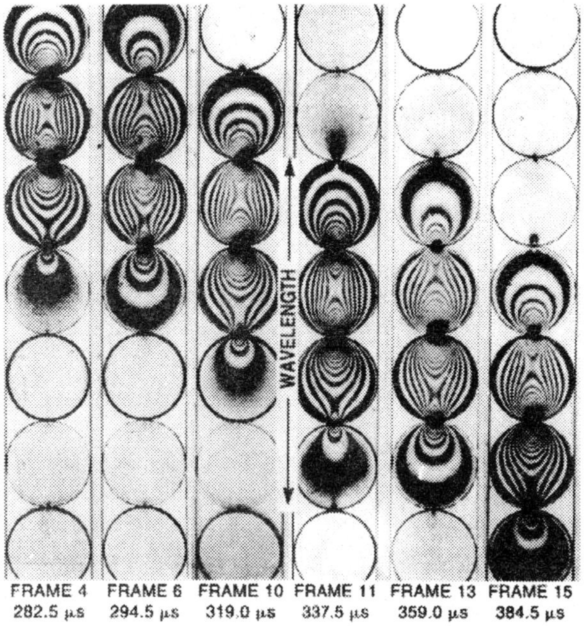

Figure 17.43 Typical isochromatic fringe patterns obtained for dynamic load transfer through a single chain assembly of disks [77]

Figure 17.44 A sequence of isochromatic fringes near a dynamically propagating crack in Homalite-100 sheet [64]

17.9.2 Digital Image Correlation

The Digital Image Correlation (DIC) technique, one of the most recent non-contact methods for analyzing full-field shape, motion and deformation, has been developed and improved during the past two decades. This technique involves recording real time deformation images with digital photography. The analysis of these images is performed by using a mapping of predefined points to obtain full field shape, motion and deformation measurements.

There are two types of DIC systems: two-dimensional systems which incorporate one camera and three-dimensional systems with two cameras. For the two-dimensional system, the post loading images are compared to the reference (zero load) image to obtain the deformation. For this system, images must be taken at a position that is perpendicular to the direction of the deformation. Only the in-plane deformation (u and v) of the surface can be measured with this approach. Any out-of-plane deformation (w) markedly affects the accuracy of the measurement. Employing a three-dimensional camera system eliminates this problem. The working mechanism of a three-dimensional system is similar to how our eyes function. Two cameras capture two images from different angles at the same instant. By correlating these two images, one can obtain the three dimensional shape of the surface. Correlating this deformed shape to a reference (zero-load) shape gives full-field in-plane and out-of-plane deformations. The details of the DIC technique are described in Chapter15.

By using high-speed cameras, the DIC method can be expanded to analyze mechanical phenomena, such as impact, blast and high-frequency vibration. Recent developments in the digital high-speed photography has significantly improved the accuracy of DIC method and simplified the experimental arrangement, consequently, bringing more attention to the DIC technique.

There are several high speed cameras currently available for dynamic experiments. Of these cameras three are most popular. The first type makes use of a single lens and multiple CCD receivers, such as the DRS IMACON 200. The different CCD receivers capture the images at different times to achieve high-speed imaging. The DRS IMACON 200 currently holds the world record of frame-rate (up to 200 million frames per second) while maintaining excellent image resolution (up to 1360×1024 pixels). Because of the difference between CCDs, the stability of the images required for DIC cannot be guaranteed. The second disadvantage is that the number of frames taken is very limited. For example, the DRS IMACON 200 can only capture a total of 16 frames.

The second type of digital high-speed camera uses a single lens and a single IS-CCD high-speed receiver, such as SHIMADZU HPV-1. It can capture 100 frames at a frame-rate up to 1 million fps while having a fixed image resolution of 312×260 pixels. This type of camera solves the image stability problem. However, the number of frames are still limited.

The third type of digital high-speed camera uses a single lens and a single CMOS receiver, such as the PHOTRON SA1. There is a relation between the framing rate and the image resolution. As the frame rate increases, the image resolution reduces. One significant advantage of this type of camera is that it incorporates an expandable inner memory. Therefore, it can record frames for longer periods of time. For example, the PHOTRON SA1, with an 8GB internal memory, can capture images at a frame-rate of 20,000 fps with an image resolution of 512×512 pixels with a 1 second time duration.

An illustration of an experimental arrangement for photographing a dynamic image of a speckle pattern is shown in Fig. 17.45. Because the real surface of the specimen is difficult to image, a speckle pattern with good contrast is usually painted on the surface of the specimen (Fig. 17.45). A sequences of real images taken by two Photron SA1 cameras and the correlated out-of-plane back face deformation for of a panel during a shock tube blast test is presented in Fig. 17.46.

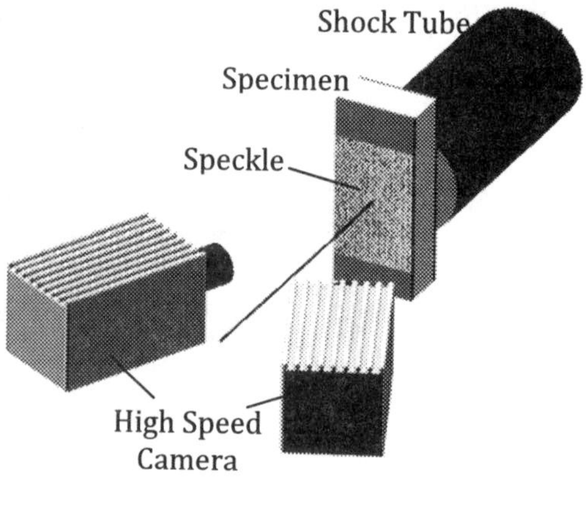

Figure 17.45 Dynamic digital image correlation setup

Other optical methods that are used for dynamic measurements include dynamic moiré, coherent gradient sensing, caustics, etc.

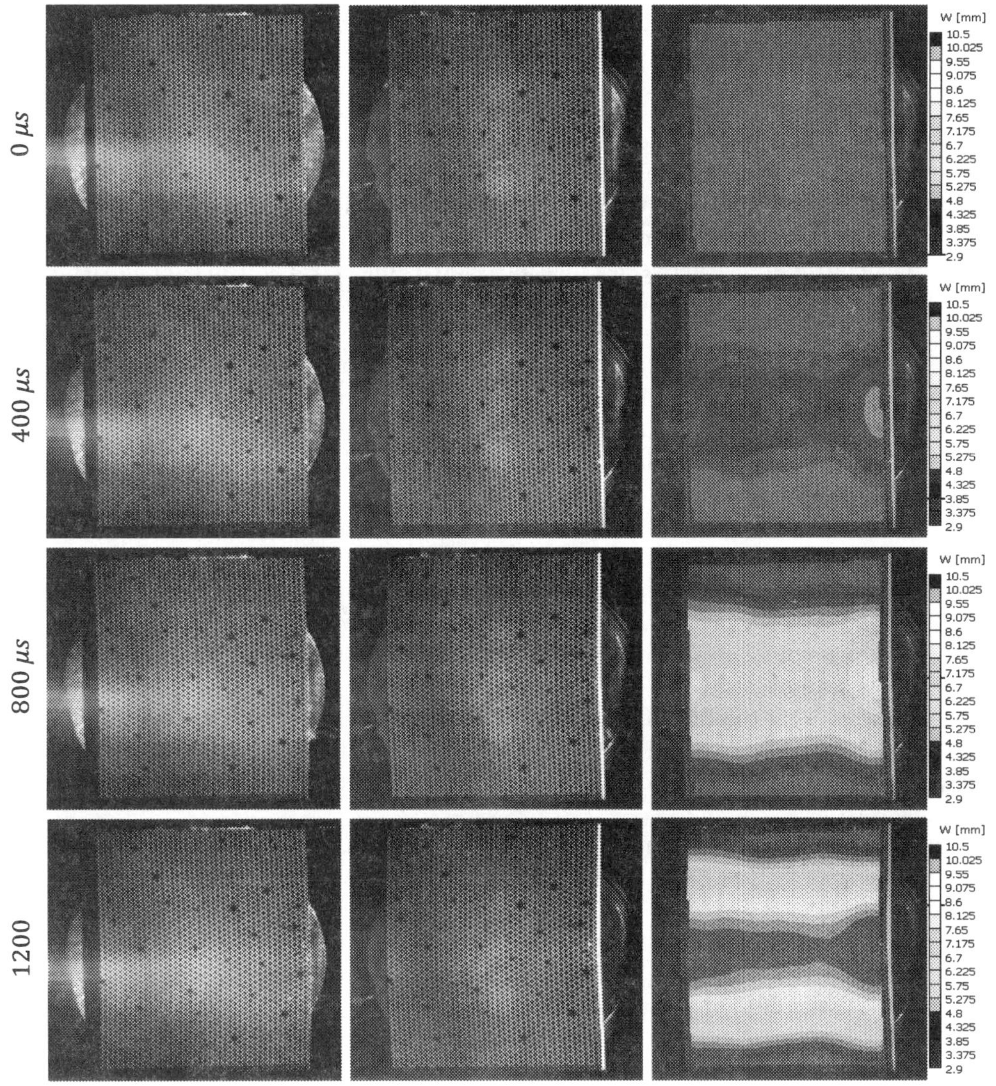

(a) Left camera (b) Right camera (c) Out-of-plane deflection

Figure 17.46 Digital images and correlation results

EXERCISES

17.1 How many parts does the servo-hydraulic system consist of? What are they?
17.2 What is the test strain rate range of the high-speed servo-hydraulic system?
17.3 Which part of the servo-hydraulic system will induce the inertia effects on the test results? How do they affect the test results?
17.4 How many parts are involved in the drop weight impact test system? What are they?
17.5 What is the test strain rate range of the drop weight impact test system?
17.6 If the initial release height of the Charpy-V notch pendulum is 1.8 m and the highest height that the pendulum reaches is 0.8 m after it has struck the specimen, how much energy is absorbed by the specimen? Assume the mass of the pendulum is 10 kg.
17.7 What is the main difference between a Charpy V-notch impact test and an Izod impact test?
17.8 What is the experimental procedure of a Taylor impact test?
17.9 How many parameters can we control in the Taylor impact test? What are they?
17.10 From Fig. 17.22, the polycarbonate cylinder has a final total length of 70 mm and a final un-deformed length of 44 mm. The initial total length is 76.2 mm and the impact velocity is 187 m/s. What is the dynamic yield strength of the polycarbonate?
17.11 Discuss the advantages of using the split Hopkinson pressure bar experiment?
17.12 What conditions are required for a valid SHPB test?
17.13 If the diameter of the specimen is 8 mm and the Poisson's ratio of the materials is 0.3, what is the proper length of the specimen for a SHPB test?
17.14 If the transmitter bar of a SHPB is a hollow bar with cross section area A_{bh}, could you follow the analysis for the solid transmitter bar from Eqs. (17.42) ~ (17.56) to derive new equations for the nominal stress, nominal strain and nominal strain rate? Assume the incident bar is still a solid bar with cross section area A_b.
17.15 How could we prevent rotation of a flyer plate during a plate impact test?
17.16 Based on the analysis in Fig. 17.39 and 17.40, and assuming the flyer plate has low wave impedance and the target plate has high wave impedance, compare the magnitude of u_1 and $u_0/2$? Which one is larger?
17.17 Prepare a detailed memorandum describing a high-speed camera that you are recommending to your manager to purchase. The application for the camera involves the following experimental methods:

(a) Dynamic photoelasticity
(b) Dynamic moiré
(c) Dynamic digital image correlation
(d) Dynamic caustics

BIBLIOGRAPHY

1. Sharpe, W.N.: Handbook of Experimental Solid Mechanics, Springer, 2008
2. Younkin, G.W.: Industrial Servo Control Systems – Fundamentals and Applications, second edition, revised and expanded, Marcel Dekker, Inc, New York, Basel, 2003.
3. Walters, R.B.: Hydraulic and Electro-Hydraulic Control Systems, Elsevier Science Publishing Co., Inc, New York, 1991.
4. Bkguelin, Ph., M. Barbezat, and H.H. Kausch: "Mechanical Characterization of Polymers and Composites with a Servohydraulic High-Speed Tensile Tester", Journal de Physique III France, Vol. 1, pp. 1867-1880, 1991.
5. Lim, J.H.: "High speed tensile test of automotive steel sheets at the intermediate strain rate", POSCO Technical Report, Vol. 10, pp. 116-122, 2007.
6. Silva, F.A., D. Zhu, B. Mobasher, C. Soranakomb, and R.D.T. Filhoc: "High Speed Tensile Behavior of Sisal Fiber Cement Composites", Materials Science and Engineering A, doi:10.1016/j.msea.2009.08.013, 2009.
7. Axel Physical Testing Services: "High Strain Rate Experiments," http://www.axelproducts.com/pages/highstrainrate.html.
8. Instron Testing Machines for Tensile, Fatigue, Impact & Hardness Testing: "VHS 8800 High Strain Rate Systems," <http://www.instron.us /wa/product/VHS-8800-High-Strain-Rate-Systems.aspx>.
9. MTS Systems: "High Rate Testing Systems," http://www.mts.com/stellent/groups/public/documents/library/dev_004212.pdf.
10. Yu, J., X. Wang, Z. Wei, and E. Wang: "Deformation and Failure Mechanism of Dynamically Loaded Sandwich Beams with Aluminum-Foam Core", International Journal of Impact Engineering, Vol. 28, pp. 331-347, 2003.
11. Gustin, J., A. Joneson, M. Mahinfalah, and J. Stone: "Low Velocity Impact of Combination Kevlar/Carbon Fiber Sandwich Composites", Composite Structures, Vol. 69, pp. 396–406, 2005.
12. Instron Testing Machines for Tensile, Fatigue, Impact & Hardness Testing: "Dynatype 9200 Series Impact Test Machines," <http://www.instron.us/wa/products/impact/series_9200.aspx>.
13. Schubel, P.M., J. Luo, and I.M. Daniel: "Low Velocity Impact Behavior of Composite Sandwich Panels, Composites": Part A, Vol. 36, pp. 1389–1396, 2005.
14. Hosur, M.V., M. Adbullah, and S. Jeelani: "Studies on the Low-Velocity Impact Response of Woven Hybrid Composites", Composite Structures, Vol. 67, pp. 253–262, 2005.
15. Anderson, T., and E. Madenci: "Experimental Investigation of Low-Velocity Impact Characteristics of Sandwich Composites", Composite Structures, Vol. 50, pp. 239-247, 2000.
16. Bhuiyan, M.A., M.V. Hosur, and S. Jeelani: "Low-Velocity Impact Response of Sandwich Composites with Nanophased Foam Core and Biaxial (±45º) Braided Face Sheets", Composites: Part B, Vol. 40, pp. 561–571, 2009.
17. Imatek: "Drop Weight Tear Testers," http://www.imatek.co.uk/product-dwtt.php, 2010.
18. ASTM-E-23, "Standard Test Methods for Notched Bar Impact Testing of Metallic Materials".
19. Jacobs, J., and T. Kilduff: Engineering Materials Technology - Structures, Processing, Properties, and Selection, Pearson Prentice Hall, ISBN 9780130481856, 2004.
20. Callister, W. D. Jr.: Materials Science and Engineering: An Introduction, John Wiley and Sons, Inc, New York, NY, 1994.
21. Lucon, E.: "Influence of Striking Edge Radius (2 vs 8 mm) on Instrumented Charpy Data and Absorbed Energies", International Journal of Fracture, Vol. 153, pp. 1-14, 2008.
22. Walter+bai Ag: "Series PH 25-50 Joule," <http://www.walterbai.com/Universal_Material_Testing/Impact_Testing_Machines/Impact_Pendulum_Tester_Series_PH_25-50_Joule.html>.
23. DIC Corporation: <Http://www.dic.co.jp/en/products/planac/mechanical.html>.
24. Instron Testing Machines for Tensile, Fatigue, Impact & Hardness Testing: "Instron Pendulum Testers," <http://www.instron.us/wa/product/Instron-Pendulums.aspx>.
25. Learning Space: "Introduction to Structural Integrity," <http://openlearn.open.ac.uk/file.php/3258/formats/T357_1_rss.xml>.

26. Taylor, G.I.: "The Testing of Materials at High Rates of Loading", Journal Institution Civil Engineers, Vol. 26, pp. 486–519, 1946.
27. Taylor, G.I.: "The Use of Flat-Ended Projectiles for Determining Dynamic Yield Stress I. Theoretical Considerations". Proceedings of the Royal Society of London. Series A, Mathematical and Physical Sciences, Vol. 194, pp. 289-299, 1948.
28. House, J.W.: "Taylor impact testing", Report from Air Force Armament Laboratory, AFATL-TR-89-41, 1989.
29. Forde, L.C., W.G. Proud, and S.M. Walley: "Symmetrical Taylor Impact Studies of Copper", Proceedings of the Royal Society A, Vol. 465, pp. 769–790, 2009.
30. Shin, H., S. Park, and S. Kim,: "Deformation Behavior of Polymeric Materials by Taylor Impact", International Journal of Modern Physics B, Vol. 22, pp. 1235-1242, 2008.
31. Sarva, S., A.D. Mulliken, and M.C. Boyce: "Mechanics of Taylor Impact Testing of Polycarbonate", International Journal of Solids and Structures, Vol. 44, pp. 2381–2400, 2007.
32. Wang, B., J. Zhang, and G. Lu: "Taylor Impact Test for Ductile Porous Materials—Part 2: Experiments", International Journal of Impact Engineering, Vol. 28, pp. 499–511, 2003.
33. Allen, D.J., W.K. Rule, and S.E. Jones: "Optimizing Material Strength Constants Numerically Extracted from Taylor Impact Data", Experimental Mechanics, Vol. 37, pp. 333-338, 1997.
34. Lopatnikov, S.L., B.A. Gama, M.J. Haque, C. Krauthauser, J.W.G. Jr, M. Guden, and I.W. Hall: "Dynamics of Metal Foam Deformation during Taylor Cylinder–Hopkinson Bar Impact Experiment", Composite Structures, Vol. 61, pp. 61–71, 2003.
35. Hopkinson, B.: "A Method of Measuring the Pressure Produced in the Detonation of High Explosive or by the Impact of Bullets", Philosophical Transactions of the Royal Society of London. Series A, Vol. 213, pp. 437-456, 1914.
36. Kolsky, H.: "An Investigation of the Mechanical Properties of Materials at Very High Rates of Loading", Proceedings of the Physical Society. Section B, Vol. 62, pp. 676-700, 1949.
37. Zhao, H., G. Gary, and J.R. Klepaczko: "On the Use of a Viscoelastic Split Hopkinson Pressure Bar", International Journal of Impact Engineering, Vol. 19, pp. 319-330, 1997.
38. Chen, W., B. Zhang, and M.J. Forrestal: "A Split Hopkinson Bar Technique for Low-Impedance Materials", Experimental Mechanics, Vol. 39, pp. 81-85, 1999.
39. Ninan, L., J. Tsai, and C.T. Sun: "Use of Split Hopkinson Pressure Bar for Testing Off-Axis Composites", International Journal of Impact Engineering, Vol. 25, pp. 291-313, 2001.
40. Huh, H., W.J. Kang, and S.S. Han: "A Tension Split Hopkinson Bar for Investigating the Dynamic Behavior of Sheet Metals", Experimental Mechanics, Vol. 42, pp. 8-17, 2002.
41. Hartley, K.A., J. Duffy, and R.H. Hawley: "The Torsional Kolsky (Split-Hopkinson) Bar", ASM Metal Handbook, 8,9th ed., American Society of Metals, Materials Park, OH, pp. 218–228, 1985.
42. Staab, G.: "A Direct-Tension Split Hopkinson Bar for High Strain-Rate Testing", Experimental Mechanics, Vol. 31, pp. 232-235, 1991.
43. Wang L.: Foundation of Stress Waves, second edition, National Defense Industry Press, Beijing, P. R. China, 2005.
44. Field, J.E., S.M. Walley, W.G. Proud, H.T. Goldrein, and C.R. Siviour: "Review of Experimental Techniques for High Rate Deformation and Shock Studies", International Journal of Impact Engineering, Vol. 30, pp.725-775, 2004.
45. Duvall, G.E.: Shock Wave in Condensed Media. Physics of High Energy Density, Academic Press, New York, 1971.
46. Marsh, S.P.: LASL Shock Hugoniot Data, University of California Press, Berkeley, CA, 1980.
47. Dunn, J.E., R. Fosdick, and M. Slemrod: editors, Shock Induced Transitions and Phase Structures in General Media, Berlin, Springer, 1993.
48. Grady, D.E.: "The Spall Strength of Condensed Matter", Journal of Mechanics and Physics of Solids, Vol. 36, pp. 353-384, 1988.
49. Meyers, M.A.: Dynamic Behavior of Materials, Wiley, New York, 1994.
50. Gray, G.T.: "Shock Wave Testing of Ductile Materials", ASM Handbook, American Society of Metals, Materials Park, OH, Vol. 8, pp. 462-476, 2000.

51. Graham, R.A., F.W. Neilson, and W.B. Benedick: "Piezoelectric Current Form Shock-Loaded Quartz – a Sub-Microsecond Stress Gage", Journal of Applied Physics, Vol. 36, pp. 1775-1800, 1965.
52. Obara, T., N.K. Bourne, and Y. Mebar: The Construction and Calibration of an Inexpensive PVDF Stress Gage for Fast Pressure Measurements", Measurement Science and Technology, Vol. 6, pp. 345-348, 1995.
53. Barker, L.M.: "The Development of the VISAR and Its Use in Shock Compression Science", Shock Compression of Condensed Matter, AIP Conference Proceedings, Vol. 505, pp. 11-18, 2000.
54. Voloshin, A.S., and C.P. Burger: "Half Fringe Photoelasticity – a New Approach to Whole Field Stress Analysis", Experimental Mechanics, Vol. 23, pp. 304-314, 1983.
55. Shukla, A., M.H. Sadd, and H. Mei: "Experimental and Computational Modeling of Wave Propagation in Antigranulocytes Materials", Experimental Mechanics, Vol. 30, pp. 377-381, 1990.
56. Wells, A.A., and D. Post: "The Dynamic Stress Distribution Surrounding a Running Crack: a Photoelastic Analysis", Proceedings of the Society for Experimental Stress Analysis, Vol. 16, pp. 69-96, 1958.
57. Sutton, M.A., J.J. Orteu, and H. Schreier: Image Correlation for Shape, Motion and Deformation Measurements – Basic Concepts, Theory and Applications, Springer, 2009.
58. Lee, D., H. Tippur, and M. Kirugulige: "Experimental Study of Dynamic Crack Growth in Unidirectional Graphite/Epoxy Composites Using Digital Image Correlation Method and High-Speed Photography", Journal of Composite Materials, Vol. 43, pp. 2081-2108, 2009.
59. Li, E.B., A.K. Tieua, and W.Y.D. Yuen: "Application of Digital Image Correlation Technique to Dynamic Measurement of the Velocity Field in the Deformation Zone in Cold Rolling", Optics and Lasers in Engineering, Vol. 39, pp. 479-488,2003.
60. Wang, E., and A. Shukla: "Blast Response of Sandwich Composites Using Digital Image Correlation Technique", 9th International Conference on Sandwich Structures, California Institute of Technology, Pasadena, California, USA, June 14-16, 2010.
61. Barthelat, F., Z. Wu, B.C. Prorok, and H.D. Espinosa: "Dynamic Torsion Testing of Nanocrystalline Coatings Using High-Speed Photography and Digital Image Correlation", Experimental Mechanics, Vol. 43, pp. 331-340, 2003.
62. Jain, N., and A. Shukla: "Asymptotic Analysis and Reflection Photoelasticity for the Study of Transient Crack Propagation in Graded Materials", Journal of mechanics of Materials and Structures, Vol. 2, pp. 595-612, 2007.
63. Shukla, A.: Practical Fracture Mechanics in Design, 2nd edition, Mercel and Dekker, 2005.
64. Shukla, A.: Dynamic Fracture Mechanics, World Scientific Publishing Co. Pte. Ltd, 2006.
65. Bruck, H.A., S.R. McNeill, M.A. Sutton, and W.H. Peters: "Digital Image Correlation Using Newton-Raphson Method of Partial Differential Correction", Experimental Mechanics, Vol. 29, pp. 261-267, 1989.
66. House, J.W., J.C. Lewis, P.P. Gillis, and L.L. Wilson: "Estimation of Flow Stress under High Rate Plastic Deformation", International Journal of Impact Engineering, Vol. 16, pp. 189-200, 1995.
67. Jones, S.E., J.A. Drinkard, W.K. Rule, and L.L. Wilson: "An Elementary Theory for the Taylor Impact Test", International Journal of Impact Engineering, Vol. 21, pp. 1-13, 1998.
68. Barton, C.S., and E.G. Volterra: "Dynamics of Plastics and Rubber-like Materials", Final Report, Rensselaer Polytechnic Institute, for ONR, June 30, 1946.
69. Lee, O.S., and M.S. Kim: "Dynamic Material Property Characterization by Using Split Hopkinson Pressure Bar (SHPB) Technique", Nuclear Engineering and Design, Vol. 226, pp. 119-125, 2003.
70. Nail, N.K., A. Asmelash, V.R. Kavala, and V. Ch: "Interlaminar Shear Properties of Polymer Matrix Composites: Strain Rate Effect", Mechanics of Materials, Vol. 39, pp. 1043-1052, 2007.
71. Shukla, A., G. Ravichandran, and Y. Rajapakse: Dynamic Failure of Materials and Structures, Springer, 2009.
72. Jain, N., and A. Shukla: "Mixed Mode Dynamic Fracture in Particulate Reinforced Functionally Graded Materials", Experimental Mechanics, Vol. 46, pp. 137-154, 2006.
73. Singh, R., A. Shukla, and H. Zervas: "Explosively Generated Pulse Propagation Through Particles Containing Natural Cracks", Mechanics of Materials, Vol. 23, pp. 255-270, 1996.

74. Chen, W., F. Lu, and M. Chen: "Tension and Compression Tests of Two Polymers under Quasi-Static and Dynamic Loading", Polymer Testing, Vol. 21, pp. 113-121, 2002.
75. Ramesh, K.T.: "On the Localization of Shearing Deformations in Tungsten Heavy Alloys", Mechanics of Materials, Vol. 17, pp. 165-173, 1994.
76. Rosenberg, Z., Y. Meybar, and D. Yaziv: "Measurement of the Hugoniot Curve of Ti-6Al-4V with Commercial Manganin Gages", Journal of Physics D: Applied Physics, Vol. 14, pp. 261-266, 1981.
77. Singh, R., A. Shukla and H. Zervas: "Effect of Flaws on the Stress Wave Propagation in Particulate Aggregates: Near and Far Field Observations," International Journal of Solids and Structures, Vol. 32, pp. 2523-2546, 1995.
78. Dally, J. W.: Ph.D. Dissertation, Illinois Institute of Technology, 1958.

CHAPTER 18

NANOSCALE MEASUREMENTS

18.1 INTRODUCTION

Over the last decade, a vast amount of fundamental work has been performed in developing and understanding experimental measurements on the nanoscale. Interest in nanoscale measurements is driven by the demand in industry to understand the mechanical behavior of materials at the nanoscale and to improve the design of nanoscale devices. The applications of such devices are widespread and can be found in a variety of fields. These applications include the medical and automotive industries, defense systems, information technology, structural health and environmental monitoring as well as many unexpected applications. On the macroscopic scale, a number of standardized techniques have been developed to measure material properties, such as the elastic modulus, yield strength, fatigue strength, and fracture toughness: whereas on the nanoscale, such measurements pose great challenges. The field of mechanical testing and measurements on the nanoscale has been rapidly adapting existing techniques, as well as developing new techniques to meet new challenges.

The rapid development of microelectromechanical systems (MEMS) and the emergence of the new class of bulky nanocrystalline materials led to the exploration and exploitation of size effects in electrical, mechanical, and thermal domains [1]. It was quickly realized that materials exhibit different properties on the microscale than those observed on the macroscopic scale [2]. The problem is further complicated by the fact that mechanical properties of microscale and nanoscale materials are significantly affected by the fabrication processes, and are very sensitive to the influence of interfaces and adjoining materials [3]. Thus, the critical importance of mechanical testing at microscale (nanoscale) is evident.

The basic measurements required in conducting experiments to determine tensile, bending, and torsion properties, at the nanoscale, do not change from those required to determine properties at the macroscale. Instead, methods used for specimen preparation and gripping as well as the way forces or displacements are applied and measured change. This chapter presents four different techniques that are capable of measuring nanoscale deformation at a high resolution. It also includes a description of the equipment, preparation of the specimens, loading fixtures and sample results.

18.2 SCANNING ELECTRON MICROSCOPE

A typical scanning electron microscope (SEM), shown in Fig. 18.1, is used to generate and control an electron beam that sweeps the specimen in a raster scan pattern. The electrons are generated by thermionic emission at a heated tungsten filament, which has a V-shaped tip about 200 μm in diameter. The filament is maintained at a high negative voltage during operation. The electrons emitted are accelerated to the anode (ground) with acceleration voltages that are adjustable from about 1 to 50 kV.

Conventional electromagnetic lenses are positioned in the SEM column to focus the electron beam by the interaction of the magnetic field of the electronic lens with the moving electrons. The condenser and objective lens are used to reduce the beam diameter by a factor of 1,000 or more, to the final spot diameter of 5 to 20 nm when it reaches the specimen.

The objective lens is used to focus the electron beam on the specimen at different working distances, which usually range from about 10 to 40 mm. The spot size is minimized by reducing the working distance to improve resolution. However, the depth of field is increased by increasing the working distance, which produces a smaller divergence angle of the beam. The depth of field is also dependent on the beam limiting aperture used in the SEM as indicated in Table 18.1.

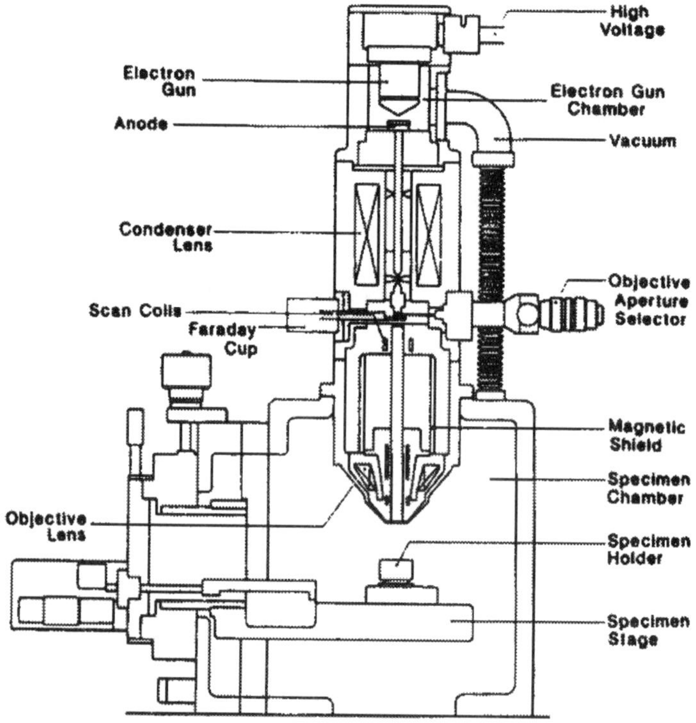

Figure 18.1 Schematic diagram of elements in the vacuum column of an SEM

Table 18.1
Typical depth of field with a working distance of 10 mm

Magnification	Aperture Diameter (μm)			
	50	100	200	500
Depth of Field (μm)				
10^1	8000	4000	2000	800
10^2	800	400	200	80
10^3	80	40	20	8
10^4	8	4	2	0.8

The diameter of the e-beam is of critical importance when writing either line or dot patterns to produce a high resolution moiré grating. The variables which affect the beam diameter include the filament material, the accelerating voltage, and the probe current. The probe current is the electrical current transmitted as the electron beam travels from the filament to the specimen. To minimize beam diameter, low probe currents (10 to 50 pA) are employed with higher accelerating voltages (20-30 kV). The filament material is generally tungsten, although filaments from lanthanum hexaboride produce a tighter beam. The beam diameter ranges from about 10 to 20 nm as the probe current increases from 10 to 50 pA with an accelerating voltage of 30 kV. The beam diameter increases as the accelerating voltage is decreased.

It is important to note the difference between the filament current and the probe current. The filament current, which is about 200 to 240 μA, heats the tungsten filament and produces the electrons in the gun. The probe current is the incident current produced by the electron beam striking the specimen. The probe current is adjusted with the condenser lenses (for a given aperture). Increasing the magnification of a condenser lens increases the divergence angle, and a larger portion of the electron beam is blocked by the beam-limiting aperture reducing the current in the beam. A beam of electrons with a diameter in the range of 10 to 20 nm can be produced in a typical SEM. Moreover, by controlling the scanning coils, this beam can be directed over the surface of the specimen to trace any specified pattern on an x-y plane.

The electrons generated in a SEM interact with the atoms that make up the sample producing signals that contain information about the sample's surface topography, composition and other properties such as electrical conductivity. A wide range of magnifications are possible in SEM, from about 10 times (about equivalent to that of a powerful hand-lens) to more than 500,000 times, which is about 250 times the magnification limit of the best light microscopes.

18.2.1 Sample Preparation

For conventional imaging in the SEM, specimens must be electrically conductive, at least at the surface, and electrically grounded to prevent the accumulation of electrostatic charge on the surface. Metal objects require little special preparation for SEM except for cleaning and mounting on a specimen stub. Nonconductive specimens tend to accumulate charge when scanned by the electron beam, especially in secondary electron imaging mode, which causes scanning faults and other image artifacts. They are therefore usually coated with an ultra-thin coating of an electrically-conducting material, commonly gold, deposited on the sample either by low vacuum sputter coating or by high vacuum evaporation. Conductive materials in current use for specimen coating include gold, gold/palladium alloy, platinum, osmium, iridium, tungsten, chromium and graphite. The coating prevents the accumulation of static electric charge on the specimen during electron irradiation.

Two important reasons for coating, even when there is more than enough specimen conductivity to prevent charge accumulation, are to maximize the signal and improve the spatial resolution, especially with samples of low atomic number. Usually, the signal increases with atomic number, especially for backscattered electron imaging. The improvement in resolution arises because in a material with a low atomic number such as carbon, the electron beam can penetrate several micrometers below the surface, generating signals from an interaction volume much larger than the beam diameter, reducing spatial resolution. Coatings with a high atomic number material, such as gold, maximizes secondary electron yield from within a surface layer a few nm thick and suppresses secondary electrons generated at greater depths. Coatings provide a signal that is predominantly derived from locations closer to the beam and closer to the specimen surface than would be the case in an uncoated material having a low atomic number.

18.2.2 Applications

The SEM enables the observation of materials in macro- and submicron ranges. The instrument is capable of generating three-dimensional images for analysis of topographic features. These features can be used for measuring displacements before, during and after a mechanical experiment is conducted in the SEM chamber. A good example of such measurements was presented by Masayuki Arai [4]. He examined the properties of a freestanding CoNiCrAlY thin film that was detached chemically from a coated substrate. A tensile test was performed using a specially fabricated loading fixture, which was inserted in the vacuum chamber of the SEM. Many of these miniature loading systems for mechanical testing, which fit into an SEM chamber, are now commercially available. One of these loading systems, used for observing the effects of tensile stress on a sample in an SEM, is shown in Fig. 18.2. The loading heads move in opposite directions, which minimizes the movement of the area of interest. The load and displacement sensors measure the applied load and the resulting extension of the specimen. The data from the sensors is then used to generate a stress-strain curve. While the SEM images were used by Arai to study the development and propagation of the crack (illustrated in Fig. 18.3), the data from the sensors on the substage were used to prepare the stress-strain graph that is shown in Fig. 18.4.

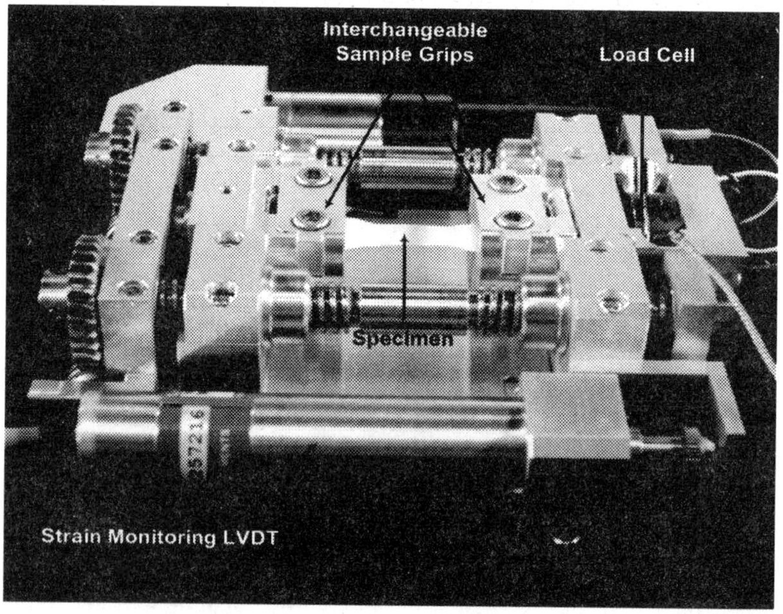

Figure 18.2 Tensile Substage used for insitu SEM tests (*courtsey MTI/Fullam*)

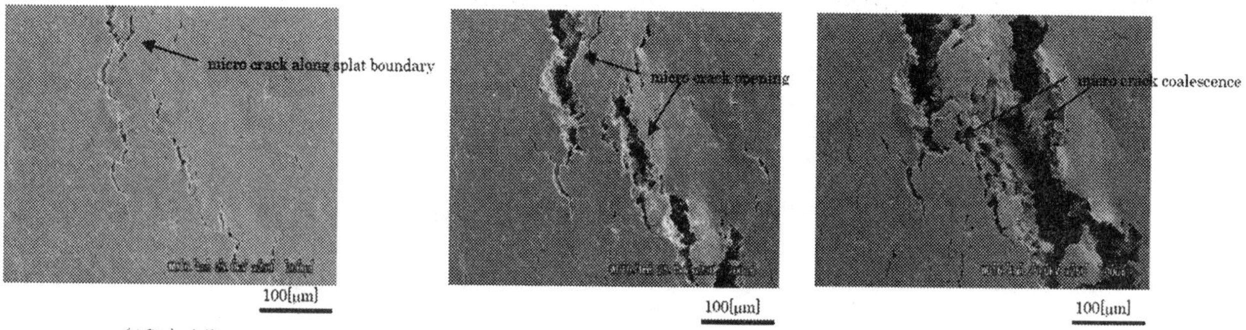

(a) Strain=1.4% (b) Strain=2.0% (c) Strain=2.5%

Figure 18.3 Continuous observation results using CoNiCrAlY tensile test at 1013K in an SEM. (4)

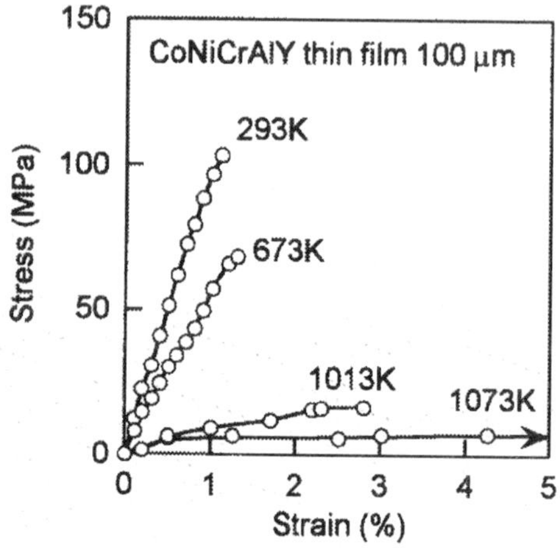

Figure 18.4 Typical stress versus strain curves (4)

The SEM can also be used to explore the surface topology of a specimen after an experiment is complete. Dawicke and Sutton [5] used SEM images to improve their evaluation of crack tip opening angles. According to their findings, a distinct change in the macroscopic appearance of the fracture surface occurs when the loading changes from fatigue cycling to stable tearing for both low and high fatigue stress experiments. SEM photographs of a typical fracture surface from a low cycle fatigue stress experiment subjected to stable tearing followed by additional fatigue cycles, are shown in Fig. 18.5. In the SEM photographs, the stable tearing regions appear light and the fatigue regions appear dark. The interface between the dark (fatigue) and light (stable tearing) region denotes the crack-front profile before the initiation of stable tearing. They also digitized the fracture surface elevation as a function of crack length and the through-thickness position using SEM images.

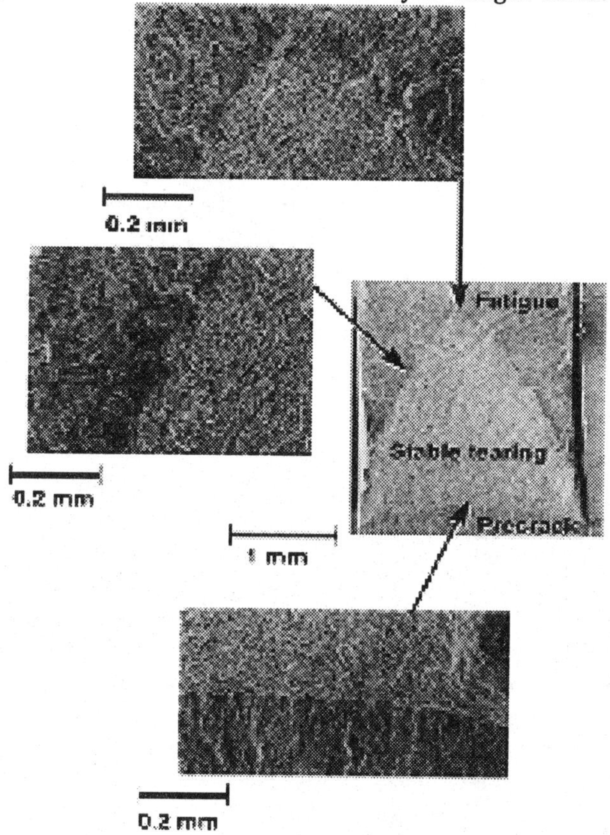

Figure 18.5 Scanning-electron-microscope photographs of the fracture surface of a M(T) specimen. [5]

An SEM has also been used by Dally and Read [6] to write very high frequency line and dot gratings (10,000 lines/mm) on homogenous and heterogeneous materials which are suitable for making moiré applications and nanoscale measurements. The details of this technique are given in Chapter 13.

18.3 TRANSMISSION ELECTRON MICROSCOPE

A schematic of a typical transmission electron microscope (TEM) is shown in Fig. 18.6. There are certain similarities between TEM and SEM, but the principle of operation of a TEM differs from a SEM. Both use electron beams, but unlike a SEM where the beam is reflected from the specimen surface, in a TEM the beam penetrates the specimen and impinges on an image plane after magnification with a system of magnetic lenses. The electrons in TEM are emitted by an electron gun, commonly fitted with a heated tungsten filament cathode as the electron source. The filament is held at a relatively high potential compared to the grounded anode located below the specimen. Typically, the electron beam is accelerated by a voltage ranging between 40 to 400 keV. The beam is focused by magnetic lenses, and transmitted through the specimen that is reasonably transparent to the transmitted electrons. In applications where the specimen is not transparent to electrons, replicas of the specimen surface are made using plastic or carbon that is transparent to electrons. When the electron beam emerges from the specimen, it carries information about the structure of the specimen that is magnified by the magnetic objective lens system in the microscope. The spatial variation in this information is observed by projecting the magnified electron image onto a fluorescent screen. The image is recorded by a CCD camera for further analysis.

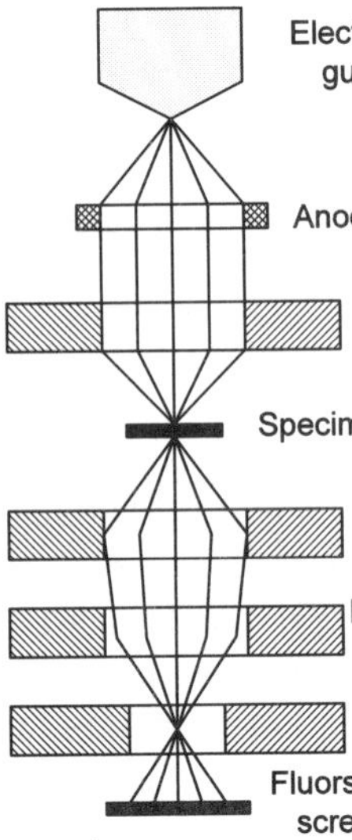

Figure 18.6 Schematic diagram of a typical Transmission Electron Microscope.

The main components in a TEM include an electron emission source, a vacuum system, a series of electromagnetic lenses and an image recording system. A specimen insertion port that allows the insertion into, motion within, and removal of specimens from the beam path is incorporated in its design. The main component of TEM, the electron gun, is comprised of the filament, a biasing circuit, a Wehnelt cap and an extraction anode. When the filament is connected to the negative side of its power supply, electrons can be "pumped" from the electron gun to the anode plate and the TEM column, thus completing the circuit. By proper design of the magnetic lenses, the electrons that exit the filament in a diverging cone are forced into a converging pattern. This electron beam is captured by the electron lenses that act similarly to optical lenses, by focusing parallel rays at some constant focal length. These electron lenses operate either electrostatically or magnetically; however, the majority of electron lenses for TEMs utilize electromagnetic coils to generate the effect of a convex lens.

Magnifications varying from 210 to 300,000 can be obtained in a TEM. The specimens must be reasonably transparent to electrons, must have sufficient local variations in thickness and/or density to provide adequate image contrast and must be small enough to fit within the specimen chamber. These requirements almost always require preparation of a carbon or plastic replica of the surface. The intensity of the images obtained in TEM increases with the accelerating potential and decreases as the specimen or replica thickness is reduced. The contrast of the image can be improved by using a smaller objective lens aperture. However, smaller lens apertures reduce the resolution of the images. These microscopes have resolution on the order of 25 to 50 A°, and are superior to the SEM where the resolution ranges from 70 to 100 A°. On the other hand, TEM images do not provide three-dimensional effects which can be achieved with an SEM image. Also, the fidelity achieved in representing the specimen surface features is better in the SEM in comparison to the TEM. In both the TEM and the SEM, the specimen is always directly probed.

18.3.1 Sample Preparation

Due to the specific requirements described above, sample preparation for a TEM examination is a complex procedure. The specimen thickness should not be more than about one hundred nanometers, as the electron beam in TEM interacts readily with the sample. Quality TEM samples have a thickness of the order of a few tens of nanometers. TEM specimen preparation is also specific to the material being analyzed and depends on the desired results from the analysis. Hence, there are many generic techniques used for the preparation of the specimen for a TEM.

Materials, such as powders or nanotubes that have dimensions small enough to permit passage of the electrons, are deposited onto a support grid or film to prepare the specimen. For other materials, the specimen is prepared as a thin foil, or etched using appropriate chemical agents. In solid mechanics applications involving fractography, the replica technique is used to transpose topographic information from the surface on high fidelity thin films that are easy to handle and to transport to the TEM specimen stage. These replicas are produced by either forming a plastic film on the specimen's surface, depositing a carbon film by vapor deposition or by forming an oxide film by chemical treatment. These replicas are sometimes shadowed to improve visualization of actual topographic shapes. The details of these replica techniques and a more detailed description of TEM are provided in [7].

18.3.2 Applications

TEM images have been used to verify dispersion of nano-sized particles in various polymer matrices. Proper dispersion of these particles is essential for improving the mechanical and transport properties of nanocomposites. Any agglomeration of particles leads to deterioration of properties. A TEM image of a nanocomposite fabricated with Al_2O_3 particles in a polyester matrix, which display good dispersion characteristics, is shown in Fig. 18.7.

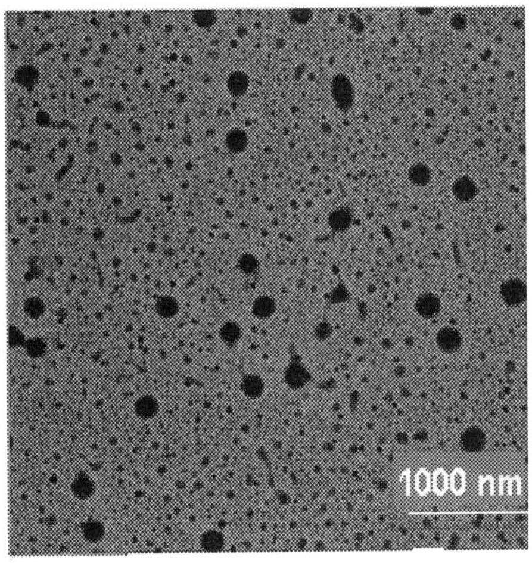

Figure 18.7 TEM image showing nano-sized aluminum oxide particles dispersed in polyester matrix.

TEM imaging is frequently used in fractography to examine fracture surfaces. Unlike the SEM that enables direct examination of the specimen surface, high fidelity replicas of the surface are used in TEM studies. Replicas have many advantages as compared to studying actual specimens, particularly when the specimen sizes are large. An example of a TEM image depicting details of a fatigue fracture surface is shown in Fig. 18.8.

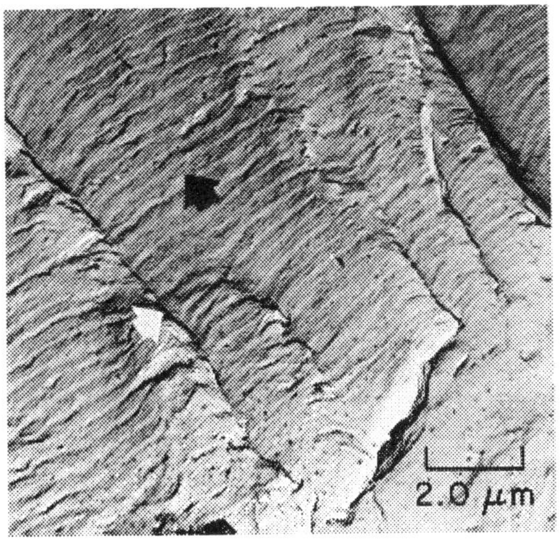

Figure 18.8 A TEM image showing fatigue striations [8].

18.4 ATOMIC FORCE MICROSCOPE

Atomic force microscopy is one of the latest methods of probing a specimen surface to perform various types of physical and chemical measurements. The AFM technique, first developed in 1986, is a derivative of the scanning tunneling microscopy. The AFM is a type of very high-resolution scanning probe microscopy, with demonstrated resolution of fractions of a nanometer. In simple terms, the AFM consists of a cantilever with a sharp tip (probe) at its end. This probe is used to scan the specimen surface, as illustrated in Fig. 18.9. When the cantilever tip is positioned a few angstroms away from the specimen surface, long range interatomic interactions prevail giving rise to interaction forces that are described by:

$$F = -\frac{C}{r^n} \qquad n \geq 7 \tag{18.1}$$

where the negative sign denotes force of attraction, r is the distance between molecules, n describes the decay length and C is a constant.

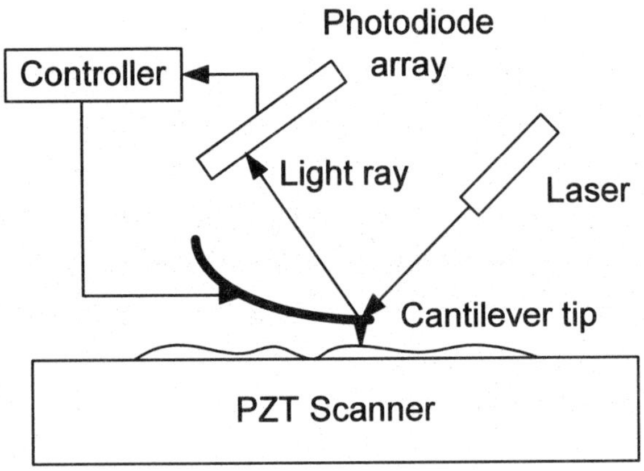

Figure 18.9 Block diagram of an atomic force microscope

Depending on the value of n in Eq. (18.1) and the sign of the equation, the interaction forces between the cantilever tip and the specimen are classified either as short range or long range forces. For very short distances, that are much less than 1 nm, short range forces are significant and the value of n is between 9 and 16. These short range forces are highly repulsive. For larger separations (5-10 nm), n is either 7 or 8 and the forces are considered long range. These long range forces are attractive and are classified as van der Waals forces. There is a separation zone where the short and long range forces are in competition with each other. There are several models that describe these interaction forces and the most common and simple model is the Leonard-Jones potential model shown in Fig. 18.10.

In the AFM, because the probe tip radius is of finite size and the ensuing interaction is between particles rather than molecules, Eq. (18.1) is modified. For example with a conical AFM tip, the interaction force is given by:

$$F = -\frac{HR\tan^2\theta}{d^2} \tag{18.2}$$

where H is the Hamaker constant, d is the distance between the surface and the AFM tip, θ is the half cone angle and R is the tip's radius.

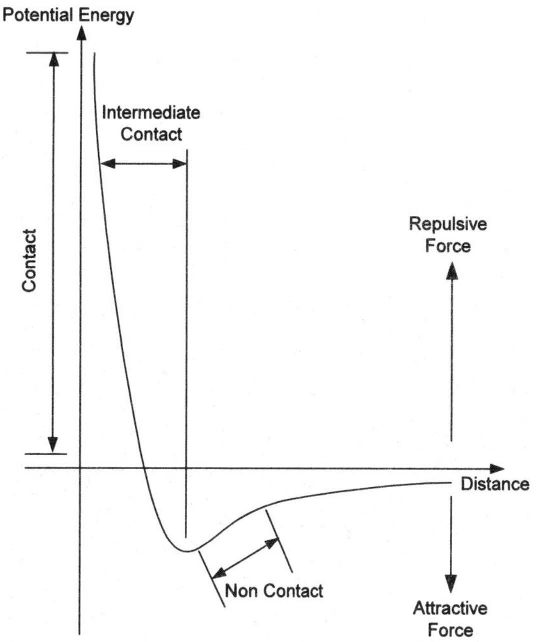

Figure 18.10 Lennard–Jones potential function showing the nature of forces with AFM tip distance.

The principle of operation of an AFM is very simple when the tip is only a few Angstrom away from the specimen surface and short range repulsive forces prevail. When the AFM cantilever tip is brought into proximity of a sample surface these forces deflect the cantilever. The resulting deflection is measured and related to the surface topology or other physical or chemical properties. AFMs can be used to measure various types of forces, such as mechanical contact force, van der Waals forces, capillary forces, chemical bonding, electrostatic forces, and magnetic forces. When the AFM is operated in the non-contact force regime, as illustrated in Fig. 18.10, then the RMS amplitude of the cantilever oscillating near its damped resonance frequency is used as a feedback signal to detect changes in distance between the cantilever tip and the sample. From a solid mechanics perspective, the most important forces are mechanical contact forces and van der Waals forces. Graphs of the measured cantilever force as a function of displacement are used to study the elastic/plastic mechanical properties of the surface. The AFM can also be used in conjunction with the digital image correlation techniques to obtain full field strain measurements on the surface with nm resolutions. Typical results from such applications are shown later in this section. AFMs can also be used to measure transport mechanical properties such as thermal conductivity.

18.4.1 Components of Atomic Force Microscope

A schematic illustration of the major components of an AFM is shown in Fig. 18.11. A piezoelectric scanner with the cantilever tip scans the specimen surface. These scanners are generally constructed by combining independently operated piezoelectric elements that can precisely manipulate samples for scanning in three dimensions. The force signal from the scanner is converted from an analog to a digital signal before processing by the control electronics. The digital signal is then transmitted to a computer where it is stored in memory. In contact AFM, illustrated in Fig. 18.10, the deflection of the cantilever beam mounted on the scanner is proportional to the force exerted on the specimen. This deflection is accurately measured by using a simple optical arrangement, which consists of a laser diode that reflects light from a reflecting surface on the cantilever beam, as shown in Fig. 18.11. The reflected light is collected by a photo sensitive detector consisting of a system of photodiodes. Differential light detected by the photodiode system is directly proportional to the deflection of the cantilever beam. Because long beam paths are used, the deflection of the cantilever beam is optically magnified resulting in very accurate measurements. Because of this geometrical signal amplification, deflections less than 10 nm can be accurately measured. In some applications, silicon cantilever beams have piezo-resistive elements deposited on them, which serve as strain gages with output signals proportional to deflection (see Chapter 7). These silicon cantilever beams are used to achieve scan speeds that are 5-10 times higher than conventional cantilever beams, but they are practical only for very smooth and flat sample surfaces. Interferometric techniques have also been used to measure cantilever deflections [9] but this technique is not usually employed in commercially available AFM instruments.

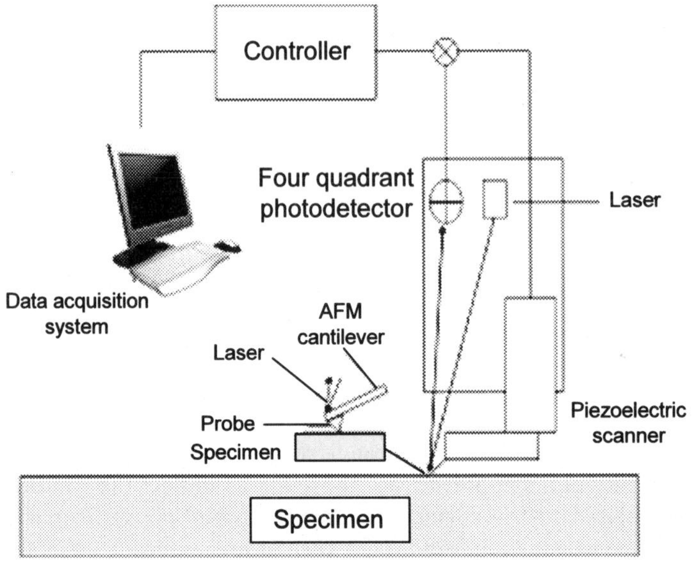

Figure 18.11 Schematic diagram of an AFM

18.4.2 Cantilever Geometry and Deflection Relations

The measurements made by AFMs are dependent on the cantilever beam geometry and its mechanical properties. The two commonly used cantilever beam geometries are shown in Fig. 18.12. These are rectangular and triangular V shaped cantilevers that are commercially available with various stiffness and different resonant frequencies. The cantilever probe supports are selected depending on the type of materials (soft to stiff) under examination. These cantilever beams also come with different tip geometries to obtain profiles from surfaces with different degrees of roughness. The primary difference between rectangular and triangular cantilevers is in their torsional stiffness. The triangular cantilevers are stiffer when torsional loads are applied about their axis of symmetry and they are used for making measurements on surfaces with strong adhesive forces or for imaging in contact AFM. The rectangular beams have significantly less torsional stiffness and are useful in making tribological and in-plane force measurements.

The stiffness or spring constant of the cantilever beams is required to determine the force measurement. This spring constant can be calculated analytically, obtained through numerical simulations, or, preferably, by experimental calibration. For a rectangular cantilever beam the spring constant is derived in standard elasticity books [10] and is given by:

$$k = \frac{Ewt^3}{4L^3} \tag{18.3}$$

where E is the elastic modulus and the other symbols are defined in Fig. 18.12.

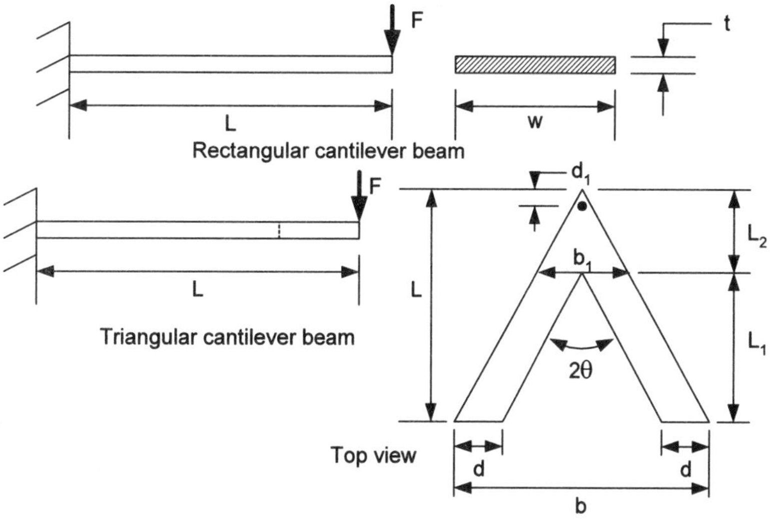

Figure 18.12 Schematic drawing of the geometry of rectangular and triangular shaped cantilever beams.

The spring constant for the triangular cantilever beam has been obtained by Cappella and Dietler [11] and is given by:

$$k = \frac{Et^3 db_1}{2b_1(L^3 - L_2^3) + 6dL_2^3} \tag{18.4}$$

where E is the elastic modulus and the beam dimensions are defined in Fig. 18.12.

The determination of the torsional spring constants is more complex and the derivations are more difficult. Sader [12] has outlined detailed derivations of these equations, and Cappella and Dietler [11] describe calibration techniques. Considering the number of different relations described in the literature for the spring constants, calibration to establish the spring constant for a specific cantilever beam is recommended.

These cantilevers are manufactured by surface and bulk micromachining methods, and are typically made from silicon, glass, or silicon nitride with a tip radius of curvature on the order of a few to tens of nanometers.

18.4.3 Imaging Methods

The imaging methods used by an AFM depend on the surface conditions and the type of material being probed. An AFM can be used either in a contact or non-contact mode.

Contact Mode

As the name implies, this method atomic force microscopy is performed in contact where the overall force is repulsive. The force exerted on the cantilever beam is measured at every point as the probe is scanned over the specimen's surface. These repulsive forces measured in the contact mode are of the order of 10 to 100 nano-newtons. The tip forces and the close proximity of the tip to the surface cause large stresses on the tip and also on the specimen that could result in tip wear and/or surface deformation. To avoid surface deformation, the contact mode AFM is generally used on specimens with hard surfaces, where the hardness is similar to the tip material. Furthermore, contact mode AFM is used with operational constraints—either under constant force or constant cantilever beam deflection. For measurements on rough surfaces the cantilever beam's deflection is held constant while the tip scans predetermined locations on the surface. As the surface roughness changes, the force acting on the tip of the cantilever beam also changes. To accommodate this change in force, the height of the PZT actuator controlling the scanner is adjusted to maintain constant force, i.e. cantilever deflection. The changes in the motion of the actuator are used to reconstruct the surface features.

For making measurements on flat surfaces with relatively small roughness—of the order of a few nanometers—the AFM is used so that the cantilever height remains constant. This implies that the tip deflection will change from point to point and, thus, provide information for reconstruction of the surface topography.

Non Contact Mode

This method of surface mapping takes advantage of van der Waals forces for generating the surface topology. Because these forces are very small (below 100 pico-newtons), the static cantilever beam deflections are difficult to distinguish from those due to random thermal vibrations (noise). To circumvent this problem, the cantilever beam is vibrated using an alternating current to excite the beam. The cantilever beam is oscillated close to its fundamental resonance frequency or a harmonic thereof. During measurement, tip-sample interaction forces modify the oscillation amplitude, phase and resonance frequency with respect to the external (reference) signal. These differences provide information about the sample's characteristics. Schemes for dynamic mode operation include frequency modulation and, the more common, amplitude modulation.

For certain applications, the AFM is sometimes used in intermittent contact mode. In this mode, the AFM image is produced by bringing the oscillating tip in close proximity to the specimen's surface where repulsive forces are maintained.

18.4.4 Applications

The AFM is used to determine the surface force—cantilever beam deflection curves. The information thus obtained can be quantitatively used to study mechanical behavior at the nanometer length scale. Ioannis Chasiotis and his colleagues [13] have successfully used the AFM to measure surface displacements and strains on thin films and composite materials when subjected to load. Using digital image analysis, they have successfully evaluated in-plane deformations with 1-2 nm resolution. Because the AFM provides information on surface features, it can be used to calculate strains if the surface features are used as markers. If the surface roughness features are orders of magnitude smaller than the specimen thickness, they can be used to determine strain by imaging them before and during the loading. The mechanical loading on the specimen is provided by special fixtures built to accommodate imaging by an AFM. In this technique, the local deformations are very small and sub-pixel resolution is required for data analysis. Sub-pixel resolution is accomplished by integrating the digital image correlation technique (see Chapter 15) with AFM measurements. For details on these measurements see references [13-20]. This combined technique with measurements by AFM and DIC has many advantages. It provides multi-scale imaging capability, nanoscale resolution for both hard and soft materials and direct correlation between the material microstructure and imposed deformations.

An excellent example of measuring mechanical properties using an AFM was demonstrated in reference [14]. A polycrystalline thin film with a circular hole of diameter 6 μm was subjected to uniaxial loading under an AFM. Then AFM images were recorded in a small region adjacent to this hole. Digital image correlation technique was applied in this region to obtain the in-plane strains. Using an inverse method, the displacement field was employed to calculate the elastic modulus and Poisson's ratio of the material. A schematic of the specimen and the DIC results are shown in Fig. 18.13.

The AFM has also been used to study fracture toughness and sub critical crack growth in thin films. Results from fracture testing of a 2 μm thick free standing pre-cracked polysilicon film are shown in Fig 18.14 [15]. DIC data in conjunction with finite element modeling was used in the calculation of mode I stress intensity factors in this study.

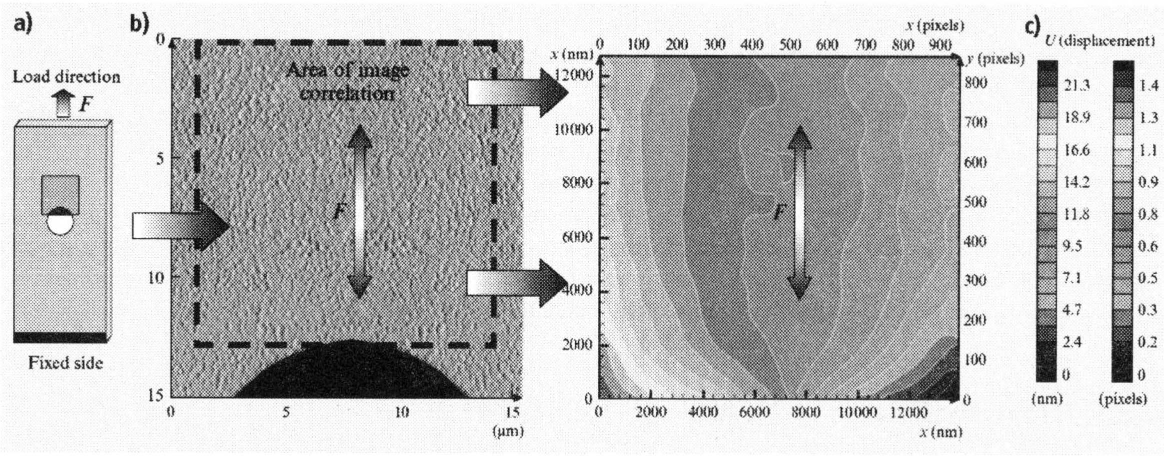

Figure 18.13 (a) Images of a freestanding 2μm thick polysilicon film with a central hole subjected to uniaxial tension. (b) AFM image of the area next to the circular hole. (c) Transverse displacement field in the marked area [14].

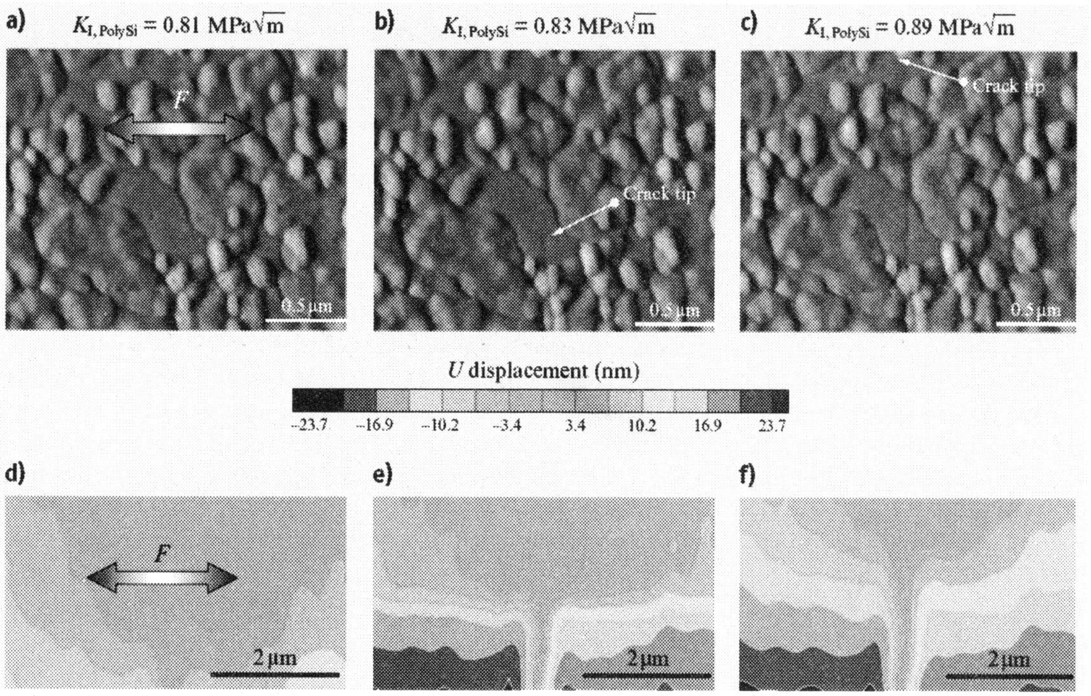

Figure 18.14 AFM micrographs of sub-critical crack growth in a freestanding polysilicon film and the corresponding displacement fields recorded at different values of macroscopically applied stress. (a) The crack tip is below the field of view at the specimen centerline, (b) crack tip location after the first step of crack growth, (c) crack tip location after second crack growth with the crack arrested at a grain boundary. The displacement contours (d, e, f) correspond to AFM images (a, b, c) respectively [15].

The AFM can also be used in conjunction with digital moiré to measure the displacement and strain fields. Liu and Chen [21] modified the existing AFM to use the moiré pattern generated by the interference between the specimen grating and the virtual reference grating formed by digital image processes. For more information on moiré methods see Chapter 13. The overlapped image is filtered by a 2-D wavelet transformation to obtain clear interference moiré patterns. From moiré patterns, the displacement and strain fields can be analyzed. Digital AFM moiré method is very sensitive and is easy to use in realizing nanoscale measurements.

The AFM has many advantages over electron microscopes. It provides a true three dimensional surface profile unlike the SEM, which provides a two dimensional image of the surface. The AFM provides higher resolution than SEM and does not require a vacuum environment for its operation. The AFM can operate under normal ambient conditions and under liquid environments. These operating conditions make it ideally suited for testing biological specimens. There are some disadvantages in using an AFM in comparison to a SEM. The area that an AFM can scan is much smaller (about 100 times) than a SEM. The AFM is also tip sensitive and is much slower than SEM.

18.5 NANOINDENTATION MEASUREMENTS

The nanoindentation method has become a popular technique in measuring local mechanical properties in a variety of materials, especially those in which one of the dimensions approaches the nanoscale, such as thin films and nanostructured materials. As the name implies, in the nanoindentation technique, nano-sized indentations are made on the surface of a flat solid material. During the indentation process, the force-depth curve for the penetration of an indenter tip into a flat solid is recorded. This information is then used along with simple analytical models to determine the modulus of elasticity and the hardness H of a given material (thin film, coating, etc.). The technique is based on the hardness testing methods developed in the past. As in traditional hardness testing, this technique also employs many different indenter shapes (e.g. Vickers, Berkovich, conical, etc.) to create nano-sized indentations. The choice of these indenter shapes, to some extent, depends on the material being tested. The nanoindentation technique can be used to measure elastic and plastic properties of materials including fracture.

A sketch of the parameters involved in a typical indentation experiment is shown in Fig. 18.15. As the indenter comes in contact with the specimen, the force on the indenter begins to increase. The indenter is pressed into the sample until a maximum force or a specified depth of penetration is achieved. The force on the indenter is then held constant for an operator defined time (generally a few seconds). After this period, the indenter is removed from the specimen at nearly the same rate as the indentation, until the applied force reduces to about 10% of the peak force. Then the indenter is idled at this location for a pre-defined time. This part of the test aids in determining the amount of the measured deflection which is due to the thermal expansion and contraction of the equipments and/or sample. This deflection is also called the "thermal drift". Finally, the indenter is completely removed from the test specimen. A sketch of a typical force-time curve is shown in Fig. 18.16, and a sketch of a typical force-depth curve is shown in Fig. 18.17. These records are used in estimating the mechanical properties of the material.

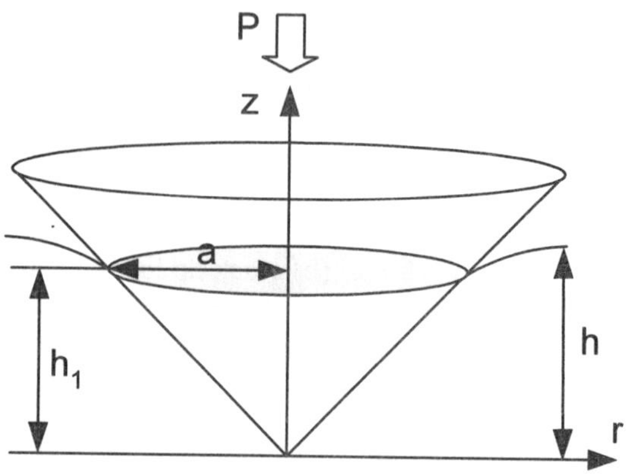

Figure 18.15 Schematic diagram of an indenter geometry.

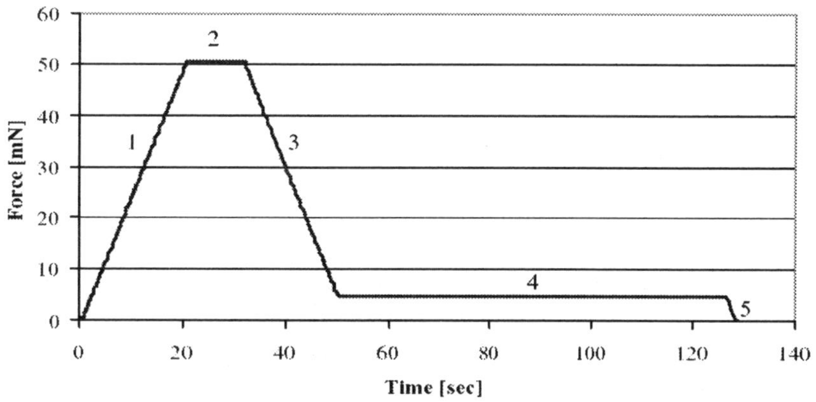

Figure 18.16 A typical force time graph from an indentation experiment [22]

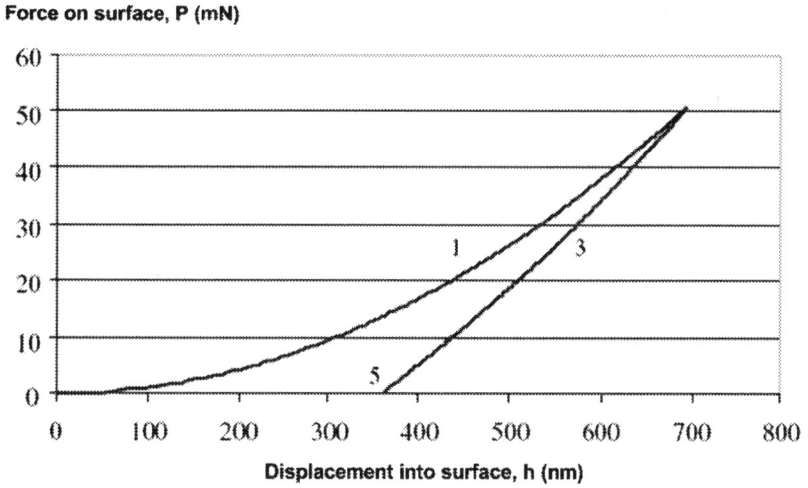

Figure 18.17 Force - displacement curve from an indentation experiment [22]

The analytical procedure for calculating the hardness and the modulus of elasticity are described below:

Following the experiment, the hardness H can be calculated by using:

$$H = P/A \tag{18.5}$$

where P is the force exerted by the indenter on the sample and A is the projected contact area normal to the direction of indentation.

The elastic modulus is calculated from:

$$E = \left(1-\nu^2\right)\left[\frac{1}{E_r} - \frac{1-\nu_i^{\,2}}{E_i}\right]^{-1} \tag{18. 6}$$

$$E_r = \frac{\sqrt{\pi}}{2}\frac{S}{\sqrt{A}} \tag{18.7}$$

where S is defined as the elastic contact stiffness and ν and ν_i are the Poisson's ratio of the test and indenter materials, respectively. E_i is the elastic modulus of the indenter material.

To use equations (18.5) and (18.7), it is necessary to calculate the values of the contact area A and the contact stiffness S. The contact area A is a function of the contact depth and is given by:

$$A = f(h_c) \tag{18.8}$$

For a Berkovich indenter, the area function is defined as:

$$A = 24.56 h_c^2 \tag{18.9}$$

For a spherical indenter the area function is defined as:

$$A = 2\pi R h_c^2 \tag{18.10}$$

where R is the tip radius. The area function for other types of indenters can be found in Hay [22].

To calculate the value of area functions, it is necessary to first determine the contact depth h_c. The contact depth for Berkovitch and spherical indenters is given by

$$h_c = h - 0.75P/S \tag{18.11}$$

The contact stiffness, S, used in equations (18.7) and (18.11), is obtained from the force-displacement (P-h) graph when the indenter is first retracted from the sample (elastic region). Generally, this is accomplished by fitting the unloading curve by an equation of the form:

$$P = B(h - h_f)^m \tag{18.12}$$

where B, h_f and m are fitting parameters.

Equation (18.12) is next analytically differentiated to obtain the slope of the force-displacement curve, which is evaluated at the maximum depth to obtain the value of stiffness S.

$$S = \left.\frac{dP}{dh}\right|_{h=h_{max}} = Bm(h_{max} - h_f)^{m-1} \tag{18.13}$$

This procedure gives all the information needed to calculate hardness and elastic modulus using equations (18.5) and (18.6).

There are many studies reported in literature [23, 24] that provide properties of different materials determined using the nanoindentation method. An excellent example of these properties measured with a broad range of materials (soft to very hard) is given in [23, 24] and illustrated in Fig. 18.18. These measurements on naturally occurring materials can be exploited to design novel materials with beneficial properties.

The nanoindentation technique is also used to evaluate quasi-static fracture toughness of brittle materials [25]. The popularity of this technique has grown because it does not require elaborate specimen machining and the loading process is relatively simple. As the indenter is pressed into the brittle material, radial cracks emanate from the corners of the contact shape as shown in Fig. 18.19. These radial cracks can be related to the fracture toughness of the brittle material through a simple model developed by Lawn and his co-workers [26]. The stress intensity factor K_r for the material when indented using Vicker's indenter is given by:

$$K_r = \xi_r \left(\frac{E}{H}\right)^{0.5} \frac{P}{c^{1.5}} = \chi_r \frac{P}{c^{1.5}} \qquad (18.14)$$

where c is the radial crack radius, P is the peak indentation load and χ_r is a dimensionless constant that depends on the specific material-indenter system [27].

Because the cracks arrest in these tests, K_r is the measurement of the crack arrest toughness which is slightly lower than the crack initiation toughness K_{Ic}. These fracture toughness measurements have been refined in recent years by incorporating additional instrumentation like acoustic emission during the indentation process [28]. Other researchers have applied hard films on softer substrates and have used the cracks in the film to predict fracture toughness [29].

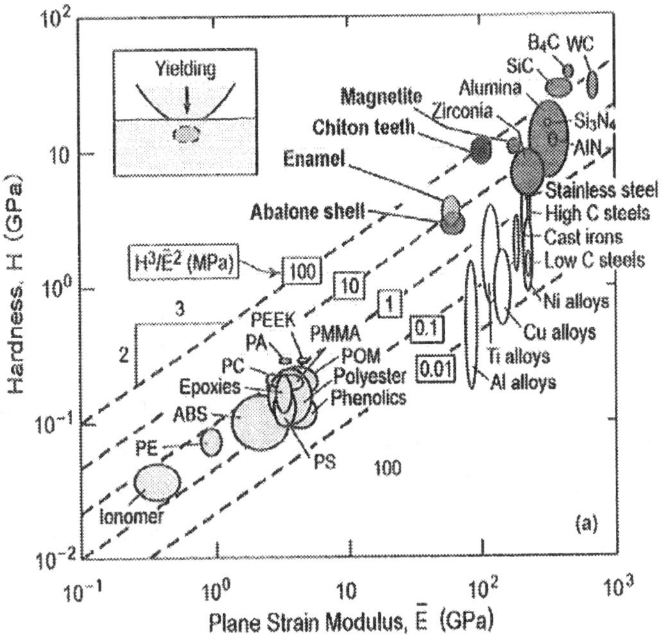

Figure 18.18: Hardness versus plane strain modulus for many materials [23]

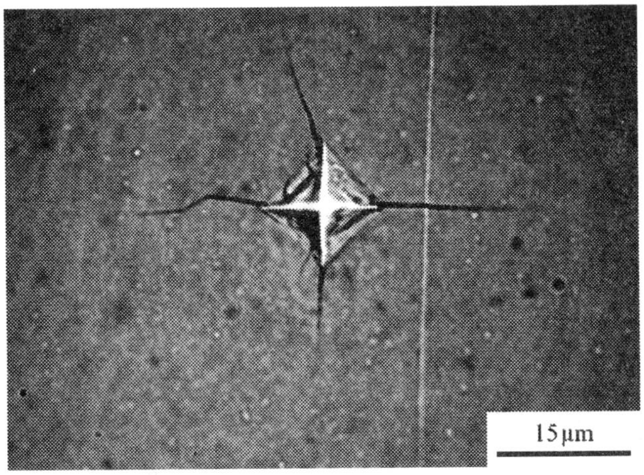

Figure 18.19: Vickers indentation in silicon, showing radical cracks [33]

During conventional nanoindentation experiments, one is unable to view the deformations in the material during the indentation process. On many occasions, there are discontinuities in the loading and unloading curves which are related to discrete events occurring in the material during loading. Coupling nano-mechanical testing with direct observations using a Scanning Electron Microscope [30], Transmission Electron Microscope [31] and an Atomic Force Microscope [32], allows the user to obtain detailed deformation information during the loading-unloading process. There are several commercially available nanoindentation systems compatible with SEM, TEM and AFM.

AFM nanoindentation is a very promising approach to study the relationship between local microstructure and material properties [32]. While AFM-based nanoindentation experiments involve a slightly inclined tip trajectory during indenting, this technique allows a unique combination of high-resolution imaging, a composition mapping and local mechanical studies with forces as small as nano-Newtons and spatial resolution down to 1 nm. Balikov et. al. performed AFM-based nanoindentation experiments on a set of polymer materials with microscopic moduli ranging from 1 MPa to 10 GPa. Some of his results are shown in Fig. 18.20.

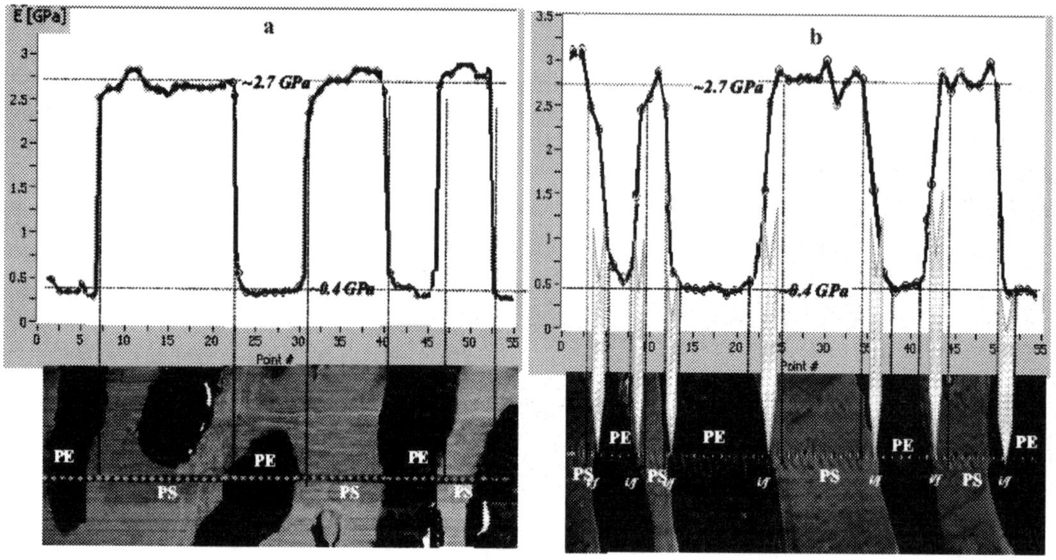

Figure 18.20 A combination of phase images (bottom) and the elastic modulus profiles (top) measured across matrix and several domains in the locations with the leftover indents [32].

EXERCISES

1. Briefly describe the importance of nanoscale measurements.
2. Explain the construction and working of a scanning electron microscope (SEM).
3. Why are the materials coated with a layer of conductive coating before a SEM image is taken?
4. Compare the working of a scanning electron microscope to a transmission electron microscope (TEM).
5. Briefly describe the limitations of TEM.
6. Explain the working of an atomic force microscope (AFM).
7. Describe the procedure for conducting a nanoindentation experiment. Include the procedure for interpreting the data collected during the experiment.
8. Prepare a memorandum for your manager recommending a commercially available nanoindentation instrument.

BIBLIOGRAPHY

1. Judy, J.: Microelectromechanical systems (MEMS): fabrication, design and application, Smart Mater Struct, pp 1115-1134, 2001.
2. Vinci, R. P., and J. J. Vlassak: Mechanical behavior of thin films, Ann. Rev. Mater. Sci., pp 431-462, 1996.
3. Nix, W. D.: Mechanical properties of thin films, Transactions. Metall. Trans., pp 2217-2245, 1989.
4. Arai, M.: In-Situ SEM observation of Deformation and Fracture Process in Plasma sprayed CoNiAlY thin film, J. Solid Mech. And Mat. Eng., pp 1389-1398, 2008.
5. Dawicke, D. S., and M. A. Sutton: CTOA and Crack-tunneling measurements in thin sheet 2024-T3 Aluminum Alloy. Experimental Mechanics, pp 357-368, 1994.
6. Dally, J. W. and D. T. Read: Electron Beam Moiré, Experimental Mechanics, Vol. 33, pp. 270-277, 1993.
7. Mills, K.: Metals Handbook, Volume 12, Fractography, American Society for Metals, Vol. 12, 1987.
8. Fellows, J.A.: Fractography and atlas of fractographs, Metals Handbook, American Society for Metals, Vol. 9, 1974.
9. Morris, V. J., A. P. Gunning and A. R. Kirby: Atomic Force Microscopy for Biologists, Imperial College Press, London, 2004.
10. Timoshenko, S. P., and J. N. Goodier: Theory of Elasticity, 4th ed., McGraw Hill, New York, 1980.
11. Cappela, B., and G. Dietler: Force-distance curves by atomic force microscopy, Surf. Sci. Rep. pp 1-104, 1999.
12. Sader, J. E.: Parallel beam approximation for V-shaped atomic force microscope cantilevers, Rev. Sci. Instrument., pp 4583-4587, 1995.
13. Chasiotis, I., S. Cho, and K. Jonnalagadda: Fracture toughness and crack growth in polycrystalline Silicon, J. Appl. Mech., pp 714-722, 2006.
14. Cho, S., J. Cárdenas-García, and I. Chasiotis: Measurement of nano-displacements and elastic properties of MEMS via the microscopic hole method. Sens. Actuators A Physical, pp 163-171, 2005.
15. Cho, S., K. Jonnalagadda, and I. Chasiotis: Mode I and mixed mode fracture of polysilicon for MEMS, Fatigue Fract., Eng. Mater. Struct., pp 21-31, 2007.
16. Israelachvili, J.: Intermolecular and Surface Forces, 2nd ed., Academic, San Diego, 1987.
17. Burnham, N. A., R. J. Colton, and H. M. Pollock: Interpretation of force curves in force microscopy, Nanotechnology, pp 64-80, 1993.
18. Chasiotis, I., and W. G. Knauss: A new microtensile tester for the study of MEMS materials with the aid of atomic force microscopy, Exp. Mech., pp 51-57, 2002.
19. Cho, S. W., I. Chasiotis, T. A. Friedman, J. Sullivan: Direct measurements of Young's modulus, Poisson's ratio and failure properties of ta-C MEMS, J. Micromech. Microeng., pp 728-735, 2005.
20. Cho, S. W., and I. Chasiotis: Elastic properties and representative volume element of polycrystalline silicon for MEMS, Exp. Mech., pp 37-49, 2007.
21. Liu, C. M., and L. W. Chen: Digital Atomic Force Microscope Moiré method, Ultramicroscopy, pp 173-181, 2004.
22. Hay, J.: Introduction to Instrumented Indentation Testing, Experimental Techniques, pp 66-72, 2009.
23. Weaver, J. C. et. al.: Analysis of an ultra hard magnetic biomineral in chiton radular teeth, Materials Today, Vol. 13, pp. 4-14, 2010.
24. Zok, F. W. and A. Miserez: Acta Mater., Vol. 55, pp. 6365-6371, 2007.
25. Lawn, B. R. and T. R. Wilshaw: Fracture of brittle solids, Cambridge University Press, 1975.
26. Lawn, B. R., A. G. Evans and D. B. Marshall: Elastic/plastic indentation damage in ceramics: The medial/radial crack system, Journal American Ceramic Society, Vol. 63, pp. 574-581, 1980.
27. Anstis, G. R., P. Chantikul, B. R. Lawn and D. b. Marshall: A critical evaluation of indentation techniques for measuring fracture toughness I, Direct crack measurements, Journal American Ceramic Society, Vol. 64, pp. 533-542, 1981.
28. Bahr. D. F. J. W. Hoehn, N. R. Moody and W. W. Gerberich: Adhesion and acoustic emission analysis of failures in nitride film with a metal interlayer, Acta Mater. Vol. 55, pp. 5163-5175, 1997.

29. Pang, M. and D. F. Bahr: Thin film fracture during nanoindentation of a titanium oxide film-titanium system, Journal of Materials Research, Vol. 16, pp. 2634-2643, 2001.
30. Nowak, Julia et.al.: In situ nanoindentation in the SEM, Materials Today, Vol. 13, pp. 16-17, 2010.
31. Ye, J., R. Mishra, A. Pelton and A. Minor: Direct Observation of the NiTi martensitic phase transformation in nanoscale volume, Materials Today, Vol. 13, pp. 18-27, 2010.
32. Belikov, S., S. Magonov, N. Erina, L. Huang, C. Su, A. Rice, C. Meyer, C. Prater, V. Ginzburg, G. Meyers, R. McIntyre, and H. Lakrout: Theoretical modeling and implementation of elastic modulus measurement at the nanoscale using atomic force microscope, J Physics: Conference Series, pp 1303-1307, 2007.
33. Bahr, D. F. and D. J. Morris, Nanoindentation: Localized probes of mechanical behavior of materials, Handbook of Experimental Solid Mechanics, Editor W. N. Sharpe, Springer.
34. Peng, B., M. Locascio, P. Zapol, S. Li, S. Mielke, G. Schatz, and H. D. Espinosa: Measurements of near-ultimate strength for multiwalled carbon nanotubes and irradiation-induced crosslinking improvements, Nature Nanotechnology, Vol. 3, No. 10, pp. 626-631, 2008.
35. Zhu, Y., and H. D. Espinosa: An electromechanical material testing system for insitu electron microscopy and applications, Proceedings of the National Academy of Sciences of the USA, Vol. 102, pp. 14503-14508, 2005.
36. Espinosa, H. D., B.C. Prorok, and M. Fischer: A Methodology for Determining Mechanical Properties of Freestanding Thin Films and MEMS Materials, Journal of the Mechanics and Physics of Solids, Vol. 51, No. 1, pp. 47-67, 2003.
37. Ke, C. H. and H. D. Espinosa: Feedback Controlled Nano-cantilever Device, Applied Physics Letters, Vol. 85, pp. 681-683, 2004.

PART V

MEASUREMENT

AND

ANALYSIS METHODS

CHAPTER 19

DIGITAL RECORDING SYSTEMS

19.1 INTRODUCTION

During the past four decades enormous progress has been made in developing digital instrumentation. Indeed, the combination of analog transducers with digital processing, accomplished through analog to digital conversion has placed new dimensions on both engineering analysis and process control because of the many advantages associated with digital processing of analog data. Equally important has been the integration of the personal computer into instrument systems. Circuit cards, which plug into slots on PCs, are commercially available that combine signal conditioning circuits, analog to digital conversion, and storage of data into memory. Software such as LabVIEW enables the engineer to access the data to perform either an analysis or to control a process.

Measurements of physical quantities such as force, torque, strain, acceleration, etc. are analog measurements. An analog instrument system, illustrated in Fig. 19.1, is used to obtain an analog signal v_0, which is proportional to the quantity that is being monitored. The analog system interfaces with a digital system through an analog to digital (A/D) interface. The key element, which represents the interface between the analog acquisition system and the digital processing system, is an analog to digital converter (ADC). This ADC takes the voltage v_0 from the acquisition system and converts it to an equivalent digital code. Once digitized, the signal (i.e. a digital code) can be displayed, processed, stored or transmitted. The arrangement of the combined analog and digital system, shown in Fig. 19.1, depends on the purpose of the system. A simple system involving the direct display of a single quantity will incorporate a transducer, a power supply, a number of integrated circuits (chips) and a numerical display[1]. A more involved digital system will include a computer for real-time processing, disk drives for storage of raw and processed data, a monitors for display, printers for hard copy and data transmission lines to off-site facilities.

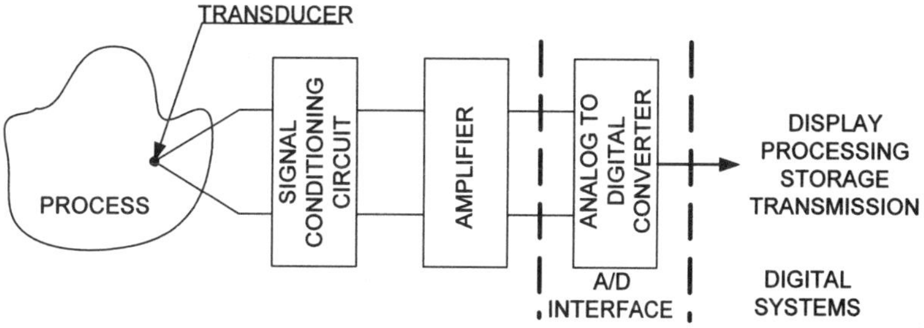

Figure 19.1 An analog/digital instrumentation system.

[1] These simple systems are often housed in a small hand held package that is powered with batteries, which enable it to be used either in the laboratory or the field.

19.2 DIGITAL CODES

Digital systems contain many logic gates, (AND, OR and NOT gates), which act like switches that can be turned on or off. Because the logic gates have only two states (on or off), digital words consist of binary elements called bits, which are either 0 for off or 1 for on. Consider a digital word consisting of a four-bit array, say 1011. The "1" at the extreme left is the most significant bit (MSB) and the "1" at the extreme right is the least significant bit (LSB). With a binary code, the least significant bit has a weight of 2^0, the next bit has a weight of 2^1, the next 2^2 and the most significant bit m has a weight 2^{m-1}. When all four bits are zero (i.e. 0000), the equivalent count is 0. When all four bits are 1 (i.e. 1111), the equivalent count is $2^3 + 2^2 + 2^1 + 2^0 = 8 + 4 + 2 + 1 = 15$. A complete listing of a 4-bit binary code is presented in Table 19.1.

Table 19.1
Equivalent count for a 4-bit binary code.

MSB[a]	Bit 2	Bit 3	LSB[a]	MSB		Bit 2		Bit 3		LSB		Count
0	0	0	0	0	+	0	+	0	+	0	=	0
0	0	0	1	0	+	0	+	0	+	2^0	=	1
0	0	1	0	0	+	0	+	2^1	+	0	=	2
0	0	1	1	0	+	0	+	2^1	+	2^0	=	3
0	1	0	0	0	+	2^2	+	0	+	0	=	4
0	1	0	1	0	+	2^2	+	0	+	2^0	=	5
0	1	1	0	0	+	2^2	+	2^1	+	0	=	6
0	1	1	1	0	+	2^2	+	2^1	+	2^0	=	7
1	0	0	0	2^3	+	0	+	0	+	0	=	8
1	0	0	1	2^3	+	0	+	0	+	2^0	=	9
1	0	1	0	2^3	+	0	+	2^1	+	0	=	10
1	0	1	1	2^3	+	0	+	2^1	+	2^0	=	11
1	1	0	0	2^3	+	2^2	+	0	+	0	=	12
1	1	0	1	2^3	+	2^2	+	0	+	2^0	=	13
1	1	1	0	2^3	+	2^2	+	2^1	+	0	=	14
1	1	1	1	2^3	+	2^2	+	2^1	+	2^0	=	15

[a] Where MSB and LSB are the most and least significant bits, respectively.

It is evident from Table 19.1 that a 4-bit binary word permits a count of 2^4 or 16, arranged from 0 to 15. The maximum count C is determined from:

$$C = (2^n - 1) \tag{19.1}$$

where n is the number of bits in the digital word. It is clear that C = 255 for an 8-bit word and 1023 for a 10-bit word.

The least significant bit represents the resolution R of a digital count containing n bits, that can be written as:

$$R = \frac{2^0}{2^n - 1} = \frac{1}{2^n - 1} = \frac{1}{C} \tag{19.2}$$

This result indicates that the resolution that can be achieved with logic gates arranged to yield an 8-bit digital word (a byte) is (1/255) or 0.392 percent of full scale. A 12-bit word provides a resolution of 1 part in 4095, which corresponds to a resolution error of 0.024%.

Resolution is an important concept in digital instrumentation because it defines the number of bits required to achieve a specified error limit in a measurement or for the conversion of an analog voltage into a digital count representing that voltage. Values for C, R and resolution error E_R as a function of the number of binary bits n are presented in Table 19.2. Clearly an 8-bit digital word that provides resolution to within ± 1 count out of full-scale count of 255 will limit the resolution error to 0.39%.

Table 19.2
Count C, Resolution R and Error E_R as a function of the number of binary bits n.

n	C	R(ppm)[a]	E_R (%)
4	15	66666	6.6666
5	31	32258	3.2258
6	63	15873	1.5873
7	127	7874	0.7874
8	255	3922	0.3922
9	511	1957	0.1957
10	1023	978	0.0978
11	2047	489	0.0489
12	4095	244	0.0244
13	8191	122	0.0122
14	16383	61	0.0061
15	32767	31	0.0031
16	65535	15	0.0015

[a] PPM is parts per million.

19.3 CONVERSION PROCESSES

The analog/digital interface, illustrated in Fig. 19.1, shows an analog to digital converter (ADC) that transforms the analog voltage into a digital count. The ADC is a one-way device converting only from an analog signal to a digital code. To convert from a digital code to an analog voltage requires a digital to analog converter (DAC), which is also a one-way device. Because the DAC and ADC are the key functional elements in analog/digital instrument systems, they will be described in detail.

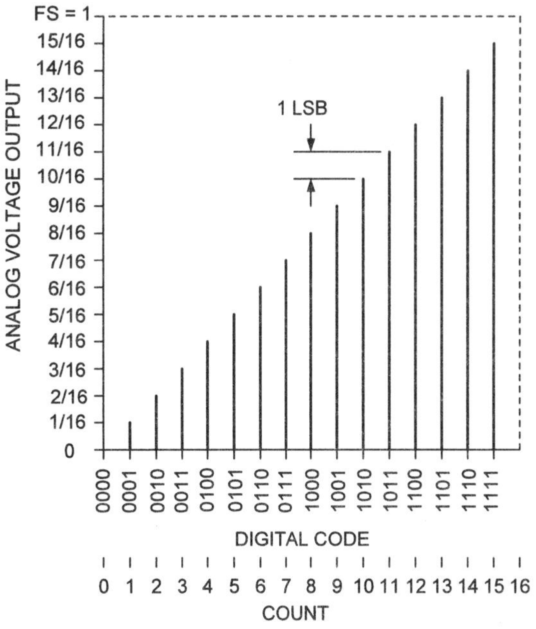

Figure 19.2 Relation between digital count and analog voltage for a digital to analog converter (DAC).

First, consider a 4-bit DAC where the input is a digital code ranging from 0000 to 1111 as listed in Table 19.1. The digital input (the independent variable) is plotted along the abscissa of Fig. 19.2 and the analog voltage output from the DAC, ranging from 0 to 15/16 of full scale, is shown along the ordinate. While the full-scale value of 16 is not available as a digital input, it represents the reference quantity to which the analog voltage is normalized. For example, let 10 V be the full scale voltage (FSV), then the digital code 1000 will give, under ideal conditions, an analog voltage of (8/16)(10) = 5 V.

Next, consider an ADC with the analog voltage being the independent variable shown in Fig. 19.3. Because all analog voltages between zero and full scale can exist, they must be quantized by dividing the complete range of voltage into sub ranges. If FSV is the full-scale analog voltage input, the quantization increment is equal to FSV × LSB where the LSB = 2^{-n}. For a 4-bit ADC, FSV × LSB = (1/16) FSV or 0.0625 FSV. All analog voltages within a given quantization increment are represented by the same digital code. The illustration in Fig. 19.3 shows that the digital code corresponds to the mid point in each increment. The quantizing process, which replaces a linear analog function with a stair case digital representation, results in a quantization uncertainty of ± ½ LSB and a quantization error that is shown at the top of Fig. 19.3. The average value of the quantization error is zero; however, if it is assumed that it is equally probable that v/FSV takes any value within the quantization increment, it can be shown that the standard deviation σ of a number of measurements is given by:

$$\sigma = \frac{1}{2\sqrt{3}}\text{LSB} = \frac{2^{-(n+1)}}{\sqrt{3}} \tag{19.3}$$

Clearly, this statistic indicates that the effective resolution of the ADC is much better than the usually specified resolution given by Eq. (19.3). Converters of either type (DAC or ADC) are not ideal and errors due to offset, gain and scale factor can occur. These three types of conversion errors are illustrated graphically in Fig. 19.4 for both the DAC and ADC.

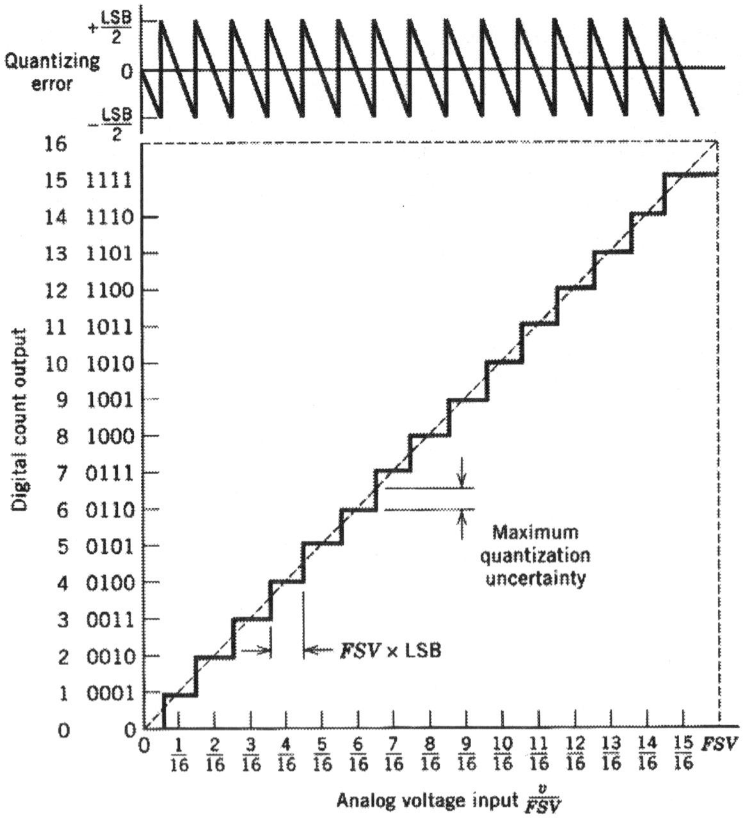

Figure 19.3 The relation between analog voltage input and digital count output for a 4-bit analog to digital converter (ADC).

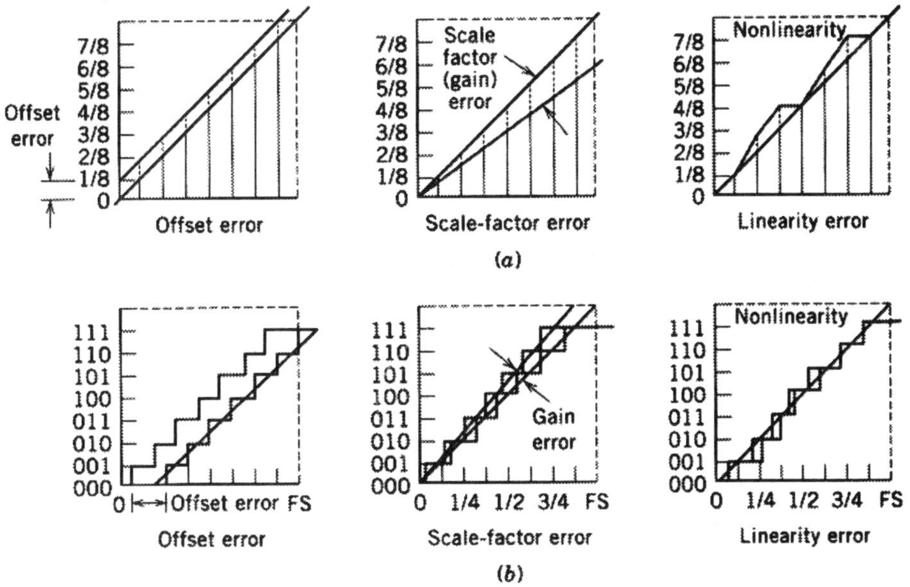

Figure 19.4 Typical sources of error in (a) digital-to-analog converters; (b) analog-to-digital converters.

19.4 DIGITAL TO ANALOG CONVERTERS

Many different circuits are employed in the design of digital to analog converters (DAC). It is beyond the scope of this book to cover the more sophisticated circuits; instead, a simple circuit will be described which shows the essential features involved in the digital to analog conversion process. The circuit for a four-bit DAC is illustrated in Fig. 19.5. A voltage reference v_R is connected to a set of precision resistors with a series of switches. These switches are gates in a digital logic circuit with 0 representing an open switch and 1 representing a closed switch. The resistors are binary weighted which means that resistance is doubled for each higher switch or bit so that:

$$R_n = 2^n R_f \tag{a}$$

where R_n is the resistance of the n_{th} bit and R_f is the feedback resistance on the operational amplifier.

When the switches are closed, a binary weighted current I_n flows to the summing bus. This current is given by:

$$I_n = \frac{v_R}{R_n} = \frac{v_R}{2^n R_f} \tag{19.4}$$

The operational amplifier converts the currents to voltages and provides a low-impedance output. The analog output voltage v_0 is given by:

$$v_0 = -R_f \sum_{n=1}^{k} I_n \tag{19.5}$$

where I_n is summed only if the switch n is closed.

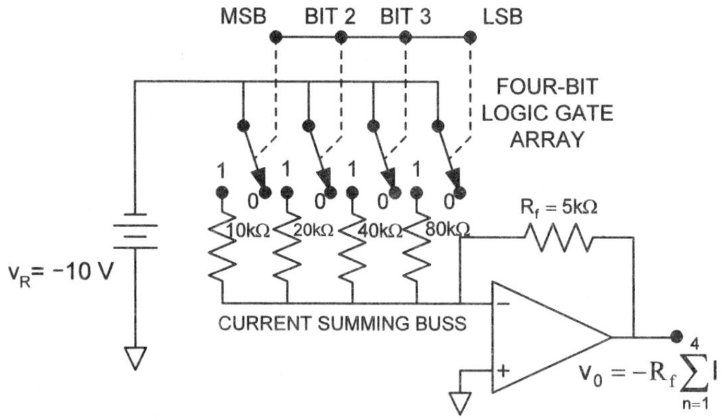

Figure 19.5 Schematic diagram of a 4-bit digital-to-analog converter.

Consider as an example the digital code 1011 (equivalent to 11) for the circuit shown in Fig. 19.5 with R_f = 5 kΩ and v_s = – 10 V. From Eq. (19.4) it is clear that I_1 = –1 mA, I_2 = 0, I_3 = – ¼ mA and I_4 = – 1/8 mA. Summing these currents and multiplying by R_f gives v_0 = 6.875 V which is the same as (11/16) of the full-scale (reference) voltage.

Commercial DACs are more complex than the schematic shown in Fig. 19.5 because they contain more bits (8, 12 and 16 are common) and have several regulated voltages, integrating circuits for switching and on-chip registers. The large number of bits is serviced with a system of parallel conductors called an input bus. The voltage on each conductor in the bus is either high or low and activates each of the switches (gates) to provide the digital code as input to the device. The analog voltage output v_0 is constant with respect to time as long as the digital code is held at the same value on the input.

In many digital systems there are several functions occurring together and proper sequencing of these different functions is mandatory if a common bus is used in the data distribution system. In these applications, the DAC is preceded by a register as indicated in Fig. 19.6. The register is a memory device where the input data may be stored and held. With a common data bus, which serves several digital devices, only select signals are to be converted by the DAC. The other signals are to be ignored as they are intended for different devices. Sorting the signals from the bus is accomplished with a strobe that is activated when the DAC has been addressed and given a signal to write. The strobe signal enables the register to read the data on the bus during the period of this signal and the new data, a digital code, replaces the old data in an update. The register is then latched and the updated data is held in memory. This new digital data is then converted to an analog voltage in the DAC and the analog voltage is held constant until the strobe signal initiates the next update.

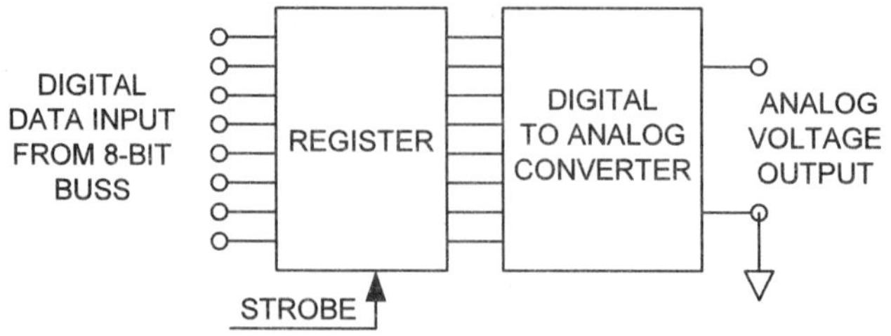

Figure 19.6 A digital-to-analog converter with a register to control data flow.

19.5 ANALOG TO DIGITAL CONVERTERS

Conversion of analog signals to digital code is extremely important in any instrument system involving digital processing of analog signals from signal conditioning circuits. Of the systems available for analog to digital conversion, three of the most common will be described—the successive-approximation method, the integration method and the flash-conversion method.

19.5.1 Successive Approximation Method

The method of successive approximation is illustrated in Fig. 19.7 where a bias voltage v_b is shown as a close approximation to an unknown analog voltage v_u, which is given by:

$$v_b = v_u + \frac{1}{2^{n+1}} FSV \tag{19.6}$$

The term $\left(\frac{1}{2^{n+1}}\right)FSV$ in Eq. (19.6) is added to the unknown voltage to place the DAC output in the center of a quantization increment. In the case illustrated in Fig. 19.7, a 4-bit DAC is employed to convert a fixed analog voltage v_u = (10/16)FSV. The bias voltage is given by Eq. (19.6) as:

$$v_b = \left(\frac{10}{16} + \frac{1}{2^5}\right)FSV = \frac{21}{32}FSV \tag{a}$$

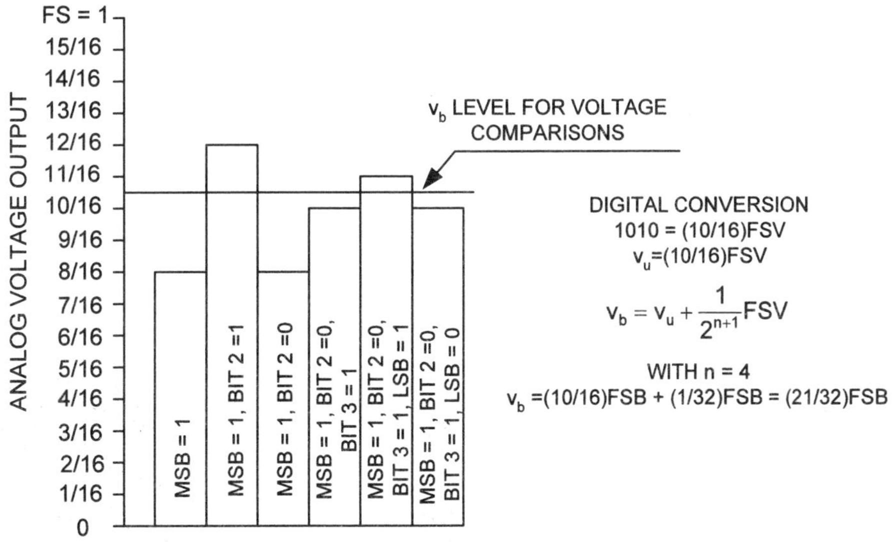

Figure 19.7 Voltage in a successive approximation conversion of an analog signal to a digital code.

The bias voltage v_b is compared to a sequence of precise voltages generated by an input controlled DAC. The digital input to the DAC is successively adjusted until the known output v_0 from the DAC compares closely with v_u.

At the start of the conversion process, the input to the DAC is set at 1000 (i.e. the most significant bit is on and all other bits are off). The analog voltage v_0 = (8/16)FSV and a voltage comparison shows $v_0 < v_b$. By turning bits on and off and making voltage comparisons after each change, the output voltage v_0 approaches the bias voltage as shown in Fig. 19.7. It should be noted that this method of successive approximations is analogous to weighing an unknown mass on a balance by using a set of n binary weights.

The input of the DAC, in this case 1010, is transferred to the output register of the ADC. In this simple example, the conversion process was exact because the unknown voltage v_u was selected at a 4-bit binary value (10/16)FSV. In general, an uncertainty in the conversion will occur when the unknown voltage differs from a binary value. To determine the uncertainty with an ADC involving n approximations consider:

$$v_0 = v_b - \frac{1}{2^n} FSV \tag{19.7}$$

Substituting Eq. (19.6) into Eq. (19.7) gives:

$$v_0 = v_u - \frac{1}{2^{n+1}} FSV \tag{19.8}$$

The relative difference between v_u and v_0 referred to the full-scale voltage is given by:

$$\frac{v_u - v_0}{FSV} = \frac{1}{2^{n+1}} \tag{19.9}$$

where n is the number of bits used in the approximation.

Equation (19.9) shows that analog to digital conversion results in a maximum uncertainty that is equivalent to (1/2)LSB. The quantization error, shown in Fig. 19.3, varies between ± (1/2)LSB and is bounded by the maximum uncertainty.

In this example, the analog input voltage did not change during the conversion process. In fact, accurate conversion cannot be accomplished when the input voltage changes with time. To avoid problems associated with fluctuating voltages during the conversion period, the ADC utilizes an input device that samples and holds the signal constant for the time required for conversion. When the conversion is complete, the 8-, 12- or 16-bit digital code is transferred to a register and the conversion process is repeated. Conversion rates depend on the number of bits, design of the circuit and the speed of the transistors used in switching. In 2003, a moderate-cost 12-bit ADC typically requires 10 μs to convert an analog signal. For a single channel application, this conversion time gives a data acquisition rate of 100,000 samples/s.

19.5.2 Integration Method

Analog to digital conversion by integration is based on counting clock pulses. A typical circuit for a dual slope ADC is shown in Fig. 19.8a. At the start of a conversion, the unknown input voltage v_u is applied together with a reference voltage v_R to a summing amplifier that gives an output voltage:

$$v_a = -\tfrac{1}{2}(v_u + v_R) \tag{a}$$

This voltage is imposed on an integrator that integrates va with respect to time to obtain:

$$v_i = \frac{(v_u + v_R)t^*}{2RC} \tag{19.10}$$

where t* is a fixed time of integration, as shown in Fig. 19.8a. Upon completion of the integration at time t*, a switch on the input of the integrator is activated which disconnects the summing amplifier and connects the reference voltage v_R to the integrator, as shown in Fig. 19.8b. The output voltage of the integrator then decreases with a slope of:

$$\frac{\Delta v_i}{\Delta t} = -\frac{v_R}{RC} \tag{19.11}$$

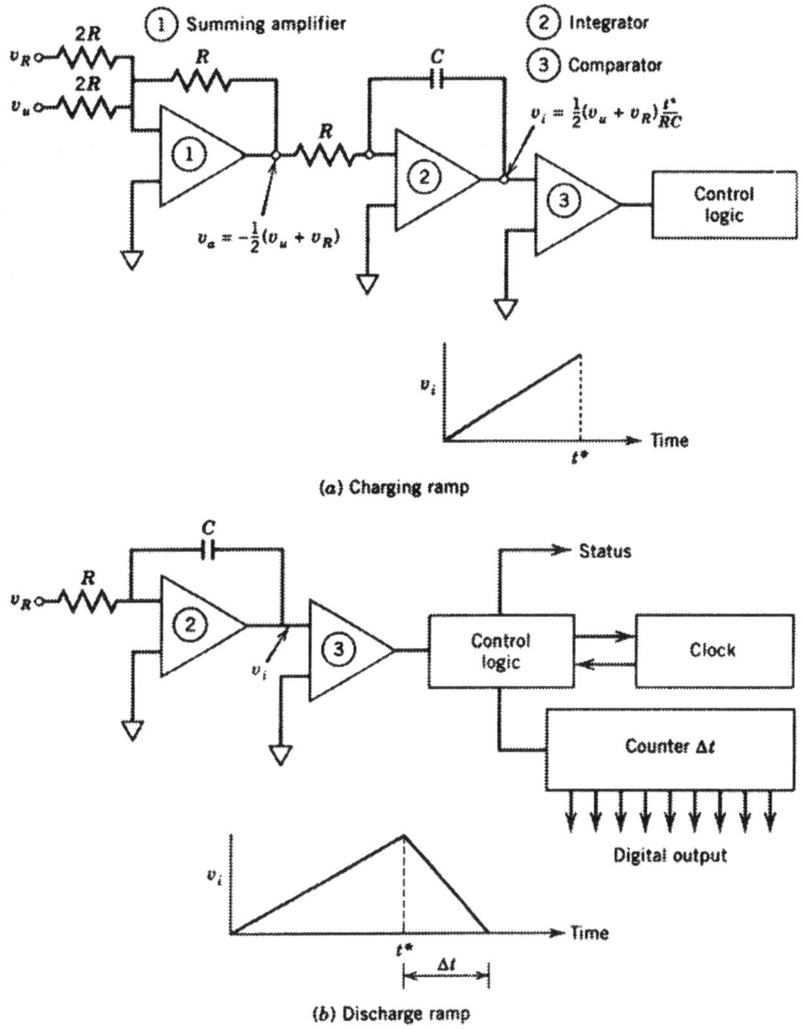

Figure 19.8 Schematic circuit diagrams for an analog to digital converter using dual slope integration.

A comparator monitors the output voltage v_i and issues a signal to the control logic when v_i goes to zero. This zero crossing occurs when

$$\frac{(v_u + v_R)t^*}{2RC} = \frac{v_R \Delta t}{RC} \tag{b}$$

Reducing Eq. (b) yields:

$$\frac{\Delta t}{t^*} = \frac{1}{2}\left(\frac{v_u}{v_R} + 1\right) \tag{19.12}$$

It is clear from Eq. (19.12) that $\Delta t/t^*$, a proportional count of clock pulses, is related to v_u/v_R. If a counter is started by a switch on the integrator, the counter will give a binary number representing the unknown voltage v_u.

The integration method for analog to digital conversion has several advantages. It is extremely accurate as its output is independent of R, C and the clock frequency because these quantities affect both the up and down ramps in the same way. The influence of noise on the unknown signal is markedly attenuated because the electrical noise, which occurs at high frequency, is averaged toward zero during the integration period t*. The primary disadvantage of the integration method is the speed of conversion which is less than ½ t* conversions per second. In order to attenuate the effect of 60 Hz noise, t* must be greater than 16.67 ms; therefore, the speed of conversion must be less than 30 samples/s. This conversion rate is too slow for large high-performance data-acquisition systems, but it is satisfactory for digital voltmeters and smaller lower cost recording systems.

The conversion speed of the ADC determines the frequency of the unknown analog signal that can be measured. To determine this frequency let the input signal to an ADC be a sinusoid with a frequency f given by:

$$v_u = \frac{v_u^*}{2} \sin 2\pi f t \qquad (19.13)$$

where $v_u^*/2$ is the amplitude = FSV/2.

The maximum rate of change of this input is obtained by differentiating Eq. (19.13) and letting cos $2\pi ft$ = 1 to obtain:

$$\left(\frac{dv_u}{dt}\right)_{max} = \pi f v_u^* \qquad (19.14)$$

The term $(dv_u/dt)_{max}$ is called the slew rate of the signal. If the ADC is to convert the signal into a digital code of n bits to within 1 LSB, then the change in input voltage Δv must be limited to Δv < (LSB × FSV) during the conversion time T. It is clear from Eq. (19.14) that:

$$\Delta v_{max} = \pi f v_u^* \, T < (LBS \times FSV) \qquad (19.15)$$

Solving Eq. (19.15) for the frequency limit f gives:

$$f < \frac{LSB \times FSV}{v_u^*}\left(\frac{1}{\pi T}\right) \qquad (c)$$

Because FSV = v_u^*, Eq. (c) reduces to:

$$f < \frac{2^{-n}}{\pi T} \qquad (19.16)$$

If the signal is unipolar, n is the number of bits; however, if the signal is bipolar, an additional bit is necessary to provide the sign. For this reason, a unipolar signal can be sampled with twice the frequency of a bipolar signal. For example, a 10-bit ADC converting at 20 readings/s can monitor a unipolar signal with a frequency f < 0.0062 Hz. This signal with v_u^* = 10 V corresponds to a maximum slew rate of 0.195 V/s in an ADC with FSV = 10V.

19.5.3 Parallel or Flash Conversion Method

Parallel or flash analog to digital conversion is the fastest but most expensive method for designing ADCs. The flash converter, illustrated in Fig. 19.9, employs (2^n – 1) voltage comparators arranged in parallel. Each comparator is connected to the unknown voltage v_u. The reference voltage is applied to a binary resistance ladder so that the reference voltage applied to a given comparator is 1 LSB higher than the reference voltage applied to the lower adjacent comparator.

When an analog signal is presented to the comparator bank, all the comparators with $v_R^* < v_u$ will go high and those with $v_R^* > v_u$ will stay low. Because they are latching type comparators, they hold high or low until they are down loaded to a system of digital logic gates that convert the parallel comparator word into a binary coded word.

The illustration shown in Fig. 19.9 is deceptively simple because an 8-bit parallel ADC contains $2^8 - 1 =$ 255 latching comparators and resistances and about 1000 logic gates to convert the output to binary code. Also, the accuracy is improved by placing a sample-and-hold amplifier before the ADC so that the input voltage does not change over the period required to operate the comparators.

Parallel comparators have improved in performance at reduced cost in the past decade with the availability of very-large-scale integrated circuits that accommodate all of the components on a single chip of silicon. Flash type ADCs are employed on very-high-speed digital systems such as digital oscilloscopes. Conversions in these applications are made at 5 to 20 G samples/s, which gives sampling times of 0.05 to 0.2 ns.

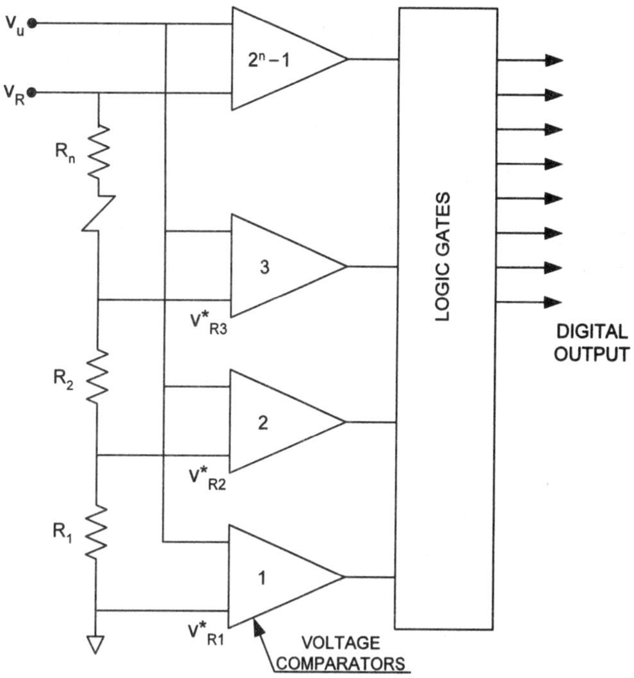

Figure 19.9 Schematic diagram for a flash analog-to-digital converter.

19.6 DIGITAL VOLTMETERS

Digital voltmeters (DVM) offer many advantages over analog-type meters, such as speed in reading, increased accuracy, better resolution and the capability of automatic operation. Digital voltmeters display the measurement with easy to read numerals, as shown in Fig. 19.10, rather than as a pointer deflection on a continuous scale as with analog meters. Digital multimeters measure current, resistance and ac and dc voltages. The DVM may also be used with a multiplexer and a printer to provide a simple but reliable automatic data-logging system.

The number of full digits in the display determines the range of a DVM. For example, a four-digit DVM can record a count to 9999. If the full scale of the DVM is set at 1 V, the count of 9999 provided by the four digits would register a reading of 0.9999 V. Some DVMs are equipped with partial digits to extend the range. The partial digit can only display a limited range of numbers. While zero and one are common for the 1/2 or most significant digit, some newer models are capable of displaying partial digits of 2, 3 or 19. The partial digit extends the resolution of the DVM. For example, consider use of a four-digit DVM for measuring 10.123 V. Because only four digits are available, the meter set on the 10-V scale would read 10.12 V. The last digit (3) would be truncated and lost. If a 4-½ digit DVM is used for the same measurement, the extra partial digit permits 100 percent over-ranging and a maximum count of 19999. The display of the 4-½ digit meter, would show 5 numbers (10.123).

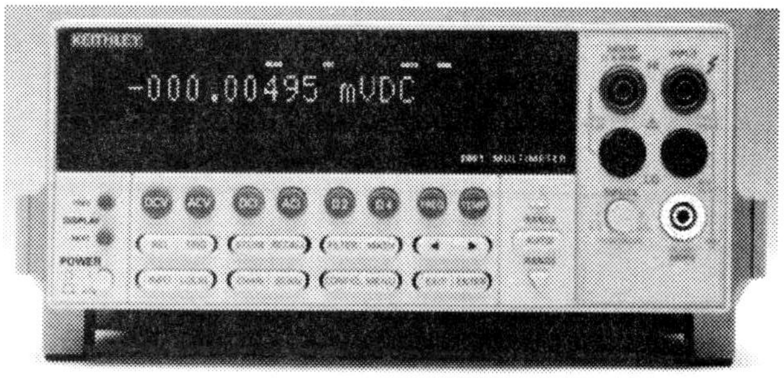

Figure 19.10 A 7-½ digital multimeter. Courtesy of Keithley Instruments, Inc.

Over-ranging may be expressed as a percentage of full scale. For instance, a four-digit DVM with 100 percent overrange displays a maximum reading of 19999. Similarly, with a 200 percent overrange, the maximum display is 29999. In some instances, the overrange capability of the DVM is expressed in terms of the specified range. The four-digit DVM with 100 percent overrange, which has a maximum display of 19999, could be specified with full-scale ranges of 2, 20, 200 V, etc. and an overrange specification is not necessary.

Resolution of a DVM is determined by the maximum count that is displayed. For example, a 4-digit DVM with a maximum count of 10,000 has a resolution of 1 part in 10,000 or 100 parts per million (PPM). The sensitivity of a DVM is the smallest increment of voltage that can be detected and is determined by multiplying the lowest full-scale range by the resolution. Therefore, a four-digit DVM with a 100-mV lowest full-scale range has a sensitivity of $(1/10{,}000) \times 100$ mV = 0.01 mV.

Accuracy of a DVM is usually expressed as ± x percent of the reading ± N digits. A typical value for a 5-½ digit instrument operating on a 2 V range is x = 0.0011% and N = ± 2. Accuracy will depend strongly on calibration and instrument stability. A modern DVM utilizes electronic calibration where calibration constants are stored in non-volatile memory and calibration is accomplished without internal adjustments. Improved stability, which reduces fluctuations in the last digit, is the key to enhanced accuracy. Hermetically sealed resistance networks in the circuits have greatly improved stability in more modern designs.

The input to the multimeter may be an ac voltage, a dc voltage, a current or a resistance; however, in all cases, the input is immediately converted to a dc voltage. The signal is then amplified with variable-gain amplifiers. Their gain is automatically adjusted by control logic so that the voltage applied to the analog to digital converter (ADC) is within specifications avoiding an overload condition.

The ADC changes the dc voltage input to a proportional clock count by using the dual-slope integration technique illustrated in Fig. 19.11. There are three different operations in the dual-slope integration technique for A/D conversion. First, during auto zero, the potential at the integrator output is zeroed for a fixed time, say 100 ms. Second, the dc input is integrated with respect to time for a fixed period, say 100 ms. The output of the integrator is a linear ramp with respect to time as shown in Fig. 19.11. At the end of the run up, the dc input voltage is disconnected from the integrator and the third operation, run down, is initiated. Run-down time may vary from zero to, say 200 ms, and will depend on the charge developed on the integrating capacitor during run up. Since the discharge rate is fixed during run down, the larger the charge on the integrating capacitor, the longer the discharge time. Because both run up and run down produce slopes on the voltage-time trace, this conversion method from voltage to time is called dual-slope integration.

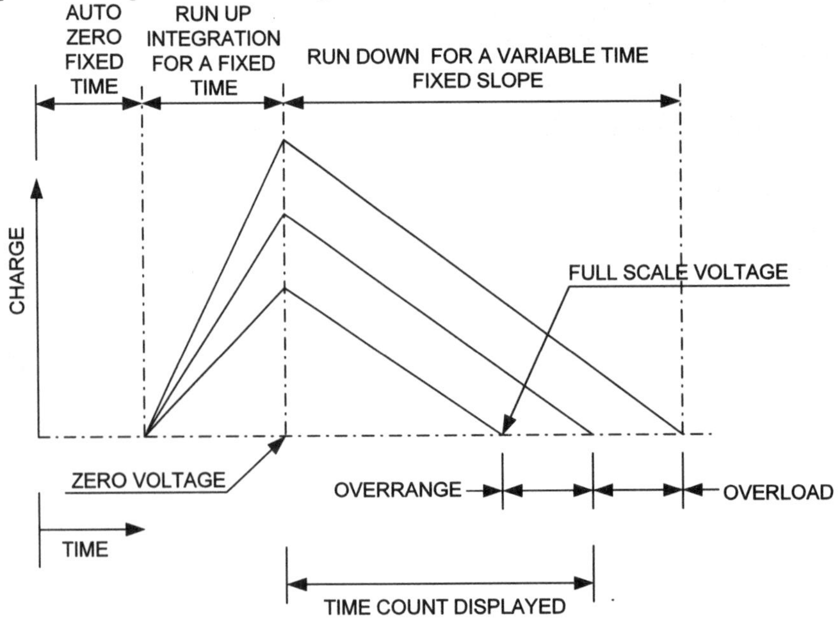

Figure 19.11 Dual slope integration technique used in most DVMs.

A counter is started at the beginning of a run down and operates until the output voltage from the integrator crosses zero. The accumulated time is proportional to the dc voltage applied to the integrator. This time count is then displayed as the voltage. Polarity, range and function information that are provided by the controller are also displayed.

The characteristics of a digital voltmeter may be altered by changing the number of digits, the time interval for integrator run up and the frequency of the clock. A professional hand held multimeter with 3-½ digits and a maximum count of 1999 is designed with five different ranges: ± 200.0 mV, ± 2.000 V, ± 20.00 V, ± 200.0 V and ± 1500 V. The highest sensitivity is 100 μV on the 200-mV range. Accuracy is ± 0.5 percent of the reading plus one digit. The clock frequency is 200 kHz and the integration time is 100 ms. The reading rate is usually about 2.5 readings per second, depending on the input.

System DVMs are more complex than the hand held or small bench type DVMs because the former are provided with a microprocessor and local memory to facilitate interfacing with other components of an automated data-processing system. A typical data processing system incorporating a system type DVM includes a scanner or multiplexer for switching input voltages into the DVM for analog to digital conversion and a bus which is compatible with a personal computer (PC). The memory in the PC (RAM or disk) is used to store the data that is acquired. Reduction, manipulation and analysis of the data are performed according to programmed instructions and results are often presented in easily understood graphical form on a PC monitor or from a digital printer.

System multimeters are higher performance devices than bench multimeters. The number of digits is usually increased to 6-½ or 8-½, which increases the count and improves the resolution to 1 or 0.01 PPM, respectively. Clock frequencies are increased and advanced conversion techniques[2] are employed to give a reading rate of 100,000 readings per second. Microprocessors are added to control the different measurements and to control the interface. System multimeters are available with either IEEE-488 or RS-232 bus structures. Also, the internal microprocessor has modest computing capability that permits one to add, subtract, multiply and divide as well as store and compare numerical information.

19.7 DATA-LOGGING SYSTEMS

A basic data-logging system consists of a scanner or multiplexer, a high quality digital voltmeter and a computer to store and analyze the data recorded. Such a system can be employed to record the output from a large number of transducers (hundreds) at a sampling rate, which depends on the capability of the DVM and the resolution required. Conversion rates for a modern DVM, which utilizes a multi-slope analog-to-digital converter, is shown in Table 19.3.

TABLE 19.3
Characteristics of a System Type DVM
Agilent Model 3458A

Number of Digits	Readings per second	Resolution (ppm)
4 - ½	1×10^5	33.3
5 - ½	5×10^4	3.33
6 - ½	6×10^3	0.333
7 - ½	60	0.033
8 - ½	6	0.003

*The number of digits, ranging from 4-½ to 8-½, is selectable with the model 3458A. The ½ digit on this model may read 0, 1 or 2.

---

[2] The advanced conversion techniques alleviate the constraint that the integration time $t^* \geq 16.17$ ms for integrating voltmeters.

Because the DVMs are relatively fast, a system controller is needed to direct the scanner to each new channel, to control the integration time for the DVM and to transfer the output from the DVM to the computer. The system controller is a microprocessor that uses two separate busses; one for data transfer and the other for memory addressing. The software, which directs the operation of the controller, is stored in read-only memories (ROMs) and a random-access memory (RAM). The system operating programs are permanently stored in the ROMs, which are programmed during the manufacturing of the instrument. The input by the operator, individual channel parameters and other program routines is entered through a keyboard and stored in RAM.

The scanner contains a bank of switches (usually three pole) that serve to switch two leads and the shield from the input cable to the integrating digital voltmeter. In most cases, high-speed solid-state switching devices are employed. The system controller directs the scanner operation. The modes of operation available include single-channel recording, single scan of all channels, continuous scan and periodic scan. In the single-channel mode, a preselected channel is continuously monitored at the reading rate of the system. In the single-scan mode, the scanner makes a single sequential sweep through a preselected group of channels. The continuous-scan mode is identical to the single-scan mode, except that the system automatically resets and recycles on completion of the previous scan. The periodic scan is simply a single scan that is initiated at preselected time intervals such as 1, 5, 15, 30 or 60 min. The scanner also provides a signal for the visual display of the channel number and a code signal to the controller to identify the transducer being monitored.

The transducer signal is switched through the scanner[3] to a high-quality integrating digital voltmeter that serves as the ADC. The speed of operation depends primarily on the capabilities of the DVM and the resolution required. As is evident in Table 19.3, high reading rates are possible even with a 5-½ digit ADC that provides a resolution of 3.33 PPM.

The output from a data logger is displayed with a digital panel meter that indicates the voltage output, its polarity and the channel number. The output of most data logging systems is recorded on either local disk memories associated with a PC or workstation or a remote disk memory associated with a central computing center. Some of the more expensive system DVMs contain on board memory to facilitate data storage. One of the principal advantages of a data-logging system is the capability for processing the data in essentially real time with an on-line computer.

19.8 PC BASED DATA-ACQUISITION SYSTEMS

Recently, relatively low cost circuit boards have become available that contain many of the elements found in higher cost more elaborate data-acquisition systems. Some data-acquisition boards (DAQs), such as the one shown in Fig. 19.12, are designed to interface with one or more sensors such as strain gages, resistance temperature detectors (RTDs), thermocouples, thermistors, etc. The sensor support contained on these circuit cards includes sensor excitation, linearization, cold reference compensation and conversion of the output to engineering units.

Figure 19.12 A circuit card that plugs into a USB port on a data personal computer.

[3] In some systems the DVM is incorporated in the scanner to provide an integrated system.

The evolution from stand alone data acquisition systems to PC based data acquisition system has occurred because of increased reliability of PCs and high capacity hard drives at reduced cost. The integration of data acquisition with the PC also affords the opportunity to process and analyze the data more effectively in less time and with less effort. Significant software developments, to enable the control of the signals on the circuit card and to interface with the PC, have also reduced the difficulty in employing these systems.

A block diagram of a data acquisition (DAQ) circuit card is illustrated in Fig. 19.13. On the left is an input/output (I/O) connector used to transfer both analog and digital signals both to and from the circuit card. The number of pins available in the connector varies depending upon the type and purpose of the circuit card employed. A typical configuration with 68 pins provides the conductors for the input analog signals, the analog trigger signals, the counter pulses, digital input or output signals and analog signals used in process control. The input signals are processed on the card and the output signals are used to drive analog devices. A digital buss that transfers the digital output signals from the DAQ to the host PC or other digital instrument is located on the far right hand side of Fig. 19.13.

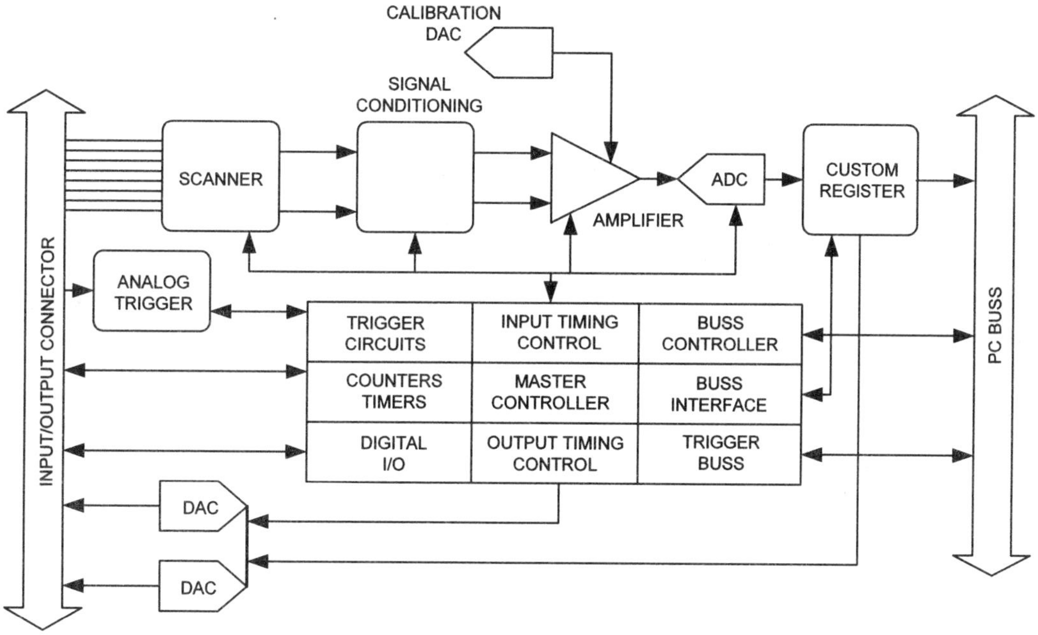

Figure 19.13 A block diagram illustrating the various operations performed by a DAQ.

These circuit cards typically contain three sections. The first section acquires the analog signals, performs the signal conditioning for the compatible sensors, multiplexes to the appropriate sensor and then amplifies the signal with a programmable gain. The multiplexing (scanning) is performed at very high speed by employing solid state switching devices. The amplifiers are very high quality with extremely rapid settling times (usually less than 5 μs) that are consistent with the high sampling rates at which the systems operate.

The second section on the DAQ performs A/D conversion. A typical card utilizes[4] either a 12-bit or 16-bit ADC with sampling rates varying from 100 kS/s to 10 MS/s. The digital output from the ADC is transferred to a custom designed register that serves an interface to the buss contained in the host PC. The output from this register is transferred to the PC's buss upon receiving a command from the master controller.

---

[4] See for example the numerous DAQs offered by National Instruments on their website www.ni.com.

The third section incorporates a microcomputer with on-board memory that is used in processing data by performing tasks such as linearization, reference junction compensation and engineering unit conversion. The microprocessor also provides the logic to scan the sensors, adjust the amplifier gain and transfer the data to the standard bus registers. The standard bus interface incorporates drivers and receivers to facilitate communication with a PC-type host computer. This section also contains a number of high-resolution counters that are used to control the timing of the signal acquisition and the subsequent processing of the signal through the scanners, amplifier and DAC. Finally, both analog and digital trigger circuits are contained in the controller to enable the operator to initiate and terminate the data acquisition process.

An advantage of utilizing a DAQ in making measurements is their capability to minimize error due to temperature changes. All recording instruments tend to drift with changes in temperature and this drift produces error. High quality DAQs are designed with premium components that are relatively stable over a wide range of temperature. They also incorporate compensating components that tend to cancel temperature induced voltage and calibration changes. Well-designed DAQs have proper placed shielding and appropriately placed ground planes in the circuit card to minimize electrical noise generation in the data acquisition system. High quality DAQs are equipped with the highest-grade ADCs. These ADCs incorporate circuits that enable self-calibration thus eliminating linear errors. Non-linear errors are eliminated through the use of premium components that remain stable with time and temperature.

The card is programmed from the PC host computer and the digitized data is transferred from the card to the memory of the PC using the standard bus in the host computer. Note that an external bus structure is not required for data transmission when the entire data-acquisition system is contained within the PC. At this point all further processing and preparation of graphics is performed on the host computer using commercially available software.

The software employed in acquiring and analyzing data is equally important in the design of an effective measurement system. Typically data acquisition software includes drivers and application codes. Drivers are a set of commands unique to the sensors or actuators involved in the measurement or in the control of a process. The interface to the sensors or actuators is compatible with the driver commands. The software also enables the operator to display and analyze the data in essentially real time. A popular software program used in measurements and process control is LabVIEW (Laboratory Virtual Instrument Engineering Workbench) is described later in Chapter 8.

19.9 SYSTEMS FOR MEASURING RAPIDLY VARYING SIGNALS

Measuring transient phenomena, where the signal from the transducer is a rapid changing function of time, is the most difficult and most expensive measurement in experimental work. Frequency response or bandwidth is the dominant characteristic required of the recording instrument in dynamic measurements and usually accuracy and economy are sacrificed in order to improve the response capabilities. Two different recorders are used in modern instrument systems[5] to record transient voltages—digital acquisition systems (DAQs), described in previous sections and the oscilloscope.

The DAQ's sampling rate controls the integrity of the record of the transient signal as will be described later in Section 19.10 when aliasing is discussed. Sampling rates for DAQs vary from about 100 kS/s to 10 MS/s, which enable them to be used to capture many of the transient signals associated with mechanical phenomena.

---

[5] The oscillograph and the magnetic tape recorder that were used in the past to record signals with relatively low frequencies have been replaced in most applications with digital instruments. An oscillograph utilizes a galvanometer, which incorporates a highly refined D'Arsonval movement, to drive a pen or a light beam over a moving strip of chart paper. In a magnetic tape recorder, the dynamic signal is stored on magnetic tape for later playback and display on either an oscillograph or an oscilloscope.

Oscilloscopes operate with a higher frequency response (bandwidth) than DAQs because they use higher performance electronic components. There are several different types of oscilloscopes including: analog, digital storage (DSO), digital phosphor (DPO) and digital sampling[6]. Digital oscilloscopes can display any transient signal within its range of operation with clarity and stability. As with DAQs, the bandwidth of the digital oscilloscope is limited by its sampling rate, although these sampling rates are usually one or two orders of magnitude higher than those found in a typical DAQ.

19.9.1 Analog Oscilloscopes[7]

The cathode-ray-tube oscilloscope is a voltage-measuring instrument that is capable of recording extremely high-frequency signals (up to about 1 GHz). The cathode-ray tube (CRT), which is the most important component in an oscilloscope, is an evacuated tube in which electrons are produced, controlled and used to provide a voltage-time record of a transient signal. The electrons produced by heating a cathode are collected, accelerated and focused onto the face of the tube with a grid and a series of hollow anodes. The impinging stream of electrons forms a bright point of light on a fluorescent screen at the inside face of the tube. Voltages are applied to horizontal and vertical deflection plates in the CRT (see Fig 19.14) to deflect the stream of electrons and thus move the point of light over the face of the tube. It is this ability to deflect the stream of electrons that enables the CRT to act as a dynamic voltmeter with an essentially zero inertia recording system. The speed of writing is controlled by the intensity of the electron stream and the characteristics of the luminous phosphors used to coat the face of the CRT.

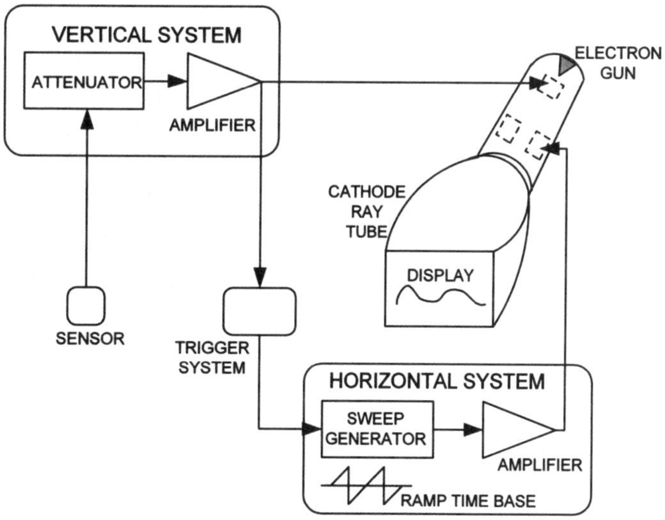

Figure 19.14 Bock diagram of the principle elements of an analog oscilloscope.

An oscilloscope can be used to record a voltage y as a function of time, or it can be used to simultaneously record two unknown voltages x and y. The block diagram of an oscilloscope, presented in Fig. 19.14, shows the input signal from a sensor and the connections to the deflection plates in the CRT. The y and the x inputs are connected to the vertical and horizontal deflection plates through amplifiers. Because the sensitivity of the CRT is relatively low (approximately 100 V are required on the deflection plates to deflect the beam of electrons 25 mm on the face of the tube), high gain amplifiers are used to increase the voltage of the input signals.

When the oscilloscope is used as a y-t recorder, the input to the horizontal amplifier is switched to a sawtooth generator. The sawtooth generator produces a voltage-time output, having the form of a ramp function where the voltage increases uniformly with time from zero to a maximum and then

[6] Digital sampling oscilloscopes that are used to capture extremely high frequency repetitive signal will not be described in this textbook. The bandwidth capability of a typical sampling oscilloscope is about 50 GHz, which is well above the frequency of interest to most engineers working in the mechanical world.

[7] Analog oscilloscopes are still available in many laboratories; however, they are being replaced with digital instruments, which offer many advantages in capturing and processing high-speed transient signals.

almost instantaneously returns to zero so that the process can be repeated. When this ramp function is imposed on the horizontal deflection plates, it causes the electron beam to sweep from left to right across the face of the tube. When the voltage from the sawtooth generator goes to zero, the electron beam is returned almost instantaneously to its starting point. The frequency of the sawtooth generator can be varied to give different sweep rates. The available sweep rates depend on the bandwidth of the oscilloscope. For high frequency oscilloscopes with a bandwidth of 1 GHz the sweep rate can be varied from 200 ps/div to 0.2 s/div in calibrated steps. Because the horizontal dimension of the face of the CRT is divided into 10 divisions (usually a division equals 10 mm), the observation time is ten times the sweep rate.

Since the observation time is usually relatively short, the horizontal sweep must be synchronized with the event to ensure that a recording of the signal from the transducer is made at the correct time. Three different triggering modes are used to synchronize the oscilloscope with the event: the external trigger, the line trigger and the internal trigger. Trigger signals from any one of these three sources activate the sawtooth generator and initiate the horizontal sweep. The external trigger requires an independent triggering pulse from an external source usually associated with the dynamic event being studied.

The line trigger utilizes the signal from the power line to activate the sawtooth generator. Because the line-trigger signal is repetitive at 60 Hz, the horizontal sweep triggers 60 times each second; therefore, the trace on the CRT appears continuous. The line trigger is quite useful when the oscilloscope is used to measure periodic waveforms that exhibit a fundamental frequency of 60 Hz.

The internal trigger makes use of the y input signal to activate the sweep. The level of the trigger signal required to initiate the sawtooth generator can be adjusted to very low levels; consequently, only a small region of the record is lost in measuring a transient pulse. If the input signal is repetitive, the frequency of the sawtooth generator can be adjusted to some multiple of the frequency of the input signal. The sawtooth generator is then synchronized with the input signal and the trace appears stationary on the CRT screen.

The trace on the screen of the CRT is produced when the electron beam impinges on a phosphor coating on the inside face of the tube. The light produced by fluorescent of the phosphor has a degree of persistence that enhances the visual observation of the trace. Permanent records of the traces can be made with special-purpose oscilloscope cameras that attach directly to the mainframe of the oscilloscope.

The amplifier used in an oscilloscope is quite important because it controls the sensitivity, bandwidth, rise time, input impedance and the number of channels that can be recorded with the instrument. Because of the importance of the amplifier to the operational characteristics of the oscilloscope, many analog instruments are designed to accept plug-in-type amplifiers. This arrangement permits the amplifier to be changed quickly and easily to alter the characteristics of the oscilloscope.

Bandwidth is defined as the frequency interval over which signals are recorded with less than 3-dB loss compared to midband performance. Because modern amplifiers perform very well at low frequencies (down to dc), bandwidth refers to the highest frequency that can be recorded with an error less than 3 dB (29.3%). Bandwidth f_{bw} and rise time t_r are related by:

$$f_{bw}\, t_r = 0.35 \tag{19.17}$$

Because good practice dictates utilization of a vertical amplifier capable of responding five times as fast as the rise time of the applied signal, Eq. (19.17) is modified in practice to:

$$f_{bw}\, t_r = 1.70 \tag{19.18}$$

It is apparent from Eq. (19.18) that an amplifier-oscilloscope combination with a bandwidth of 100 MHz is capable of recording signals with a rise time of approximately 17 ps.

Reciprocal sensitivity, S_R, refers to the voltage of the input signal required to produce a specified vertical deflection on the CRT. The sensitivity of a typical amplifier used to control the y deflection ranges from 1 mV/div to 5 V/div in calibrated steps. Higher sensitivity amplifiers are available with S_R equal to 10 μV/div to 50 mV/div; however, the bandwidth is reduced as sensitivity is increased. A sensitivity-bandwidth tradeoff occurs in the selection of an amplifier and for lower frequency measurements it is usually advisable to specify the minimum bandwidth required and to work with higher sensitivity lower noise amplifiers.

The input amplifier also controls the number of traces displayed on the oscilloscope screen. In some models, an electronic switch is housed in the amplifier that alternately connects two input signals to the vertical deflection system in the CRT. The principal advantages of using this feature to produce a dual-trace oscilloscope are lower cost and better comparison capabilities. Both of these advantages are due to the fact that only one horizontal amplifier and one pair of deflection plates is used in making both traces. However, high-speed transient events are difficult to record in this manner because a significant variation might occur on one channel while the beam is tracing on the other channel. Whenever two nonrecurring signals of very short duration must be recorded together, dual-beam oscilloscopes are employed. A dual-beam oscilloscope has independent deflection plates within the CRT for each beam and employs independent horizontal and vertical amplifiers for each beam. The dual-beam system is superior to the dual-trace system, because it can display two signals separately and simultaneously; however, it is more costly.

The modular oscilloscope with plug-in amplifiers and time bases provides a versatile instrument that can be adapted to a wide number of applications in a multipurpose laboratory. For more specific applications, portable oscilloscopes without plug-in amplifiers may be more suitable, because they are lower in cost and simpler to operate. Also, many portable oscilloscopes are battery powered, small and lightweight, which permits their application in remote-site field measurements.

19.9.2 Digital Storage Oscilloscopes

A digital storage oscilloscope (DSO) is identical to an analog oscilloscope except for:

1. The manipulation of the input signal prior to display on the cathode-ray tube.
2. The permanent-storage capabilities of the instrument.
3. The signal-processing capabilities of its microprocessor.

A block diagram showing the essential elements in a DSO is presented in Fig. 19.15. It is clear from this figure that the input signal is amplified and then converted to digital form in an ADC. A clock in the controller determines the sampling rate for the ADC. Next in this serial arrangement is a DEMUX that divides the data stream for subsequent processing. The digital values, called sample points are stored in an acquisition memory. Usually many data points represent a waveform. The number of data points used to establish a waveform is called the record length. A trigger system similar to that found in an analog oscilloscope initiates the sampling process.

The signal path in a DSO passes through a microprocessor that processes the digital data, coordinates display activities and manages the front panel controls. The signal then is stored in the display memory before being displayed. From the display memory the digital signal is transferred to a DAC, where it is reconverted to an analog voltage for display on the CRT.

A photograph of a modern digital oscilloscope is shown in Fig. 19.16.

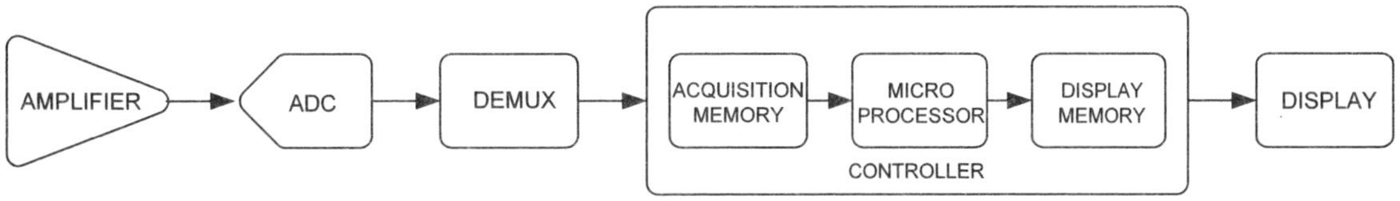

Figure 19.15 Serial processing employed in a digital storage oscilloscope.

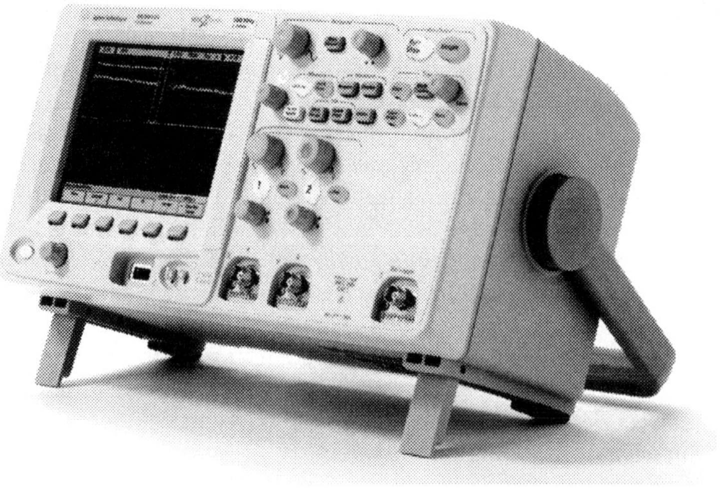

Figure 19.16 A modern digital oscilloscope. Courtesy of Agilent Technologies.

Because the data are stored, in addition to being displayed, operation of the digital oscilloscope differs from operation of an analog oscilloscope. The display on the CRT of the digital oscilloscope is a series of points produced by the electron beam at locations controlled by the data in storage. The data collected during the sweep of the oscilloscope is in memory and can be recalled and analyzed either within the digital oscilloscope or by downloading to a host computer.

Several features establish the capabilities of a digital storage oscilloscope. First, the sampling rate and the bandwidth are important in recording transient signals. Bandwidth requirements described previously under analog oscilloscopes also apply to digital oscilloscopes. An additional factor is the sampling rate that determines the time interval between data points. For example, a digital oscilloscope with a 1.0 GS/s rate can sample, hold, convert and store a data point in 1.0 ns.

Second, the size of the memory is important because its size controls the record length that can be recorded. The record length is the number of data points (samples) that can be stored in the DSO's memory. Because of memory limitations, an oscilloscope can store only a specified number of samples; consequently, the waveform duration t_d is inversely proportional to the oscilloscope's sampling rate as given below:

$$t_d = RL/SR \tag{19.19}$$

where RL is the record length and SR is the sampling rate.

The number of data points needed to adequately analyze a waveform depends on the application. If the waveform is a stable sinusoidal, a hundred data points are usually sufficient. However, if the analysis pertains to a complex data stream, several thousand data points may be required to adequately record the event. The waveform duration t_d that can be covered with a DSO is dependent on the choice of sampling rates and the duration can vary widely (500,000/1 is common).

A final feature of importance is whether the A/D conversion method is designed to measure single-shot events or repetitive signals. Repetitive signals are easier to measure as sampling can be repeated on the second and subsequent waveforms to give instruments with apparent sampling rates that are an order of magnitude higher than the real sampling rates. The delayed sequential sampling technique that is used to increase the number of data points (samples) that define a repetitive waveform is illustrated in Fig. 19.17. For pulse measurements, the signal occurs once and only once and repetitive measurements cannot be used to increase the sampling rate.

In operation, A/D conversion takes place continuously at a prescribed sampling rate with the words going to storage until the data acquisition memory (DAM) is full. The address for each data word stored in the DAM is proportional to the time when the data were taken. After a sweep, when the DAM is full, the data is discarded, unless a trigger signal is received during the sweep, and the conversion process continues. If a trigger signal occurs during the sweep, the data in the DAM is transferred to the display memory. This data is then processed and displayed as a voltage-time trace on the CRT.

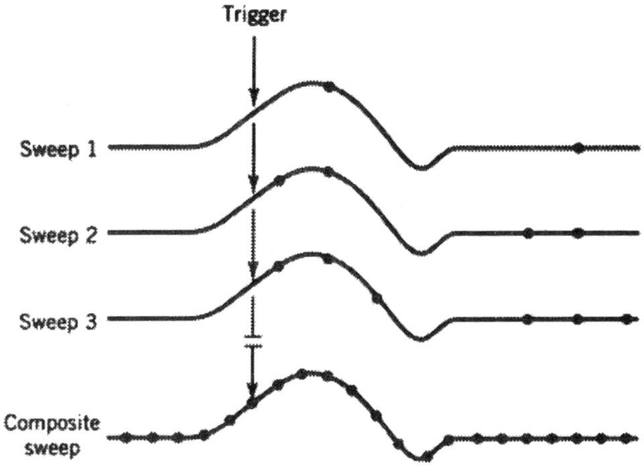

Figure 19.17 By varying the time between the trigger and the ADC, different data points can be recorded on each repetition (sweep).

The fact that the input signal has been stored offers many advantages for data display and data processing. The data are displayed on the CRT in a repetitive manner so that traces from even one-shot transient events appear stationary. The trace can also be manipulated by expanding either the horizontal or vertical scales or both. This expansion feature permits a small region of the record to be enlarged and examined in detail. Readout of the data from the trace is also much easier and more accurate with digital oscilloscopes. A pair of marker lines called cursors (one vertical and the other horizontal) can be positioned anywhere on the screen. The procedure is to position the vertical line at a time on the trace when a reading of the voltage is required. The horizontal marker (or cross hair) automatically positions itself on the trace. The coordinates of the cross hair, intersection with the trace, are presented as a numerical display on the screen.

Modern digital oscilloscopes are usually equipped with a microprocessor that provides different on-board signal analysis features that often include:

1. Pulse characterization—rise time, fall time, base line and top line width, overshoot, period, frequency, rms, mean, standard deviation, duty cycle, etc.
2. Frequency analysis—power, phase and magnitude spectrum.
3. Spectrum analysis—100 to 50,000 point fast Fourier transforms (FFT).
4. Math package—add, subtract, multiply, integrate and differentiate.
5. Smoothing —1, 3, 5, 7 or 9 point.
6. Counter—average frequency and event crossings.
7. Display Control—x zoom, x position, y gain, and y offset.
8. Plotting Display in either dots or vectors.
9. Mass storage to floppy disk, hard disk or nonvolatile memory.

If additional processing is required, the data can be downloaded to a host computer for final data analysis.

Early models of DSOs were introduced in 1972. Initially, their performance was limited due to the low-bandwidth capability; however, marked improvements in ADCs, microprocessors and high-speed, high-capacity memory chips have greatly enhanced speed of conversion, improved resolution and expanded the amount of data, which can be stored. With these improvements, the DSO is superior in every respect to an analog oscilloscope. Digital methods of recording have in most cases replaced analog methods because costs are competitive and performance is superior.

19.9.3 Digital Phosphor Oscilloscopes

The digital phosphor oscilloscope (DPO) utilizes a parallel architecture to enhance the probability of observing rare transient events that may occur in a data stream. The parallel architecture of a DPO is illustrated in Fig. 19.18. Reference to this figure shows that the first stage in a DPO is similar to that of an analog oscilloscope because the signal is amplified. The second stage is similar to that of a DSO because it is represented by a ADC to convert the signal from analog to digital format. However, the DPO differs significantly from the DSO following analog-to-digital conversion.

Figure 19.18 Parallel processing architecture of a digital phosphor oscilloscope.

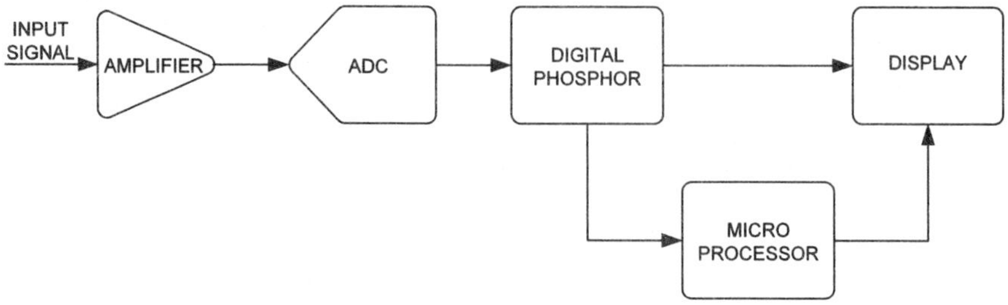

For any type of oscilloscope there is always a dead time during which the instrument processes the most recently acquired data, resets the electronics and waits for the next trigger signal to initiate a new sweep. During this dead time, the oscilloscope is ignoring ongoing signal activity. For this reason, the probability of observing an infrequent event decreases as the dead time interval increases.

A DSO uses a serial processes to store waveforms in the data acquisition memory. The speed of the microprocessor represents a bottleneck in the process because it limits the waveform capture rate. In the DPO instruments, the digitized waveform is transferred as a raster signal to a digital phosphor database 30 times each second. This signal is then transmitted to the display as illustrated in Fig. 19.18. This direct rasterization of the waveform together with the direct copy to the display removes the data-processing bottleneck that is inherent in serial oscilloscope architectures. By temporarily bypassing the microprocessor, the DPO provides signal details that are captured in real-time. The microprocessor in the DPO operates in parallel with this system for display management, measurement automation and instrument control and does not affect the acquisition speed of the DPO.

Digital phosphor is quite different from the chemical phosphor that has been used for many decades on CRT tubes. Digital phosphor is an electronic database with separate "cells" of information for each pixel on the display. Each time a waveform is captured, it is mapped onto the array of cells in the digital phosphor database. In the mapping operation, some cells are activated by the presence of a waveform signal while others are not. For repetitive signals the intensity of the activated cells accumulates, which enables the display to show intensified waveform signals that are proportional to the frequency of occurrence of the waveform at each point on the y-t traces.

DPOs combine the advantages of analog and digital oscilloscopes. They can be employed to observe high and low frequencies, repetitive waveforms, transient signals and signal variations in real time. A DPO provides the Z (intensity) axis in real time that cannot be achieved with the electronics employed in DSOs.

19.10 ALIASING

Digital data-acquisition systems contain an ADC that converts an analog signal to a digital signal at a specified sampling rate. This sampling rate is extremely important in dynamic measurements where high-frequency analog signals are being processed. For a well-defined representation of a dynamic waveform, the analog signal should be determined with a digital data point (sample) taken 10 or more times during the period of the waveform. This concept of using 10 samples to define a sine wave is illustrated in Fig. 19.19. As the number of digital data points decreases, the definition of the type of waveform and its characteristics degrade to the point where the digital representation can be misleading. When the sampling frequency f_s is:

$$f_s \leq 2f \qquad (19.20)$$

the waveform with frequency f takes on a false identity. Nyquist sampling theory (beyond the scope of this textbook) is the basis for Eq. (19.20). The minimum sampling frequency is $f_s = 2f^*$ and the maximum conversion time is $T_s^* = 1/(2f^*)$ where f^* is called the Nyquist frequency and T_s^* is the Nyquist interval.

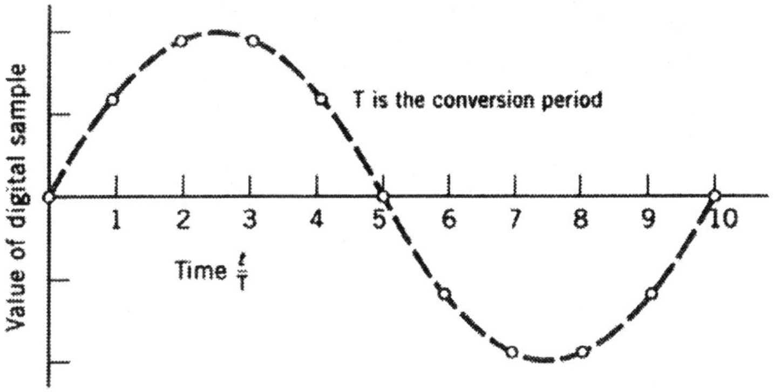

Figure 19.19 Digital representation of a sinusoidal waveform that illustrates the definition achieved with ten sampling points.

If the frequency of the analog signal $f \geq f_s/2$, the sampling process is inadequate and the output from the ADC gives a false low-frequency waveform, called an alias, which differs from the true analog signal. To illustrate aliasing, consider a sinusoidal analog signal with a frequency f_1 given by:

$$v_1(t) = \cos(2\pi f_1 t) \qquad (19.21)$$

If the ADC has a sampling frequency f_s, the times at which the signal is sampled is given by:

$$t = nT_s = n/f_s \qquad n = 0, 1, 2, \ldots \qquad (19.22)$$

Substituting Eq. (19.22) into Eq. (19.21) gives the sampled voltages as:

$$v_1(nT_s) = \cos\left(\frac{2\pi n f_1}{f_s}\right) \qquad (19.23)$$

Next, consider a second sinusoidal signal with a frequency f_2 which is greater than a cutoff frequency f_c so that:

$$f_2 = 2mf_c \pm f_1 \qquad m = 1, 2, \ldots. \qquad (19.24)$$

If the ADC used with both signals is the same, then $v_2(nT_s)$ is obtained by interchanging f_2 for f_1 in Eq. (19.21) to give:

$$v_2(nT_s) = \cos\left(\frac{2\pi n(2mf_c \pm f_1)}{f_s}\right) \qquad (19.25)$$

Let the cutoff frequency be given by:

$$f_c = f_s/2 \qquad (19.26)$$

as indicated by sampling theory. Substituting Eq. (19.26) into Eq. (19.25) gives:

$$v_2(nT_s) = \cos\left(2\pi mm \pm \frac{2\pi nf_1}{f_s}\right) = \cos\left(\frac{2\pi nf_1}{f_s}\right) \qquad (19.27)$$

A comparison of Eqs. (19.27) and (19.23) shows that v_1 and v_2 are identical at each sampling time (nT_s) and it is impossible to distinguish the signal amplitudes between the two frequencies f_1 and f_2 . For example if f_c = 200 Hz, then f_s = 400 Hz and an alias signal with a frequency f_1 = 50 Hz occurs whenever the input signal has frequencies f_2 of 350, 450, 750, 850, etc. The relation showing the alias frequency is obtained by substituting Eq. (19.26) into Eq. (19.24) to obtain:

$$\pm f_1 = f_2 - mf_s \qquad \text{if} \qquad f_2 \geq f_s/2 \qquad m = 1, 2, \ldots \qquad (19.28)$$

where f_1 is the frequency of the alias signal and f_2 is the frequency of the input signal.

The results of Eq. (19.28) are illustrated in Fig. 19.20 for the example when $f_s = (3/2)f_2$. Note that the recorded digital signal exhibits the alias frequency $f_1 = f_2/2$.

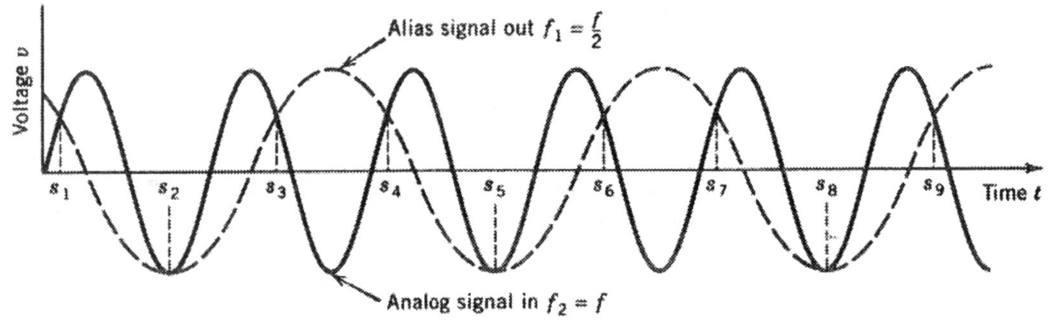

Figure 19.20 Effect of aliasing when a sinusoid is sampled at a frequency $f_s = (3/2)\, f_2$.

Aliasing can be avoided if the sampling frequency exceeds twice the maximum frequency in the analog signal.

$$f_s = 2f_c > 2f_2 \qquad (19.29)$$

This relation is essentially the same as Eq. (19.26), which defines the cutoff frequency in terms of the sampling frequency. Clearly, if the maximum frequency in the analog signal does not exceed the cutoff frequency, the identity of Eqs. (19.23) and (19.27) cannot be established, and aliasing will not occur.

19.10.1 Anti-aliasing Filters

Commercial instruments avoid the aliasing problem by using anti-aliasing analog filters to reduce the high frequency components in the analog signals. These filters have a frequency response function that exhibits a relatively sharp signal attenuation beginning at $f = 0.4\, f_s$, as shown in Fig. 19.21. The analog signal is attenuated 40 dB at the Nyquist frequency $f_s^* = f_s/2$. Because 40 dB is equivalent to a signal transmission of only 1%, the filter reduces the signal component that produces aliasing to an insignificant amount.

Unfortunately, anti-aliasing filters can severely distort transient signals with high-frequency components. The possibility of distortion on the one hand and the need to prevent aliasing of the signal on the other implies that care must be exercised when selecting the sampling rate of the ADC for a specific high frequency measurement.

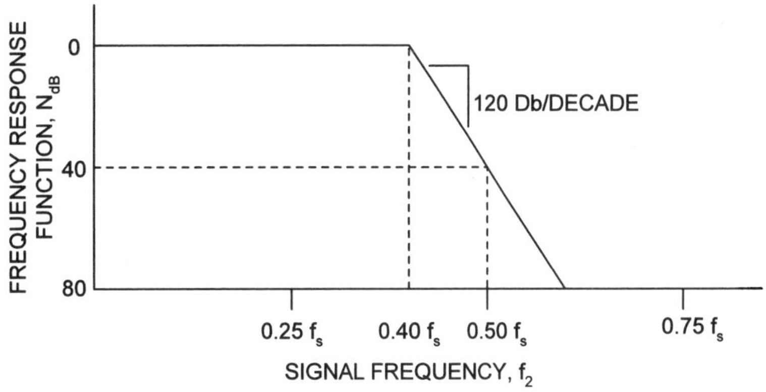

Figure 19.21 Frequency response function for an ideal analog anti-aliasing filter for an ADC with a sampling rate of f_s.

19.11 SUMMARY

Digital systems that employ analog to digital converters (ADCs) or digital to analog converters (DACs) are widely used in instrument systems for engineering analysis and for process control. The principal advantage of digital systems is the ability to store data and to process that data by using low cost commercially available storage devices, computers, controllers and significant amounts of effective application software.

Analog to digital conversion is the key element in systems utilized for data acquisition and for digital instruments such as digital voltmeters, data acquisition systems and digital oscilloscopes. Features such as sampling rate, resolution, accuracy and cost are used to compare ADCs. For high resolution and low cost, ADCs usually employ the integration method for conversion; however, the sampling rate is relatively low. For very high sampling rates, the flash method of conversion is used with more limited resolution (8 bits) and significantly higher cost. The method of successive approximation is used in ADCs with moderate sampling rates, good resolution and intermediate cost.

Digital to analog conversion is also employed in process control. After digital processing of data according to programmed instructions, the system automatically determines if the process must be modified. For example, assume the process being controlled is a curing oven and the temperature is to be increased. The digital signal for a specified temperature increase is converted to an analog signal using a DAC. This signal is then amplified and used to activate a temperature controller that increases the current flowing through its heating elements.

Finally, developments in digital devices are continuing at a rapid rate as the technology used in manufacturing integrated circuits permits further reductions in device size. Continued improvements will lead to higher sampling rates, more bits, higher resolution and lower costs for digital hardware. Software written for measurement and control is currently improving each year, particularly with PC-based data-acquisition systems.

EXERCISES

19.1 Prepare a block diagram showing a combined analog-digital instrumentation system to measure pressure for an application involving:
(a) engineering analysis (b) process control
19.2 Prepare a table showing the maximum count C as a function of the number of bits n. Let n equal from 4 to 32 in steps of 4.
19.3 Add a column showing the resolution R to the table in Exercise 19.2.
19.4 Add a column showing the resolution error E_R (%) to the table in Exercise 19.2.
19.5 If you select an instrument with 8-bit logic circuits, what is the resolution error that can be expected.
19.6 Write an engineering brief describing an A/D converter (ADC).
19.7 Write an engineering brief describing a D/A converter (DAC).
19.8 Describe the difference between resolution error and quantizing error. Which is the most important?
19.9 Prepare a graph showing the standard deviation of a number of measurements of v/FSV as n increases from 4 to 32 in steps of 4.
19.10 Prepare a graph showing C versus v/FSV that demonstrates offset error.
(a) for a D/A converter (b) for an A/D converter.
19.11 Prepare a graph showing C versus v/FSV that demonstrates scale factor error.
(a) for a D/A converter (b) for an A/D converter.
19.12 Prepare a graph showing C versus v/FSV that demonstrates linearity error
(a) for a D/A converter (b) for an A/D converter.
19.13 For the 4-bit DAC shown in Fig. 19.5, determine the output voltage v_0 for the following digital codes.
(a) 1101 (c) 0110 (e) 1001
(b) 1010 (d) 0101 (f) 1110
19.14 Describe in a brief paragraph a register and indicate some of its uses in a digital system.
19.15 What is a strobe signal and how is it used in a D/A converter? Why is it necessary?
19.16 What are the three common systems used in designing A/D converters? List the advantages and disadvantages of each system.
19.17 Prepare an illustration, similar to Fig. 19.7, that demonstrates A to D conversion by the method of successive approximations if the fixed analog input voltage v_u is:
(a) (1/8) FSV (c) (3/4) FSV
(b) (7/16) FSV (d) (13/16) FSV.
19.18 Explain why the conversions of Exercise 19.17 were all exact.
19.19 Because A to D conversions requires some time for switching and comparing how do we avoid errors due to voltage fluctuations during the conversion period?
19.20 Using Fig. 19.7 as a guide, describe the operation of a successive approximation A/D converter. Indicate the purpose of each block element and each input or output signal.
19.21 Begin with Eq. (19.10) and verify Eq. (19.12).
19.22 Using Fig. 19.8 as a guide, describe the operation of a dual slope integrating A/D converter. Indicate the purpose of each block element, component and each input or output signal.
19.23 Verify Eq. (19.16) beginning with Eq. (19.13).
19.24 For a 12-bit unipolar dual slope integrating A/D converter capable of 20 S/s, determine the frequency limit.
19.25 If the A/D converter of Exercise 19.24 were bipolar determine the frequency limit.
19.26 Determine the slew rate of the A/D converter in Exercise 19.24 if the FSV is:
(a) 1 V (c) 5 V
(b) 2 V (d) 10 V
19.27 Using Fig. 19.9 as a guide, describe the operation of a flash type A/D converter.

19.28 Describe instruments that employ the flash type A/D converter. What is the sampling rate employed in these instruments? How are the sampling rate, bandwidth and sampling times related?

19.29 Why is a plug-in data acquisition board so cost effective?

19.30 A five-digit DVM is capable of what maximum count if it has

(a) 0 over ranging (b) 100% over ranging (c) 200% over ranging

19.31 You are to measure a voltage v_i with a 5-½ digit DVM capable of 100 % over ranging. If the meter is specified with an accuracy of ± 0.002% and ± 2 counts, determine the maximum and minimum readings anticipated if v_i is:

(a) 1.80000 V (b) 2.50000 V (c) 9.99996 V

19.32 Determine the error in each of the three cases of Exercise 19.31.

19.33 The purchasing department of a state agency asks you to write a specification so that they can procure bids for a system multimeter. Prepare this specification.

19.34 Explain the use of ROM and RAM memory incorporated into a data logging system.

19.35 Describe the scanner employed in a data logging system.

19.36 Compare digital and analog oscilloscopes and cite the advantages and disadvantages of each.

19.37 Prepare a graph showing observation time in a digital oscilloscope as a function of sampling rate. Use memory size in words as a parameter and let the sampling rate vary from 10 S/s to 1 GS/s. Let the memory size be 1000, 2000, 5000, 10,000 and 20,000 words.

19.38 In measuring periodic signals, it is possible to increase the apparent sampling rate of a digital oscilloscope. Explain how this is accomplished.

19.39 Explain how data preceding the trigger time can be recovered on a digital oscilloscope.

19.40 What are the common on-board signal analysis features found on digital oscilloscopes equipped with a microprocessor?

19.41 Verify Eq. (19.28).

19.42 Prepare a graph, similar to the one shown in Fig. 19.20, showing the analog and the alias signals if:

	f_c (Hz)	f_c (Hz)	f_c (Hz)
(a)	200	400	850
(b)	300	600	1,500
(c)	1,000	2,000	5,000
(d)	5,000	10,000	30,000

19.43 What are the characteristics of an anti-aliasing filter?

19.44 List the advantages and disadvantages of using an anti-aliasing filter.

BIBLIOGRAPHY

1. Ahmed, H. and P. J. Spreadbury: Analogue and Digital Electronics for Engineers, 2nd Edition, Cambridge University Press, New York, 1984.
2. Stone, H. S.: Microcomputer Interfacing, Addison Wesley, Boston, MA, 1988.
3. Bibbero, R. J. and D. M. Stern: Microprocessor Systems: Interfacing and Applications, John Wiley & Sons, New York, 1982.
4. Floyd, Thomas L.: Digital Fundamentals, 8th Edition, Prentice-Hall, Englewood Cliffs, NJ 2003.
5. Hall, D. V.: Digital Circuits and Systems, McGraw Hill, New York, 1989.
6. Tocci, L. and F. Ambrosio: Microprocessors and Microcomputers: Hardware and Software, 6th Edition, Prentice-Hall, Englewood Cliffs, NJ, 2003.
7. Hoeschele, D. F.: Analog-to-Digital/Digital-to-Analog Conversion Techniques, 2nd Edition, John Wiley & Sons, New York, NY, 1994.
8. Beards, D.: Analog Digital Electronics, 2nd Edition, Prentice-Hall, Englewood Cliffs, NJ, 1996.
9. Spencer, C. D.: Digital Design for Computer Data Acquisition, Cambridge University Press, New York, 1990.
10. Kaplan, D. M. and C. G. White: Hands-On Electronics, Cambridge University Press, New York, 2003.
11. Smith, R.: Electronics: Circuits and Devices, 3rd edition, John Wiley & Sons, New York, 1987.
12. Dally, J. W., Riley W. F. and K. G. McConnell: Instrumentation for Engineering Measurements, 2nd Edition, John Wiley & Sons, New York, 1993.
13. Figliola, R. S. and D. E. Beasley, Theory and Design for Mechanical Measurements, 3rd Edition, John Wiley & Sons, New York, 2000.

CHAPTER 20

STATISTICAL ANALYSIS OF EXPERIMENTAL DATA

20.1 INTRODUCTION

Experimental measurements of quantities such as pressure, temperature, length, force, stress or strain will always exhibit some variation if the measurements are repeated a number of times with precise instruments. This variability, which is fundamental to all measuring systems, is due to two different causes. First, the quantity being measured may exhibit significant variation. For example, in a materials study to determine fatigue life at a specified stress level, large differences in the number of cycles to failure are noted when a number of specimens are tested. This variation is inherent in the fatigue process and is observed in all fatigue life measurements. Second, the measuring system, which includes the transducer, signal conditioning equipment, A/D converter, recording instrument, and an operator may introduce error in the measurement. This error may be systematic or random, depending upon its source. An instrument operated out of calibration produces a systematic error, whereas, reading errors due to interpolation on a chart are random. The accumulation of random errors in a measuring system produces a variation that must be examined in relation to the magnitude of the quantity being measured.

The data obtained from repeated measurements represent an array of readings not an exact result. Maximum information can be extracted from such an array of readings by employing statistical methods. The first step in the statistical treatment of data is to establish the distribution. A graphical representation of the distribution is usually the most useful form for initial evaluation. Next, the statistical distribution is characterized with a measure of its central value, such as the mean, the median, or the mode. Finally, the spread or dispersion of the distribution is determined in terms of the variance or the standard deviation.

With elementary statistical methods, the experimentalist can reduce a large amount of data to a very compact and useful form by defining the type of distribution, establishing the single value that best represents the central value of the distribution (mean), and determining the variation from the mean value (standard deviation). Summarizing data in this manner is the most meaningful form of presentation for application to design problems or for communication to others who need the results of the experiments.

The treatment of statistical methods presented in this chapter is relatively brief; therefore, only the most commonly employed techniques for representing and interpreting data are presented. A formal course in statistics, which covers these techniques in much greater detail as well as many other useful techniques, should be included in the program of study of all engineering students.

20.2 CHARACTERIZING STATISTICAL DISTRIBUTIONS

For purposes of this discussion, consider that an experiment has been conducted n times to determine the ultimate tensile strength of a fully tempered beryllium copper alloy. The data obtained represent a sample of size n from an infinite population of all possible measurements that could have been made. The simplest way to present these data is to list the strength measurements in order of increasing magnitude, as shown in Table 20.1.

Table 20.1
The ultimate tensile strength of beryllium copper, listed in order of increasing magnitude

Sample number	Strength ksi (MPa)	Sample number	Strength ksi (MPa)
1	170.5 (1175)	21	176.2 (1215)
2	171.9 (1185)	22	176.2 (1215)
3	172.6 (1190)	23	176.4 (1217)
4	173.0 (1193)	24	176.6 (1218)
5	173.4 (1196)	25	176.7 (1219)
6	173.7 (1198)	26	176.9 (1220)
7	174.2 (1201)	27	176.9 (1220)
8	174.4 (1203)	28	177.2 (1222)
9	174.5 (1203)	29	177.3 (1223)
10	174.8 (1206)	30	177.4 (1223)
11	174.9 (1206)	31	177.7 (1226)
12	175.0 (1207)	32	177.8 (1226)
13	175.4 (1210)	33	178.0 (1228)
14	175.5 (1210)	34	178.1 (1228)
15	175.6 (1211)	35	178.3 (1230)
16	175.6 (1211)	36	178.4 (1230)
17	175.8 (1212)	37	179.0 (1236)
18	175.9 (1213)	38	179.7 (1239)
19	176.0 (1214)	39	180.1 (1242)
20	176.1 (1215)	40	181.6 (1252)

These data can be arranged into seven groups to give a frequency distribution as shown in Table 20.2. The advantage of representing data in a frequency distribution is that the central tendency is more clearly illustrated.

Table 20.2
Frequency distribution of ultimate tensile strength

Group intervals ksi (MPa)	Observations in the group	Relative frequency	Cumulative frequency
169.0-170.9 (1166-1178)	1	0.025	0.025
171.0-172.9 (1179-1192)	2	0.050	0.075
173.0-174.9 (1193-1206)	8	0.200	0.275
175.0-176.9 (1207-1220)	16	0.400	0.675
177.0-178.9 (1221-1234)	9	0.225	0.900
179.0-180.9 (1235-1248)	3	0.075	0.975
181.0-182.9 (1249-1261)	1	0.025	1.000
Total	40		

20.2.1 Graphical Representations of the Distribution

The shape of the distribution function representing the ultimate tensile strength of beryllium copper is indicated by the data groupings of Table 20.2. A graphical presentation of this group data, known as a **histogram**, is shown in Fig. 20.1. The histogram method of presentation shows the central tendency and variability of the distribution much more clearly than the tabular method of presentation of Table 20.2. Superimposed on the histogram is a curve showing the relative frequency of the occurrence of a group of measurements. Note that the points for the relative frequency are plotted at the midpoint of the group interval.

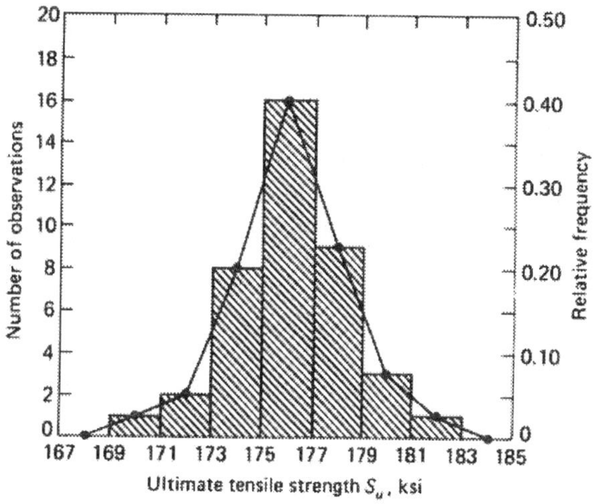

Figure 20.1 Histogram with superimposed relative-frequency diagram.

A cumulative frequency-diagram, shown in Fig. 20.2, is another way of representing the ultimate-strength data from the experiments. The cumulative frequency is the number of readings having a value less than a specified value of the quantity being measured (ultimate strength) divided by the total number of measurements. As indicated in Table 20.2, the cumulative frequency is the running sum of the relative frequencies. When the graph of cumulative frequency versus the quantity being measured is prepared, the end value for the group intervals is used to position the point along the abscissa.

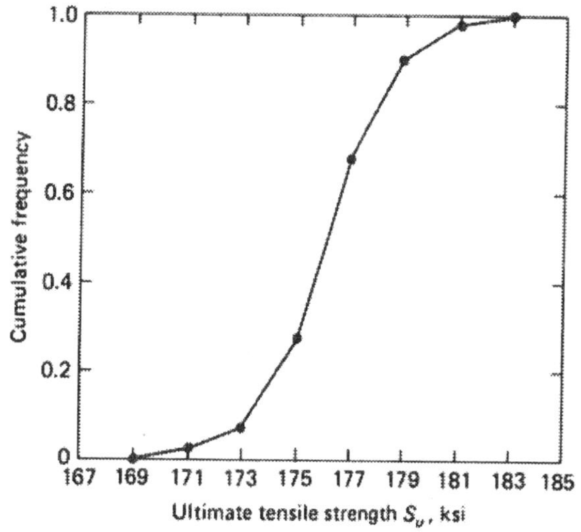

Figure 20.2 Cumulative frequency diagram.

20.2.2 Measures of Central Tendency

While histograms or frequency distributions are used to provide a visual representation of a distribution, numerical measures are used to define the characteristics of the distribution. One basic characteristic is the central tendency of the data. The most commonly employed measure of the central tendency of a distribution of data is the sample mean $\overline{x}$, which is defined as:

$$\overline{x} = \sum_{i=1}^{n} \frac{x_i}{n} \tag{20.1}$$

where x_i is the i th value of the quantity being measured and n is the total number of measurements.

Because of time and costs involved in conducting tests, the number of measurements is usually limited; therefore, the sample mean $\overline{x}$ is only an estimate of the true arithmetic mean μ of the population. It is shown later that $\bar{x}$ approaches μ as the number of measurements increases. The mean value of the ultimate-strength data presented in Table 20.1 is $\bar{x}$ = 176.1 ksi (1215 MPa).

The median and mode are also measures of central tendency. The median is the central value in a group of ordered data. For example, in an ordered set of 41 readings, the 21st reading represents the median value with 20 readings lower than the median and 20 readings higher than the median. In instances when an even number of readings are taken, the median is obtained by averaging the two middle values. For example, in an ordered set of 40 readings, the median is the average of the 20th and 21st readings. Thus, for the ultimate tensile strength data presented in Table 20.1, the median is ½ (176.1 + 176.2) = 176.15 ksi (1215 MPa).

The mode is the most frequent value of the data; therefore, it is located at the peak of the relative-frequency curve. In Fig. 20.1, the peak of the relative probability curve occurs at an ultimate tensile strength S_u = 176.0 ksi (1214 MPa); therefore, this value is the mode of the data set presented in Table 20.1.

It is evident that a typical set of data may give different values for the three measures of central tendency. There are two reasons for this difference. First, the population from which the samples were drawn may not be Gaussian where the three measures are expected to coincide. Second, even if the population is Gaussian, the number of measurements n is usually small and deviations due to a small sample size are to be expected.

20.2.3 Measures of Dispersion

It is possible for two different distributions of data to have the same mean but different dispersions, as shown in the relative-frequency diagrams of Fig. 20.3. Different measures of dispersion are the range, the mean deviation, the variance, and the standard deviation. The standard deviation S_x is the most popular and is defined as:

$$S_x = \left[\sum_{i=1}^{n} \frac{(x_i - \bar{x})^2}{n-1}\right]^{1/2} \qquad (20.2)$$

Because the sample size n is small, the standard deviation S_x of the sample represents an estimate of the true standard deviation σ of the population. Computation of S_x and $\bar{x}$ from a data sample is easily performed with most scientific type calculators.

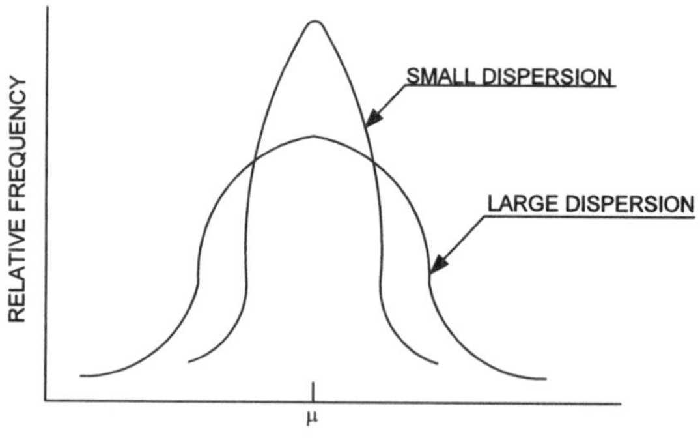

Figure 20.3 Relative frequency diagrams with large and small dispersions.

Expressions for the other measures of dispersion, namely, range R, mean deviation d_x, and variance S_x^2 are given by:

$$R = x_L - x_s \qquad (20.3)$$

$$d_x = \sum_{i=1}^{n} \frac{|x_i - \bar{x}|}{n} \qquad (20.4)$$

$$S_x^2 = \sum_{i=1}^{n} \frac{(x_i - \bar{x})^2}{n-1} \qquad (20.5)$$

where x_L is the largest value of the quantity in the distribution and x_s is the smallest value.

Equation (20.4) indicates that the deviation of each reading from the mean is determined and summed. The average of the n deviations is the mean deviation. The absolute value of the difference ($x_i - \bar{x}$) must be used in the summing process to avoid cancellation of positive and negative deviations. The variance of the population σ^2 is estimated by S_x^2 where the denominator (n – 1) in Eqs. (20.2) and (20.5) serves to reduce error introduced by approximating the true mean μ with the estimate of the mean $\bar{x}$. As the sample size n is increased the estimates of $\bar{x}$, S_x, and S_x^2 improve as shown in the discussion of Section 20.4. Variance is an important measure of dispersion because it is used in defining the normal distribution function.

Finally, a measure known as the coefficient of variation C_v is used to express the standard deviation S_x as a percentage of the mean $\bar{x}$. Thus:

$$C_v = \frac{S_x}{\bar{x}}(100) \tag{20.6}$$

The coefficient of variation represents a normalized parameter that indicates the variability of the data in relation to its mean.

20.3 STATISTICAL DISTRIBUTION FUNCTIONS

As the sample size is increased, it is possible in tabulating the data to increase the number of group intervals and to decrease their width. The corresponding relative-frequency diagram, similar to the one illustrated in Fig. 20.1, will approach a smooth curve (a theoretical distribution curve) known as a **distribution function.**

A number of different distribution functions are used in statistical analyses. The best-known and most widely used distribution in experimental mechanics is the Gaussian or normal distribution. This distribution is extremely important because it describes random errors in measurements and variations observed in strength determinations. Other useful distributions include binomial, exponential, hypergeometric, chi-square χ^2, F, Gumbel, Poisson, Student's t, and Weibull distributions. The reader is referred to references [1-5] for a complete description of these distributions. Emphasis here will be on Gaussian and Weibull distribution functions because of their wide range of application in experimental mechanics.

20.3.1 Gaussian Distribution

The Gaussian or normal distribution function, as represented by a normalized relative-frequency diagram, is shown in Fig. 20.4. The Gaussian distribution is completely defined by two parameters; the mean μ and the standard deviation σ. The equation for the relative frequency f in terms of these two parameters is given by:

$$f(z) = \frac{1}{\sqrt{2\pi}} e^{-\left(z^2/2\right)} \tag{20.7}$$

where

$$z = \frac{\bar{x} - \mu}{\sigma} \tag{20.8}$$

Experimental data (with finite sample sizes) can be analyzed to obtain $\bar{x}$ as an estimate of μ and S_x as an estimate of σ. This procedure permits the experimentalist to use data drawn from small samples to represent the entire population.

The method for predicting population properties from a Gaussian (normal) distribution function utilizes the normalized relative-frequency diagram shown in Fig. 20.4. The area A under the entire curve is given by Eq. (20.7) as:

$$A = \frac{1}{\sqrt{2\pi}} \int_{-\infty}^{\infty} e^{-\left(z^2/2\right)} dz = 1 \qquad (20.9)$$

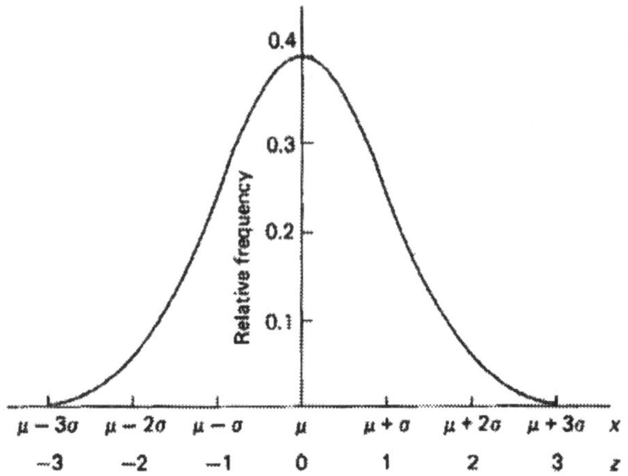

Figure 20.4 The normal or Gaussian distribution function.

Equation (20.9) implies that the population has a value z between $-\infty$ and $+\infty$ and that the probability of making a single observation from the population with a value $-\infty \leq z \leq +\infty$ is 100%. While the previous statement may appear trivial and obvious, it serves to illustrate the concept of using the area under the normalized relative-frequency curve to determine the probability p of observing a measurement within a specific interval. Figure 20.5 shows graphically, with the shaded area under the curve, the probability that a measurement will occur within the interval between z_1 and z_2. Thus, from Eq. (20.7) it is evident that:

$$p(z_1, z_2) = \int_{z_1}^{z_2} f(z)dz = \frac{1}{\sqrt{2\pi}} \int_{z_1}^{z_2} e^{-\left(z^2/2\right)} dz \qquad (20.10)$$

Evaluation of Eq. (20.10) is most easily accomplished by using tables that list the areas under the normalized relative-frequency curve as a function of z. Table 20.3 lists one-side areas between limits of $z_1 = 0$ and z_2 for the normal distribution function.

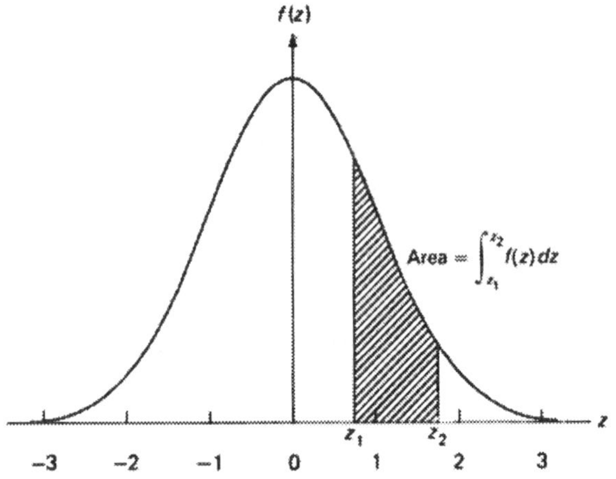

Figure 20.5 Probability of a measurement of x between limits of z_1 and z_2.
The total area under the curve f(z) is 1.

Because the distribution function is symmetric about z = 0, this one-sided table is sufficient for all evaluations of the probability. For example, A(−1,0) = A(0,+1) leads to the following determinations:

$$A(-1,+1) = p(-1,+1) = 0.3413 + 0.3413 = 0.6826$$
$$A(-2,+2) = p(-2,+2) = 0.4772 + 0.4772 = 0.9544$$

$$A(-3,+3) = p(-3,+3) = 0.49865 + 0.49865 = 0.9973$$
$$A(-1,+2) = p(-1,+2) = 0.3413 + 0.4772 = 0.8185$$

Table 20.3
Areas under the normal distribution curve from $z_1 = 0$ to z_2 (one side)

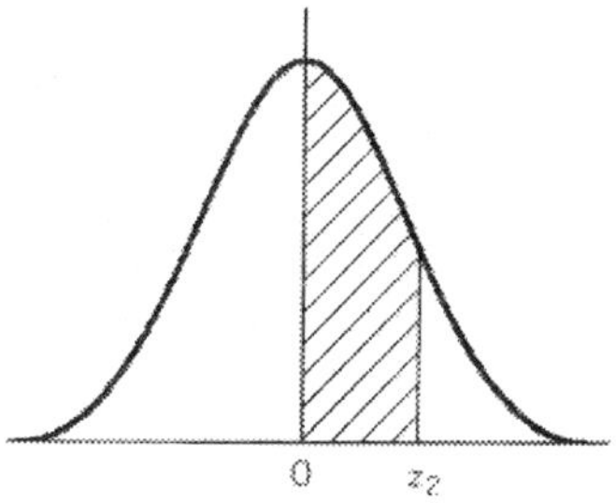

$z_2 = \frac{x-\bar{x}}{S_x}$	0.00	0.01	0.02	0.03	0.04	0.05	0.06	0.07	0.08	0.09
0.0	0.0000	0.0040	0.0080	0.0120	0.0160	0.0199	0.0239	0.0279	0.0319	0.0359
0.1	0.0398	0.0438	0.0478	0.0517	0.0557	0.0596	0.0636	0.0675	0.0714	0.0753
0.2	0.0793	0.0832	0.0871	0.0910	0.0948	0.0987	0.1026	0.1064	0.1103	0.1141
0.3	0.1179	0.1217	0.1255	0.1293	0.1331	0.1368	0.1406	0.1443	0.1480	0.1517
0.4	0.1554	0.1591	0.1628	0.1664	0.1700	0.1736	0.1772	0.1808	0.1844	0.1879
0.5	0.1915	0.1950	0.1985	0.2019	0.2054	0.2088	0.2123	0.2157	0.2190	0.2224
0.6	0.2257	0.2291	0.2324	0.2357	0.2389	0.2422	0.2454	0.2486	0.2517	0.2549
0.7	0.2580	0.2611	0.2642	0.2673	0.2704	0.2734	0.2764	0.2794	0.2823	0.2852
0.8	0.2881	0.2910	0.2939	0.2967	0.2995	0.3023	0.3051	0.3078	0.3106	0.3233
0.9	0.3159	0.3186	0.3212	0.3238	0.3264	0.3289	0.3315	0.3340	0.3365	0.3389
1.0	0.3413	0.3438	0.3461	0.3485	0.3508	0.3531	0.3554	0.3577	0.3599	0.3621
1.1	0.3643	0.3665	0.3686	0.3708	0.3729	0.3749	0.3770	0.3790	0.3810	0.3830
1.2	0.3849	0.3869	0.3888	0.3907	0.3925	0.3944	0.3962	0.3980	0.3997	0.4015
1.3	0.4032	0.4049	0.4066	0.4082	0.4099	0.4115	0.4131	0.4147	0.4162	0.4177
1.4	0.4192	0.4207	0.4222	0.4236	0.4251	0.4265	0.4279	0.4292	0.4306	0.4319
1.5	0.4332	0.4345	0.4357	0.4370	0.4382	0.4394	0.4406	0.4418	0.4429	0.4441
1.6	0.4452	0.4463	0.4474	0.4484	0.4495	0.4505	0.4515	0.4525	0.4535	0.4545
1.7	0.4554	0.4564	0.4573	0.4582	0.4591	0.4599	0.4608	0.4616	0.4625	0.4633
1.8	0.4641	0.4649	0.4656	0.4664	0.4671	0.4678	0.4686	0.4693	0.4699	0.4706
1.9	0.4713	0.4719	0.4726	0.4732	0.4738	0.4744	0.4750	0.4758	0.4761	0.4767
2.0	0.4772	0.4778	0.4783	0.4788	0.4793	0.4799	0.4803	0.4808	0.4812	0.4817
2.1	0.4821	0.4826	0.4830	0.4834	0.4838	0.4842	0.4846	0.4850	0.4854	0.4857
2.2	0.4861	0.4864	0.4868	0.4871	0.4875	0.4878	0.4881	0.4884	0.4887	0.4890
2.3	0.4893	0.4896	0.4898	0.4901	0.4904	0.4906	0.4909	0.4911	0.4913	0.4916
2.4	0.4918	0.4920	0.4922	0.4925	0.4927	0.4929	0.4931	0.4932	0.4934	0.4936
2.5	0.4938	0.4940	0.4941	0.4943	0.4945	0.4946	0.4948	0.4949	0.4951	0.4952
2.6	0.4953	0.4955	0.4956	0.4957	0.4959	0.4960	0.4961	0.4962	0.4963	0.4964
2.7	0.4965	0.4966	0.4967	0.4968	0.4969	0.4970	0.4971	0.4972	0.4973	0.4974
2.8	0.4974	0.4975	0.4976	0.4977	0.4977	0.4978	1.4979	0.4979	0.4980	0.4981
2.9	0.4981	0.4982	0.4982	0.4983	0.4984	0.4984	0.4985	0.4985	0.4986	0.4986
3.0	0.49865	0.4987	0.4987	0.4988	0.4988	0.4988	0.4989	0.4989	0.4989	0.4990

Because the normal distribution function has been well characterized, predictions can be made regarding the probability of a specific strength value or measurement error. For example, one may anticipate that 68.3% of the data will fall between limits of $\bar{x} \pm 1.0\ S_x$, 95.4% between limits of $\bar{x} \pm 2.0\ S_x$, and 99.7% between limits of $\bar{x} \pm 3.0\ S_x$. Also, 81.9% of the data should fall between limits of $\bar{x} - 1.0\ S_x$ and $\bar{x} + 2.0\ S_x$.

Table 20.4
Areas under the normal distribution curve from z_1 to $z_2 \Rightarrow \infty$ (one side)

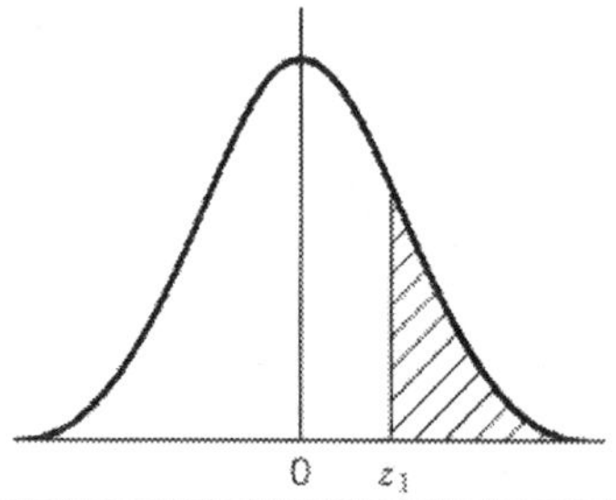

$z_1 = \frac{x - \bar{x}}{S_x}$	0.00	0.01	0.02	0.03	0.04	0.05	0.06	0.07	0.08	0.09
0.0	0.5000	0.4960	0.4920	0.4880	0.4840	0.4801	0.4761	0.4721	0.4681	0.4641
0.1	0.4602	0.4562	0.4522	0.4483	0.4443	0.4404	0.4364	0.4325	0.4286	0.4247
0.2	0.4207	0.4168	0.4129	0.4090	0.4052	0.4013	0.3974	0.3936	0.3897	0.3859
0.3	0.3821	0.3783	0.3745	0.3707	0.3669	0.3632	0.3594	0.3557	0.3520	0.3483
0.4	0.3446	0.3409	0.3372	0.3336	0.3300	0.3264	0.3228	0.3192	0.3156	0.3121
0.5	0.3085	0.3050	0.3015	0.2981	0.2946	0.2912	0.2877	0.2843	0.2810	0.2776
0.6	0.2743	0.2709	0.2676	0.2643	0.2611	0.2578	0.2546	0.2514	0.2483	0.2451
0.7	0.2430	0.2389	0.2358	0.2327	0.2296	0.2266	0.2236	0.2206	0.2177	0.2148
0.8	0.2119	0.2090	0.2061	0.2033	0.2005	0.1977	0.1949	0.1922	0.1894	0.1867
0.9	0.1841	0.1814	0.1788	0.1762	0.1736	0.1711	0.1685	0.1660	0.1635	0.1611
1.0	0.1587	0.1562	0.1539	0.1515	0.1492	0.1469	0.1446	0.1423	0.1401	0.1379
1.1	0.1357	0.1335	0.1314	0.1292	0.1271	0.1251	0.1230	0.1210	0.1190	0.1170
1.2	0.1151	0.1131	0.1112	0.1093	0.1075	0.1056	0.1038	0.1020	0.1003	0.0985
1.3	0.0968	0.0951	0.0934	0.0918	0.0901	0.0885	0.0869	0.0853	0.0838	0.0823
1.4	0.0808	0.0793	0.0778	0.0764	0.0749	0.0735	0.0721	0.0708	0.0694	0.0681
1.5	0.0668	0.0655	0.0643	0.0630	0.0618	0.0606	0.0594	0.0582	0.0571	0.0559
1.6	0.0548	0.0537	0.0526	0.0516	0.0505	0.0495	0.0485	0.0475	0.0465	0.0455
1.7	0.0446	0.0436	0.0427	0.0418	0.0409	0.0401	0.0392	0.0384	0.0375	0.0367
1.8	0.0359	0.0351	0.0344	0.0336	0.0329	0.0322	0.0314	0.0307	0.0301	0.0294
1.9	0.0287	0.0281	0.0274	0.0268	0.0262	0.0256	0.0250	0.0244	0.0239	0.0233
2.0	0.0228	0.0222	0.0217	0.0212	0.0207	0.0202	0.0197	0.0192	0.0188	0.0183
2.1	0.0179	0.0174	0.0170	0.0166	0.0162	0.0158	0.0154	0.0150	0.0146	0.0143
2.2	0.0139	0.0136	0.0132	0.0129	0.0125	0.0122	0.0119	0.0116	0.0113	0.0110
2.3	0.0107	0.0104	0.0102	0.00990	0.00964	0.00939	0.00914	0.00889	0.00866	0.00840
2.4	0.00820	0.00798	0.00776	0.00755	0.00734	0.00714	0.00695	0.00676	0.00657	0.00639
2.5	0.00621	0.00604	0.00587	0.00570	0.00554	0.00539	0.00523	0.00508	0.00494	0.00480
2.6	0.00466	0.00453	0.00440	0.00427	0.00415	0.00402	0.00391	0.00379	0.00368	0.00357
2.7	0.00347	0.00336	0.00326	0.00317	0.00307	0.00298	0.00288	0.00280	0.00272	0.00264
2.8	0.00256	0.00248	0.00240	0.00233	0.00226	0.00219	0.00212	0.00205	0.00199	0.00193
2.9	0.00187	0.00181	0.00175	0.00169	0.00164	0.00159	0.00154	0.00149	0.00144	0.00139

In many problems, the probability of a single sample exceeding a specified value z_2 must be determined. It is possible to determine this probability by using Table 20.3 together with the fact that the area under the entire curve is unity (A = 1); however, Table 20.4, which lists one-sided areas between limits of $z_1 = z$ and $z_2 \Rightarrow \infty$, yields the results more directly.

The use of Tables 20.3 and 20.4 can be illustrated by considering the ultimate-tensile-strength data presented in Table 20.1. By using Eqs. (20.1) and 20.2), it is easy to establish estimates for the mean $\bar{x}$ and standard deviation S_x as $\bar{x}$ = 176.1 ksi (1215 MPa) and S_x = 2.25 ksi (15.5 MPa). The values of $\bar{x}$ and S_x characterize the population from which the data of Table 20.1 were drawn. It is possible to establish the probability that the ultimate tensile strength of a single specimen drawn randomly from the population will be between specified limits (by using Table 20.3), or that the ultimate tensile strength of a single sample will not be above or below a specified value (by using Table 20.4). For example, one determines the probability that a single sample will exhibit an ultimate tensile strength between 175 and 178 ksi by computing z_1 and z_2 and using Table 20.3. Thus:

$$z_1 = \frac{175 - 176.1}{2.25} = -0.489 \qquad z_2 = \frac{178 - 176.1}{2.25} = 0.844$$

$$p(-0.489, 0.844) = A(-0.489, 0) + A(0, 0.844) = 0.1875 + 0.3006 = 0.4981$$

This simple calculation shows that the probability of obtaining an ultimate tensile strength between 175 and 178 ksi from a single specimen is 49.8%. The probability of the ultimate tensile strength of a single specimen being less than 173 ksi is determined by computing z_1 and using Table 20.4. Thus:

$$z_1 = \frac{173 - 176.1}{2.25} = -1.37$$

$$p(-\infty, -1.37) = A(-\infty, -1.37) = A(1.37, \infty) = 0.0853$$

Thus, the probability of drawing a single sample with an ultimate tensile strength less than 173 ksi is 8.5%.

20.3.2 Weibull Distribution

In investigations of the strength of materials due to brittle fracture, of crack-initiation toughness, or of fatigue life, researchers often find that the Weibull distribution provides a more suitable approach to the statistical analysis of the available data. The Weibull distribution function p(x) is defined as:

$$\begin{aligned} p(x) &= 1 - e^{-[(x - x_0)/b]^2} && \text{for } x > x_0 \\ p(x) &= 0 && \text{for } x < x_0 \end{aligned} \qquad (20.11)$$

where x_0, b, and m are the three parameters which define this distribution function In studies of strength, p(x) is taken as the probability of failure when a stress x is placed on the specimen. The parameter x_0 is the **zero strength** since p(x) = 0 for $x < x_0$. The constants b and m are known as the **scale parameter** and the **Weibull slope parameter (modulus)**, respectively.

Four Weibull distribution curves are presented in Fig. 20.6 for the case where x_0 = 3, b = 10, and m = 2, 5, 10, and 20. These curves illustrate two important features of the Weibull distribution. First, there is a threshold strength x_0 and if the applied stress is less than x_0, the probability of failure is zero. Second, the Weibull distribution curves are not symmetric, and the distortion in the S-shaped curves is controlled by the Weibull slope parameter m. Application of the Weibull distribution to predict failure rates of one percent or less of the population is particularly important in engineering projects where reliabilities of 99% or greater are required.

To utilize the Weibull distribution requires knowledge of the Weibull parameters. In experimental investigations, it is necessary to conduct experiments and obtain a relatively large data set to accurately determine x_0, b, and m. Consider as an illustration, Weibull's own work in statistically characterizing the fiber strength of Indian cotton. In this example, an unusually large sample (n = 3,000) was studied by measuring the load to fracture (in grams) for each fiber. The strength data obtained was placed in sequential order with the lowest value corresponding to k = 1 first and the largest value corresponding to k = 3,000 last. The probability of failure p(x) at a load x is then determined from:

$$p = \frac{k}{n+1} \tag{20.12}$$

where k is the order number of the sequenced data and n is the total sample size.

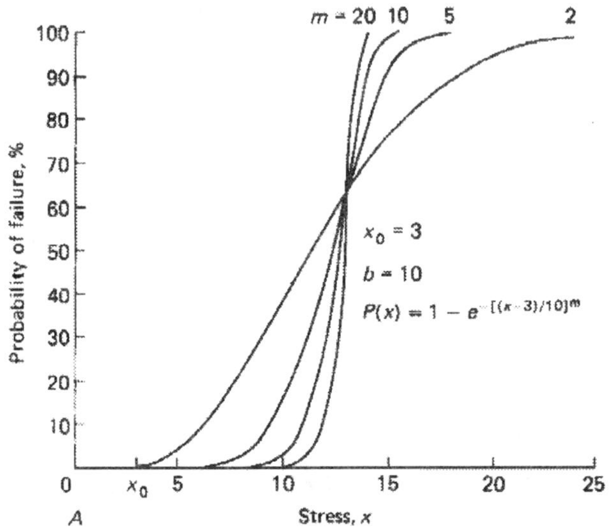

Figure 20.6 The Weibull distribution function.

At this stage it is possible to prepare a graph of probability of failure P(x) as a function of strength x to obtain a curve similar to that shown in Fig. 20.6. However, to determine the Weibull parameters x_0, b, and m requires additional conditioning of the data. From Eq. 20.11, it is evident that:

$$e^{[(x-x_o)/b]^m} = [1 - p(x)]^{-1} \tag{20.13}$$

Taking the natural log of both sides of Eq. (20.13) yields:

$$\left[\frac{(x-x_o)}{b}\right]^m = \ln[1-p(x)]^{-1} \tag{20.14}$$

Taking $\log_{10}$ of both sides of Eq. (20.14) gives a relation for the slope parameter m. Thus:

$$m = \frac{\log_{10} \ln[1-p(x)]^{-1}}{\log_{10}(x-x_o) - \log_{10} b} \tag{20.15}$$

The numerator of Eq. (20.15) is the reduced variate $y = \log_{10} \ln[1 - p(x)]^{-1}$ used for the ordinate in preparing a graph of the conditioned data as indicated in Fig. 20.7. Note that y is a function of p alone and for this reason both the p and y scales can be displayed on the ordinates (see Fig. 20.7). The lead term in the denominator of Eq. (20.15) is the reduced variate $x = \log_{10} (x - x_0)$ used for the abscissa in Fig. (20.7).

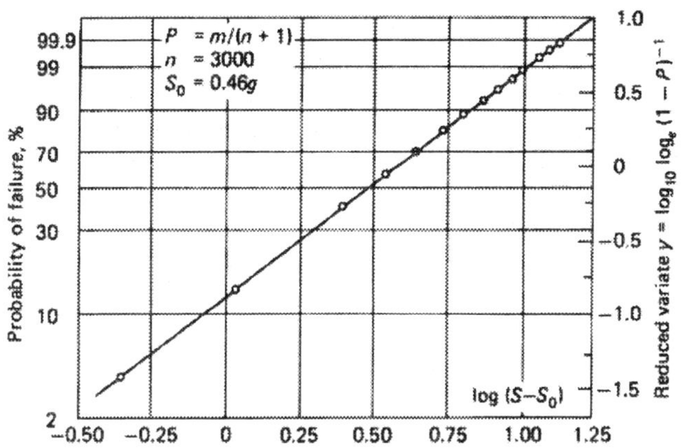

Figure 20.7 Fiber strength of Indian cotton shown in graphical format with Weibull's reduced variate (from data by Weibull).

In the Weibull example, the threshold strength x_0 was adjusted to 0.46 grams so that the data would fall on a straight line when plotted against the reduced x and y variates. The constant b is determined from the condition that:

$$\log_{10} b = \log_{10}(x - x_0) \qquad \text{when } y = 0 \tag{20.16}$$

Note from Fig. 20.7 that y = 0 when $\log_{10}(x - x_0) = 0.54$ which gives b = 0.54. Finally m is given by the slope of the straight line when the data is plotted in terms of the reduced variates x and y. In this example problem, m = 1.48.

20.4 CONFIDENCE INTERVALS FOR PREDICTIONS

When experimental data are represented with a normal distribution by using estimates of the mean $\bar{x}$ and standard deviation S_x and predictions are made about the occurrence of certain measurements, questions arise concerning the confidence that can be placed on either the estimates or the predictions. One cannot be totally confident in the predictions or estimates because of the effects of sampling error. Sampling error can be illustrated by drawing a series of samples (each containing n measurements) from the same population and determining several estimates of the mean $\bar{x}_1$, $\bar{x}_2$, $\bar{x}_3$, ... A variation in $\bar{x}$ will occur, but fortunately, this variation can also be characterized by a normal distribution function, as shown in Fig. 20.8. The mean of the x and $\bar{x}$ distributions is the same; however, the standard deviation of the $\bar{x}$ distribution $S_{\bar{x}}$ (sometimes referred to as the **standard error**) is less than S_x because:

$$S_{\bar{x}} = \frac{S_x}{\sqrt{n}} \tag{20.17}$$

When the standard deviation of the population of $\bar{x}$'s is known, it is possible to place confidence limits on the determination of the true population mean μ from a sample of size n, provided n is large (n > 25). The confidence interval within which the true population mean μ is located is given by the expression:

$$(\bar{x} - z\,S_{\bar{x}}) < \mu < [\bar{x} + z\,S_{\bar{x}}] \tag{20.18}$$

where $\bar{x} - zS_{\bar{x}}$ is the lower confidence limit and $\bar{x} + z\,S_{\bar{x}}$ is the upper confidence limit.

The width of the confidence interval depends upon the confidence level required. For instance, if z = 3 in Eq. (20.18), a relatively wide confidence interval exists; therefore, the probability that the population mean μ will be located within the confidence interval is high (99.7%). As the width of the confidence interval decreases, the probability that the population mean μ will fall within the interval decreases. Commonly used confidence levels and their associated intervals are shown in Table 20.5.

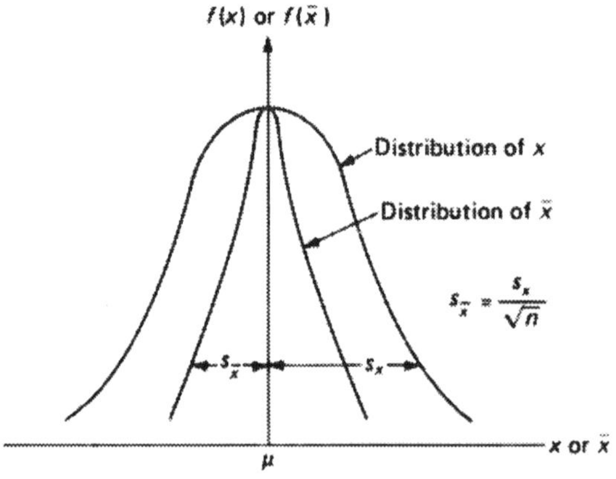

Figure 20.8 Normal Distribution of individual measurements of the quantity x and of the mean $\overline{x}$ from samples of size n.

Table 20.5
Confidence interval variation with confidence level interval = $\overline{x} + z\, S_{\overline{x}}$

Confidence level,%	z	Confidence level,%	z
99.9	3.30	90.0	1.65
99.7	3.00	80.0	1.28
99.0	2.57	68.3	1.00
95.0	1.96	60.0	0.84

When the sample size is very small ($n < 20$), the standard deviation S_x does not provide a reliable estimate of the standard deviation μ of the population and Eq. (20.18) should not be employed. The bias introduced by small sample size can be removed by modifying Eq. (20.18) to read as:

$$(\overline{x} - t(a)\, S_{\overline{x}}) < \mu < [\overline{x} + t(\alpha)\, S_{\overline{x}}] \qquad (20.19)$$

where $t(\alpha)$ is the statistic known as **Student's t**, and α is the level of significance (the probability of exceeding a given value of t).

The Student t distribution is defined by a relative frequency equation f(t), which can be expressed as:

$$f(t) = F_0 \left(1 + \frac{t^2}{\nu}\right)^{(\nu+1)/2} \qquad (20.20)$$

where F_0 is the relative frequency at $t = 0$ required to make the total area under the f(t) curve equal to unity and ν is the number of degrees of freedom.

The distribution function f(t) is shown in Fig. 20.9 for several different degrees of freedom ν. The degrees of freedom equal the number of independent measurements employed in the determination. It is evident that as ν becomes large, Student's t distribution approaches the normal distribution. One-side areas for the t distribution are listed in Table 20.6 and illustrated in Fig. 20.10.

The term $t(\alpha) S_{\overline{x}}$ in Eq. (20.19) represents the measure from the estimated mean $\overline{x}$ to one or the other of the confidence limits. This term may be used to estimate the sample size required to produce an estimate of the mean $\overline{x}$ with a specified reliability. Noting that one-half the band width of the confidence interval is $\delta = t(\alpha) S_{\overline{x}}$ and using Eq. (20.17), it is apparent that the sample size is given by:

$$n = \left[\frac{t(\alpha) S_{\overline{x}}}{\delta} \right]^2 \tag{20.21}$$

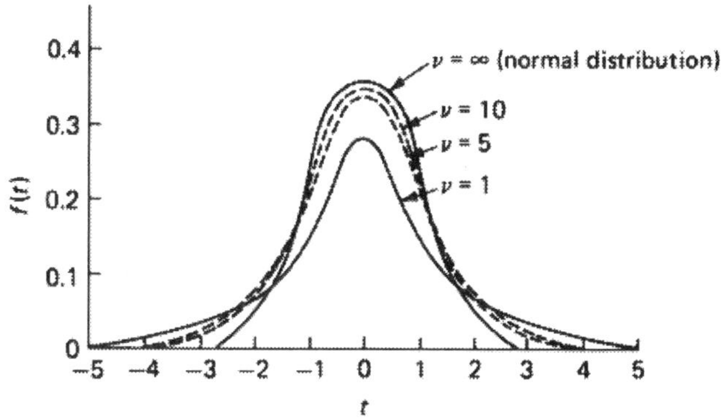

Figure 20.9 Student's t distribution for several degrees of freedom ν.

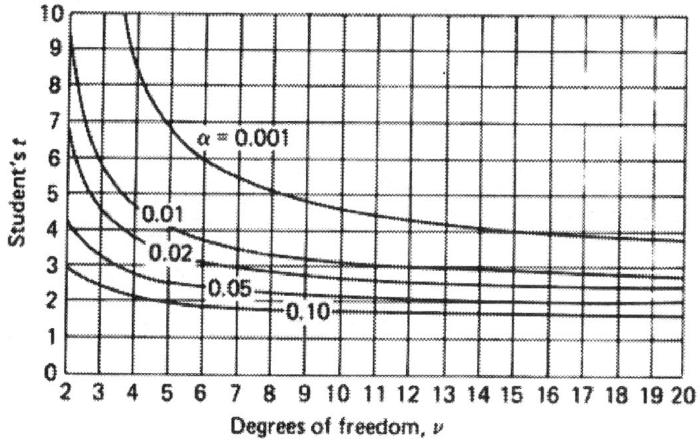

Figure 20.10 Student's t statistic as a function of the degrees of freedom ν with α as the probability of exceeding t as a parameter.

The use of Eq. (20.21) can be illustrated by considering the data in Table 20.1, where S_x = 2.25 ksi and $\overline{x}$ = 176.1 ksi. If this estimate of μ is to be accurate to ± 1% with a reliability of 99% then:

$$\delta = (0.01)(176.1) = 1.76 \text{ ksi}$$

Since t(α) depends on n, a trial-and-error solution is needed to establish the sample size n needed to satisfy the specifications. For the data of Table 20.1, n = 40: therefore ν = 39 and t(α) = t(0.995) = 2.71 from Table 20.6. The value t(α) = t(0.995) is used since 0.5% of the distribution must be excluded on each end of the curve to give a two-sided area corresponding to a reliability of 99%. Substituting into Eq. (20.21) yields:

$$n = \left[\frac{2.71(2.25)}{1.76} \right]^2 = 12.00$$

This result indicates that a much smaller sample than 40 will be sufficient. Next try n = 12, ν = 11, and t(α) = 3.11; then:

$$n = \left[\frac{3.11(2.25)}{1.76}\right]^2 = 15.80$$

Finally, with n = 15, ν = 14, and t(α) = 2.98; then:

Table 20.6
Student's t distribution for ν degrees of freedom showing t(α) as a function of area A (one side)

	Confidence level, α									
ν	0.995	0.99	0.975	0.95	0.90	0.80	0.75	0.70	0.60	0.55
1	63.66	31.82	12.71	6.31	3.08	1.376	1.000	0.727	0.325	0.158
2	9.92	6.96	4.30	2.92	1.89	1.061	0.816	0.617	0.289	0.142
3	5.84	4.54	3.18	2.35	1.64	0.978	0.765	0.584	0.277	0.137
4	4.60	3.75	2.78	2.13	1.53	0.941	0.741	0.569	0.271	0.134
5	4.03	3.36	2.57	2.02	1.48	0.920	0.727	0.559	0.267	0.132
6	3.71	3.14	2.45	1.94	1.44	0.906	0.718	0.553	0.265	0.131
7	3.50	3.00	2.36	1.90	1.42	0.896	0.711	0.549	0.263	0.130
8	3.36	2.90	2.31	1.86	1.40	0.889	0.706	0.546	0.262	0.130
9	3.25	2.82	2.26	1.83	1.38	0.883	0.703	0.543	0.261	0.129
10	3.17	2.76	2.23	1.81	1.37	0.879	0.700	0.542	0.260	0.129
11	3.11	2.72	2.20	1.80	1.36	0.876	0.697	0.540	0.260	0.129
12	3.06	2.68	2.18	1.78	1.36	0.873	0.695	0.539	0.259	0.128
13	3.01	2.65	2.16	1.77	1.35	0.870	0.694	0.538	0.259	0.128
14	2.98	2.62	2.14	1.76	1.34	0.868	0.692	0.537	0.258	0.128
15	2.95	2.60	2.13	1.75	1.34	0.866	0.691	0.536	0.258	0.128
16	2.92	2.58	2.12	1.75	1.34	0.865	0.690	0.535	0.258	0.128
17	2.90	2.57	2.11	1.74	1.33	0.863	0.689	0.534	0.257	0.128
18	2.88	2.55	2.10	1.73	1.33	0.862	0.688	0.534	0.257	0.127
19	2.86	2.54	2.09	1.73	1.33	0.861	0.688	0.533	0.257	0.127
20	2.84	2.53	2.09	1.72	1.32	0.860	0.687	0.533	0.257	0.127
21	2.83	2.52	2.08	1.72	1.32	0.859	0.686	0.532	0.257	0.127
22	2.82	2.51	2.07	1.72	1.32	0.858	0.686	0.532	0.256	0.127
23	2.81	2.50	2.07	1.71	1.32	0.858	0.685	0.532	0.256	0.127
24	2.80	2.49	2.06	1.71	1.32	0.857	0.685	0.531	0.256	0.127
25	2.79	2.48	2.06	1.71	1.32	0.856	0.684	0.531	0.256	0.127
26	2.78	2.48	2.06	1.71	1.32	0.856	0.684	0.531	0.256	0.127
27	2.77	2.47	2.05	1.70	1.31	0.855	0.684	0.531	0.256	0.127
28	2.76	2.47	2.05	1.70	1.31	0.855	0.683	0.530	0.256	0.127
29	2.76	2.46	2.04	1.70	1.31	0.854	0.683	0.530	0.256	0.127
30	2.75	2.46	2.04	1.70	1.31	0.854	0.683	0.530	0.256	0.127
40	2.70	2.42	2.02	1.68	1.30	0.851	0.681	0.529	0.255	0.126
60	2.66	2.39	2.00	1.67	1.30	0.848	0.679	0.527	0.254	0.126
120	2.62	2.36	1.98	1.66	1.29	0.845	0.677	0.526	0.254	0.126
∞	2.58	2.33	1.96	1.65	1.28	0.842	0.674	0.524	0.253	0.126

$$n = \left[\frac{2.98(2.25)}{1.76}\right]^2 = 14.50$$

Thus, a sample size of 15 would be sufficient to ensure an accuracy of ± 1% with a confidence level of 99%. The sample size of 40 listed in Table 20.1 is too large for the degree of accuracy and confidence level specified. This simple example illustrates how sample size can be reduced and cost savings affected by using statistical methods.

20.5 COMPARISON OF MEANS

Because the Student's t distribution compensates for the effect of small sample bias and converges to the normal distribution in large samples, it is a very useful statistic in engineering applications. A second important application utilizes the t distribution as the basis for a test to determine if the difference between two means is significant or due to random variation. For example, consider the yield strength of a steel determined with a sample size of n_1 = 20 which gives $\bar{x}_1$ = 78.4 ksi, and $S_{\bar{x}1}$ = 6.04 ksi. Suppose now that a second sample from another supplier is tested to determine the yield strength and the results are n_2 = 25, $\bar{x}_1$ = 81.6 ksi, and $S_{\bar{x}2}$ = 5.56 ksi. Is the steel from the second supplier superior in terms of yield strength? The standard deviation of the difference in means $S_{(\bar{x}2-\bar{x}1)}$ can be expressed as:

$$S^2_{(\bar{x}2-\bar{x}1)} = S_p^2\left(\frac{1}{n_1} + \frac{1}{n_2}\right) = S_p^2\,\frac{n_1 + n_2}{n_1 n_2} \tag{20.22}$$

where S_p^2 is the pooled variance that can be expressed as:

$$S_p^2 = \frac{(n_1 - 1)S^2_{x1} + (n_2 - 1)S^2_{x2}}{n_1 + n_2 - 2} \tag{20.23}$$

The statistic t can be computed from the expression:

$$t = \frac{|\bar{x}_2 - \bar{x}_1|}{S_{(\bar{x}2-\bar{x}1)}} \tag{20.24}$$

A comparison of the value of t determined from Eq. (20.24) with a value of t(α) obtained from Table 20.6 provides a statistical basis for deciding whether the difference in means is real or due to random variations. The value of t(α) to be used depends upon the degrees of freedom $\nu = n_1 + n_2 - 2$ and the level of significance required. Levels of significance commonly employed are 5% and 1%. The 5% level of significance means that the probability of a random variation being taken for a real difference is only 5%. Comparisons at the 1% level of significance are 99% certain; however, in such a strong test, real differences can often be attributed to random error.

In the example being considered, Eq. (20.23) yields S_p^2 = 33.37 ksi, Eq. (20.22) yields $S^2_{(\bar{x}2-\bar{x}1)}$ = 3.00 ksi, and Eq. (20.24) yields t = 1.848. For a 5% level of significance test with ν = 43 and α = 0.05 (the comparison is one-sided, since the t test is for superiority), Table 20.6 indicates that t(α) = 1.68. Because t > t(α), it can be concluded with a 95% level of confidence that the yield strength of the steel from the second supplier was higher than the yield strength of steel from the first supplier.

20.6 STATISTICAL SAFETY FACTOR

In experimental mechanics, it is often necessary to determine the stresses acting on a component and its strength in order to predict whether failure will occur or if the component is safe. The prediction can be a difficult if both the stress σ_{ij} , and the strength S_y, are variables since failure will occur only in the region of overlap of the two distribution functions as shown in Fig. 20.11.

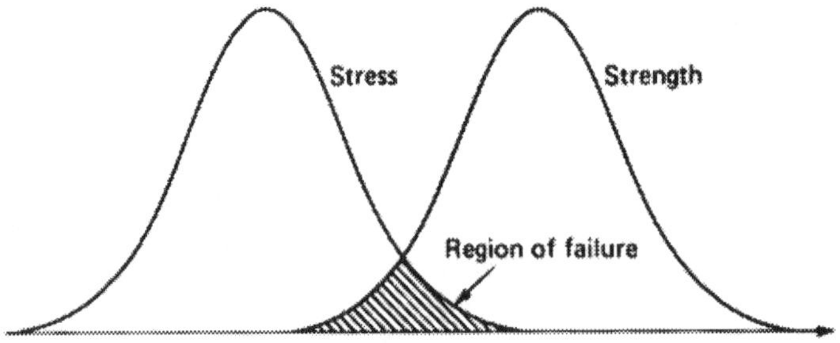

Figure 20.11 Superimposed normal distribution curves for strength and stress showing the region of failure.

To determine the probability of failure, or conversely, the reliability, the statistic z_R is computed by using the equation:

$$z_R = \frac{\bar{x}_S - \bar{x}_\sigma}{S_{S-\sigma}} \tag{20.25}$$

where

$$S_{S-\sigma} = \sqrt{S_S^2 - S_\sigma^2} \tag{20.26}$$

and the subscripts S and σ refer to strength and stress, respectively.

The reliability associated with the value of z_R determined from Eq. (20.25) may be determined from a table showing the area $A(z_R)$ under a standard normal distribution curve by using:

$$R = 0.5 + A(z_R) \tag{20.27}$$

Typical values of R as a function of the statistic z_R are given in Table 20.7.

Table 20.7
Reliability R as a function of the statistic z_R

R, %	z_R	R, %	z_R
50	0	99.9	3.091
90	1.288	99.99	3.719
95	1.645	99.999	4.265
99	2.326	99.9999	4.753

The reliability determined in this manner incorporates a safety factor of 1. If a safety factor of N is to be specified together with a reliability, then Eqs. (20.25) and (20.26) are rewritten to give a modified relation for z_R as:

$$z_R = \frac{\bar{x}_S - N\,\bar{x}_\sigma}{\sqrt{S_S^2 + S_\sigma^2}} \tag{20.28}$$

this can be rearranged to give the safety factor N as:

$$N = \frac{1}{\overline{x}_\sigma}\left(\overline{x}_S - z_R\sqrt{S_S^2 + S_\sigma^2}\right) \tag{20.29}$$

20.7 STATISTICAL CONDITIONING OF DATA

Previously it was indicated that measurement error can be characterized by a normal distribution function and that the standard deviation of the estimated mean $S_{\overline{x}}$ can be reduced by increasing the number of measurements. In most situations, sampling cost places an upper limit on the number of measurements to be made. Also, it must be remembered that systematic error is not a random variable; therefore, statistical procedures cannot serve as a substitute for precise accurately calibrated, and properly zeroed measuring instruments.

One area where statistical procedures can be used very effectively to condition experimental data is with the erroneous data point resulting from a measuring or recording mistake. Often, this data point appears questionable when compared with the other data collected, and the experimentalist must decide whether the deviation of the data point is due to a mistake (hence to be rejected) or due to some unusual but real condition (hence to be retained). A statistical procedure known as **Chauvenet's criterion** provides a consistent basis for making the decision to reject or retain such a point from a sample containing several readings.

Application of Chauvenet's criterion requires computation of a deviation ratio DR for each data point, followed by comparison with a standard deviation ratio DR_0. The standard deviation ratio DR_0 is a statistic that depends on the number of measurements, while the deviation ratio DR for a point is defined as:

$$DR = \frac{x_i - \overline{x}}{S_x} \tag{20.30}$$

The data point is rejected when $DR > DR_0$ and retained when $DR \leq DR_0$. Values for the standard deviation ratio DR_0 are listed in Table 20.8.

Table 20.8
Deviation ratio DR_0 used for statistical conditioning of data

Number of measurements n	Deviation ratio DR_0	Number of measurements n	Deviation ratio DR_0
4	1.54	25	2.33
5	1.65	50	2.57
7	1.80	100	2.81
10	1.96	300	3.14
15	2.13	500	3.29

If the statistical test of Eq. (20.30) indicates that a single data point in a sequence of n data points should be rejected, then the data point should be removed from the sequence and the mean $\overline{x}$ and the standard deviation S_x should be recalculated. Chauvenet's method can be applied only once to reject a data point that is questionable from a sequence of points. If several data points indicate that $DR > DR_0$, then it is likely that the instrumentation system is inadequate or that the process being investigated is extremely variable.

20.8 REGRESSION ANALYSIS

Many experiments involve the measurement of one dependent variable, say y, which may depend upon one or more independent variables, $x_1, x_2, \ldots\ldots, x_k$. Regression analysis provides a statistical approach for conditioning the data obtained from experiments where two or more related quantities are measured.

20.8.1 Linear Regression Analysis

Suppose measurements are made of two quantities that describe the behavior of a process exhibiting variation. Let y be the dependent variable and x the independent variable. Because the process exhibits variation, there is not a unique relationship between x and y and the data, when plotted, exhibit scatter, as illustrated in Fig. 20.12. Frequently, the relation between x and y that most closely represents the data, even with the scatter, is a linear function. Thus:

$$Y_i = mx_i + b \tag{20.31}$$

where Y_i is the predicted value of the dependent variable y_i for a given value of the independent variable x_i. A statistical procedure used to fit a straight line through scattered data points is called the least-squares method. With the least-squares method, the slope m and the intercept b in Eq. (20.31) are selected to minimize the sum of the squared deviations of the data points from the straight line shown in Fig. 20.12. In utilizing the least-squares method, it is assumed that the independent variable x is free of measurement error and the quantity

$$\Delta^2 = \sum (y_i - Y_i)^2 \tag{20.32}$$

is minimized at fixed values of x. After substituting Eq. (20.31) into Eq. (20.32), the minimization process of Δ^2 implies that:

$$\begin{aligned} \frac{\partial \Delta^2}{\partial b} &= \frac{\partial}{\partial b}\sum (y_i - mx - b)^2 = 0 \\ \frac{\partial \Delta^2}{\partial m} &= \frac{\partial}{\partial m}\sum (y_i - mx - b)^2 = 0 \end{aligned} \tag{a}$$

Differentiating yields:

$$\begin{aligned} 2\sum (y_i - mx - b)(-x) &= 0 \\ 2\sum (y_i - mx - b)(-1) &= 0 \end{aligned} \tag{b}$$

Solving Eqs. (b) for m and b yields:

$$m = \frac{\sum x \sum y - n\sum xy}{\left(\sum x\right)^2 - n\sum x^2} \quad \text{and} \quad b = \frac{\sum y - m\sum x}{n} \tag{20.33}$$

where n is the number of data points. The slope m and intercept b define a straight line through the scattered data points such that Δ^2 is minimized.

In any regression analysis it is important to establish the correlation between x and y. Equation (20.31) does not predict the exact values that were measured, because of the variation in the process. To illustrate, assume that the independent quantity x is fixed at a value x_1 and that a sequence of measurements is made of the dependent quantity y. The data obtained would give a distribution of y, as illustrated in Fig. 20.13. The dispersion of the distribution of y is a measure of the correlation. When the dispersion is small, the correlation is good and the regression analysis is effective in describing the variation in y. If the dispersion is large, the correlation is poor and the regression analysis may not be adequate to describe the variation in y.

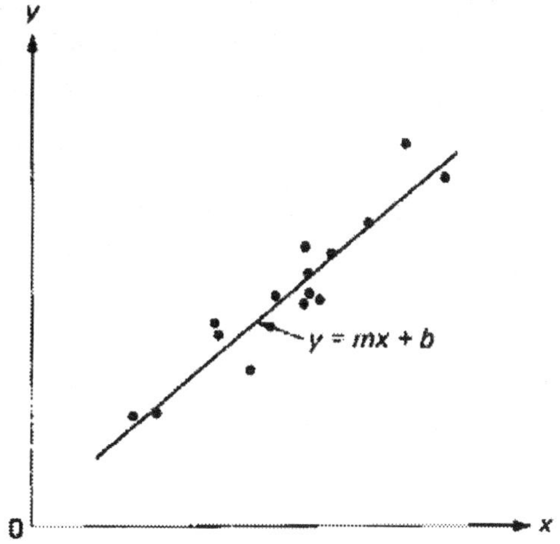

Figure 20.12 Linear regression analysis is used to fit a least squares line through scattered data points.

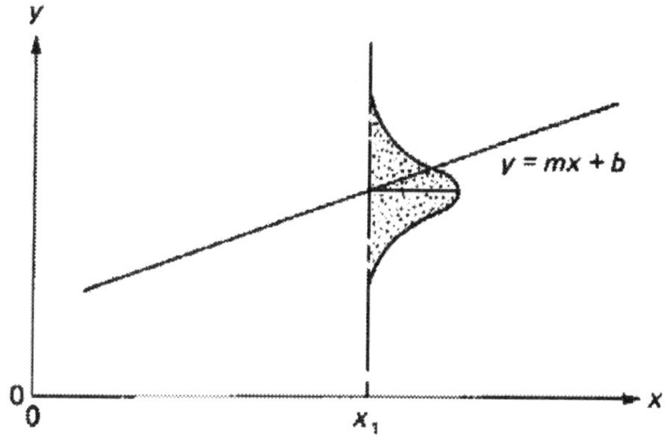

Figure 20.13 Distribution of y at a fixed value of x super-imposed on the linear-regression graph.

The adequacy of regression analysis can be evaluated by determining a correlation coefficient R^2 that is given by the following expression:

$$R^2 = 1 - \frac{n-1}{n-2}\left[\frac{\{y^2\} - m\{xy\}}{\{y^2\}}\right] \tag{20.34}$$

where $\{y^2\} = \Sigma y^2 - (\Sigma y)^2/n$) and $\{xy\} = \Sigma xy - (\Sigma x)(\Sigma y)/n$

When the value of the correlation coefficient $R^2 = 1$, perfect correlation exists between y and x. If R^2 equals zero, no correlation exists and the variations observed in y are due to random fluctuations and not changes in x. Because random variations in y exist, a value of $R^2 = 1$ is not obtained even if y(x) is linear. To interpret correlation coefficients $0 < R^2 < 1$, the data in Table 20.9 is used to establish the probability of obtaining a given R^2 due to random variations in y.

Table 20.9
Probability of obtaining a correlation coefficient R^2 due to random variations in y.

	Probability			
n	0.10	0.05	0.02	0.01
5	0.805	0.878	0.934	0.959
6	0.729	0.811	0.882	0.917
7	0.669	0.754	0.833	0.874
8	0.621	0.707	0.789	0.834
10	0.549	0.632	0.716	0.765
15	0.441	0.514	0.592	0.641
20	0.378	0.444	0.516	0.561
30	0.307	0.362	0.423	0.464
40	0.264	0.312	0.367	0.403
60	0.219	0.259	0.306	0.337
80	0.188	0.223	0.263	0.291
100	0.168	0.199	0.235	0.259

As an example, consider a regression analysis with n = 15 which gives $R^2 = 0.65$ as determined by Eq. 20.34. Reference to Table 20.9 indicates that the probability of obtaining $R^2 = 0.65$ due to random variations is slightly less than one percent. Thus one can be 99% certain that the regression analysis represents a true correlation between y and x.

20.8.2 Multi-Variate Regression

Many experiments involve measurements of a dependent variable y, which depends upon several independent variables x_1, x_2, x_3, ..., etc. It is possible to represent y as a function of x_1, x_2, x_3, ..., by employing the multi-variate regression equation:

$$Y_i = a + b_1x_1 + b_2x_2 + \ldots + b_kx_k \tag{20.35}$$

The regression coefficients a, b_1, b_2, ..., b_k are determined by using the method of least squares in a manner similar to that employed for linear regression analysis where the quantity $\Delta^2 = \Sigma(y_i - Y_i)^2$ is minimized. Substituting Eq. (20.32) into Eq. (20.35) yields:

$$\Delta^2 = \Sigma(y_i - a - b_1x_1 - b_2x_2 - \ldots - b_kx_k)^2 \tag{20.36}$$

Differentiating yields:

$$\begin{aligned}
\frac{\partial \Delta^2}{\partial a} &= 2\left[\sum (y_i - a - b_1x_1 - b_2x_2 - \ldots - b_kx_k)(-1)\right] = 0 \\
\frac{\partial \Delta^2}{\partial b_1} &= 2\left[\sum (y_i - a - b_1x_1 - b_2x_2 - \ldots - b_kx_k)(-x_1)\right] = 0 \\
\frac{\partial \Delta^2}{\partial b_2} &= 2\left[\sum (y_i - a - b_1x_1 - b_2x_2 - \ldots - b_kx_k)(-x_2)\right] = 0 \\
&\ldots\ldots\ldots\ldots\ldots\ldots\ldots\ldots\ldots\ldots \\
\frac{\partial \Delta^2}{\partial b_k} &= 2\left[\sum (y_i - a - b_1x_1 - b_2x_2 - \ldots - b_kx_k)(-x_k)\right] = 0
\end{aligned} \tag{20.37}$$

Equations (20.37) lead to the following set of k + 1 equations, which can be solved for the unknown regression coefficients a, b_1, b_2, ..., b_k.

$$\begin{aligned}
an + b_1\sum x_1 + b_2\sum x_2 + \ldots\ldots b_k\sum x_k &= \sum y_i \\
a\sum x_1 + b_1\sum x_1^2 + b_2\sum x_1x_2 + \ldots\ldots b_k\sum x_1x_k &= \sum y_ix_1 \\
a\sum x_2 + b_1\sum x_1x_2 + b_2\sum x_2^2 + \ldots\ldots b_k\sum x_2x_k &= \sum y_ix_2 \\
\ldots\ldots\ldots\ldots\ldots\ldots\ldots\ldots\ldots\ldots\ldots\ldots \\
a\sum x_k + b_1\sum x_1x_k + b_2\sum x_2x_k + \ldots\ldots b_k\sum x_k^2 &= \sum y_ix_k
\end{aligned} \tag{20.38}$$

The correlation coefficient R^2 is again used to determine the degree of association between the dependent and independent variables. For multiple regression equations, the correlation coefficient R^2 is given as:

$$R^2 = 1 - \frac{n-1}{n-k}\left[\frac{\{y^2\} - b_1\{yx_1\} - \{yx_2\} - \ldots\ldots - \{yx_k\}}{\{y^2\}}\right] \tag{20.39}$$

where

$$\{yx_k\} = \sum yx_k - \frac{\left(\sum y\right)\left(\sum x_k\right)}{n} \quad \text{and} \quad \{y^2\} = \sum y^2 - \frac{\left(\sum y\right)^2}{n}$$

This analysis is for linear, noninteracting, independent variables; however, the analysis can be extended to include cases where the regression equations would have higher-order and cross-product terms. The nonlinear terms can enter the regression equation in an additive manner and are treated as extra variables. With well-established computer routines for regression analysis, the set of (k + 1) simultaneous equations given by Eqs. (20.38) can be solved quickly and inexpensively. No significant difficulties are encountered in adding extra terms to account for nonlinearities and interactions.

20.8.3 Field Applications of Least-Square Methods

The least-squares method is an important mathematical process used in regression analysis to obtain regression coefficients. Sanford showed that the least-squares method could be extended to field analysis of data obtained with optical techniques (photoelasticity, moiré, holography, etc.). With these optical methods, a fringe order N, related to a field quantity such as stress, strain, or displacement, can be measured at a large number of points over a field (x, y). The applications require an analytical representation of the field quantities as a function of position (x, y) over the field. Several important problems including calibration, fracture mechanics, and contact stresses have analytical solutions where coefficients in the governing equations require experimental data for complete evaluation. Two examples will be described which introduce both the linear and nonlinear least-squares method applied over a field (x, y).

Linear Least-Squares Method

Consider a calibration model in photoelasticity and write the equation for the fringe order N(x, y) as:

$$N(x, y) = \frac{h}{f_\sigma}G(x, y) + E(x, y) \tag{a}$$

where G(x,y) is the analytical representation of the difference of the principal stresses ($\sigma_1 - \sigma_2$) in the calibration model, h is the model thickness, f_σ is the material fringe value and E(x,y) is the residual birefringence.

Assume a linear distribution for E(x,y) which can be expressed as:

$$E(x,y) = Ax + By + C \tag{b}$$

For any selected point (x_k,y_k) in the field where N_k is determined:

$$N(x,y) = \frac{h}{f_\sigma} G(x,y) + Ax_k + By_k + C \tag{20.40}$$

Note that Eq. (20.40) is linear in terms of the unknowns (h/f_σ), A, B, and C. For m selected data points, with m > 4, an overdeterministic system of linear equations results from Eq. (20.40). This system of equations can be expressed in matrix form as:

$$[N] = [a][w]$$

where

$$[N] = \begin{bmatrix} N_1 \\ N_2 \\ \cdots \\ \cdots \\ N_m \end{bmatrix} \qquad [a] = \begin{bmatrix} G_1 & x_1 & y_1 & 1 \\ G_2 & x_2 & y_2 & 1 \\ \cdots & \cdots & \cdots & \cdots \\ \cdots & \cdots & \cdots & \cdots \\ G_m & x_m & y_m & 1 \end{bmatrix} \qquad [w] = \begin{bmatrix} h/f_\sigma \\ A \\ B \\ C \end{bmatrix}$$

The solution of the set of m equations for the unknowns h/f_σ, A, B, and C can be achieved in a least-squares sense through the use of matrix methods. Note that:

$$[a]^T [N] = [c][w]$$

where

$$[c] = [a]^T [a]$$

and that:

$$[w] = [c]^{-1} [a]^T [N]$$

where $[c]^{-1}$ is the inverse of **[c]**. Solution of the matrix **[w]** gives the column elements which are the unknowns. This form of solution is easy to accomplish on a small computer which can be programmed to perform the matrix manipulations.

The matrix algebra outlined above is equivalent to minimizing the cumulative error E which is:

$$\mathcal{E} = \sum_{k=1}^{m} \left[\frac{h}{f_\sigma} G(x_k, y_k) + Ax_k + By_k + C - N_k \right]^2 \tag{20.41}$$

The matrix operations apply the least-squares criteria which require that:

$$\frac{\partial \mathcal{E}}{\partial (h/f_\sigma)} = \frac{\partial \mathcal{E}}{\partial A} = \frac{\partial \mathcal{E}}{\partial B} = \frac{\partial \mathcal{E}}{\partial C} = 0 \tag{20.42}$$

The advantage of this statistical approach to calibration of model materials in optical arrangements is the use of full field data to reduce errors due to discrepancies in either the model materials or the optical systems.

Nonlinear Least-Squares Method

In the preceding section a linear least-squares method provided a direct approach to improving the accuracy of calibration with a single-step computation of an overdeterministic set of linear equations. In other experiments involving either the determination of unknowns arising in stresses near a crack tip or contact stresses near a concentrated load, the governing equations are nonlinear in terms of the unknown quantities. In these cases, the procedure to be followed involves linearizing the governing equations, applying the least-squares criteria to the linearized equations, and finally iterating to converge to an accurate solution for the unknowns.

To illustrate this statistical approach, consider a photoelastic experiment that yields an isochromatic fringe pattern near the tip of a crack in a specimen subjected to mixed-mode loading. In this example, there are three unknowns K_I, K_{II}, and σ_{0x} which are related to the experimentally determined fringe orders N_k at positions (r_k, θ_k). The governing equation for this mixed-mode fracture problem is:

$$\begin{aligned}\left(\frac{Nf_\sigma}{h}\right)^2 &= \frac{1}{2\pi r}\left[(K_I \sin\theta + 2K_{II}\cos\theta)^2 + (K_{II}\sin\theta)^2\right] \\ &+ \frac{2\sigma_{0x}}{\sqrt{2\pi r}}\sin\frac{\theta}{2}\left[K_I \sin\theta(1+2\cos\theta) + K_{II}\left(1+2\cos^2\theta+\cos\theta\right)\right] \\ &+ \sigma_{0x}^2\end{aligned} \tag{20.43}$$

Equation (20.43) can be solved in an overdeterministic sense, by forming the function $f(K_I, K_{II}, \sigma_{0x})$ as:

$$\begin{aligned}f_k(K_I, K_{II}, \sigma_{0x}) &= \frac{1}{2\pi r_k}\left[(K_I \sin\theta_k + 2K_{II}\cos\theta_k)^2 + (K_{II}\sin\theta_k)^2\right] \\ &+ \frac{2\sigma_{0x}}{\sqrt{2\pi r_k}}\sin\frac{\theta_k}{2}\left[K_I \sin\theta_k(1+2\cos\theta_k) + K_{II}\left(1+2\cos^2\theta_k+\cos\theta_k\right)\right] \\ &+ \sigma_{0x}^2 - \left(\frac{N_k f_\sigma}{h}\right)^2 = 0\end{aligned} \tag{20.44}$$

where k = 1, 2, 3, ... m and (r_k,θ_k) are coordinates defining a point on an isochromatic fringe of order N_k. A Taylor series expansion of Eq. (20.44) yields:

$$(f_k)_{i+1} = (f_k)_i + \left(\frac{\partial f_k}{\partial K_I}\right)_i \Delta K_I + \left(\frac{\partial f_k}{\partial K_{II}}\right)_i \Delta K_{II} + \left(\frac{\partial f_k}{\partial \sigma_{0x}}\right)_i \Delta\sigma_{0x} \tag{20.45}$$

where i refers to the i th iteration step and ΔK_I, ΔK_{II}, and $\Delta\sigma_{0x}$ are corrections to the previous estimate of ΔK_I, ΔK_{II}, and $\Delta\sigma_{0x}$. It is evident from Eq. (20.45) that corrections should be made to drive $f(K_I, K_{II}, \sigma_{0x})$ toward zero. This fact leads to the iterative equation:

$$\left(\frac{\partial f_k}{\partial K_I}\right)_i \Delta K_I + \left(\frac{\partial f_k}{\partial K_{II}}\right)_i \Delta K_{II} + \left(\frac{\partial f_k}{\partial \sigma_{0x}}\right)_i \Delta\sigma_{0x} = -(f_k)_i \tag{20.46}$$

In matrix form the set of m equations represented by Eq. (20.46) can be written as:

$$[f] = [a][\Delta K] \tag{20.47}$$

where:

$$[f] = \begin{bmatrix} f_1 \\ f_2 \\ \cdots \\ \cdots \\ f_m \end{bmatrix} \qquad [a] = \begin{bmatrix} \partial f_1/\partial K_I & \partial f_1/\partial K_{II} & \partial f_1/\partial \sigma_{0x} \\ \partial f_2/\partial K_I & \partial f_2/\partial K_{II} & \partial f_2/\partial \sigma_{0x} \\ \cdots & \cdots & \cdots \\ \cdots & \cdots & \cdots \\ \partial f_m/\partial K_I & \partial f_m/\partial K_{II} & \partial f_m/\partial \sigma_{0x} \end{bmatrix} \qquad [\Delta K] = \begin{bmatrix} \Delta K_I \\ \Delta K_{II} \\ \Delta \sigma_{0x} \end{bmatrix}$$

The least-squares minimization process is accomplished by multiplying, from the left, both sides of Eq. (20.47) by the transpose of matrix **[a]**, to give:

$$[a]^T [f] = [a]^T [a][\Delta K]$$

or

$$[d] = [c][\Delta K]$$

where

$$[d] = [a]^T [f]$$

$$[c] = [a]^T [a]$$

Finally, the correction terms are:

$$[\Delta K] = [c]^{-1} [d] \tag{20.48}$$

The solution of Eq. (20.48) gives ΔK_I, ΔK_{II}, and $\Delta\sigma_{0x}$ which are used to correct initial estimates of K_I, K_{II}, and σ_{0x} and obtain a better fit of the function $f_k(K_I, K_{II}, \sigma_{0x})$ to m data points.

The procedure is executed on a small computer programmed using MATLAB. One starts by assuming initial values for K_I, K_{II}, and σ_{0x}. Then, the elements of the matrices **[f]** and **[a]** are computed for each of the m data points. The correction matrix **[ΔK]** is then computed from Eq. (20.48) and finally the estimates of the unknowns are corrected by noting that:

$$(K_I)_{i+1} = (K_I)_i + \Delta K_I$$

$$(K_{II})_{i+1} = (K_{II})_i + \Delta K_{II} \tag{20.49}$$

$$(\sigma_{0x})_{i+1} = (\sigma_{0x})_i + \Delta\sigma_{0x}$$

The procedure is repeated until each element in the correction matrix **[ΔK]** becomes acceptably small. As convergence is quite rapid, the number of iterations required for accurate estimates of the unknowns is usually small.

20.9 CHI-SQUARE TESTING

The chi-square χ^2 test is used in statistics to verify the use of a specific distribution function to represent the population from which a set of data has been obtained. The chi-square statistic χ^2 is defined as:

$$\chi^2 = \sum_{i=1}^{k} \left[\frac{(n_o - n_e)^2}{n_e} \right] \tag{20.50}$$

where n_o is the actual number of observations in the i th group interval and n_e is the expected number of observations in the i th group interval based on the specified distribution and k is the total number of group intervals.

The value of χ^2 is computed to determine how closely the data fits the assumed statistical distribution. If $\chi^2 = 0$, the match is perfect. Values of $\chi^2 > 0$ indicate the possibility that the data are not represented by the specified distribution. The probability p that the value of χ^2 is due to random variation is illustrated in Fig. 20.14. The degree of freedom is defined as:

$$\nu = n - k \qquad (20.51)$$

where n is the number of observations and k is the number of conditions imposed on the distribution.

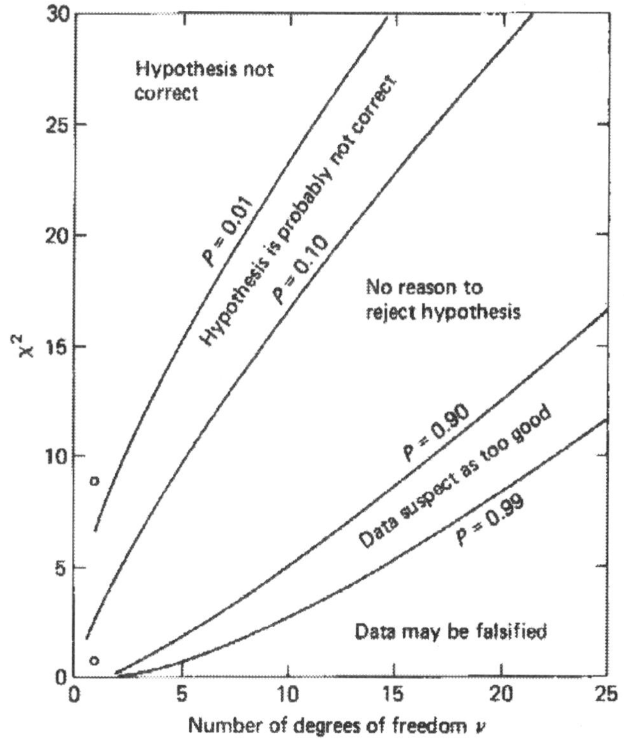

Figure 20.14 Probability of χ^2 values exceeding those shown as a function of the number of degrees of freedom.

As an example of the χ^2 test, consider the ultimate tensile strength data presented in Table 20.10 and judge the adequacy of representing the ultimate tensile strength with a normal probability distribution described with $\bar{x}$ = 176.1 ksi and S_x = 2.25 ksi. By using the properties of a normal distribution function, the number of specimens expected to fall in any strength group can be computed. The observed number of specimens in Table 20.1 exhibiting ultimate tensile strengths within each of seven group intervals, together with the computed number of specimens in a normal distribution in the same group intervals, are listed in Table 20.10. The computation of the χ^2 value ($\chi^2 = 1.785$) is also illustrated in the table. The number of groups n = 7. Because the two distribution parameters $\bar{x}$ and S_x were determined by using these data, k = 2; therefore, the number of degrees of freedom is $\nu = n - k = 7 - 2 = 5$.

Table 20.10
Chi-Squared χ^2 computation for grouped ultimate tensile strength data.

Group interval	Number observed	Number expected	$(n_o - n_e)^2/n_e$
0-170.9	1	0.468	0.604
171-172.9	2	2.904	0.281
173-174.9	8	9.096	0.132
175-176.9	16	13.748	0.369
177-178.9	9	9.836	0.071
179-180.9	3	3.360	0.039
180-∞	1	0.588	0.289
			$1.785 = \chi^2$

Plotting these results ($\nu = 5$ and $\chi^2 = 1.785$) in Fig. 20.14 shows that the point falls in the region where there is no reason to expect that the hypothesis in not correct. The hypothesis is to represent the tensile strength with a normal probability function. The χ^2 test does not prove the validity of this hypothesis, but instead fails to disprove it.

The lines dividing the χ^2 - ν graph of Fig. 20.14 into 5 different regions are based on probabilities of obtaining χ^2 values greater than the values shown by the curves. For example, the line dividing the regions "no reason to reject hypothesis" and "hypothesis probably is not correct" has been selected at a probability level of 10%. Thus, there is only one chance in 10 that data drawn from a population correctly represented by the hypothesis would give a χ^2 value exceeding that specified by the $p > 0.10$ curve. The hypothesis rejected region is defined with the $p > 0.01$ curve indicating only one chance in 100 of obtaining a χ^2 value exceeding those shown by this curve.

The χ^2 function can also be used to question if the data have been adjusted. Probability levels of 0.90 and 0.99 have been used to define regions where "data suspect as too good" and "data may be falsified". For the latter classification there are 99 chances out of 100 that the χ^2 value will exceed that determined by a χ^2 analysis of the data.

The χ^2 statistic can also be used in contingency testing where the sample is classified under one of two categories—pass or fail. Consider, for example, an inspection procedure with a particular type of strain gage where 10% of the gages are rejected due to etching imperfections in the grid. In an effort to reduce this rejection rate, the manufacturer has introduced new clean-room techniques that are expected to improve the quality of the grids. On the first lot of 2000 gages, the failure rate was reduced to 8%. Is this reduced failure rate due to chance variation, or have the new clean-room techniques improve the manufacturing process? A χ^2 test can establish the probability of the improvement being the result of random variation. The computation of χ^2 for this example is illustrated in Table 20.11.

Table 20.11
Observed and expected inspection results

Group Interval	Number Observed	Number Expected	$(n_o - n_e)^2/n_e$
Passed	1840	1800	0.89
Failed	160	200	8.00
			$8.89 = \chi^2$

Plotting the results from Table 20.11 on Fig. 20.14 after noting that $\nu = 1$ shows that $\chi^2 = 8.89$ falls into the region "hypothesis is not correct". In this case the hypothesis was—there has been no improvement. The χ^2 test has shown that there is less than one chance in a hundred of the improvement in rejection rate (8% instead of 10%) being due to random variables. Thus, one can conclude, with confidence, that the new clean-room techniques were effective in improving yield.

20.10 ERROR PROPAGATION

Previous discussions of error have been limited to error arising in the measurement of a single quantity: however, in many engineering applications, several quantities are measured (each with its associated error) and another quantity is predicted on the basis of these measurements. For example, the volume V of a cylinder could be predicted on the basis of measurements of two quantities (diameter D and length L). Thus, errors in the measurements of diameter and length will propagate through the governing mathematical formula $V = \pi DL/4$ to the quantity (volume, in this case) being predicted. Because the propagation of error depends upon the form of the mathematical expression being used to predict the reported quantity, standard deviations for several different mathematical operations are listed below.
For addition and/or subtraction of quantities ($y = x_1 \pm x_2 \pm .... \pm x_n$), the standard deviation $S_{\bar{y}}$ of the mean $\bar{y}$ of the projected quantity y is given by:

$$S_{\bar{y}} = \sqrt{S_{\bar{x}1}^2 + S_{\bar{x}2}^2 + ... + S_{\bar{x}n}^2} \tag{20.52}$$

For multiplication of quantities ($y = x_1x_2 ....x_n$), the standard deviation $S_{\bar{y}}$ is given by:

$$S_{\bar{y}} = (\bar{x}_1\bar{x}_2....\bar{x}_n)\sqrt{\frac{S_{\bar{x}1}^2}{\bar{x}_1^2} + \frac{S_{\bar{x}2}^2}{\bar{x}_2^2} + ... + \frac{S_{\bar{x}n}^2}{\bar{x}_n^2}} \tag{20.53}$$

For division of quantities ($y = x_1/x_2$), the standard deviation $S_{\bar{y}}$ is given by:

$$S_{\bar{y}} = \frac{\bar{x}_1}{\bar{x}_2}\sqrt{\frac{S_{\bar{x}1}^2}{\bar{x}_1^2} + \frac{S_{\bar{x}2}^2}{\bar{x}_2^2}} \tag{20.54}$$

For calculations of the form ($y = x_1^k$), the standard deviation $S_{\bar{y}}$ is given by:

$$S_{\bar{y}} = k\bar{x}_1^{k-1}S_{\bar{x}1} \tag{20.55}$$

For calculations of the form ($y = x^{1/k}$), the standard deviation $S_{\bar{y}}$ is given by:

$$S_{\bar{y}} = \frac{\bar{x}_1^{1/k}}{k\bar{x}_1}S_{\bar{x}1} \tag{20.56}$$

Consider, for example, a rectangular rod where independent measurements of its width, thickness, and length have yielded $\bar{x}_1 = 2.0$ with $S_{\bar{x}1} = 0.005$, $\bar{x}_2 = 0.5$ with $S_{\bar{x}2} = 0.002$, and $\bar{x}_3 = 16.5$ with $S_{\bar{x}3} = 0.040$, where all dimensions are in inches. The volume of the bar is:

$$V = \bar{x}_1\,\bar{x}_2\,\bar{x}_3$$

The standard error of the volume can be determined by using Eq. (20.53). Thus:

$$S_{\bar{y}} = (\bar{x}_1\bar{x}_2....\bar{x}_n)\sqrt{\frac{S_{\bar{x}1}^2}{\bar{x}_1^2}+\frac{S_{\bar{x}2}^2}{\bar{x}_2^2}+.......+\frac{S_{\bar{x}n}^2}{\bar{x}_n^2}}$$

$$= (2.0)(0.5)(16.5)\sqrt{\frac{(0.005)^2}{(2.0)^2}+\frac{(0.002)^2}{(0.5)^2}+\frac{(0.040)^2}{(16.5)^2}}$$

$$= 0.0875 \text{ in.}$$

This determination of $S_{\bar{y}}$ for the volume of the bar can be used together with the properties of a normal probability distribution to predict the number of bars with volumes within specific limits.

The method of computing the standard error of a quantity $S_{\bar{y}}$ as given by Eqs. (20.52) to (20.56), which are based on the properties of the normal probability distribution function, should be used where possible. However, in many engineering applications, the number of measurements that can be made is small; therefore, the data $\bar{x}_1$, $\bar{x}_2$, ..., $\bar{x}_n$ and $S_{\bar{x}1}$, $S_{\bar{x}2}$, ..., $S_{\bar{x}n}$ needed for statistical based estimates of the error are not available. In these instances, error estimates can still be made but the results are less reliable.

A second method of estimating error when data are limited is based on the chain rule of differential calculus. For example, consider a quantity y that is a function of several variables:

$$y = f(x_1, x_2, ... , x_n) \qquad (20.57)$$

Differentiating yields:

$$dy = \frac{\partial y}{\partial x_1}dx_1 + \frac{\partial y}{\partial x_2}dx_2 + ... + \frac{\partial y}{\partial x_n}dx_n \qquad (20.58)$$

In Eq. (20.58), dy is the error in y and dx_1, dx_2, ... , dx_n are errors involved in the measurements of x_1, x_2, ... x_n. The partial derivatives $\partial y/\partial x_1$, $\partial y/\partial x_2$, ... , $\partial y/\partial x_n$ can be determined exactly from Eq. (20.57). Frequently, the errors dx_1, dx_2, ... , dx_n are estimates based on the experience and judgment of the test engineer. An estimate of the maximum possible error can be obtained by summing the individual error terms in Eq. (20.58). Thus:

$$dy\Big|_{max} = \left|\frac{\partial y}{\partial x_1}dx_1\right| + \left|\frac{\partial y}{\partial x_2}dx_2\right| + ... + \left|\frac{\partial y}{\partial x_n}dx_n\right| \qquad (20.59)$$

Equation (20.59) gives a worst case estimate of error, because the maximum errors dx_1, dx_2, ... , dx_n are assumed to occur simultaneously and with the same sign.

A more realistic equation for estimating error is obtained by squaring both sides of Eq 20.58 to give:

$$(dy)^2 = \sum_{i=1}^{n}\left(\frac{\partial y}{\partial x_i}\right)^2 (dx_i)^2 + \sum_{i=1,j=1}^{n}\left(\frac{\partial y}{\partial x_i}\right)\left(\frac{\partial y}{\partial x_j}\right)dx_i dx_j \qquad (20.60)$$

where $i \neq j$.

If the errors dx_i are independent and symmetrical with regard to positive and negative values then the cross product terms will tend to cancel and Eq. (20.60) reduces to:

$$dy = \sqrt{\left(\frac{\partial y}{\partial x_1}dx_1\right)^2 + \left(\frac{\partial y}{\partial x_2}dx_2\right)^2 + \ldots\ldots + \left(\frac{\partial y}{\partial x_n}dx_n\right)^2} \tag{20.61}$$

20.11 SUMMARY

Statistical methods are extremely important in engineering, because they provide a means for representing large amounts of data in a concise form, which is easily interpreted and understood. Usually, the data are represented with a statistical distribution function that can be characterized by a measure of central tendency (the mean $\bar{x}$) and a measure of dispersion (the standard deviation S_x). A normal or Gaussian probability distribution is by far the most commonly employed; however, in some cases, other distribution functions may have to be employed to adequately represent the data.

The most significant advantage resulting from use of a probability distribution function in engineering applications is the ability to predict the occurrence of an event based on a relatively small sample. The effects of sampling error are accounted for by placing confidence limits on the predictions and establishing the associated confidence levels. Sampling error can be controlled if the sample size is adequate. Use of Student's t distribution function, which characterizes sampling error, provides a basis for determining sample size consistent with specified levels of confidence. The Student t distribution also permits a comparison to be made of two means to determine whether the observed difference is significant or whether it is due to random variation.

Statistical methods can also be employed to condition data and to eliminate an erroneous data point (one) from a series of measurements. This is a useful technique that improves the data base by providing strong evidence when something unanticipated is affecting an experiment.

Regression analysis can be used effectively to interpret data when the behavior of one quantity y depends upon variations in one or more independent quantities $x_1, x_2, \ldots, x_n$. Even though the functional relationship between quantities exhibiting variation remains unknown, it can be characterized statistically. Regression analysis provides a method to fit a straight line or a curve through a series of scattered data points on a graph. The adequacy of the regression analysis can be evaluated by determining a correlation coefficient. Methods for extending regression analysis to multivariate functions exist. In principle, these methods are identical to linear regression analysis; however, the analysis becomes much more complex. The increase in complexity is not a concern, because computer subroutines are available that solve the tedious equations and provide the results in a convenient format.

Many probability functions are used in statistical analyses to represent data and predict population properties. Once a probability function has been selected to represent a population, any series of measurements can be subjected to a chi-squared (χ^2) test to check the validity of the assumed function. Accurate predictions can be made only if the proper probability function has been selected.

Finally, statistical methods for accessing error propagation were discussed. These methods provide a means for determining error in a quantity of interest y based on measurements of related quantities $x_1, x_2, \ldots, x_n$ and the functional relationship $y = f(x_1, x_2, \ldots, x_n)$.

EXERCISES

20.1 Ten measurements of the fracture strength (ksi) of an aluminum alloy are:

25.0	25.2	24.9	25.5	24.6
24.8	25.2	25.0	24.8	25.0

Determine the mean, median, and mode which represent the central tendency of this data.

20.2 Verify the mean value $\bar{x}$ of the data listed in Table 20.1.

20.3 Determine the range R, mean deviation d_x and variance S^2_x for the data given in Exercise 20.1.

20.4 Determine the range R, mean deviation d_x and variance S^2_x for the data listed in Table 20.1.

20.5 Find the coefficient of variation for the data given in Exercise 20.1.

20.6 Find the coefficient of variation for the data listed in Table 20.1.

20.7 Consider a Gaussian population with a mean $\mu = 100$ and a standard deviation $\sigma = 10$. Determine the probability of selecting a single sample with a value in the interval between:

(a) 75 – 80 (b) 98 – 102 (c) 92 - 97
(d) 115 – 123 (e) greater than 125

Also determine the percent of data which will probably be within limits of:

(a) $\bar{x} \pm 2.5\ S_x$ (b) $\bar{x} - 1.0\ S_x$ and $\bar{x} + 1.5\ S_x$
(c) $\bar{x} - 1.5\ S_x$ and $\bar{x} + 1.0\ S_x$ (d) $\bar{x} \pm 0.55\ S_x$

20.8 A manufacturing process yields aluminum rods with a mean yield strength of 35,000 psi and a standard deviation of 1000 psi. A customer places a very large order for rods with a minimum yield strength of 32,000 psi. Prepare a letter for submission to the customer that describes the yield strength to be expected and outline your firm's procedures for assuring that this quality level will be achieved and maintained.

20.9 Determine the Weibull distribution function corresponding to the parameters $x_0 = 5$, $b = 5$ and $m = 10$. Plot P(x) for $0 < x < 50$.

20.10 For the Weibull distribution of Exercise 20.9, predict the probability of failure if $x = 6.5$.

20.11 A Weibull distribution function describing the strength of a brittle ceramic uses $x_0 = 3$ ksi, $b = 2$ ksi, and $n = 3$. Compute the expected values of the ten lowest strengths measured if a total of 400 specimens were tested.

20.12 Repeat 20.11 but let $b = 3$ ksi.

20.13 In calibrating a brittle coating 12 calibration beams are tested to determine that $\bar{x} = 470\ \mu\varepsilon$ and $S_x = 60\ \mu\varepsilon$. Determine the standard distribution of the mean $S_{\bar{x}}$. Give the confidence level associated with the statement that the true mean μ of the calibration was between 460 and 480 $\mu\varepsilon$.

20.14 In the calibration test of the brittle coating in Exercise 20.13 the number of beams is increased from 12 to 40. The new results gives $\bar{x} = 465\ \mu\varepsilon$ and $S_x = 70\ \mu\varepsilon$. Determine $S_{\bar{x}}$ and give confidence limits on the statement that the true threshold strain is between 455 and 475 $\mu\varepsilon$.

20.15 Use Eq. (20.20) with two degrees of freedom and evaluate f(t) as a function of t. Plot your results on Fig. 20.9. Select F_0 so that:

$$\int_{-\infty}^{\infty} f(t)dt = 1$$

20.16 Compare the results of Exercise 20.15 with the distribution function from a normal distribution.

20.17 Determine the sample size necessary to insure that the average strength of a material exceeds 70 MPa if the sample is drawn from a population with an estimated mean of 65 MPa and an estimated standard deviation of 2 MPa. Use a confidence level of 5%.

20.18 An inspection laboratory samples two large shipments of dowel pins by measuring both length and diameter. For shipment A, the sample size was 40, the mean diameter was 6.12 mm, the mean length was 25.3 mm, the estimated standard deviation on diameter was 0.022 mm, and the estimated standard deviation on length was 0.140 mm. For shipment B, the sample size was 60, the mean diameter was 6.04 mm, the mean length was 25.05 mm, the estimated standard deviation on diameter was 0.034 mm, and the estimated standard deviation on length was 0.203 mm.

(a) Are the two shipments of dowel pins the same?
(b) What is the level of confidence in your prediction?
(c) Would it be safe to mix the two shipments of pins? Explain.

20.19 Repeat Exercise 20.18 for the following two shipments of dowel pins.

	Shipment A	**Shipment B**
Number	20	10
Diameter	$\bar{x}$ = 6.05 mm, S_x = 0.03 mm	$\bar{x}$ = 5.98 mm, S_x = 0.04 mm
Length	$\bar{x}$ = 24.9 mm, S_x = 0.22 mm	$\bar{x}$ = 25.4 mm, S_x = 0.18 mm

20.20 Fatigue tests of a component under simulated load conditions indicated that the mean failure load for the specified life was $\bar{F}$ = 4115 N with an estimated standard deviation of 340 N. In service this component will be subjected to an average load of 2800 N but this load may vary and the estimated standard deviation is 700 N. Determine the reliability of the component over the specified life.

20.21 For the data in Exercise 20.20, determine the safety factor for a reliability of (a) 90%, (b) 95%, and (c) 99%.

20.22 The following sequence of measurements of atmospheric pressure (millimeters of mercury) was obtained with a barometer.

764.3	764.6	764.4	765.2	764.5
764.5	765.7	765.4	764.8	765.3
765.2	764.9	764.6	765.1	764.6

Determine the mean and the standard deviation after employing Chauvenet's criterion to statistically condition the data.

20.23 Determine the slope m and the intercept b for a linear regression equation y = mx + b and the correlation coefficient R^2 for the data listed below:

x	1	2	3	4	5	6	7	8	9	10
y	2.1	3.9	6.4	8.2	9.5	12.0	13.6	16.7	17.9	19.4

20.24 The drying of a coating containing a solvent is a diffusion controlled process which can be described by the equation $C = ae^{-mt}$, where C is the concentration (% solvent), t is the time (seconds), and a and m are diffusion constants. Use linear regression methods to determine a and m for the following data which was obtained by weighing a sample of coating during drying.

t	100	200	300	400	500	700	1000
C	3.40	1.40	0.561	0.195	0.067	0.0097	0.0004

20.25 Determine the regression coefficients a, b_1, b_2, and b_3 and the correlation coefficient R^2 for the following data set.

y	x_1	x_2	x_3
6.8	1.0	2.0	1.0
7.8	1.5	2.0	1.5
7.9	2.0	2.0	2.0
8.0	2.5	3.0	1.0
8.3	3.0	3.0	1.5
8.4	3.5	3.0	2.0
8.5	4.0	4.0	1.0
8.6	4.5	4.0	1.5
8.9	5.0	4.0	2.0
9.1	5.5	5.0	1.0
9.3	6.0	5.0	1.5
9.6	6.5	5.0	2.0

20.26 Write a computer program which utilizes field data taken from a calibration disk in photoelasticity. The program should accept 20 data points N(x,y), the model thickness h, and analytic functions G(x,y) and E(x,y). The output [see Eq. (20.40)] should give f_σ, A, B, and C.

20.27 Write a computer program to implement the non-linear least squares method. Use the mixed-mode relations given in Eq. (20.43) as the example. The program input is field data N(r,θ) and calibration data f_σ/h. The output is K_I, K_{II}, and σ_{0x}.

20.28 A die-casting operation produces bearing housings with a rejection rate of 4% when the machine is operated over an 8-h shift to produce a total output of 3200 housings. The method of die cooling was changed in an attempt to reduce the rejection rate. After 2 h of operation under the new cooling conditions, 775 acceptable castings and 25 rejects had been produced.

(a) Did the change in the process improve the output?
(b) How certain are you of your answer?

20.29 The stress σ_x at a point on the free surface of a structure or machine component can be expressed in terms of the normal strains ε_x and ε_y measured with electrical resistance strain gages as:

$$\sigma_x = \frac{E}{1-\nu^2}\left(\varepsilon_x + \nu\varepsilon_y\right)$$

If ε_x and ε_y are measured within ± 2% and E and ν are measured within ± 5%, estimate the error in σ_x.

20.30 A gear-shaft assembly consists of a shaft with a shoulder, a bearing, a sleeve, a gear, a second sleeve, and a nut. Dimensional tolerances for each of these components are listed below:

Component	Tolerance, mm
Shoulder	0.050
Bearing	0.025
Front sleeve	0.100
Gear	0.050
Second sleeve	0.100
Nut	0.100

(a) Determine the anticipated tolerance of the series assembly.
(b) What will be the frequency of occurrence of the tolerance of Part (a)?

20.31 Estimate the error in determining the weight of a cylindrical rod if the dimensional measurements are accurate to 0.5% and the specific weight is accurate to 0.1%.

20.32 Estimate the error in determining strain from two displacement measurements if the displacements are measured with an accuracy of 2% and the positions x_1 and x_2 each are measured with an accuracy of 1%.

BIBLIOGRAPHY

1. Bethea, R. M. and R. R. Rhinehart: Applied Engineering Statistics, Dekker, New York, 1991.
2. Bethea, R. M., B. S. Duran, and T. L. Boullion: Statistical Methods for Engineers and Scientists, 2nd Edition, Dekker, New York, 1985.
3. Blackwell, D.: Basic Statistics, McGraw-Hill, New York, 1969.
4. Bragg, G. M.: Principles of Experimentation and Measurement, Prentice-Hall, Englewood Cliffs, NJ, 1974.
5. Chou, Y.: Probability and Statistics for Decision Making, Holt, Rinehart & Winston, New York, 1972.
6. Snedecor, G. W. and W. G. Cochran: Statistical Methods, 8th Edition, Iowa State University Press, Ames, IA, 1989.
7. Davies, O. L. and P. L. Goldsmith: Statistical Methods in Research and Production, Hafner, New York, 1972.
8. McCall, C. H. Jr.: Sampling and Statistics Handbook for Research, Iowa State University Press, Ames, IA., 1982.
9. Weibull, W.: Fatigue Testing and Analysis of Results, Pergamon Press, New York, 1961.
10. Young, H. D.: Statistical Treatment of Experimental Data, McGraw-Hill, New York, 1962.
11. Zehna, P. W.: Introductory Statistics, Prindle, Weber & Schmidt, Boston, 1974.
12. Zehna, P. W.: Probability Distributions and Statistics, Allyn and Bacon, Boston, 1970.

INDEX